FOURTH EDITION

The Extraordinary Chemistry of Ordinary Things

CARL H. SNYDER
University of Miami

JOHN WILEY & SONS, INC.

TO SARAH

A joyful combination of atoms, ions, and molecules.

Acquisitions Editor	Jennifer Yee
Marketing Manager	Bob Smith
Senior Production Editor	Elizabeth Swain
Text and Cover Designer	Madelyn Lesure
Illustration Editor	Anna Melhorn
Photo Editor	Lisa Gee
Front Cover Photo	Steve Lupton/CORBIS.
Back Cover Photo	Alfred Pasieka/Peter Arnold, Inc.

This book was set in Times Ten by Preparé and printed and bound by Von Hoffmann.
The cover was printed by Von Hoffmann.

This book is printed on acid free paper. ∞

Library of Congress Cataloging-in-Publication Data:

Snyder Carl H.
The extraordinary chemistry of ordinary things / Carl H. Snyder.-- 4th ed.
p. cm.
Includes bibliographical references and index.
ISBN 0-471-41575-8 (cloth : acid-free paper)
1. Chemistry. I. Title.

QD33.2 .S58 2002
540--dc21 2002190821

Printed in the United States of America

10 9 8 7 6 5 4 3 2

PREFACE

The Central Theme

This 4th edition of *The Extraordinary Chemistry of Ordinary Things* continues the central thesis of the earlier editions: Since we live our daily lives immersed in substances that are composed of chemicals and that owe their properties and behavior to these chemicals—our foods, fuels, consumer goods, environmental materials, even the physical materials of our own bodies—the most effective way to begin an examination of chemistry is through these goods and substances of our everyday world.

As in the earlier editions, our survey of chemistry takes place

- in the context of chemistry as an experimental science;
- in the context of the ordinary things of our everyday lives, and a few that aren't so ordinary;
- in the context of the larger realm of science, drawing on chemical principles to illustrate the workings of science as a whole; and
- in the context of the need for an understanding of science and its relationship to our society so that all of us, scientist and nonscientist alike, can make reasoned judgments on societal issues that are affected by the processes and fruits of science in general and chemistry in particular.

New to This Edition

- The flexibility of free-standing chapters that can be covered in a variety of sequences;
- Expanded coverage of environmental topics;
- The addition of web questions/activities for every chapter;
- A new chapter covering energy, Chapter 8: Working With Chemistry–Food, Fuel, and Energy
- A new chapter covering biochemistry, Chapter 18: The Chemistry of Heredity–Molecular Genetics
- Revised end-of-section and end-of-chapter questions

Focus, Content, and Organization

Several important changes distinguish this revision from earlier editions. In its early chapters this 4th edition focuses more sharply on energy—its sources, uses, and impact on society—and introduces a new chapter devoted entirely to the topic. Another new chapter, later in the text, introduces biochemistry with an examination of the chemical basis of heredity and genetics.

As in earlier editions, this new organization is adaptable to either a one- or a two-term course. Of the 23 chapters, the first 12 cover topics generally regarded as fundamental to the science of chemistry. The remaining 11 deal with applications of chemistry to the more common goods and processes of our everyday world.

The first three chapters are introductory. They focus on atoms, ions, molecules, elements, compounds, bonding, and the periodic table. Next come five chapters covering chemical aspects of work, energy, and nuclear, fossil and biological fuels. Chapter 4 takes us through a chronological narrative spanning the half-century from the discovery of radioactivity though the development and use of the atomic bomb. Chapter 5 examines peaceful uses of nuclear reactions including the generation of commercial electrical power, medical applications, and the dating of ancient objects. Chapters 6 and 7 introduce the chemistry of carbon compounds and the role of petroleum hydrocarbons as a source of energy. Chapter 8, new to this edition, caps these energy-related themes as it combines and unites chemical aspects of energy that had been dispersed throughout several chapters of earlier editions. It joins scientific and societal issues involved in the expanding use of energy, with special emphasis on the greenhouse effect. The next four chapters round out the foundations of chemistry with the mole concept and chemical calculations; acids and bases; redox; and the states of matter and the gas laws.

Following this set of 12 chapters comes a group of 11 that view the foundations of chemistry through their applications to everyday life. An examination of detergents in Chapter 13 provides insights into surface chemistry, while the environmental focus of Chapter 14 emphasizes quantitative aspects of pollution. The following three chapters on macronutrients examine the chemistry of esters (as the triglycerides of fats and oils), introduce stereochemistry and optical activity (through carbohydrates), and survey amino acids, polypeptides, and proteins. This provides a foundation for Chapter 18, a new chapter that introduces biochemistry through the chemistry of heredity, DNA, and the genetic code.

The last five chapters consider chemical aspects of vitamins, minerals and food additives; safety and hazards with emphasis on chemical poisons and toxins; plastics and polymers; cosmetics and personal care; and medicines and drugs.

A valuable feature of this approach is its flexibility. Except for the first three chapters, the set of two dealing with nuclear chemistry, and another set of two covering first proteins and then molecular genetics, the chapters of this 4th edition are largely free-standing, capable of being covered in a variety of sequences. Copious cross-referencing of Chapters and Sections throughout the text adds to this quality.

Questions, Exercises, and Examples

As in earlier editions, virtually every section is followed by a question designed to induce immediate reflection and review. Exercises at the end of each chapter are divided into three categories: (1) review, for a straightforward reexamination of the factual content of the chapter; (2) mathematical, to emphasize the mathematical aspects of chemistry; and (3) thought-provoking. Exercises in this last category sometimes have no "right" answer but are intended to stimulate thought about the interconnection of chemistry, society, and individual values. Answers are provided for virtually all the even-numbered problems of the first two categories.

The number of both the end-of-section questions and end-of-chapter exercises has been increased substantially from the 3rd edition, by about 10% and 23% respectively. Another change from the 3rd edition has been the replacement of reprints of articles featuring the news of science (which followed the index) with questions designed to stimulate an expansion of coverage through personal research on the Web. These appear at the end of each chapter and range from five to 10 questions per chapter.

As in the 3rd edition, many of the chapters, especially the earlier ones, contain worked examples to ease the way through some of the more difficult concepts. Other characteristics of the presentation include the introduction of definitions, concepts, symbols, and the like, largely on a need-to-know basis. It seems to make more sense to explain and describe the world about us as we encounter it, rather than to define and categorize ideas well before they find useful applications. Etymologies are also included. Students often find it's easier to learn technical terms if they know where the words came from. Other preferences and pedagogical tools I'm fond of, perhaps subconsciously, show up here and there as well, beneficially I hope.

A Personal View: Chemistry as an Experimental Science

Many of us teach chemistry as we know it to be, with an understanding shaped by many years, even decades, of study. We see chemistry as a coherent, rational whole, and we transmit this model of the chemical universe to our students. Yet I have found, and I suspect many others would agree, that merely transmitting this model is insufficient and unsatisfying to both the teacher and the student. It's important to describe not only the coherent model of the universe that the science of chemistry presents, but also to demonstrate why we are forced to accept this particular model as preferable, more useful, and more intellectually satisfying than any other.

I use the word "forced" because the model of the world that chemistry gives us is one that we are required to accept. We are forced to this particular model by our contact with physical reality, by our tests of physical reality, by the questions we frame as we test this real universe experimentally, and by the answers we receive from our experimental tests. Chemistry is, above and beyond all else, an experimental science.

We are forced to mold the universe into one particular intellectual construct because of our commitment to the scientific method and its experimental approach to knowledge. To teach chemistry, I am convinced, requires teaching the broad outlines of the scientific method, explicitly or implicitly. We have no choice, for example, but to acknowledge that atoms, ions, molecules, and chemical bonds do actually exist. But why are we forced to this view of the world? This is what students must come to understand if they are to learn chemistry in its richest context: why we are forced to see the physical universe as we do.

We accept the reality of atoms and all the other structures and concepts of chemistry because we have no other rational choice. Our experimental tests of our universe, through the scientific method, lead us to them and only to them. Let us then give our students real, demonstrable evidence that what we tell them is, indeed, true chemistry, real chemistry. Let us show them in lectures and in textbooks that what we tell them is true not because we say it is, but because they see it is.

Descriptions of Activities and Introductory Observations

In the spirit of the preceding section, all but a few chapters begin with a description of an activity or observation intended to lead to insights into chemistry. These are not to be mistaken for traditional laboratory experiments. This is a textbook, not a laboratory manual. If carried out at all, these activities should be carried out only by or under the direct guidance and supervision of an experienced, professional chemist, in an institutional, academic setting, and with proper regard for the safety of both the environment and all persons involved.

In light of this, the following comments should be kept in mind: The descriptions of these activities are generalized. They are not presented with the details, cautions, or warnings that are found in laboratory manuals. Instructions for the disposal of waste material are not included in the descriptions of the activities, yet all wastes, whatever their origin, must be disposed of properly to avoid hazards to the environment and/or to individuals. Furthermore, although many of these descriptions refer to common consumer products, the misuse of a consumer product, or use in ways or under circumstances for which it was not intended, can be hazardous.

To state this in a more formal sense: Except for the activities described in this text in association with a stated warning, hazard or caution, the author believes the activities to be safe and to offer valuable educational experience. However, any and all duplication or performance of these activities is conducted at one's own risk. The author does not warrant or guarantee the safety of individuals performing these activities. It cannot be assumed that all necessary warning and precautionary measures are contained in this text, or that other or additional information or measures may not be required. The author hereby disclaims any liability for any loss or damage claimed to have resulted from or been related in any way to the activities, regardless of the form of action.

Supplements

An innovative package of supplements to accompany *The Extraordinary Chemistry of Ordinary Things*, 4th edition, is available to assist both the instructor and the student.

1. **Study Guide,** by Ann Ratcliffe of the University of Northern Colorado and David Dever, formerly of Macon College. This Guide is an invaluable tool for the student, containing unusual, illustrative scenarios as well as the more traditional study-guide features such as chapter overviews and solutions to the in-text questions. Also included are worked-out solutions to the problems in the text along with additional exercises of the same nature and level of difficulty.
2. **Laboratory Manual,** by Bruce Richardson of Highline Community College and Thomas Chasteen of Sam Houston State University. Twenty-five laboratory exercises are included in this Manual, all written in a clear, concise, and unintimidating fashion. The themes emphasized in the Laboratory Manual closely parallel those of the text, incorporating experiments with both consumer and environmental applications.
3. **Instructor's Manual,** by John Thompson of Texas A&M University–Kingsville. Chapter overviews contain learning objectives, chapter outlines, discussion and critical thinking questions, postscripts/chapter lead-ins, key terms, and additional class demonstrations for each chapter in the text.
4. **Test Bank.** Written by the text author, the Test Bank contains over 1,000 multiple-choice questions.
5. **Computerized Test Bank.** IBM and Macintosh versions of the entire Test Bank are available with full editing features to help you customize your tests.
6. **Instructor's resource CD-ROM** contains PowerPoint-rendered text illustrations and selected photos for classroom projection.
7. **E-grade** is an on-line quizzing and homework management program that allows students to do practice tests and email homework assignments directly to the professor.
8. **Web Site.** The web site (www.wiley.com/college/snyder) for the text includes on-line quizzing, links to other outside on-line resources.

Acknowledgments

My thanks for their contributions to making this 4th edition a reality go to many people, especially four to whom I owe special debts.

Principally, I want to thank Jennifer Yee of John Wiley & Sons. Through her role as Project Editor, she played a determining role in bringing this edition to life. Her special combination of enthusiasm, energy, professionalism, and grace was instrumental in moving it along to completion.

As in the 1st edition, I again thank Harry P. Schultz, my departmental chairman at the University of Miami during the creation of the first edition of this textbook. After my initial suggestion, years before, that we introduce a new course for nonscience students, Harry gave me unreserved support, encouragement, and recognition. This book came into being largely because of Harry's repeated urgings that I put all my class notes onto paper and eventually into print. Without Harry neither the course nor the book would exist.

For this 4th edition, I'm grateful to Professor Peter Luykx of the University of Miami Biology Department for inspiring the interest and understanding that led to the introduction of biochemistry through an examination of molecular genetics. Peter's elegant descriptions of the connection between the abstract molecular architecture of DNA and RNA and the biological characteristics—the physical flesh and bones—of living organisms were invaluable in the creation of this chapter.

And I again thank my wife, Jean, for patience, understanding, support, and unfailing good cheer during the long process of revision, and for forbearance during all those times when my physical being was sitting a few feet away at home, but my mind, spirit, psyche, and all else were somewhere in another universe, working on these pages.

Additional thanks go to several others at John Wiley & Sons, for special contributions: David Harris, Associate Publisher, for his vision in bringing this edition into existence; Lisa Gee, Photo Editor, for uncovering the striking new photographs that enhance the text; Anna Melhorn, Illustration Coordinator, for ensuring the accuracy and quality of chemical structures, drawings, and related graphics; Elizabeth Swain, Senior Production Editor, for maintaining an energetic and efficient production process; and Robert Smith, Marketing Manager, for distributing the news of this important new edition.

Finally, I want to thank the reviewers for their contributions and help in bringing this edition into existence. Their suggestions and guidance, consistently both positive and useful, were invaluable.

Scott A. Davis, *Mansfield University*
Peter Zurbach, *St. Joseph's University*
Anthony Dribben, *Mississippi College*
Elena J. Colicelli, *College of Saint Elizabeth*
Joseph Tausta, *State University of New York–College at Oneonta*
Pamela Kerrigan, *Manhattan College, College of Mount St. Vincent*
Jacqueline Waldman, *Goucher College*
Raj Khanna, *University of Maryland*
Akbar Salam, *Texas Tech University*
John S. Thompson, *Texas A&M University-Kingsville*
Xiche Hu, *University of Toledo*

GUIDE TO THE INTRODUCTORY OBSERVATIONS

The following is a complete list of the chapter-opening activities, observations, and descriptions, with the concepts they illustrate. They help set the stage for examining a principal topic of the chapter and its role in our everyday lives.

Chapter 1 **Enlightenment From a Flashlight** See what the differences in the ways salt and sugar conduct (or don't conduct) electricity in water tell us about their composition.

Chapter 2 **Atoms and Paper Clips** Gain insight into the modern view of an atom by using a pile of paper clips.

Chapter 3 **Scouring Pads and Kitchen Magnets** Demonstrate the changes in the properties of an element as it enters into a compound, by using a kitchen scouring pad.

Chapter 4 **The Bomb** The first detonation of an atomic bomb capped a half century of discovery and application of the secrets of the atomic nucleus. Learn what happened when the bomb exploded on July 16, 1945.

Chapter 5 **Static at Stagg Field: The First Atomic Pile** The energy of disintegrating nuclei was harnessed in a controlled manner on December 2, 1942, inside an unused football stadium in Chicago. Discover how this was accomplished through the construction and activation of an atomic pile.

Chapter 6 **A Candle Burning in a Beaker: Energy From Hydrocarbons** Observe how the chemistry of a burning candle releases the power of hydrocarbons.

Chapter 7 **Petroleum and Strong Tea** See how valuable products are obtained from petroleum by examining the distillation of pure water from strong tea.

Chapter 8 **Warming Up With Work** Demonstrate the principle that doing work generates heat, simply by rubbing your hands together.

Chapter 9 **The Glass Where Pollution Begins** Understand how we count chemical particles, what concentrations are, and why quantitative measures of pollution are important, using drinking glasses, a ruler, a marking pen, and some water, salt and sugar.

Chapter 10 **Breath with the Strength of Red Cabbage** Learn how a simple breath test using red cabbage, household ammonia, vinegar, and water can reveal the acidity of exhaled breath.

Chapter 11 **Galvanized Tacks, Drugstore Iodine, and Household Bleach** See how galvanized tacks, iodine, and household bleach produce color changes as a result of electron transfers.

Chapters 12 **Squeezing Air Out of a Bottle** Demonstrate the properties of gases through a parlor trick that involves the illusion of squeezing the air out of a bottle with your bare hands.

Chapter 13 **With Nerves as Steady as a Chemical Bond** Examine the chemical phenomenon of surface tension with a stunt that seems to demonstrate that your hands are steadier than anyone else's. The secret lies in the toothpick you use.

Chapter 14 **How to Generate Air Pollution in Your Own Kitchen** See where the acidity of acid rain comes from using red cabbage leaves, a pot of water, matches, a candle, and a paper towel.

Chapter 15 **Vegetable Oil Chemistry** Learn why the effect of two different vegetable oils on drugstore iodine reveals differences in their degrees of unsaturation.

Chapter 16 **How to Tell a Potato from an Apple, the Hard Way** Distinguish between a potato and an apple using only iodine and your sense of sight.

Chapter 17 **How to Turn an Egg White White** Turn an egg white white without cooking it to demonstrate the process of denaturation.

Chapter 18 **Pineapple Gelatin** Learn how a pineapple plant protects itself against predatory insects as you compare a dish of pineapple gelatin made with chunks of canned pineapple and one containing pieces of the fresh fruit.

Chapter 19 **The Power of Vitamin C** Visualize Vitamin C's chemical prowess by examining its ease of oxidation, using Vitamin C tablets, iodine, and household bleach.

Chapter 20 **Spreading Poison All Around** You deliberately introduce a toxic chemical into the environment every time you spray an insecticide into the air. Consider the consequences and decide whether you should be concerned about them.

Chapter 21 **Plastics, Water, and Rubbing Alcohol** Separate plastics of different densities using a mixture of rubbing alcohol and water.

Chapter 22 **The Wave of Chemistry** See how a home permanent changes the shape of protein molecules in hair.

Chapter 23 **Perception, Reality, and Chemicals** Test your own perceptions of reality with some unusual graphics.

BRIEF CONTENTS

CONTENTS

CHAPTER 1

An Introduction to Chemistry

TURNING ON THE LIGHT

Science—to know and to understand the universe.

Enlightenment from a Flashlight

Table salt (sodium chloride) and table sugar (sucrose).

Our study of chemistry begins with two of our most common consumer products, chemicals that most of us use virtually every day: table salt and table sugar. Table salt, known chemically as *sodium chloride*, and table sugar, *sucrose*, are both white, crystalline solids. It's almost impossible for most of us to tell them apart except by taste.

There's another important way they differ, though, and that is through their electrical behavior in water. The difference in how each of these consumer products conducts—or doesn't conduct—electricity when it's dissolved in water tells us a great deal about their composition.

We can observe the difference with a simple flashlight, some salt, sugar, a bit of warm water, and a sponge or absorbent towel. (This demonstration works best with fresh batteries and a new flashlight, with clean electrical contacts.) In a small glass dissolve a tablespoon of sodium chloride (table salt) in about a quarter of a cup of warm water. Do the same with a tablespoon of sucrose (table sugar) in another small glass. Now wet a sponge or a piece of an absorbent towel with some warm water. Turn the flashlight on and unscrew the back of the flashlight.

An **electric circuit** is the path the electrons follow. **Electrons** are extremely small particles that carry a negative electrical charge.

An **electric current** is a flow of electrons.

Naturally, the flashlight goes off when you remove the back. The spring that you see on the back of the flashlight does two jobs. It pushes the batteries firmly up against the bulb and it serves as part of the flashlight's **electric circuit**, which is simply the path the **electrons** follow within the flashlight. Electrons are extremely small particles that carry a negative electrical charge. We'll have more to say about them in Chapters 2 and 3. For the moment, all we need to know about electrons is that their movement along a circuit produces an **electric current**. Electrons leave the base of the batteries, travel to the bulb over metal parts within the flashlight, move through the bulb's filament—the small, narrow wire in a light bulb that heats up and emits light as electrons flow through it—and then back to the batteries (Fig. 1.1).

The spring we've just removed is part of this circuit. It acts as the path that carries the electrons from the bottom of the nearest battery to the inside wall of the flashlight. When we unscrew the back of the flashlight, we interrupt the path, so the light goes

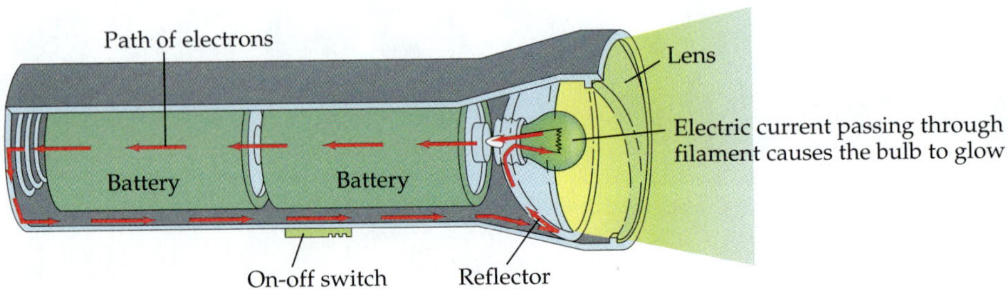

Figure 1.1 The electrical circuit of a flashlight.

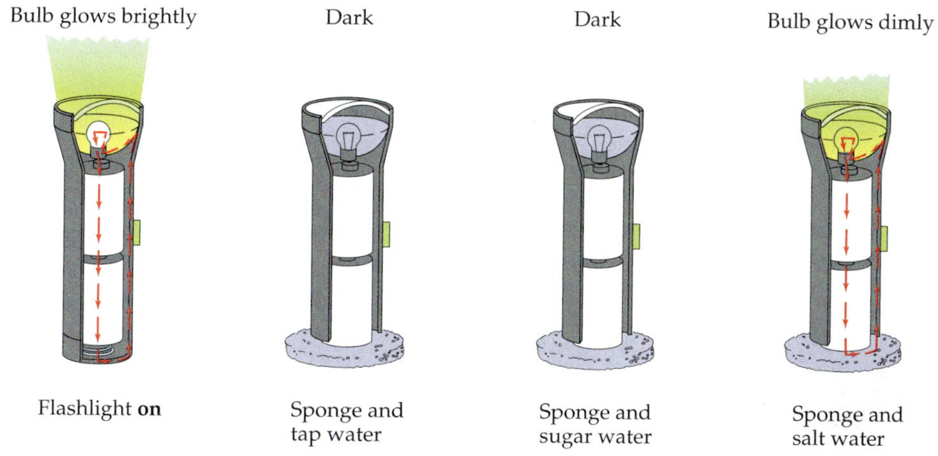

Figure 1.2 The electrical conductivity of tap water, sugar water, and salt water.

out. We're now going to try to make the light go back on by replacing the metal spring with something else.

First we'll try plain tap water. With the bulb pointed upward (to keep water from running down into the flashlight), press the wet sponge or towel firmly against the back of the exposed battery. Push hard against both the back of the battery and the inside wall of the flashlight so that you make good contact with both. Watch the bulb as you make contact. You'll see that the flashlight bulb remains dark since water itself is a poor conductor of electricity. With tap water, not enough electrons flow across the wet sponge or towel to heat the filament to glowing. Remove the batteries, clean and dry them and the inside of the flashlight, and reinsert the batteries.

Now squeeze the tap water out of the sponge and wet it with the warm sugar water solution. Once again push the sponge or towel into the back of the flashlight so that it forms a tight bridge between the back of the battery and the inside wall of the flashlight. The bulb still fails to glow. We're forced to conclude that sugar water is no better at conducting an electric current than plain tap water. Again, remove and dry the batteries, dry the flashlight, and reinsert the batteries.

Finally, repeat the entire process with the salt water. If you watch the bulb, in dim light, you'll see it glow faintly for a few moments just as you first press the sponge or towel against the back of the battery and the inside wall of the flashlight. This must mean that while sugar doesn't add to water's ability to conduct a current, salt does. Figure 1.2 sums up these observations.

Since the voltage of the flashlight batteries is so very low, this demonstration of the difference in electrical behavior between salt and sugar is simple and safe enough for you to carry out (and does no damage to the flashlight or batteries as long as they are cleaned and dried thoroughly at the end). ■

1.1 Light Bulbs, Salt, and Sugar

A much more dramatic (*and much more dangerous*) demonstration would use pure water rather than tap water, a common household light bulb, and 120 volts of house current, along with table salt and table sugar.

WARNING...Don't actually try to do this next version of the demonstration yourself! Attempting it would involve a potentially lethal combination of bare wires and house current—120 volts—which must not be handled by anyone who isn't professionally skilled in working with electricity.

Figure 1.3 An electric light bulb. Pure water is a very poor conductor of electricity, as is a solution of sucrose (table sugar) in water. A solution of sodium chloride (table salt) in water is a good conductor of electricity.

We know that if we plug a light bulb into an electric outlet, the bulb lights up (Fig. 1.3). The same principles are at work here as in the flashlight. If someone skilled in working with the high voltages of house current were to change the circuit by cutting one of the wires and then bending the free ends into a beaker of pure water, the bulb would remain dark. As we saw with the flashlight, pure water is a poor conductor of electricity. With water in the circuit, too little current could pass through the bulb's filament to make it glow. Replacing the pure water with a solution of sucrose (table sugar) in water would make no difference whatever. The light bulb would remain dark and we would conclude, as we did with the flashlight, that a solution of sucrose in water is no better a conductor of electricity than is pure water itself. If the bent wires were dipped into a solution of sodium chloride (table salt), though, the bulb would glow brightly.

The remarkable difference between sucrose and sodium chloride, visible with the flashlight batteries and dramatically apparent with 120 volts of house current, is that sodium chloride conducts electricity when it's dissolved in water but sucrose doesn't. Substances that conduct electricity when dissolved in water—or when melted—are called **electrolytes**. Those that don't are **nonelectrolytes**. Sodium chloride, then, is an electrolyte, while sucrose is a nonelectrolyte. This difference in electrical conductivity between salt and sucrose arises from differences in the chemical compositions of these two everyday substances. More exactly, it comes from differences in the kinds of forces that hold matter together in the two.

An **electrolyte** is a substance that conducts electricity when it is dissolved in water or when it is melted.

A **nonelectrolyte** does not conduct electricity when it is dissolved in water or when it is melted.

To understand what these forces are and how they differ in sucrose and in sodium chloride, we'll soon examine the *atoms*, *ions*, *molecules*, and *chemical bonds* that make up the world around us (and that make up us as well) and that hold it all together. For the moment it's enough to know that

- Atoms, which were once thought to be the ultimate, indivisible particles that make up all matter, are among the fundamental particles of the science of chemistry. By **matter** we mean simply all the different kinds of substances that make up the material things of the universe.
- Molecules are groups of two or more atoms held together by the forces of chemical bonds.
- Ions are atoms or groups of atoms that carry a positive or negative electrical charge.

Matter consists of all the different kinds of substances that make up the material things of the universe.

Figure 1.4 Gold atoms and a dollar bill.

In the first few chapters of this book we'll examine these particles in more detail, we'll define them more precisely, and we'll learn how they are held together in the substances about us. We'll find, among other things, that atoms are extraordinarily small particles. Strung along in a straight line like beads on a necklace, for example, it would take 500,000,000 atoms of gold to stretch across the length of a dollar bill (Fig. 1.4).

After we have laid this foundation, we'll move on, in later chapters, to some of the ordinary (and sometimes not so very ordinary) things that these particles form and make work. Before we do, though, we'll examine briefly just what chemistry is and how we are going to explore it in the chapters that lie ahead. Then we'll return to our investigation of sugar, salt, water, and the light bulb to learn why table salt makes the bulb light up, but sugar doesn't.

■ QUESTION

What do you think would be the result of placing the wires into a solution made up of a *mixture* of equal parts of sucrose and sodium chloride?

1.2 The Extraordinary Chemistry of Ordinary Things

Chemistry is the branch of science devoted to the study of matter, its composition, its properties, and the changes it undergoes. Chemistry, then, studies the material substance of the universe, the stuff we can hold, throw, feel, weigh, smell, see, touch, and taste.

Chemistry is the branch of science that studies the composition and properties of matter and the changes that matter undergoes.

In this book we'll examine chemistry largely through the properties and the compositions of ordinary things, some as ordinary as the water, table salt, and table sugar we started with, some a bit less common or less obvious but nonetheless important to the way we live today or perhaps will live tomorrow. Since all the materials of our everyday lives, including the food and the consumer products that form both our necessities and our luxuries, are made up of ordinary matter, the study of their composition, properties, and changes is, in fact, the study of chemistry itself.

As we examine this science of chemistry we'll come across many more demonstrations or experiments like the one involving sugar water and salt water that begins this chapter. Perhaps their most important function is to emphasize repeatedly that chemistry is an experimental science. All that we know about the material universe comes from observations drawn from experiments no different in spirit from these opening demonstrations. Our attempts to explain observations drawn from these and more complex and sophisticated experiments lead us to a better, more complete understanding of the universe itself, as we'll see in the Perspective that concludes this chapter.

Energy is the ability to do work.

Several major themes occur repeatedly throughout this study. One is that chemical reactions provide us with the energy that drives our society and that provides fuel for our bodies, through processes ranging from the combustion of the hydrocarbons that make up gasoline to the oxidation of the carbohydrates, fats, and proteins of our foods. **Energy** itself is simply the ability to do work. Without the power of chemistry, expressed literally as the energy that these chemical reactions release, both our lives and our society (at least as we now know it) would come to an end.

Chemicals give us energy, form the clothes and other materials of our everyday lives, and provide the substance of our physical bodies.

Another theme is that chemicals provide us with the bulk, physical structures of our everyday lives, from the plastics that form our consumer goods and that wrap them in convenient packages to the substances that form, shape, and wrap our own bodies: our bones, organs, and skin. Every bit of whatever material substance we see, touch, or use is made of chemicals, and so are we.

A third theme reveals the importance and the structure of chemical particles themselves, from the smallest ions, atoms, and molecules to the largest and most complex molecules of our bodies and of the world about us. Water, a simple molecule consisting of one oxygen and two hydrogen atoms, forms more than half the weight of our bodies; no living thing can exist without it. Molecular oxygen, a union of two small oxygen atoms, provides life to the entire animal kingdom. Atmospheric ozone, with a molecular structure consisting of three oxygen atoms, shields and protects living things from deadly doses of ultraviolet solar radiation. At the other end of this spectrum are the huge, intricate, precisely designed molecules of life, including the DNA that provides us with our genetic heritage.

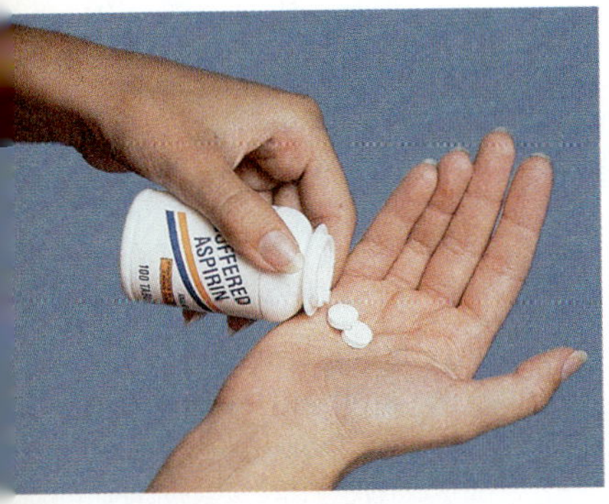

Aspirin, in recommended doses, can relieve pain, swelling, and fever. In overdoses it can cause nausea, bleeding, even death.

Our final theme is that chemicals are good or bad, beneficial or harmful, only in the ways we use them. Chemicals can cause illness and death, and yet, as judiciously chosen food additives for example, they can also protect us against microorganisms that cause illness and death. The chemicals of medications and pharmaceuticals can alleviate pain and suffering and cure illnesses, but used incorrectly or excessively they can produce the very suffering they were designed to relieve. The sturdy and durable synthetic plastics that have long and useful lives in consumer products can become persistent, degradation–resistant components of our trash. And the by-products of the chemical reactions that provide energy and materials to today's world can themselves become pollutants that foul our air, land, and water, as we'll see in Chapter 14. Our wise use of chemicals depends both on our own good judgment and on our clear understanding of what these chemicals of our world are and what they can do both for us and to us. In a real and broader sense the study of chemistry is the study of our society itself. Examining either without an understanding of the other leaves a void of ignorance about the modern world.

■ QUESTION

Describe a circumstance in which water, H_2O, is (a) a life-sustaining or a life-saving substance; (b) a hazardous or life-threatening substance.

1.3 Ions: Electricity in Motion and at Rest

Now, with our understanding of what chemistry is and how we are going to examine it, we can return to our investigation of table salt and table sugar and learn how our observations of their properties help us understand them and the world about us.

We've seen that the light bulb of Figure 1.3 glows as electricity passes through its filament. An electric current moves through the wires of the circuit in the form of electrons, which are extremely small particles that carry a negative electrical charge.

Since the bulb remains dark when a beaker of pure water is put into the circuit, we can conclude that water itself doesn't conduct electricity very well. Actually, we rarely find pure water in the world around us. Most samples of water we encounter contain dissolved salts, much like the table salt of Figure 1.3, and therefore most of the water we find in our everyday lives does conduct electricity. Dissolved salts are partly responsible for the common observation that (impure) water often conducts electricity, and sometimes very well indeed.

Sucrose doesn't change water's ability to conduct electricity, but sodium chloride does. In fact, a solution of sodium chloride in water (or molten sodium chloride, at a very high temperature) is very effective at conducting an electric current. Clearly, adding the sodium chloride introduces something into the water that allows electricity to flow from one wire to the other. Adding sucrose does not.

The simplest explanation for all this and other, related observations is that sodium chloride is made up of electrically charged particles that can move about in water and can transport an electric current through water much as electrons transport current through the wires of an electrical circuit. On the other hand, we have to conclude that sucrose is *not* made up of electrically charged particles (Figure 1.5).

These small particles of sodium chloride and other electrolytes, each bearing a negative or positive electric charge, are called **ions**. As we saw in Section 1.1, *an ion is an atom or a group of atoms that carries an electrical charge.* To understand more fully what ions are—and how it is that sodium chloride can carry an electric current although sucrose cannot—we have to understand what atoms are (and, later, what molecules are). We'll have more to say about atoms and ions in Chapter 2.

An **ion** is an atom or a group of atoms that carries an electrical charge.

Figure 1.5 The effect of sucrose and of sodium chloride on water's ability to conduct an electric current.

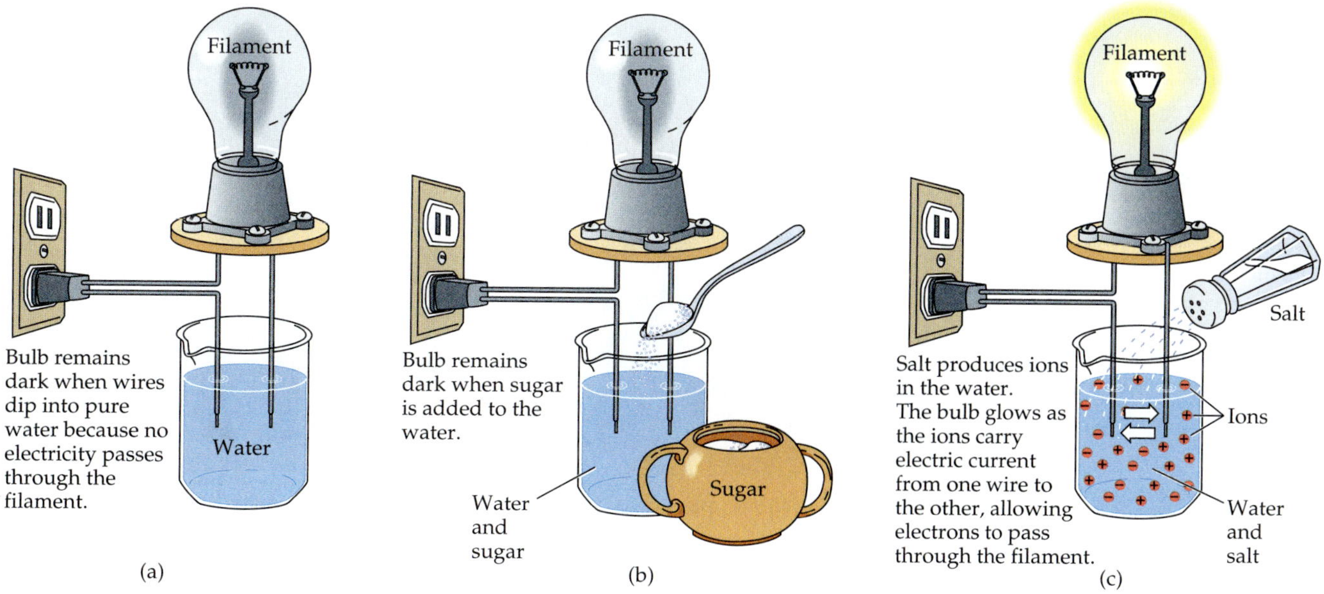

An **anion** is a negatively charged ion; a **cation** is a positively charged ion.

Ion is a term derived from a Greek word meaning "to go." In 1834 the English physicist Michael Faraday used the word to describe chemical particles that move to one electrical pole or another. He divided ions into two categories: **anions** are the negatively charged chemical particles that move to the positive electrical pole (the anode); **cations** are the positively charged chemical particles that move to the negative electrical pole (the cathode). We still use the same definitions today, but now we recognize that ions can remain at rest, too. The sodium chloride of our crystalline table salt, for example, is made up entirely of ions that remain quietly in the salt shaker.

An **ionic bond** is a chemical bond resulting from the mutual attraction of oppositely charged ions.

Sodium chloride, we now know, is composed of sodium cations and chloride anions held together in crystalline table salt by the mutual attraction of opposite electrical charges. The positively charged sodium cations and the negatively charged chloride anions are held close to each other through the action of **ionic bonds**, whose strength comes from this mutual attraction of opposite electrical charges. Together, the two *elements*, sodium and chlorine, form the *compound* sodium chloride.

■ QUESTION

Would you expect to find many ions in pure water? Explain.

1.4 Elements and Compounds

Elements are the fundamental substances of chemistry, composed of the atoms described in Section 1.1. We'll define what an element is in Chapter 2, after we learn more about the structure of atoms. For the moment we'll describe what elements are and how they behave.

Today we recognize the existence of 113 different elements, ranging from the common or well-known calcium, carbon, chlorine, gold, helium, hydrogen, iron, nitrogen, oxygen, phosphorus, sodium, sulfur, uranium, and zinc, to uncommon examples such as astatine, hafnium, and ununquadium. Neither these nor any other

The elements gold and silver.

Elements mercury, copper, and carbon.

Sodium chloride, hydrogen peroxide, water, and sucrose are some of the more common compounds of our everyday world.

element can be decomposed or converted to a simpler substance by any form of energy we deal with in our everyday lives. Neither heat, light, electricity, sound, magnetism, nor any other common form of energy, no matter how intense, can produce either of these changes in an element. Nor can one element be transformed into another element, a process known as *transmutation*, except under conditions far removed from our common experience, such as at extraordinarily high temperatures, close to those found at the surface of the sun, and in some other unusual ways as well. We'll examine some of these in Chapter 4.

In combination with other elements they (at least the more common elements) form our water, air, food, clothing, homes, automobiles, medicines, and our own bodies. Table 1.1 describes some of the elements that make up the human body.

By combining with each other in precise, well–defined ratios, two or more elements can form a pure substance known as a **compound**. The table salt, sugar,

A **compound** is a pure substance formed by the chemical combination of two or more different elements in a specific ratio.

Table 1.1 Representative Elements of the Human Body

Element	Grams in the Body of a 60-kg (132-lb) Person (Approximate)	Bodily Location and Function	Dietary Source
Calcium	1300	More that 99% of the body's calcium is in the bones and teeth	Milk and milk products including cheese and ice cream
Chlorine	90	As chloride ion, it is the principal cellular anion; when combined with hydrogen ion, it forms the hydrochloric acid of gastric juices	Sodium chloride (table salt)
Cobalt	Trace	Component of vitamin B_{12}	Widely distributed (as a component of vitamin B_{12}) in meat, especially liver, kidney, and heart; also in clams, oysters, milk, and milk products
Copper	Trace	Component of many enzymes	Widely distributed, especially in nuts, shellfish, kidneys, and liver
Iodine	Trace	Thyroid; necessary for normal functioning of the thyroid	Seafood, iodized table salt
Iron	2	Red blood cells; a component of hemoglobin	Beef, liver, dried fruits, whole-grain and enriched cereal products; and egg yolk
Magnesium	20	Second most abundant cation in body cells (after potassium); over half the body's magnesium is in the bones	Green, leafy vegetables
Phosphorus	690	Component of ATP, a cellular energy-releasing agent	Meat, eggs, milk, and milk products
Potassium	200	Most abundant cation in body cells; regulates water balance in cells	Widely distributed in a large variety of foods
Sodium	60	Most abundant cation in body fluids outside cells; regulates water balance in the body	Sodium chloride; widely distributed in a large variety of foods
Zinc	Trace	Occurs in bones and many enzymes	Seafood, especially oysters; meat, liver, eggs, milk, and whole-grain products

and water of our investigation with the light bulb are examples of compounds. Table salt (sodium chloride) is a pure substance formed from a combination of the two elements sodium and chlorine, with both elements present in a specific, well–defined, fixed ratio to each other. Table sugar (sucrose) is a pure substance formed from a combination of the three elements carbon, hydrogen, and oxygen in a distinct ratio. Water forms when hydrogen and oxygen combine, again in a fixed proportion. Hydrogen peroxide, a bleaching agent and an antiseptic, also consists of a combination of hydrogen and oxygen in a specific ratio, but one different from the ratio of these two elements in water.

Any compound can be decomposed into its individual elements, some by the action of heat or light, some by the effect of an electric current, and some through reaction with still another compound or with an element. Water, for example, decomposes into hydrogen and oxygen when an electric current passes through it. Heating sucrose decomposes it into another compound (water) and the element carbon.

It's worth noting that when elements react with one another the ratio of their weights determines the nature of the product. For example, if exactly 23.0 g of sodium and 35.5 g of chlorine are allowed to react with each other, they form 58.5 g of pure sodium chloride, with neither sodium nor chlorine left over. As long as this weight ratio (23.0/35.5) is maintained, only pure sodium chloride results. But if either sodium or chlorine is present in excess, the product is a mixture of sodium chloride and the element that's in excess. We'll see why this is so in Chapter 9.

For an illustration of the importance of the ratios of reacting substances, consider the following example.

EXAMPLE

Reacting in Ratios

Suppose we allow 10.0 g of sodium to react with an equal weight of chlorine. What is the composition of the product?

■ We know that a ratio of 23.0 g of sodium to 35.5 g of chlorine produces pure sodium chloride and that if either sodium or chlorine is present in excess, the product is a mixture of sodium chloride and the element that is present in excess. Since 35.5 g of chlorine reacts with a *smaller* weight of sodium (23.0 g) to produce sodium chloride, it's clear that with equal weights of the two there's an excess of sodium. Our problem, then, is to calculate just how much excess sodium is present. To find this value we multiply the 10.0 g of chlorine we're given by the ratio 23.0 g sodium/35.5 g chlorine:

$$10.0\text{ g chlorine} \times \frac{23.0\text{ g sodium}}{35.5\text{ g chlorine}} = 6.5\text{ g sodium}$$

This means that to maintain the ratio of 23.0 g sodium to 35.5 g chlorine, 6.5 g of sodium must react with the 10.0 g of chlorine provided in this illustration. The reaction consumes all of the chlorine present (10.0 g) and 6.5 g of the original 10.0 g of sodium to produce 16.5 g of sodium chloride (from the 10.0 g of chlorine and 6.5 g of sodium), with

10.0 g	sodium present originally
−6.5 g	sodium reacting with chlorine
3.5 g	sodium remaining

The product, then, is composed of 16.5 g sodium chloride (from the combination of 10.0 g chlorine and 6.5 g sodium) and the remaining 3.5 g of sodium. (You might want to consult the appendices at the end of the book for help with this example and others throughout the text.)

■ QUESTION

What results from the reaction of 15.0 g of chlorine with 5.0 g of sodium?

1.5 Light Bulbs, Salt, and Sugar Revisited

Let's pause here to review what we've just done and seen. We began our study of chemistry with the observation that pure water is a very poor conductor of electricity. We've seen that dissolving *table sugar* in water doesn't improve its ability to conduct an electric current but that dissolving *table salt* in water does increase this ability, quite dramatically, as demonstrated by the light bulb of Figure 1.3.

Looking back on the opening demonstration we can recognize that we used table salt, table sugar, some water, and a flashlight to answer a sequence of questions:

- Does ordinary *tap water* conduct an electric current?
- Does adding *table sugar* to water produce any change in water's ability to conduct an electric current?
- Does adding *table salt* to water produce any change in water's ability to conduct an electric current?

We found the answers to these questions by observing the results of a few experiments. To help organize these results we set up two categories of substances:

- *electrolytes*, which (like table salt) *do* improve water's ability to conduct an electric current, and
- *nonelectrolytes*, which (like table sugar) *don't* improve water's ability to conduct an electric current.

"Electrolytes" and "nonelectrolytes" are simply terms that help us *organize* some of the material we used and the results we observed. In order to *interpret* our observations, to *explain* how and why table salt acts as an electrolyte, we went a step further. More exactly, people much like us who made similar observations long ago—Michael Faraday, for example—went further. They gained new insight into how the universe is constructed and how it operates by reasoning (from their own observations and those of others) that when we dissolve an electrolyte in water it must produce particles that can move through the water, and that can carry electricity from one place to another. In our opening demonstration these particles carried electricity between the bottom of the battery and the side of the flashlight.

To reflect the ability of these particles to move about in water and other substances these scientists turned to the Greek language and called them *ions*. Since ions can transport electricity from one place to another they must themselves carry electrical charges. Thus, we now have *anions*, the ions that carry negative electrical charges, and *cations*, the positively charged ions.

In reviewing our opening demonstration and its implications, we have just gotten our first glimpse of the *scientific method*, the process by which *science* works. We'll examine just what science is and how it operates in a bit more detail as we close this opening chapter.

■ QUESTION

How would you go about determining whether an aspirin tablet contains any electrolytes among its ingredients?

PERSPECTIVE

Science—Understanding the Universe

Science itself is a way of knowing and understanding the universe we live in. (The word *science* comes to us from the Latin *scire*, to know.) Science often operates by

- Asking questions of the universe by means of experiments and similar tests of the physical world (just as we asked, implicitly, "Does pure water conduct electricity?" when we carried out the investigation of Fig. 1.2).
- Observing the way these questions are answered (as we observed that the light bulb does not glow when the connection is made by water alone).
- Asking additional questions that are generated by the answers to our earlier question (as we asked, again implicitly, "Does adding sucrose or sodium chloride to the water improve its electrical conductivity?").
- Interpreting the answers to our questions—what we have observed—to help us understand how the universe operates and increase our understanding of it (as we inferred from our observations of the light bulb that adding sodium chloride to the water introduces something into the water, something we call "ions," that allows electricity to flow from one point to another, but that adding sucrose does not introduce these ions).
- Communicating our observations and interpretations to others so that they can examine what we have done, repeat and confirm our own observations, perhaps suggest alternative explanations for what we have observed, ask questions of their own based on our observations and interpretations, and thereby continue the entire process.

This general procedure provides us with a never-ending supply of questions to be asked and results in a continuously refined interpretation of the universe we live in. Its fruits are the hypotheses and theories we use to explain our world. A *hypothesis* is a shrewd but tentative explanation of a relatively small set of observations; a *theory* is more firmly grounded interpretation, based on a larger set of confirmed observations and generally accepted by a large number of people. It's important to understand, though, that even a universally accepted theory supported by a great number and variety of observations may have to be modified, revised, or even abandoned completely as the result of even a single new observation. In this sense the ultimate value of science is that it allows us to explain what we continue to observe and experience in the world about us, through sets of descriptions (which we may call theories) that are relatively simple, that are consistent with one another, that

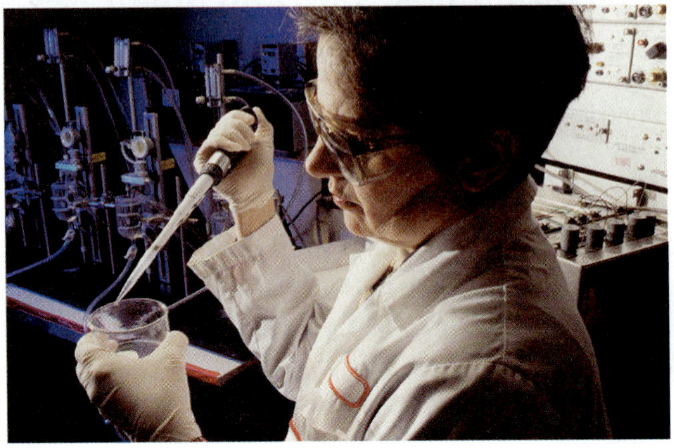

Science: A scientist prepares to investigate fundamental properties of a chemical in solution.

cover large numbers of observations and experiments, and that are generally accepted by most of us.

Taken as a whole, this method of learning is known as the **scientific method**. The specific steps of the scientific method may vary from time to time and from investigation to investigation, but they always involve asking a question of the universe, determining the answer through an experiment or other test, and then using the results of the experiment to refine our knowledge of the universe, often with the generation of still other questions. The quality of the investigation and of our resulting understanding of the universe depends on the cleverness of the questions we ask, the skill with which we carry out the experiments, and our ability to convert the results of the experiments into an ever more sophisticated understanding of ourselves and all that surrounds us.

As important as it is to understand what science is and how it operates, it's equally important to understand the difference between science and *technology*. Although the two terms are often used together, as "science and technology," and sometimes (incorrectly) interchangeably, they refer to two quite different activities. Science, as we've seen, is a way of understanding how the universe operates. **Technology** comes from a Greek word *tekhne*, meaning art, craft, or skill. While science gives us the ability to know and to understand the physical universe, technology provides the art, skill, or craft to apply the knowledge gained through science. Throughout this study of chemistry we'll see repeatedly that the knowledge

Technology: A cellular telephone brings the applications of science into our daily lives.

gained through science works to our benefit or to our detriment through the technology we use to put it to work. We'll see in Chapters 4 and 5, for example, that the knowledge of how atoms are constructed and how they behave can be applied through one form of technology to build powerful weapons of destruction or through another technology to detect and cure disease. ■

Science is a way of knowing and understanding the universe.

The **scientific method** is the process by which science operates.

Technology is the art, skill, or craft used to apply the knowledge provided by science.

EXERCISES

For Review

1. Following are a statement containing a number of blanks and a list of words and phrases. The number of words equals the number of blanks within the statement, and all but two of the words fit correctly into these blanks. Fill in the blanks of the statement with those words that do fit, then complete the statement by filling in the two remaining blanks with correct words (not in the list) in place of the two words that don't fit.

> __________ are substances that conduct electricity when dissolved in __________ or when __________. They consist of __________ (__________ charged ions) and __________ (positively charged ions). Common table salt, __________, is an example of an electrolyte, while table sugar, __________, is a __________. That is, table sugar does not __________ when dissolved in water. We conclude that table sugar is not composed of __________.

anions	cations	chemistry
compound	electrolytes	ions
melted	nonelectrolyte	sodium chloride
sucrose	water	

2. Complete this statement using the same conditions as described for the preceding exercise.

> __________, which is the study of the composition and properties of matter, and of the changes that it undergoes, is a branch of __________, which itself provides us with a way of knowing and understanding the universe we live in. In the operation of the __________ we ask questions of the universe through tests and __________. By observing the results that we get, we can formulate additional questions, perform additional experiments, and finally develop a tentative explanation of what we have learned. If this tentative explanation or __________ is confirmed by others and becomes widely accepted it becomes a __________ and helps us understand better the world about us.

chemistry	experiments	hypothesis
science	scientific method	theory

3. Match each item in group A with one in group B.

Group A	Group B
______ a. chloride anion	1. assembly of atoms
______ b. compound	2. chemical particle that carries either a positive or a negative electrical charge
______ c. electrolyte	3. conducts electricity when it is dissolved in water or when it is molten
______ d. ion	4. element
______ e. ionic bond	5. negatively charged ion
______ f. molecule	6. positively charged ion
______ g. sodium cation	7. pure substance formed by combination of two or more elements in a specific ratio
______ h. sulfur	8. results from the attraction of oppositely charged ions

4. What characteristic determines whether or not a small particle is be classified as (a) an *ion*, (b) a *cation*, (c) an *anion*?

5. What characteristic determines whether or not a substance can be classified as an *electrolyte* or a *nonelectrolyte*?

6. What is the distinction between an *element* and a *compound*?

7. a. Name three elements.

b. Name three different compounds, each of which contains at least one of these elements.

8. What kind of chemical bond holds the sodium cations and chloride anions in place, close to each other, in a crystal of table salt?

9. Name five elements found in the human body, describe their functions in the body, and identify a good dietary source for each of these elements.

10. In your own words, describe briefly the steps of the scientific method.

11. What is the difference between a *hypothesis* and a *theory*?

12. What is the difference between *science* and *technology*?

13. In what way are water and hydrogen peroxide similar in chemical composition? In what way are they different? (See Section 1.4.)

A Little Arithmetic and Other Quantitative Puzzles

14. Suppose you carry out a large number of tests as follows. In each experiment you use 10.0 g of sodium. In the first experiment you allow the 10 g of sodium to react with 0.1 g of chlorine; in the second you allow 10 g of sodium to react with 0.2 g of chlorine; in the third, with 0.3 g of chlorine; and so forth. In each subsequent test you allow 10 g of sodium to react with an additional 0.1 g of chlorine until the series ends with 10 g of sodium and 20.0 g of chlorine. Describe qualitatively how the product(s) of the individual reactions change(s) as the series of experiments progresses from 0.1 g of chlorine to 20.0 g of chlorine.

15. How much chlorine would you have to use if you wanted 10.0 g of sodium to react with it to produce pure sodium chloride, with neither excess sodium nor excess chlorine left over?

16. Water is composed of hydrogen and oxygen in the ratio 1 g hydrogen to 8 g oxygen. Hydrogen peroxide is composed of hydrogen and oxygen in the ratio 1 g hydrogen to 16 g oxygen (Section 1.4). What would result from the reaction of a mixture of 1 g hydrogen and 12 g oxygen?

Think, Speculate, Reflect, and Ponder

17. Which of the following are suitable subjects for the science of chemistry? (a) determining the composition of rocks; (b) analyzing the effects of a day's experiences on the nature of dreams; (c) designing a low-calorie sweetener that is also an electrolyte; (d) preparing a plastic that decomposes readily to environmentally harmless products; (e) defining the difference between music and random noise; (f) determining how the body converts food into skin, hair, and other body tissue; (g) looking for a correlation between hyperactivity in children and seasonal changes in daylight.

18. Suppose you discovered a new substance and weren't sure whether you should classify it as a new element or a new compound. How would you go about determining whether you had discovered a new element or a new compound?

19. The following summarizes the results we described for the light bulb, water, sodium chloride, and sucrose early in this chapter:

Condition of the Light Bulb	
pure water only	dim or dark
water and sucrose	dim or dark
water and sodium chloride	bright

What would you have concluded for each of the following results:

A. Condition of the Light Bulb	
pure water only	dim or dark
water and sucrose	bright
water and sodium chloride	dim or dark

B. Condition of the Light Bulb	
pure water only	dim or dark
water and sucrose	bright
water and sodium chloride	bright

C. Condition of the Light Bulb	
pure water only	bright
water and sucrose	bright
water and sodium chloride	bright

20. Why was it necessary to test the electrical conductivity of pure water before adding sodium chloride and sucrose?

21. Suppose water itself were a very good conductor of electricity. How might you then determine whether table sugar, table salt, or any other substance is an electrolyte or a nonelectrolyte?

22. Would you expect seawater to be a good conductor of electricity? Explain.

23. Is it possible to determine whether *air* is an electrolyte? If your answer is *yes*, explain how you might do it; if *no* explain why not.

24. Several centuries ago, it was generally believed that heavy objects fall faster than lighter objects, that the more an object weighs the faster it falls. According to one legend, the Italian scientist Galileo Galilei, who lived from 1564 to 1642, corrected this error by dropping two spheres of different weight simultaneously from a high point on the Leaning Tower of Pisa and demonstrating that they reach the ground at the same time. Describe what Galileo supposedly did in terms of the scientific method. What question did he ask of the universe? What test did he use to obtain an answer? What did he observe? How did he interpret this observation?

25. The old saying "A watched pot never boils" comes from the observation that a pot of water seems to take longer to come to a boil if you are watching it than if you aren't. Describe how you could use the scientific method to learn whether water really does take longer to boil if you are watching it. If water takes just as long to boil whether or not you are watching it, how could you use the scientific method to determine whether water actually does seem to take longer if you are, indeed, watching it, and just how much longer it seems to take? Describe what questions you would ask, how you would go about finding experimental tests that would give you answers to these questions, and how you would interpret the answers.

26. Describe another popular belief or perception, similar to the one about the watched pot or falling objects, that you could investigate by application of the scientific method.

27. Give an example of a benefit brought to us through the operation of *science*. Do the same for a benefit brought through the operation of *technology*.

28. Water can be considered to be a "good" chemical in the sense that we cannot live without it. Yet it is "bad" when floods destroy property or when someone drowns. Similarly, aspirin is "good" in the sense that it relieves headaches, yet it is "bad" in that children have mistaken it for candy, taken overdoses, and died.

a. Give examples of two other chemicals that are ordinarily beneficial yet can produce undesirable, even deadly effects.

b. Give examples of two chemicals that are ordinarily thought of as hazardous or harmful, yet in the right circumstances can be beneficial.

Web Questions

Use the Web to answer or accomplish each of the following. Report your Web source for each response.

1. Describe a scientific contribution made by Michael Faraday, in addition to his work with ions and electrolytes.

2. The terms "chemistry," "energy," "matter," "technology," and "science" are all defined in this chapter. Present alternative definitions derived from Web sources.

3. Show how application of the steps of the scientific method led to a scientific discovery made: (a) during the 19th century; (b) during the past 10 years.

4. Illustrate each of the themes discussed in Section 1.2 with examples obtained from the Web.

5. Find when each of the elements of Table 1.1 was discovered and who made the discovery. If the date or the discoverer is unknown, describe early sources and uses for the element.

Atoms and Elements

THE BUILDING BLOCKS OF CHEMISTRY

Bars and nuggets of gold, an element.

Atoms and Paper Clips

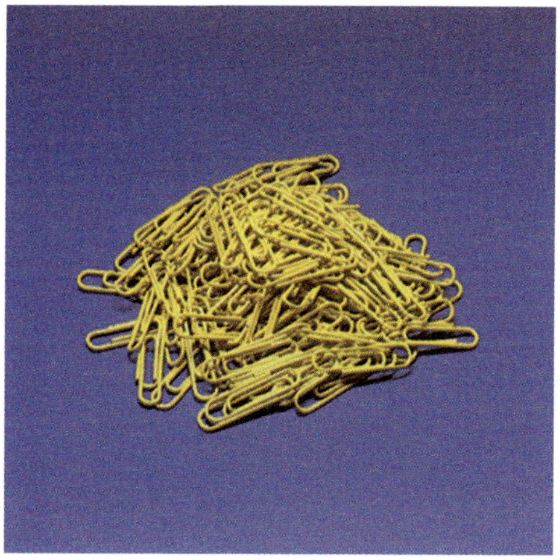

To gain some insight into the modern view of an atom, try this demonstration. It draws an analogy between a pile of paper clips and a small amount of one of the 113 known elements, a bar of gold for example. Although it's only a simple analogy, one that can't be stretched very far, it does illustrate an important point about atoms.

Place a pile of 15 to 20 paper clips on a level surface and imagine a bar of pure gold sitting next to them (Fig. 2.1). Now divide the pile of paper clips roughly in half. Subdivide one of the new, smaller piles in half and repeat the process again and again until you are down to a "pile" that consists of a single paper clip. In our analogy, that single paper clip represents an "atom" of paper clips, the smallest part of the original pile of paper clips that you can still identify as a paper clip. In dividing and subdividing the original pile of paper clips into ever smaller piles, you finally came to the smallest part of the original pile that you can still identify as a paper clip. By analogy, that simple paper clip represents an "atom."

Now imagine that you perform the same operation with the bar of gold. Picture yourself dividing the (imaginary) bar in half again and again, as you did with the paper clips, until you finally reach the smallest particle of the bar that you can identify as the element gold. Like the single paper clip you eventually reached, that smallest particle—the smallest one you can still identify as gold—is an atom of gold. This time, though, it's an authentic atom.

We can't carry this analogy very far. The paper clip that remains has all the properties we expect of a paper clip, but even if we could see the atom of gold, which has a

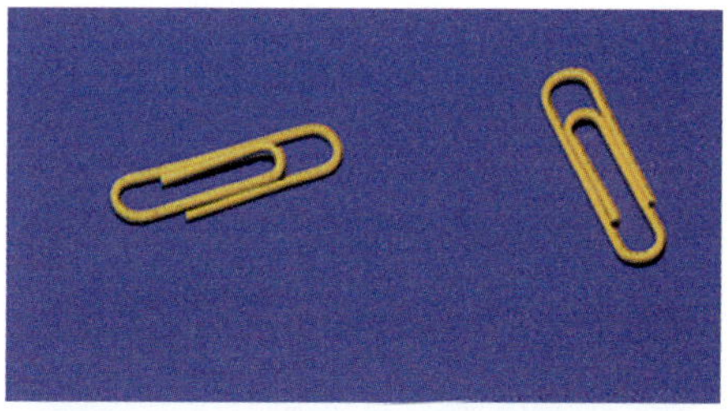

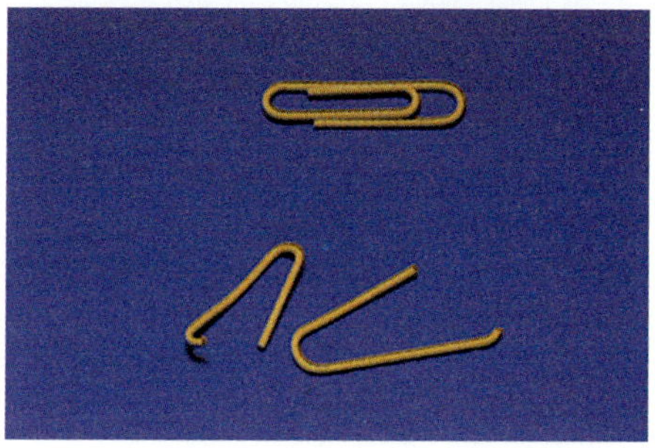

Piles of paper clips, individual paper clips, and fragments of a paper clip. The fragments no longer represent a paper clip.

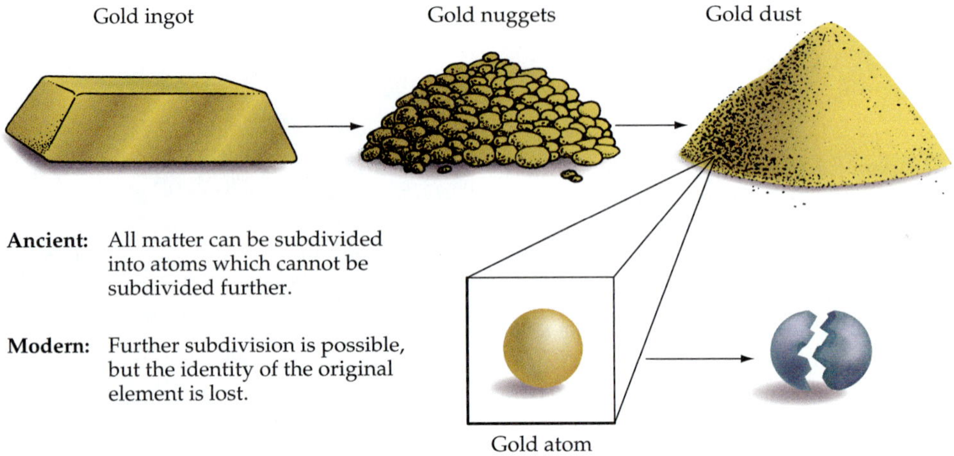

Figure 2.1 In a parallel to the demonstration with the paper clips, repeatedly subdividing a piece of gold produces smaller and smaller groups of atoms. Dividing a single gold atom into two parts produces fragments that no longer represent the element gold.

diameter about 1/500,000,000th the length of a dollar bill, as we saw in Figure 1.4, we wouldn't expect it to have many of the properties of the shiny, yellow metal we started with. Nonetheless, as we'll see in Section 2.5, you would still be able to characterize this ultimate particle of the bar as a bit of the element gold.

Now for the final part of the analogy. Suppose you cut the paper clip in half or twist it and break it into two pieces, as in Figure 2.1. Each of these pieces might have some sort of use. Perhaps you can even devise sensible names for them. Whatever their value, though, and whatever names you might assign to them, it's impossible to identify either of the fragments as a paper clip.

Similarly, we can divide a portion of an element into smaller and smaller pieces until we reach a single atom of that particular element. As with the paper clip, we can split an atom into fragments, but whatever the uses and properties of these fragments and whatever names we may give them, we cannot identify any of them as the element from which the atom originally came. ■

2.1 Democritus and a Bar of Gold

The ancient Greeks, who gave us the word *ion* for a moving electrical particle, also gave us the word for the structure that is both the smallest particle of an element and the fundamental particle of chemistry, the *atom*. The word comes from the Greek *atomos*, meaning *indivisible* or an *indivisible particle*. One of the Greek philosophers of antiquity, Democritus held that infinitely small, indivisible, and eternal particles constitute the essence of all matter and give substances their particular properties. Democritus and his followers believed that any particular piece of matter can be divided and subdivided down to its ultimate particles, its *atomos*, and no further.

Both philosophy and science have advanced quite a bit since the time of Democritus. Although we now know that atoms have extremely small but finite sizes and can indeed be divided or split, Democritus was nonetheless right in one limited but very real sense: Chemists now recognize atoms as the smallest particles of the 113 known elements that make up our entire universe. We can subdivide any quantity of any element as much as we wish and still have that unique element, until we reach the atoms that compose it. Once we split an atom, we can no longer identify the element from which it came. An **atom**, then, is *the smallest electrically neutral particle of an element that we can identify as that element.*

An **atom** is the smallest electrically neutral particle of an element that can be identified as that element.

2.2 The Size and Abundance of Atoms

Atoms themselves seem almost as small as Democritus believed them to be. They are far too small to be seen with even the most powerful optical microscope, although we can produce images of them with other kinds of microscopes. A gold atom, for example, is about 3×10^{-8} cm in diameter and has a mass of about 3.3×10^{-22} g. (For an explanation of this form of writing numbers, called *exponential notation* or *scientific notation*, please turn to Appendix A, A Short Guide to Exponential Notation. Generally we can use *mass* and *weight* interchangeably. In the next section we'll see how the two terms differ.)

In decimal notation, a gold atom has a diameter of about 0.00000003 cm and has a mass of about 0.00000000000000000000033 g. Compared with atoms of other elements, the gold atom is about average. Other atoms range in size from those of helium, a little over half the diameter of a gold atom, to atoms even larger than those of cesium, which has almost twice gold's diameter (Fig. 2.2).

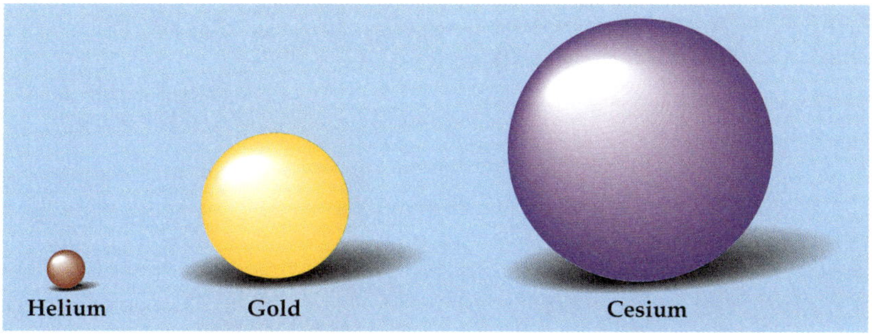

Figure 2.2 Relative sizes of helium, gold, and cesium atoms.

The following example gives us a sense of just how small an atom is.

EXAMPLE

Bill of Gold

What is the mass of a sheet of gold atoms that's the size of a dollar bill and just one atom deep?

■ A dollar bill is 15.7 cm long and 6.6 cm wide. If, as we've already seen, it takes about half a billion spherical gold atoms to stretch across the length of a dollar bill, it takes a proportionally smaller number of atoms to run along its width:

$$500{,}000{,}000 \text{ gold atoms} \times \frac{6.6 \text{ cm (the bill's width)}}{15.7 \text{ cm (the bill's length)}}$$
$$= 210{,}000{,}000 \text{ gold atoms along the width}$$

If we assume that the atoms are lined up on the dollar bill in 210,000,000 rows (or 2.1×10^8 rows) of 500,000,000 (or 5.0×10^8) atoms each, then we can calculate that there are

$$2.1 \times 10^8 \text{ rows} \times \frac{5.0 \times 10^8 \text{ atoms of gold}}{\text{row}} = 10.5 \times 10^{16} \text{ atoms of gold}$$

present in the rectangle. Since, as we've just seen, each gold atom weighs 3.3×10^{-22} g, the entire rectangle of gold weighs

$$10.5 \times 10^{16} \text{ atoms} \times \frac{3.3 \times 10^{-22} \text{ g}}{\text{atom}} = 35 \times 10^{-6} \text{ g, or } 0.000035 \text{ g}$$

For comparison, a dollar bill itself weighs very nearly 1.0 g.

Clusters of gold atoms organized into a map of the Western Hemisphere. The images are seen through a scanning tunneling microscope (STM).

As for the abundance of the various elements, in terms of the actual numbers of atoms present, hydrogen is by far the most plentiful element both in the universe (Fig. 2.3) and in our bodies (Fig. 2.4). The four elements hydrogen, oxygen, carbon, and nitrogen make up 96% of our weight and over 99% of all the atoms of our bodies (Fig. 2.5).

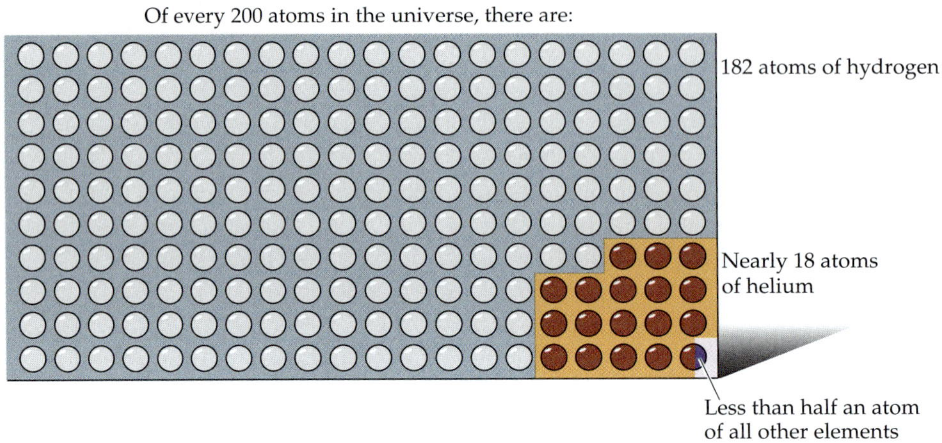

Figure 2.3 Distribution of the elements in the universe.

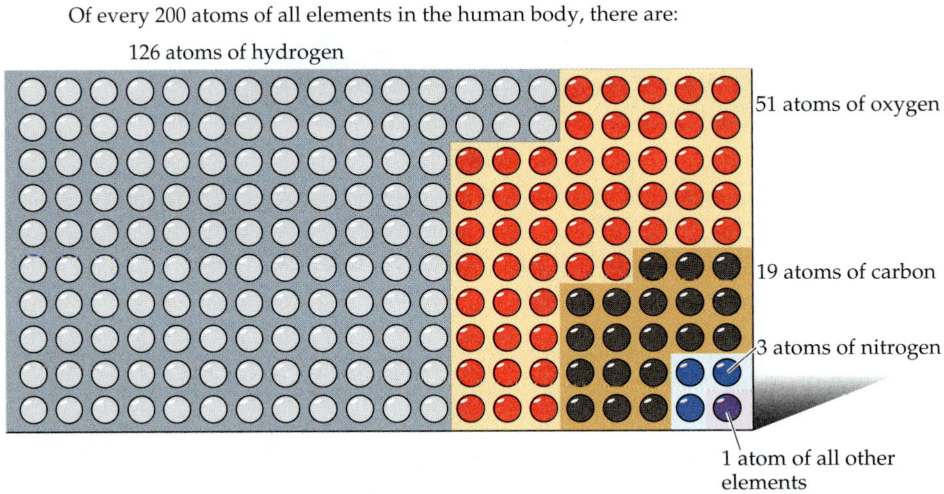

Figure 2.4 Distribution of the elements in the human body.

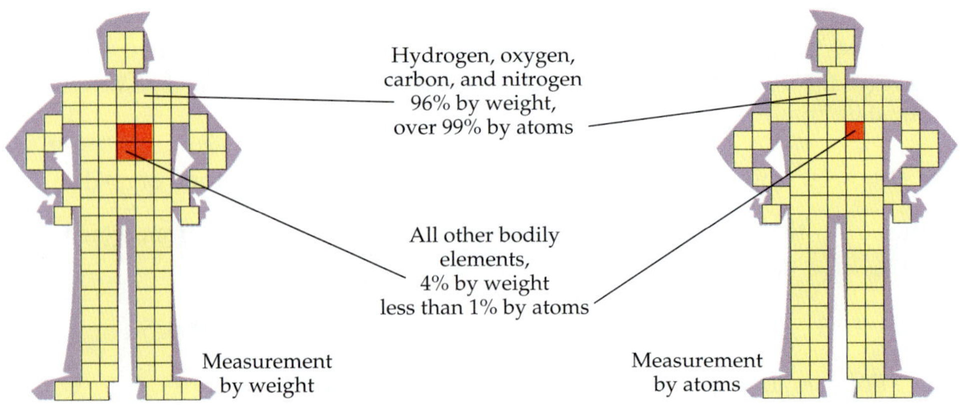

Figure 2.5 Composition of the human body.

■ QUESTION

In this Section, the mass of a gold atom is given as about 3.3×10^{-22} g. A more precise value is 3.27×10^{-22} g. Given this more precise value, how many gold atoms are there in 197 g of gold? You might find the discussions of Appendices A and C useful. The Example "One Gold Atom" of Section 9.3 reveals the significance of this exercise.

2.3 Mass and Weight

In describing the size of atoms and their abundance in our bodies, we have used two terms, mass and weight, that might seem to be completely interchangeable. They aren't, even though the difference between the two is virtually insignificant in the practical activities of our everyday world. To be precise, though, mass is a fundamental property of matter that is unaffected by its location; the **mass** of a body is defined as its resistance to acceleration no matter where in the universe it may be. Weight, on the other hand, results from the force of gravity. A body's **weight** depends on where it is, on what gravitational force is acting on it. (Appendix B, The Metric System, contains a discussion of units of mass and weight in the metric system and in the English system.)

Mass is observed as a body's resistance to acceleration.

Weight results from the pull of gravity.

A person who weighs 60 kg (132 lb) on earth, for example, would weigh only 10 kg (22 lb) on the moon, where the gravitational force is only one-sixth that of the earth, and would weigh nothing at all in an orbiting spaceship (Fig. 2.6). Yet that person's mass remains the same on the moon and in the spaceship as it is on earth. (*Mass*, a body's resistance to acceleration, is sometimes defined more simply as the *quantity of matter* present.)

Figure 2.6 Weight depends on the size of the gravitational force. Mass can be defined as a body's resistance to acceleration.

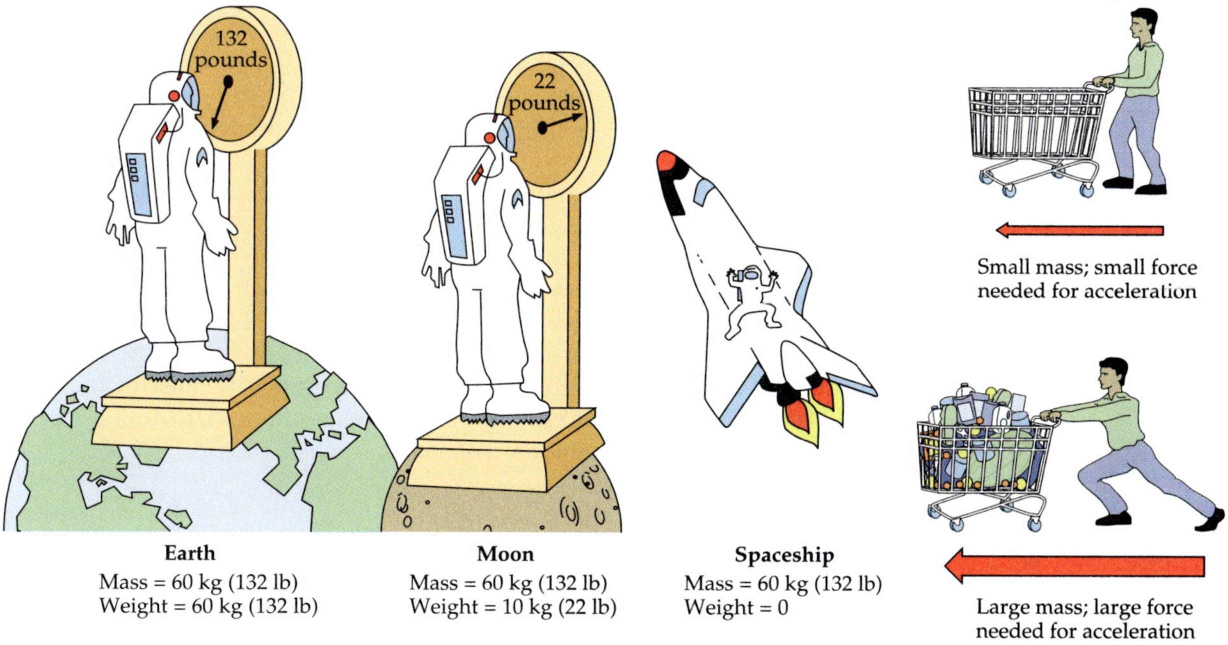

The astronauts have mass but no weight.

At any rate, since all the activities of our daily lives take place on or near the surface of the earth, with insignificant variations in the effects of gravity, we can think of mass and weight as equivalent to each other, even in chemistry. We speak of atomic *weights*, for example, and then define them in terms of *mass*, as we'll see in the next section.

■ **QUESTION**

Suppose you were an astronaut floating weightless at one end of an orbiting spaceship. An astronaut at the other end of the ship wants two screwdrivers you have in your hands. One of the tools has a mass of 50 g; the other has a mass of 100 g. (a) What does each of the screwdrivers weigh in the orbiting spaceship? (b) Would you have to push one with more force than you would use on the other to get them moving with the same speed toward the other astronaut or would you be able to use the same force on each?

2.4 Subatomic Particles: Protons, Neutrons, and Electrons

The **nucleus** is the positively charged central core of an atom.

A **proton** is a subatomic particle with a mass of 1 amu and a charge of 1+. A **neutron** is a subatomic particle with a mass of 1 amu and no electrical charge. An **electron** is a subatomic particle with negligible mass and a charge of 1−.

The **atomic mass unit**, or **amu**, is one-twelfth the mass of the most common kind of carbon atom.

Protons, neutrons, and electrons are the three subatomic particles that form an atom and determine its properties. Structurally, all atoms consist of two parts:

- the **nucleus**, a positively charged central core that holds protons and (except for the great majority of hydrogen atoms) neutrons, and
- surrounding shells of negatively charged electrons (Fig. 2.7).

Small as atoms are, the protons, neutrons, and electrons that compose them are even smaller. A **proton** has a mass of about 1.673×10^{-24} g and carries an electrical charge of 1+, a unit positive charge. With a mass of 1.675×10^{-24} g, a **neutron** has a slightly greater mass than a proton. Unlike the proton, the neutron carries no electrical charge. An **electron**'s mass is about 9.11×10^{-28} g, which we'll round off and write as 0.0009×10^{-24} g for a more direct comparison with the proton and neutron. The electron bears a charge of 1−, a unit negative charge. It's often convenient to use p or p^+ to represent a proton, n or n^0 to represent a neutron, and e or e^- for an electron.

Using these numerical values, expressed in grams, for the weights of subatomic particles and atoms soon becomes cumbersome. It's much more convenient to describe the masses of subatomic particles and of the various atoms by a unit called the **atomic mass unit**, or **amu**, which is defined as exactly one-twelfth the mass of the most common kind of carbon atom (Fig. 2.8). This gives the proton a mass of 1.007 amu and the neutron a mass of 1.009 amu. For most ordinary uses we can round both of these off to 1 amu. The mass of the electron (0.0005 amu) is so small in comparison with the other particles that we can consider it as zero. Table 2.1 summarizes the characteristics of these three subatomic particles.

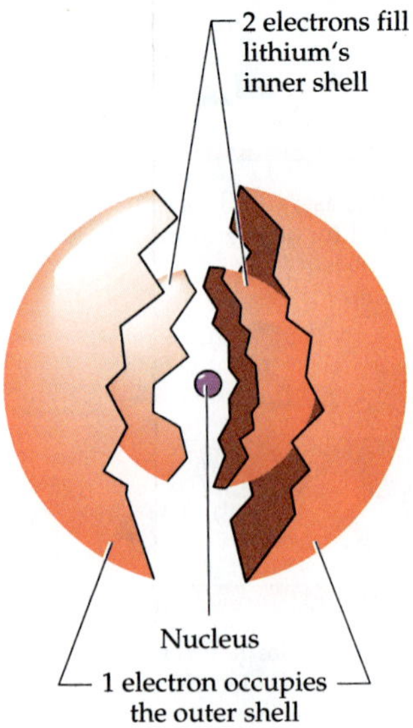

Figure 2.7 The nucleus and electron shells of a lithium atom, a typical small atom.

Figure 2.8 The atomic mass unit, amu.

Table 2.1 Subatomic Particles

	g	amu	Location in Atom	Charge	Symbol
Neutron	1.67×10^{-24}	1	Nucleus	0	n, n^0
Proton	1.67×10^{-24}	1	Nucleus	1+	p, p^+
Electron	0.0009×10^{-24}	0	Outside the nucleus	1−	e, e^-

■ QUESTION

(a) The nucleus of a hydrogen atom contains only a single proton. What is the mass of the hydrogen atom in atomic mass units? (b) The nucleus of a fluorine atom contains 9 protons and 10 neutrons. What is the mass of a fluorine atom in amu?

2.5 Atoms, from *A* to *Z*

Atoms of hydrogen are the simplest of all atoms and provide a good starting point for a study of atomic structure. As we saw in Section 2.2, hydrogen atoms are the most abundant of all atoms in the universe as a whole, as well as in our own bodies. They make up two-thirds of all atoms in water, but just over 11% of the water's weight. Through a process we'll describe near the end of Chapter 4, hydrogen atoms provide the energy of the sun.

The overwhelming majority of all hydrogen atoms consist of just one proton and one electron. The proton forms the nucleus of the atom, and the electron occupies a spherical shell or envelope surrounding the nucleus (Fig. 2.9). Since the single negative charge of hydrogen's lone electron exactly balances the single positive charge of its single proton, the hydrogen atom itself has no net electrical charge. This illustrates a general rule: *In all atoms of all elements, the number of electrons surrounding the nucleus exactly equals the number of protons within the nucleus.* All atoms of all elements, then, are electrically neutral.

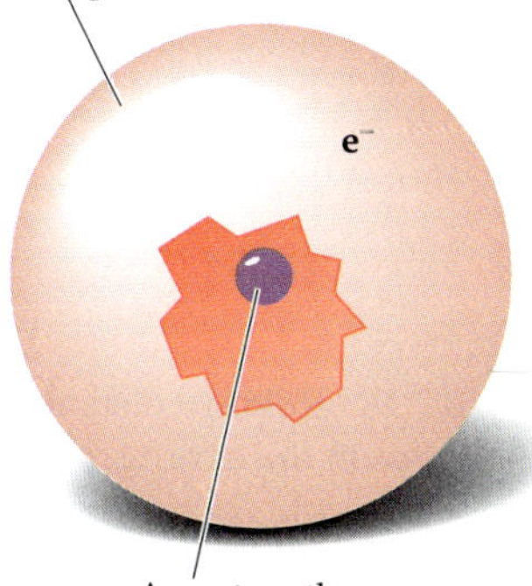

Figure 2.9 The structure of the hydrogen atom.

Because the mass of an electron is negligible in comparison with the mass of both a proton and a neutron, virtually the entire mass (99.95%) of an atom lies in its nucleus. Since both protons and neutrons have masses of very nearly

Filled with the flammable element hydrogen for buoyancy, the German airship *Hindenburg* burned in May 1937. The hydrogen ignited as the airship approached its dock in Lakehurst, New Jersey, after a transatlantic crossing. Thirty-six people died as a result.

1 amu, we need only add up all the protons and all the neutrons in an atomic nucleus to get a value for the mass of an atom. This sum of an atom's protons and neutrons is known as its **mass number**, represented by the symbol A. For a hydrogen atom with a nucleus consisting of a single proton, the mass number is 1. That is, $A = 1$.

The **mass number** of an atom is the sum of the protons and neutrons in its nucleus.

To gain some sense of the difference in mass between the proton and the electron and of the enormous distance between the two in the hydrogen atom, we can represent its nucleus, the single proton, as an adult of average weight sitting in an empty field. The atom's electron could then appear as a small bird, perhaps a common sparrow, flying around the person at a distance of 2 miles (Fig. 2.10).

In addition to mass numbers, one more atomic value is important to us: the **atomic number**, represented by the symbol Z. The atomic number Z is simply *the total number of protons in an atomic nucleus*. With one proton in its nucleus, hydrogen's atomic number is 1. For hydrogen, $Z = 1$. Atomic numbers are particularly important in chemistry because *all atoms of the same element have the same atomic number*. Conversely, *all atoms of any specific atomic number are atoms of the same element*. This gives us a convenient definition of an **element** as a substance whose atoms all have the same atomic number.

The **atomic number** is the sum of all of an atom's protons. An **element** is a substance whose atoms all have the same atomic number.

As a chemical symbol for hydrogen, we use the capital letter H, which can represent either the element itself or a single hydrogen atom, depending on the context. Chemically, then, H can represent a single proton surrounded by a spherical

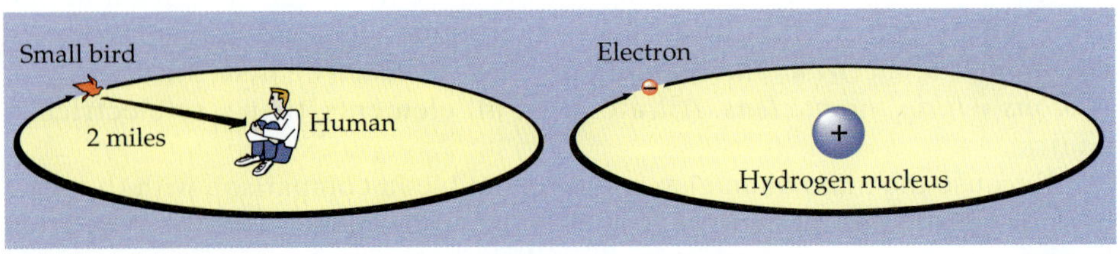

Figure 2.10 Relative distances and masses in the hydrogen atom.

shell containing a single electron as shown in Figure 2.9. If we wish to designate not only an atom of a particular element but its mass number and atomic number as well, we write the mass number (A) at the upper left of the atomic symbol and the atomic number (Z) at the lower left. For a hydrogen atom with a mass number of 1 and atomic number of 1, we write

$$\begin{array}{r} \text{mass number } (A)\searrow \\ \text{elemental symbol} \longrightarrow {}^{1}_{1}\text{H} \\ \text{atomic number } (Z)\nearrow \end{array}$$

With the definitions of mass number and atomic number in mind, we can easily determine the number of neutrons in an atomic nucleus by subtracting the atomic number from the mass number: $A - Z =$ the number of neutrons.

As an illustration, a common atom of the element fluorine has a mass number of 19 and an atomic number of 9. Thus its nucleus is made up of 9 protons and 10 neutrons ($19 - 9 = 10$). Because all atoms must be electrically neutral, we know that there are as many electrons surrounding the fluorine atom's nucleus as there are protons within its nucleus, or 9 of each. Moreover, since the chemical symbol for fluorine is F, we can represent the atom as

$${}^{19}_{9}\text{F}$$

EXAMPLE

How Many Neutrons?

How many neutrons are there in the nucleus of an aluminum atom of mass number 27? (The atomic number of aluminum is 13.)

■ With an atomic number of 13, $Z = 13$ for aluminum. Since the mass number of this particular atom of aluminum is 27, $A = 27$. Knowing that the number of neutrons $= A - Z$,

$$\text{neutrons} = A - Z$$

$$\text{neutrons} = 27 - 13$$

$$\text{neutrons} = 14$$

The nucleus of this aluminum atom contains 14 neutrons.

■ **QUESTION**

The most common kind of lithium atom, atomic number 3, has a mass number of 7. How many protons are there in the nucleus of this lithium atom? How many neutrons? How many electrons surround the nucleus? Given that the chemical symbol for lithium is Li, show A, Z, and the symbol as we did in the example of fluorine.

2.6 Isotopes: Deuterium and Tritium

As we've seen, the atomic nucleus can contain neutrons as well as protons. An atom with one proton and also a single neutron in its nucleus is an atom of hydrogen with a mass number of 2. It *must* be an atom of the element hydrogen because with one proton in its nucleus its atomic number is 1, and all atoms with an atomic number of 1 belong to the element hydrogen. But its mass number must be 2 since both the proton and the neutron contribute 1 amu each to the total mass.

Isotopes are atoms of the same element with different mass numbers.

Atoms that have the same atomic number (and therefore belong to the same element) but that differ in mass number are called **isotopes**. To state this a bit

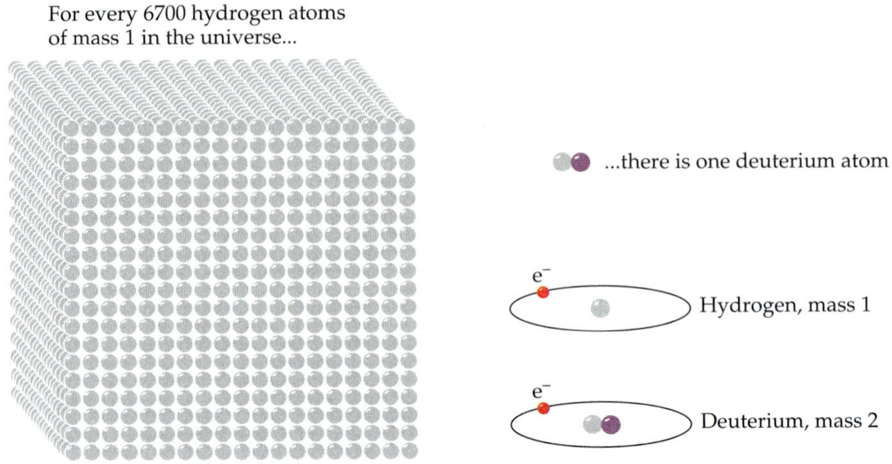

Figure 2.11 Relative abundance of deuterium.

differently, isotopes of any particular element all have the same number of protons (and therefore the same atomic number) but carry different numbers of neutrons in their nuclei (and therefore have different mass numbers). To help differentiate between these two isotopes of the element hydrogen, the isotope of mass number 2 is called *deuterium*. Deuterium is the so-called "heavy hydrogen" used in the construction of the hydrogen bomb (Chapter 4).

Naturally occurring deuterium is extremely rare; there's only about one deuterium atom for every 6700 hydrogen atoms of mass number 1 that occur in nature (Fig. 2.11). Because the ratio of the two isotopes overwhelmingly favors the atom of mass number 1, the word *hydrogen* commonly refers either to the naturally occurring mixture of the two or simply to the isotope of mass number 1. Where confusion can occur, the term *protium* is used for the isotope of mass number 1. Furthermore, the symbol D represents specifically an atom of deuterium.

Naturally occurring hydrogen consists almost exclusively of the two isotopes protium and deuterium. But it's possible to manufacture a third isotope, *tritium*, by adding a second neutron to the nucleus. Tritium, with a nucleus containing one proton and two neutrons, has a mass number of 3 and an atomic number of 1. Tritium is used along with deuterium to produce the explosive force of the hydrogen bomb. The hydrogen of the universe consists of about 99.985% protium, 0.015% deuterium, and just a trace of tritium.

■ QUESTION

As we've just seen, an atom with a nucleus consisting of *one proton* and *two neutrons* is tritium, an isotope of hydrogen. Is an atom with a nucleus consisting of *one neutron* and *two protons* still another isotope of hydrogen? Explain.

2.7 Building up the Elements: Hydrogen through Neon

Although adding a neutron to an atomic nucleus increases its mass number and thereby generates a different isotope of the same element, adding a proton produces an entirely different element. (In actual practice it isn't nearly as easy to add a proton to a nucleus as it is to add a neutron. In any case, what concerns us here isn't the specific procedure we might use for the addition, but rather the consequence of adding a proton to an atomic nucleus.) Adding protons, as we have

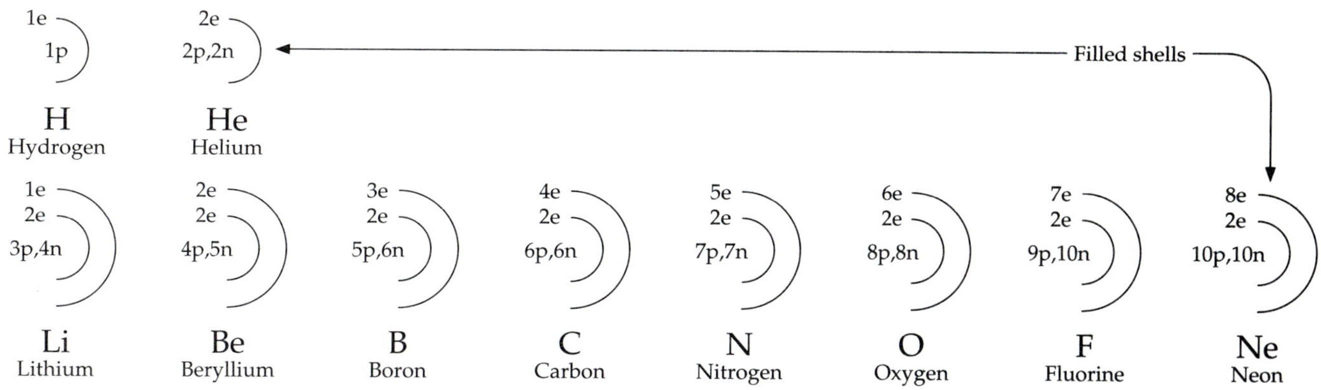

Figure 2.12 Names, symbols, electron structures, and compositions of nuclei of the first 10 elements.

seen, increases atomic numbers as well as mass numbers. Adding one proton to a hydrogen nucleus, for example, produces an atom of the element *helium*. Virtually all the helium atoms in the universe have two neutrons in their nuclei as well as two protons, so a helium atom's mass number is 4 and its atomic number is 2. With two positively charged protons in its nucleus, a helium atom has two negatively charged electrons in the surrounding shell. These two electrons completely fill this particular shell; no more electrons can enter it. We're familiar with helium as a gas used to fill balloons. Since helium is less dense than air, helium-filled balloons tend to rise upward into the atmosphere.

Helium-filled balloons.

A third proton produces *lithium*, atomic number 3 and (for the most common isotope) mass number 7. (Lithium is a metal used in small, long-lasting batteries that power digital watches, calculators, and similar electronic equipment.) Adding the proton to the nucleus requires adding a third electron to maintain electrical neutrality. Since the shell containing the first two electrons is now full and can hold no more electrons, the third electron goes into a second shell, larger than the first and concentric with it, as shown in Figure 2.7. Each of these electron shells occupied by the electrons that surround the nucleus is called a **quantum shell** and receives a *quantum number*: 1 for the shell closest to the nucleus (filled by two electrons), 2 for the next shell (which can hold a maximum of eight electrons), 3 for the next, and so forth. We use the term **electron structure** to indicate the distribution of electrons in the quantum shells surrounding a nucleus.

Each of the electron shells surrounding an atomic nucleus is a **quantum shell**. The **electron structure** of an atom refers to the distribution of electrons in its quantum shells.

Continuing the addition of protons to the nuclei and of electrons to the surrounding shells forms, in sequence, *beryllium*, *boron*, *carbon*, *nitrogen*, *oxygen*, *fluorine*, and *neon* and completes the series of the first 10 elements. Figure 2.12 shows the names, chemical symbols, electron structures, and compositions of the nuclei for these. We'll learn more about them in the concluding Perspective.

■ QUESTION

How many quantum shells does the sodium atom, atomic number 11, have? (We've just noted that the second quantum shell can hold a maximum of eight electrons.)

2.8 More Electron Structures: Sodium through Calcium

In the neon atom the tenth proton is balanced by a tenth electron, which enters the second shell and completes that shell with its full complement of eight electrons. (Remember, there are two or more electrons in the first quantum shell of every

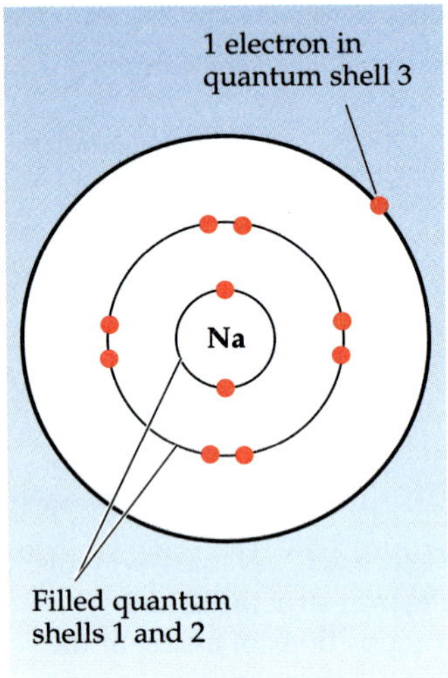

Figure 2.13 Sodium atom.

element except hydrogen.) Formation of *sodium* by the addition of an eleventh proton and a counterbalancing eleventh electron places that new electron in the *third* quantum shell (Fig. 2.13). Adding more protons to the nucleus and more electrons to the third quantum shell produces, in succession, *magnesium*, *aluminum*, *silicon*, *phosphorus*, *sulfur*, *chlorine*, and *argon*.

Adding a proton to the argon nucleus and another electron, the nineteenth, produces *potassium* and begins filling the fourth quantum shell. *Calcium*, with atomic number 20, has two electrons in the first quantum shell, eight in the second, eight in the third, and two in the partially filled fourth. The first 20 elements are listed in Table 2.2 with their chemical symbols and electron structures.

Table 2.2 Electron Structures of the First 20 Elements

			Quantum Number of Shell			
Element	Symbol	Atomic Number	1	2	3	4
Hydrogen	H	1	1			
Helium	He	2	2			
Lithium	Li	3	2	1		
Beryllium	Be	4	2	2		
Boron	B	5	2	3		
Carbon	C	6	2	4		
Nitrogen	N	7	2	5		
Oxygen	O	8	2	6		
Fluorine	F	9	2	7		
Neon	Ne	10	2	8		
Sodium	Na	11	2	8	1	
Magnesium	Mg	12	2	8	2	
Aluminum	Al	13	2	8	3	
Silicon	Si	14	2	8	4	
Phosphorus	P	15	2	8	5	
Sulfur	S	16	2	8	6	
Chlorine	Cl	17	2	8	7	
Argon	Ar	18	2	8	8	
Potassium	K	19	2	8	8	1
Calcium	Ca	20	2	8	8	2

■ QUESTION

Among the first 20 elements, which hold electrons only in completely filled quantum shells? (These are among a group of elements we'll meet again when we examine the periodic table in Section 3.7.)

2.9 Atoms and Paper Clips Revisited

In the opening demonstration we drew an analogy between atoms and paper clips. We saw that in some ways an individual paper clip is related to a handful of paper clips in the same way an individual atom is related to a few grams of an element. Each is the smallest unit that we can identify as a member of the group it represents. We'll now extend this analogy a bit by applying it to atomic weights and isotopes.

Like elements, paper clips vary quite a bit in their characteristics. There are triangular, rectangular, and the familiar oval clips; large clips and small clips; metal clips and plastic clips; and clips of all colors (Fig. 2.14). Although there probably aren't as many different kinds of paper clips as there are elements, the variety of clips that are available offers a rough parallel to the variety of the chemical elements.

For example, we can use a small, metal paper clip with rounded ends or one of the very same kind but a bit larger (and therefore a bit heavier). In a general sort of way, the two clips of Figure 2.15 could illustrate isotopes, both representing the same kind of paper clip but one heavier than the other. (The heavier clip is larger, too, but that's not part of the analogy; isotopic atoms are about the same size.)

If we drop one of these larger and heavier clips into a pile of 6700 of the smaller clips, we see a parallel to the distribution of hydrogen atoms in the universe, with one heavier deuterium for every 6700 lighter protiums. The addition of the heavier clip raises the average weight of all the paper clips just a little above the weight of one of the more plentiful (smaller) clips. Similarly, the average atomic weight of all the hydrogen in the universe is just a bit greater than the mass of a protium atom.

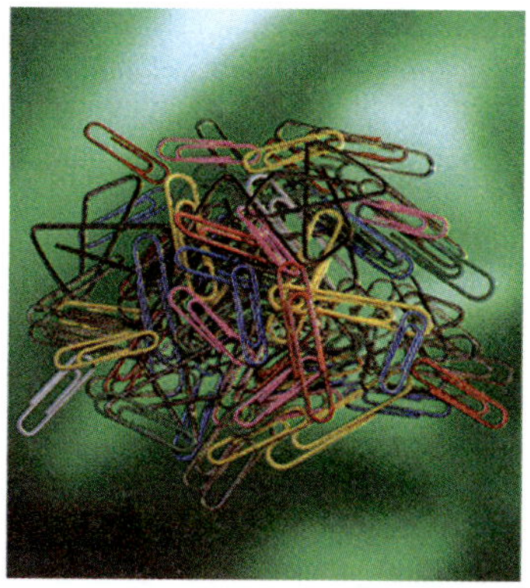

Figure 2.14 The many varieties of paper clips reflect the varieties of elements and their atoms.

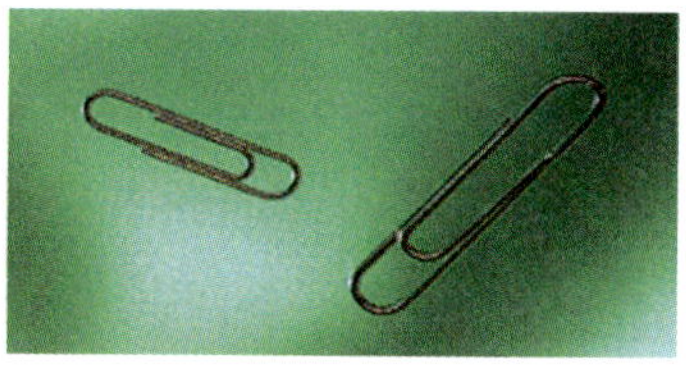

Figure 2.15 Two paper clips of the same kind but different sizes can be used as models for atomic isotopes. Atomic isotopes represent the same element, but have different masses.

■ **QUESTION**

To carry the analogy with paper clips a bit further, suppose you had a pile of 100 small paper clips that weigh exactly 0.100 g each. (a) What's the total weight of the entire pile? (b) What's the average weight of a paper clip in the pile? Now suppose you drop one more paper clip into the pile, a large paper clip that weighs exactly 0.200 g. You now have a pile of 101 paper clips: 100 small and 1 large. (c) What's the total weight of the entire pile now? (d) What's the *average* weight of a paper clip in the pile? (You might review the material of Appendix D, Significant Figures, in answering this question.)

PERSPECTIVE

The First 10 Elements

The first 10 elements, those with atomic numbers from 1 to 10, are about as diverse a group of substances as we can imagine. We'll take a quick look at some of their more interesting facets and then, in Chapter 3, examine the ways in which many of these and other elements combine to form some of the ordinary and extraordinary substances of our world.

Hydrogen, the first in this series, was originally identified as a distinct substance in 1766 by Henry Cavendish, a British scientist. In examining this newly identified gas, Cavendish was struck by its ability to form water when it burns in air. This observation led the French chemist Antoine Lavoisier, a member of the aristocracy and a contemporary of Cavendish, to suggest the name hydrogen, which comes from Greek words meaning *produces water*. (In 1794, shortly after he made this suggestion, Lavoisier was executed at the guillotine by the leaders of the French Revolution.)

The isotope deuterium was first identified in 1931 by the American chemist Harold C. Urey, who received the 1934 Nobel Prize in Chemistry for this discovery. In a real sense hydrogen is one of the most important of all the elements: Both our own bodies and the universe as a whole contain more atoms of hydrogen than of all the other elements combined.

Helium, the second element, is a gas of extraordinarily low reactivity. It's used to dilute gaseous anesthetics during surgery and to inflate lighter-than-air blimps and balloons. One of its most important properties is its inability to burn; helium is completely *nonflammable*. In one sense helium has the most unusual history of all the elements: It was found on the sun before it was discovered on the earth. In 1868 analysis of light radiating from the sun led scientists to conclude that a previously unreported element existed on the sun. The British astronomer J. Norman Lockyer called the newly discovered element helium, which he took from the Greek word for the sun, *helios*.

Lithium is a reactive metal that forms hydrogen gas on contact with water. First isolated in 1818, it's used commercially in long-lasting batteries (Section 2.7) and medically in the compound *lithium carbonate* to moderate the swings of manic depression. The element's name comes from the Greek word for stone, *lithos*, and reflects its original isolation from a mineral, *petalite*.

Like lithium, beryllium was also discovered (in 1797) in a mineral, in this case *beryl*. A hard but very toxic metal, beryllium has few uses in consumer goods. Combined with copper, it increases the strength of various specialized tools, springs, and electrical components; because of its low absorption of X-rays, it's used to make the casings of X-ray tubes. It's also used in the construction of spacecraft and nuclear reactors (Chapter 5).

Boron is a semimetallic element, which means that if it's isolated under certain conditions, it's a solid with a dull, metallic shine; under other conditions it's a black powder that's very much unlike any metal. Although boron was first prepared in a very crude form by the French chemist Joseph Louis Gay-Lussac and a co-worker in 1808, the pure element wasn't obtained until 1909. Pure boron has few if any uses. Consumers use boron largely as a component of the antiseptic *boric acid*. The origin of the element's name is uncertain.

Carbon is a nonmetallic element found in every compound (except water) that's important to sustaining life and health. Because of its widespread occurrence in living things, the chemistry of carbon is called "organic chemistry." (We'll examine this in greater detail in Chapter 6.) The element itself exists in four distinct forms: *charcoal*, the black material formed when wood and other living or once-living materials are heated to high temperatures; *graphite*, a slippery form of carbon that's used as a solid lubricant and that forms the "lead" of pencils; *diamond*, the precious gem; and *fullerenes*, a form of carbon discovered in 1985 in which five- and six-membered rings of carbon unite to form large spheres, the most common of which contain 60 carbon atoms. Fullerenes are currently under intense study for their properties and potential uses. Carbon itself has been used by humans probably since the discovery of fire and the burning of wood. The word "carbon" comes from a Latin term for coal or charcoal.

Nitrogen, a gas that makes up about 78% of the air we breathe, was discovered independently in the early 1770s by two British chemists, Joseph Black and Joseph Priestly, and a Swedish chemist, Karl Wilhelm Scheele. Its name reflects its occurrence in *nitre*, known also as *potassium nitrate* or *saltpeter*, a component of gunpowder. Nitrogen is found in an enormous number of compounds important to life and good health, including medications, agricultural fertilizers, and the vitamins and proteins of our foods.

Oxygen, which makes up almost 21% of our atmosphere, was also discovered by Priestly and Scheele, again independently and again in the early 1770s. Like nitrogen, oxygen occurs in a large variety of important compounds, including the proteins, carbohydrates, and fats and oils of our foods. The name "oxygen" comes from Greek words and reflects its presence in many acidic or sharp-tasting substances.

Fluorine, a gas, is one of the most reactive elements known. Although it was recognized as an element in the early 1800s, its ability to react with virtually anything it touches delayed the preparation of sizable quantities of pure fluorine until 1886, when Henri Moissan, a French chemist, first isolated it. Compounds of fluorine known as *fluorides* are important for hardening the enamel of teeth. It's also a component of Teflon, a compound used as a coating on machine parts and kitchen utensils to make their surfaces nearly friction-free. Fluorine's name is taken from a group of minerals in which it occurs.

Neon, the tenth element of our series, is a gas as inert as helium. The British chemist William Ramsay and his assistant M. Travers discovered the element in 1898 by isolating it from air and named the new element neon from a Greek word meaning *new*. Its principal consumer application lies in generating the glow of neon signs.

All of these are elements since, using the description of Section 1.4, they can't be produced from any other substances (that don't already contain them), nor can they be decomposed or broken down to any other substances by any of the common forces of our everyday world. Next, in Chapter 3, we'll examine how these and other elements combine to form compounds. ■

EXERCISES

1. Following are a statement containing a number of blanks and a list of words and phrases. The number of words equals the number of blanks within the statement, and all but two of the words fit correctly into these blanks. Fill in the blanks of the statement with those words that do fit, then complete the statement by filling in the two remaining blanks with correct words (not in the list) in place of the two words that don't fit.

> __________ are the smallest electrically neutral particles of an __________, a fundamental substance of chemistry and of the world, that can be identified as that element. All atoms consist of two parts: (1) a __________ that contains positively charged __________ and (except for __________, an __________________ of hydrogen) electrically neutral __________; and (2) __________, which lie in __________ surrounding the nucleus. The number of protons in the nucleus defines an atom's __________ and is represented by the symbol __________. The sum of the protons and neutrons in the nucleus determines the atom's __________ and is represented by __________. As the atomic number of the atom increases, the number of electrons in the shells also increases so that the number of electrons always equals the number of protons and the atom remains electrically __________. The first quantum shell can contain a maximum of __________ electrons; the second a maximum of __________ electrons.

A	negative
atomic number	neutral
atoms	neutrons
deuterium	nucleus
eight	protons
element	quantum shells
isotope	two
mass number	*Z*

2. Name an element that is

a. the major gas of the air we breathe

b. used in small, long-lasting batteries

c. used to harden the enamel of teeth

d. present in graphite, diamonds, petroleum, and all living things

3. Of the first 10 elements, which are metals?

4. Exercise 3 contains the phrase "the first 10 elements." What does this refer to?

5. How many elements are currently known to exist?

6. In terms of the number of atoms present, hydrogen is the most abundant element in the universe and also in our bodies. What is the second most abundant element in the universe, again in terms of the number of atoms present? What is the second most abundant element in our bodies?

7. Given the atomic number and the mass number of an atom, how do we determine the number of protons in the nucleus? The number of neutrons in the nucleus? The number of electrons in the surrounding shells?

8. Do an atom of protium and an atom of deuterium represent atoms of two different elements? Explain.

9. (a) Name the three isotopes of the element whose atomic number is 1. (b) What collective name do we give to a mixture of these three isotopes when they are present in the same ratios as in the universe as a whole?

10. What is a name used for the isotope of hydrogen indicated by ${}^{3}_{1}H$?

11. What chemical symbol is used as the equivalent of ${}^{2}_{1}H$?

12. Name and give the chemical symbols for the elements with the first 10 atomic numbers.

13. Atoms of what element are used to define the atomic mass unit?

14. Can any isotope of any element have a mass number of zero? Explain.

15. (a) What would remain if we removed an electron from a hydrogen atom? What charge would this particle bear? What would its mass be? (b) How do you think we could convert a sodium atom into one of the sodium cations discussed in Chapter 1?

16. Write the chemical symbol, showing the atomic number and the mass number, for atoms of two different elements each of which has two neutrons in its nucleus.

17. Name:

a. three elements that have only a single electron in their outermost quantum shell

b. two elements that have exactly two electrons in their outermost quantum shell

c. three elements with filled outermost quantum shells

d. one element that has only a single electron in its innermost quantum shell

18. The nucleus of ${}^{13}_{6}C$ and the nucleus of ${}^{14}_{7}N$ contain the same number of __________________ but different numbers of

________________. The nucleus of $^{14}_{6}C$ and the nucleus of $^{13}_{6}C$ contain the same number of ________________ but different numbers of ________________.

19. (a) Which element was found on the sun before it was found on the earth? (b) Which element has a name that indicates its presence in many acids or sour-tasting substances? (c) How many years passed between the identification of hydrogen as an element and the discovery of its isotope deuterium? (d) Which of the first 10 elements are gases? (e) Which is the most reactive of the first 10 elements?

20. What do *diamonds* and *fullerenes* have in common?

A Little Arithmetic and Other Quantitative Puzzles

21. Give the number of protons and the number of neutrons in the nucleus of each of the following atoms and the number of electrons in their first and second quantum shells: (a) $^{10}_{5}B$; (b) $^{14}_{6}C$; (c) $^{3}_{2}He$; (d) $^{40}_{19}K$; (e) $^{16}_{8}O$.

22. Section 2.5 tells us that hydrogen atoms make up two-thirds of all atoms in water but just over 11% of the water's weight. Given that there are twice as many hydrogen atoms in water as there are oxygen atoms and that virtually all the hydrogen atoms in water are $^{1}_{1}H$ and virtually all the oxygen atoms are $^{16}_{8}O$, how do you explain this apparent discrepancy?

23. What is the weight of a sheet of gold atoms 1 atom thick and with the same dimensions as this page? You can answer this question by measuring the width and length of this page and carrying out a calculation similar to the one in the example of Section 2.2.

24. Using 3×10^{-8} cm as the diameter of a gold atom and 3.3×10^{-22} g as its weight, calculate the weight of 1 cm^3 of gold. (You can start by finding how many gold atoms fit on a line 1 cm long; see Section 2.2.) How does your calculated value compare with the measured density of gold, 19.3 g/cm^3? Suggest some factors that might account for the difference.

Think, Speculate, Reflect, and Ponder

25. (a) If you split the nucleus of a gold atom into two parts, would you now have two smaller gold atoms? Explain. (b) If you combined the nuclei of two gold atom into a single nucleus, would you now have one, larger gold atom? Explain.

26. a. Lithium, sodium, and potassium are all metals that react with water to liberate hydrogen gas. What, if anything, do atoms of each of these metals have in common?

b. Helium is an unreactive gas and neon is a gas of extremely low reactivity. What, if anything, do their atoms have in common?

27. Suppose we could combine one electron with one proton to form a single, new subatomic particle. What mass, in amu, would the resulting subatomic particle have? What electrical charge would it carry? What known subatomic particle would the resulting particle be equivalent to?

28. Suppose someone discovered a particle that consisted of a single neutron surrounded by a shell containing a single electron. Would you classify this as an atom? Would you classify it as an ion? Would it represent a new element? Explain your answers.

29. a. What is the difference between an object's mass and its weight?

b. A bag of table sugar (sucrose) weighs 10 kg on the moon. What is its weight on the earth?

c. A bag of table salt (sodium chloride) has a mass of 5 kg on the moon. What is its mass on the earth?

d. A piece of charcoal has a mass of 1 kg on the earth. What is its weight on the moon?

30. You have two spheres, one made of lead and one made of cork. They are standing next to each other at some spot on the surface of the earth. At that location each weighs 1 kg. Compare their masses at that location. Does one have a greater mass than the other? If so, which has the greater mass? Now move both spheres to one particular location on the surface of the moon and again compare their masses. Does one have a greater mass? If so, which one? Does one weigh more than the other on the moon? If so, which? Finally, leave the cork sphere on the moon and move the lead sphere back to its original location on earth. Again, compare both their masses and their weights.

31. Suppose you discovered how to produce an isotope of hydrogen containing *three* neutrons in its nucleus.

a. Write its elemental symbol, showing its atomic number and its mass number.

b. What name might you give this new isotope to distinguish it from protium, deuterium, and tritium.

c. What new letter-symbol would you give this new isotope to distinguish it from D and T, but without repeating the symbol of another element.

d. Explain or give a reason for your responses to parts *b* and *c*.

32. The description of the atom that we have used in this chapter resembles in some ways our own solar system. What part of our solar system corresponds to the nucleus of an atom? What part corresponds to the electron shells surrounding the nucleus? What are some of the other similarities between the structure of the atom, as described in this chapter, and our own solar system? What are some of the more obvious differences?

Web Questions

Use the Web to answer or accomplish each of the following. Report your Web source for each response.

1. The modern view of the atom originated with the work of John Dalton. Write a brief biography of John Dalton, including the dates of his birth and death, his nationality, background, education, and a description of his view of the atom and how he arrived at it.

2. What is the principal factor affecting a planet's gravitational force? If a person weighs 60 kg on earth, what would that same person weigh on Mercury, Venus, Mars, Saturn, and Jupiter?

3. Who discovered each of the following? How and when was the discovery made? (a) the electron, (b) the neutron, (c) the proton?

4. The Perspective summarizes the occurrence, history, properties, etc., of the first 10 elements, those with atomic numbers 1–10. Write similar short paragraphs for the next five elements, those with atomic numbers 11–15.

5. Henry Moseley, Isaac Newton, and Harold Urey each made a major contribution to the topics discussed in this chapter. Identify the topic associated with each of these three scientists and describe the scientist's contribution.

Chemical Bonding

IONIC AND COVALENT COMPOUNDS

Periodicity of the seasons, seen in a tree.

Scouring Pads and Kitchen Magnets

Figure 3.1 The attraction that iron has for a magnet isn't carried over into the iron oxides it forms as it rusts.

In this chapter we examine how and why elements combine to form compounds. We learn, too, that their characteristics can change as they join each other through chemical bonds. Dangerous or poisonous elements can form benign compounds, and reasonably harmless elements can form toxic, corrosive, or otherwise hazardous compounds. An element seldom carries more than a few of its properties, if any at all, into the compounds it forms.

We see this in our daily lives. Carbon is the element that makes up the charcoal of the briquettes we use in backyard grills. It's a brittle, black solid that rubs off onto whatever touches it. Another element, oxygen of the air, is a gas that supports life and is consumed by anything that burns. When we light charcoal briquettes, the elemental carbon of the briquette and the elemental oxygen of the air combine to produce heat and the compound *carbon dioxide*. Unlike carbon, carbon dioxide is a colorless gas. Unlike oxygen, carbon dioxide doesn't support life and won't keep a fire going. Indeed, carbon dioxide is widely used to extinguish fires. The characteristics of the carbon and of the oxygen are transformed as they combine to form carbon dioxide.

We can see this same sort of thing happening with the metal scouring pads we use for cleaning pots and pans. These pads are composed largely of iron that's drawn out into fine filaments and woven into pads that are often impregnated with a detergent. As wet pads of this sort stand in the open air, the iron in the filaments reacts slowly with the oxygen of the air to form the reddish-brown oxides of iron we call rust.

Try this simple but effective demonstration of the change in properties of an element—iron, in this case—as it enters into a compound. Elemental iron is attracted to magnets. Touch a magnet to an iron scouring pad and feel the attraction they have for each other. You can use one of the small, strong magnets that holds notes, coupons, and so forth to the sides of refrigerators.

Now wet the scouring pad with a little vinegar or salt water and let it stand in the open and rust. As the days pass and the rust accumulates, the attraction between the pad and the magnet drops. Elemental iron is attracted to magnets, but it doesn't carry this characteristic into its compounds. The oxides of iron that form rust aren't attracted to magnets (Fig. 3.1).

Iron, oxygen, and carbon are elements we find easily in their free state in the world about us. In this chapter we'll concentrate our attention on two elements we don't often see in the free state, sodium and chlorine. We'll focus our attention on them and on the compound they form when they react with each other: sodium chloride, our common table salt.

3.1 Table Salt Revisited

At the beginning of Chapter 1 we entered our study of chemistry with a demonstration of a remarkable difference between two common chemicals, sodium chloride (table salt) and sucrose (table sugar). We saw that pure water is a very poor conductor of electricity and that although adding sucrose to water does nothing to improve its conductivity, adding sodium chloride does increase water's ability to conduct an electric current dramatically, causing the light bulb to burn brightly. That observation led us to classify substances as either electrolytes or nonelectrolytes and introduced us to the chemical particles known as ions.

Now that we've learned something of elements and compounds, and of the ions and other chemical particles that make up our world, we'll return to sodium chloride, this time to contrast its properties to those of two *un*common chemicals, elemental *sodium* and elemental *chlorine*. In the discussions that follow, we'll use the term *elemental* to refer to pure elements themselves as they stand alone, uncombined with any other element, either in a compound or any other way. With this definition, "elemental sodium" refers to pure, metallic sodium and "elemental chlorine" refers to pure chlorine gas, the forms in which we ordinarily find the pure elements themselves. Thus, although the element sodium is present in both the metal (as sodium atoms) and the compound sodium chloride (as sodium cations), only a piece of the metal itself consists of *elemental* sodium. Similarly, the element chlorine is present in both elemental, gaseous chlorine and table salt, but only the gas itself consists of *elemental* chlorine.

As we proceed in this chapter, we'll examine atoms and ions in more detail; we'll learn how atoms are converted into ions; and we'll learn about the two different kinds of bonds, *ionic* and *covalent*, that hold chemical particles together in the multitude of substances we find and use in our daily lives. What's more, we'll learn that although the *sodium cation* of our table salt and the *sodium atom* of metallic, elemental sodium differ only in the number of electrons present in their most remote quantum shells (and, as a result, in the net electrical charge on each), this apparently small difference produces remarkable differences in their properties. And while the words "chlor*ide*," as in "sodium chloride," and "chlor*ine*," as in "chlorine gas," sound very much alike, the two chemical particles they represent could hardly be more different in the ways they behave.

■ QUESTION

The element of atomic number 53 is iodine, with the chemical symbol I. Iodine is an antiseptic available in commercial preparations in drugstores; potassium iodide is added to "iodized" table salt to protect against the condition known as "goiter" (Section 19.2). In which would you find (a) elemental iodine, (b) the I^- anion, and (c) the element iodine?

3.2 Sodium and Chlorine

We find elemental chlorine only rarely and elemental sodium not at all in the common substances of our everyday world, and for a very good reason: They are both extremely reactive and can be highly dangerous substances if they aren't used with care. Chlorine is capable of destroying living tissue and can be a health hazard and even deadly if it is inhaled in more than trace amounts. Because it kills bacteria, chlorine is used in small quantities to eliminate infectious organisms from both swimming pools and public drinking water. It's also a very effective household bleaching agent. You can smell the odor of traces of elemental chlorine in the air

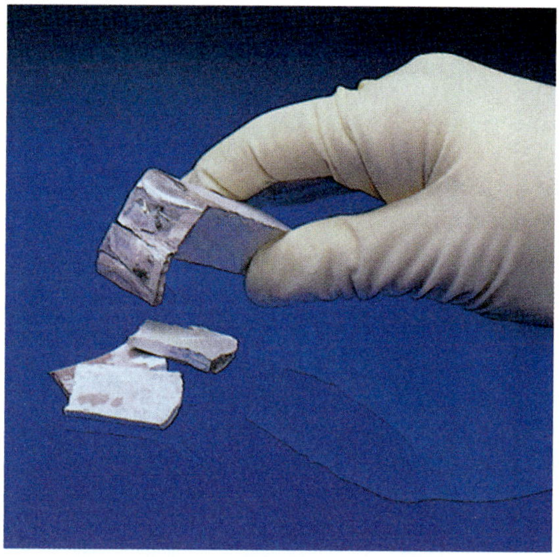

(Left) Elemental sodium, a metallic solid.
(Right) A flask containing elemental chlorine, a yellow gas.

near swimming pools and around open bottles of liquid household bleach, and you can taste it in the municipal drinking water of some regions. Released into the atmosphere in large volumes, it can kill plants and any animal that breathes it.

Chlorine was the first of a series of poisonous gases to be used as weapons in World War I. On April 22, 1915, in a battle near the Belgian city of Ypres, German troops released large quantities of the gas from cylinders stored in their trenches. Wind carried the greenish-yellow gas across the battlefield to positions held by French and other Allied troops. Coupled with a vigorous artillery attack, the chlorine forced the Allied forces to retreat. The gas attack was so successful that chlorine and other poisonous and irritating gases were used repeatedly during the rest of the war.

These various applications illustrate the problem with trying to label chemicals as "good" or "bad" or as "beneficial" or "harmful" (Section 1.2). The same elemental chlorine that kills when it's used as a war gas saves lives when it's used to purify drinking water. It makes our lives safer, easier, and more pleasant as it rids swimming pools of bacteria and takes stains out of clothing. Released into the atmosphere in large quantities, it has served as a war gas; released judiciously into swimming pools and drinking water, it helps prevent epidemics. The qualities of "good" and "bad" aren't properties of the chlorine itself, only of how we ourselves use it.

Sodium reacts explosively with water.

As for elemental sodium, the hazards of even small amounts of this shiny metal far outweigh any benefits it might provide. Elemental sodium reacts with water to produce highly flammable hydrogen gas. A piece of metallic sodium about the size of a walnut can produce disastrous results if it comes in contact with water. Elemental sodium has no uses among our consumer products.

■ QUESTION

What compound forms when highly reactive, elemental sodium comes into contact with highly reactive, elemental chlorine?

3.3 Chemical Particles: Risks and Benefits

Sodium chloride, the compound that forms when elemental sodium reacts with elemental chlorine, is a relatively harmless electrolyte that many of us use daily to modify the taste of food. While it isn't completely without hazard—it's implicated

The vigorous reaction of sodium with chlorine.

Common table salt, the product of the reaction of sodium with chlorine.

as a contributing factor in high blood pressure, or *hypertension*, and when eaten in large quantities it can be deadly, especially to small animals, children, and infants—common table salt simply isn't in the same class as elemental sodium or chlorine. Yet it is composed of the same elements. Why both elemental sodium and chlorine should be so hazardous and yet the sodium and chloride ions of their compound—sodium chloride—be so benign stems partly from the difference between an ion and an atom.

This example shouldn't lead us to assume that when elements themselves are hazardous their compounds will necessarily be harmless. The reverse can easily be true. For example, elemental nitrogen and oxygen, the principal gases of the air we inhale many times each minute, are themselves ordinarily entirely harmless to us, yet several of their compounds are quite hazardous. One of them, *nitrogen dioxide*, a compound containing twice as many oxygen atoms as nitrogen atoms, is partly responsible for the damage of acid rain (Section 14.5) and can produce a potentially fatal inflammation of the lungs if inhaled. The only generalization we can make is that no generalization is possible. There's no necessary connection between a hazard of an element and the chemical form it's in. Neither elements, compounds, atoms, ions, nor molecules are necessarily filled with either risks or benefits to us. In any case, both the risks and benefits we derive from them depend, as we have already seen, on how we use them.

■ **QUESTION**

Name an element that is harmless or beneficial to us in its elemental form. Name a different element that can be hazardous in its elemental form. Now switch things around and name a hazardous compound of the harmless or beneficial element and a harmless or beneficial compound of the hazardous element.

3.4 Valence Electrons

Several factors affect the properties of any particular atom or ion, including its atomic number, mass number, and electronic structure. Plutonium (Pu), with an atomic number of 94, may well be the most deadly of all the elements, regardless of mass number or electronic structure (Section 5.3). And while water, a compound of hydrogen and oxygen, is a necessity of life, drinking water in which deuterium atoms (${}^{2}_{1}H$; mass number 2) have replaced the protium atoms (${}^{1}_{1}H$; mass number 1; Section 2.6) can be fatal. Yet even as atomic number and mass number affect the properties of a chemical particle, electronic structure also plays a major role in its behavior.

The electron structures of sodium and chlorine, for example, are largely responsible for sodium chloride's ability to conduct an electric current. In considering the effects of electron structure we'll focus on the electrons in the outermost electron shell, which is known as the **valence shell**. It is these **valence electrons** that are so important, since the number of these valence electrons determines, to a large extent, the properties and the reactivity of any particular element.

Valence electrons are the electrons of the **valence shell**, which is the outermost electron shell of an atom.

As we examine the connection between the number of valence electrons and the behavior of an element or its atoms, we'll use a simplified way of showing electronic structures called **Lewis structures** or **electron dot structures**. Named for Gilbert N. Lewis (1875–1946), an American chemist who contributed much to our knowledge of valence electrons, these structures show only the chemical symbol of the element and its valence electrons, which are arranged in paired or unpaired dots around the symbol. Table 3.1 shows the Lewis structures of the first 20 elements. (Sections 3.7 and 3.8 will explain a good bit about the arrangement of the elements in the table.) For many purposes Lewis structures give us all the information we need. We'll use these Lewis structures as we examine *periodicity* and its application to both valence electrons and the properties of elements.

The **Lewis structure**, or **electron dot structure**, shows simply the elemental symbol and the valence electrons, arranged in paired or unpaired dots.

Table 3.1 Lewis Structures of the First 20 Elements

H·							He:
Li·	·Be·	·Ḃ·	·Ċ· (4 dots)	·N: (5 dots)	·O: (6 dots)	·F: (7 dots)	:Ne: (8 dots)
Na·	·Mg·	·Al· (3 dots)	·Si· (4 dots)	·P: (5 dots)	·S: (6 dots)	·Cl: (7 dots)	:Ar: (8 dots)
K·	·Ca·						

■ QUESTION

Which elements of Table 3.1 have only one valence electron each? Which have seven? Eight?

3.5 Periodicity

When characteristics repeat themselves again and again over time we say that they recur *periodically*, or that they are *periodic*. The seasons are periodic. By and large, summer's days are the warmest of the year; days are warmest, periodically, during the summer. Plants bloom, periodically, in the spring. Farm harvests come, periodically, during the fall. The sun also rises periodically.

If we examine the properties and behavior of all the elements (with the possible exception of hydrogen, which is the first in the sequence of atomic numbers and something of a chemical maverick), we find that their properties repeat periodically, much as days grow warm or flowers bloom periodically from one year to the next. The properties of the elements repeat themselves, but their repetition occurs over irregular sequences of atomic numbers rather than over regular periods of time. We also find a periodicity in their Lewis structures.

Table 2.2, for example, shows that neon (atomic number 10) and argon (atomic number 18) both hold eight valence electrons. Like neon (Perspective, Chapter 2), argon is a virtually inert gas. The chemical reactions of these two elements are few and rare. Similarly, krypton (36) and xenon (54) hold eight valence electrons and, like the others, are gases with extremely little reactivity. We would be correct in guessing that a valence shell containing eight electrons marks an element with extraordinarily little chemical reactivity, or none at all, and that *any atom that has (or acquires) a valence shell containing exactly eight electrons often is (or becomes) quite inert.*

From a slightly different viewpoint we can say that any atom or ion that has or acquires the same electronic structure (or Lewis structure) as one of these virtually inert gases drops sharply in its reactivity. This generalization gives us the **octet rule**: *Atoms often react so as to obtain exactly eight electrons in their valence shells.* In acquiring an octet of electrons the atom loses much of its chemical reactivity.

The **octet rule** states that atoms often react so as to obtain exactly eight electrons in their valence shells.

(It's important to recognize that helium, like hydrogen, is a bit anomalous. Helium, an unreactive gas with *two* valence electrons, doesn't fit this generalization. Nonetheless, the generalization is still partly valid since both lithium and beryllium often react so as to acquire the same electronic configuration as the inert helium.)

We can extend our observations about periodicity: Elements with the same number of valence electrons tend to behave in the same way. Lithium (atomic number 3), sodium (11), potassium (19), rubidium (37), and cesium (55), for example, have one valence electron each. All are metals and all are reactive, even though some of these metals react with certain substances much more vigorously than do others of the series. For example, each of these metals reacts with water at its own characteristic rate to produce highly flammable hydrogen gas. That's why, as we saw in Section 3.2, sodium ignites or explodes when it comes into contact with water. Sodium, like the rest of the metals with only one valence electron, reacts with water to liberate the hydrogen of the water in the form of elemental hydrogen gas. With sodium and all the other alkali metals except lithium, the heat released in the reaction often causes the hydrogen to ignite or explode.

■ QUESTION

Name two elements you would expect to show the same kind of chemical reactivity as magnesium. What is the basis for your choices?

3.6 Atomic Weights and the Genius of Dimitri Mendeleev

Dimitri Mendeleev devised the first comprehensive periodic table of the elements.

Recognizing repetitions in properties similar to those we have just examined, Dimitri Mendeleev, professor of chemistry at the Technological Institute of St. Petersburg, Russia, devised an arrangement of the elements known as the periodic table. The youngest of 17 children, Mendeleev was born in Siberia in 1834. In his later years, during the last decade of the nineteenth century, he became the director of Russia's Bureau of Weights and Measures.

An exceptional teacher, extremely popular with his students at the Institute, Mendeleev was preparing in 1869 to write a chemistry textbook. As he was reviewing the properties of the 63 elements known at the time—today we recognize

113 (Section 1.4)—he realized that by organizing them in order of increasing *atomic weight* he produced, on a much greater scale, the same sort of periodic repetition of properties we have just described.

Because of their importance, we pause here to examine atomic weights briefly. An element's **atomic weight** is simply the average of the masses of all of its isotopes, weighted for their individual abundances. For example, when we mix in the small amount of deuterium (^2_1H) present in the universe along with all the protium (^1_1H), the average weight of all atoms of atomic number 1 becomes slightly larger than 1 (the mass of protium alone). Thus, the atomic weight of hydrogen, the weighted average of all its isotopes (neglecting the trace of tritium present in the universe), is 1.008 amu.

The **atomic weight** of an element is the average of the masses of all of its isotopes, weighted for the abundance of each.

E X A M P L E

Calculating Atomic Weight

Calculate the atomic weight of chlorine from a knowledge of the abundances of its most common isotopes. The isotope of mass number 35 makes up 75.77% of all chlorine atoms, while the isotope of mass number 37 accounts for 24.23% (the remainder) of chlorine atoms.

■ To obtain the contribution of each isotope, weighted for its abundance, we multiply the mass of each isotope by the fraction of the total that each isotope represents. The isotope of mass 35 (75.77% of the total) thus contributes

$$0.7577 \times 35 \text{ amu} = 26.52 \text{ amu}$$

to the weight-averaged atomic weight of all atoms of the element chlorine. The isotope of mass 37 (24.23% of the total) contributes

$$0.2423 \times 37 \text{ amu} = 8.96 \text{ amu}$$

The weight-averaged atomic weight of chlorine is the sum of these two contributions:

$$\begin{array}{r} 26.52 \text{ amu} \\ +8.96 \text{ amu} \\ \hline 35.48 \text{ amu} \end{array}$$

which we can round off to 35.5 amu. For a review of the atomic mass unit (amu), refer to Section 2.4.

Although others had also noted the same sort of periodic repetition of properties even before Mendeleev's inspiration, none had pursued the idea as vigorously or in as much detail as the Russian chemist. Mendeleev organized the known elements into a table in which, with a few exceptions, atomic weights increased regularly in the horizontal rows, from left to right, and elements with similar properties fell neatly into placc in the table's vertical columns. To maintain a consistent periodicity in elemental properties, he was forced to leave some of the boxes empty. He predicted that they would be filled by as yet unknown elements. Within 15 years three of the gaps were, indeed, filled through the discoveries of the elements gallium, scandium, and germanium, all of which proved to have the very properties he predicted from the locations of the vacancies. What's more, atomic weights of some of the elements Mendeleev had to place (apparently) out of the normal order in his table were later found to be incorrect. Newer, better values showed that he was, indeed, correct in his placement of those particular elements and that they did not violate a sequence of (now-corrected) atomic weights. It was the measured values of their atomic weights (as they were known at the time) that were in error, not Mendeleev's idea of periodicity.

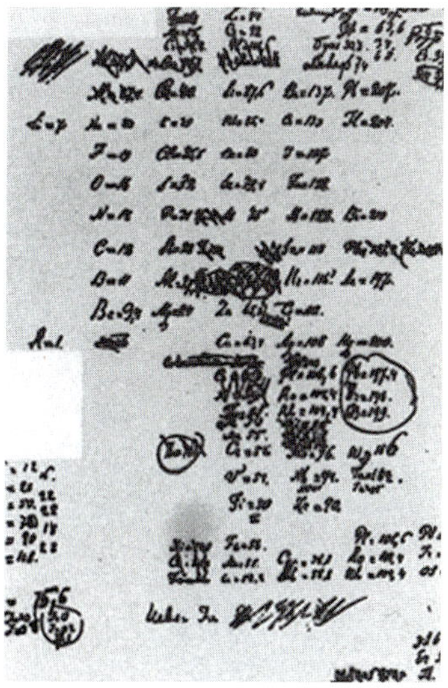

A portion of one of Mendeleev's handwritten drafts of the periodic table.

■ QUESTION

The atomic weight of lithium (Li) is 6.94. The lithium isotope of mass number 7 is the most abundant of all lithium's isotopes, making up about 92.5% of all lithium atoms. What is likely to be the mass number of the second most abundant isotope of lithium? Explain.

3.7 The Periodic Table

In the modern version of the periodic table (Fig. 3.2) the elements are organized in a sequence of increasing atomic *number* rather than atomic *weight*. Mendeleev, who lived from 1834 to 1907, had only atomic weights to work with; the concept of atomic numbers was not devised until 1913. In most modern versions of the table, the chemical symbol of each element appears in the center of the squares, with the atomic number above the symbol and the atomic weight below.

Now we're in a position to connect the number of valence electrons of an element (Section 3.4) with the element's position in the periodic table. We now know that each of the table's columns contains elements with the same numbers of valence electrons and that this is the basis of their chemical similarities. In recognition of these similarities we call all the elements in any individual column a **family** of elements. The leftmost family, consisting of lithium, sodium, potassium, rubidium, cesium, and francium, each with one valence electron, is the *alkali metal* family.

A **family** of elements consists of all the elements in any column of the periodic table. The elements in each row of the periodic table represent a **period**.

Each row of the periodic table represents a **period**. As we move from left to right within each period, valence shells of the atoms fill with electrons, with one additional electron for each step to the right. Just to the right of the alkali metal family lies the column of the *alkaline earth* metals. Each of these elements has two valence electrons. The rightmost column, composed of gases with eight valence electrons—two in the case of helium—and with little or no chemical reactivity, is the *inert gas* or *noble gas* family, so-called because of their almost complete lack of chemical association with the more common elements. Just to its left is a family of *halogens* consisting of fluorine, chlorine, bromine, iodine, and astatine. (*Halogen* comes from the Greek *hals*, meaning salt or sea. The halogens readily form *salts*, which we'll discuss further in Chapter 10.) Elemental fluorine is extremely reactive; in combination with carbon it forms the plastic Teflon (Perspective, Chapter 2). Along with its use as a bleach and disinfectant, chlorine is a component of synthetic rubber, of plastics such as *polyvinyl chloride* (PVC), and of various pharmaceuticals. Bromine and iodine are also used in the manufacture of pharmaceuticals.

Because it concisely furnishes information about chemical symbols, atomic weights, atomic numbers, valence electrons, chemical families, and, to those famil-

Figure 3.2 A modern version of the periodic table of the elements. Elements 113 and 115–118 have received names and symbols although they have not yet been observed.

Group / Period	1	2		3	4	5	6	7	8	9	10	11	12	13	14	15	16	17	18
1	1 H																		2 He
2	3 Li	4 Be												5 B	6 C	7 N	8 O	9 F	10 Ne
3	11 Na	12 Mg												13 Al	14 Si	15 P	16 S	17 Cl	18 Ar
4	19 K	20 Ca		21 Sc	22 Ti	23 V	24 Cr	25 Mn	26 Fe	27 Co	28 Ni	29 Cu	30 Zn	31 Ga	32 Ge	33 As	34 Se	35 Br	36 Kr
5	37 Rb	38 Sr		39 Y	40 Zr	41 Nb	42 Mo	43 Tc	44 Ru	45 Rh	46 Pd	47 Ag	48 Cd	49 In	50 Sn	51 Sb	52 Te	53 I	54 Xe
6	55 Cs	56 Ba	*	71 Lu	72 Hf	73 Ta	74 W	75 Re	76 Os	77 Ir	78 Pt	79 Au	80 Hg	81 Tl	82 Pb	83 Bi	84 Po	85 At	86 Rn
7	87 Fr	88 Ra	**	103 Lr	104 Rf	105 Db	106 Sg	107 Bh	108 Hs	109 Mt	110 Uun	111 Uuu	112 Uub	113 Uut	114 Uuq	115 Uup	116 Uuh	117 Uus	118 Uuo
*Lanthanoids			*	57 La	58 Ce	59 Pr	60 Nd	61 Pm	62 Sm	63 Eu	64 Gd	65 Tb	66 Dy	67 Ho	68 Er	69 Tm	70 Yb		
**Actinoids			**	89 Ac	90 Th	91 Pa	92 U	93 Np	94 Pu	95 Am	96 Cm	97 Bk	98 Cf	99 Es	100 Fm	101 Md	102 No		

iar with its use, the reactivities of the elements, the periodic table provides one of the most valuable reference works in the entire field of chemistry. Inside the back cover of this book you will find a list of the data of the periodic table, alphabetized by the names of the elements, and a list of the names and symbols of the elements, alphabetized by the symbols.

■ QUESTION

What are the name and the chemical symbol of the alkaline-earth metal with the smallest atomic number?

3.8 Valence Shells and Chemical Reactivity

Combining a knowledge of the periodic table and of the reactions of the chemical elements produces the useful generalization we first saw in Section 3.5: *Atoms often react so as to obtain exactly eight electrons in their valence shells*. As a general rule, atoms or ions with an octet of valence electrons are far more stable, or unreactive, than those without this octet. Since the valence shells of the inert gases (with the exception of helium) already have an octet, these elements have virtually no tendency to react in any way with any other elements.

Each alkali metal, though, has a single electron in its valence shell. Losing that one valence electron to some other element leaves the alkali-metal atom with a valence shell (the shell just below) identical to that of the noble gas of one lower atomic number. The halogens, such as chlorine, have valence shells containing seven electrons. Adding one electron to the valence shell of a halogen gives each of the halogens the same electron structure as its adjacent inert gas.

(At this point we might ask why the sodium atom can't *gain seven* electrons to form an octet and why a chlorine atom can't *lose seven* electrons, leaving the next lower shell with its octet of electrons. We'll answer these questions in the next section.)

We'd be correct in expecting that metallic, elemental sodium would react with gaseous, elemental chlorine with the transfer of sodium's lone valence electron into chlorine's valence shell, giving both sodium and chlorine octets. Our expectation is not only correct; it is *explosively* correct: Sodium and chlorine react violently, with the rapid release of a great deal of energy, to produce the very stable, very inert household commodity, sodium chloride, our common table salt (Fig. 3.3).

One important use for the periodic table now becomes clear. Understanding its construction allows us to begin predicting what elements will react with each other and what kinds of compounds they can form. Since each member of the alkali-metal family contains only a single electron in its valence shell, and since each halogen can acquire an octet by adding a single valence electron, any of the alkali metals can react

Figure 3.3 The reaction of sodium and chlorine to produce sodium chloride. Full electronic structures are shown at the top; Lewis structures are at the bottom.

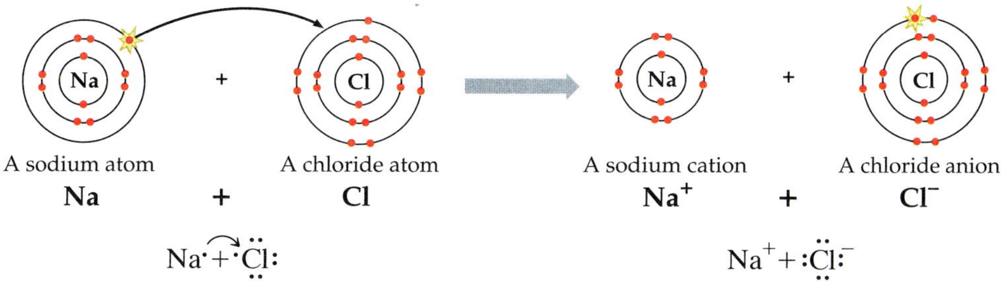

with any of the halogens to form a compound by the simple transfer of a single electron from the valence shell of the alkali metal atom to the valence shell of the halogen. Potassium, for example, can react with any of the halogens to form a compound similar in many ways to NaCl. The results are KF, KCl, KBr, and KI. (Astatine, At, is a very rare element, and although it is a member of the halogen family, we virtually never come across astatine or any of its compounds.) As a parallel illustration, chlorine reacts not only with sodium, but with the other alkali metals as well, to form LiCl, NaCl, KCl, RbCl, and CsCl. (Francium, Fr, and its compounds are about as rare as astatine and are easily ignored.)

■ QUESTION

(a) How many electrons must a magnesium atom lose to acquire a valence shell with eight electrons? (b) How many electrons must oxygen gain to form an octet? (c) Would you expect magnesium and oxygen to react with each other to form a chemical compound? Why?

3.9 From Sodium Atoms to Sodium Cations, from Chlorine Atoms to Chloride Anions

Ionic compounds are compounds composed of ions.

As we've just seen, the transfer of one or more electrons between atoms converts the atoms into ions and results in the formation of **ionic compounds**, which are compounds composed of ions. Sodium chloride is just such an ionic compound. It forms when elemental sodium and elemental chlorine react with each other and in other ways as well. (The suffix *-ide* of *chloride* tells us that we are dealing with an anion. Fluor*ide*, chlor*ide*, brom*ide*, and iod*ide* are the names of the hal*ide* anions. The *-ine* suffix applies to the names of the elements and to the elemental forms themselves, as in fluor*ine*, chlor*ine*, etc.)

A sodium *atom* contains 11 protons within its nucleus and 11 electrons in the surrounding shells. The negative charges of these 11 electrons counter the 11 positive charges of the protons. As shown in the complete electronic structures of Figure 3.3, the loss of the single electron from its valence shell leaves the sodium atom with only 10 electrons (in two quantum shells) to counterbalance the 11 nuclear protons. This leaves sodium with a net excess of one proton and thus a net electrical charge of 1+. On reacting with chlorine, then, the sodium atom loses one electron and is converted to a sodium cation, Na^+. We might ask, as we did in the parenthetical portion of the preceding section, why the sodium atom can't just as well *gain seven* electrons and thereby complete an octet. If it did, the resulting ion would contain an excess of seven electrons over protons and thus bear an electrical charge of 7−. Since like electrical charges repel each other (and opposite electrical charges attract), the accumulation of seven negative charges on the same small particle would produce a very high-energy anion that would react quickly to lose some of its excess electrons. Losing a single electron and forming the Na^+ cation, with a single positive charge, produces a much more stable particle than the one that would result from a gain of seven electrons.

The Cl *atom* contains 17 protons in its nucleus and an equal number of electrons in three quantum shells. As the valence shell, which holds seven electrons in the Cl atom, completes the octet with an eighth electron acquired from some other atom, the Cl gains a net excess of one electron over its nuclear protons and is transformed from a chlorine atom into a chloride anion, Cl^-. Just as the *gain* of seven electrons would produce a highly reactive anion in the case of sodium, the *loss* of seven electrons by the chlorine would produce an unstable cation with a charge of 7+.

The electronic structure of the sodium cation (two electrons in the first quantum shell and eight in the second) resembles the electronic structure of what inert or noble gas element? The electronic structure of the chloride anion resembles that of what inert or noble gas element?

■ QUESTION

3.10 The Ionic Bonds of Table Salt

The crystals of the sodium chloride we shake out of a salt shaker are made up of equal numbers of sodium cations and chloride anions. *Crystals* are the well-defined, solid shapes that pure substances often form as their ions or molecules arrange themselves into a precise, regular, three-dimensional order. Crystals of sodium chloride result as the oppositely charged sodium cations and chloride anions attract each other and become organized into an extensive, orderly arrangement known as a **crystal lattice**. Figure 3.4 shows a part of this crystal lattice of NaCl. As this lattice extends outward in three dimensions, it forms the well-defined shape of the crystals that appear, much enlarged, in Figure 3.5.

A **crystal lattice** is the orderly, three-dimensional arrangement of the chemical particles that make up a crystal.

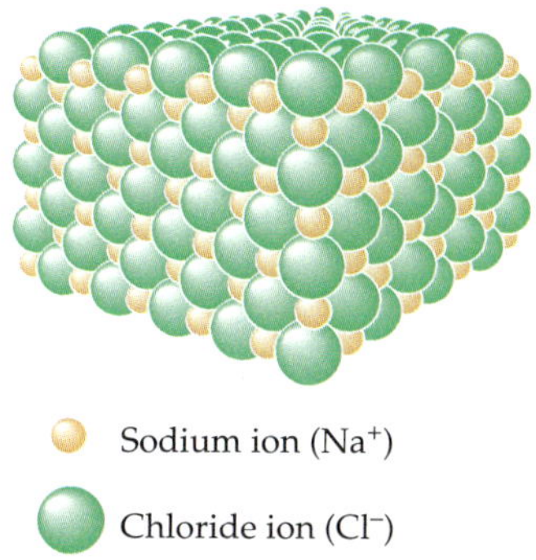

Figure 3.4 Part of a sodium chloride crystal.

The force that holds the ions in place within the lattice comes from the mutual attraction of the oppositely charged cations and anions. It's called an **ionic bond**, and it results in the formation of ionic compounds. (The formula NaCl does not tell us that sodium chloride is an ionic compound. To emphasize that fact we can write the formula as Na^+Cl^-.) When sodium chloride dissolves in water, the cations and anions separate from the lattice and from each other, enter the water solution, and provide the ions that carry an electric current between the poles in Figure 1.5. Crystals of table sugar (sucrose) consist of electrically uncharged molecules held in place in the lattice by forces much weaker than the ionic bonds of sodium chloride. When molecules of sucrose separate from their crystal lattice, they don't provide ions and thus they don't conduct a current (Fig. 1.5). That's why the light bulb glows brightly when the electrolyte sodium chloride is added to the water but not when the nonelectrolyte sucrose is added.

An **ionic bond** results from the attraction of oppositely charged ions.

What do we call the structures, shown enlarged in Figure 3.6, that form as water crystallizes in the atmosphere in cold weather?

■ QUESTION

Figure 3.5 Crystals of sodium chloride, an ionic compound and an electrolyte.

Figure 3.6 Crystals of water support a popular sport.

3.11 Chemical Formulas

Sucrose is a nonelectrolyte because it isn't made up of ions and doesn't provide any means for transporting an electric current across the water between the two wires of Figure 1.5. Sucrose, water, and many other substances are made up of *molecules*, which we described briefly in Section 1.1 as groups of two or more atoms held together by chemical bonds. While the ions of a crystalline ionic compound lie next to each other in a lattice that extends to the edges of the crystal, the atoms of a molecule are bound cohesively into a unit—the molecule—that has a well-defined size and shape and that exists as that same unit whether it's packed into a crystal or dissolved in a liquid like water or in a gas such as the air we breathe.

Water itself is made up of molecules that consist of two hydrogen atoms and one oxygen bound firmly to each other through chemical bonds. These three atoms exist as a molecular unit of specific size and shape no matter whether the water is in the form of ice, liquid water, or steam. To show the numbers of hydrogen and oxygen atoms in a water molecule, we write the chemical formula of water as H_2O, with the subscript 2 indicating the presence of two hydrogen atoms and an implied subscript 1 after the oxygen, which indicates only a single oxygen. Sucrose, too, is made up of molecules, but it contains hydrogens and oxygens in numbers much larger than in water, and carbons as well. Sucrose molecules have the formula $C_{12}H_{22}O_{11}$, which indicates that 12 carbon atoms, 22 hydrogens, and 11 oxygens make up a molecule of sucrose. We'll discuss the forces that hold molecules together as a third kind of chemical particle (in addition to atoms and ions) in Section 3.13, after we examine chemical formulas in more detail.

A **chemical formula** is a sequence of chemical symbols and subscripts that shows the elements present in a compound and the ratio of their ions in a lattice or the actual number of their atoms in a molecule.

The **chemical formula** of a compound provides us with two pieces of information:

1. It tells us what elements are present.
2. It gives us the ratio of their ions or the actual numbers of atoms of each element present in a molecule.

The chemical formula for sodium chloride (NaCl), for example, tells us that both sodium and chlorine are present and that they are present in a one-to-one ratio. The chemical formula H_2O tells us that two hydrogen atoms and one oxygen atom make up a water molecule. Both sodium chloride and water are **binary compounds**, made up of two different elements.

A **binary compound** is made up of two elements.

Two general rules are helpful for writing chemical formulas:

1. *For compounds made up of ions, whether the compounds are binary or contain more than two elements, write the cation first, then the anion.* The cations of these compounds usually come from the alkali-metal or the alkaline-earth families of the periodic table (Section 3.7). The anions often come from the halogen family or the family beginning with oxygen. For these, name the cation first and then the anion, with the suffix *-ide* for the anion. For example, CaF_2 is *calcium fluoride*, Li_2O is *lithium oxide*, and SrS is *strontium sulfide*.
2. *For molecules containing carbon (Chapter 6), write carbon first, hydrogen second, and then the other elements present in alphabetical order.* (The names of many of these compounds don't follow any simple rules.) Examples are $CHCl_3$ (chloroform, an industrial solvent), $C_{12}H_{22}O_{11}$ (sucrose), $C_{16}H_{13}ClN_2O$ (the relaxant Valium, known chemically as diazepam), and $C_6H_8N_2O_2S$ (sulfanilamide, the first of the sulfa drugs).

EXAMPLE

Cations Come First

Name the compounds Cs_2O and $BaCl_2$.

■ Since Cs is written first in the formula Cs_2O, it represents the cation *cesium* (see back-cover end papers). Then comes O, representing the anion of the element oxygen and providing the word *oxide*. The full name of Cs_2O is *cesium oxide.*

For $BaCl_2$ we have the cation *barium* and the anion of the element chlorine, *chloride*. The name of $BaCl_2$ is *barium chloride.*

■ **QUESTION**

Name (a) MgI_2, (b) BeS, and (c) K_2O. Write the formula of a molecule of urea, a compound that we excrete to rid our bodies of excess nitrogen and that is composed of one carbon, one oxygen, two nitrogen, and four hydrogen atoms.

3.12 Valence and Chemical Formulas

The ratio of ions in a binary, ionic compound (and therefore its chemical formula as well) depends on the valence of the elements it contains. We've seen that sodium loses and chlorine gains one electron each, so they combine in a one-to-one ratio in NaCl.

Unlike sodium, the element calcium has *two* electrons in its valence shell. In its reaction with chlorine, each calcium atom transfers *two* valence electrons to chlorine atoms. Since each chlorine atom still needs only a single electron to complete its octet, one calcium atom combines with *two* chlorine atoms to form the ionic compound *calcium chloride* (Fig. 3.7). To show the association of *two* chloride anions with *one* calcium cation we write the chemical formula of calcium chloride as $CaCl_2$.

Combinations with this ratio of cation and anion can occur between any member of the alkaline-earth family and any halogen. A magnesium atom can combine with two iodine atoms, for example, to form *magnesium iodide*, MgI_2, and beryllium and fluorine can combine to form *beryllium fluoride*, BeF_2.

Similarly, one oxygen atom can *accept* both electrons from a single atom or one from each of two individual atoms. In the question at the end of Section 3.8, for example, we saw that a single oxygen atom can accept two electrons from the valence shell of a single magnesium atom to form *magnesium oxide*, MgO. In the same way an oxygen atom and a calcium atom can combine to form the compound *calcium oxide*, CaO. In these examples both the magnesium atom and the calcium atom lose two electrons from their valence shells to form a magnesium cation, Mg^{2+}, and a calcium ion, Ca^{2+}. In each case the oxygen acquires two electrons to form an oxide anion, O^{2-}. By a similar process two sodium atoms can react with one oxygen to produce *sodium oxide*, Na_2O.

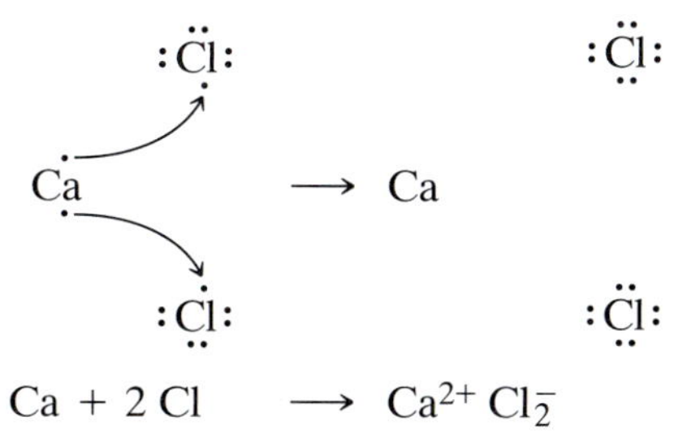

Figure 3.7 The reaction of calcium and chlorine.

The subscripts reflect the requirement that the net electrical charge of all the ions combined must be zero. In practice this means that the total number of positive charges provided by the cation(s) must equal the total number of negative charges brought by the anion(s). This occurs when

$$\text{cation subscript} \times \text{cation charge} = \text{anion subscript} \times \text{anion charge}$$

Notice that we ignore the sign (+ or −) of the charge here and concern ourselves only with the numerical value of the charge. The simplest case occurs when the electrical charges of the cation and the anion are equal, as in NaCl, MgO, and CaO. If the charges are equal, the subscripts must be equal, so we simply use (an implied) 1 for each.

The value of an ion's charge, regardless of its sign, also defines the *valence* of the ion. Here we use the term *valence* to describe what we might call the "combining power" of an element. ("Valence" comes from the Latin word *valentia*, meaning *vigor* or *capacity*.) With the ability to lose or accept a single electron, both sodium and chlorine form ions with a valence of 1; calcium's valence is 2. In this same vein, since magnesium can lose two electrons from its valence shell, magnesium has a valence of 2. Oxygen, which needs two electrons to complete its valence shell, also has a valence of 2.

EXAMPLE

Charges Must Equal

Write the chemical formulas of calcium chloride, sodium oxide, and aluminum oxide.

■ In the first case the charge on the calcium cation is 2+, while the charge on the chloride anion is 1−. Here we have

$$1 \text{ (the } Ca^{2+} \text{ subscript)} \times 2 \text{ (the } Ca^{2+} \text{ charge)} = 2 \text{ (the } Cl^{-} \text{ subscript)} \times 1 \text{ (the } Cl^{-} \text{ charge)}$$

$$2 = 2$$

The formula is $CaCl_2$.

For the compound of sodium and oxygen,

$$2 \text{ (the } Na^{+} \text{ subscript)} \times 1 \text{ (the } Na^{+} \text{ charge)} = 1 \text{ (the } O^{2-} \text{ subscript)} \times 2 \text{ (the } O^{2-} \text{ charge)}$$

$$2 = 2$$

Once again, 2 = 2 and now the formula is Na_2O.

When aluminum reacts with oxygen, the aluminum loses three valence electrons (rather than gaining 5) and acquires a charge of 3+. Oxygen gains two valence electrons and acquires a charge of 2−. For aluminum oxide,

$$2 \text{ (the } Al^{3+} \text{ subscript)} \times 3 \text{ (the } Al^{3+} \text{ charge)} = 3 \text{ (the } O^{2-} \text{ subscript)} \times 2 \text{ (the } O^{2-} \text{ charge)}$$

$$6 = 6$$

and we have Al_2O_3.

In each of these formulas the product of (valence × subscript) of one ion equals the product of (valence × subscript) of the other. Knowing the formula and the valence of one of the ions allows us to calculate the valence of the other. Normally, the valence of hydrogen, the alkali metals, and the halogens is 1, and the valence of the alkaline earths and oxygen is 2.

EXAMPLE

Calculating Valence

Calculate the valence of lead in the compound PbO_2.

■ For lead dioxide, PbO_2, a compound used in the lead-acid batteries of automobiles (Chapter 11), we can calculate the valence of the lead as follows:

1 (the implied Pb subscript) × (the valence of Pb)

= 2 (the O subscript) × 2 (the valence of O)

1 × (the valence of Pb) = 4

The valence of the lead in PbO_2, then, is 4. Since the oxygen carries a negative charge, the Pb ion must exist as Pb^{4+}.

■ **QUESTION**

What is the valence of (a) platinum in PtO_2, a compound used in the catalytic converters of automobiles (Chapter 7); (b) copper in Cu_2O, a compound used as a red pigment in glass and ceramics; and (c) mercury in HgS, a pigment used in plastics, paper, rubber, and other commodities?

3.13 Water, a Covalent Compound

As we saw in our work with the light bulb at the opening of Chapter 1, neither sucrose nor water is a good conductor of electricity. It follows that neither one can be composed primarily of ions. Water, whose molecules are much simpler than those of sucrose, offers a good example of chemical bonding in nonionic compounds.

As shown by the formula H_2O, water's molecules are formed of two hydrogens and one oxygen. Since water is not a good electrolyte, some force other than ionic attraction must bond the hydrogens to the oxygen. That force is the **covalent bond**; we will now see how it forms.

A **covalent bond** consists of a pair of electrons shared between two atoms.

Hydrogen does not release its lone valence electron to oxygen as easily as sodium loses its valence electron to chlorine. As a result, no full transfer of electrons from one atom to another takes place in the formation of water, so ions don't form as they do in sodium chloride. Instead, the oxygen and the hydrogens of water *share* their valence electrons to acquire valence shells resembling those of the inert gases. As shown in Figure 3.8, two hydrogens (each with one valence electron) and one oxygen (with its six valence electrons) combine through the sharing of two pairs of electrons. Each hydrogen, by sharing one of oxygen's electrons, acquires a valence shell of two electrons, which resembles helium's. The lone oxygen, by sharing one electron from each of the two hydrogens, acquires a valence shell resembling neon's. The elements of the water molecule, then, are held together by the sharing of two pairs of electrons. Each pair consists of one electron from the valence shell of a hydrogen and one electron from the valence shell of the oxygen.

A *shared pair of electrons constitutes a covalent bond*, which can be represented as a pair of dots (H:O:H) or, more often, as a dash (H—O—H.) Unlike

Figure 3.8 Covalent bonding in H_2O.

A **molecule** is a discrete chemical structure held together by covalent bonds.

crystals of an ionic compound, which are aggregates of enormous numbers of cations and anions held together by electrostatic forces (Fig. 3.4), **molecules** are discrete, often relatively small chemical structures held together by covalent bonds (Fig. 3.9). Molecules have distinct sizes and shapes. In contrast, ionic compounds, as we saw in Section 3.10, are formed of ions that exist in orderly lattices extending in all directions to the surfaces of their individual crystals.

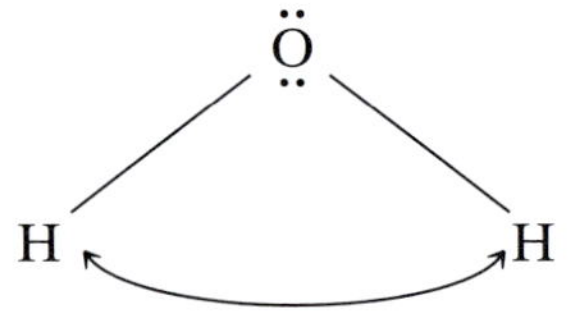

Bond angle, 104.5°

Figure 3.9 The structure of the water molecule.

Covalent compounds consist of discrete molecules; ionic compounds consist of vast crystal lattices. And while the chemical formula of an ionic compound shows the ratio of its ions, the chemical formula of a covalent compound shows the actual numbers of atoms that make up each of its molecules. The chemical formula of a covalent compound is often called its **molecular formula**. Knowing the molecular formula of a compound, we can calculate its **molecular weight**, which is simply the sum of the atomic weights of all of the atoms in each of its molecules. The molecular weight of water, for example, is simply the sum of the atomic weights of its two hydrogen atoms and one oxygen atom. Using the atomic weights of the periodic table (Fig. 3.2) to the nearest full atomic mass unit, we get $(2 \times 1) + 16$, or 18, as the molecular weight of water.

A **molecular formula** is the chemical formula of a covalent compound.

The **molecular weight** of a compound is the sum of the atomic weights of all of the atoms in each of a compound's molecules.

Generally, different elements that obtain stable electronic structures by the gain or loss of one or two electrons and that lie far from each other in the periodic table, form ionic bonds with each other. Other elements, including those of the same chemical family, occupying the same vertical column, generally form covalent bonds with each other. Hydrogen, which we called a maverick element in Section 3.5, usually forms covalent bonds with other elements. These are, of course, no more than useful generalizations. Exceptions do occur, some of them quite important.

Of all molecules, hydrogen, H_2, is the simplest. It's made up of two hydrogen atoms, each sharing its single electron with the other and each thereby forming a covalent bond (Fig. 3.10). With two atoms composing each of its molecules, hydrogen is *diatomic*. Other diatomic molecules include those of elemental nitrogen, N_2, oxygen, O_2, and all the halogens, F_2, Cl_2, Br_2, and I_2.

$$\mathrm{H} + \mathrm{H} \longrightarrow \mathrm{H{-}H}$$
$$\mathrm{H^{\cdot}} + \mathrm{{}^{\cdot}H} \longrightarrow \mathrm{H{:}H}$$

Figure 3.10 Formation of a molecule of H_2.

■ QUESTION

A molecule of propane, the fuel of liquid propane gas supplied by tanks to outdoor grills, trailers, and many homes, has the structure

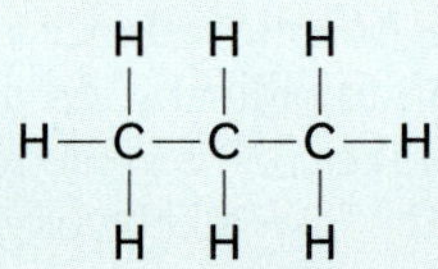

(a) What is the total number of covalent bonds in propane? (b) What is the total number of shared electrons in propane? (c) What is the molecular formula of propane?

3.14 Hydrogen Halides and the Halogens: Polar Covalent Bonds

We've seen that a hydrogen atom does not release its electron to the oxygen of water, but rather shares it in a covalent bond. Similarly, the hydrogen of a *hydrogen halide*, a compound of hydrogen and a halogen, does not release its electron to the halogen atom, but shares it instead. Figure 3.11 shows the result for *hydrogen chloride*. With this sharing, both the hydrogen and the halogen atom have obtained the electronic structure of an inert gas. We can write the resulting molecule as HCl or, to emphasize the covalent bond, H—Cl.

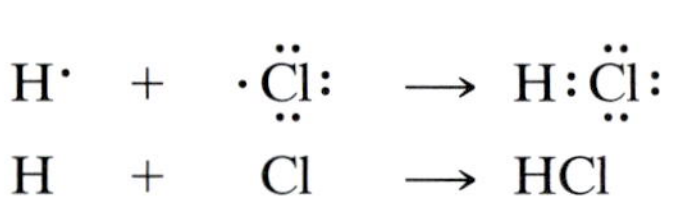

Figure 3.11 Formation of a molecule of HCl from a hydrogen atom and a chlorine atom.

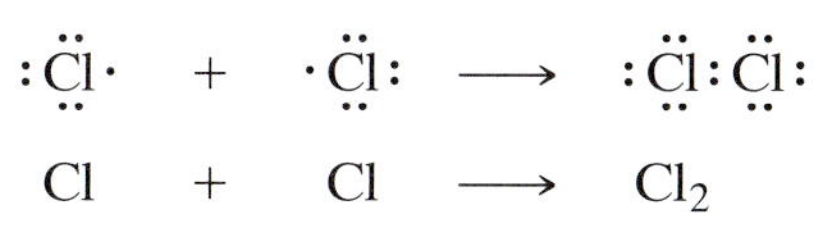

Figure 3.12 Two chlorine atoms combine to form a chlorine molecule.

Something similar results when two halogen atoms react with each other. Chlorine atoms, for example, lack a single electron for an octet. The sharing of two electrons, one from the valence shell of each of two different chlorine atoms, gives each an octet and results in the formation of a covalent bond. The result is a diatomic chlorine molecule, Cl_2 or Cl—Cl (Fig. 3.12).

In their pure states, then, both the hydrogen halides and the halogens exist as diatomic, covalent molecules. But while molecules of Cl_2 and HCl (and H_2 as well) are all held together by covalent bonds, the bond that forms H—Cl is a little different from those of Cl—Cl and H—H. Just as elements differ in their chemical reactivities and other properties, they differ in their attraction for shared electrons. Generally, the elements that lie to the upper right of the periodic table (except for the inert gases) exert a greater attraction for shared electrons than do most of the other elements, especially those that lie toward the lower left. This ability to attract the electrons of a covalent bond is called an atom's **electronegativity**.

An atom's **electronegativity** is a measure of its ability to attract the electrons of a covalent bond toward itself.

Chlorine has a greater electronegativity than hydrogen, so the chlorine of the H—Cl bond pulls the shared electrons toward itself more strongly than does the hydrogen. The result is a **polar covalent bond**, in which the electrons of the bond lie quite a bit closer to the chlorine nucleus than to the hydrogen nucleus. This shift in the position of the bonding electrons produces a **dipole**, in which the chlorine's end of the molecule bears a slightly negative charge. The charge isn't as large as the full negative charge we find on an electron, but it's a small negative charge nonetheless. The hydrogen's end of the molecule bears a small positive charge. These less-than-full positive and negative charges are indicated by the use of the small Greek letter δ (Fig. 3.13). Naturally, all the atoms of any particular element have the same electronegativity, so the shared electrons of Cl—Cl and H—H lie exactly midway between the two atoms. These molecules have no dipole.

A **polar covalent bond** is formed when the shared electrons of a covalent bond lie closer to one of the atoms than to the other.

A **dipole** results when there is a separation of positive and negative charges in the molecule.

There's a useful analogy for all this. Imagine a tug of war between two teams, with a pair of electrons on the rope, midway between the two. If the teams are equally matched (have identical electronegativities), the electrons remain at that midway point and form a nonpolar covalent bond. If one of the teams is just a bit stronger than the other (has a greater electronegativity), the electrons lie a little closer to the stronger team, resulting in a polar covalent bond. But if one of the teams completely overpowers the other, it carries the electrons completely over to its side, which results in an ionic bond (Fig. 3.14).

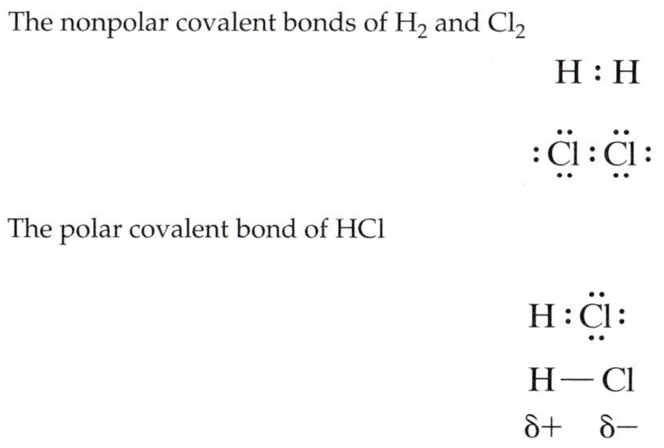

Figure 3.13 Nonpolar and polar covalent bonds.

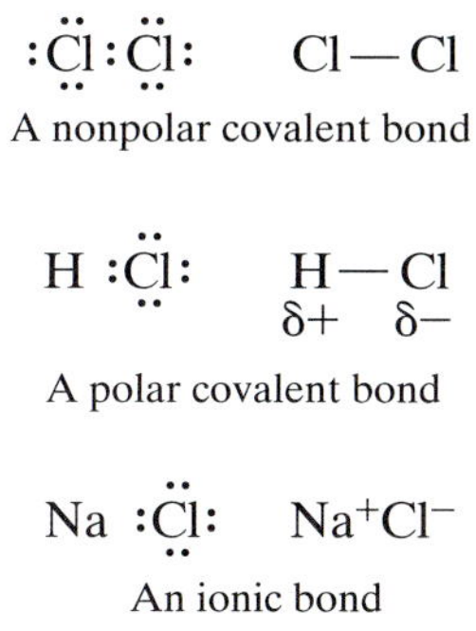

Figure 3.14 Covalent and ionic bonds.

■ QUESTION

Which of the following would you expect to contain polar covalent bonds?

(a) F—F (b) 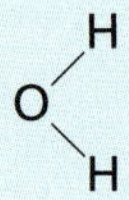 (c) O_2 (d) 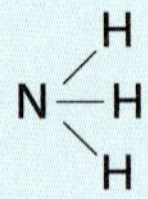 (e) Br_2 (f) I—Br

3.15 Ionization, from Molecules to Ions

Things are not quite as simple as the previous sections might suggest. Sodium chloride, a fine electrolyte, is indeed an ionic compound, and sucrose is a nonelectrolyte that does indeed exist as covalent molecules (of very complex structure) both in its pure, solid, crystalline state and also when it's dissolved in water.

But water itself is a very weak electrolyte. Actually, pure water *does* conduct electricity, but only very poorly. Although pure, liquid water is composed almost (but not quite) entirely of covalent H_2O molecules, it does nonetheless contain a very small concentration of ions. What's more, while the hydrogen halides are, indeed, covalent compounds, composed of molecules in which the hydrogen is attached to the halogen by a polar covalent bond, if you dissolve a hydrogen halide, such as HCl, in water you'll find that it is a fine electrolyte and, like the sodium chloride of Chapter 1, carries an electric current through water quite nicely. In this section we'll examine how covalent compounds can become electrolytes.

$$H:\underset{\cdot\cdot}{\overset{\cdot\cdot}{O}}:H \longrightarrow H^+ + :\underset{\cdot\cdot}{\overset{\cdot\cdot}{O}}:H^-$$

$$H_2O \longrightarrow H^+ + OH^-$$

Hydrogen cation   Hydroxide anion

Figure 3.15 The ionization of water to a hydrogen cation (a proton) and a hydroxide anion.

Although hydrogen is less efficient than sodium at releasing its valence electron, full electron transfers do take place to a very small extent between hydrogen and oxygen, and an occasional covalent water molecule, H_2O, can be transformed into an association of a cationic hydrogen ion, H^+, and a *hydroxide ion*, OH^-. This occurs as the hydrogen nucleus, a proton, leaves the electron pair that bonded it to the oxygen, with the oxygen keeping all of its electrons and also remaining bonded to the second hydrogen of the water molecule (Fig. 3.15).

Ionization occurs when a covalent molecule is converted into ions.

The conversion of a covalent molecule into ions is known as **ionization**. In pure water ionization to a hydrogen cation and a hydroxide anion occurs to only a very small extent, which explains why water is such a poor conductor of electricity. At room temperature there is at any given moment only one pair of H^+ and OH^- ions for every 556,000,000 water molecules. For comparison, to produce the same ratio of Na^+ and Cl^- ions (one pair for every 5.56×10^8 water molecules), we could add 5 g of sodium chloride (one teaspoon of table salt) to 850,000 liters of water. That's about 225,000 gallons of water, enough to cover the entire playing surface of a football field to a depth of about 7.5 in.

$$H_2O \rightleftharpoons H^+ + OH^-$$

Figure 3.16 The reversible ionization of a water molecule.

With both the ionization and the reverse reaction—the recombination of ions into water molecules—occurring readily and rapidly throughout the liquid, there is always a dynamic *equilibrium* of the two processes that keeps the concentration of the ions constant at one pair of ions for every 5.56×10^8 water molecules. This equilibrium is symbolized chemically by two arrows. One shows the ionization as the separation of a water molecule into a proton and a hydroxide ion. The other, reversed in direction, shows the recombination (Fig. 3.16).

$$H\ :\underset{\cdot\cdot}{\overset{\cdot\cdot}{Cl}}: \longrightarrow H^+ + :\underset{\cdot\cdot}{\overset{\cdot\cdot}{Cl}}:^-$$

$$H—Cl \longrightarrow H^+ + Cl^-$$

Figure 3.17 The ionization of a molecule of H—Cl to a hydrogen ion and a chloride anion.

While the hydrogen halides are all covalent compounds that exist as gases under ordinary conditions, all but hydrogen fluoride ionize completely when they dissolve in water. (HF ionizes only slightly in water, producing only a few H^+ and F^- ions.) Hydrogen chloride (HCl), hydrogen bromide (HBr), and hydrogen iodide (HI) all ionize completely in water, each producing a hydrogen ion (H^+) and the associated halide ion (Fig. 3.17).

Which would conduct an electric current more effectively, a solution of 0.1 g of HBr in 100 mL of water, or a solution of 0.1 g of HF in 100 mL of water? Why?

■ QUESTION

3.16 Scouring Pads and Kitchen Magnets Revisited

In the opening demonstration we saw that the metallic iron of a scouring pad can combine with the gaseous oxygen of air to form what we commonly call rust. This brown granular material is usually a complex mixture of substances, but most of the iron in rust is combined with oxygen in the form of a material called *ferric oxide*, Fe_2O_3. As we saw, the rust that's formed has properties different from those of either of the elements that compose it. It's a solid rather than a gas like oxygen, and it isn't attracted to a magnet, as elemental iron is. We now know that the rust forms through interchanges of valence electrons between the iron and the oxygen, and that the driving force for this interchange is the formation of electronic structures resembling those of the noble gases. As a result of this interchange a new material forms, with new properties.

Given that the valence of the oxygen in ferric oxide is 2−, what is the valence of the iron?

■ QUESTION

PERSPECTIVE

How Do We Know?

With this chapter we have completed the foundation for our continuing study of chemistry. It's a good point to stop, reflect for a moment, and ask, "How do we know?" How do we know that the hydrogen halides are all formed of covalent molecules, and that HCl, HBr, and HI ionize completely when they dissolve in water? How do we know that at room temperature there is at any given moment only one pair of H^+ and OH^- ions for every 556,000,000 water molecules (Section 3.14)? How do we know, for that matter, that atoms exist, that they have nuclei at their cores and electrons in their shells, and that there are such things as protons and neutrons? How do we know any of this?

The answer is that we ask experimental questions of nature and we interpret the answers so that they make as much sense as possible. That, in essence, sums up the scientific method we examined in Chapter 1. At its heart, *chemistry is an experimental science*. There's no way in the world to know that ordinary table salt will cause a light bulb to light up and that table sugar won't, unless someone tries it. The results of that experiment and others like it tell us all we know of table salt and everything else. There's no other way to learn about our physical universe except through the experimental questions we ask and the answers we get by asking them.

In the remainder of this book we're going to learn of the results of experimental questions others have asked, and of the answers they got and of the theories that these answers eventually coalesced into. Sometimes we'll follow the reasoning that leads to our current view of the world, as it exists today. Occasionally we'll examine the experimental tests themselves that lead to this view. In every case, though, what we learn comes from real tests, real experiments, real questions that people through the centuries have asked, and the hypotheses and theories that have grown from them.

In the next few chapters we'll examine more closely the chemistry of atoms, their nuclei, and the electrons surrounding the nuclei. We'll examine further the covalent bonds they form as well as many examples of the covalent molecules that are formed by these bonds. Initially, we'll examine all this with special emphasis on one of our major themes: how we obtain, store, and use chemical energy. We'll begin with the energy of the atomic nucleus and how we use it now, and how we may be able to use it even more effectively in the future to provide plentiful energy, to cure illness and save lives, and to learn more about our own past. ■

EXERCISES

1. Following are a statement containing a number of blanks and a list of words and phrases. The number of words equals the number of blanks within the statement, and all but two of the words fit correctly into these blanks. Fill in the blanks of the statement with those words that do fit, then complete the statement by filling in the two remaining blanks with correct words (not in the list) in place of the two words that don't fit.

In its modern form the __________ is an organization of all the known elements arranged in order of increasing __________. Elements that lie in the same __________ have similar __________, have the same number of __________, and belong to the same chemical __________. The table shows the element's chemical __________, atomic number, and __________, which is the average mass of all the __________ of the element, weighted for the __________ of each. The table is particularly useful for predicting the outcome of chemical reactions between __________ since an atom of any element tends to react so as to convert its electronic structure to that of a nearby __________. When atoms react by a complete __________ of electrons, the products are __________. When they react by __________ pairs of electrons, the result is a __________. While ionic compounds exist as crystals made up of extensive __________ of ions, covalent compounds are composed of discrete __________. If the atoms that form the covalent bonds have significantly different __________, the bond that forms is a __________.

abundance
atomic number
atomic weight
column
covalent bond
crystals
electronegativities
elements
halogens
ions
isotopes
lattices
molecules
noble gas
periodic table
polar covalent bond
properties
symbol
transfer
valence electrons

2. Define, illustrate, or explain each of the following:

a. alkaline earth metal
b. binary compound
c. chemical family
d. chemical formula
e. covalent bond
f. halogen
g. ionization
h. molecular formula
i. molecular weight
j. molecule
k. valence electron

3. Identify the elements and the compounds among the following: (a) oxygen, (b) water, (c) chlorine, (d) sucrose, (e) hydrogen, (f) sodium chloride, (g) sulfur, (h) propane, (i) argon, (j) carbon.

4. Identify each of the following as consisting principally of molecules, principally of ions, or principally of atoms: (a) water, (b) carbon, (c) sodium chloride, (d) propane, (e) lithium iodide, (f) sucrose, (g) potassium.

5. Write chemical formulas for (a) lithium bromide, (b) calcium sulfide, (c) sodium sulfide, (d) beryllium oxide, (e) the chlorine molecule, and (f) the compound formed on reaction of silicon and hydrogen.

6. Write the chemical formulas of all the ionic compounds that could result from reaction of a mixture of lithium and potassium with a mixture of fluorine and bromine.

7. Name a commercial product that contains (a) elemental chlorine, (b) the iodide ion, (c) both the element nitrogen and the element chlorine, (d) the element fluorine, (e) the chloride ion, (f) the element sulfur, (g) elemental iodine.

8. Which of the following contain (or are a form of) *elemental* carbon? Which contain *the element* carbon? (Reviewing the Perspective of Chapter 2 may be helpful.) (a) carbon dioxide, (b) graphite, (c) water, (d) diamond.

9. Where might you find (a) elemental oxygen, (b) elemental gold, (c) the element carbon, (d) sodium ions, (e) the element hydrogen, and (f) the elements carbon, hydrogen, nitrogen, and oxygen in a single compound?

10. What element has the following:

a. an atomic weight of almost exactly 14
b. an atomic number of 16
c. one electron in its first quantum shell
d. two quantum shells, both of which are filled with electrons
e. the electron structure 2 8 2
f. a valence shell filled with a total of two electrons
g. a total of three quantum shells, with two electrons in its valence shell
h. a total of two quantum shells, with three electrons in its valence shell
i. twice as many electrons in its second shell as in its first

11. Each of the following pairs of elements reacts to form a binary compound. Which form an ionic compound and which form a covalent compound?

a. sulfur and oxygen
b. hydrogen and fluorine
c. barium and chlorine
d. sodium and bromine
e. nitrogen and oxygen
f. hydrogen and carbon
g. lithium and iodine
h. magnesium and oxygen

12. Write the chemical symbols of the following elements: (a) aluminum, (b) argon, (c) calcium, (d) carbon, (e) cesium, (f) fluorine, (g) gold, (h) hydrogen, (i) lead, (j) manganese.

13. Name the elements that the following chemical symbols represent: (a) Ag, (b) Be, (c) He, (d) I, (e) Kr, (f) Mg, (g) Na, (h) Si, (i) Zn.

14. Write the Lewis structures of (a) a chlorine atom, (b) a fluorine atom, (c) a fluoride anion, (d) a sodium atom, (e) a magnesium atom, (f) a carbon atom, (g) an aluminum atom,

(h) a boron atom, (i) a molecule of HCl, (j) a water molecule, (k) a molecule of hydrogen peroxide, H_2O_2.

15. How many electrons would an oxygen atom have to lose to obtain the same electronic structure as an inert gas? Why doesn't an oxygen atom lose these electrons, rather than acquire two, when it reacts with other elements?

16. a. What are two properties that all elements in the same column of the periodic table as helium have in common?

b. What are two properties that all elements in the same column of the periodic table as lithium have in common?

17. What would you expect to happen if you place a small piece of potassium metal into water?

18. What change occurs in the valence shells of elements as we move from left to right in any given row of the periodic table?

19. What change occurs in the valence shells of elements as we move down a column in the periodic table?

20. Why do elements in the same column of the periodic table show similar chemical behavior?

21. What is the valence of sulfur in the gas *sulfur dioxide*, SO_2? In the gas *sulfur trioxide*, SO_3?

22. The element oxygen can exist in two molecular forms. In its most common form, it exists as a diatomic molecule, O_2, but it can also exist as ozone, in which three atoms of oxygen combine to form a covalent molecule. Write the molecular formula of ozone.

23. Like water, hydrogen peroxide (Fig. 3.18) ionizes slightly to produce a hydrogen cation and an anion. What is the chemical structure of the anion resulting from the ionization of H_2O_2? Show the structure in two ways: (1) with pairs of electrons represented by pairs of dots and (2) with covalent bonds represented by dashes.

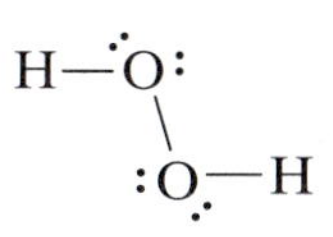

Figure 3.18 The hydrogen peroxide molecule, H_2O_2.

24. Pure hydrogen chloride is a gas under ordinary conditions. Molecules of pure, gaseous hydrogen chloride consist of a hydrogen atom covalently bonded to a chlorine atom. When hydrogen chloride gas dissolves in water, the resulting solution easily conducts an electric current. As the hydrogen chloride molecules enter the water, what process occurs to generate an electrolyte?

A Little Arithmetic and Other Quantitative Puzzles

25. Give the valence of the cation in each of the following compounds: (a) SrF_2, (b) $ZnCl_2$, (c) $AlCl_3$, (d) CuO, (e) Fe_2O_3.

26. What is the total number of electrons being shared among all the covalent bonds of a molecule of methane, CH_4?

27. What would you write as the molecular formula for the nitrogen dioxide of Section 3.3? (*Di-* is a prefix meaning "two.") What is the valence of the nitrogen of this molecule?

28. Bromine's atomic weight is 79.9. Two isotopes make up virtually all the bromine in the universe. One, with a mass number of 79, makes up 50.69% of all the bromine atoms. What is the mass number of the other isotope?

29. The permanganate anion, MnO_4^-, bears a single negative charge. Knowing that oxygen acquires an octet by acquiring two electrons, what do you calculate as the valence of the manganese atom in this ion?

30. What is the valence of neon?

Think, Speculate, Reflect, and Ponder

31. Many compounds, such as carbon dioxide, CO_2, methane, CH_4, water, H_2O, and sucrose, $C_{12}H_{22}O_{11}$, exist as molecules. Hydrogen gas, H_2, also exists as molecules. Does this mean that the gas H_2 is a compound? Explain.

32. The covalent compound acetylene, the fuel of the oxyacetylene torch used by welders, has the molecular formula C_2H_2. The covalent compound benzene, a commercial solvent, has the molecular formula C_6H_6. Each of these compounds contains carbon and hydrogen atoms in a one–to–one ratio. Would it be correct to write the chemical formula of each as CH? Explain your answer.

33. Elemental oxygen exists as a diatomic gas, O_2. How many electrons do the two oxygen atoms have to share between them to form this diatomic molecule? How many covalent bonds unite the two oxygen atoms of the O_2 molecule?

34. Arranging the elements in order of increasing atomic number also places the elements in order of increasing atomic weight, with a few exceptions. One of these appears in the sequence of tellurium (Te) and iodine (I). The atomic weight of tellurium (atomic number 52) is 127.6. The atomic weight of iodine (atomic number 53) is 126.9. How do you account for this?

35. Hydrogen chloride, HCl, exists as a gas composed of covalent molecules. If we dissolve this gas in water, the water conducts an electric current well. If we dissolve HCl in benzene, a solvent of molecular formula C_6H_6, the solution does not conduct an electric current. (Pure benzene does not conduct an electric current either.) What do you conclude from this observation?

36. The element carbon, as we will see later, forms the basis for all life as we know it. Science fiction writers sometimes speculate on the properties of different forms of life based on an element other than carbon. What element do you think they usually choose, and why? (Referring to Fig. 3.2 may help you answer this question.)

37. Some forms of the periodic table show hydrogen twice, once at the top of the column of the alkali metals and once again at the top of the column of the halogens. Suggest a reason for putting hydrogen into both families. (*Hint:* Consider the structure of the hydrogen atom and its similarities to the other elements of each of these families.)

38. Silver chloride, AgCl, is not soluble in water. Explain how you might be able to determine whether this is an ionic or a covalent compound. (*Hint:* Refer to Section 1.1.)

39. Are there any exceptions to the octet rule? If your answer is no, explain why there aren't any. If your answer is yes, describe one and explain why it is an exception.

Web Questions

Use the Web to answer or accomplish each of the following. Report your Web source for each response.

1. Section 3.6 contains the sentence "Although others had also noted the same sort of periodic repetition of properties even before Mendeleev's inspiration, none had pursued the idea as vigorously or in as much detail as the Russian chemist." Identify two other scientists who reported a periodic repetition of properties among elements before Mendeleev devised his periodic table. What repetitions did they find? Among what elements?

2. Figure 3.2 represents one of several modern versions of the periodic table. Find two other versions, one of which represents a three-dimensional periodic table.

3. Identify three different compounds each of which has the molecular formula $C_6H_{12}0_6$. Describe briefly some similarities and some differences among them.

4. Find the molecular formulas of the pain-relievers aspirin and ibuprofen; the antibiotics penicillin and erythromycin; the sweeteners aspartame and saccharin.

5. Write a brief summary of the life of Linus Pauling. Include descriptions of his contributions to the topics of this chapter.

Discovering the Secrets of the Nucleus

FROM A PHOTOGRAPHIC MYSTERY TO THE ATOMIC BOMB

Tracks of subatomic particles in a magnetic field.

The Bomb

The world's first atomic explosion, July 16, 1945, at Alamogordo, New Mexico.

At 5:30 A.M., during a lull in a storm in the predawn darkness of July 16, 1945, near Alamogordo, New Mexico, the world's first atomic bomb was detonated. About 6 kg ($13\frac{1}{4}$ lb) of plutonium exploded with the force of 20,000 tons of TNT. The resulting fireball, 10,000 times hotter than the sun, lit up the sky. An observer watching unprotected 20 miles away was temporarily blinded. People saw the light or felt the blast for hundreds of miles. It produced a cloud rising 8 miles, shaped like a mushroom. The hundred-foot steel tower holding the bomb vaporized and the desert sands below it melted to glass for half a mile. The power of the nucleus had been unleashed.

In this chapter we'll trace the events of just under 50 years that led from the discovery, in 1896, of a puzzling image that formed on a sealed photographic plate to the construction and detonation, in 1945, of the world's first nuclear weapon. As we follow this story of the discovery of the secrets of the nucleus, we'll learn how and why nuclear energy was first applied in the creation of a weapon of massive destruction.

Following this, in Chapter 5, we'll turn from warfare to the peaceful uses of atomic energy. We'll learn how transformations of the nucleus are used to produce commercial electric power, to cure illness and save lives, and to reveal something of the history of our society and of our planet. We'll also examine some of the real, potential, and imagined hazards inherent in today's applications.

4.1 The Discovery of Radioactivity: How the Unexpected Exposure of a Photographic Plate Led to Two Historic Nobel Prizes

Antoine Henri Becquerel. In 1903 he shared a Nobel Prize with Marie and Pierre Curie for the discovery of radioactivity.

In 1896 Antoine Henri Becquerel, a professor at the École Polytechnique, Paris, and the son and grandson of physics professors, made the unexpected observation that begins our story. Becquerel's discovery occurred while he was looking for a (nonexistent) connection between X-rays, which had been discovered just the year before, and *phosphorescence*, a phenomenon that causes certain substances to glow visibly for a short time after they have been exposed to some forms of radiation, such as ultraviolet light.

Becquerel thought (incorrectly) that the generation of X-rays might somehow be connected to phosphorescence, that phosphorescing substances might emit X-rays as well as visible light. To test this hypothesis, he conducted a series of experiments based on the observation that crystals of uranium compounds phosphoresce after they have been exposed to sunlight. Becquerel placed a crystal of a uranium compound on a photographic plate well wrapped in black paper. He then placed the combination of the crystal and the wrapped plate in bright sunlight so that the ultraviolet component of the sunlight would induce phosphorescence.

If the phosphorescing crystal also emitted X-rays, he reasoned, these rays would penetrate the black paper and expose the photographic plate.

Sure enough, when Becquerel developed the plate he found that it contained an exposed spot corresponding to the position of the crystal. Since he had already shown that a similarly wrapped plate placed in the bright sunlight *without* the uranium compound on top of it would remain unexposed, he concluded (correctly) that the crystal of the uranium compound was responsible for the exposed spot and (incorrectly) that the crystal emitted X-rays as it glowed with the visible light of phosphorescence.

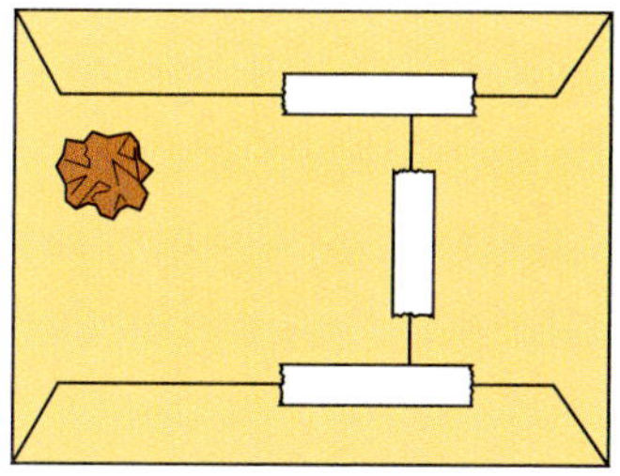

Placing a uranium compound on a wrapped sheet of film produces...

...an exposed spot on the developed negative.

Figure 4.1

The true source of the photographic image was revealed when a period of cloudy weather forced Becquerel to suspend his research. Without sunlight to induce the phosphorescence (and, Becquerel thought, the accompanying generation of X-rays), he stored the crystal of the uranium compound in his desk drawer, near a well-wrapped, unexposed photographic plate that he also kept there. A few days later, as he was developing other plates used in his work, Becquerel also developed this wrapped and presumably unexposed plate. To his amazement, it showed a distinct spot that indicated intense exposure. *Something* had penetrated the black paper and had caused a strong image to form on the plate even though the photographic emulsion had never been exposed to phosphorescence or to X-rays or to any other known form of radiation.

With additional experiments Becquerel soon found that this mysterious, penetrating radiation originated in the element uranium itself and had no connection whatever with ultraviolet light, X-rays, or phosphorescence. Almost two years after the original discovery, Marie Sklodowska Curie, a student of Becquerel's (and often referred to more simply as Madame Curie) took up the study of this strange form of radiation. She gave it the name **radioactivity**, described many of its properties, showed that it was also emitted from elements other than uranium, and proposed that its origin lay in changes that occur within atoms themselves. It was this radioactivity that had penetrated through the black wrapping paper to form an image on the plate in Becquerel's desk drawer (Fig. 4.1). In continuing studies of radioactivity she and her husband Pierre, who had joined her in her work, soon discovered two additional radioactive elements, *radium* and *polonium*. Later work confirmed Marie Curie's ideas on the origin of radioactivity. Today we recognize that this form of radiation results from the spontaneous emission of radiant energy and/or high-energy particles from the nucleus itself.

Radioactivity is the spontaneous emission of radiant energy and/or high-energy particles from the nucleus of an atom.

In 1903 Marie Curie, a native of Poland, became the first woman to receive a Nobel Prize as she, Pierre, and Becquerel shared the award in physics for the discovery of radioactivity. In 1911 she again made history, this time by becoming the first person ever to receive the Nobel Prize a second time. This second prize was awarded in the field of chemistry for her discovery of radium and polonium. Pierre, who might well have shared the honor with her, had been killed five years earlier when he was struck by a horse-drawn cart in the streets of Paris. Further recognition of their achievements came in the naming of an element—curium (Cm), atomic number 96—and a unit of radioactivity—the curie (Sec. 4.5)—in honor of Pierre and Marie Sklodowska Curie.

(In 1935 the Curies' daughter, Irène Joliot-Curie, also received the Nobel Prize. For producing radioactive isotopes of nickel, phosphorus, and silicon by artificial means, Irène and her husband Frédéric Joliot-Curie shared the award in chemistry.)

Marie Sklodowska Curie with her daughter, Irène. Irène later shared a Nobel Prize with her husband, Frédéric Joliot-Curie.

The story of the discovery of radioactivity provides a fine example of *serendipity*, which is the ability to make fortunate or happy discoveries quite unexpectedly. It was entirely by chance that the Paris weather happened to turn cloudy during

Becquerel's investigations; lucky that he stored the uranium crystal and the photographic plate in the same drawer; and fortunate that he decided to develop the plate, even though he might have assumed that it had remained unexposed. And yet Becquerel's serendipitous discovery depended on more than just luck. He had to be thoroughly prepared intellectually for the work he was doing in order to grasp the significance of the mysterious spot on the developed plate and to conclude that he had discovered a new form of radiation. What's more, he had to use the scientific method to confirm his discovery and to investigate its implications with additional experiments. We'll see other examples of serendipitous discoveries as we progress through our study of chemistry.

(The word *serendipity* itself was coined in 1754 by the English politician and writer Horace Walpole. It refers to the ability to make accidental discoveries similar to those of the characters of the children's story *The Three Princes of Serendip*.)

■ QUESTION

What *incorrect* assumption did Becquerel make in beginning his investigation of uranium, sunlight, and photographic paper? What *correct* conclusion did he reach when he found spots on paper that was in contact with uranium exposed to the sun? What *incorrect* conclusion did he reach when he made this observation? What result caused him to turn the entire project over to Marie Sklodowska Curie for further investigation?

4.2 Radioactivity and Radioactive Decay

In 1899, three years after the discovery of radioactivity, the British physicist Ernest Rutherford showed that its rays consist of more than one form of radiation. He identified two kinds, which he differentiated by letters of the Greek alphabet: (1) the *α rays* (*alpha rays*), easily stopped by paper, metal foil, or even a few centimeters of air (as well as by skin and lightweight clothing), and (2) the more penetrating *β rays* (*beta rays*). It takes a sheet of metal a few millimeters thick, a block of dense wood, or heavy clothing to stop β rays. A third component of the radiation, *γ rays* (*gamma rays*), was discovered a short time later by the French scientist P. V. Villard. Even more deeply penetrating than either the α or β rays, γ rays pass through most substances with ease. Blocking them requires a shield made of many centimeters of lead or an even thicker wall of concrete (Fig. 4.2).

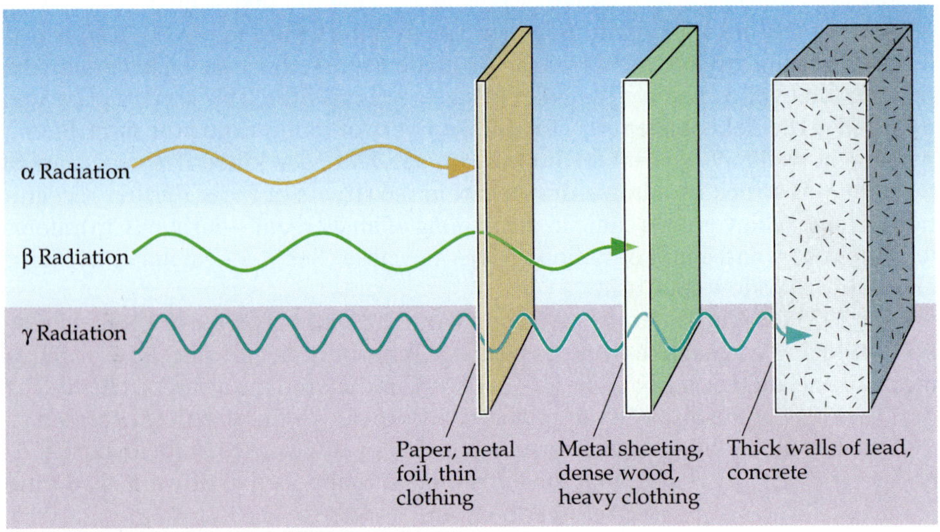

Figure 4.2 The penetrating power of radiation.

Ernest Rutherford was born in New Zealand and did much of his work with radioactivity, first at McGill University, in Canada, and then in Manchester, England. His studies gave us our modern view of the atom as a chemical particle consisting of a dense, positively charged central nucleus surrounded by clouds of negatively charged electrons. In 1908 Rutherford received the Nobel Prize in chemistry for his contributions to our understanding of radioactivity.

As a result of the work done by Rutherford and his students and co-workers, we now know that the radiation discovered by Becquerel comes directly from the nuclei of the atoms of radioactive isotopes. We also know that as the radioactive atoms emit these rays, their atomic numbers and mass numbers often change. (Sometimes only the atomic number changes, sometimes only the mass number, sometimes both, and sometimes, but far less often, neither. We'll see an example of this last case in Chapter 5.)

Since each atomic number corresponds to a specific element (recall that the atomic number of hydrogen is 1; helium, 2; lithium, 3; etc.), radioactivity often results in the transformation of atoms of one element into atoms of (one or more) other elements. Furthermore, if the newly formed nuclei are themselves radioactive they, too, emit radiation and are themselves transformed into still other nuclei until, eventually, the final product of a series of transformations is a *stable* or *nonradioactive* nucleus. The entire and sometimes intricate path followed in these transformations, from one radioactive atom to another and eventually to a stable atom, is called a chain or sequence of *radioactive decay*. As we've just seen, the radiation that accompanies this decay consists of α, β, and γ rays.

Ernest Rutherford discovered α rays and β rays. He received a Nobel Prize in 1908 for his contributions to our understanding of radioactivity.

■ QUESTION

Which component of radioactivity would be stopped by this single page? Which two components would be stopped (completely or almost completely) by this entire book? Which component would pass, almost undiminished, through both this single page and this entire book?

4.3 Nuclear Notation

To follow the events of even the simplest path of radioactive decay and the changes that occur in the nucleus of an atom as a result of its radioactivity, and to understand the nature of α, β, and γ rays, we examine here a convenient form of notation that shows both the atomic number and the mass number of an isotope or a subatomic particle. With the particle's mass number written to the top left of its symbol, and its atomic number to the bottom left, we can see at a glance its mass and its total nuclear charge.

For subatomic particles that carry an electrical charge it's also useful to show the charge at the upper right of the symbol. Occasionally, we'll use only the symbol and its charge or the symbol alone. With this system (Fig. 4.3) the common isotope of hydrogen, protium (Sec. 2.6), is written $^{1}_{1}H$, and the rarer deuterium (often written simply as D) is $^{2}_{1}H$. Neither of these two naturally occurring isotopes of hydrogen is radioactive. Tritium, a hydrogen isotope that is radioactive and occurs in nature only in extremely small quantities, is $^{3}_{1}H$. As another example, virtually all the helium in the universe occurs as the stable isotope $^{4}_{2}He$, with a mass number of 4 and atomic number 2. Fluorine occurs almost exclusively as a stable isotope of mass number 19 and atomic number 9, $^{19}_{9}F$. Somewhat simpler forms of notation, such as fluorine-19 or F-19, emphasize the mass number of a particular isotope.

39 Mass number
19 Atomic number
1+ Charge

Protium	$^{1}_{1}H$
Deuterium	$^{2}_{1}H$
Tritium	$^{3}_{1}H$
Helium atom	$^{4}_{2}He$
Helium nucleus	$^{4}_{2}He^{2+}$
Uranium–235	$^{235}_{92}U$
Potassium atom	$^{39}_{19}K$
Potassium cation	$^{39}_{19}K^{1+}$

Figure 4.3 Notation showing atomic number, mass number, and charge.

EXAMPLE

Symbolic Gold

Almost all atoms of gold (Au) have a mass number of 197. Write the chemical symbol for this most abundant isotope of gold.

■ Referring to the periodic table, or to a table of elements and their atomic numbers, we find that the atomic number of gold is 79. Writing 79 as a subscript to the left of the chemical symbol for gold, and writing the mass number (197) as a superscript to the left of the symbol, we have $^{197}_{79}Au$.

A **radioisotope** is an isotope of an element that emits radioactivity.

Like hydrogen, many other elements can exist both in the form of stable isotopes and also as naturally occurring or manufactured radioactive isotopes, known as **radioisotopes**. About 98.9% of all carbon atoms, for example, exist as the stable isotope $^{12}_{6}C$, with another stable isotope, $^{13}_{6}C$, making up virtually all of the remainder. The existence of a trace of a naturally occurring radioisotope $^{14}_{6}C$, also known as *carbon-14*, *C-14*, or *radiocarbon*, allows us to date ancient objects, as we will see in Chapter 5.

■ **QUESTION**

How many neutrons are in the nuclei of each of the stable isotopes of carbon, $^{12}_{6}C$ and $^{13}_{6}C$, and in the radioisotope $^{14}_{6}C$? (For help, see Sec. 2.5.)

4.4 α Particles, β Particles, and γ Rays

This same notation helps describe α, β, and γ radiation. The α rays that Rutherford discovered are composed of streams of high-energy helium nuclei—combinations of two protons and two neutrons—given off during radioactive decay and traveling at about 5% to 7% the speed of light. In effect, these rays consist of large numbers of fast-moving helium atoms stripped of their two surrounding electrons, or $^{4}_{2}He^{2+}$. Other ways of writing them are $^{4}_{2}\alpha$ and simply α.

The α rays penetrate into and through matter only very poorly. With their two positive charges and their mass of 4 amu, the helium nuclei that make up these rays simply can't make much headway through the enormous numbers of atoms that make up any bit of matter. Frequent collisions with atoms cause the rays to be stopped easily, even by a thin sheet of paper, as we saw in Section 4.2. In addition, since a nucleus that emits an **α particle** loses two protons and two neutrons, the atom losing the α particle drops by two units in atomic number and by 4 amu in mass.

Alpha particles are high-energy helium nuclei emitted by radioactive nuclei and traveling at 5% to 7% the speed of light.

Beta radiation consists of a stream of high-energy electrons traveling in a range of velocities, up to about 90% the speed of light. Each of the high-energy electrons that make up the β rays carries a charge of 1− and has virtually no mass. Unlike the α particle, the β particle isn't equivalent to the nucleus of any atom. As a result, we can't use atomic numbers and mass numbers (as we defined them in Sec. 2.5) to describe this particle. Nonetheless, we can translate the β particle's lack of protons and neutrons into a zero at the upper left of the symbol, where we would ordinarily write a mass number. In place of an atomic number we'll write 1− to the lower left to show the charge. This 1−, which represents the electrical charge on the β particle, serves a function similar to the atomic number, which represents the number of protons and therefore the positive charge on an atomic nucleus. For β rays (and the particles that form them) we can now write $^{0}_{1-}e$, $^{0}_{1-}\beta$, β^{-}, or simply β.

With their high speed, single negative charge, and extremely small mass, the high-energy **β particles** pass through matter much more easily than do α particles and are stopped only by heavy clothing or thick walls. They have about 100 times the penetrating power of α particles.

Beta particles are high-energy electrons emitted by radioactive nuclei and moving at up to 90% the speed of light.

The loss of a single β particle produces an *increase* of one unit in the atomic number of the nucleus. This follows from the requirement that the creation or destruction of net electrical charge can never occur in a chemical reaction, or in any other process for that matter. No positive (or negative) charge can appear or vanish unless the opposite charge also appears or vanishes at the same time. The appearance or disappearance of charges occurs through the formation of pairs of ions or of subatomic particles bearing opposite electrical charges, or their combination and destruction or neutralization.

With this in mind we can see that if the nucleus of an atom *loses* a β particle, which carries a single unit of *negative* charge, then the nucleus must balance this loss of one unit of negative charge by simultaneously *increasing* in *positive* charge, again by one unit. For a physical picture of the ejection of a β particle from the nucleus, we can think of a neutron as a combination of a proton and an electron. Since the electron's mass is negligible compared with the mass of a proton or a neutron (Sec. 2.4), combining an electron with a proton (or removing an electron from a neutron) would hardly affect the mass of the particle. Moreover, as a union of one positive and one negative charge the combination would be electrically neutral, as a neutron indeed is. In effect, then, the loss of a β particle converts a neutron to a proton without substantially changing the total mass of the nucleus but nonetheless increasing its atomic number by one unit.

Like visible light, radio and television waves, and X-rays, the **γ rays** that constitute the third member of this set of radioactive emissions are a form of electromagnetic radiation. Because of their high energies (even greater than the energies carried by X-rays) and their ability to penetrate deeply into matter, they can do considerable biological damage, as we'll see in Chapter 5. Without either mass or charge, γ rays are written $^{0}_{0}\gamma$ or simply γ.

Gamma rays are a form of high-energy electromagnetic radiation. They have no mass and carry no electrical charge.

The physical characteristics of this kind of radiation help explain its very great penetrating power. Traveling with high energy at the speed of light and without either mass or charge, γ rays penetrate matter easily. Complete protection from their effects requires the use of thick lead or concrete shields. They have about 100 times the penetrating power of β rays.

X-ray examination of luggage at a security station.

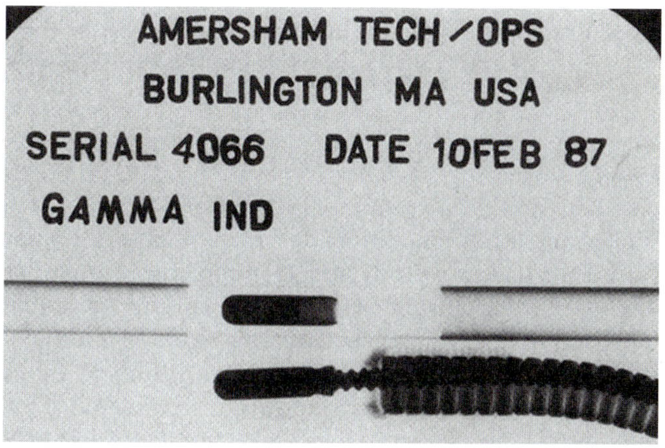

Gamma ray analysis of a fitting for a medical device shows it to be free of flaws.

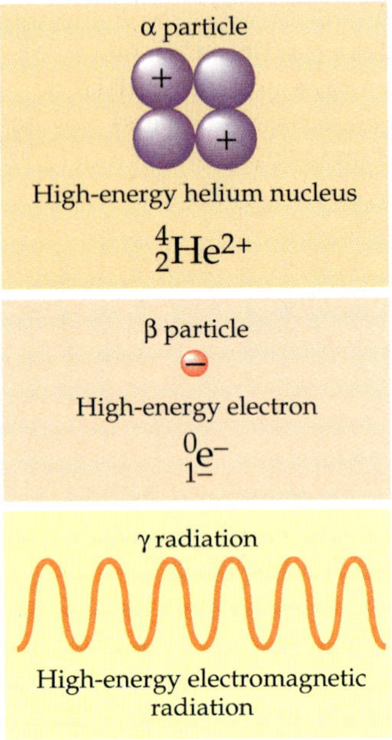

Figure 4.4 The components of α rays, β rays, and γ rays.

Ionizing radiation is any form of radiation capable of converting electrically neutral matter into ions.

In an interesting commercial application, γ rays are used to detect flaws in metal parts and structures in much the same way as X-rays are used to detect and diagnose fractures in bones or cavities in teeth, or to examine luggage at security stations. A sample of the radioisotope cobalt-60, enclosed in a protective capsule, is placed in or near the metal part under examination. The γ rays from the radioisotope pass through the metal and strike a sheet of photographic film or an electronic detector on the other side. Examining the image produced on the film or on a video screen reveals flaws, much as medical and dental X-rays show the interior of a bone or a tooth or the contents of luggage.

When high-energy α or β particles or γ rays collide with electrically neutral matter, they can knock electrons away from atoms or molecules and generate pairs of ions. Along with other forms of radiation that are powerful enough to do this, including cosmic rays and X-rays, these components of radioactivity are known as **ionizing radiation**. Figure 4.4 and Table 4.1 sum up their characteristics.

Table 4.1 The Ionizing Radiation of Radioactivity

Radiation	Component	Symbols	Velocity	Penetrating Power
α rays	Helium nuclei	$^{4}_{2}He^{2+}$, $^{4}_{2}\alpha$, α	5% to 7% of the speed of light	Low
β rays	Electrons	$^{\ 0}_{-1}e$, $^{\ 0}_{-1}\beta$, β^{-}, β	Varies; up to 90% of the speed of light	Moderate
γ rays	Electromagnetic waves	$^{0}_{0}\gamma$, γ	Speed of light	High

■ QUESTION

How does an atom's atomic number change when its nucleus loses a(n): (a) α particle, (b) β particle, (c) γ ray? How does that atom's mass number change with the loss of each of these?

4.5 Naturally Occurring Radioactive Decay: From Tritium to Helium, from Carbon to Nitrogen, from Uranium to Lead

We can now use our knowledge of nuclear radiation and its symbols to examine a few representative illustrations of naturally occurring radioactive decay. As we do, we'll use nuclear notation to write the kinds of equations that will help explain, both here and in the next chapter, how the mysteries of the nucleus were explored, how the atomic bomb was built, and how we now use nuclear reactions in medicine, in commerce, and in the study of history.

Tritium, the simplest of all the radioisotopes, decays by *β emission* (the loss of a β particle) to form $^{3}_{2}He$, a stable isotope of helium:

$$^{3}_{1}H \longrightarrow {}^{3}_{2}He + {}^{\ 0}_{-1}\beta$$

When the tritium nucleus loses the electron, one of the neutrons of its nucleus is transformed into a proton. Although its mass number remains unchanged at 3, the atom increases by one unit in atomic number and becomes the element helium

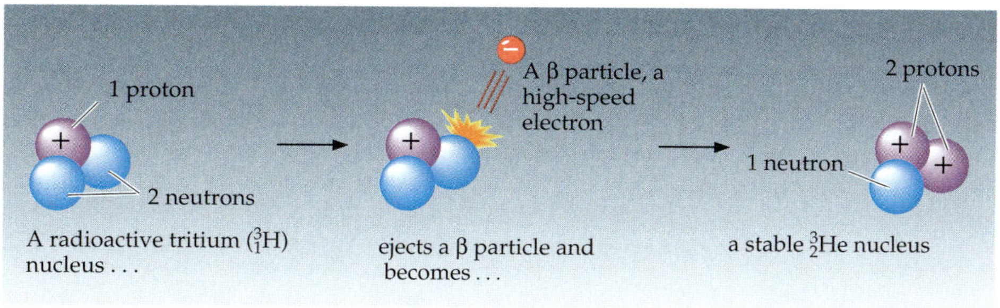

Figure 4.5 Radioactive decay of tritium.

(Fig. 4.5). Tritium itself is a key component of the hydrogen bomb. The combination of tritium with deuterium at extremely high temperatures provides the explosive force of one version of the hydrogen bomb (Sec. 4.14).

Decay of carbon-14 also occurs with loss of a β particle. The result

$$^{14}_{6}C \longrightarrow {}^{14}_{7}N + {}^{\ 0}_{1-}\beta$$

in this case is the transformation of a radioactive carbon atom into the most common isotope of nitrogen (Fig. 4.6). As in the case of tritium's decay, the mass of the nucleus remains unchanged while its atomic number increases by one unit. The radioactive decay of carbon-14 is useful for establishing the dates of ancient objects, as we'll see in detail in Section 5.11.

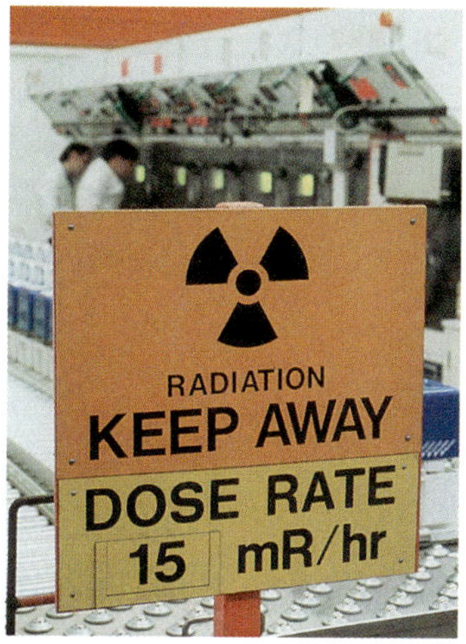

The universal trident symbol for radioactivity.

Uranium, which formed an image on the photographic plate in Becquerel's desk drawer and led him and the Curies to radioactivity, also provided the explosive force of one of the first atomic bombs. Its most common isotope, $^{238}_{92}U$, makes up over 99% of the uranium found in nature—mostly in *pitchblende*, an ore of the earth's crust—and represents the radioisotope of highest mass and highest atomic number of any found (in more than trace amounts) in nature. The radioactive decay of U-238 to thorium by loss of an α particle and emission of γ radiation represents the first step in one particular sequence of radioactive decay. Since the γ radiation that accompanies uranium's decay has neither mass nor charge and therefore doesn't affect the mass number or atomic number of any of the nuclei produced, it's often omitted from the equations. The equation for the decay of uranium to thorium is usually written as

$$^{238}_{92}U \longrightarrow {}^{234}_{90}Th + {}^{4}_{2}He$$

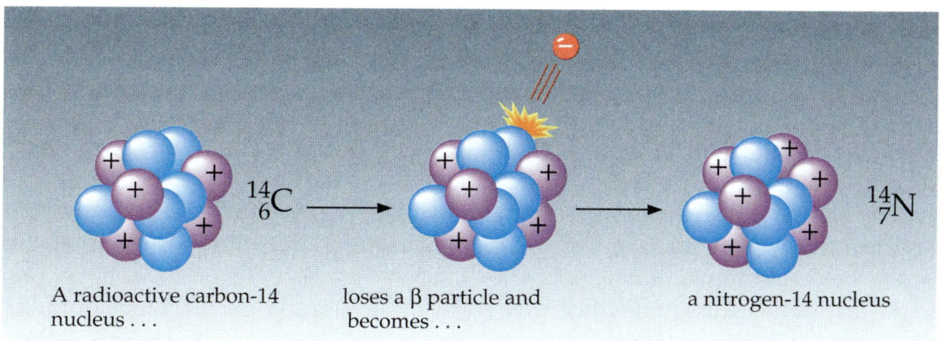

Figure 4.6 Radioactive decay of carbon-14.

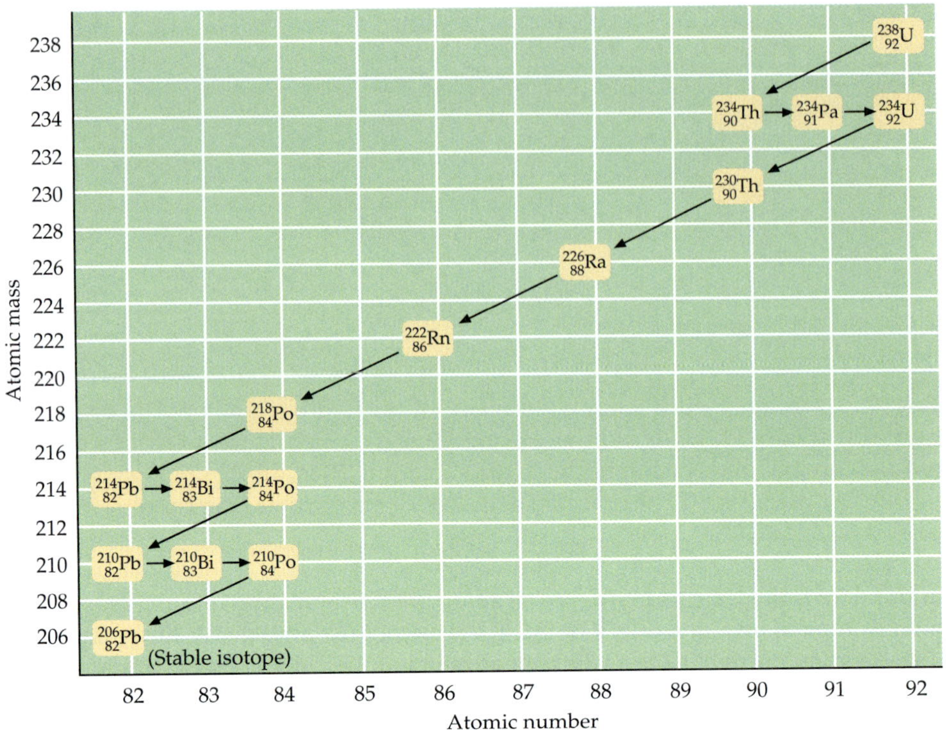

Figure 4.7 The sequence of radioactive decay from $^{238}_{92}U$ to $^{206}_{82}Pb$.

Notice that:

- the sum of the mass numbers to the left of the arrow, the 238 of the U-238, equals the sum of the mass numbers to the right of the arrow, 234 + 4.
- the sum of the atomic numbers to the left of the arrow, the 92 of the uranium, equals the sum of the atomic numbers to the right of the arrow, 90 + 2.

After 13 additional steps the entire chain finally comes to an end with the formation of a stable isotope of lead, $^{206}_{82}Pb$. The entire sequence appears in Figure 4.7.

Notice in Figure 4.7 that the radioactive element *radon* (Rn) is a transient part of this decay chain, forming from the radioactive decay of radium (Ra) and decaying, in turn, to polonium (Po). Radon itself is a gaseous element, a member of the inert family of elements of the periodic table.

There's concern that gaseous radon formed by the decay of very small amounts of radioisotopes that occur naturally in some kinds of rocks and soils may seep upward through the ground, enter homes and other buildings, and present an indoor pollution hazard. Ordinarily, we might expect that any radon gas we inhale would leave our lungs when we exhale. But radioactive decay of radon as it remains inside our lungs, during the moment between inhaling and exhaling, would produce minuscule amounts of polonium and other radioisotopes. Any traces of these radioactive elements that might remain within our bodies could do considerable biological damage through ionizing radiation (Chapter 5). There's still much dispute about whether radon actually constitutes a significant indoor hazard. Nevertheless, the simple possibility of harm from indoor radon has spurred the design and manufacture of various radon detectors for use in homes and other buildings.

The following example illustrates some of the steps in the sequence of radioactive decay shown in Figure 4.7.

EXAMPLE

Subatomic Loss

What subatomic particle is lost from Ra-226 as it decays to Rn-222?

■ We can write the reaction that occurs here as

$$^{226}_{88}\text{Ra} \longrightarrow {}^{222}_{86}\text{Rn} + \text{subatomic particle}$$

Since the mass number drops by 4 and the atomic number drops by 2 as radium-226 is transformed into radon 222, the subatomic particle that is lost must be a combination of two protons and two neutrons. Thus, it must be a helium nucleus, which is the same in this case as saying that it is an α particle. The complete nuclear reaction is

$$^{226}_{88}\text{Ra} \longrightarrow {}^{222}_{86}\text{Rn} + \alpha$$

What subatomic particle is lost as radon-222 is converted to polonium-218?

Again, the equation

$$^{222}_{86}\text{Rn} \longrightarrow {}^{218}_{84}\text{Po} + \text{subatomic particle}$$

shows us that a particle with an atomic number of 2 and a mass number of 4 is lost. It must once again be a helium nucleus. The complete equation is

$$^{222}_{86}\text{Rn} \longrightarrow {}^{218}_{84}\text{Po} + \alpha$$

What subatomic particle is lost as bismuth-210 is converted to polonium-210 in the next-to-last step of the sequence shown in Figure 4.7?

Here the equation is

$$^{210}_{83}\text{Bi} \longrightarrow {}^{210}_{84}\text{Po} + \text{subatomic particle}$$

Since there is no change in the mass number and an *increase* by 1 in atomic number, an electron must have been lost from the nucleus as a β particle. The equation is therefore

$$^{210}_{83}\text{Bi} \longrightarrow {}^{210}_{84}\text{Po} + \beta$$

EXAMPLE

Curium's First Step

In its radioactive decay, curium-245 (Cm-245) loses an α particle. What radioisotope results from the decay of Cm-245?

■ Reference to the periodic table shows that curium's atomic number is 96. Since an α particle has a mass number of 4 and an atomic number of 2, curium-245 loses 4 units of mass and two units of atomic number. The result is an isotope with a mass number of 241 (245 − 4) and an atomic number of 94 (96 − 2). The element with an atomic number of 94 is plutonium (Pu).

Thus the first step in the chain of radioactive decay of curium-245 produces plutonium-241:

$$^{245}_{96}\text{Cm} \longrightarrow {}^{241}_{94}\text{Pu} + {}^{4}_{2}\alpha$$

Other naturally occurring radioisotopes decay by paths different from the one shown in Figure 4.7. Uranium-235, for example, decays initially to thorium-231, by α emission, and then, through an additional 13 steps, to the stable lead-207. After a total of 15 steps, radioactive curium-245 ends up as a stable isotope of bismuth, Bi-209.

(The *curie*, the unit of radioactivity mentioned in Section 4.1, represents the rate of decay of a radioisotope: 1 curie = 3.7×10^{10} radioactive disintegrations per second.)

■ QUESTION

What subatomic particle is lost by $^{210}_{84}Po$ as it is converted into $^{206}_{82}Po$ in the final step of the decay sequence of U-238? Write the reaction for this step.

4.6 Rutherford's Transmutation of Nitrogen

Although scientists could observe the changes that naturally occurring radioisotopes undergo, as Becquerel and the Curies had done, until 1919 no one had as yet planned and carried out an artificial nuclear transformation that would convert a *stable* isotope of one element into another element. In that year Ernest Rutherford (Sec. 4.2) published a report of the first artificial transformation of one element to another. By bombarding nitrogen-14, a stable (and the most common) isotope of nitrogen, with α particles emitted by radium, Rutherford transformed the nitrogen atom into an atom of oxygen. In the process, the nitrogen nucleus absorbs the two protons and two neutrons of the α particle and then loses a proton, all with a net gain of 3 in mass number and 1 in atomic number. The entire procedure results in a **transmutation**, the conversion of one element into another:

Transmutation is the conversion of one element into another element.

$$^{14}_{7}N + ^{4}_{2}He \longrightarrow ^{17}_{8}O + ^{1}_{1}H$$

As we saw for the decay of uranium-238 (Sec. 4.5), the sum of the mass numbers to the left of the arrow equals the sum of the mass numbers to the right, and the sum of the atomic numbers to the left equals the sum of the atomic numbers to the right. With his conversion of nitrogen into oxygen, Rutherford had achieved the age-old dream of the transmutation of the elements.

One of the goals of *alchemy*, an ancient, mystical practice that preceded the science of chemistry and in many respects prepared its way, was the transmutation of one elemental substance into another. The alchemists sought especially to convert a common, inexpensive metal such as lead into precious gold. Through nuclear transformations much like Rutherford's, the alchemical dream has by now been realized, but without fulfilling the promise of the riches the ancients sought. Not only is expensive platinum (rather than lead) used as the starting metal for the formation of gold, but the cost of the highly purified platinum needed for the modern transmutation is far greater than the value of the gold that's generated.

■ QUESTION

The conversion of platinum into gold has been achieved by bombarding platinum-198 with neutrons to produce platinum-199. This isotope, in turn, decays to gold-199 with the loss of a subatomic particle:

$$^{198}_{78}Pt + ^{1}_{0}n \longrightarrow ^{199}_{78}Pt$$

$$^{199}_{78}Pt \longrightarrow ^{199}_{79}Au + \text{a subatomic particle}$$

What subatomic particle is lost by the platinum as it becomes an atom of gold?

4.7 A Nuclear Wonderland

As the 20th century progressed, science probed ever more deeply into the structure and behavior of the atom, continuously stripping away its secrets. Today the nucleus stands as a strange and complex world, seemingly no less bizarre than the one Alice found when she fell down the rabbit hole of Lewis Carroll's *Alice's Adventures in Wonderland*.

It's certainly not a world of fiction and imagination like Alice's. The nucleus is, after all, the very real foundation of the real matter of our everyday world. It's a world, though, that seems to operate by a very different set of rules.

One of its most striking features is the ability of large numbers of protons (all positively charged) to remain closely packed in the stable clusters of the nucleus. In our more familiar world such a compact assembly of similarly charged particles would fly apart, propelled by the forces of electrical repulsion (Sec. 3.9). Yet most nuclei are stable; they follow laws of nature that operate most effectively in the nucleus and that we simply don't find at work in our more familiar, everyday world.

There are still other strange facets to the nuclear world. Each charged subatomic particle (protons and electrons, for example) has a counterpart in another particle that is identical in every respect but carries the opposite electrical charge. **Positrons**, which are identical to electrons but carry positive charges, appear briefly as the result of radioactive decay or high-energy collisions at the nuclear level and then vanish in still other transformations. When a positron meets an electron, the pair disappears in a burst of gamma radiation. We'll meet the positron once again in Section 4.14.

A **positron** is a subatomic particle that carries a charge of 1+ but is otherwise identical to an electron.

Quarks are fundamental particles that compose larger subatomic particles such as protons and neutrons.

The combination of a positron and an *antiproton* (identical with a proton except that it carries a negative charge) forms a particle with a negatively charged nucleus and a surrounding, positively charged positron. This laboratory-created atom of *antihydrogen*, an example of *antimatter*, exists for only about 40 billionths of a second before it collides with ordinary matter and disappears in a burst of energy.

Another strange set of particles, the **quarks**, are among the truly fundamental particles of the universe. First proposed in 1964 to make sense of the large and growing number of subatomic particles being discovered, quarks remained hypothetical particles until a set of experiments begun in 1967 demonstrated their existence. For discovering quarks, Jerome I. Friedman and Henry W. Kendall of the Massachusetts Institute of Technology and Richard E. Taylor of the Stanford Linear Accelerator Center shared the 1990 Nobel Prize in physics.

Jerome I. Friedman. In 1990 he shared a Nobel Prize with Richard E. Taylor and Henry W. Kendall for their experimental demonstration of the existence of quarks.

Named for a word coined by the Irish writer James Joyce in his book *Finnegan's Wake*, quarks are particles that combine in various ways to form still other subatomic particles, including protons and neutrons. It takes a set of three quarks, for example, to form a proton or a neutron. In the work that demonstrated the existence of quarks, Friedman, Kendall, and Taylor also found experimental evidence for still another strange particle, the **gluon**, which had been proposed years earlier to account for the "glue" that holds the (then-hypothetical) quarks together in their small clusters. The whimsy evident in the word *quark* and in its origin, as well as in gluon, appears repeatedly in terms used to describe quarks and their properties, classifications such as *color*, *flavor*, *charm*, and *beauty*.

With one exception, the stranger particles and properties of the nucleus—its extraordinary behavior and its newly found particles with their unfamiliar colors, flavors, charms, and beauties—are beyond the scope of our examination. That single, important exception concerns the effect of neutrons on the stability of the nucleus. It is an effect that has led to the production of new, artificially produced radioisotopes, valuable in medical diagnosis and treatment, and it led as well to the construction of the atomic bomb.

Gluons are subatomic particles that hold the quarks in their clusters.

■ QUESTION

In what ways does the structure of an atom of *antihydrogen* differ from the structure of an atom of *hydrogen*? In what ways are the two atoms similar?

4.8 Nuclear Fission

James Chadwick received the Nobel Prize in physics in 1935 for his discovery of the neutron.

The stability of any particular nucleus depends, among other things, on the ratio of its neutrons to its protons. The ratio of greatest stability starts at about 1 : 1 (one neutron per proton) for elements with low atomic numbers, such as helium, and rises to about 1.5 : 1 for stable isotopes of mercury and lead. The most abundant (stable) isotope of lead, for example, $^{208}_{82}Pb$, contains 126 neutrons and 82 protons, which amounts to 1.54 neutrons for every proton.

With a few exceptions, nuclei with ratios at or very near the ideal are stable, but those that lie beyond a narrow range are apt to decay. Thus, protium and deuterium are stable isotopes of hydrogen, while tritium, with its ratio of two neutrons to one proton, is radioactive. Carbon-12 and carbon-13 are stable isotopes of carbon but carbon-14, with a higher ratio than either of the others, is radioactive. Adding a neutron to the nucleus of platinum-198 causes it to decay by β emission to gold-199 (see the question at the end of Section 4.6). One relatively simple way to destabilize a nucleus, then, is to increase its content of neutrons.

Although Rutherford had proposed the neutron as a nuclear particle as early as 1920, its existence wasn't observed experimentally until 12 years later, in 1932. (The neutron's discoverer, British physicist James Chadwick, received the Nobel Prize for physics in 1935.) In 1938, six years after the discovery of the neutron, an investigation of its effect on the nucleus of a uranium atom led to a new kind of nuclear transformation and to what would soon become an entirely new source of energy. The world was about to be changed.

The critical event occurred in 1938, almost exactly one year before the outbreak of World War II. Until then all of the changes either observed or produced artificially in atomic nuclei involved additions and/or losses of small particles: α particles, β particles, neutrons, and protons. What's more, while some of the nuclear transformations released a bit of energy, others consumed more energy than they produced. All this changed in a laboratory in Berlin late in 1938. There Otto Hahn, a German radiochemist who would receive the 1944 Nobel Prize in chemistry for his discovery, bombarded uranium with neutrons. Among the products that he and a co-worker, Fritz Strassmann, isolated from the experiment was an isotope of barium, an atom with a little less than two-thirds the protons of a uranium atom and less than two-thirds uranium's mass number. This was an astonishing discovery, not easily explained on the basis of what was then known about the behavior of atomic nuclei.

Lise Meitner interpreted Otto Hahn's experimental observations as confirmation that he had split a uranium nucleus.

What had been dislodged from the uranium nucleus by increasing its ratio of neutrons to protons ever so slightly was not a few small, subatomic particles, not a couple of α particles or β particles or a proton or two, but fully a third of the entire nucleus. Something remarkable had happened to the nucleus. Once again, serendipity was at work.

Nuclear fission is the splitting of the atomic nucleus into two or more large fragments.

Uncertain about how to interpret the generation of barium by the addition of neutrons to the uranium nucleus but convinced of its significance, Hahn and Strassmann sent word of their discovery to Lise Meitner, a physicist who had worked with Hahn in earlier studies of radioactivity. Born in Austria, Meitner had fled to Denmark when the Nazis took power. Working with her nephew, Otto Frisch, she concluded that, rather than knocking some small subatomic particle out of the uranium nucleus, the neutrons had actually cleaved the nucleus of an atom of uranium into two or more large fragments, a transformation without precedent in all the previous studies of the atom. (A few weeks later Hahn and Strassmann came to the same conclusion as well.) It was Lise Meitner who coined the term **nuclear fission** for the splitting of an atomic nucleus.

■ QUESTION

When an atom of uranium-235 is bombarded with neutrons, one of the many fission reactions it can undergo produces barium and an additional element (as well as energy and additional neutrons) but no α particles or β particles. With this in mind, and with reference to the periodic table, name the additional element produced in this particular mode of fission.

4.9 A Chain Reaction, a Critical Mass

Soon it became clear that fission differs in one very important way from the other nuclear reactions known at the time. Fission releases enormous amounts of energy, far more than that released by the loss of a few small particles from a radioactive nucleus. With the remarkable discovery of nuclear fission and recognition of the amount of energy it might release, and with the coming of World War II, research on the atomic nucleus accelerated. If a *sustained* sequence of fission reactions could be maintained, consuming an entire package of fissionable material in an instant, the result might be explosive!

Investigations quickly revealed that the most efficient uranium fission comes from the addition of a neutron to the isotope of mass 235, *uranium-235*, or *U-235*, rather than the more common U-238. More than 99% of all naturally occurring uranium is U-238; less than three-quarters of 1% occurs as the more fissionable U-235.

Uranium-235, a source of nuclear power.

The cleavage of a U-235 nucleus can occur in many different ways and can produce any of numerous sets of products. It can, for example, produce tellurium, zirconium, and two neutrons (and energy),

$$^{235}_{92}U + ^{1}_{0}n \longrightarrow ^{137}_{52}Te + ^{97}_{40}Zr + 2\,^{1}_{0}n$$

or barium, krypton, and three neutrons (and, again, energy),

$$^{235}_{92}U + ^{1}_{0}n \longrightarrow ^{140}_{56}Ba + ^{93}_{36}Kr + 3\,^{1}_{0}n$$

or still other sets of products. Some 200 different isotopes representing 35 different elements result from the fission of U-235 atoms.

In addition to these various isotopes, the fission of U-235 also releases several neutrons, as shown in the preceding equations. Any one (or more) of these could penetrate into another U-235 nucleus and continue the chain of energy-releasing fission reactions or start a new branch. Averaging over all the sets of products that result from the fission of U-235, one fission reaction produces about 2.5 neutrons. If each released neutron could cause the fission of another nucleus, every cleavage of *two* U-235 nuclei could release enough neutrons to split an additional *five*.

In this way the fission of a single U-235 atom could begin a cascading **chain reaction** that could consume all the U-235 present and release energy instantaneously in amounts never before achieved by humans. (The slow release of energy, even in very large amounts, can be tapped to obtain electrical, mechanical, or other kinds of power. The instantaneous release of immense amounts of energy is explosive.)

A **chain reaction** is a continuing series of nuclear fissions that occur when neutrons released in the fission of one atom cause the fission of additional atoms, which in turn release still more neutrons and produce still more fissions, and so on.

Thoughts quickly turned toward the new war and to the building of an atomic bomb that would convert the sudden release of energy from a rapid chain reaction into the immense explosive force of a devastating weapon. What was needed was a **critical mass** of U-235. This is the minimum mass of fissionable material needed to produce a self-sustaining chain reaction. The critical mass is reached when the mass of fissionable material becomes large enough to ensure that the released neutrons are, indeed, absorbed by other fissionable nuclei and that these

A **critical mass** is the minimum mass of fissionable material needed to sustain a chain reaction.

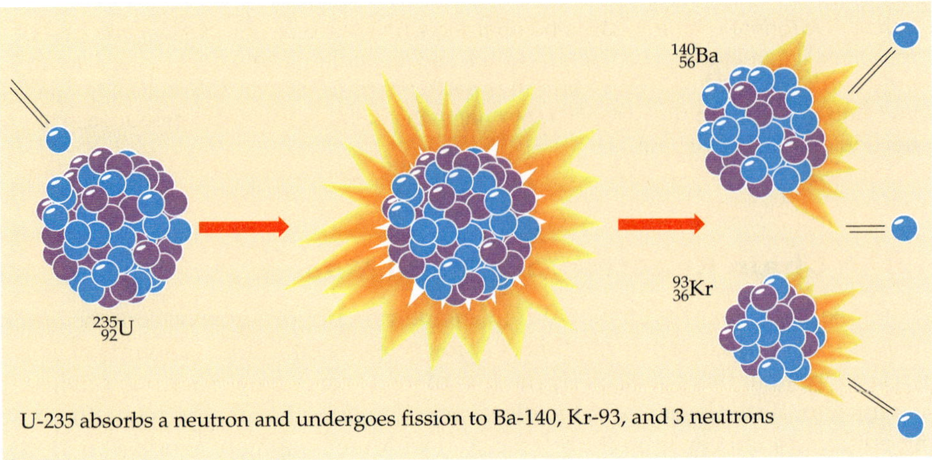

Figure 4.8 A typical fission reaction of U-235.

released neutrons produce a continuing chain of fission reactions and thereby generate and sustain an energy-releasing chain reaction. Figures 4.8 and 4.9 show a typical fission of a U-235 atom and the resulting chain reaction.

Uranium-235 isn't alone in its ability to begin and sustain a chain reaction. Plutonium-239, a radioisotope that occurs in nature only in traces, can also generate a chain similar to that of U-235. The fission of Pu-239 is a bit more efficient and releases about 20% more energy than the fission of U-235.

Figure 4.9 Schematic diagram of the cascading effect of a typical chain reaction initiated by a single neutron.

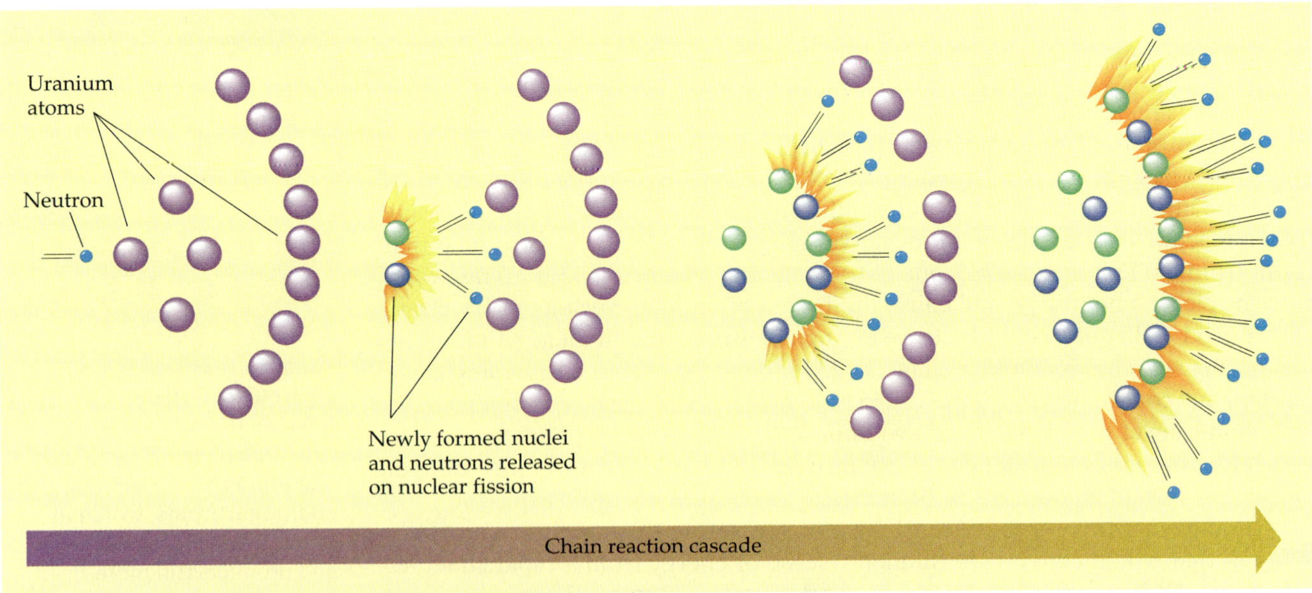

■ **QUESTION**

(a) What is the *advantage* to using U-235 rather than U-238 as the fissionable material in building a fission bomb? (b) What is the *disadvantage* to using U-235 rather than U-238 as the fissionable material?

4.10 The Manhattan Project

Spurred on by the recognition that a weapon as powerful as an atomic bomb could determine the outcome of the war, and with the fear that Germany might be making rapid progress toward the same goal, the United States set out to build the atomic weapon as quickly as possible. Complete control of the project and all the economic and political power to complete the work were given to the U.S. Army Corps of Engineers. Initially located in a New York office under the code name of the Manhattan Engineering District, the entire operation quickly became known more simply as the *Manhattan Project*. To provide the nuclear explosive, several secret production facilities were constructed, the most successful of which were at Oak Ridge, Tennessee, and Hanford, Washington. The scientific center that actually designed and assembled the bomb itself was located at Los Alamos, New Mexico.

At Oak Ridge common uranium, a mixture of isotopes, was converted to its gaseous fluoride, UF_6, *uranium hexafluoride*. The desired isotopic ${}^{235}_{92}UF_6$ was separated from the more plentiful ${}^{238}_{92}UF_6$ by taking advantage of subtle differences in the physical properties of the gases through a process known as *gaseous diffusion* (Fig. 4.10). The hexafluoride of the U-235 is 3 amu lighter in mass than that of the more common U-238, and its molecules move about a bit more rapidly than those containing the heavier isotope. As a result the lighter ${}^{235}_{92}UF_6$ diffuses through extremely small openings of a porous barrier a little more quickly than the heavier ${}^{238}_{92}UF_6$ (Fig. 4.10). Reconverting the UF_6 enriched in ${}^{235}_{92}UF_6$ into the metal produces uranium containing a high concentration of U-235.

Because of the importance of the bomb, and the wartime urgency in its construction, and because the building of the atomic bomb was an entirely new event in human history, with little or no certain knowledge or established techniques that would ensure success, several different routes were followed simultaneously. With several paths taken at the same time, the failure of any one need not doom the project. As an alternative to U-235 as the explosive, for example, Pu-239 was manufactured at the Hanford plant. There, fissionable Pu-239 was generated by a controlled reaction that begins with the absorption of neutrons by U-238. This became the first large-scale synthetic transformation of one element into another. The alchemists' dream had become reality on a massive, industrial scale.

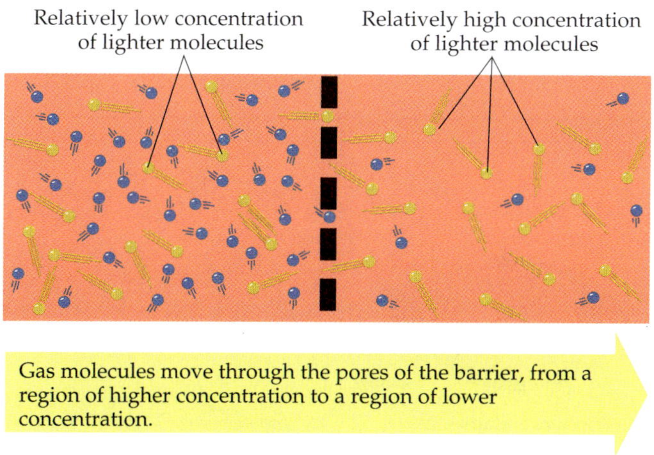

Figure 4.10 Enrichment by gaseous diffusion.

J. Robert Oppenheimer and Leslie Groves at the remains of the tower used in the test of the first atomic bomb. Oppenheimer was the scientist who led the team that designed and assembled the bomb; Groves, an army general, was overall director of the Manhattan Project.

The first atomic bombs. *Top*: The plutonium bomb, which measured 3.25 m (10.7 ft) long and 1.5 m (5 ft) in diameter, was given the code name "Fat Man." *Bottom*: The uranium bomb, 3 m (10 ft) long and 0.7 m (2.3 ft) in diameter, was called "Little Boy."

With production of fissionable uranium and plutonium achieved, the bomb itself could now be constructed. It would have to contain fissionable material dispersed in the form of one or more subcritical masses. In the absence of a single critical mass no chain reaction could occur, so the bomb could not explode. The design of a bomb, then, hung on the mechanical act of converting one or more subcritical masses of fissionable material into a single, critical, explosive mass instantaneously, and with enough force to keep it all together long enough to produce a nuclear explosion.

Because of differences in the way uranium and plutonium behave, two different designs were developed. For the uranium bomb, code named Little Boy, a conventional explosive would be used to fire a subcritical uranium "bullet" into a subcritical ring of uranium. As the bullet filled the hole of the ring, an explosive critical mass would be formed. For the plutonium bomb, named Fat Man, a sphere of plutonium with a hollow core, something resembling a cantaloupe, would be crushed, or *imploded*, forming a nugget of critical mass through the force of a surrounding explosive (Fig. 4.11).

For both bombs the neutrons needed to start the chain reaction would be supplied by an *initiator* formed by the combination of beryllium and radioactive polonium-210. Alpha particles from this radioisotope of polonium would knock the needed neutrons out of the beryllium. [It was this same combination, beryllium and polonium-210, that first revealed the existence of neutrons to James Chadwick (Sec. 4.8).] The mechanical act of forming the critical mass in each bomb would also be used to bring the polonium and beryllium together. No more than an estimated 10 neutrons would be needed to begin the chain reaction for the detonation of the plutonium bomb.

Figure 4.11 The operation of fission bombs.

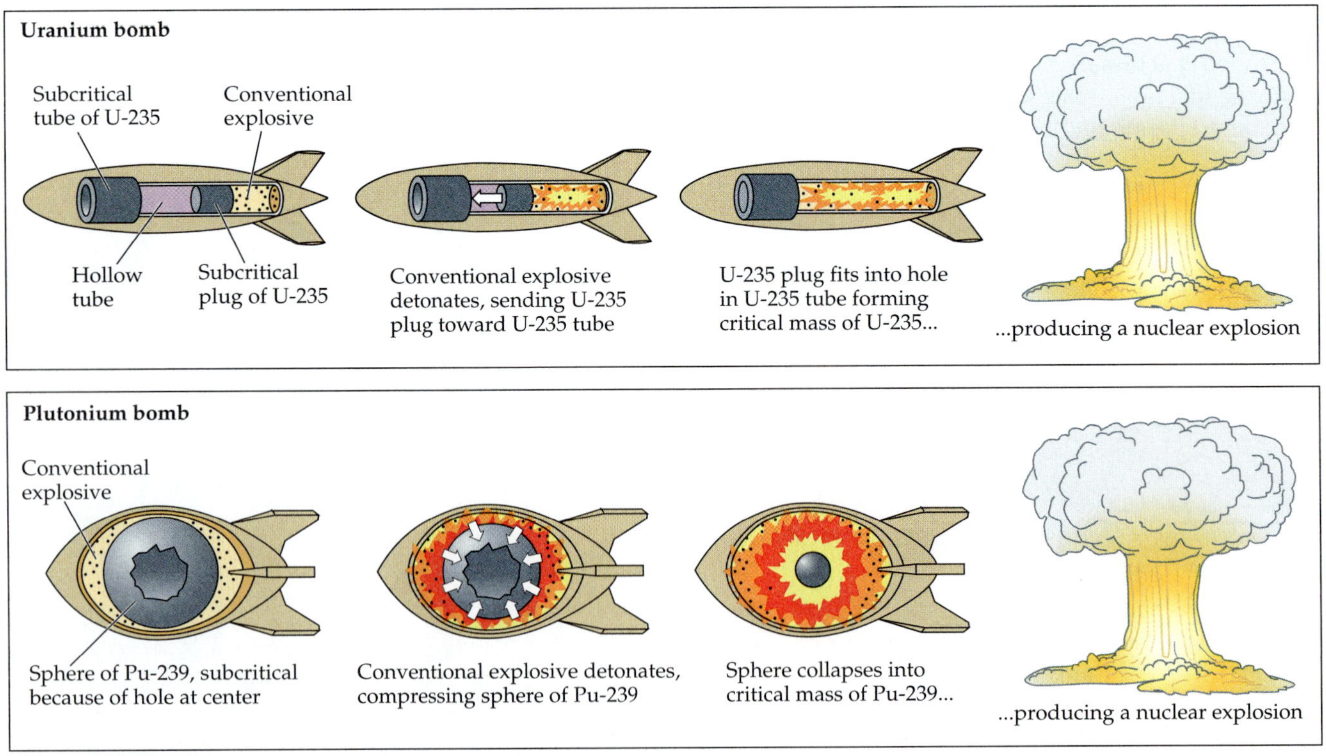

Although the design of the uranium bomb was simple enough and seemed certain to produce an explosion, the plutonium bomb was another matter. Because Fat Man would not detonate unless the implosion occurred with perfect symmetry, to form a perfectly spherical critical mass of plutonium, a test of a finished bomb was necessary. In the summer of 1945 a plutonium bomb was constructed at the Los Alamos laboratories and detonated in desert testing grounds near Alamogordo, New Mexico, with results described at the beginning of this chapter. Following this successful test, a second bomb (Little Boy, the untested uranium bomb) and a third (Fat Man, another plutonium bomb) were exploded above the Japanese cities of Hiroshima and then Nagasaki on the mornings of August 6 and 9, destroying the cities and killing or wounding an estimated 200,000 people. Victims who survived the initial blast and its immediate aftermath suffered for decades from delayed effects of the radiation. The choices of Hiroshima and Nagasaki as targets rested on a combination of factors, including their military importance, the visibility of targets, prevailing weather conditions, and, not least, that they had as yet been spared from attack; subsequent estimates of the destructive force of the atomic bombs would not be complicated by damage caused by conventional bombs.

Remains of a building after the explosion of the uranium bomb at Hiroshima, August 6, 1945.

The bombs were history's first atomic weapons. They produced unimaginable destruction and deaths, both immediately from the blast and fireball and lingering from the lethal radiation. The uranium bomb dropped on Hiroshima exploded with a power equivalent to 13,500 lb of TNT. Destruction of the city was virtually total for a distance of over a mile from the center of the blast; temperatures of more than 3000°C incinerated buildings and people within 2 miles of the center. Seeing the blast and the mushroom cloud from the cockpit of the *Enola Gay*, the B-29 that dropped the bomb over Hiroshima, the copilot of the flight wrote in his journal, "My God, what have we done?" Within a few days Japan surrendered. World War II had ended.

We turn now to the source of this immense power and, in the next chapter, to how we have harnessed it for peace.

■ QUESTION

How was a critical mass achieved in the detonation of the U-235 bomb? In the detonation of the Pu-239 bomb?

4.11 Energy, Mass, and Albert Einstein

The unimaginable power released by nuclear reactions, whether through the explosion of an atomic bomb in warfare or by the peaceful generation of electricity within the nuclear reactor of a power plant, comes from the conversion of mass itself directly into energy. In the chemical changes we examined in Chapter 1 and in the chemical reactions that are the subjects of the remaining chapters—reactions that involve the sharing and the transfer of valence electrons—we invariably observe the operation of two fundamental laws of chemistry. One, the **Law of Conservation of Mass**, recognizes that mass can neither be created nor destroyed as a result of chemical transformations. In a parallel fashion, the **Law of Conservation of Energy** holds that energy can neither be created nor destroyed as a result of chemical transformations. The first of these laws requires that there must be exactly as much matter (and therefore exactly as many atoms) among the combined products of a chemical reaction as in its combined reactants. Similarly, the second requires that the sum of all the energy present in the products (including any that is liberated as the reaction progresses) must equal the

The **Law of Conservation of Mass** states that mass can neither be created nor destroyed as a result of chemical transformations.

The **Law of Conservation of Energy** states that energy can neither be created nor destroyed as a result of chemical transformations.

sum of all the energy in the reactants (including any that is added to produce the reaction). In brief, although matter can be converted from one substance (a reactant) to another (a product), it can neither be created nor destroyed; and although energy can be transferred or changed from one form to another, it can neither be created nor destroyed. These laws hold, to the limits of our ability to detect changes in mass and energy, as long as the nuclei of atoms remain intact.

But in reactions that take place at the subatomic level, radioactivity and nuclear fission, for example, we can observe both the formation and disappearance of matter and the formation and disappearance of energy as matter and energy are interconverted, one into the other. We can observe the conversion of matter into energy and the formation of matter from energy.

Albert Einstein. He derived the equation that relates mass and energy.

Albert Einstein, born in Germany in 1879, naturalized as an American citizen in 1940, and generally recognized as the most brilliant theoretical physicist of the 20th century, was the first to describe these interchanges in mathematical terms. He found that matter and energy are themselves equivalent and interconvertible in the mathematics of their behavior. He showed that a specific quantity of mass is equivalent to a specific quantity of energy, calculated through the equation $E = mc^2$. In Einstein's equation E represents energy, m represents mass, and c is the speed of light in a vacuum. For his work in this and other areas of physics, particularly on a phenomenon known as the photoelectric effect, Einstein received the Nobel Prize for physics in 1921.

(It was Einstein who, not quite a month before the outbreak of World War II and at the urging of fellow scientists, wrote a letter to President Roosevelt revealing to him the possibility of producing an atomic bomb. He took no part in the actual creation of the bomb.)

■ QUESTION

Using the Law of Conservation of Mass and the Law of Conservation of Energy, show why nuclear fission might not be viewed as a chemical reaction.

4.12 The Matter of the Missing Mass: Mass Defect and Binding Energy

Knowing that matter and energy are equivalent and that a specific quantity of mass is equivalent to a specific quantity of energy, we're now in a position to find the source of the energy that binds the protons and neutrons into an atom's compact, dense nucleus and, in the next section, to examine the source of the energy released by fission.

We'll look at binding energy first, using an atom of U-235 as an illustration.

EXAMPLE

Total Mass

Calculate the mass of an atom of uranium-235.

■ We might expect the total mass of an atom of U-235 to be the sum of the masses of all of its subatomic particles—its protons, neutrons, and electrons. Using the values of Section 2.4 for protons, neutrons, and electrons

$$\text{proton} = 1.007 \text{ amu}$$

$$\text{neutron} = 1.009 \text{ amu}$$

$$\text{electron} = 0.0005 \text{ amu}$$

we can calculate the mass of an atom of U-235. With the periodic table or a table of atomic weights and atomic numbers, we find that the atomic number of uranium is 92. Thus, there must be 92 protons in the nucleus and 92 electrons in the surrounding shells. Furthermore, since its mass number is 235, the total, combined number of protons and neutrons in the nucleus must be 235. With 92 protons in the nucleus, the number of neutrons in U-235 is

$$\text{neutrons} = 235 - 92 = 143$$

The subatomic particles that make up an atom of U-235 are thus

92 protons
92 electrons
143 neutrons

With a mass of 1.007 amu each, the protons contribute

$$92 \text{ protons} \times 1.007 \text{ amu/proton} = 92.64 \text{ amu}$$

With a mass of 0.0005 each, the electrons contribute

$$92 \text{ electrons} \times 0.0005 \text{ amu/electron} = 0.0460 \text{ amu}$$

With a mass of 1.009 each, the neutrons contribute

$$143 \text{ neutrons} \times 1.009 \text{ amu/neutron} = 144.29 \text{ amu}$$

Combining these individual contributions, we find that the calculated mass of the U-235 atom is

protons	92.64 amu
electrons	0.05 amu
neutrons	144.29 amu
	236.98 amu

We can round this off to 237.0 amu for the calculated mass of the U-235 atom.

As we've just seen, the calculated mass of the U-235 atom arrived at by adding up the masses of all the subatomic particles that compose it is 237.0 amu. Yet the actual mass of an atom of U-235 *as measured experimentally* (and extremely accurately) is only 235.043924 amu, which we can round off here to 235.0 amu. The difference between the calculated mass of all the protons, neutrons, and electrons present in an atom and the actual, *measured* mass is known as the atom's **mass defect**. The mass defect for the U-235 atom, then, is about 2.0 amu.

An atom's **mass defect** is the difference between the mass of the atom as a whole and the mass of all the individual protons, neutrons, and electrons that compose it.

Where did the missing mass go? It has been converted from *mass* into *energy*, the very energy that binds all of the protons and neutrons together to form a compact, coherent nucleus. This **binding energy**—the cement that holds the nucleus together—comes from the conversion into energy of a very small fraction of the masses of the subatomic particles that compose the atom. For U-235 the binding energy is the energy equivalent of 2.0 amu, the difference between the sum of the masses of its individual particles and the mass of the whole atom itself. An atom's *binding energy* is the energy equivalent of its *mass defect.*

The **binding energy** of an atom, the energy that holds the nucleus together as a coherent whole, is the energy equivalent of its mass defect.

Similar calculations and measurements for atoms of other elements show that the mass defect per atomic mass unit (and therefore the binding energy per atomic mass unit) rises sharply as mass numbers increase, reaches its maximum at mass numbers of 50 to 60, and then drops slowly as mass numbers increase further.

■ QUESTION

Calculate the mass defect of an atom of Pu-239. The measured mass of an atom of Pu-239 is 239.05 amu.

4.13 The Energy of Uranium Fission

Now we can turn to the source of the energy released by nuclear fission, once again using the equivalence of mass and energy. We use the neutron-induced fission of U-235 to barium, krypton, and three neutrons as an illustration (Sec. 4.9):

$$^{235}_{92}U + ^{1}_{0}n \longrightarrow ^{140}_{56}Ba + ^{93}_{36}Kr + 3\,^{1}_{0}n$$

Using accurately measured masses of the neutron and the atoms that take part in this reaction

Particle	Precise Mass (amu)
the neutron	1.009
U-235	235.044
Ba-140	139.911
Kr-93	92.931

we can calculate the total mass of all the particles that enter into this reaction (one U-235 atom and one neutron) and all the particles that result from the fission (one Ba-140 atom, one Kr-93 atom, and three neutrons), and the difference between the two sets.

EXAMPLE

Gone Fission

Calculate the difference in mass between the reactants and the products in the fission of U-235 just shown.

■ The reactants and their masses are

Particle	Mass
1 U-235 atom	235.044
1 neutron	1.009
Total	236.053 amu

The products and their masses are

Particle	Mass
1 Ba-140 atom	139.911
1 Kr-93 atom	92.931
3 neutrons	3.027
Total	235.869 amu

As a result of this particular fission reaction, there is a loss of mass equal to

total mass of reactants	236.053
total mass of products	−235.869
mass lost	0.184 amu

This loss of 0.184 amu is the amount of mass that is converted into energy in the fission reaction of a single U-235 atom. Released instantaneously from a large mass of uranium, it's the source of the explosive power of the uranium bomb; released slowly, under the controlled conditions of a nuclear power plant, it's a source of commercial electricity.

From a different perspective, a loss of a little less than a tenth of a percent of the total mass of the reactants occurs in this particular U-235 fission:

the mass lost in fission

$$\frac{0.184 \text{ amu}}{236.053 \text{ amu}} \times 100 = 0.078\%, \quad \text{the portion of the mass converted into energy}$$

the combined mass of reactants: a U-235 nucleus and a neutron

If it were to proceed through this nuclear reaction exclusively, the fission of a kilogram of U-235, for example, would convert about 0.78 g of matter into energy. Naturally, this energy could be used to do useful work just as the energy of burning oil, gasoline, coal, or wood or of falling water is used. The energy released by nuclear reactions is no different from the energy released by any of these.

Although the conversion of about a tenth of a percent of a mass of uranium into energy might seem to be an inefficient way to produce energy, for a bomb or for a power plant, Einstein's equation shows otherwise. By use of the equation $E = mc^2$, it's possible to calculate the amount of work that can be accomplished by the energy equivalent to this lost matter. Table 4.2 presents a comparison of the lengths of time a 100-watt light bulb could be kept burning by this and other representative sources of energy and includes a comparison with the amount of energy we get from gasoline and from the biological metabolism of ordinary table sugar. In Chapter 5 we'll examine the advantages and disadvantages of nuclear fission as a commercial source of energy.

Table 4.2 Energy Production in Terms of a 100-Watt Light Bulb

Energy Source	Approximate Amount of Time the Energy Would Keep a 100-Watt Light Bulb Burning
Complete conversion of 1 g of matter into energy	29,000 years
Fission of 1 kg of U-235	23,000 years
Fission of 1 g of U-235	23 years
Efficient burning of 1 g of gasoline	8 minutes
Nutritional energy that 1 g of sugar (about a fifth of a teaspoon) provides to the human body	Slightly less than 3 minutes

■ **QUESTION**

Table 4.2 shows that the efficient burning of 1 g of gasoline could keep a 100-watt light bulb burning for 8 minutes. How long would the energy obtained by the complete conversion of the gram of gasoline directly into energy keep the bulb lit?

4.14 Nuclear Fusion: Another Bomb, with the Power of the Sun

Nuclear fusion is the process by which several nuclei of small mass combine to form a single nucleus of larger mass.

We've seen that in *nuclear fission* the nuclei of certain isotopes of relatively large mass split apart into smaller fragments, releasing energy in the process through the conversion of matter into energy. In **nuclear fusion**, a transformation resembling the *reverse* of fission, several atoms of small mass come together at extremely high temperatures, 10 million to 100 million degrees Celsius, to form larger nuclei. As in fission, nuclear fusion also results in the conversion of matter into energy. The fusion of four hydrogen nuclei to a helium nucleus and two positrons (Sec. 4.7) produces the energy of the sun. As in all other reactions, this fusion reaction occurs with the preservation of electrical neutrality: The two positive charges of the ejected positrons are balanced by the two negative charges of two electrons lost from valence shells. (The four hydrogen atoms taking part in the fusion reaction contain a total of four valence electrons; the single helium atom produced has only two. Thus, two electrons are lost from valence shells as a result of the fusion.)

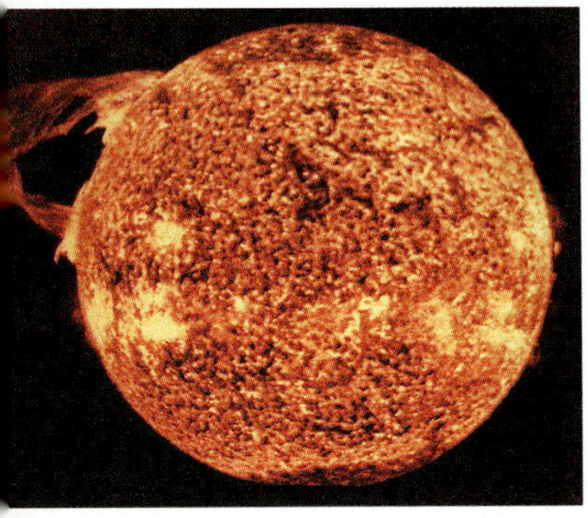

A flare ejected from the surface of the sun. The sun's energy comes from a fusion reaction that converts hydrogen to helium.

$$4\,{}^{1}_{1}\text{H} \longrightarrow {}^{4}_{2}\text{He} + 2\,{}^{0}_{1+}\text{e}^{+} + 2\,{}^{0}_{1-}\text{e}^{-}$$

The oppositely charged positrons and electrons combine with each other; each pair of these subatomic particles is transformed into a pair of γ rays (Sec. 4.7):

$${}^{0}_{1+}\text{e}^{+} + {}^{0}_{1-}\text{e}^{-} \longrightarrow 2\gamma$$

The overall fusion reaction of the sun becomes

$$4\,{}^{1}_{1}\text{H} \longrightarrow {}^{4}_{2}\text{He} + 4\gamma$$

EXAMPLE

Gone Fusion

In Section 4.13 we saw that the *fission* of 1 kg of U-235 converts about 0.78 g of matter into energy. How much matter is converted to energy by the *fusion* of 1 kg of hydrogen, or more specifically, protium? The experimentally determined mass of a helium-4 atom is 4.0026 amu; for an atom of protium, it's 1.0078 amu.

■ We've just seen that 4 protium atoms fuse to form one atom of He-4. The mass of 4 protium atoms is

$$4 \times 1.0078 \text{ amu} = 4.0312 \text{ amu}$$

Since the measured mass of a He-4 atom is 4.0026 amu, the amount of mass converted into energy is

mass of 4 protium atoms	4.0312 amu
mass of 1 He-4 atom	4.0026 amu
mass converted into energy	0.0286 amu

On a percentage basis, the amount of mass converted into energy is

$$\frac{0.0286 \text{ amu}}{4.0312 \text{ amu}} \times 100 = 0.709\% \quad \text{of the mass of the 4 protium atoms}$$

Applying this percentage to 1 kg of protium, we get

$$0.709\% \times 1000 \text{ g} = 7.09 \text{ g of matter}$$

Gram for gram, the fusion of protium converts about 7.1/0.78, or about 9 times as much mass into energy as does the fission of uranium-235.

Clearly, a bomb powered by a fusion reaction would be far more powerful than the fission bombs dropped on Hiroshima and Nagasaki. As a practical matter, though, the fusion of four protium atoms occurs too slowly to produce an explosion. Instead, the more rapid combination of a deuterium atom with a tritium atom provides the power of one kind of *fusion bomb*, or *hydrogen bomb*. (Although fast enough to produce an explosion, this reaction isn't quite as efficient in its conversion of matter into energy as is the combination of four protium atoms. Nevertheless, the fusion of deuterium and tritium still produces more power per gram of material than does uranium or plutonium fission.)

$$^{2}_{1}\text{H} + ^{3}_{1}\text{H} \longrightarrow ^{4}_{2}\text{He} + ^{1}_{0}\text{n}$$

The detonation of a small fission bomb within the larger fusion bomb produces the temperatures of tens of millions of degrees needed for the fusion reaction to begin.

■ QUESTION

The fusion bomb just described uses both deuterium ($^{2}_{1}$H) and tritium ($^{3}_{1}$H) as fuel. The deuterium can be obtained by concentrating quantities of the isotope that occur naturally in the water molecules of seawater. The tritium needed for the reaction is produced by wrapping the fission detonator with a compound of $^{6}_{3}$Li. This isotope of lithium reacts with neutrons generated by the fission explosion to produce a tritium atom and another particle:

$$^{6}_{3}\text{Li} + ^{1}_{0}\text{n} \longrightarrow ^{3}_{1}\text{H} + ?$$

What is the additional particle produced in this reaction? Use the masses and the atomic numbers of the neutron and the lithium and hydrogen isotopes given in the equation to arrive at your answer. (*Hint:* Review Sec. 4.5.)

PERSPECTIVE

Mental Images of Our Physical World

In this chapter we began the story of atomic energy with the serendipity of Antoine Henri Becquerel as he examined an unexpected image on a photographic plate in 1896. We traced the story of atomic energy from the discovery of radioactivity to the development of a bomb that explodes and incinerates with the power of the sun. At one level or another the following chapters continue to explore several of the themes introduced here.

The next chapter carries Becquerel's discovery from its first application in the weapons of war to the peaceful pursuits of today's world. We'll see how we now apply the effects of nuclear changes to the commercial production of electricity, the healing arts of medicine, and the study of history. Following that, in Chapter 8, we'll examine energy itself, both nuclear and non-nuclear, and learn how the reactions taking place on the sun even now—the same sort that provide the power of the hydrogen bomb—furnish the energy of our everyday world and of our own bodies.

But in another sense one of the themes developed in this chapter is reflected in all the chapters that follow. It's the theme that the world of chemical structures—atoms and molecules—is a bit different from the world of our everyday lives, just as the world of *Alice's Adventures in Wonderland* (Sec. 4.7) differs from the familiar world of the earth's surface.

In the world we know so well, we expect like electrical charges to repel each other and unlike charges to attract. Yet in the world of the atom the positively charged protons are packed closely together in the nucleus while negatively charged electrons lie in shells far from the positively charged nucleus. Similarly, in our more familiar world we put the mass of physical bodies into one category, and the energy of work they do into another. Yet we learned in this chapter that mass and energy are interconvertible, that mass can literally vanish, to be replaced by its equivalent in energy.

In another sense, chemistry asks us to look at the physical structures of our world not with our eyes but with our minds. As it explains the behavior of the matter that makes up our world, chemistry invokes structures—atoms and molecules—too small to be seen with even the most powerful optical microscopes, structures that can be made visible only with such uncommon devices as the scanning tunneling microscope (Sec. 2.2). In this sense, the study of chemistry asks us to use our imagination to give shape and structure to the nucleus and its surrounding electrons and the bonds they form.

In its descriptions of the material of our world, what follows asks us to use letters of the English alphabet to represent the nuclei and electrons shells of atoms that are invisible to our eyes; numerical subscripts to represent clusters of these atoms; and straight lines to represent the clouds of electrons that hold these atoms into the well-formed structures we call molecules. The same kind of mental activity that allows us to visualize the structure and behavior of atoms and the conversion of matter directly into energy (and the reverse) will convert the letters, lines, and subscripts that follow into the incredibly small particles that form us and everything else in our everyday world. ■

EXERCISES

For Review

1. Following are a statement containing a number of blanks and a list of words and phrases. The number of words equals the number of blanks within the statement, and all but two of the words fit correctly into these blanks. Fill in the blanks of the statement with those words that do fit, then complete the statement by filling in the two remaining blanks with correct words (not in the list) in place of the two words that don't fit.

Antoine Henri Becquerel discovered the phenomenon of __________, which occurs when the nucleus of a __________ emits an __________, a __________, and/or a __________. Each of these forms of radiation can cause the ionization of matter so each is an example of __________. When an atomic nucleus loses either an α particle, which is a high-energy __________, or a β particle, which is a high-energy __________, the nucleus undergoes a change in its atomic number and becomes transformed into a nucleus of a different element in a process known as __________. If the newly formed isotope is also radioactive, it too emits radiation and is transformed into still another element. Eventually, after a series of these transformations, a nonradioactive __________ is formed, and this __________ comes to an end.

α particle	neutron
β particle	proton
γ ray	radioactivity
chain of radioactive decay	radioisotope
electron	stable isotope
ionizing radiation	

2. Repeat the instructions of Exercise 1 for the following statement and list of words and phrases.

The first atomic weapons—the bombs tested near Alamogordo, New Mexico, and those dropped on Hiroshima and Nagasaki—derived their explosive power from the conversion of ____________________ into __________ through the process of __________. The first atomic bomb detonated was made of __________, an artificially produced isotope that does not occur naturally (in more than trace amounts). The bomb dropped on Hiroshima contained __________, which was separated from its more common isotope, __________, by the process of __________. Each of these weapons was detonated by setting off a conventional explosion that compressed subcritical portions of the metal into a __________ and set off a __________, which produced the nuclear blast. Still another kind of nuclear weapon, which harnesses the power of the sun, relies on __________ for its energy. In one form, __________ and __________ nuclei come together at very high temperatures to produce __________ and release nuclear power.

critical mass	nuclear fission
deuterium	nuclear fusion
energy	plutonium-239
helium	tritium
hydrogen	uranium-235
initiator	uranium-238
matter	

3. Each of the following scientists or groups of scientists in the left-hand column received a Nobel Prize for contributing to our understanding of the structure and properties of the atomic nucleus. Match each of these individuals or groups with one of their major accomplishments (often, but not in every case, the one for which the prize was awarded), shown in the right-hand column.

______ a. A. H. Becquerel	1. derived the equation $E = mc^2$
______ b. James Chadwick	2. discovered α and β rays
______ c. Marie Curie	3. discovered the neutron
______ d. Albert Einstein	4. discovered nuclear fission
______ e. Otto Hahn	5. discovered radioactivity
______ f. Ernest Rutherford	6. discovered radium and polonium

4. Place the following events in chronological order:

a. detonation of the first atomic bomb
b. discovery of α, β, and γ rays
c. discovery of the neutron
d. discovery of nuclear fission
e. discovery of radioactivity
f. the first planned, successful transmutation of an element

5. In the process of unlocking the secrets of the nucleus and in the story of the atomic bomb, what important event or process is associated with each of the following cities?

a. Alamogordo, New Mexico
b. Berlin
c. Hanford, Washington
d. Los Alamos, New Mexico
e. Oak Ridge, Tennessee
f. Paris

6. a. What element was the first one found to be radioactive?
b. What element was the first one found to undergo nuclear fission?
c. What radioisotope was the first to be produced artificially on an industrial scale?
d. Why was the element of part (c) produced in such large quantities?
e. What element is produced on the sun through a nuclear fusion reaction?
f. What isotope forms the final, stable product of the chain of radioactive decay that starts with U-238?
g. What element forms when carbon-14 undergoes radioactive decay?
h. What element forms initially on radioactive decay of radon?
i. What element served as the starting material for the first planned, successful, transmutation of one element into another?
j. What element was produced as a result of the first planned, successful transmutation?

7. What is *phosphorescence* and what event revealed to A. H. Becquerel that the mysterious spot on his photographic plates did *not* originate in the phosphorescence of uranium?

8. What one characteristic is common to all end products of all possible chains of radioactive decay?

9. How does the ratio of neutrons to protons in an atomic nucleus affect the stability of the nucleus?

10. A *neutron bomb* is a proposed nuclear weapon that would explode with very little force (and therefore do little physical damage to buildings and equipment) but would release immense amounts of neutrons as ionizing radiation that would disable and kill troops and civilians. Which would you expect to penetrate matter more effectively: (a) a beam of high-speed neutrons or a beam of γ radiation? (b) a beam of high-speed neutrons or a beam of α radiation? Give reasons for your choices.

11. (a) What results from the collision of an electron with a positron? (b) Explain why this is an example of the conversion of mass into energy.

12. Protons and neutrons are the particles that make up the atomic nucleus. What particles make up protons and neutrons?

13. Describe a commercial use for γ radiation that is *not* a medical application.

14. What does the *implosion* accomplish in the plutonium bomb?

15. In the initiator of both the uranium and plutonium bombs, what function does the polonium-210 serve? What function does the beryllium serve?

16. In what way did Ernest Rutherford bring the dreams of the alchemists to reality?

A Little Arithmetic and Other Quantitative Puzzles

17. Using the form of notation shown in Figure 4.3 and (if needed) data obtained from the periodic table or lists of atomic symbols, atomic numbers, and mass numbers, write complete symbols for (a) radon-222, (b) carbon-13, (c) the most common isotope of lithium, (d) the isotope of chlorine of mass number 37, (e) an atom containing 43 protons and 56 neutrons, (f) a helium nucleus, (g) a proton, (h) a γ ray, (i) a β particle, (j) a chloride ion, and (k) a sodium ion.

18. (a) When a radioactive nucleus ejects an α particle, both the atomic number and the mass number decrease. By what quantity does each decrease? (b) When a radioactive nucleus ejects a β particle, only one of these (atomic number or mass number) changes. Which one? Does it increase or decrease? By what quantity?

19. (a) What effect does ejection of a γ ray have on atomic number? On mass number? (b) What effect does ejection of a positron have on atomic number? On mass number?

20. Tritium is a radioactive isotope of hydrogen. Explain why tritium *cannot* decay with the loss of an α particle.

21. What form of radioactivity, if any, occurs with an *increase* in the mass number of the radioactive nucleus? Explain.

22. The experimentally measured mass of an atom is always less than the mass calculated as the sum of the masses of all the protons, neutrons, and electrons that compose it. Why?

23. If U-238 absorbed *one* neutron and the resulting U-239 then underwent fission to generate *three* neutrons and two (and only two) mutually equivalent nuclear particles that were equal in their mass numbers and atomic numbers, what isotope of what element would these two new atomic particles represent? Do you think they would be stable isotopes or radioisotopes? Explain.

24. Per unit of atomic mass, Fe-56, the most common isotope of iron, has one of the greatest mass defects and one of the highest binding energies of all atoms.

a. Given that the experimentally measured atomic mass of this isotope of iron is 55.9349, calculate its mass defect. The atomic number of iron is 26.

b. What is the mass defect of Fe-56 per atomic mass unit?

c. What is the mass defect of U-235 per atomic mass unit? (See Sec. 4.12.)

d. Which has the greater binding energy per atomic mass unit, Fe-56 or U-235?

25. Complete the following equations with the symbol for the atom or particle represented by the question mark. Show the mass numbers and atomic numbers of the isotopes formed or the symbols of the subatomic particles:

a. $^{210}_{84}\text{Po} \longrightarrow ? + \alpha$

b. $^{234}_{91}\text{Pa} \longrightarrow ? + \beta$

c. $^{131}_{53}\text{Pa} \longrightarrow ^{131}_{54}\text{Xe} + ?$

d. $^{230}_{90}\text{Th} \longrightarrow ^{226}_{88}\text{Ra} + ?$

e. $? \longrightarrow ^{239}_{94}\text{Pu} + \alpha$

f. $? \longrightarrow ^{236}_{92}\text{U} + \beta$

26. Why was a source of neutrons needed in the detonation of the atomic bombs?

Think, Speculate, Reflect, and Ponder

27. Describe one difference between the way the laws of the physical universe operate within the nucleus and the way they operate in our everyday world.

28. Describe Becquerel's use of the scientific method as he sought to establish a connection between phosphorescence and X-rays. How did he explain the observation of an exposed spot on the photographic film that had been wrapped in black paper and kept in a drawer with the crystal of the uranium compound, out of contact with sunlight? How did this observation affect the idea that a phosphorescing substance emits X-rays along with visible light? If Becquerel wished to continue his investigation of a connection between phosphorescence and X-ray emission, what test(s) might he have used next?

29. Suppose that instead of using crystals of a uranium compound in his investigations of phosphorescence, Becquerel had used crystals of a compound that was *not* radioactive but that *did* nonetheless phosphoresce. With your own knowledge that phosphorescence and the generation of X-rays are not connected, what do you think Becquerel would have observed? What do you think his conclusion would have been?

30. Was Columbus's discovery of the American continent an example of serendipity? If the American continent did not exist and Columbus had indeed reached the Orient by sailing westward, as he had expected, would that have been an example of serendipity? Explain your answers.

31. Give an example of a discovery (other than those discussed in this chapter) made through serendipity. The example you use may come from outside the field of science.

32. Contrast and compare the use of serendipity and the use of the scientific method in making discoveries. Can the two work in common in making discoveries or must the use of one exclude the use of the other?

33. In the operation of the scientific method the questions that scientists ask of the universe (in their experiments) are almost always based on something already known of the universe. The questions are chosen carefully, so that there is a good chance that the answers (the results of the experiments) will yield interesting or important information. With this in mind, why do you think that Otto Hahn chose to bombard uranium nuclei with neutrons rather than with α or β particles?

34. It has been suggested that in areas with certain kinds of soil and underground rock and mineral deposits, it may be more hazardous to live in a house that is well sealed against drafts than in one with loose-fitting doors and windows that allow a continual flow of air into and out of the house. Suggest a reason why this may be so.

35. Explain why radioactivity is hazardous to humans.

36. Radioactivity does considerable biological damage and can be harmful to living things. Of the three forms of radiation discussed in this section, α, β, and γ rays, which would be the most hazardous to a person standing a few feet from a radioactive substance that emits all three of these forms of radiation? Which would be the least hazardous?

37. The first significant application of newly discovered nuclear energy was in the construction of an atomic bomb. Suggest a reason why this was so.

38. Nuclear weapons tests in space, in the atmosphere, and in the ocean are prohibited by international treaty, but underground tests are still permitted. Why are underground tests of nuclear weapons considered to be different from the other tests?

39. Detonations of high explosives such as dynamite and TNT are used for peaceful purposes, in demolition and in construction, for example, as well as in weapons of war. Suggest and describe peaceful uses for detonations of fission or fusion nuclear explosives.

40. As we saw in Exercise 10, a neutron bomb produces a very small blast when it explodes but releases very large numbers of neutrons. Would you classify a neutron bomb as a more humane weapon or a less humane weapon than a uranium bomb of the power that exploded over Hiroshima? Explain.

41. Chemical reactions release and absorb far less energy (per atom involved) than the nuclear reactions of fission and fusion. Using this observation as a basis, suggest a reason why the Law of Conservation of Mass *might not* be valid.

42. As we saw in Section 4.8, Otto Hahn and Fritz Strassmann carried out certain experiments on uranium late in 1938 in Berlin, and Lise Meitner identified the results of these experiments as newly discovered nuclear fission, a process that could release incredible amounts of energy. In September 1939, Germany invaded Poland and World War II began. Not quite a month before the outbreak of war, believing that Germany was working toward the development of an atomic bomb, several scientists persuaded Albert Einstein to write a letter to President Roosevelt, as we saw in Section 4.11, revealing the possibility that such a bomb might be built. Assume that you are a nuclear scientist living in August 1939. You are aware of the possibility of building a bomb of such devastating power, and you are also aware of the events taking place in Europe and of the implications they hold for our form of society. Would you have joined other scientists in urging Einstein to write his letter or would you have urged that he not write it? Explain your answer.

43. Assuming that the nations of the world will continue to maintain military forces and arm themselves with weapons of war, do you think that nuclear weapons ought to be an accepted part of a nation's arsenal of weapons? Do you think that all nations ought to renounce the use of nuclear weapons, destroy any and all that they now possess, and refrain from building or helping others to build more? If you think that nations ought to maintain nuclear weapons, do you think these weapons ought to be considered as conventional weapons, like bullets, shells, high-explosive bombs, and mines, or that they ought to be considered to be unconventional weapons, like nerve gas, poison gas, and the biological weapons of germ warfare? Give your reasons for answering as you do.

44. One of the motives for dropping atomic bombs on Japan at the end of World War II was to save the lives of American troops who might otherwise have had to invade the home islands of Japan. There is still dispute about whether such an invasion would have been necessary. Regardless of the likelihood that such an invasion might have been required, suppose that one would indeed have been necessary and that the lives of American troops would have been lost as a result. Suppose you had been President of the United States in 1945 and that the best military estimates were that dropping the bombs would save the lives of 1 *million* U.S. troops. Knowing what you do now of the effects of the bombs that dropped on Hiroshima and Nagasaki, would you have authorized their use? Suppose that the best estimates were that Japan was lightly defended and that our troops were so well prepared that no more than *10* U.S. lives were likely to be lost. Under those conditions, would you have authorized the use of nuclear weapons? If you would have authorized their use at the level of 1 million casualties, but not at a level of 10, can you cite a specific number of casualties that would have been a dividing line between your decision to authorize and your refusal to authorize the use of the atomic bombs? Explain your answers.

Web Questions

Use the Web to answer or accomplish each of the following. Report your Web source for each response.

1. Section 4.1 contains the sentence "In 1903 Marie Curie, a native of Poland, became the first woman to receive a Nobel Prize as she, Pierre, and Becquerel shared the award in physics for the discovery of radioactivity." In what year was the Nobel Prize in physics first awarded? Who received the Prize, and for what discovery? How was the award of that first Prize in physics connected to the events described in this chapter?

2. Find an example of a scientific discovery or advance, in addition to those described in this chapter, that came about as a result of serendipity.

3. Write a brief biography of Ernest Rutherford emphasizing his scientific achievements and describing several ways in which he advanced our understanding of the atom and the particles that compose it.

4. Write a brief description of Paul Villard's discovery of gamma rays.

5. Identify Murray Gell-Mann and describe his contribution to our knowledge or understanding of one of the topics of this chapter.

6. Section 4.11 contains the sentence "It was Einstein who, not quite a month before the outbreak of World War II and at the urging of fellow scientists, wrote a letter to President Roosevelt revealing to him the possibility of producing an atomic bomb." Describe the circumstances surrounding the writing of this letter.

7. Identify each of the following, describe how each was involved in the creation of the atomic bomb, and describe a scientific contribution of each that is unrelated to the atomic bomb: (a) Leo Szilard, (b) Edward Teller, (c) John von Neumann, (d) Eugene Wigner.

8. What were some of the arguments used to justify the use of atomic bombs at the close of World War II? What were some of the criticisms of their use? What alternatives to dropping them on occupied cities were proposed before their actual use?

9. What is the current state of nuclear armaments among the world's nations? What publicly available information now exists about the kinds and numbers of nuclear weapons that each nation now holds? What international agreements are now in effect concerning the construction and/or testing of nuclear weapons? What agreements, if any, are now under consideration or review?

Harnessing the Secrets of the Nucleus

NUCLEAR ENERGY, NUCLEAR MEDICINE, AND A NUCLEAR CALENDAR

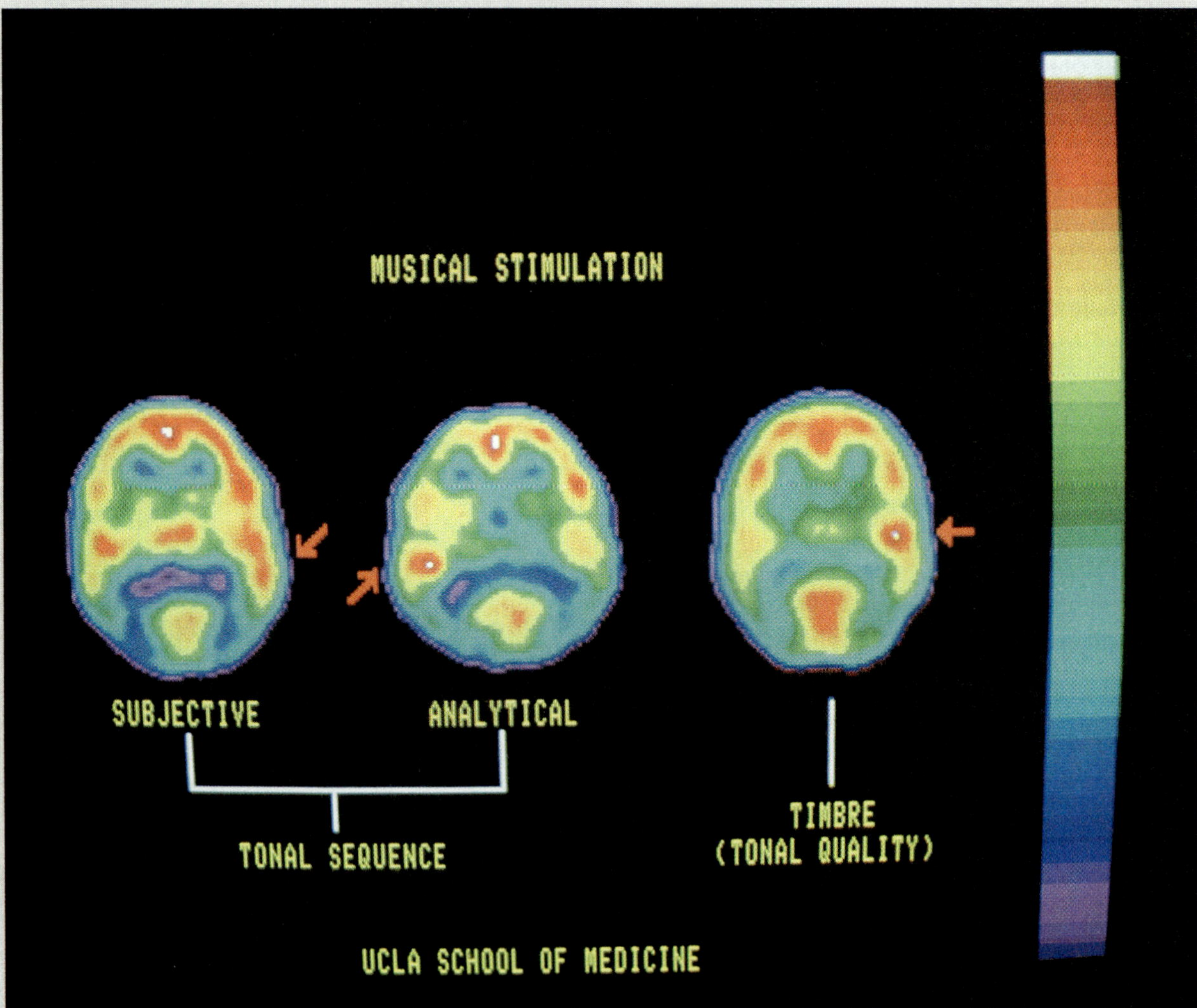

Positron emission tomography reveals the effects of musical stimulation on the brain.

Static at Stagg Field: The First Atomic Pile

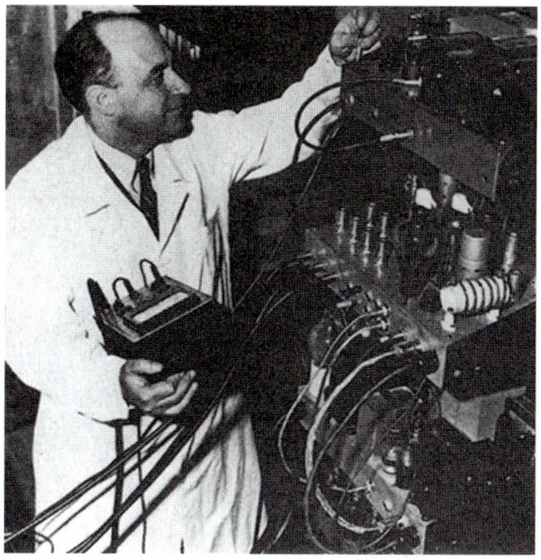

Enrico Fermi built the first atomic pile and produced the first controlled chain reaction on December 2, 1942.

If the spectacular blast that dwarfed the sun above the barren sands of New Mexico in the summer of 1945 heralded the use of nuclear power for war, then the birth of peaceful nuclear power came more quietly on a cold winter's day in 1942, with a noise no greater than loud static, in a squash court within a football stadium in the populous city of Chicago.

It happened like this. The bomb's designers knew in the early days of World War II that detonation would occur only if a sustained chain reaction could be generated as the critical mass formed. It would be the chain reaction that would cause the explosive, instantaneous fission of all the nuclei. But in 1942 no one had ever seen or produced a sustained chain reaction. No one was quite certain that it would even occur when the critical mass formed. To produce a *controlled* chain reaction, to demonstrate that such a chain reaction could be made to occur without restraint within the bomb itself, the physicist Enrico Fermi designed and built the first *atomic pile* (also known as a nuclear pile) in a doubles squash court under the stands of Stagg Field, the University of Chicago's unused football stadium. Born in Italy, Fermi won the Nobel Prize in physics in 1938 for his work with radioactivity and for finding a way to slow the speed of neutrons emitted in nuclear reactions, an achievement critical to harnessing nuclear power. With the rise of fascism in Italy, Fermi came to the United States, became a citizen, and continued his work.

The pile that Fermi built was roughly cubical, 20 feet high, with a base about 20×24 feet. It consisted of a lattice of uranium and uranium oxide cores and bricks of highly purified graphite, a form of carbon. Rapidly moving neutrons slow down considerably as they pass through graphite and certain other materials, which are known as *moderators* because of their ability to slow or moderate the speed of neutrons passing through them. Slowing the speed of the neutrons increases the chance that some of them will be absorbed into a uranium-235 nucleus and produce fission. (Slower moving neutrons are more often absorbed by these nuclei than are faster ones, which tend to bounce off the nucleus in a process called *neutron scattering*.) Thus neutrons produced in one fuel rod must leave it, pass through a moderator, slow down, and then enter another fuel rod where they are absorbed by nuclei and continue the chain reaction.

On December 2, 1942, 16 days after construction had begun, Fermi decided that the 432-ton pile was ready for a trial. Located throughout the lattice were 10 rods of cadmium, an element that absorbs large quantities of neutrons. These cadmium rods, which prevented any chain reaction from starting, would be removed in stages to get the neutrons flowing and the pile running.

Would the test fizzle? Would it blow up—pile, squash court, stadium, and all—right there in Chicago? No one could be entirely certain. At about 2:00 p.m. on the afternoon of December 2, after preliminary testing earlier in the morning, the critical phase began. Nine of the 10 cadmium rods were removed from the pile and were stationed nearby, set to be thrust back into the pile if needed. One attendant held an axe to a rope suspending

A depiction of the dawn of nuclear power as the first chain reaction begins beneath Stagg Field, Chicago.

one of the rods above the pile, ready to drop it back into the pile in an emergency. Three other assistants stood above the pile with bottles of a cadmium solution. If all else failed they would douse the pile with cadmium, smothering any runaway reaction.

As the single, remaining cadmium rod was withdrawn slowly, a few inches at a time, radiation detectors stationed around the pile, one of them emitting audible clicks, began responding. With each stage of removal the clicks grew more frequent. As the rod was removed, the level of response rose, in step with the rod. The clicking soon became a continuous static. Finally, just before 4:00 in the afternoon, at a point predicted by Fermi, the meters began responding out of proportion to the withdrawal of the rod. The neutrons were cascading freely; a controlled chain reaction had begun. After it had proceeded successfully for 4.5 minutes, the central cadmium rod was driven back into the pile and the experiment was shut down.

It had worked. A chain reaction had been started, had been allowed to continue for just under 5 minutes, and had been stopped. The bomb could be built. What's more, the chain reaction had been controlled. Because the power of the nucleus could be released slowly, moderately, under controlled conditions, it could be used as a source of energy for peace as well as war.

In this chapter we'll see how the slow release of nuclear energy from atomic piles, similar to the one Fermi used, is used for the production of electricity on a commercial scale. Then we'll examine some of the applications of radioactivity in medicine and, still later, we'll see how an unusual property of naturally occurring radioisotopes allows us to determine the age of ancient objects.

5.1 Nuclear Power: The Promises and the Problems

Released slowly, the energy of nuclear fission can provide electricity in much the same way as the energy of burning oil, gas, or coal: by furnishing the power that turns the shaft of an electric generator. Any force that can turn a generator's shaft—the action of a human arm, the force of the wind or of falling water, or the power of steam generated by the heat of burning oil, gas, or coal, or by an atomic pile—can produce electricity.

To the list of more conventional fuels we can now add nuclear fuel, especially as it generates heat in an atomic pile. Ideally, nuclear fusion would be more desirable than fission for producing electrical energy. Fusion would be more efficient (Sec. 4.14) and would generate fewer radioactive wastes (Sec. 5.5). In practice, though, there's no known material that can withstand the high temperatures needed to maintain a fusion reaction, so there's no way at present to construct an enclosure for this kind of nuclear furnace. One hope lies in constricting the fusion mixture with a powerful magnetic field, pulling the extraordinarily hot gases away from contact with the interior surface of their container. This has not yet proved practical for a continuous fusion reaction. At present it appears that a realistic, commercially practical fusion reactor will cost billions of dollars to produce and lies many decades in the future.

Today's practical nuclear power plant consists of a fission reactor built around a nuclear pile similar to Fermi's. The heat generated by nuclear fission within the pile converts water into steam, which strikes the blades of a turbine attached to the shaft of an electric generator. The pile serves as the core of the reactor. Movable control rods, often made of boron or cadmium, slide into and out of the pile to control the rate of heating and energy production. These control rods maintain the neutron flow through the fissionable material at a level high enough to sustain useful energy production but low enough to prevent overheating.

About two-thirds of the nuclear power plants currently operating in the United States are *pressurized-water reactors (PWR)*, in which circulating, pressurized water carries the pile's heat to the water (in another container) that generates the steam that turns the turbine (Fig. 5.1). The remainder are *boiling-water reactors (BWR)*, which dispense with the circulating water and immerse the pile directly into the container of water that's converted to steam.

The pile designed and built by Enrico Fermi in 1942 was intended for research rather than the production of power. During the 4.5 minutes of the first test, the pile generated only about enough energy to light the bulb of a flashlight. The practical, large-scale generation of electricity by nuclear fission began almost exactly nine years later—in December 1951—when the National Reactor Testing Station in Idaho began supplying power to about 50 nearby homes.

Initially, in the middle of the 20th century, the promise of nuclear energy was one of cheap, plentiful electricity produced by a fuel that would free us from the

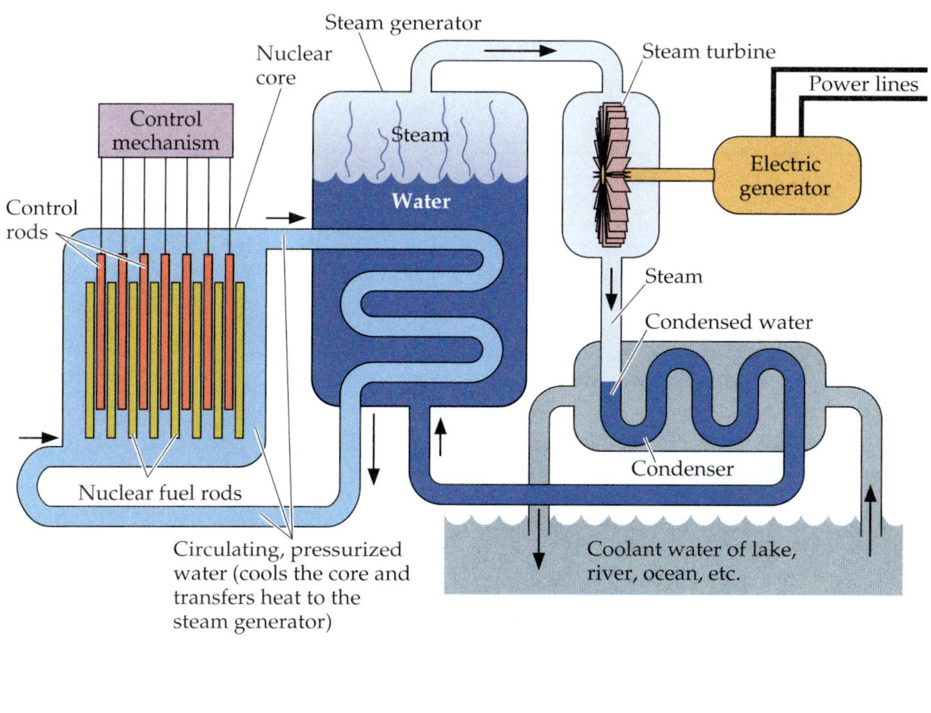

Figure 5.1 Schematic diagram of a nuclear power plant.

Cooling towers of a nuclear power plant.

smoke, smog, and other pollutants of coal, oil, and similar fuels. With the application of a bit of scientific ingenuity, it was thought, our supply of nuclear fuel could provide us with centuries of plentiful, inexpensive, pollution-free energy.

The introduction of nuclear energy started well in mid-century United States, but it soon gave way to problems and fears that have stunted the growth of commercial nuclear power. By 1970, 18 nuclear plants were operating in the United States. The number rose rapidly, reaching 108 by 1988, but it has changed little since then, dropping slightly to 104 in 2000 as a result of shutdowns. (In 2000 a total of 438 nuclear power plants were in operation throughout the world, with another 31 planned or under construction.) The percentage of commercial electricity produced from nuclear power in the United States rose from a minuscule 1.4% in 1970 to 11% in 1980, then to about 20% in 1988, a level that has remained virtually unchanged.

Table 5.1 presents recent data for commercial electric power production by nuclear energy in several industrialized nations. In the United States, only coal, which generates a little over half of all commercial electricity, surpasses nuclear power in the production of electric power.

The dream of cheap, plentiful nuclear power, though far from dead, has grown dim over the past few decades. Several factors have contributed to the fading of the dream. Among those we will consider here are (1) the real and imagined dangers of nuclear power plants themselves; (2) limitations in the available supply of fissionable fuels; (3) hazards involved in the disposal of nuclear waste; and (4) the financial costs of nuclear power.

Table 5.1 Nuclear Power Production in 2000

Country	Nuclear Power Plants in Operation	Percentage of Electric Power Produced
United States	104	20
France	59	76
Japan	53	34
United Kingdom	35	22
Russia	29	15
Germany	19	31

■ QUESTION

(a) What percentage of the world's nuclear power plants are located in the United States? (b) Rank the countries of Table 5.1 in terms of the percentage of electric power produced from nuclear plants. Where in this ranking does the United States stand?

5.2 Problem 1. Nuclear Power Plants: Real and Imagined Dangers

Perhaps the most obvious and yet least realistic of all the fears is that a nuclear power plant might explode in a nuclear fireball. This concern is completely groundless. A nuclear explosion requires the almost instantaneous release of nuclear energy as the fissionable material is compacted into an explosive critical mass. A fission pile, on the other hand, is designed to produce a slow and continuous release of energy. Because of the low concentration of U-235 in the nuclear fuel, there just isn't enough U-235 present to form an explosive mass of fissionable material: The fuel rods of nuclear piles contain only about 2 to 3% U-235; nuclear weapons require U-235 with a purity on the order of 90%. In the absence of such a mass of purified U-235, neither a weapon nor a fission pile can produce a nuclear explosion. (The uranium that occurs naturally, in the earth's crust, contains about 0.72% U-235; over 99% is U-238; see Sec. 4.9.)

Concerns about other kinds of accidents are more realistic. These include, among others, the accidental overheating of the core and the occurrence of non-nuclear fires or explosions that might release radioactive material. In December 1952, almost exactly one year after the first distribution of nuclear-generated electricity to consumers by the Idaho plant (Sec. 5.1), the first major accident occurred at a nuclear generator. Four control rods were accidentally removed from the core of a pile near Ottawa, Canada. The result was a partial *meltdown*, a situation in which uncontrolled heating leads to the literal melting of the reactor and the destruction of the building that contains it. Although millions of gallons of water within the reactor became contaminated with radioactive material, there were no reported injuries. Five years later a fire in a reactor at Seascale, on the Irish Sea in northwestern England, led to the release of radioactive material into the countryside with the eventual radiation-related deaths of an estimated 39 people.

The most serious accident in the United States occurred on March 28, 1979, in a plant at Three Mile Island, near Harrisburg, Pennsylvania. A combination of equipment failures and human errors led to a loss of coolant. The core overheated to perhaps 2000°C, producing a partial meltdown in which much of the nuclear fuel melted. The accident resulted in the flooding of the building surrounding the reactor, the *containment building*, with large quantities of radioactive water and the release of a small amount of radioactive gas into the atmosphere.

Worldwide, by far the most spectacular and most damaging accident produced explosions and a fire in Unit 4 of a four-unit nuclear reactor complex at Chernobyl, Ukraine, early in the morning of April 26, 1986. (At the time of the explosion, Ukraine was one of the member states of the Soviet Union.) A combination of faulty reactor design and a series of human errors, including the failure to follow prescribed safety procedures, resulted in rapid, catastrophic overheating and rupture of the nuclear pile.

There were two explosions. Flash overheating of the reactor core produced an almost instantaneous, explosive vaporization of coolant water into steam, rupturing the core, tossing its 1000-ton cover, and blowing the roof off its housing. Seconds later the rapid vaporization of nuclear fuel blew the core apart, spewing its radioactive contents up, out of the building, into the atmosphere and the surrounding environment.

The accident released huge amounts of radioactive smoke and debris from the reactor. Radioactive particles traveled high into the atmosphere, moving first northwestward to Finland and Sweden—where analysis of radioisotopes in the upper atmosphere first revealed the nuclear accident—and then eastward and westward into Europe and Asia, with small amounts reaching the Western Hemisphere. The explosion and ensuing release of radiation caused two deaths immediately at the plant and 28 more within a few days or weeks.

After the immediate, physical shock of the disaster, and beyond Chernobyl itself, the greatest danger came from the radioactive debris released by the accident

The nuclear power plant at Chernobyl, after the accident of April 16, 1986.

Fallout is fine radioactive debris that is released into the atmosphere by nuclear explosions and accidents and that settles to the earth.

in the form of fine particles. Whether formed in the atmosphere by a nuclear weapon or thrown into the atmosphere by a ground-level accident like the one at Chernobyl, this radioactive dust disperses and settles to the earth as radioactive **fallout**. The danger posed by the fallout forced the evacuation of about 116,000 people from the region surrounding Chernobyl. Nearby fields of crops had to be destroyed, and even the bark of trees growing as far as 40 miles away had to be scraped free of radioactivity. Even 15 years later the heavy machinery used to clean up the debris, the trucks, cranes, helicopters, and bulldozers abandoned near the site, were still emitting hazardous radioactivity. Except for certain areas in and near Chernobyl that were scoured of radioactivity, everything within 18 miles of the reactor was still off-limits to humans.

Beyond residual radioactivity at the site itself, the major hazard comes from radioactive components of the fallout that enter the food supply, especially iodine-131. Fallout containing this radioisotope settles on grasslands and fields and enters the human food chain through the meat and milk of the animals that feed on contaminated vegetation. The thyroid, a gland located in the neck, is particularly sensitive to I-131 since most of the iodine contained in our food and drink concentrates in this gland. Children who drink milk containing I-131—produced by cows that graze on grass contaminated by fallout—are especially at risk of thyroid cancer. In the 15 years following the explosion the rate of thyroid cancer among those living in the surrounding area and older than 18 years was 4.5 times the rate for the 15 years preceding the blast; for younger residents the rate was about 34 times that of the earlier period.

Of the four nuclear plants at Chernobyl, Unit 4 was destroyed in the 1986 accident; Unit 2 stopped operating after a fire in 1991; Unit 1 was closed in 1996, 10 years after the explosion at Unit 4; Unit 3, the last of the nuclear reactors at Chernobyl, was closed in December 2000. Chernobyl is no longer a center of nuclear power production.

Despite the devastation at Chernobyl and its consequences, it's important to remember that the damage was not caused by either a fission or a fusion reaction. The real danger was not one of a nuclear blast, but rather of the accidental release of radioactive material into the environment.

■ QUESTION

What is the most serious form of damage that could occur if a natural disaster such as a hurricane, a tornado, or an earthquake struck a nuclear power plant? Explain.

5.3 Problem 2. Available Supplies of Fissionable Fuels: Is the Breeder Reactor a Solution?

Even if the perfect fission reactor could be built, completely immune to any form of failure or accident, major problems would still exist in the

- supply of economical, fissionable fuel;
- disposal of nuclear wastes; and
- overall economics of energy production.

We'll examine first the supply of nuclear fuel. We've already seen that U-235, the most efficient of the fissionable isotopes of uranium, constitutes less than three-quarters of one percent of the naturally occurring uranium ore deposits (Secs. 4.9 and 5.2). At the current rate of energy production, estimates of the time it would take to exhaust the world's supply of this isotope—at least the supply that can be extracted from the ore at a reasonable cost—range from less than 50 years to about a century.

One way to extend the supply of nuclear fuel by a factor of 100 or more is through a **breeder reactor**, which actually produces more fissionable material than it consumes. Several different types of breeder reactors are technically feasible, each using a different type of nuclear fuel and each operating through a different set of nuclear reactions. In one kind, plutonium-239, manufactured in ways similar to those used in building the first atomic bombs (Secs. 4.9 and 4.10), serves as the nuclear fuel in the core. In addition to the Pu-239 this reactor also contains rods or pellets of U-238, the common and relatively abundant but ineffective isotope of uranium. Neutrons emitted by fission of the Pu-239 (or by some other spontaneous process) strike the U-238 and convert the uranium, through a short segment of a decay chain, to Pu-239, as shown in the following three reactions.

A **breeder reactor** is a nuclear reactor that produces more fissionable fuel than it consumes.

In the first reaction of the sequence, a neutron enters a U-238 nucleus and converts it to the isotopic U-239:

$$^{238}_{92}\text{U} + {}^{1}_{0}\text{n} \longrightarrow {}^{239}_{92}\text{U}$$

In the next step, the uranium-239 (the product of the first reaction) loses a β particle from its nucleus to produce neptunium-239 (Np-239). With an increase in the atomic number (and therefore the positive charge) of the nucleus, but no change in the number of electrons surrounding the nucleus, the neptunium is produced as a cation (Sec. 1.3). This newly formed Np-239 cation captures an electron from some outside source. As the newly captured electron enters the shells surrounding the nucleus, it neutralizes the positive charge of the cation and transforms the Np-239 cation into an electrically neutral Np-239 atom. In the entire process, then, the *net* change is the loss of an electron from the nucleus of the U-239 atom into the surrounding electron shells. Although the electron that ends up in the outer shells isn't necessarily the same one that left the nucleus as a β particle, we write the set of equations as if it were:

$$^{239}_{92}\text{U} \longrightarrow {}^{239}_{93}\text{Np}^{+} + {}^{0}_{1-}e^{-} \longrightarrow {}^{239}_{93}\text{Np}$$

Finally, the Np-239 loses a β particle in a similar manner to form Pu-239. The entire process produces energy and at the same time "breeds" fissionable Pu-239 from U-238:

$$^{239}_{93}\text{Np} \longrightarrow {}^{239}_{94}\text{Pu}^{+} + {}^{0}_{1-}e^{-} \longrightarrow {}^{239}_{94}\text{Pu}$$

With the conversion of an inefficient isotope such as U-238 into the much more efficient Pu-239 through breeding, the useful supply of fissionable fuels can be made to last thousands of years. Because this type of reactor uses faster moving neutrons than those of the U-235 reactor and also uses liquid sodium (which exists as a solid metal under ordinary conditions) as a coolant rather than high-pressure water, it's sometimes called a *liquid metal fast breeder (LMFB) reactor*.

Breeder reactors offer some important advantages in energy production, including a long operating life and low operating expenses. But because of their relatively high initial costs as well as some important technical and political factors, very few now exist throughout the world. Virtually all breeder reactors operate on an experimental rather than a commercial scale. One of the major political concerns centers on the ease with which the generated plutonium can be converted into a fission bomb.

The widespread use of breeder reactors could extend nuclear weapons to countries that do not now have them and perhaps even to terrorist organizations. Another problem lies in the extraordinarily great health hazard presented by plutonium, which emits α particles as it decays. When the element enters the body it tends to settle in the lungs and other organs and in the bone marrow. The accidental ingestion of even traces of plutonium presents a serious cancer hazard; as little as 1 microgram (1 μg, or 10^{-6} g) in the lungs may be enough to induce cancer.

■ QUESTION

Describe one advantage of a breeder reactor over a conventional nuclear reactor. Describe one disadvantage.

5.4 Problem 3. The Disposal of Nuclear Wastes: The Persistence of Radioactivity

Disposal of radioactive wastes by burial in a shallow pit.

Solutions to the first two problems of Section 5.1, safety and the supply of fissionable resources, still leave the third and perhaps most troubling problem of all: What is to be done with the radioactive wastes generated by nuclear power plants? The most serious concerns about radioactive material come not so much from fears of an accidental nuclear explosion but rather from its ionizing radiation, from the α, β, and γ rays given off by decaying radioisotopes.

As we saw in Section 4.9, the fission of a U-235 nucleus can occur in many different ways with the production of a great variety of isotopes, almost all of which are radioactive. We've also seen, in Section 4.5, that the spontaneous, radioactive decay of any particular radioisotope (with the loss of a small nuclear particle) can be part of a decay chain of several steps that represent the appearance and decay of still other radioisotopes. It follows, then, that as a nuclear pile operates it generates an enormously large number of radioactive isotopes, each emitting its own hazardous radiation. The combination of all these by-products of power generation, including residual nuclear fuel whose radioactivity, while still hazardous, has dropped below useful levels, constitutes the waste matter of nuclear power generation. The safe disposal of these radioactive wastes represents the most serious of all the technical problems presented by commercial nuclear power.

Eventually, of course, the radiation dies out and the hazard disappears. As each chain of decay comes to an end in a stable isotope (as we saw in Section 4.5 for the decay of U-238 to Pb-206), every radioisotope that now exists in the universe will become transmuted into a stable, nonradioactive atom. This doesn't mean, though, that one day, very far in the future, radioactivity will vanish from the universe. Natural processes, such as the one that converts plentiful, nonradioactive nitrogen-14 into radiocarbon (which we'll examine in Section 5.11), serve to replenish the supply of some radioisotopes.

Nevertheless, the by-products of nuclear power generation are certainly radioactive, and some of the radioisotopes among them remain dangerously radioactive for a very long time. Until all radioactive nuclear wastes undergo spontaneous transformations all the way down the chain to the stable nuclei at the end, they present an environmental hazard. The actual amount of time required for the transformation depends on the *half-life* of each radioisotope.

■ QUESTION

What is the ultimate fate of every radioactive atom now in existence?

5.5 More About Problem 3. The Half-Life of a Radioisotope and the Permanent Storage of Nuclear Wastes

The **half-life** of a radioisotope is the time it takes for exactly half of any given quantity of the isotope to decay.

The **half-life** of a radioactive isotope represents the length of time it takes for exactly *half* of any quantity of that isotope to decay. Since each isotope decays at its own particular rate, each has its own, specific half-life. Iodine-131, for example, has a half-life of 8 days. This means that if we start with 100 g of $^{131}_{53}I$, only 50 g of I-131 will be left after 8 days; 25 g will remain after another 8 days (a total of 16 days); 12.5 g will be left after another 8 days (a total of 24 days); and so on (Table 5.2). During this time the lost iodine is transformed into other isotopes in its decay chain.

The half-life of C-14 is 5730 years; for U-235 it's 7×10^8 years; for Kr-93, one of the many products of U-235 fission (Sec. 4.9), it's 1.3 seconds. These and other representative half-lives appear in Table 5.3.

If we note two general characteristics of half-lives we can understand a bit more clearly the problems associated with nuclear waste disposal. First, regardless of the actual numerical value of a half-life, we can't consider any particular radioisotope to have vanished (for all practical purposes) until at least 10 half-lives have passed. As we can see from the data in Table 5.4 and Figure 5.2, half (50%)

Table 5.2 Radioactive Loss of I-131

Total Elapsed Time (days)	Number of Days Since Previous Measurement	Remaining I-131 (g)
0	0	100.0
8	8	50.0
16	8	25.0
24	8	12.5
32	8	6.2
40	8	3.1
48	8	1.6
56	8	0.8
64	8	0.4
72	8	0.2
80	8	0.1

Table 5.3 Typical Half-Lives

Radioisotope	Half-Life
Kr-93	1.3 seconds
U-239	23.5 minutes
Tc-99*m*	6 hours
Np-239	2.4 days
I-131	8 days
Ba-140	12.8 days
C-14	5730 years
Pu-239	2.4×10^4 years
U-235	7×10^8 years
K-40	1.25×10^9 years
U-238	4.5×10^9 years

of any specific radioisotope remains after one half-life has passed; half of this (25% of the original quantity) remains after a second half-life; and so on. After 10 half-lives just under 0.1% of the original material remains, over 99.9% having decayed into other materials. The first general rule is that, regardless of the length of the half-life, *10 half-lives must pass before the residual radioactivity becomes negligible.*

To illustrate the importance of half-lives, let's return for a moment to the discovery of nuclear fission in the laboratory of Otto Hahn (Sec. 4.8) and to one of the possible fission reactions that may have taken place there.

E X A M P L E

(Half) Life Goes On

Recall from Section 4.8 that after having bombarded uranium with neutrons, Otto Hahn found barium among the products. Assuming that the barium-producing reaction that occurred in his laboratory was the one shown in Figure 4.8,

$$^{235}_{92}U + ^{1}_{0}n \longrightarrow ^{140}_{56}Ba + ^{93}_{36}Kr + 3\,^{1}_{0}n$$

and assuming that it took Hahn about a week to carry out the experiment and to analyze the products, do you think he might have found the krypton produced in this reaction, as well as the barium? (Both Ba-140 and Kr-93 are radioisotopes.)

■ Referring to Table 5.3, we find that the half-life of Ba-140 is 12.8 days and the half-life of Kr-93 is 1.3 seconds. We'll also refer to Table 5.4 and Figure 5.2, which show the percentage of a radioisotope that remains after successive half-lives.

One week is much less than one half-life of Ba-140 (which is 12.8 days), so we shouldn't be surprised that a majority (more than half) of the Ba-140 generated in the neutron bombardment remained and could be detected within a reasonable time for Hahn to have completed his work, perhaps a week.

Table 5.4 The Disappearance of a Radioisotope

Half-Lives Expired	Percent of Radioisotope Remaining (to the nearest 0.001%)
0	100.
1	50.
2	25.
3	12.500
4	6.250
5	3.125
6	1.562
7	0.781
8	0.391
9	0.195
10	0.098

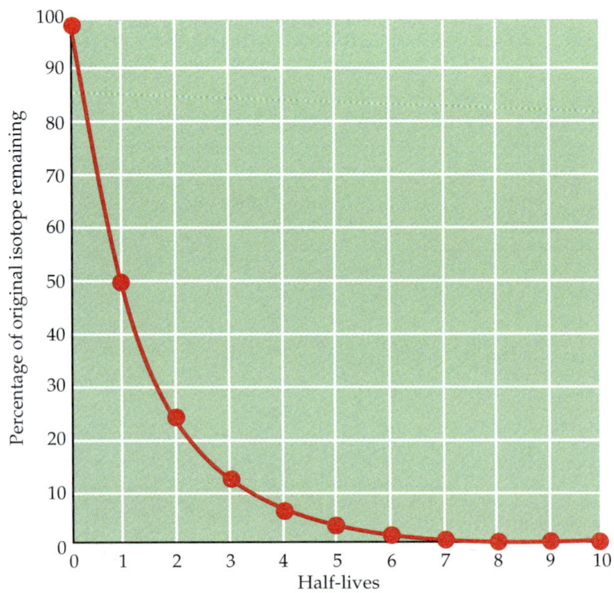

Figure 5.2 Graphical representation of the disappearance of a radioisotope.

Yet with a half-life of just 1.3 seconds (Table 5.3), more than 99.9% of any Kr-93 formed would have vanished in just 10 × 1.3 seconds, or 13 seconds. (This period corresponds to 10 half-lives of the radioisotope.) It's unlikely, then, that Hahn would have detected any of the Kr-93 that formed in this reaction.

The second important characteristic of radioactivity is that there's no known way to change the rate of the kinds of radioactive decay we have examined in this chapter. The half-lives of these radioisotopes can't be increased or decreased by temperature, pressure, magnetic fields, or any other known physical phenomenon. Since we can't increase the rate of their decay (so as to convert them quickly to stable or nonradioactive isotopes), large quantities of radioactive nuclear wastes will be with us and with those who come after us, for generations almost beyond measure. To protect ourselves and future generations from their radiation, a safe way must be found to store these wastes so that they remain out of contact with our environment until they no longer pose a threat or until we devise other ways to dispose of them.

These radioactive wastes include a variety of materials from several different kinds of sources, including:

- medical diagnostic and therapeutic materials, including pharmaceuticals and disposable equipment such as syringes, film, and containers;
- contaminated gloves, clothing, and instruments used in working with radioisotopes in hospitals, nuclear power plants, academic and commercial laboratories, and similar locations;
- exhausted fuel from nuclear power plants; and
- uranium and plutonium extracted from dismantled nuclear weapons.

Regardless of their origin, nuclear wastes can be divided into the two categories of (1) weakly radioactive (*low-level*) wastes and (2) more highly radioactive (*high-level*) and therefore more dangerous materials. Through the Low-Level Waste Policy Act of 1980 and the Nuclear Waste Policy Act of 1982 and subsequent amendments to each, Congress provided for the storage of both low-level and high-level wastes. Low-level wastes are either stored where they are generated or shipped to several federally approved burial sites throughout the United States.

High-level wastes are another matter. In its Act of 1982 Congress directed that high-level wastes be buried deep underground. In a 1987 amendment, Congress was more specific. It authorized the U.S. Department of Energy to determine whether Yucca Mountain, a dry, isolated site about 100 miles northwest of Las Vegas, Nevada, might be a suitable site for long-term storage of high-level radioactive wastes.

In plans currently under study these wastes would be sealed in glass blocks, packed into secure, corrosion-resistant containers, and stored in tunnels dug 1000 feet under the surface of Yucca Mountain. There it's believed they might lie some 1000 feet above the subterranean water table, undisturbed for at least 10,000 years. It's important that the containers remain far from water and perfectly dry since the high-level radioactivity of their contents keeps them warm. No matter how corrosion resistant the containers might seem by our current standards, contact of the warm containers with moisture over years, decades, centuries, and millennia could lead to eventual corrosion and leakage of their radioactive contents into the environment, and especially into underground water supplies.

Construction of a tunnel that will be used for burial of radioactive wastes deep within Yucca Mountain, Nevada.

With approval of the Yucca Mountain site by various governmental bodies, construction could begin by 2005 and storage in 2010. Until then, high-level wastes must be stored on the grounds of nuclear power plants themselves and at sites that had been used for the production of nuclear weapons.

In addition to isolation through burial, another solution to the problem of high-level wastes is *reprocessing*, in which the relatively low levels of uranium and plutonium remaining in exhausted nuclear fuel are extracted, concentrated, and recycled into new, regenerated fuel packets. Plants in France, Great Britain, and Russia currently reprocess nuclear fuel from nuclear generators in their own countries as well as from generators in Germany and Japan. The United States discontinued nuclear reprocessing in the 1970s for a variety of reasons, including the possibility that (as with breeder reactors, Sec. 5.4) regenerated nuclear fuel might be stolen or otherwise diverted into the manufacture of nuclear weapons.

■ QUESTION

If each of the radioisotopes of Table 5.3 were stored at the Yucca Mountain site, which would still be present after 10,000 years at a level of 10% or more of the original amount?

5.6 Problem 4. The Cost of Nuclear Power

The fourth problem, the economic cost of nuclear power, fluctuates with changes in costs of the more traditional fuels. The economics of nuclear power include all the expenses of building and maintaining the plant, obtaining the nuclear fuel, adhering to safety standards, and disposing of the nuclear wastes. But there are costs uniquely associated with the more common fuels as well, including the costs of the coal, petroleum, and natural gas used as fuel, the pollution controls needed to clean the products of combustion—the smoke, ash, sulfur dioxide, and other pollutants—before they enter the atmosphere, and more subtle environmental costs as well (Ch. 14). (One of the environmental benefits of nuclear power is that it does not generate atmospheric pollutants.)

The cost of electricity produced by nuclear power plants must be compared with the cost of power generated by other means. When coal, petroleum, and natural gas are readily available and inexpensive, and when pollution control technology is recognized as efficient and effective in protecting the atmosphere, the economics of nuclear power can become the most significant restraint in its application to the production of electric power. But with high prices for traditional fuels and mounting concerns about atmospheric pollution, nuclear power can become appealing economically.

■ QUESTION

(a) What is one cost of electric power production by nuclear power that is not a factor in the use of coal, petroleum, or natural gas? (b) What is one cost of electric power production by these fuels that is not a factor in the use of nuclear power?

5.7 The Power to Kill and ...

With all the problems involved in generating electrical power from nuclear energy, the most valuable use of nuclear reactions in today's society may well lie in their medical applications: their ability to cure disease and to save lives. So that we can understand more clearly how and why radiation has the power to cure and to prolong lives, we'll first return briefly to the other side of nuclear radiation, the danger that radioactivity poses to living things.

We saw in Chapter 4 that radioisotopes emit ionizing radiation consisting of α particles, β particles, and γ rays. As this ionizing radiation passes through matter, it collides with molecules to form ions that can react with still other chemical particles to form new and unusual products. The result of the passage of ionizing radiation through matter, then, is the transformation of molecular forms originally present into other, often undesirable products. Enough damage of this sort in a living system can produce illness and death.

In general, the damage done to living things by ionizing radiation depends on both the radiation's ability to penetrate tissue—its *penetrating power*—and its ability to produce disruptive ionizations within that tissue—its *ionizing power*. Alpha particles, with a relatively large mass (4 amu) and carrying two positive charges, have enough mass and charge to do considerable damage to whatever atoms and molecules they collide with. They have considerable ionizing power. But they're so big that they're easily stopped by frequent collisions with other chemical particles. They have little penetrating power and simply don't travel very far. Yet they do a considerable amount of damage in the short distances they do travel.

A β particle, with a much smaller mass and carrying a single negative charge, has less ionizing power than an α particle, but being smaller than the α particle, it has less frequent collisions so it travels farther. Over comparable distances β particles do less damage than α particles. Gamma rays, with neither mass nor charge, have considerably more penetrating power, but with little ionizing power they do less damage than α particles or β particles over comparable distances of travel.

As an analogy, you might think of a narrow alleyway cluttered with cars parked here and there, trash bins strewn around, pushcarts blocking the way, and assorted other obstacles on the street and in the gutters. A large, multiwheel tractor-trailer truck (the α particle) trying to bully its way through the alley probably

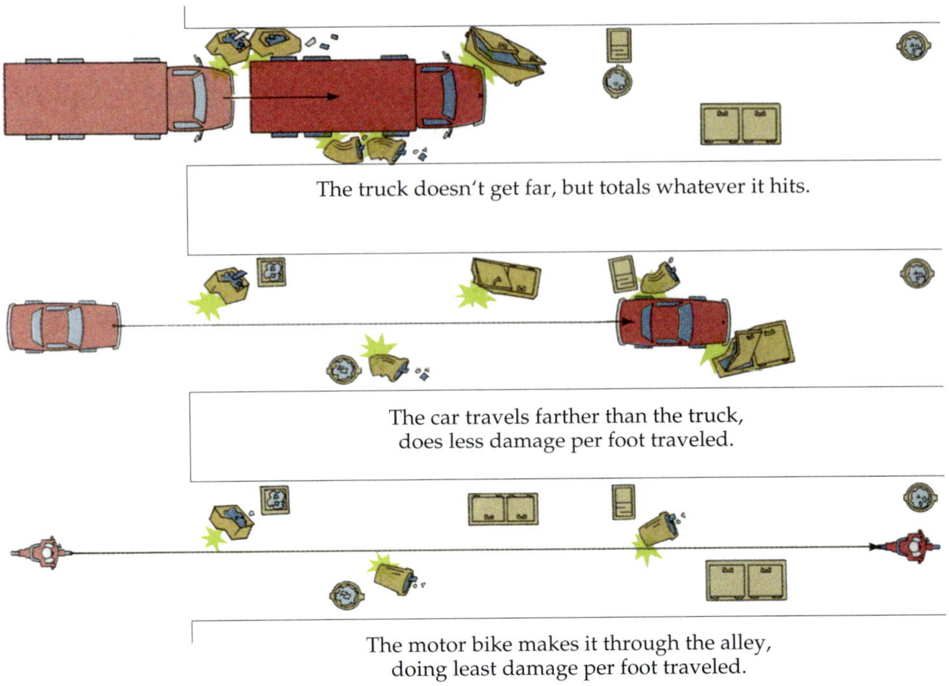

Figure 5.3 Ionizing power and penetrating power: an analogy. The truck represents an α particle, the car represents a β particle, and the motor bike represents a γ ray.

wouldn't get very far, but it would do a lot of damage over the short distance it traveled. A small sports car (the β particle) speeding recklessly along might get farther along in the alley with fewer collisions and less overall damage, but it still wouldn't make it all the way through. Someone on a motor bike (the γ ray) would be more likely to get all the way through the clutter to the other end of the alley, causing only minor damage (Fig. 5.3).

Although no analogy is perfect, this one provides a useful comparison of the penetrating power and ionizing power of α particles, β particles, and γ rays. In fact, we can carry it one step further. While a single motor bike wouldn't do much damage to whatever is in the alleyway, hordes of motorbikes racing through could destroy everything. Similarly, while small doses of γ rays might be harmless or even beneficial, as we'll see in the next section, massive doses can kill.

As ionizing radiation passes through living matter, much like the vehicles in the crowded alleyway, it encounters and collides with a variety of molecules, ranging from simple water molecules to the incredibly elaborate structures of the enzymes of life and the DNA and RNA of cell reproduction and heredity (Ch. 18). The chemical processes necessary for life and for successful reproduction depend on the precise interactions of all of these substances, each with its own characteristic and intricate molecular structure. Large doses of ionizing radiation, with their capacity to transform the finely detailed structures of life into a confusion of new and utterly useless or even harmful molecular forms, can bring the processes to a halt. When this damage is extensive enough to cause illness or death of the living body receiving the radiation, it is called **somatic damage**. The more immediate results of somatic damage range from reduced white cell counts in the blood, through such symptoms as fatigue and nausea, to painful death. Delayed effects appear as damage to bodily organs (the spleen, for example), glands (such as the thyroid), and bone marrow, and through the development of leukemia and other cancers. As with chemical

Somatic damage is injury to a living body, causing illness or death.

poisons and other things that injure the body, the extent of the damage depends on the nature of the radiation, the size of the dose, and the fitness of the individual.

Smaller amounts of radiation, too small to produce perceptible damage to the living organism receiving the radiation, can nonetheless disorganize the molecules of heredity in the cells of the reproductive system, those DNA structures that carry genetic information from one generation to the next. Such **genetic damage** can leave an individual animal, for example, in apparently good physical condition but with subtle reproductive damage that causes offspring to be born with physical and/or mental defects or heightened susceptibility to diseases. An important distinction between somatic damage and genetic damage is that somatic damage affects the individual receiving the radiation; genetic damage produces illness and physical and mental defects in generations yet to come.

Genetic damage is damage that is transmitted to future generations.

Genetic damage caused by radiation. The deformed pig was born the year following the Chernobyl disaster, at a radiation-contaminated farm 40 miles from the power plant.

Although genetic damage from radiation has been observed in laboratory animals for more than 75 years, indications that similar damage might occur in humans have appeared only recently. Analysis of DNA in families living near Chernobyl suggests that radiation-induced genetic damage in parents may have been passed on to children born years after the explosion. Although the link between radiation and genetic damage is well established for laboratory animals, it's still tenuous for humans.

While there's no reason to doubt that damage of this sort can occur in humans just as it does in laboratory animals, there may be several good reasons why it has yet to be demonstrated conclusively. One of these is that large-scale exposure to artificial or human-generated radiation is still a relatively recent phenomenon, at least on the time scale of human generations. Perhaps not enough time has passed to allow the phenomenon to appear. Moreover, the total number of children born to all the survivors of large-scale radiation exposure is only now, after Chernobyl, beginning to grow large enough to allow the effects of artificially induced genetic damage to show up clearly against a background of genetic changes that occur normally. (We'll discuss radiation from natural sources in Sec. 5.13.) Finally, analytical techniques continue to improve, allowing correlations to be made today that might have escaped notice years ago.

■ QUESTION

Name and describe two types of biological damage caused by ionizing radiation.

5.8 ... The Power to Cure

We can turn now to the brighter side of radiation—its capacity to heal and to cure. These beneficial effects of radioactivity depend on both the nature of the radiation itself and the chemistry of the element producing it. The radiation emitted by specific isotopes can be used medically in two distinct ways: (1) in diagnosis, to generate visual images of organs and glands, as an aid in detecting tumors for example, and (2) in therapy, for treating cancers and other disorders of the body.

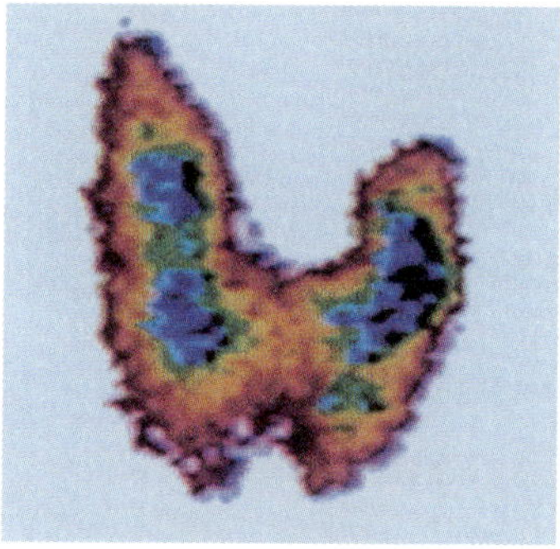

An image of a thyroid gland obtained through the use of radioactive iodine.

As an illustration, the thyroid gland accumulates much of the iodine that enters the body through food and drink, and it uses this element in regulating our growth and metabolism (Sec. 5.2). (Adding potassium iodide, KI, to "iodized" table salt helps prevent enlargement of the thyroid, a condition known as *goiter*; see Chapter 19.) Images of the gland, useful in diagnosing metabolic irregularities, can be obtained from the radiation emitted by several different radioisotopes of iodine, usually introduced into the body as NaI. Iodine-131, which emits both β and γ rays and has a half-life of eight days, is widely used in these examinations of the thyroid.

Preparations for generating technetium-99*m*. The radioisotope will be used as a source of gamma rays in medical diagnostic procedures.

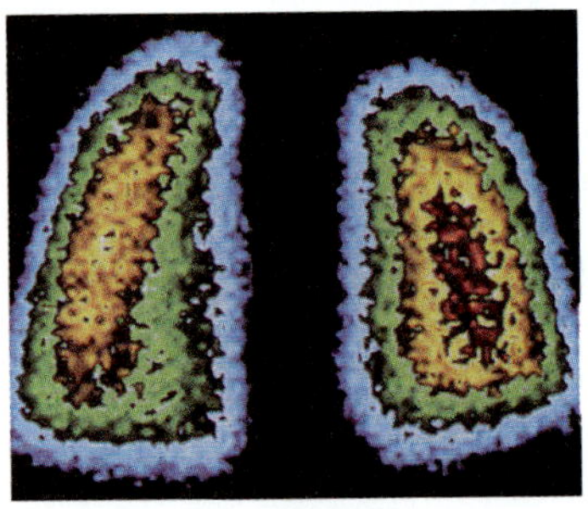

Images of human lungs obtained from a γ-ray scan. A compound containing radioactive technetium-99*m* was the source of the radiation.

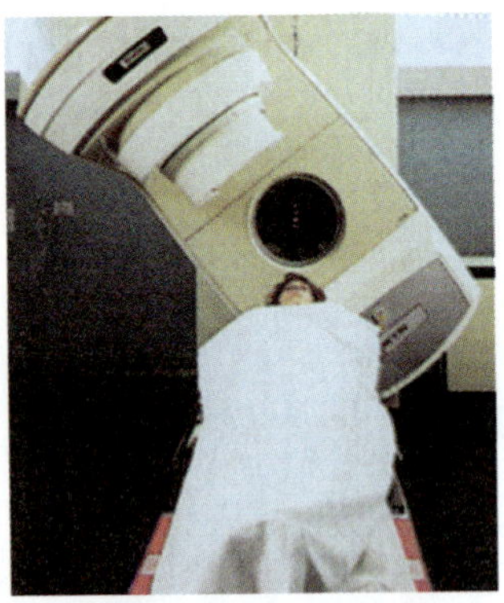

A cancer patient receiving radiation therapy.

Generally, the procedure involves administering a small dose of the radioactive NaI to the patient—who drinks a solution of the sodium iodide in water—and, after a brief period to allow the thyroid to accumulate the radioisotope, detecting the radiation coming out of the thyroid and translating its patterns into an image of the gland. Because of the great sensitivity of the instruments used in detecting and analyzing the radiation, images can be obtained from quantities of radioiodine far too small to cause any harm to the gland or to the rest of the body. In larger amounts this radioisotope of iodine can produce enough radiation to kill controlled numbers of thyroid cells, thereby repressing the excessive formation of thyroid hormones and alleviating or curing a condition known as *hyperthyroidism*, a condition in which an overactive thyroid causes weight loss (regardless of appetite), a rapid heartbeat, general irritability, and an apparent enlargement of the eyeballs.

Technetium-99*m* ($^{99m}_{43}Tc$) is the radioisotope most widely used in medical diagnosis. The *m* indicates that it is a *metastable* (unstable) isotope that decays with loss of γ rays alone and therefore without any change in its atomic mass or number:

$$^{99m}_{43}Tc \longrightarrow {}^{99}_{43}Tc + \gamma$$

Incorporated into appropriate chemical compounds, technetium-99*m* can be introduced into the body so that it accumulates in various organs, including the heart, kidneys, liver, and lungs, and in glands such as the thyroid. Since it emits only γ rays, which penetrate tissue and exit the body as effectively as they enter it, technetium-99*m* is particularly useful in generating diagnostic images. This radioisotope, which has a half-life of 6 hours, can be used in quantities small enough to avoid chemical poisoning or radiation damage. It's their ability to leave the body efficiently, with few ionizing collisions, that makes γ rays particularly useful in diagnostic medicine.

[Historically, technetium was the first new element to be produced artificially. Unknown before 1937, the element had been sought for many years to fill one of the gaps in the periodic table (Sec. 3.6). In 1937 two scientists at the University of California, Berkeley, discovered technetium in a sample of molybdenum that had been bombarded with deuterium. Its name comes from the Greek word *tekhnetos*, meaning synthetic or artificial.]

Radioactivity originating outside the body is also useful in medical therapy. Because cancer cells divide more rapidly than normal cells and are more active metabolically, they are also more susceptible to damage by ionizing radiation. Cancers located deep within the body are sometimes treated with sharply focused beams of γ rays emitted by an external source of cobalt-60. In *cobalt radiation therapy* the cobalt source is swung around the body in a circular arc with the tumor at its center. The tumor, at the beam's focus, receives the γ radiation continually and, therefore, in a large cumulative dose; the rays pass only briefly through the surrounding tissues. With this procedure the γ rays kill cancer cells selectively while sparing healthy ones.

■ **QUESTION**

How is each of the following used in medical diagnosis and/or therapy? (a) I-131; (b) Tc-99*m*; (c) Co-60.

5.9 Positron Emission Tomography

In addition to providing therapy (by killing off cancerous cells selectively, for example) and yielding structural images of bodily organs, γ rays are also useful for examining the actual operation of an organ through a diagnostic technique known as **positron emission tomography**, or **PET**.

Positron emission tomography, or **PET,** produces images of planes, within an organ, generated through the analysis of γ rays emitted by collisions of positrons and electrons.

PET operates through the emission of positrons by nuclei with unstable neutron-proton ratios. As we saw in Section 4.7, a positron is a subatomic particle indistinguishable from an electron except for its positive charge. We also saw, in Section 4.8, that the stability of a nucleus depends partly on the ratio of its neutrons to its protons. Certain radioisotopes, especially those containing relatively few neutrons, emit positrons, which disappear in a burst of γ radiation when they collide with electrons. (The electrons vanish as well.) Whenever a nucleus emits a positron, one of the protons within the nucleus is transformed into a neutron and thereby improves the ratio. To keep track of these nuclear transformations we can think of a proton as a combination of a neutron and a positron. Looking at it in this way, the loss of a positron from the combination (the proton) leaves the neutron behind, with the net conversion of one proton to one neutron.

(Recall that in Section 4.4 we followed β-particle emission by viewing a neutron as a combination of a proton and an electron. Although these combinations may seem to run counter to each other and against common sense, we've already seen in Section 4.7 that the behavior of the subatomic world seldom follows the ordinary logic of the everyday world about us. In any event, β-particle emission *does* convert a neutron to a proton, and positron emission *does* convert a proton to a neutron.)

In PET, a positron-emitting radioisotope—most often carbon-11, nitrogen-13, oxygen-15, or fluorine-18—is incorporated into the molecular structure of a compound normally used by the organ under examination or into a closely related compound that travels along the same biological paths. The compound containing the radioisotope is administered to the patient and travels into the organ. Positrons emitted by the isotope collide with nearby electrons and disappear (along with the electrons) in bursts of energy consisting of two γ rays traveling in opposite directions. Nearby detectors and computers convert these unusual bursts into images of planes or slices of the organ and reveal the molecular activity within it. (*Tomography* comes from the Greek word *tomos*, meaning slice or section.)

In this way PET provides views of the molecular traffic within organs such as the brain and the heart and shows what's going on inside them. As an illustration, PET reveals graphically that the human brain processes the names of different kinds of objects—simple tools and common animals, for example—in different regions. It also shows the effects of various drugs and medications on an organ. In one of PET's most useful applications, an atom of fluorine-18 is attached to a molecule of glucose, the sugar used by the brain as its exclusive nutritional fuel. This tagged glucose enters the brain along with ordinary glucose and emits its positrons, which collide with electrons and emit bursts of γ rays. Analysis of images obtained in this way allows physicians to follow the path of the glucose within the brain and to diagnose and treat abnormalities of this organ.

Among the advantages of PET are the very short half-lives of the more common positron-emitting isotopes, ranging from 2 minutes for oxygen-15 to 110 minutes for fluorine-18. With isotopes of such short half-lives, the source of ionizing radiation soon disappears from the body. Figure 5.4 illustrates the positron-emitting activity of fluorine-18.

E X A M P L E

Positively Positron

Into what element is an atom of O-15 converted when it emits a positron?

■ Oxygen's atomic number is 8. There must therefore be 8 protons in oxygen's nucleus. Since positron emission converts a proton into a neutron, the atomic number must decrease by 1 unit as the positron leaves the nucleus. Thus oxygen, with its atomic number of 8, is converted into the element whose atomic number is 7, *nitrogen*.

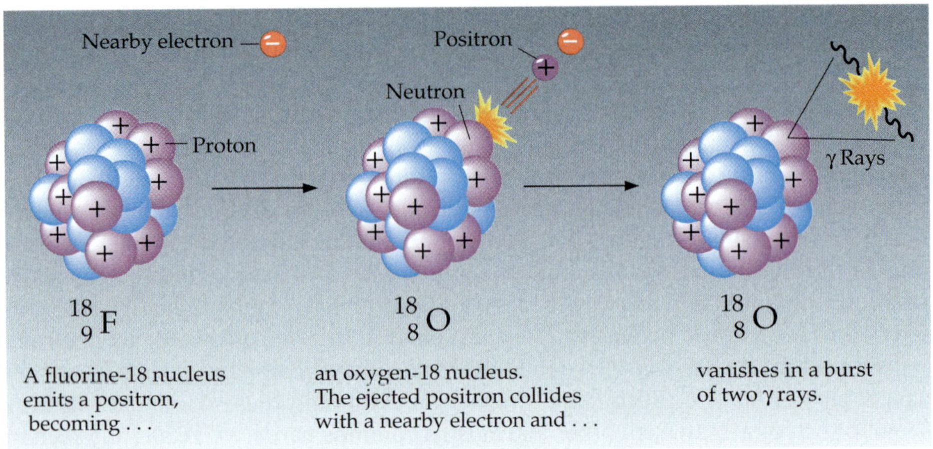

Figure 5.4 Positron emission by fluorine-18.

■ **QUESTION** Into what element is an atom of nitrogen-13 transformed when it emits a positron?

5.10 The Nuclear Calendar: Uranium and the Age of the Earth

With their individual, predictable, and immutable rates of decay, naturally occurring radioisotopes provide an important tool for dating ancient objects, ranging in age from artifacts created a few thousand years ago to bodies as old as the earth and the solar system.

We can estimate the earth's age from our knowledge of the half-lives of the radioisotopes that constitute the decay chain of U-238 (Table 5.3) and from a chemical analysis of the oldest known rocks. To understand how, suppose that the oldest sample of rock we can find contains exactly equal numbers of atoms of lead-206 (the stable isotope of lead at the end of the U-238 decay series) and of uranium-238. If we assume that all of the lead-206 came from decay of the U-238, then we have evidence that exactly one half-life of the U-238 has passed since the rock was formed. This amounts to 4.5×10^9 years (Table 5.3). If we find a ratio of three atoms of Pb-206 for every atom of U-238, then two half-lives (9.0×10^9 years) have passed, and so on. By measuring the ratio of the atoms of each of these isotopes and by making several critical assumptions (especially that all the lead atoms were formed from decay of uranium atoms), we can calculate the age of the rock containing the U-238.

EXAMPLE

Old Rocks

Analysis of a piece of meteorite shows that it contains seven Pb-206 atoms for every atom of U-238. How old is the meteorite?

■ To calculate the age of the meteorite we make the same sort of assumptions described in this section. We assume that no Pb-206 was present at the formation of the meteorite and that all the Pb-206 now present in the meteorite came from radioactive decay of the U-238.

We can now construct a table showing the number of half-lives of U-238 that have passed, and the number and ratio of U-238 and Pb-206 atoms. To simplify the arithmetic, we'll assume that we start with 64 U-238 atoms. Notice that since each U-238 atom decays eventually to one Pb-206 atom, the total number of U-238 and Pb-206 atoms must remain constant:

Half-Lives of U-238	Number of U-238 Atoms	Number of Pb-206 Atoms	Pb/U Ratio
0	64	0	0
1	32	32	1
2	16	48	3
3	8	56	7

Since the *observed* ratio is 7 atoms of Pb-206 for every U-238 atom in the sample, 3 half-lives must have passed since this sample of the meteorite was formed. As the half-life of U-238 is 4.5×10^9 years (Table 5.3), 3 half-lives are

$$3 \times 4.5 \times 10^9 \text{ years} = 13.5 \times 10^9 \text{ years}$$

If our assumptions are correct, the sample of the meteorite is 13.5×10^9 years old.

This technique, as well as others that are based on different chains of radioactive decay and that require different assumptions, give estimates of the age of the earth's surface that fall in the broad range of 3.5 billion to 4 billion years (3.5×10^9 to 4×10^9 years).

We might expect, at first, that the half-lives of all the other radioisotopes in the chain have to be used in the calculation as well, but this isn't the case at all. The U-238 decays so much more slowly than any of the other isotopes that its half-life alone reveals the age of the rock. To see why, we can use an analogy involving an hourglass with an unusual shape.

Imagine a sand-filled hourglass shaped like the one shown in Figure 5.5. The time it takes for the sand to fall from the top bulb to the bottom is determined exclusively by the width of the narrowest neck, which happens to be the first in the series. The remaining necks are much wider than the first one, so they don't affect the rate at which the sand falls. Similarly, the rate of decay of U-238 to Pb-206 is determined exclusively by the slowest step in the sequence, which again happens to be the first step.

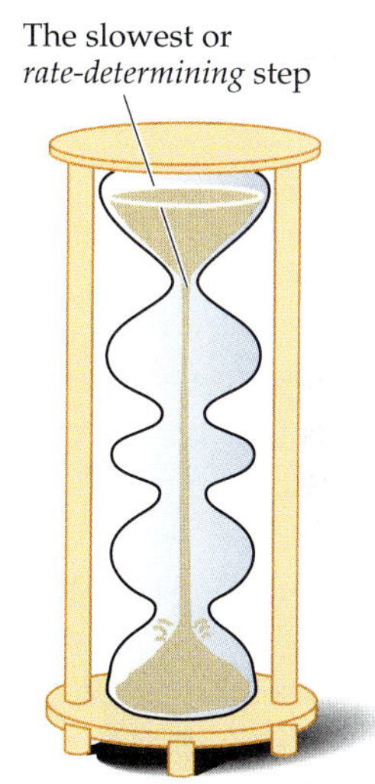

Figure 5.5 An unusual hourglass.

With this analogy, both the rate at which the sand falls through the narrowest (first) constriction and the rate of decay of the U-238 through the slowest (first) decay step determine the time it takes for each process to occur. The slowest step of a sequence that proceeds through other, much faster steps as well—the single step that determines the rate of the entire sequence—is called the **rate-determining step** of the overall process.

A rate-determining step is the slowest step in a sequence of steps that, by itself alone, determines the rate of the entire, multistep process.

■ QUESTION

What would the ratio of Pb-206 to U-238 atoms be if the sample of meteorite in the exercise were 18.0×10^9 years old?

5.11 The Nuclear Calendar: Carbon-14 and the Shroud of Turin

A variation of this dating technique allows us to determine the age of carbon-containing objects made of materials derived from plants. Unlike uranium-238, which decays without being replenished by any natural phenomenon, the radioactive isotope carbon-14 ($^{14}_{6}C$) is formed continuously by the action of cosmic radiation on the earth's atmosphere. Neutrons generated by the cosmic radiation react with atmospheric nitrogen atoms to produce carbon-14 and a proton:

$$^{1}_{0}n + ^{14}_{7}N \longrightarrow ^{14}_{6}C + ^{1}_{1}H$$

With its half-life of 5730 years, this newly formed C-14

- combines with atmospheric oxygen to form $^{14}_{6}CO_2$,
- migrates (as part of the carbon dioxide) down to the surface of the earth, and
- becomes incorporated into the cellulose and other carbohydrates of plant life.

In this way the C-14 becomes part of our cotton, paper, and wooden products and enters the bodies of all animals through the food chain. The entire process, from formation of the C-14 in the upper atmosphere to its incorporation into living plants and animals, is relatively rapid compared with its half-life.

All this time, of course, the radiocarbon continues to decay to common nitrogen by loss of a β particle:

$$^{14}_{6}C \longrightarrow ^{14}_{7}N + \beta^-$$

With the death of the plant—through its use as food or its conversion into fabric, paper, or wooden objects—the plant's incorporation of atmospheric $^{14}_{6}CO_2$ into its carbohydrates stops. Yet the C-14 already present within it continues its long decay to nitrogen, thereby continually decreasing the ratio of the radioactive C-14 to the stable isotopes of carbon, C-12 and C-13.

The carbon dating technique itself begins with a measurement of the relative amounts of C-12 and C-13 (both of which are stable, naturally occurring isotopes) and the radioactive C-14 still present in the organic material. By comparing these measured ratios with ratios normally present in the atmosphere, it's possible to estimate the amount of time that has passed since the incorporation of $^{14}_{6}CO_2$ into the organic matter ended. Because the decay of C-14 takes much longer than its conversion into the organic matter of life, and because the atmospheric C-14 seems to be replaced about as fast as it's removed, radiocarbon dating provides good estimates of the age of substances composed of organic materials.

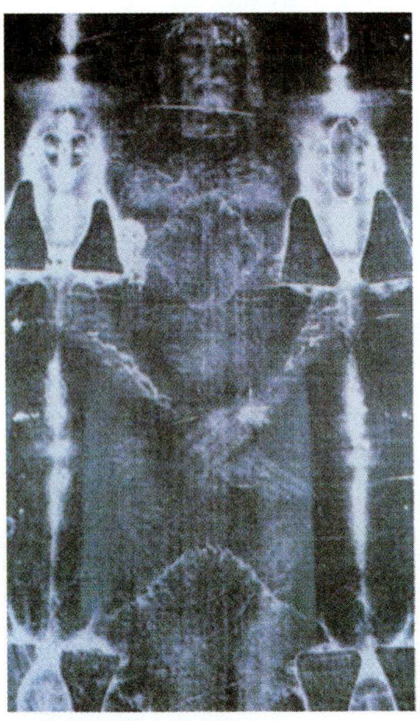

The image on the Shroud of Turin.

In 1988 this form of dating was applied to the Shroud of Turin, a 14-foot-long piece of linen in the shape of a burial cloth. The shroud first appeared in France sometime between 1350 and 1360 and was later transferred to the Italian city of Turin. Imprinted on it is the distinct image of a man. Many have believed that the image is that of Jesus and that the shroud is the actual burial cloth of Jesus, which would make the shroud approximately 2000 years old. The image was believed to have been imprinted miraculously on the shroud as it held Jesus's body. Still others have held that the image was placed on the cloth, in one way or another, by a human agent and that the cloth is more likely a hoax than an authentic relic.

Since the cloth could be the authentic shroud only if its fibers were formed *no later* than the time of Jesus's death, radiocarbon dating was used to examine its authenticity. The analysis, which was carried out in 1988 on several portions of the cloth, each about the size of a postage stamp, placed the time of the formation of the carbon-containing material of its linen as probably between 1260 and 1390, but surely later than 1200. The results of the radiocarbon dating technique require that the linen could not have been in existence at the time of Jesus's death, that it must have been fabricated centuries afterward. The radiocarbon dating thus indicates that, however the image itself may have been formed, the cloth could not have held the body of Jesus. Some, unconvinced by the radiocarbon results, maintain that pollen and other plant material of relatively recent origin may have contaminated the Shroud and may have invalidated the analytical data. Claims and counterclaims for the authenticity of the Shroud and of the radiocarbon dating have continued since the 1988 analysis.

■ QUESTION

Which of the following can be dated by radiocarbon techniques: (a) a rock; (b) a leather slipper; (c) a wooden boat; (d) a mummified body; (e) a silver spoon. Describe your reasoning.

5.12 The Source of the Static at Stagg Field: Detecting Radiation

Even with its remarkable abilities to produce ions, to destroy tissue, and to kill, as well as to diagnose and heal, there are nonetheless some important things that nuclear radiation doesn't do. It doesn't produce any taste, odor, sound, or any other kind of immediate sensation. You can't see, hear, taste, smell, or feel it. To detect nuclear radiation you have to use a device or instrument designed specifically for this purpose. Some are complex and expensive; others are as simple as a piece of film wrapped in plastic.

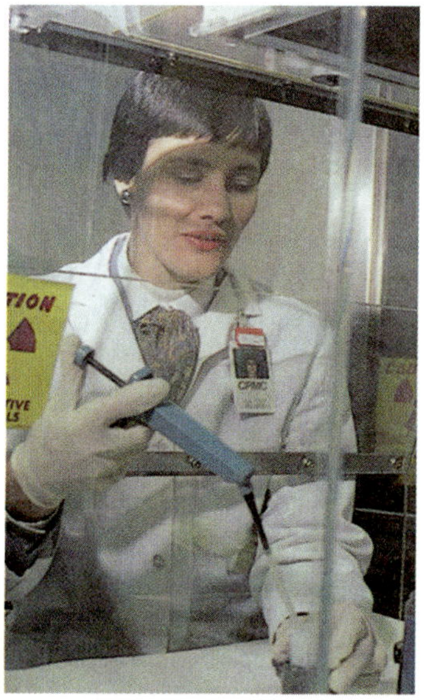

A medical worker wearing a film badge. The badge records the accumulated radiation exposure.

It was the steady clicking of one of these more complex instruments that produced the static under Stagg Field as Enrico Fermi's atomic pile began its controlled chain reaction. That historic device resembled a *Geiger counter*, which consists of a gas-filled cylinder connected to a high-voltage battery and to a sounding device and a meter. A gap between the wall of the cylinder and an electrode passing into its center serves to break an electrical circuit connecting the battery to the rest of the counter. In the absence of ionizing radiation this gap keeps the battery from activating the sounding device or the meter. Any ionizing radiation that passes through the cylinder produces a stream of ions in the gas, which temporarily completes the circuit between the cylinder's wall and the central electrode and registers as a click from a speaker. The frequency of the resulting clicks and the meter reading indicate the intensity of the radiation (Fig. 5.6). The Geiger counter is named for one of its inventors, Hans Geiger, a German scientist who had been a student of Ernest Rutherford (Sec. 4.2).

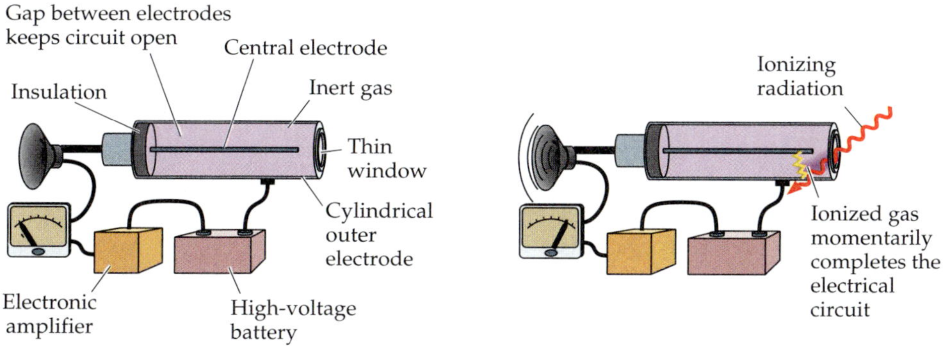

Figure 5.6 The Geiger counter.

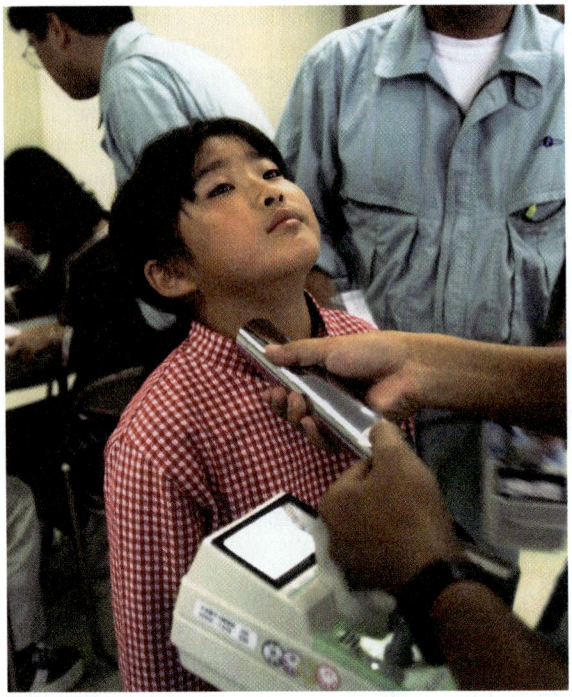

Girl being scanned with a Geiger counter after a nuclear accident.

Another kind of detector, a *scintillation counter*, uses a material that *fluoresces*, or emits an instantaneous flash of light when the invisible radiation strikes it. Zinc sulfide, ZnS, is a substance of this sort. A light-sensitive detector in the counter registers the intensity of radiation through the amount of light emitted by the fluorescing material. Both the Geiger counter and the scintillation counter are particularly useful for detecting and measuring radiation as it occurs.

Still another device, the film badge, is worn by medical personnel and others who work near radiation in order to monitor the amount of radiation they receive over a period of time, rather than at any particular moment. These badges respond to radiation much as Becquerel's sealed photographic plate did when it produced an image as it shared a desk drawer with radioactive ore (Chapter 4). Developing the film of the film badge reveals the wearer's cumulative exposure to radiation.

■ QUESTION

Which one or more of the detection devices described in this Section would you use if you wished to determine immediately whether the residue left by a spilled chemical is radioactive? Which would you use if you wanted to determine the total, cumulative amount of radiation you might be exposed to in the course of an entire month?

5.13 Ionizing Radiation in Everyday Life

Like many other things we may encounter either occasionally or in our everyday lives, our contact with nuclear radiation can be either harmful or beneficial. We've seen, for example, that radiation I-131 released by the Chernobyl explosion (Sec. 5.2) caused an increase in thyroid cancer in children, yet this same radiation from I-131 provides medically useful images of the thyroid and can cure or alleviate the medical condition of hyperthyroidism (Sec. 5.8). These are specific illustrations of the generality that ionizing radiation can cause illness and can kill, yet it can also provide us with the means to diagnose and cure illness and to save lives.

As with the chemicals of our environment, our foods and our lives, the question of harm or benefits from ionizing radiation depends on several factors, including

- the conditions of their use;
- their interaction with our bodies, collectively as human beings and personally as unique individuals; and
- the extent of our contact with them.

This last factor is quantitative; it requires some means for expressing its value. For ionizing radiation, a widely accepted unit is the **rem**, a shortened version of *roentgen-equivalent-man*.

The **rem,** a shortened form of roentgen-equivalent-man, is a unit of measurement of the biological effect of ionizing radiation on humans.

The *roentgen* itself is a unit of ionizing radiation, such as X-rays. It's named for Wilhelm Roentgen, the physicist who discovered X-rays in 1895. For this he received the first Nobel Prize given in physics, in 1901. (Roentgen's discovery of X-rays set the stage for Becquerel's discovery of radioactivity the following year; Sec. 4.1.) The additional term *equivalent-man* reflects differences in the effects of different kinds of ionizing radiation on human beings. Thus 1 rem of each of several different kinds of ionizing radiation represents the same amount of biological effect in human beings, regardless of the kind of ionizing radiation. The rem, then, is a measure of the biological effect of ionizing radiation on humans. In practice, because our exposures are so small, a more common unit is the *millirem*, or *mrem*, which equals 0.001 rem.

As a rough estimate, with plenty of variation possible, the typical resident of the United States receives an average of 360 mrem per year. Some of this comes from ionizing radiation we're exposed to in common activities (Table 5.5).

Table 5.5 Approximate Radiation Exposures

Activity	Exposure (mrem)
Lower gastrointestinal tract X-ray	500
Working inside a nuclear power plant, annually	300
Full set of dental X-rays or one chest X-ray	10–40
Round-trip, cross-country flight	2–5, depending on flight altitude and aircraft shielding
Viewing television, annually at 3 hours per day	1

In addition to exposures of this sort, we are continuously bathed in a very low level of radiation from natural sources, called **background radiation**. The bricks of our buildings and the rocks and soil under our feet, for example, contain traces of naturally occurring radioisotopes. In many localities the radioactive gas radon escapes from the earth under us and enters our homes (Sec. 4.5). What's more, high-energy cosmic rays, carrying greater energy than γ rays, constantly bombard the surface of the earth and everything on it, including us. We could shield ourselves from these forms of background radiation only by living encased in a thick shell of lead.

Background radiation is the low level of natural radiation to which we are all exposed.

But even if we should choose to live our lives in a lead shell, we would still carry within our very bodies a completely inescapable source of ionizing radiation: the

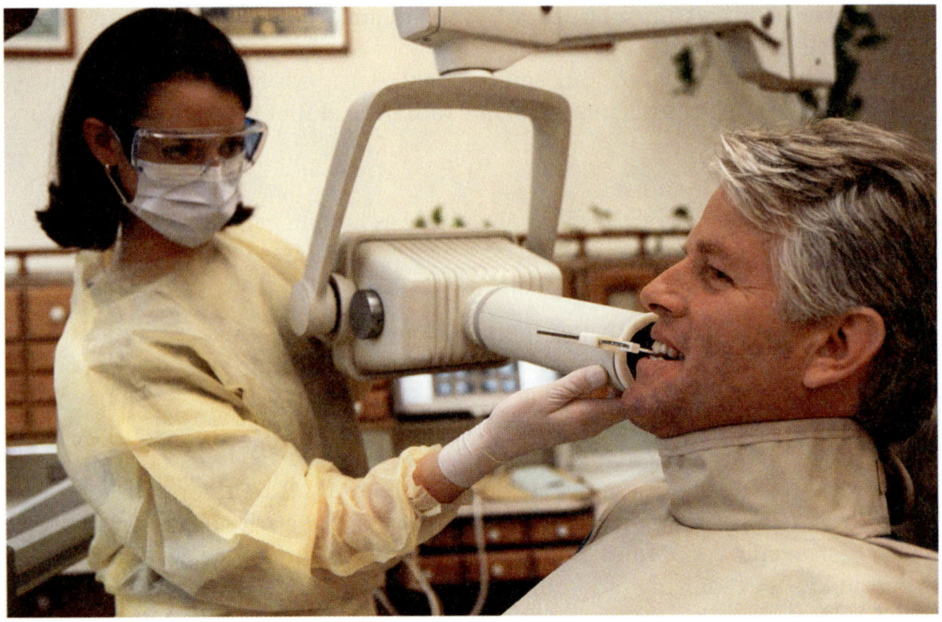

Receiving radiation from a dental X-ray.

radioisotope potassium-40. As we saw in Table 1.1, potassium is an element essential to life. It's the most abundant cation within our cells and helps regulate the water balance in our bodies. A 60-kg (132-1b) person carries around roughly 200 g of potassium at all times. We could not live without it. Yet just over 0.01% of all the potassium in the universe, including the potassium in our bodies, consists quite naturally of the radioisotope K-40, with a half-life of 1.25×10^9 years (Table 5.3). A simple calculation shows that each of us carries about 20 mg of radioactive potassium-40 within our bodies at all times. That, too, contributes to the background radiation. It's thought that what we eat, drink, and breathe brings us 25 mrem each year.

Table 5.6 sums up all this exposure, once again for the average person, with plenty of variation possible in individual cases.

Table 5.6 Radiation Sources in the United States (Average Values)

Source	Exposure (mrem)	Percent of Total
Radon	200	55
Soil, rocks, atmospheric	55	15
Medical procedures	53	15
Internal, radioisotopes within the body	39	11
Consumer products and activities	10	3
Miscellaneous	3	1

■ QUESTION

Assign the data of each of the sources of Table 5.6 into one of two categories: (a) originates in human activities; (b) natural, not dependent on human activities. What total percentage of exposure is associated with each of the two categories?

PERSPECTIVE

Energy, From the Nucleus to the World We Live In

In Chapter 1 we defined *chemistry* as *the branch of science that studies the composition and properties of matter and the changes that matter undergoes.* This definition properly focuses on the matter, the material substance, of the world about us and the universe in general. But when we look at "the changes that matter undergoes" we are also looking at energy. Virtually without exception, when there's a change in matter, there's movement of energy, into or out of the matter that's changing. We must remove energy from water (cool it) to freeze it. We must add energy to water (heat it) to boil it. When atoms split (fission) or combine (fusion), they release energy. When we burn gasoline in a car's engine or oxidize food in our own bodies, we release the energy stored in the gasoline and in the food.

As Einstein pointed out with $E = mc^2$, matter and energy are in a real sense equivalent to each other. In Section 4.12 we saw that the energy that holds a nucleus together, the atom's binding energy, is derived from the atom's mass defect. We've seen that much of the importance of nuclear transformations (and associated changes in binding energies) lies in the energy they release—instantaneously in nuclear weapons, more slowly in nuclear reactors and in medical applications. We've seen that this energy can do harm when it destroys biological molecules vital to health and life, or provide lifesaving benefits when it kills cancer cells.

The next several chapters, and those further ahead, continue examining this connection between changes in matter and the release or absorption of energy, and carry it beyond nuclear transformations to changes in the structure of molecules, especially the organic molecules of our fuels and of the food we eat. We'll define what energy itself is, learn how molecules store and release it, examine how the chemicals of our food provide us with our own biological energy, and trace the energy of our life back to its ultimate origin. ■

EXERCISES

For Review

1. Following are a statement containing a number of blanks, and a list of words and phrases. The number of words equals the number of blanks within the statement, and all but two of the words fit correctly in these blanks. Fill in the blanks of the statement with those words that do fit, then complete the statement by filling in the two remaining blanks with correct words (not in the list) in place of the two words that don't fit.

> Peaceful uses of nuclear reactions include the generation of _________, medical _________, and _________. Currently the most practical form of nuclear power plant is the _________, in which heat produced by an _________ creates steam, which turns a turbine attached to an electrical generator. Although most nuclear power plants use a uranium fuel, which is enriched in the more highly fissionable isotope _________, a _________ converts the more common, but less efficient isotope _________ into fissionable _________ as it simultaneously generates power. Someday electrical power may be produced by an environmentally cleaner _________, in which small nuclei, such as hydrogen isotopes, combine to form larger nuclei, such as helium atoms.
>
> Medically, _________ is valuable in the diagnosis and control of hyperthyroidism; _________ is used widely as a source of diagnostic γ rays.
>
> Since the _________ of radioisotopes are well known and are constants, unaffected by physical or chemical conditions, naturally occurring radioisotopes are useful in determining the age of ancient materials. The dating of objects made of wood, paper, and linen, for example, is often accomplished by analysis of their _________ content.

atomic pile	half-lives
carbon-14	iodine-131
control rods	plutonium-239
diagnosis and therapy	potassium-40
electrical power	radioisotopic dating
fission reactor	uranium-235
fusion reactor	uranium-238

2. Identify, describe, or explain each of the following:

a. background radiation	e. meltdown
b. film badge	f. PET
c. Geiger counter	g. rem
d. LMFB reactor	h. scintillation counter

3. In what way was Enrico Fermi's nuclear pile at Stagg Field important to building the first atomic bomb? In what way was it important to the development of nuclear-powered electric generating plants?

4. Describe the steps involved in the conversion of the heat produced in a nuclear pile into electricity by a commercial power plant.

5. Describe how electricity is generated through the action of steam.

6. Describe how a source of energy *other than heat* is used to generate electric power on a commercial scale.

7. What is the principal advantage of nuclear fusion over nuclear fission for generating power? What is the principal obstacle yet to be overcome in building a fusion reactor?

8. List four factors that have slowed the expected growth of nuclear power as a source of commercial electricity.

9. What would result if all the control rods of a commercial fission reactor were removed from the nuclear pile and not reinserted? What would happen if all the rods were fully inserted into the pile and allowed to remain?

10. Explain why a nuclear explosion cannot occur in a commercial fission reactor. What kinds of accidents can occur? What happens during a *meltdown*?

11. (a) Was the accident at Three Mile Island the result of a *nuclear explosion* of fissionable material? (b) Was the accident at Chernobyl the result of a *nuclear explosion* of fissionable material? Explain your answers.

12. Of the accidents at Seascale, England, Three Mile Island, Pennsylvania, and Chernobyl, Ukraine, which released the greatest amount of radioactive fallout?

13. Write the three sequential nuclear reactions that convert U-238 into Pu-239 in a breeder reactor.

14. In the conversion of U-238 to Pu-239 in the breeder reactor, another element is produced momentarily from the uranium and decays quickly to plutonium. What is this transient element produced on the path from U-238 to Pu-239?

15. In the term "liquid metal fast breeder reactor," what does the word "fast" refer to?

16. Describe two hazards associated with a breeder reactor that converts uranium to plutonium.

17. Why is it important that high-level nuclear wastes be stored in an area that is completely dry?

18. What was the first new element to be produced by artificial means? How does its name reflect its origin?

19. Describe how ionizing radiation can be used to cure or control hyperthyroidism.

20. What property of radioisotopes of iodine makes them particularly useful for diagnosis and therapy of disorders of the thyroid gland?

21. How does the emission of a positron affect the atomic number of an atom? How does it affect the mass number of an atom?

22. What form of radiation is detected by the instruments used in PET?

23. Describe how ionizing radiation can cause (a) somatic damage and (b) genetic damage.

24. Name four sources of background radiation that might affect someone standing outside in a rocky field. Name one source of background radiation to which someone would be exposed if: (a) standing in a well-sealed room with walls of thick lead; (b) buried deep within a mine.

25. List α rays, β rays, and γ rays in order of increasing ability to pass through a thick wooden wall. List them in order of increasing ability to generate ions over equivalent paths of travel within that wooden wall.

A Little Arithmetic and Other Quantitative Puzzles

26. Using the data in Table 5.3, calculate how long it would take 1 kg of Tc-99*m* to be reduced to just under 1 g of Tc-99 by the process of radioactive decay.

27. Ten grams of krypton-93 were prepared through the use of nuclear reactions. A measurement taken after a certain period passed showed that radioactive decay left a total of 0.625 g of this radioisotope. How long a period elapsed between the formation of the 10 g of krypton-93 and the measurement?

28. You have a balance that will weigh masses down to 0.1 g but gives a reading of zero for anything less than that. Using that balance, you have just measured out 100.0 g of each of the radioisotopes in Table 5.3. What weight of each of these isotopes will be left (using the same balance) after 20 years have passed?

29. Section 4.9 informs us that U-235 can undergo the following mode of fission:

$$^{235}_{92}\text{U} + {}^{1}_{0}\text{n} \longrightarrow {}^{137}_{52}\text{Te} + {}^{97}_{40}\text{Zr} + 2\,{}^{1}_{0}\text{n}$$

Suppose that as Otto Hahn bombarded uranium with neutrons, some of the fission had followed this route. Suppose further that Hahn had taken about 4 days to analyze the products of this study. Do you think he might have found any tellurium-137? Do you think he might have found any zirconium-97? (Te-137 has a half-life of 4 seconds; Zr-97 has a half-life of 16.8 hours.)

30. You have just examined a wooden utensil recovered from an ancient archeological site and have found that the ratio of C-14 to C-13 is less than 0.1% of the ratio of C-14 to C-13 measured in a branch just cut from a nearby tree. What, if anything, can you conclude about the age of the wooden utensil? What assumptions do you make in arriving at your answer?

31. In determining the age of the earth by measuring the ratio of lead-206 atoms to uranium-238 atoms in the earth's crust, we made the assumption that all of the lead-206 in the rock samples came from the radioactive decay of uranium-238. How would our calculations of the earth's age be affected if some of the lead-206 atoms were formed at the same instant the uranium-238 atoms were formed? Suppose, that is, that at least some of the lead-206 atoms that we found along with the uranium *did not* come from radioactive decay of the U-238 but were formed along with the U-238. Would the earth be older or younger than our calculations would show? Describe your reasoning.

32. Plutonium-239 has a half-life of about 24,000 years. How long will it take for 99.9% of the plutonium wastes of a breeder reactor to decay into other substances? If we assume that there are four generations of humans in each century, how many generations will this period cover?

33. Using the graph of Figure 5.2 and the data of Table 5.3, determine how long it will take for 100 g of U-235 to decay to 90 g of U-235.

34. You have 10 g of a mixture made up of exactly 5 g of I-131 and 5 g of K-40 and you also have a radiation detector that is able to detect radiation down to a level of 0.10% of the current level of radiation given off by the sample. How long do you estimate it will take for the level of radiation of the 10-g sample to drop to (a) 75% of its current level? (b) half of its current level? (c) 0.10% of its current level?

35. If the Shroud of Turin were analyzed for C-14 today, what percentage of the C-14 originally present at its formation would you expect to be present? (Figure 5.2 may be useful to you.)

36. Would it make any difference in calculations of the age of rocks, based on the isotopic ratio of U-238 and Pb-206, if the decay step with the longest half-life were at the *end* of the sequence rather than at the beginning? Use an hourglass analogy in arriving at your answer.

Think, Speculate, Reflect, and Ponder

37. What was the function of the graphite in Fermi's pile? What was the function of the cadmium?

38. How did Enrico Fermi's studies of nuclear reactions contribute to the use of atomic energy for the production of commercial power?

39. Why do health professionals working in radiology departments of hospitals wear film badges on their coats?

40. Describe the operation of each of the following radiation detection devices: (a) a Geiger counter; (b) a scintillation counter; (c) a film badge. How would you use each to estimate the *quantity* of ionizing radiation that is present? Which one of these detectors do you think is the best for determining the amount of accumulated radiation exposure a person receives over a long period, a week for example? Describe your reasoning.

41. How could you construct a crude but effective radiation detector out of common consumer materials readily available to you? Would this device measure radiation the moment it occurs or would it measure cumulative radiation, detected over a period of time?

42. What properties should a radioisotope have if it is to be administered internally in a medical diagnostic procedure?

43. You are the administrator of a major research institute. One of your research scientists presents a plan to build an instrument much like that used for cobalt radiation therapy, except that this new instrument would use an α emitter rather than a β emitter for the treatment of deep-seated tumors. What would your response be? Would you (a) promote or (b) fire the research scientist who proposed the idea? Why?

44. Of α radiation, β radiation, and γ radiation, which is most damaging to living things when it is emitted by a nearby, external source? Explain your answer. Of the three, which is most dangerous to a living thing when it is emitted by a radioisotope located within the body? Explain your answer.

45. Describe the similarities and the differences in the use of γ radiation (a) to determine the shape and size of an internal organ and to detect the presence of tumors in it; (b) to determine how well the organ is performing its biological functions, such as metabolizing a particular nutrient; (c) to kill a cancerous tumor in the organ.

46. Suggest *three* possible reasons why genetic damage by ionizing radiation has not yet been observed in humans.

47. Suppose there were to be a vote soon in your state to ban *all* forms of production of ionizing radiation by human activities, such as in the production of electricity by nuclear power plants, the production of radioisotopes for medical use, and the use of X-rays in medical and dental examinations. Would you vote for or against the proposal? Explain your answer.

48. Which one of the four factors that have slowed the growth of commercial nuclear power do you think provides the most important argument against further development of commercial nuclear power? Which one carries the least weight in arguments against further development?

49. Suppose a commercial electric utility wanted to build a nuclear power plant a few miles from where you live and your county or municipal government held a referendum to determine whether to permit the construction and operation of the plant. This would give you the opportunity to cast a vote in favor of the plant or against it. What factor(s) would sway you toward voting *for* the construction and operation of the nuclear power plant? What factor(s) would sway you toward voting *against* the plant?

50. Is there any way to detect an actual difference between the electricity that is generated by a nuclear power plant and the electricity generated by a plant run with oil? If your answer is yes, how would you go about detecting the difference? If your answer is no, how would you go about convincing someone (who believes otherwise) that you are right?

51. What's the major difference between a *pressurized water reactor* and a *boiling water reactor*? Which do you think would be less expensive to build and operate? Why? Which do you think would present less hazard of an accident or of radioactive contamination of the environment? Why?

52. Do you think that the continued production of electricity by nuclear plants justifies the accumulation of high-level nuclear wastes? Explain.

53. If you had a choice between connecting your home or apartment to electric power provided by a plant that burns imported petroleum or a plant powered by nuclear reactions, which would you choose? Why?

54. Among the various proposals suggested for the disposal of high-level nuclear wastes is the possibility of sealing them in drums and transporting them by rocket to the moon, to remain on its surface forever. What would be the advantages of such a plan? What would be the drawbacks? Would you support a plan of this sort? Explain.

55. It's possible to convert the energy of steam into electricity, and it's possible to reverse the process by using electricity to produce steam from water. It's possible to convert the energy of falling water into electricity, and it's possible to reverse the process by using electricity to pump water from a lower level to a higher level. It's possible to convert the energy of the wind into electricity with a windmill connected to a generator, and it's possible to reverse the process by using a fan to convert electricity into a stream of moving air. It's possible to convert the energy of fissioning atoms into electricity. Do you think it might be possible to reverse the process and use electricity to produce fusion? Describe your reasoning.

56. Suppose that the isotopic analysis of the Shroud of Turin had dated its origin to a few years earlier than the birth of Jesus. What conclusion(s) could be drawn from this result?

57. The ability of different kinds of ionizing radiation to penetrate human tissue varies from one form of radiation to another. Gamma rays penetrate the entire body readily, α rays are stopped by the skin, while β rays lie between these two. Yet 1 rem of any one of these forms of radiation does the same amount of damage as 1 rem of either of the other two. Explain why this is so despite the differences in penetrating ability.

Web Questions

Use the Web to answer or accomplish each of the following. Report your Web source for each response.

1. Describe the current state of plans or expectations for the use of nuclear *fusion* to produce commercial electric power.

2. What is the current condition of the nuclear power plant at Chernobyl, Ukraine? Does it currently produce electric power? Are there current plans for rebuilding it? For tearing it down? For replacing it with a new power plant?

3. Describe another reported accident at a nuclear power plant anywhere in the world, in addition to those discussed in Section 5.2.

4. Find the location, history, and current status of a breeder reactor in current use, either for experimental, research, or commercial purposes.

5. What is the current status of the proposal for the use of Yucca Mt., Nevada, for the long-term storage of nuclear wastes?

6. Identify a radioisotope other than those discussed in this chapter that is used for medical diagnosis or treatment. Describe its medical application.

7. Find a positron emission tomogram posted on the Web and describe the medical condition or study it illustrates.

8. Radioisotopes of elements other than carbon and uranium have been used for the dating of various substances. Identify an isotope of one of these other elements that has been used for dating and describe an example of its use.

9. Section 5.11 ends with the sentence, "Claims and counterclaims for the authenticity of the Shroud and of the radiocarbon dating have continued since the 1988 analysis." Summarize some of the arguments supporting and opposing the conclusion discussed in this chapter.

CHAPTER 6

Introduction to Organic Chemistry

THE POWER OF HYDROCARBONS

The burning hydrocarbons of a candle produce heat and light.

A Candle Burning in a Beaker: Energy from Hydrocarbons

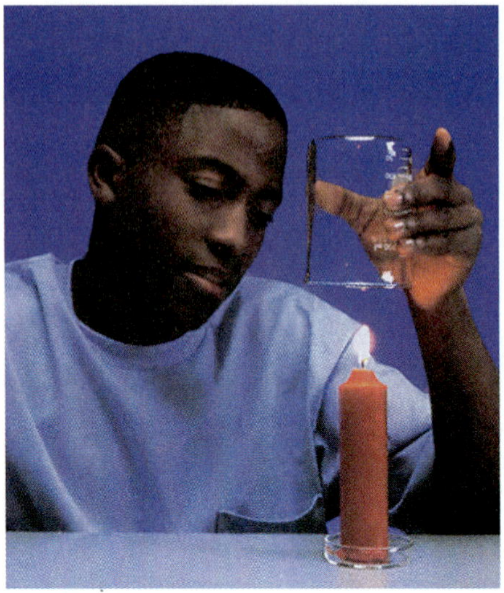

Our society gets most of its energy by burning things. We burn gasoline and diesel fuel to run our cars, trucks, and buses. We burn jet fuel for our airplanes. We burn wood, coal, heating oil, and kerosene to heat our homes and offices. We burn petroleum and natural gas to produce most of our electricity.

Much of what we burn, including gasoline, diesel fuel, jet fuel, heating oil, kerosene, and natural gas, belongs to the class of chemicals we call *hydrocarbons*. In this chapter we'll examine these hydrocarbons as representatives of a branch of chemistry known as *organic chemistry*. We'll survey their structures and names and learn about the different kinds of hydrocarbons that exist.

In the next chapter we'll see how we obtain these compounds from petroleum, how we use their energy to transport us across the earth, and how we modify their molecular structures to make them more useful in a variety of applications. First, though, here in this chapter, we'll examine the effects of some of the chemical reactions we're going to discuss. We'll do this by the simple act of lighting a candle and inserting it into a cool, inverted laboratory beaker.

WARNING...Don't try what follows with an ordinary drinking glass. The heat of the flame could cause the glass to crack or shatter. Common laboratory glassware is made of a special glass that expands and contracts very little during heating and cooling, so it's unlikely to break as the flame heats it.

If you light an ordinary household candle, which is composed largely of hydrocarbons, and insert it into a dry beaker at the temperature of a cool room (Fig. 6.1), you'll

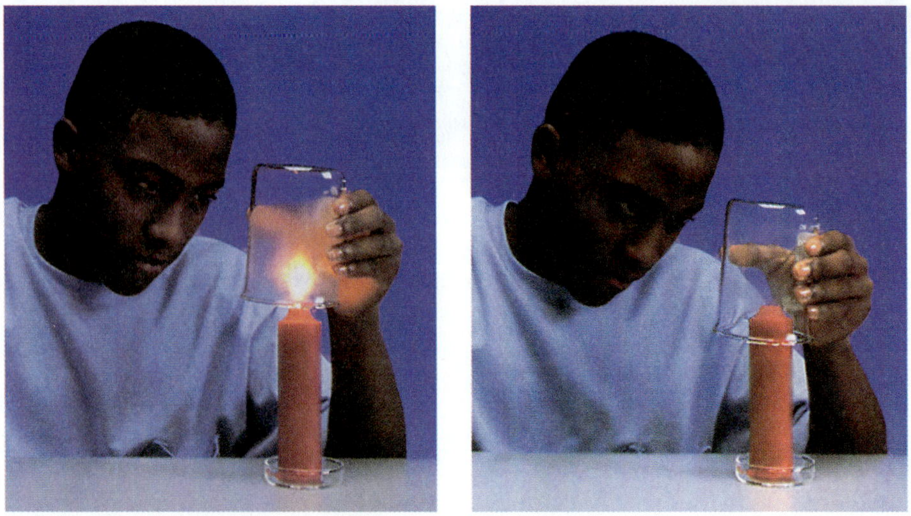

Figure 6.1 Observing the chemistry of a burning candle. As you lower a beaker over a candle flame, you first see evidence of the water that is formed on combustion of the hydrocarbons. Then, as the oxygen in the beaker is replaced by carbon dioxide, the flame dies.

find that the image of the flame grows less distinct as the glass surrounding it becomes fogged. Then you'll see the flame itself grow small and dim. If the candle is far enough inside the beaker, the flame soon dies out. If you remove the candle and run your finger around the inside of the fogged portion of the beaker after it cools, you'll find that the fogged band inside the beaker is wet.

This chapter and the next are about the chemistry that causes all this. Our examination of the chemistry of the ordinary things of our daily lives moves outward now from the atomic nuclei and the valence electrons of earlier chapters to the shared electrons of covalent bonds (Sec. 3.13) and the molecules they hold intact.

Since the compounds we'll examine, hydrocarbons, introduce us to the branch of chemistry known as organic chemistry, and since the term *organic* is itself applied in various ways in our society—we have "organic" foods, fertilizers, pesticides, gardens, fabrics, vitamins, and so on—and is sometimes poorly defined, we'll start by clarifying what we mean here by "organic." ■

■ QUESTION

A burning candle releases two different forms of energy. What are they?

6.1 What's Organic About Organic Chemistry?

All life depends on water and on the compounds of carbon. Water furnishes the fluids of life, and carbon, in covalent combination with other carbon atoms and with atoms of hydrogen, oxygen, nitrogen, sulfur, and phosphorus as well, provides the molecules of life. Carbon compounds occur in all living things. No life exists without them.

The chemistry of carbon compounds is **organic chemistry**, a term bequeathed to us by the earliest chemists. Until about 150 years ago, at the dawn of modern chemistry, organic matter—substances obtained from things that are alive or were once alive—served as the only source of these compounds. *Ethyl alcohol*, for example, an organic compound whose intoxicating effects have been known from antiquity, has long been obtained from the fermentation of grains, which gives it the alternative name *grain alcohol*. Soap, another organic substance, has been made for centuries from fats (also organic) rendered from the bodies of slaughtered animals. Organic dyes and drugs of all sorts have been extracted for centuries from great varieties of plants. The red-orange dye *henna*, used since antiquity for coloring hair and leather, comes from the lawsonia shrub; *quinine*, the first effective treatment for malaria, was originally isolated from the bark of South American trees. If you had lived as late as the first third of the 19th century and you had wanted one of these compounds or any other compound of carbon, you would have had to isolate it from organic matter. There were simply no other sources.

Organic chemistry is the chemistry of carbon compounds.

Friedrich Wöhler synthetized urea in his laboratory in 1828. With this synthesis he ended the reign of the "vital force" theory of the formation of organic compounds.

The "organic" mystique was so powerful in those years that it engendered a belief in a "vital force," which was supposedly possessed by all living things and was thought to be uniquely capable of producing the carbon compounds they contain. *Urea* illustrates the idea nicely. It's through urea, with its molecular formula CH_4N_2O, that almost all mammals excrete the unused nitrogen of proteins in their foods. Urea makes up 2–5% of human urine. First isolated in 1773, urea was considered "organic" in the sense that urea (so it was believed) could be generated only through the action of the mysterious vital force that exists only within living bodies, and not at all in the sterile glassware of the chemist's laboratory. Substances like the urea that comes from living things were supposedly different in a very mysterious sense from those obtained from nonliving sources—water, for example—which were called "inorganic."

In 1828 the German chemist Friedrich Wöhler changed all this. His preparation of urea from ammonia (NH_3) and cyanic acid (HNCO), carried out in the

"Organic" foods.

Inorganic compounds are compounds that do not contain carbon. **Inorganic chemistry** is the chemistry of compounds that do not contain carbon.

ordinary apparatus of his chemical laboratory, proved that the idea of a "vital force" was irrelevant to the development of chemistry. The urea of Wöhler's laboratory was shown to be identical in every way with the urea formed in the bodies of mammals. Indeed, since the time of Wöhler's work no difference has ever been demonstrated between the structure, properties, or behavior of a pure substance isolated from a living or once-living thing and that very same pure substance prepared in a chemist's laboratory.

By now the term *organic* has lost its mysterious aura and has become simply the category of the compounds of carbon. The meaning of *inorganic* has changed too. Today, with only a very few exceptions, **inorganic compounds** are those that do not contain carbon and **inorganic chemistry** is the chemistry of these compounds. Ironically Wöhler, the chemist who changed the meaning of "organic" with his brilliant synthesis of urea, was himself an inorganic chemist. It was Wöhler who first isolated the elements aluminum, boron, and silicon.

Today, "organic" is once again acquiring meanings that suggest the operation of mysterious vital forces, as in "organic" foods, "organic" gardening, and "organic" vitamins. Chemically, however, Wöhler's work and the ideas that evolved from it remain unchallenged. Organic chemistry is neither more nor less than the chemistry of carbon compounds.

■ **QUESTION**

In the year 1800 what characteristic was used to define a chemical as an *organic* chemical? What characteristic is used today to define a chemical as an *organic* chemical?

6.2 Methane: The Simplest Hydrocarbon

Because a carbon atom contains four valence electrons, it can form covalent bonds with as many as four other atoms at the same time (Sec. 3.13). Carbon is therefore *tetravalent*. (The *tetra* of "tetravalent" comes from a Greek word meaning four; *valent* is from a Latin word that implies capacity. Carbon has a capacity for forming four bonds. See Sec. 3.12.) Carbon atoms themselves form strong bonds to hydrogen, oxygen, nitrogen, sulfur, and the halogens, as well as to other carbon atoms. What's more, when four or more carbon atoms occur in a single hydrocarbon molecule, the carbon atoms can all lie in a single, unbroken or *unbranched* chain, with each (except the end carbons) bonded to two others, or the carbon atoms can form a *branched* sequence, much as a twig can branch out. The unbranched sequence of carbons is sometimes called a *linear* or *straight-chain* sequence:

Four carbons in an unbranched, linear, or straight-chain, sequence

Four carbons in a branched sequence

With carbon's ability to

- bond to as many as four other atoms simultaneously,
- form strong bonds to atoms of several other elements, and
- form molecules containing many and varied branches as well as molecules composed of long, unbranched strings of carbons,

organic compounds present a complexity of structures and properties as varied as life itself. More than 10 million organic compounds are now known, each with its unique molecular structure, name, and chemical and physical properties.

To bring a sense of order to this enormous number and almost incomprehensible variety of carbon compounds, chemists have organized them into **families**, each of which consists of compounds of similar molecular structures and similar properties. One of the largest of these is the family of **hydrocarbons**, composed exclusively of compounds containing just two elements: hydrogen and carbon. Even here we find it convenient to organize the large hydrocarbon family further into four smaller families:

Families of compounds are composed of compounds of similar structures and similar properties.

Hydrocarbons are compounds composed exclusively of hydrogen and carbon.

- the alkane hydrocarbons
- the alkene hydrocarbons
- the alkyne hydrocarbons
- the aromatic hydrocarbons

Methane is the simplest of all the hydrocarbons (Fig. 6.2). Its molecules consist of a single carbon and four hydrogens; it has the lowest molecular weight of all organic compounds, 16 amu. (As we'll see in Sec. 6.3, methane is one of the alkane hydrocarbons.)

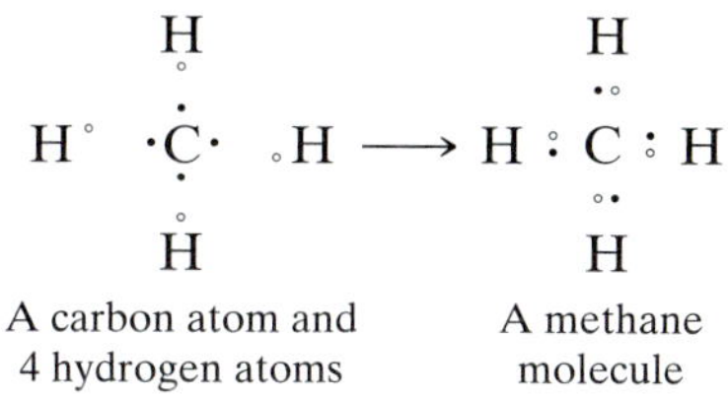

Figure 6.2 Methane, CH_4.

Four shared electron pairs bond the four hydrogens of methane to its carbon. In **stereochemistry**, which refers to *the arrangement of atoms in space*, methane is *tetrahedral*. That is, drawing straight lines from each hydrogen to its three neighboring hydrogens forms a tetrahedron, a geometric figure composed of four equilateral triangles joined along their edges. The carbon atom lies at the tetrahedron's center and the four hydrogens at its apexes, or corners, with four covalent bonds pointing outward from the central carbon atom to the four corners (Fig. 6.3). Since the geometry of the four bonds extending from the central carbon fixes the stereochemistry of the molecule, we can visualize the carbon of methane—with its valence electrons, covalent bonds, and substituents all forming a tetrahedron—as a **tetrahedral carbon**.

Stereochemistry refers to the arrangement of the atoms of a molecule in three-dimensional space.

A **tetrahedral carbon** is a carbon atom that lies at the center of a tetrahedron. It forms covalent bonds to the four atoms lying at the apexes, or corners, of the tetrahedron.

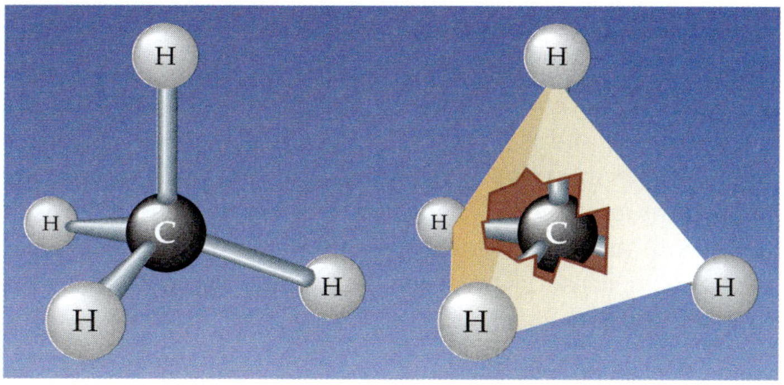

Figure 6.3 Methane and the tetrahedral carbon.

■ QUESTION

What is the total number of covalent bonds in the methane molecule? What is the total number of electrons in these covalent bonds? What is the total number of electrons in the entire methane molecule?

6.3 The Alkanes, From Methane to Decane

Methane not only represents the simplest of all the hydrocarbons, it's also the first and simplest member of the *alkanes*, one of the smaller families that exist within the larger hydrocarbon family. By examining the structures, properties, and names

A **free radical** is a chemical structure that contains an unpaired electron.

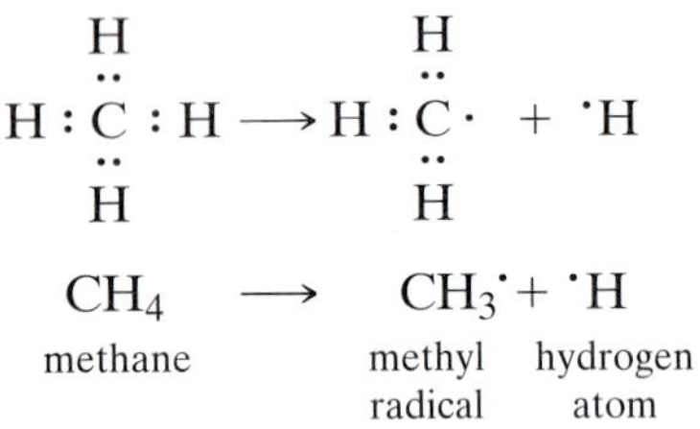

Figure 6.4 Methane and the methyl free radical.

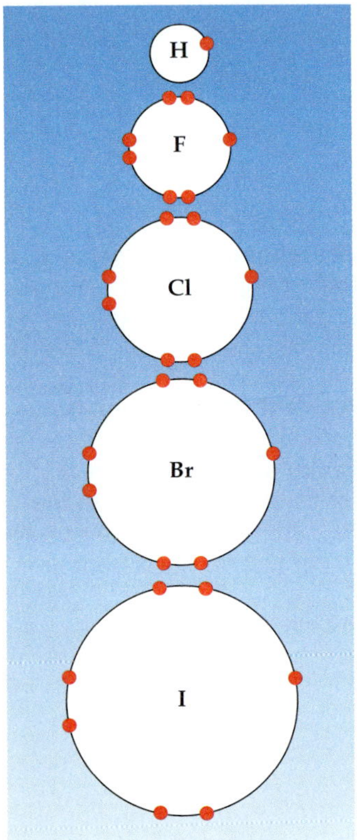

Figure 6.5 Valence shells of hydrogen and the halogens. Each atom is a free radical.

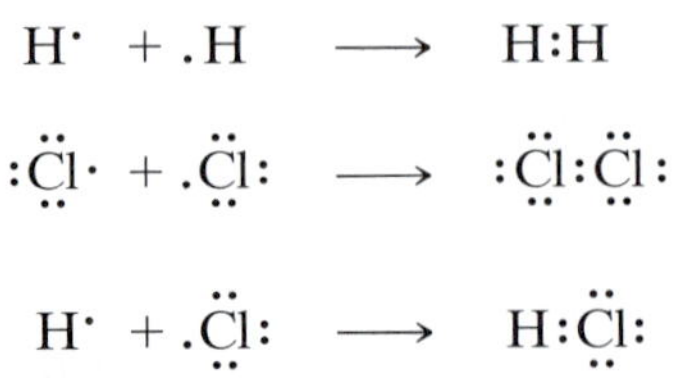

Figure 6.6 Formation of diatomic molecules of H_2, Cl_2, and HCl through pairing of electrons.

of the alkanes we'll lay the foundation for an examination of the other hydrocarbons and other kinds of organic compounds as well.

Removing a hydrogen atom from a tetrahedral methane molecule produces a *methyl* **free radical**, which is often shortened to the simpler *methyl radical* (Fig. 6.4). The distinguishing characteristic of a free radical is that it contains an unpaired electron. A detailed experimental study of atomic structures reveals that valence electrons tend to couple up in pairs. Sometimes, though, electrons remain unpaired, as though they were aloof. Any atom or group of bonded atoms that contains an unpaired electron is a free radical and reacts in ways different from atoms or groups containing only paired electrons. In the *methyl* free radical, the group of atoms that contains the unpaired electron is $CH_3^{\cdot}$. This is the covalent structure that remains after methane has lost a hydrogen atom.

A hydrogen atom with its single electron is a free radical, and so is a halogen atom, with six valence electrons paired into three sets of two electrons each and a seventh unpaired electron (Fig. 6.5). Free radicals such as these are usually highly reactive and often react so as to pair up their odd electrons with another electron, from another chemical species. The drive toward the pairing of electrons is a major force leading to the formation of covalent bonds. It results in the formation of diatomic molecules of hydrogen and the halogens and of other elements as well (Fig. 6.6).

Like other radicals, the methyl radical is also highly reactive. When two of these methyl radicals meet, they join together to pair their odd electrons into a covalent bond and thereby form a new organic compound, the alkane *ethane* (C_2H_6), which is the second member of the alkane family. To emphasize the carbon–carbon covalent bond shown in Figure 6.7, we'll write the molecular structure of ethane as $CH_3—CH_3$. With the two odd electrons of the two methyl radicals paired up, there are no free radicals in ethane.

Here, as in other molecules containing CH_3's that are covalently bonded to other atoms, we call the CH_3's *methyl groups*. Their carbons are *methyl carbons* and their hydrogens are *methyl hydrogens*. For example, we can call $CH_3—Cl$ *methyl chloride* since it consists of a methyl group covalently bonded to a chlorine. (In naming organic compounds, the *-ide* suffix, as in methyl chlor*ide*, does *not* imply that the chlorine exists as an anion.) In a similar fashion removing a hydrogen from ethane gives us the *ethyl radical*, $CH_3—CH_2^{\cdot}$. The compound $CH_3—CH_2—Br$ is *ethyl bromide* and $CH_3—CH_2—$ is an *ethyl group*. And just as methane and ethane are representatives of the more general family of alkanes, methyl groups and ethyl groups are specific examples of the more general class of *alkyl groups*. In this sense both methyl chloride and ethyl bromide are examples of *alkyl halides*. The symbol R— is used to represent the generalized alkyl group. Thus R—Cl represents methyl chloride, ethyl chloride, and any other alkyl chloride.

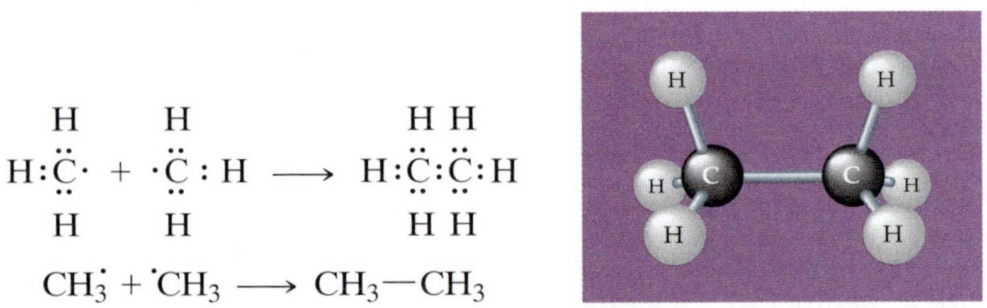

Figure 6.7 The combination of two methyl free radicals to form ethane.

Notice that the molecular formula of ethane, C_2H_6, simply shows us that each of its molecules is made up of two carbon atoms and six hydrogen atoms. The molecular structure, though, shows explicitly the organization of the atoms within the molecule. The *full* or *expanded* structure shows all the covalent bonds. The *condensed* structure omits the carbon–hydrogen bonds. It's important to remember that even though we may write the structure of ethane as $CH_3—CH_3$, the covalent bond that's shown links the two carbon atoms of the methyl groups to each other:

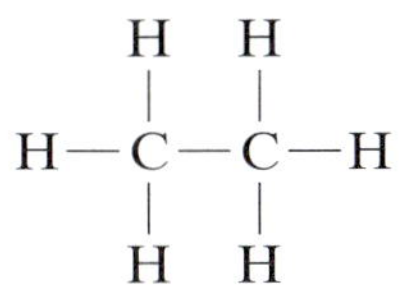

full or expanded structure of ethane

$H_3C—CH_3$

one way of writing the condensed structure, showing the C—C bond

$CH_3—CH_3$

A more common way of writing the condensed structure of ethane

Propane, C_3H_8 or $CH_3—CH_2—CH_3$, is the third member of the alkane family (Fig. 6.8). Notice that adding CH_2 (which simply represents one carbon atom and two hydrogen atoms) to the molecular formula of methane (CH_4) gives C_2H_6, the molecular formula of ethane, and adding CH_2 to ethane's molecular formula (C_2H_6) gives the molecular formula of propane, C_3H_8. Continuing this addition of CH_2 gives the molecular formulas of the entire alkane family: C_4H_{10}, C_5H_{12}, C_6H_{14}, and so on.

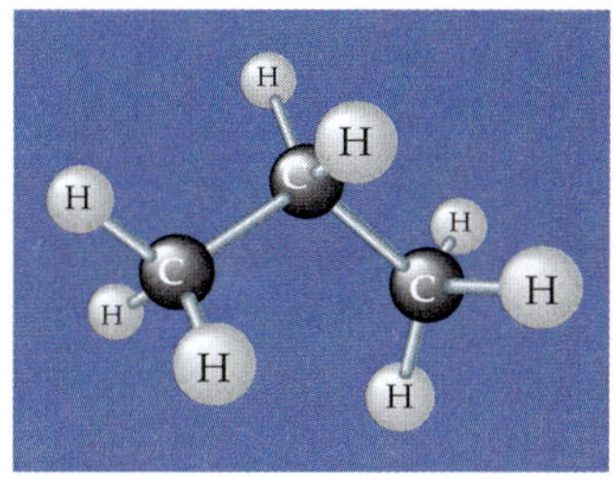

Figure 6.8 Propane.

The molecular formula of each member of the alkane series can be obtained from the *general formula* for alkanes, C_nH_{2n+2}, with n representing the number of carbons in the alkane molecule. To generate the molecular formula of any particular alkane, simply write the numerical value of n as the subscript for the carbon, then double n and add 2 to get hydrogen's subscript. This general formula defines the subset of hydrocarbons that we call the alkanes: Any compound whose molecular formula fits the general formula C_nH_{2n+2} is an **alkane**.

An **alkane** is a hydrocarbon whose molecular formula fits the general formula C_nH_{2n+2}.

■ QUESTION

Decane is an alkane with 10 carbons in each molecule. How many hydrogens does each decane molecule contain?

6.4 Alkanes: Properties and Uses

Alkanes as a class are compounds of very low toxicity. They are colorless and odorless. Methane, ethane, and propane, the lowest-molecular-weight members of the family, are all gases at ordinary pressures and temperatures. They ignite easily when mixed with air and either burn or explode, depending on their concentrations in the air.

Methane itself is the major component of the natural gas used for cooking and heating in some areas. Mixtures of about 5–15% methane in air are highly explosive. They are the dangerous *firedamp* that often forms in coal mines. Because methane and other alkanes are odorless, a small leak of natural gas itself, from a gas range for example, could remain undetected long enough to produce a flammable or explosive atmosphere in a building. Even without ignition the leak could prove deadly since the escaping gas, if it remained undetected, could displace the air in a house and asphyxiate the occupants. To prevent such catastrophes, a small amount of an intensely odorous gas, usually an organic compound of sulfur called a *mercaptan*, is added to natural gas to give the gas a strong odor and to ensure the early detection of any leak.

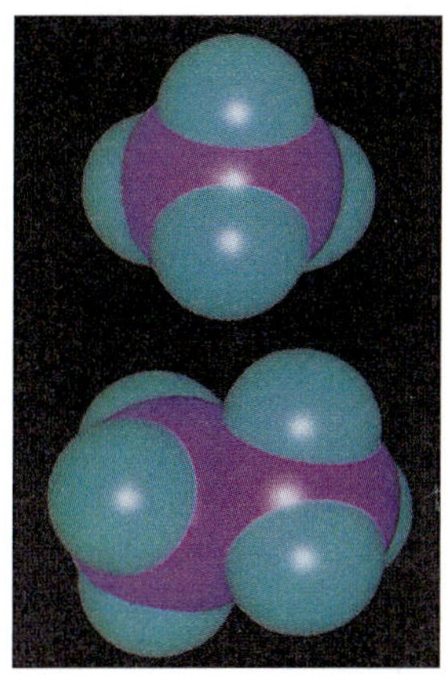

Representation of molecules of methane, CH_4 and ethane, $CH_3—CH_3$.

Gasoline, mineral oil, lubrificating oil and petroleum jelly. Consumer products composed of hydrocarbons.

As consumers we have little direct use for ethane, but many of us use a mixture of propane, butane, and other low-molecular-weight alkanes as the commercial fuel, LP gas. Several of the higher-molecular-weight alkanes, beginning with pentane, are liquids and are major components of gasoline (Chapter 7), while even larger alkanes are oily or greasy liquids or waxy solids at room temperature. The candle wax of the demonstration that began this chapter consists of these larger alkane molecules. (As an illustration of the lubricant qualities of alkanes, books and magazine articles on household hints sometimes suggest rubbing a candle on a door hinge to quiet a squeak.)

Because of their very low toxicity the higher alkanes are useful as medical lubricants. Both petroleum jelly and mineral oil, an intestinal lubricant used as a laxative, are mixtures rich in alkanes.

■ **QUESTION**

Name three consumer products, found in most drugstores or supermarkets, that consist largely or exclusively of alkanes.

6.5 The Classes of Carbons and Hydrogens

Since there's only one way to put together a covalent molecule containing four hydrogens and one tetravalent carbon, there is, necessarily, only one possible three-dimensional molecular structure for methane, the one we saw in Figure 6.3. Furthermore, since each one of the four hydrogens of methane is entirely equivalent in every way to each of the other three, removing any one of the four hydrogens and replacing it with a methyl group gives one and only one ethane (Fig. 6.9). We can make a similar statement about ethane. Since each one of the six hydrogens of ethane is entirely equivalent to each of the other five, removing any one of the six and replacing it with a methyl group gives one and only one propane (Fig. 6.10).

Figure 6.9 Forming one and only one ethane by replacing any one of the four equivalent hydrogens of methane with a methyl group.

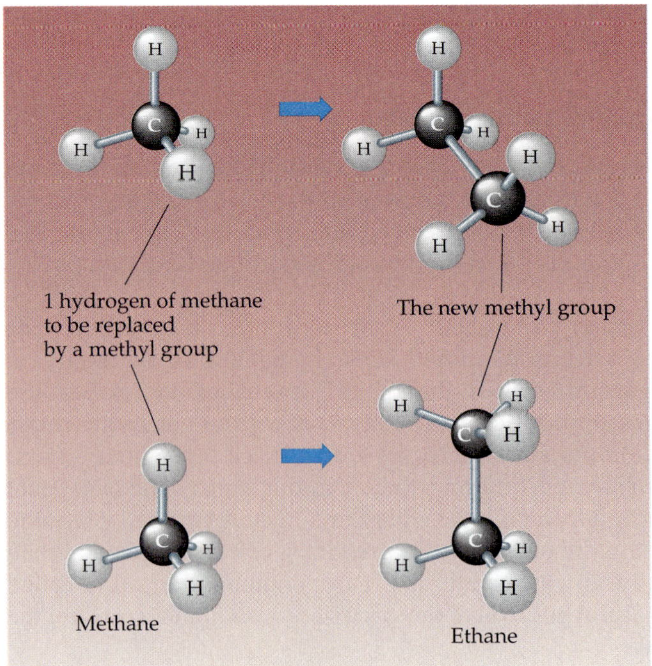

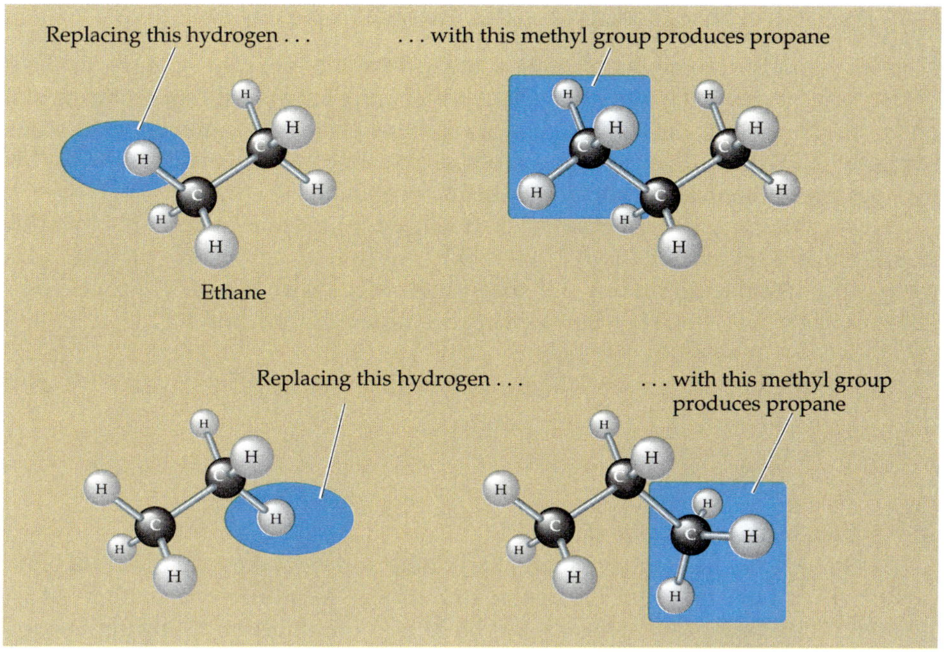

Figure 6.10 Forming one and only one propane by replacing any one of the six equivalent hydrogens of ethane with a methyl group.

With propane, though, we come to something new in alkane structures: a molecule with more than one kind of hydrogen. Each of the two terminal, methyl carbons of propane bears three hydrogens, and all six of these terminal hydrogens are mutually equivalent methyl hydrogens (Sec. 6.3). But they are all different from the two hydrogens bonded to propane's central carbon. That is, the six methyl hydrogens of the two methyl groups at the ends of the chain behave differently—have different chemical properties—from the two hydrogens of the central carbon, the $-CH_2-$ group. A $-CH_2-$ group is known as a *methylene group*. Its carbon is a *methylene carbon* and the hydrogens bonded to it are *methylene hydrogens*. Although the terms *methyl hydrogen* and *methylene hydrogen* are widely accepted by chemists and are useful for discussing the different kinds of hydrogens in a molecule, there's another, more precise way to describe them. That's by defining three different classes of carbons and the hydrogens that are bonded to them: primary carbons and hydrogens, indicated by the symbol 1°; secondary, with the symbol 2°; and tertiary, 3°.

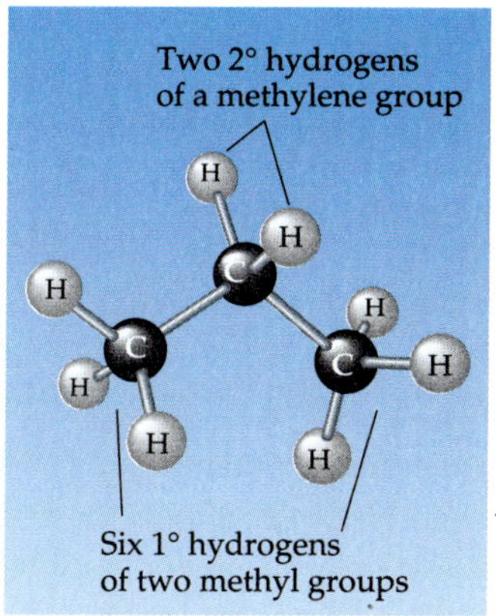

Figure 6.11 The 1° (methyl) and 2° (methylene) hydrogens of propane.

Any carbon bonded to one and only one other carbon is called a **primary carbon**, and each of the hydrogens bonded to it is a *primary hydrogen*. Thus in propane we have the two 1° carbons of the methyl groups and their six 1° hydrogens (Fig. 6.11). Any carbon bonded to just two other carbons is a **secondary carbon**, and its hydrogens are *secondary hydrogens*. The single carbon of the central methylene group in propane is a 2° carbon and it holds two 2° hydrogens. Finally, a carbon bonded to exactly three other carbons is a **tertiary carbon** and its hydrogen is a *tertiary hydrogen*. In propane there are no 3° carbons and no 3° hydrogens. [Carbons bonded to four other carbons are *quaternary* (4°), but these are rare in hydrocarbon molecules and the term is seldom used.]

A **primary** (1°) **carbon** is bonded to exactly one other carbon; a **secondary** (2°) **carbon** is bonded to exactly two other carbons; a tertiary (3°) **carbon** is bonded to exactly three other carbons.

■ **QUESTION**

How many carbons and how many hydrogens of each class (1°, 2°, and 3°) are there in $CH_3-CH_2-CH_2-CH_3$? How many of each class are there in the structure?

$$CH_3-\underset{\displaystyle CH_3}{\underset{|}{CH}}-CH_3$$

6.6 Isomerism: Butane and Isobutane

Now we can return to methane, ethane, and propane to see more clearly just how propane differs from the other two alkanes. In Figure 6.9 we saw that replacing any one of the four equivalent hydrogens of methane gives one and only one ethane. Similarly, in Figure 6.10, we saw that replacing any one of the six equivalent methyl hydrogens of ethane gives one and only one propane.

But we don't get the same sort of result with propane. It contains two different kinds of hydrogens, two 2° and six 1°, so it matters now which hydrogen we replace. Replacing one of the six 1° hydrogens with a methyl group gives $CH_3—CH_2—CH_2—CH_3$, the alkane of molecular formula C_4H_{10}, called *butane*.

Replacing one of the two 2° hydrogens with a methyl group gives

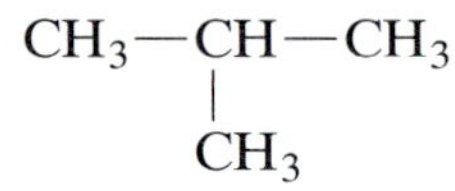

which is known as *isobutane* (Fig. 6.12).

As implied by their different names, butane and isobutane are different compounds. They have different molecular structures—with different numbers of 1°, 2°, and 3° carbons and hydrogens, as you saw in the question following Section 6.5—and different properties. Butane, for example, boils at −0.5°C, isobutane at −11.6°C. Yet, although they are different compounds, they share the same molecular formula, C_4H_{10}. Different compounds that share the same molecular formula are known as **isomers**, a word derived from the Greek words *isos*, meaning equal, and *meros*, for a part or a share. In this sense isomers "share equally" in the same molecular formula. Butane and isobutane thus provide us with an example of this phenomenon of *isomerism*.

Isomers are different compounds that have the same molecular formula.

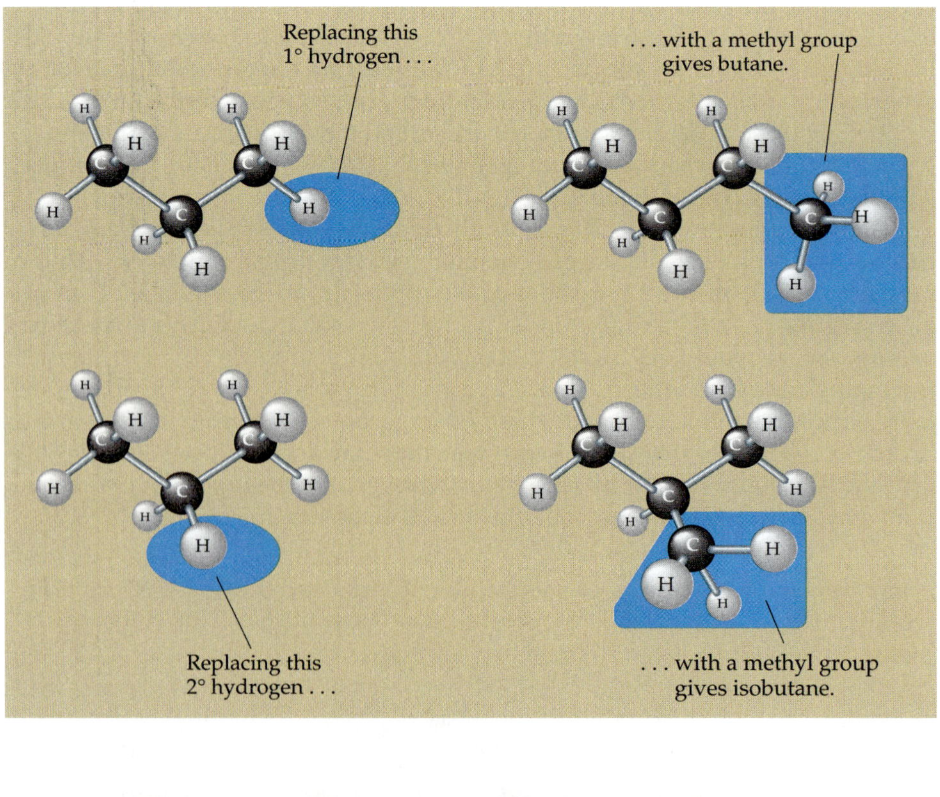

Figure 6.12 From propane to two different butanes.

Isobutane and butane also serve, respectively, as examples of compounds with a *branched chain* and an *unbranched chain*, also known as a *straight chain* (Sec. 6.2). We'll look at the straight chain of butane first. To see why its carbon chain is called a straight chain, imagine that you are standing on one of the two methyl groups that form the ends of the butane chain and you begin walking toward the methyl group at the other end:

start here . . . (↓) or here (↓)

$$CH_3{-}CH_2{-}CH_2{-}CH_3$$

butane

You could walk across the two methylene groups to the methyl group at the other end of the chain by following a direct or straight path. You wouldn't be faced with any branches as you move along the chain. This is the sense in which the carbon chain of butane is straight or unbranched.

If you started at any one of the three methyl groups of isobutane, though, you would immediately come to a fork or branch in your path:

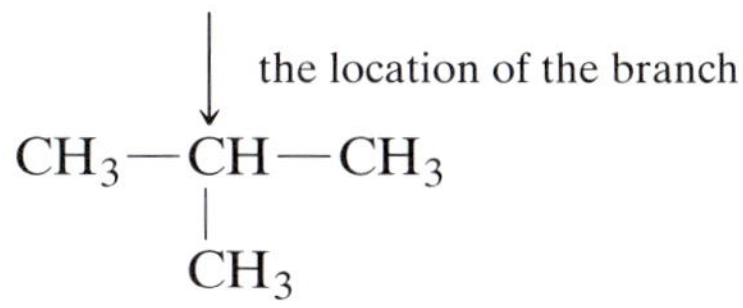

At the 3° carbon you would have to decide which one of the two remaining methyl groups you wanted to travel to, and take the appropriate branch. In this sense isobutane is a branched-chain hydrocarbon.

The multitude and structural variety of organic compounds result partly from the possibilities of isomerism. The greater the number of carbons there are in an alkane, or in any other kind of organic compound, the greater the variety of covalent structures that can be formed and the greater the number of isomers that are possible. Isomerism usually isn't possible in small molecules made up of relatively few atoms. As we've seen, only one structure is possible for methane (one carbon and four hydrogens), one for ethane (two carbons, six hydrogens), and one for propane (three carbons, eight hydrogens). Butane, though, with its four carbons and ten hydrogens, can be arranged into two different molecules, butane and isobutane. As the number of atoms in a molecule increases, so does the capacity for isomerism, and so does the number of names we must use to identify each of the isomers. In the next few sections we'll examine how we can give each of the straight-chain alkanes and each of their isomers a unique name.

■ QUESTION

Following are three isomers of C_5H_{12}. Identify each structure as branched or unbranched:

(a) $CH_3{-}CH_2{-}CH_2{-}CH_2{-}CH_3$

(b) 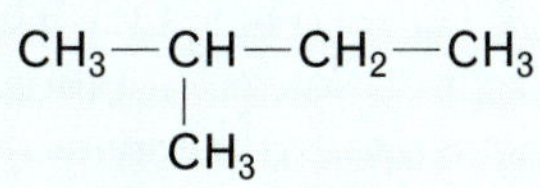

(c) 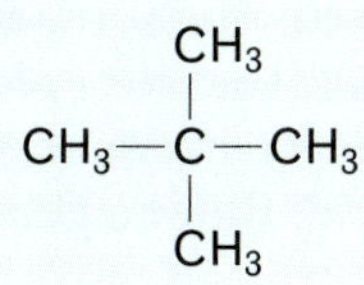

6.7 Two Alcohols, the First Fat, and Rancid Butter: How the Alkanes Got Their Names

To summarize, we've seen that the following names are used for the first five alkanes:

methane	CH_4
ethane	$CH_3—CH_3$
propane	$CH_3—CH_2—CH_3$
butane	$CH_3—CH_2—CH_2—CH_3$
isobutane	$CH_3—CH(CH_3)—CH_3$

To understand how these names came to be used and how the higher alkanes and still other kinds of organic compounds are named, we'll examine a bit of *etymology*.

Etymology is the study of origins of words; *chemical etymology* gives us a glimpse into the history of chemistry through the origins of chemical terms. As implied in the names meth*ane*, eth*ane*, prop*ane*, and but*ane*, all alkanes end in the suffix *-ane*. The root *meth-* of *methane* reflects this particular alkane's connection to *methyl alcohol*, also known as *methanol*, $CH_3—OH$. In 1661 the English natural philosopher Robert Boyle discovered methyl alcohol among the products formed as he heated wood in the absence of air so that the wood decomposed without igniting. From the time of Boyle's discovery until about 1930 his *destructive distillation* of wood was the major source of methanol and is still reflected in one of its more commonly used names, *wood alcohol*. Methyl alcohol is now made commercially from natural gas and coal.

Early in the 19th century the word *methylene*, or one of its derivatives, was applied to compounds related to wood alcohol. *Methylene* itself was coined from the Greek words *methe* (wine) and *hylē* (wood) with the intention of showing the connection between a host of similar chemical compounds and methyl alcohol, which was considered to be the "wine of wood." Thus methane, the one-carbon alkane, is related chemically to methyl alcohol, the one-carbon wine of wood.

Wood alcohol is the wine of wood only in the most poetic sense; methyl alcohol is highly toxic. (Since it attacks and destroys the optic nerve rapidly, blindness is one of the first symptoms of methanol poisoning.) Nonetheless this alcohol is a valuable industrial solvent and raw material. It's used in the manufacture of such diverse goods as plastics, jet fuels, and embalming fluid. It also shows some promise as a replacement for gasoline as a fuel for the internal combustion engine.

The etymology of the two-carbon alkane, *ethane*, is equally intricate. The *eth-* of ethane comes from the chemical name *ether* (more correctly either *diethyl ether* or *ethyl ether*) given to the very volatile and very flammable compound

$$CH_3—CH_2—O—CH_2—CH_3$$

to represent the ease with which it ignites and burns. *Ether* is taken from the Greek *aither*, meaning to kindle, to burn, or to shine. The chemical ether itself gave its name to the two-carbon alcohol from which it was first prepared, *ethyl alcohol* (or *ethanol*), $CH_3—CH_2—OH$. The name *ethane* connects the two-carbon alkane to this two-carbon source of ether.

The root *prop-* of propane, the third member of the alkane family, comes from *propionic acid*, a three-carbon acid that is the first in a series of organic acids that can be obtained from fats and oils. The word *propionic* was coined from the Greek words *protos* (first) and *pion* (fat). Propane, the three-carbon alkane, takes its name from the three-carbon acid. Butane gets it root, *but-* from the four-carbon *butyric acid*, an acid that comes from butterfat and gives rancid butter its peculiar taste and odor. *Butyrum* is the Latin word for butter.

As colorful as these roots are, with derivations that reflect the origins of the compounds they name, the rapid development of chemistry as a rigorous science demanded a more orderly approach to the naming of organic compounds. The system of nomenclature chosen as the 1800s progressed, and that is still with us today, uses Greek or Latin roots to designate the number of carbons in the alkane molecules rather than to reflect their sources or chemical genealogies. Thus the straight-chain alkane $CH_3—CH_2—CH_2—CH_2—CH_3$ is *pentane*, from the Greek *pente*, five, while the straight-chain isomer of C_6H_{14} is *hexane*, from the Greek *hex*, six. Tables 6.1 and 6.2 summarize this information for the first 10 straight-chain alkanes. Even though the names methane, ethane, propane, and butane do not reflect the number of carbons in molecules of these compounds and are thus at odds with the system used for the higher alkanes, they are fully accepted as the only correct names for these four alkanes.

Table 6.1 First 10 Straight-Chain Alkanes

Name	Molecular Formula
Methane	CH_4
Ethane	$CH_3—CH_3$
Propane	$CH_3—CH_2—CH_3$
Butane	$CH_3—CH_2—CH_2—CH_3$
Pentane	$CH_3—CH_2—CH_2—CH_2—CH_3$
Hexane	$CH_3—CH_2—CH_2—CH_2—CH_2—CH_3$
Heptane	$CH_3—CH_2—CH_2—CH_2—CH_2—CH_2—CH_3$
Octane	$CH_3—CH_2—CH_2—CH_2—CH_2—CH_2—CH_2—CH_3$
Nonane	$CH_3—CH_2—CH_2—CH_2—CH_2—CH_2—CH_2—CH_2—CH_3$
Decane	$CH_3—CH_2—CH_2—CH_2—CH_2—CH_2—CH_2—CH_2—CH_2—CH_3$

Table 6.2 Origins of the Names of the First 10 Straight-Chain Alkanes

Molecular Formula	Name of Alkane	Root	Meaning	Derivation
CH_4	Methane	meth-	wine	Greek *methe*
			wood	Greek *hylē*
C_2H_6	Ethane	eth-	to burn	Greek *aither*
C_3H_8	Propane	pro-	first	Greek *protos*
		pion	fat	Greek *pion*
C_4H_{10}	Butane	but-	butter	Latin *butyrum*
C_5H_{12}	Pentane	pent-	five	Greek *pente*
C_6H_{14}	Hexane	hex-	six	Greek *hex*
C_7H_{16}	Heptane	hept-	seven	Greek *hepta*
C_8H_{18}	Octane	oct-	eight	Greek *oktō*; Latin *octo*
C_9H_{20}	Nonane	non-	nine	Latin *novem*
$C_{10}H_{22}$	Decane	dec-	ten	Greek *deka*; Latin *decem*

One more brief example of etymology tells us something about the chemistry of the alkanes. Originally this family of hydrocarbons was called the *paraffin* family, a name that is still with us in the paraffin wax of candles, wax paper, and the sealing material of home canning and preserving. *Paraffin* comes from two Latin words: *parum*, meaning barely or too little, and *affinis*, for having affinity or, in the chemical sense, having reactivity. The word was coined in 1830 in recognition of the very low chemical reactivity of these hydrocarbons. Indeed, their only reaction of real significance to us as consumers is the one that provides us with most of our energy: hydrocarbons burn. The energy they release on combustion heats us, cools us (through air conditioning), moves our cars, airplanes and ships, and allows us to cook.

Later in this chapter we'll examine the societal consequences of using hydrocarbon combustion as a source of energy, and in Chapter 7 we'll see how the internal combustion engine converts that energy into the force of motion.

QUESTION A *pentane* molecule contains five carbons, a *decane* molecule contains 10 carbons, a *pentadecane* molecule contains 15 carbons. Write the molecular formula of the alkane *octadecane*.

6.8 The Problem with Prefixes

Using classical prefixes to indicate the number of carbon atoms works well for the straight-chain alkanes, but it doesn't solve the problem of finding names for all the isomeric alkanes. We've already seen that the prefix *iso-* is used to identify the branched-chain isomer of C_4H_{10} as isobutane. For C_5H_{12} the problem becomes a little larger since we must name three isomers:

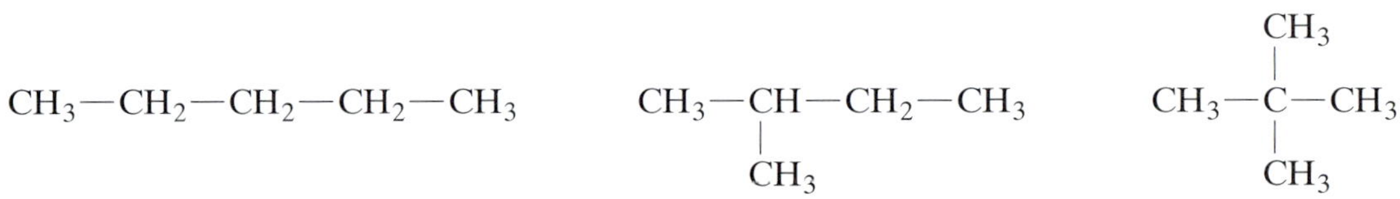

pentane   isopentane   neopentane

Pentane, isopentane, and neopentane.

We've already seen that the prefix *iso-* of isobutane and isopentane comes from the Greek word for *equal* (Sec. 6.6). The *neo-* of neopentane comes from the Greek *neos*, new. Used as a prefix, *iso-* usually refers to a chain of carbons with a single methyl group branching off at one end, as in *isohexane* and *isopropyl alcohol*:

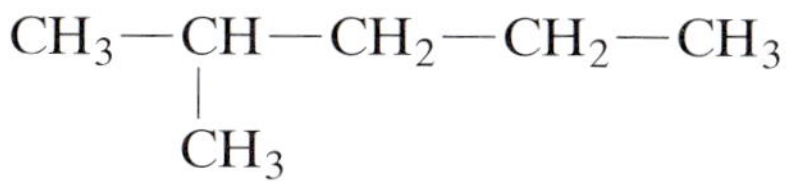

isohexane

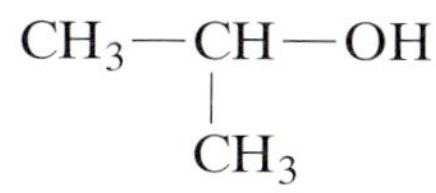

isopropyl alcohol

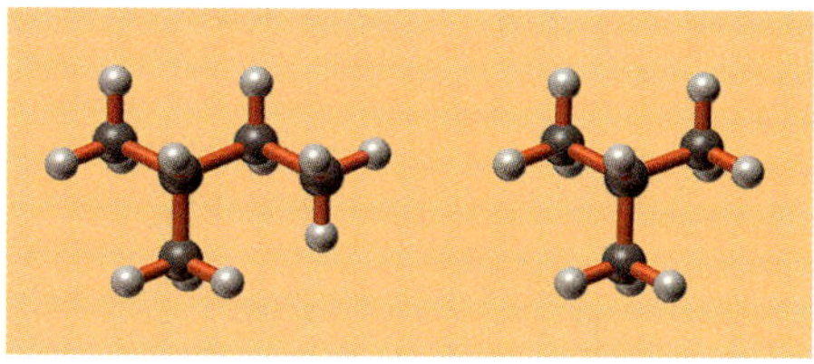

Isopentane and isobutane.

Using the combination of a classic root to define the number of carbons in an alkane and a prefix to identify the specific isomer becomes hopeless in the face of the sheer number of possible isomers that can exist. With 6 carbons and 14 hydrogens, C_6H_{14} exists as 5 different isomers; $C_{10}H_{22}$ forms 75; and $C_{20}H_{42}$ is capable of 366,319 (Table 6.3). Certainly, even if we wished to we couldn't find enough prefixes in Greek, Latin, and all other languages combined to name all possible isomers of even the first 20 alkanes. Nor could we remember them all.

Table 6.3 Numbers of Possible Isomers of the Alkanes

Carbon Content	Molecular Formula	Possible Isomers
1	CH_4	1
2	C_2H_6	1
3	C_3H_8	1
4	C_4H_{10}	2
5	C_5H_{12}	3
6	C_6H_{14}	5
7	C_7H_{16}	9
8	C_8H_{18}	18
9	C_9H_{20}	35
10	$C_{10}H_{22}$	75
15	$C_{15}H_{32}$	4,347
20	$C_{20}H_{42}$	366,319

As organic chemistry developed it became clear that the application of the classical languages to the naming of organic compounds has its limits. A simple, yet rigorous, systematic and unambiguous approach to devising names for organic molecular structures was needed. In the next section we will see how it came about.

■ QUESTION

What is the alkane of lowest molecular weight (carbon content) that contains one or more of each of the following three classes of carbons: 1°, 2°, and 3°?

6.9 IUPAC

In 1892 a convention of chemists met in Geneva, Switzerland, under the auspices of the International Union of Pure and Applied Chemistry, to simplify and systematize the naming of organic compounds. The set of rules they adopted and that is still in general use is known today as the IUPAC system, for the acronym of the host organization.

In essence, the IUPAC system operates through a series of simple rules:

1. As the parent name of the compound, use the name of the alkane corresponding to the *longest continuous chain* of carbon atoms within the molecule.
2. Locate the position of each group connected to the chain—a methyl group, for example—by numbering the chain, starting from the end that will give the position the *lower number.*
3. Indicate the total number of each set of mutually identical groups with a *prefix*. For example, use dimethyl for two methyl groups, trimethyl for three methyl groups, and so on.
4. If more than one group is present on the chain, give *each group* its individual location number.
5. Separate numbers from each other by *commas*, and separate numbers from letters by a *hyphen.*

Some examples will help. Although isobutane has a total of four carbons, the longest continuous carbon chain in the isobutane molecule contains only three carbons. So isobutane becomes *2-methylpropane* under the IUPAC system. The number 2 places the single *methyl* group at the second of a three-carbon (*propane*) chain:

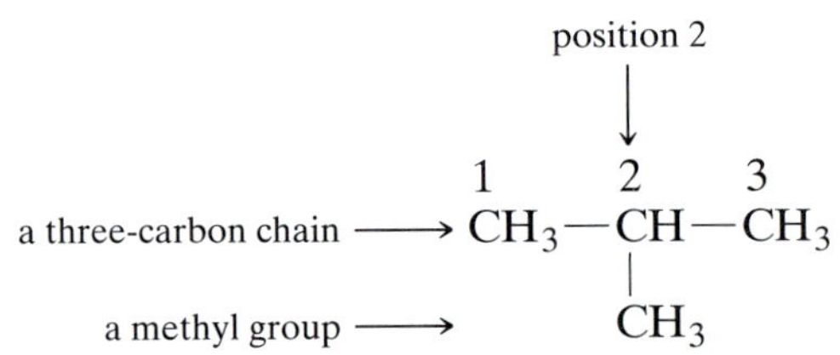

2-methylopropane
(also known as
isobutane)

The IUPAC names of some of the other alkanes we have discussed appear below:

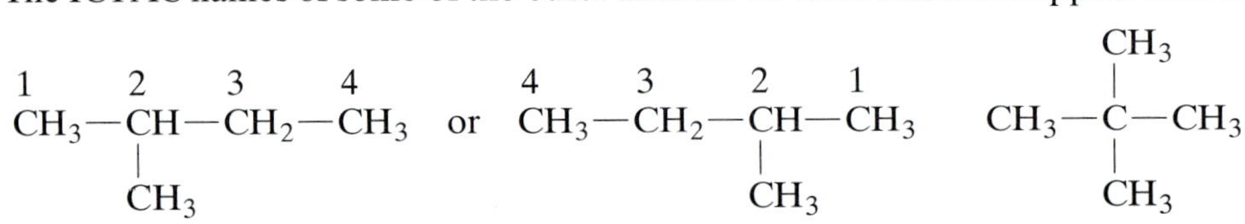

2-methylbutane
(isopentane)

2,2-dimethylbutane
(neopentane)

Notice that the IUPAC system provides a bit of useful redundancy, which minimizes mistakes: The set of numbers to the left of the hyphen must agree with the prefix to the right. For *2,2-di*methylpropane, *2,2* and *di* agree, with each indicating the presence of two groups on the chain.

The following simple example and one that's a bit more complex sum up these rules.

E X A M P L E

Isohexane

What is the IUPAC name of the C_6H_{14} isomer known commonly as isohexane?

■ In Section 6.8 we saw that the structure of isohexane is

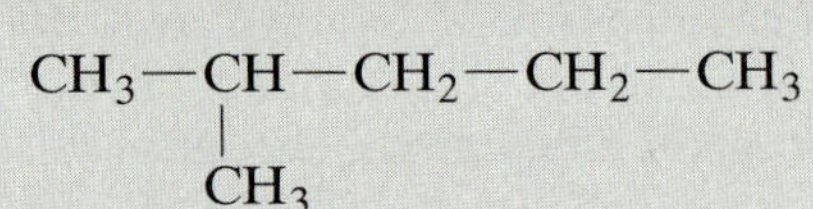

isohexane

Following rule 1, we find that the longest continuous chain of carbon atoms within the molecule consists of a string of five carbons. Since the five-carbon alkane is pentane, we name this as a *pentane.*

Under rule 2, we find that there is one methyl group on this five-carbon chain, which makes it a *methylpentane*.

Following rule 2, we number the chain so that the methyl group appears on the chain carbon bearing the lowest possible number. Here we start at the left end of the chain so that CH_3 is on carbon 2:

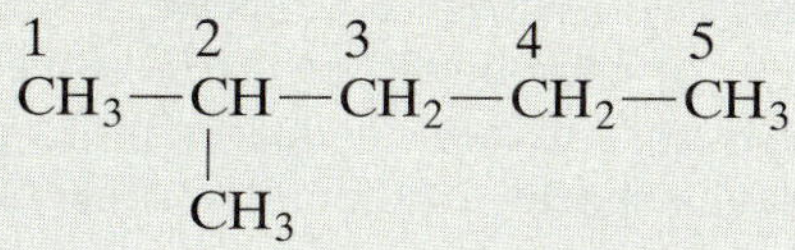

Rule 5 requires that we separate the 2 from the first letter of *methylpentane* with a hyphen. The IUPAC name of isohexane is *2-methylpentane*.

E X A M P L E

Getting Octane's Numbers

As we'll see in the next chapter, the isomer of octane that establishes a gasoline's octane number has the following molecular structure. What is its IUPAC name?

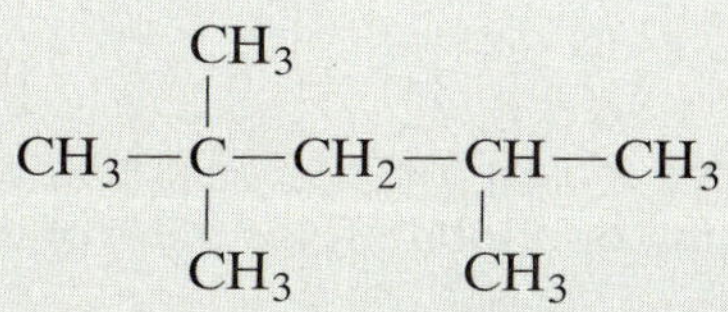

■ As in the previous examples, following rule 1, we find that the longest continuous chain of carbon atoms within the molecule consists of a string of five carbons. Since the five-carbon straight-chain alkane is pentane, we name this as a *pentane*.

Under rule 2, we find that three methyl groups exist as substituents on the chain, so we number the chain to give the carbons bonded to these methyl groups the lowest possible numbers. Here we start at the left end of the chain so that two CH_3's are on carbon 2 and one CH_3 is on carbon 4:

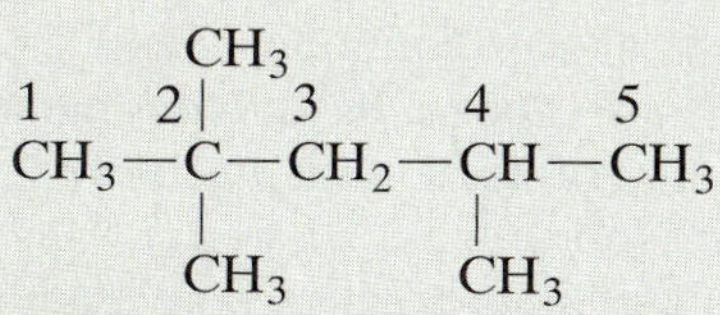

(If we had started at the right end as the structure is written above, reversing the sequence of the numbers we've just written, then there would have been only one CH_3 on carbon 2 and two CH_3's on carbon 4. The procedure we use above gives us a set of lower numbers.)

Following rule 3, we see that the five-carbon chain bears three methyl groups, so it is a *trimethylpentane*.

Rule 4 gives each of the methyl groups its own position number. Three groups require three numbers, which, following rule 5, we separate from each other by commas and from the letters by a hyphen. We now have the full name, *2,2,4-trimethylpentane*.

■ **QUESTION** Draw the molecular structures and give the IUPAC names of all five isomers of C_6H_{14}.

6.10 Beyond the Alkanes: Alkenes, Alkynes, and Cycloalkanes

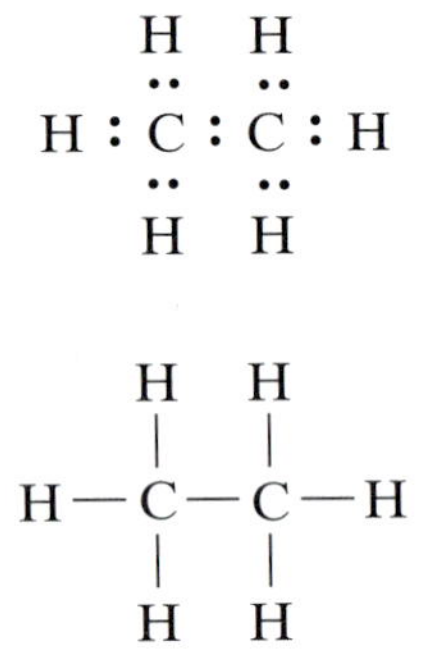

Ethane

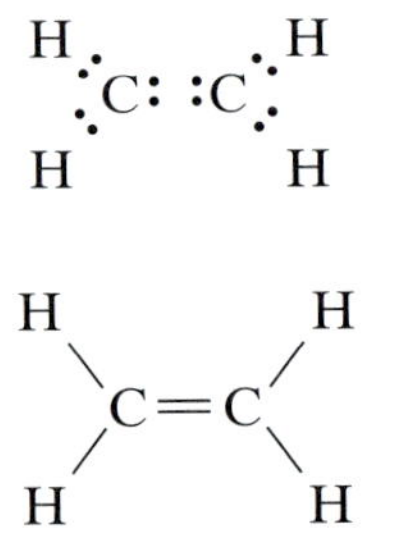

Ethene, or ethylene

Figure 6.13 The carbon–carbon single bond of ethane and the carbon–carbon double bond of ethylene.

An **alkene** is a hydrocarbon that contains a carbon–carbon double bond and fits the general formula C_nH_{2n}.

In addition to the proportions defined by the general formula of alkanes, C_nH_{2n+2}, carbons and hydrogens can combine in other ratios as well. To understand how, we can start by removing *two* hydrogens from ethane: one from each of its methyl groups. In removing these two hydrogen atoms we remove not only the proton that makes up the nucleus of each but its electron as well. These two hydrogen atoms could now unite to form a diatomic hydrogen molecule, held together by the covalent bond formed by the sharing of their electrons:

$$H^{\cdot} + {}^{\cdot}H \longrightarrow H{:}H \quad \text{(also written as H—H or } H_2\text{)}$$

We don't need to be concerned further with this molecule of hydrogen. Instead, what interests us here is the fate of the two unpaired electrons that remain on the carbons that held the hydrogens. Those two electrons come together, pair up, and form a new, second covalent carbon–carbon bond:

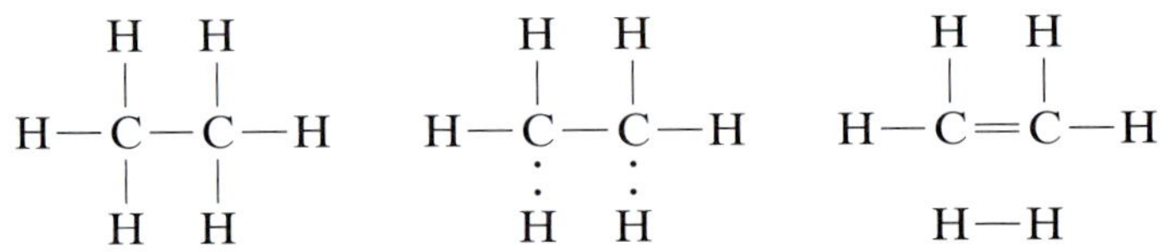

The result is the appearance of a carbon–carbon *double bond*, two covalent bonds formed through the sharing of a total of *two pairs* of electrons by neighboring carbons (Fig. 6.13).

The hydrocarbon that results from this procedure is known as *ethylene* or, in the IUPAC system, *ethene*. (Notice that the *-ane* ending of the two-carbon *ethane* is replaced by an *-ene* ending for the two-carbon *ethene*.)

Ethylene is the first compound in a family of hydrocarbons called the **alkenes**, all of which share the general formula C_nH_{2n} and each of which contains a carbon–carbon double bond. In the IUPAC system, the hydrocarbons of this family are named much like those of the alkane family, but with the *-ane* ending of the alkane names replaced by *-ene* of the alkene. While the alkane C_3H_8 is propane, the alkene of the same carbon content, C_3H_6, is *propene*, $CH_2{=}CH{-}CH_3$. (Propene is often known by a more common, non-IUPAC name, *propylene*.) The alkene family as a whole is also known by an older name, the *olefin* family.

Carrying out an operation of this sort on any two adjacent C—H bonds of any alkane produces an alkene. Notice that if we were to start with butane, we'd have a choice of removing hydrogens either from carbons 1 and 2, or from 2 and 3. Introducing a double bond between carbons 1 and 2 gives us *1–butene*. The *1* of 1-butene indicates that the double bond connects carbon atoms 1 and 2 of the chain. Since the double bond connects two consecutive atoms of the chain, using the lower of their two numbers defines the position of the double bond without any ambiguity.

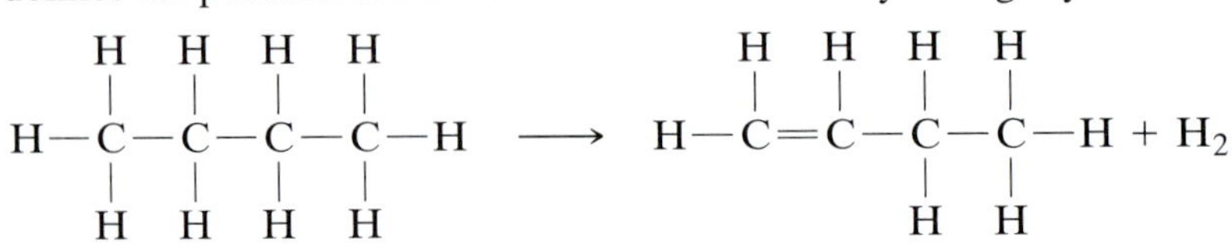

carbon number: 1 2 3 4

1-butene

$$CH_3{-}CH_2{-}CH_2{-}CH_3 \longrightarrow CH_2{=}CH{-}CH_2{-}CH_3 + H_2$$

carbon number: 1 2 3 4

1-butene

With the double bond located between carbons 2 and 3, the compound is 2–butene. Here, 2 represents the lower numbered of the two carbons joined by the double bond:

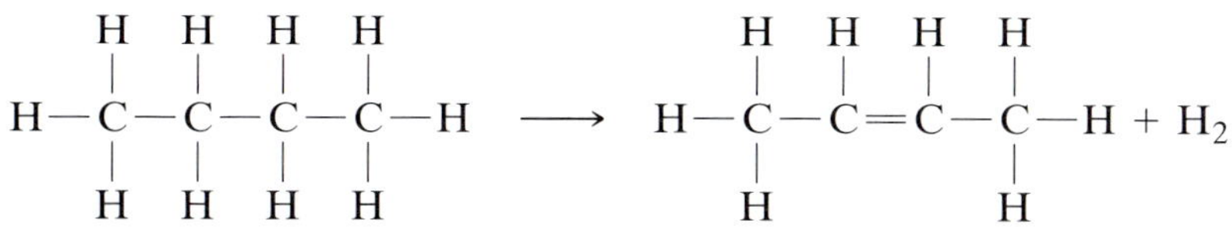

carbon number: 1 2 3 4

2-butene

$$CH_3{-}CH_2{-}CH_2{-}CH_3 \longrightarrow CH_3{-}CH{=}CH{-}CH_3 + H_2$$

carbon number: 1 2 3 4

2-butene

We can carry this a step further. Applying this same procedure to ethylene removes two hydrogens from $CH_2{=}CH_2$ and produces the *triple bond* of *acetylene*, $HC{\equiv}CH$, which is named *ethyne* in the IUPAC system (Fig. 6.14). Acetylene, with its formula C_2H_2, is the first of the **alkynes** (pronounced al-KINZ), sometimes referred to as the *acetylene* family. Like acetylene, all members of this family contain carbon–carbon triple bonds. The general formula for alkynes is C_nH_{2n-2}. In naming the members of the alkyne family by the IUPAC system, we replace the *-ane* of the corresponding alkane with *-yne*. The three-carbon member of this family, C_3H_4, is *propyne*.

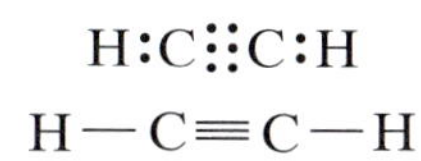

Ethyne, or acetylene

Figure 6.14 The triple bond of acetylene.

An **alkyne** is a hydrocarbon that contains a carbon–carbon triple bond and fits the general formula C_nH_{2n-2}.

E X A M P L E

Naming An Alkyne

Name the compound

$$HC{\equiv}C{-}CH_2{-}CH_3$$

■ The presence of the carbon–carbon triple bond, C≡C, shows that this is an alkyne. With four carbons in the chain, it's a butyne. As with alkenes, we have to number the carbons of the four–carbon chain so as to place the triple bond between the lower numbered carbons:

$$\overset{1}{H}\overset{2}{C}{\equiv}\overset{3}{C}{-}CH_2{-}\overset{4}{CH_3}$$

With the triple bond located between carbons 1 and 2, we name this isomer of C_4H_6 *1–butyne*.

Unlike alkanes, the first few members of the alkene and alkyne families have characteristic odors, ranging from slightly sweet to sharp and pungent, depending on their concentrations. Most of these hydrocarbons are of little direct use to consumers. Many fruits and vegetables, especially citrus fruit and tomatoes, release

Table 6.4 Alkanes, Alkenes, and Alkynes

Family	General Formula	First Member	Structure
Alkane (also known as *paraffin*)	C_nH_{2n+2}	Methane, CH_4	H HCH H
Alkene (also known as *olefin*)	C_nH_{2n}	IUPAC name: ethene Common name: ethylene	$CH_2{=}CH_2$
Alkyne (also known as *acetylene*)	C_nH_{2n-2}	IUPAC name: ethyne Common name: acetylene	$HC{\equiv}CH$

Welding with the oxyacetylene torch.

A **saturated** hydrocarbon is one that contains the maximum number of hydrogen atoms possible and that contains only single covalent bonds.

An **unsaturated** molecule is one to which hydrogen can be added.

ethylene as they ripen. This released ethylene speeds further ripening. Commercial growers sometimes use this same effect to speed the ripening of fruits and vegetables by bathing their unripened produce in ethylene gas.

Industrially, more ethylene is manufactured each year in the United States and worldwide than any other organic chemical. In 1999, U.S. firms manufactured 28 million tons of the alkene. In terms of total mass produced, ethylene accounts for about 30% of all industrially produced organic chemicals.

None of this ethylene reaches consumers directly. Most of it goes into the production of plastics, synthetic fibers, and the kinds of film we find in freezer wrappings and trash bags. (We'll have more to say about these and other plastics in Chapter 21.) With a U.S. production of almost 15 million tons, propylene ranks second to ethylene in the industrial production of organic chemicals, with much of that alkene also going into plastics.

Among the alkynes, only acetylene, the first member of the family, has extensive commercial use. Because it burns with a very hot flame, acetylene is commercially important as the fuel of the oxyacetylene torch, widely used in welding. Table 6.4 sums up the names and structures of the simplest alkanes, alkenes, and alkynes.

Alkanes are said to be **saturated**, while alkenes, alkynes, and all other organic compounds containing double or triple bonds are **unsaturated**. A compound containing a double or triple bond is considered to be unsaturated because it's possible to add molecular hydrogen to it until it will take up no more—to saturate it with hydrogen. A molecule to which we cannot add any hydrogen is considered to be saturated. Alkane molecules, then, are already saturated with as much hydrogen as they can hold (Fig. 6.15).

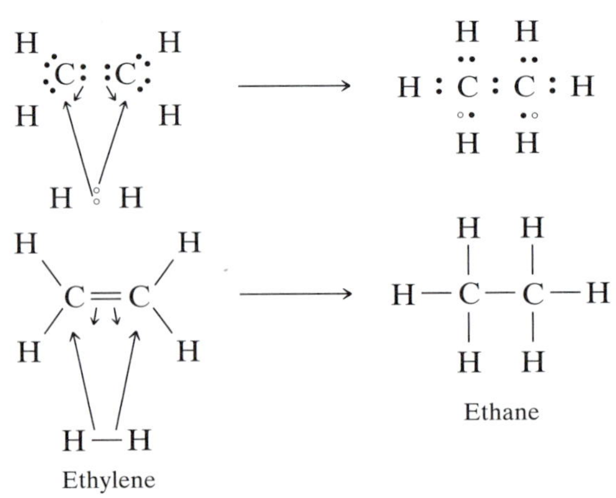

Figure 6.15 The conversion of unsaturated ethylene to saturated ethane.

Chemicals other than hydrogen can also add to unsaturated hydrocarbons, converting them to saturated compounds. These, of course, would no longer be classified as hydrocarbons since they would then contain elements other than hydrogen and carbon. In the presence of a small amount of acid, for example, water can add to an alkene to form an alcohol (Fig. 6.16).

$$\underset{\text{ethylene}}{CH_2{=}CH_2} + H_2O \xrightarrow[\text{a small amount of acid}]{\text{in the presence of}} \underset{\text{ethyl-alcohol}}{CH_3{-}CH_2{-}OH}$$

Figure 6.16 Adding water to an alkene produces an alcohol.

The terms *saturated* and *unsaturated* also apply to the fats and oils of our foods in much the same way they apply to hydrocarbons. As we'll see in Chapter 15, the unsaturated and polyunsaturated oils of foods are rich in carbon–carbon double bonds. Fats have far fewer of these double bonds, although they can still be found in small quantities in even the most highly "saturated" fat.

While all compounds fitting the general formula C_nH_{2n+2} must be alkanes since they are all fully saturated, not all compounds of the formula C_nH_{2n} are alkenes. Some are *cycloalkanes*. Each cycloalkane molecule consists of a ring or cycle of methylene groups (Sec. 6.5). *Cyclopropane*, the first member of the cycloalkane family, consists of a ring of three methylene groups, each located at an apex of an equilateral triangle. In *cyclobutane* the methylene groups form a ring nicely represented by a square, in *cyclopentane* they form a pentagon, in *cyclohexane* a hexagon, and so on (Fig. 6.17).

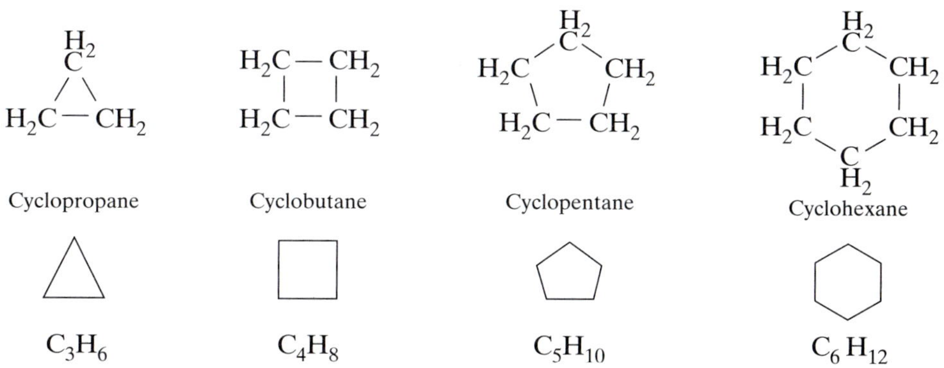

Figure 6.17 Cycloalkanes. (In the lower set of figures, only the carbon–carbon bonds are shown; the carbon atoms, hydrogen atoms, and carbon–hydrogen bonds are implied.)

■ QUESTION

The general formula C_nH_{2n+2} can be used to identify a compound as an alkane. Can the general formula C_nH_{2n} be used similarly to identify a compound as an alkene? Explain.

6.11 Aromatic Hydrocarbons

Still another family of compounds, the *aromatic hydrocarbons*, is important to a great variety of consumer products, including plastics, synthetic fibers, pharmaceuticals, gasoline, and food flavorings. All aromatic compounds contain at least one highly unsaturated ring of six carbons. The hydrocarbon *benzene* is the simplest of all of these (Fig. 6.18). The circle within the hexagonal ring of one of the structures shown in Figure 6.18 represents a peculiar facet of benzene's molecular structure and chemical reactivity. We'll see why.

Figure 6.18 Three representations of benzene.

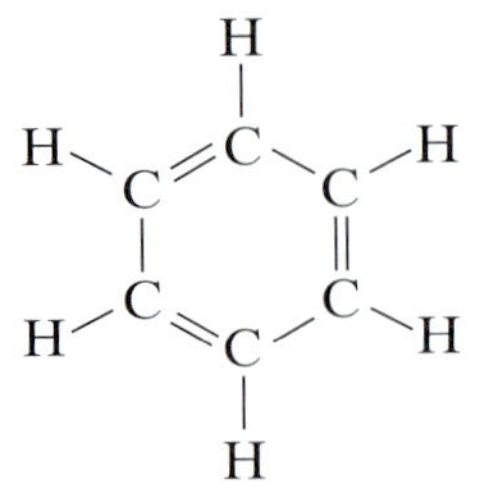

All atoms and bonds are shown.

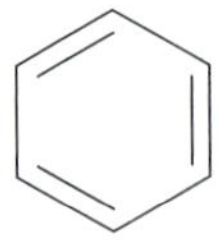

Only carbon–carbon bonds are shown. Carbon atoms, hydrogen atoms, and carbon–hydrogen bonds are implied.

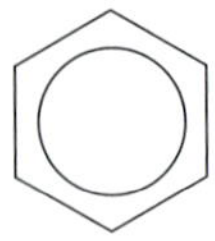

A representation showing six electrons (two from each of the three carbon–carbon double bonds) distributed evenly over the entire ring.

By drawing benzene's molecular structure with alternating double bonds (four shared electrons) and single bonds (two shared electrons) around the ring, we suggest that two different kinds of covalent bonds alternate among the carbons of the ring. In fact, though, plenty of experimental evidence shows very clearly that all the carbon–carbon bonds of the ring are mutually equivalent in every way. Every carbon–carbon bond is just like each of the others. This runs counter to the representation that shows three single bonds and three double bonds in the ring.

Moreover, it would seem from the structure containing three double bonds that benzene must be highly unsaturated, that it should add hydrogen molecules or water molecules as readily as the alkenes do, for example. But benzene adds hydrogen only under extreme conditions that are far more rigorous than those needed with alkenes and alkynes. And benzene doesn't add water molecules at all. Benzene is far less reactive than this structure would suggest. So representing benzene as a six-membered ring with three double bonds nicely positioned between adjacent pairs of carbons doesn't do justice to the compound's actual properties.

The sets of two shared electrons and four shared electrons simply aren't located as neatly between the carbons as the structure with alternating single and double bonds would suggest. Instead six electrons—two from each of the three C=C double bonds—are more accurately represented as being smeared evenly around the entire ring, distributed in equal portions among all six carbons. The circle within the hexagon of Figure 6.18 is intended to represent this smearing, this equal distribution, and to remind us that aromatic rings react in ways different from the alkenes and alkynes.

An **aromatic** compound is one that contains at least one benzene ring as part of its molecular structure.

The term *aromatic* also needs some qualification. Although benzene itself is the first member of the aromatic family of hydrocarbons, the compound has a somewhat disagreeable odor. **Aromatic**, as it's used here, simply reflects the presence of a benzene ring in the molecular structure of a compound. The classification originated in the observation that many plant-based compounds with particular pleasant odors also contain one or more benzene rings in their molecular structures. These compounds are, therefore, aromatic in the sensory perceptions they generate and in the chemical sense as well. They're classified as *aromatic* because their molecules contain one or more benzene rings, not because of their pleasant or unpleasant odors.

Among the more pleasant-smelling aromatic compounds are the organic compounds responsible for the fragrances of cinnamon (cinnamaldehyde), oil of wintergreen (methyl salicylate), and oil of anise (anethole), an oil used in the blending of perfumes and as a flavoring in foods and candies (Fig. 6.19). Though not

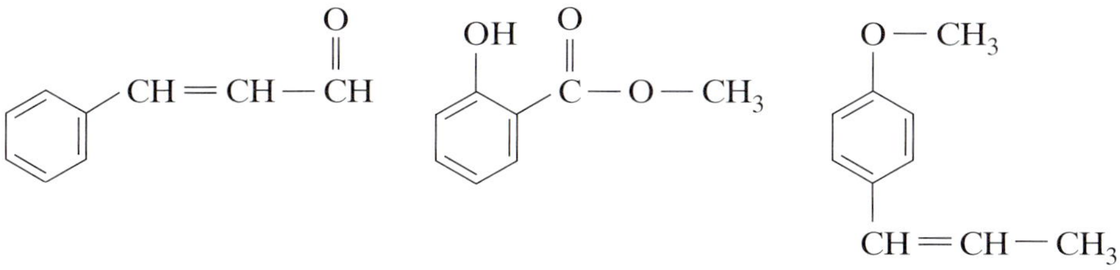

Cinnamaldehyde (occurs in cinnamon) C_9H_8O

Methyl salicylate (in oil of wintergreen) $C_8H_8O_3$

Anethole (in oil of anise) $C_{10}H_{12}O$

Figure 6.19 Aromatic compounds.

hydrocarbons (since they contain other elements in addition to hydrogen and carbon), such compounds are nonetheless considered to be aromatic chemically because of the major role the benzene ring plays in their chemical behavior. Thus the family of aromatic compounds includes both hydrocarbons and compounds that are not hydrocarbons. Table 6.5 sums up the kinds of compounds—alkanes, alkenes, alkynes, cycloalkanes, and aromatic compounds—that make up the hydrocarbon family.

Table 6.5 Families Within the Hydrocarbon Family

	Alkane	Alkene	Alkyne	Cycloalkane	Aromatic
Alternate family name	Paraffin	Olefin	Acetylene	—	—
General formula	C_nH_{2n+2}	C_nH_{2n}	C_nH_{2n-2}	C_nH_{2n}	—
First member	Methane	Ethene, ethylene	Ethyne, acetylene	Cyclopropane	Benzene
Molecular formula of first member	CH_4	C_2H_4	C_2H_2	C_3H_6	C_6H_6
Molecular structure of first member	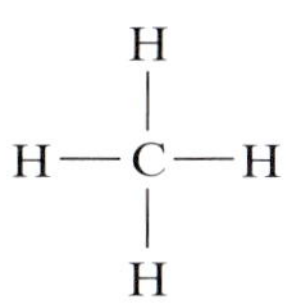	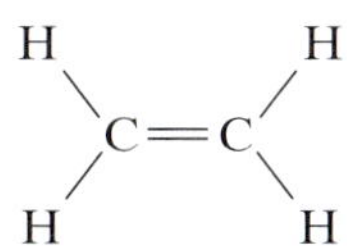	$H-C\equiv C-H$	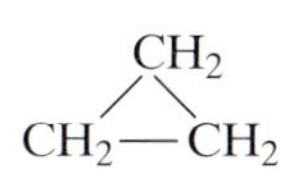	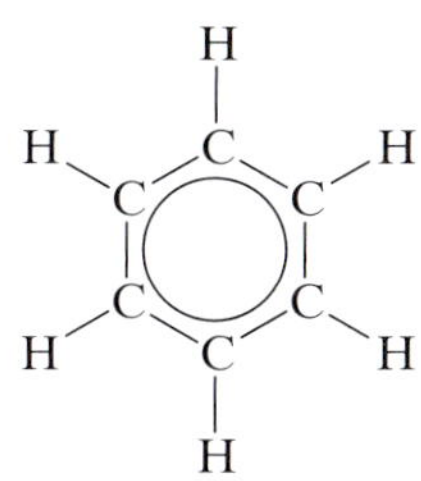

The hydrocarbon families we've been examining aren't exclusive. Any particular hydrocarbon molecule may contain any combination of double bonds, triple bonds, aromatic rings, or (if it's an alkane or a cycloalkane) none at all. Thus any particular hydrocarbon may overlap several different families. To date, the largest hydrocarbon molecule known, with a molecular formula $C_{1398}H_{1278}$, consists largely of both benzene rings and carbons held together by triple bonds. It can be expected to have properties of both alkynes and aromatic compounds.

■ QUESTION

What characteristic of molecular structure is common to all organic compounds classified as *aromatic*?

6.12 Functional Groups

Although the focus of this chapter is on hydrocarbons—compounds of carbon and hydrogen—several of the substances described here contain atoms or groups of atoms representing other elements as well. Known as *functional groups*, these include

- the —OH of the methyl and ethyl alcohols (Sec. 6.7),
- the halogens of methyl chloride and ethyl bromide (Sec. 6.3), and
- the $-\overset{\displaystyle O}{\overset{\|}{C}}H$, $-\overset{\displaystyle O}{\overset{\|}{C}}-O-CH_3$, $-O-CH_3$, and —OH groups of the aromatic compounds of Figure 6.19.

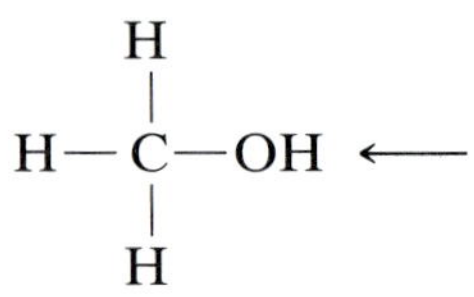

—OH, the functional group of an alcohol

methyl alcohol

A covalent combination of an oxygen and a hydrogen has replaced one of the hydrogens of methane.

A **functional group** is a small set of atoms held together by covalent bonds in a specific, characteristic arrangement that is responsible for the principal chemical and physical properties of a compound.

A **functional group** is a small set of atoms held together by covalent bonds in a specific, characteristic arrangement that is responsible for the principal chemical and physical properties of a compound. Functional groups influence melting points, boiling points, densities, and the ways in which a molecule reacts with other molecules.

As we saw in Section 6.2, it would be both foolhardy and futile to try to learn the chemistry of each of the millions upon millions of individual organic compounds now known to exist. Instead we focus our attention on the functional groups that form the most reactive portions of their molecular structures. Since all compounds that bear the same functional group undergo the same sort of chemical reactions—some faster than others in the same class, some slower, some with interesting and unusual twists—learning the chemistry of each of these relatively few functional groups allows us to generalize our knowledge to include enormous numbers of individual organic compounds, each bearing the same group. We've seen in Section 6.10, for example, that H_2 will add to ethylene to produce ethane. This is a reaction of the carbon–carbon double bond of ethylene, ethylene's functional group. Similarly, other compounds containing this same functional group will also add molecular hydrogen with similar results. Addition of H_2 to propylene and to 1-butene generates propane and butane, respectively:

$$\underset{\text{propylene}}{CH_2{=}CH{-}CH_3} + H_2 \longrightarrow \underset{\text{propane}}{CH_3{-}CH_2{-}CH_3}$$

$$\underset{\text{1-butene}}{CH_2{=}CH{-}CH_2{-}CH_3} + H_2 \longrightarrow \underset{\text{butane}}{CH_3{-}CH_2{-}CH_2{-}CH_3}$$

What's more, knowing the chemistry of each functional group allows us to predict the chemistry of new compounds as yet undiscovered or unknown to us. It's through their functional groups that compounds are organized into families and that we approach the study and practice of organic chemistry in a rational way.

Table 6.6 summarizes some of the functional groups discussed in this chapter. We'll examine compounds with these and additional functional groups as we continue our study of chemistry.

Table 6.6 Common Functional Groups

Structure	Functional Class	Representative Compound
$>C=C<$	Alkene	$CH_2=CH_2$ Ethylene
$-C\equiv C-$	Alkyne	$HC\equiv CH$ Acetylene
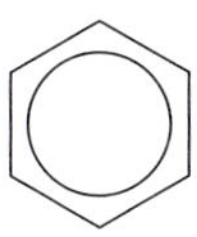	Aromatic	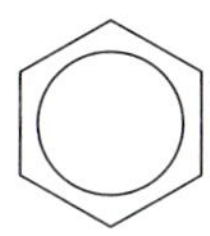 Benzene
$-\overset{O}{\overset{\parallel}{C}}-O-\overset{\mid}{\underset{\mid}{C}}-$	Ester	$CH_3-\overset{O}{\overset{\parallel}{C}}-O-CH_2-CH_3$ Ethyl acetate
$-\overset{\mid}{\underset{\mid}{C}}-OH$	Alcohol	CH_3-CH_2-OH Ethyl alcohol
$-\overset{\mid}{\underset{\mid}{C}}-X$	Alkyl halide (X = F, Cl, Br, or I)	CH_3-CH_2-Br Ethyl bromide

We've seen that the addition of H_2 to 1-butene produces butane. What product results from the addition of H_2 to 2-butene?

■ QUESTION

6.13 Candle Chemistry

Our opening demonstration—a candle burning in an inverted beaker—illustrates the reaction of hydrocarbons that's most widely used in our society: combustion. When hydrocarbons burn they consume oxygen, produce water and carbon dioxide, and liberate energy. You carry out this combustion reaction, with one kind of hydrocarbon or another, whenever you light a candle, turn on a gas stove, ignite an oil-burning furnace, or drive a car, truck, or bus that runs on gasoline or diesel fuel.

When you lit the candle in the opening demonstration, the wick started burning and melted a bit of the paraffin wax, which rose up, into the wick. The heat of the burning wick vaporized the liquid hydrocarbons; the resulting hydrocarbon vapor mixed with the surrounding air and ignited. This produced a chemical reaction in which the carbon and hydrogen atoms of the hydrocarbon molecules combined with the oxygen of the air to form carbon dioxide, CO_2, and water, H_2O, and released energy in the process. The water that formed on the inside of the beaker was a product of this chemical reaction. The released energy came to us in the form of the light and heat of the candle flame.

The explanation of how and why all this happened is the province of chemistry. We can follow the course of the chemical changes through the written chemical equation that represents the changes taking place.

6.14 Balanced Combustion Equations

Each of the hydrocarbon molecules of the paraffin wax contains a great many carbons and hydrogens. To understand how the many atoms of each molecule combine with oxygen to form water and carbon dioxide, we'll begin with a much simpler example, a charcoal briquette burning in a backyard grill. Here the fuel is the carbon itself.

The briquette burns slowly, releasing energy (heat) as the carbon atoms combine with diatomic molecules of oxygen in the surrounding air to produce carbon dioxide. Since the carbon atoms aren't bonded to hydrogen atoms, no water forms.

With plenty of oxygen available, the reaction is written as

$$C + O_2 \longrightarrow CO_2$$

It's a simple reaction, represented by a simple equation. Atomic carbon and molecular oxygen are the reactants; carbon dioxide is the product. The equation tells us something else as well. The implied *1*s in front of the C, the O_2, and the CO_2 tell us that *one* atom of carbon combines with *one* diatomic molecule of oxygen to produce *one* molecule of carbon dioxide.

This simple chemistry of a backyard grill becomes more complex, and deadly, when the charcoal burns in an enclosed space, such as a closed room. As the oxygen is consumed and becomes scarce, the chemical reaction changes. One molecule of oxygen now reacts with *two* carbon atoms to produce *two* molecules of *carbon monoxide*, CO, a dangerously toxic, odorless gas:

$$2C + O_2 \longrightarrow 2CO$$

Occasionally, on very cold nights, people living in houses with poor heating light up a charcoal grill in a closed bedroom to keep warm. That can be deadly. Not only does the glowing charcoal use up the life-sustaining oxygen, but, as the burning carbon consumes the oxygen in the enclosed room, the chemical reaction changes from one yielding CO_2 to the one that produces lethal CO. The result is often tragic.

A **balanced equation** contains the same number of atoms of each of the elements among both the reactants and the products.

It's important to recognize that both these equations are balanced. *In a **balanced equation** each atom of each element of the products, to the right of the arrow, exactly balances an atom of the same element among the reactants, to the left of the arrow* (Fig. 6.20). Equations must be balanced because of the *Law of Conservation of Mass*, which, as we saw in Section 4.11, states that *matter can neither be created nor destroyed in a chemical reaction*. In a balanced equation each atom that takes part in the reaction appears among both the reactants and the products. No atoms are created and no atoms are destroyed, so the Law of Conservation of Mass holds firm.

The balanced equation for the reaction of carbon and oxygen to produce carbon dioxide:

$$C + OO \longrightarrow OCO$$

The balanced equation for the reaction of carbon and oxygen to produce carbon monoxide:

$$\begin{matrix} C & & & & CO \\ & + & OO & \longrightarrow & \\ C & & & & CO \end{matrix}$$

Figure 6.20 Balanced equations: Each atom of each element among the products corresponds to an atom of the same element among the reactants.

E X A M P L E

Burning Hydrogen

While carbon burns in air to produce carbon dioxide but no water, hydrogen burns to produce water but no carbon dioxide. The unbalanced reaction below shows the combination of diatomic hydrogen molecules with diatomic oxygen molecules to produce water molecules.

■ Balance the following equation:

$$\textit{unbalanced:} \quad H_2 + O_2 \longrightarrow H_2O$$

Here we can start by balancing the number of oxygen atoms on both sides of the equation. This requires placing a 2 before the H_2O so that there are two oxygen atoms on both sides of the arrow.

$$\textit{unbalanced:} \quad H_2 + O_2 \longrightarrow 2H_2O$$

The reaction is still unbalanced since there are now four hydrogen atoms among the products (two in each of two water molecules) but only two among the reactants. To balance the equation we place a *2* in front of the H_2 to the left of the arrow:

$$\textit{balanced:} \quad 2H_2 + O_2 \longrightarrow 2H_2O$$

Now the equation is balanced with four hydrogens and two oxygens on each side of the arrow.

The hydrocarbon molecules of paraffin wax contain both carbon and hydrogen atoms and produce both water and carbon dioxide when they burn. Even though there's no methane in candle wax—CH_4 is a gas under ordinary conditions—we can use methane to illustrate a balanced equation for the combustion of alkanes in general, including the more complex hydrocarbons of the burning candle.

E X A M P L E

Burning Methane

Balance the chemical equation for the combustion of methane.

■ The reaction produces both water and carbon dioxide:

$$\textit{unbalanced:} \quad CH_4 + O_2 \longrightarrow CO_2 + H_2O$$

We can balance the four hydrogens of the CH_4 by showing the formation of two water molecules, which contain two hydrogens each for a total of four hydrogens:

$$\textit{unbalanced:} \quad CH_4 + O_2 \longrightarrow CO_2 + 2H_2O$$

Now there are four oxygen atoms among all the products but only two among the reactants. We need two diatomic oxygen molecules to achieve a balanced equation:

$$\textit{balanced:}\quad CH_4 + 2O_2 \longrightarrow CO_2 + 2H_2O$$

The reaction is now balanced with one carbon, four hydrogen, and four oxygen atoms among all the products and all the reactants.

EXAMPLE

One Step Further

Balance the equation for the combustion of ethane.

■ We can start as we did for methane.

$$\textit{unbalanced:}\quad C_2H_6 + O_2 \longrightarrow CO_2 + H_2O$$

Now we find two carbon atoms and six hydrogen atoms among the reactants. To account for all these atoms we can indicate two CO_2 molecules and three H_2O molecules among the products:

$$\textit{unbalanced:}\quad C_2H_6 + O_2 \longrightarrow 2CO_2 + 3H_2O$$

While this balances the carbon and hydrogen atoms among both the products and reactants, it leaves us with two oxygen atoms among the reactants and seven among the products (both CO_2 and H_2O).

Since 7 is an odd number, we might expect to need 3 1/2 molecules of diatomic oxygen to balance the seven oxygen atoms of the products. To avoid the use of half a molecule in a balanced equation, we simply double all the written and implied numbers of molecules in the entire equation.

$$\textit{balanced:}\quad 2C_2H_6 + 7O_2 \longrightarrow 4CO_2 + 6H_2O$$

The equation is balanced with 4 carbon, 12 hydrogen, and 14 oxygen atoms on both sides of the arrow.

■ **QUESTION**

A typical alkane molecule of paraffin wax contains 25 carbon atoms. Write a balanced chemical equation for its oxidation to carbon dioxide and water.

PERSPECTIVE

A Burning Candle Illuminates the Science of Chemistry

Our examination of the burning candle illustrates nicely one of the ways science works to explain the world we live in. It all starts with observations. When we inserted the flame into the inverted beaker, we saw that the flame's image blurred. Examining the inside of the beaker, we found that the blurring was caused by the formation of a band of water droplets. And when we inserted the flame farther into the beaker, we saw that the flame weakened and soon died.

Additional, more elaborate experiments would show us that the water on the beaker came from the candle itself and that the flame went out because the oxygen inside the beaker was consumed and replaced with carbon dioxide (Fig. 6.21). An analysis of the candle itself would show that it's composed of alkane hydrocarbons.

Extending these observations to interpretations of how and why the alkane hydrocarbons burn to generate water and carbon dioxide and consume the air's oxygen, we sum-

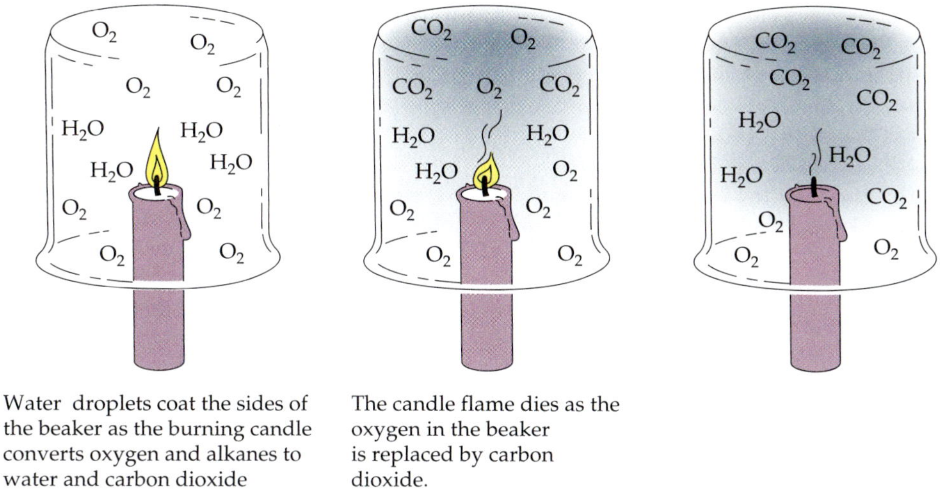

Figure 6.21
A burning candle.

marize our thoughts in the form of chemical equations. These are no more than shorthand expressions for our ideas that atoms and molecules can interact with each other to be converted from one set of chemical structures to another set, with entirely different properties.

All this takes place within the context of still other observations and still other interpretations. That is, we know that the Law of Conservation of Mass holds in all chemical reactions and that under the usual conditions of our everyday world atoms are neither created nor destroyed, nor is one element converted into another element. This leads us to express the chemical reactions in terms of balanced chemical equations, placing our explanations of what we observe on a quantitative foundation.

This applies to the entire range of material substances of the world we live in, from solids like the hard wax of our burning candle to invisible gases such as those within the beaker. In Chapter 8 we'll move beyond these material substances to the energy they consume and release as they react with each other, like the heat and light given off by the candle's flame. ■

E X E R C I S E S

For Review

1. Following are a statement containing a number of blanks, and a list of words and phrases. The number of words equals the number of blanks within the statement, and all but two of the words fit correctly into these blanks. Fill in the blanks of the statement with those words that do fit, then complete the statement by filling in the two remaining blanks with correct words (not in the list) in place of the two words that don't fit.

> Any compound containing the element carbon is an __________ compound. Those compounds containing both carbon and hydrogen, but with no other element present, are __________. Those containing carbon and hydrogen in a ratio that fits the general formula C_nH_{2n+2} are known as ______________ or, by an older system of nomenclature, ______________. The modern system of naming organic compounds, the ____________ system, came about as a result of an international gathering of chemists at Geneva, Switzerland, in 1892.
>
> Because the component atoms of all but the smallest molecules can be joined to each other in a variety of ways, most molecular formulas can represent a variety of molecules. Different compounds that share the same molecular formula are known as ____________. The molecular formula C_4H_{10}, for example, applies to two different compounds, known by their IUPAC names as ________________ and ________________.

alkanes	IUPAC
butane	organic
hydrocarbons	paraffins
isobutane	stereochemistry

2. Define, describe, or give an example of each of the following:

a. 1° carbon	f. double bond
b. alkene	g. heptane
c. alkyl group	h. saturated hydrocarbon
d. aromatic compound	i. stereochemistry
e. cycloalkane	j. unsaturated

3. Name three carbon-containing compounds that have been obtained for centuries from living or once-living substances. Describe their uses.

4. Using the chemical definition described in this chapter, identify each of the following as organic or inorganic: (a) table salt, (b) table sugar, (c) methyl alcohol, (d) oxygen gas, (e) the isotope of uranium used in nuclear reactors, (f) the dye henna, (g) urea, (h) the carbon dioxide of your exhaled breath, (i) the carbon dioxide of a carbonated soft drink.

5. Given a specific molecular structure, describe how you would determine whether it represents an alkane, an alkene, or an alkyne.

6. What do all aromatic compounds have in common?

7. Name three consumer products that are composed exclusively or largely of hydrocarbons. Describe the properties of alkanes that are most important to each of the three consumer products.

8. Name each of the following:

a. $CH_3{-}CH_2{-}CH_2{-}CH_2{-}CH_2{-}CH_3$

b. 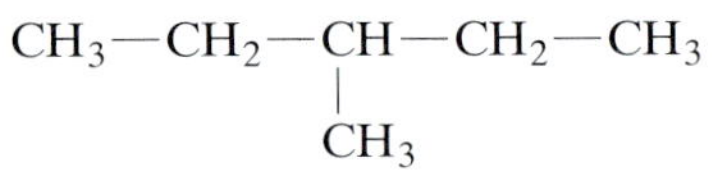

c. 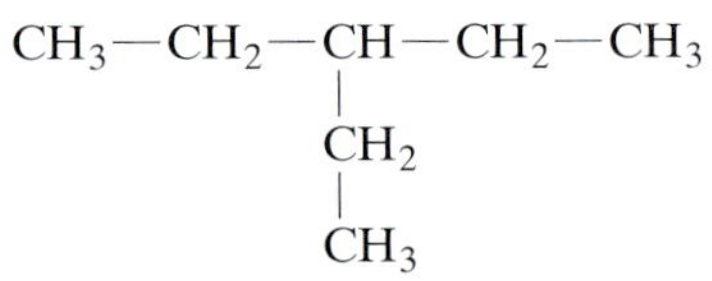

d. 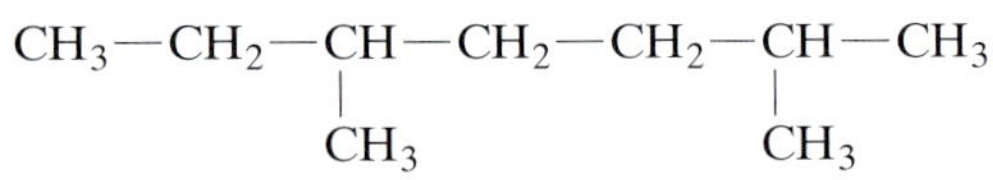

e. 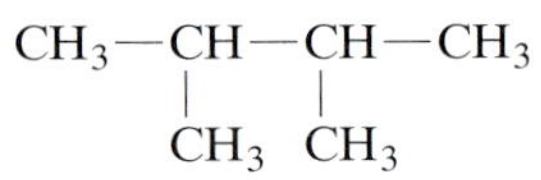

f. $CH_3{-}CH{=}CH{-}CH_2{-}CH_2{-}CH_3$

g. 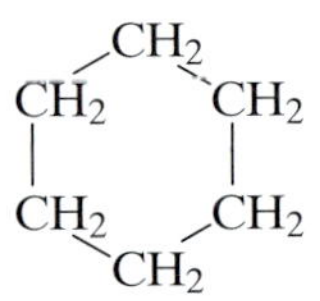

9. Identify five organic compounds that are isolated from plants. Identify one organic compound that is generated by animals or animal products.

10. What are three properties of carbon that result in the large number and great variety and complexity of organic compounds?

11. In what way do free radicals react when two of them encounter each other?

12. Match each of the following organic compounds with the property or source that is the origin of its common name:

	Compound	Property or source
________	a. methane	1. Related to the first in a series of acids found in fats
________	b. ethane	2. Related to an acid found in rancid butter
________	c. propane	3. Related to a very highly flammable compound
________	d. butane	4. Related to an alcohol obtained from wood

13. How can you determine whether any particular organic compound is a member of the alkane family?

14. Describe a property or a bit of chemistry: (a) common to the entire hydrocarbon family; (b) common to all alkenes but not to alkanes.

15. What compound is produced when: (a) diatomic hydrogen is added to ethylene, (b) water is added to ethylene?

16. Classify each of the following as to whether it is (1) a hydrocarbon or *not* a hydrocarbon, (2) an alkane or *not* an alkane, and (3) an aromatic compound or *not* an aromatic compound.

a. isobutane
b. acetylene
c. hexane
d. cyclohexane
e. carbon dioxide
f. benzene
g. $C_{17}H_{36}$
h. cinnamaldehyde

17. What's the difference between the molecular structures of ethene and ethyne?

18. Draw the molecular structure of 2-pentyne.

19. How would you determine whether a chemical equation is balanced or unbalanced?

A Little Arithmetic and Other Quantitative Puzzles

20. Write balanced equations for the combustion of (a) methane, (b) ethylene, (c) propane, (d) propyne, (e) benzene.

21. Balance each of the following equations:

a. $Na + Cl_2 \longrightarrow NaCl$
b. $H_2 + Cl_2 \longrightarrow HCl$
c. $Ca + O_2 \longrightarrow CaO$
d. $Na + O_2 \longrightarrow Na_2O$
e. $Mg + Br_2 \longrightarrow MgBr_2$

22. Draw condensed molecular structures for all the isomers of heptane, C_7H_{16}, that have exactly six 1° hydrogens.

23. The number of 1° hydrogens in a molecule must be a multiple of 3, and the number of 2° hydrogens in a molecule must be a multiple of 2. Why is this so?

24. Hydrocarbon molecules that contain two carbon–carbon double bonds are called *dienes*. What is the general molecular formula for the family of dienes? What other family of hydrocarbons has this same general molecular formula?

25. How many molecules of hydrogen would have to be added to a molecule of benzene to produce cyclohexane?

26. To what class or classes of compounds might each of the following hydrocarbons belong: (a) C_5H_{10}, (b) C_8H_{18}, (c) C_6H_{10}, (d) $C_{10}H_{20}$, (e) C_9H_{20}?

27. Draw all the isomers of C_3H_6 that can exist and name each.

28. Why can't a compound with a molecular formula of C_3H_{10} exist?

29. Can an alkane molecule contain an odd number of hydrogens? Explain.

30. The largest hydrocarbon discovered to date has the molecular formula $C_{1398}H_{1278}$ and is composed largely of aromatic rings and carbons joined by triple bonds (Sec. 6.11). Write the balanced equations for its combustion to water and carbon dioxide.

31. How many straight-chain isomers of $C_{50}H_{102}$ can there be? Explain your answer.

Think, Speculate, Reflect, and Ponder

32. *Methane* and *ethane* are both members of the *alkane* family of hydrocarbons. Describe one characteristic that methane and ethane have in common. Identify one difference between methane and ethane.

33. Draw the structure of cyclohexene.

34. Are 1-butene and cyclobutane isomers? Explain.

35. Draw the structure of two isomers of C_4H_6 that contain a triple bond. Draw one isomer that contains a ring.

36. Draw the structure and give the IUPAC name of the alkane of lowest molecular weight that contains a 4° (quaternary) carbon.

37. Name the following two compounds (Sec. 6.3 may provide help for part b):

a. 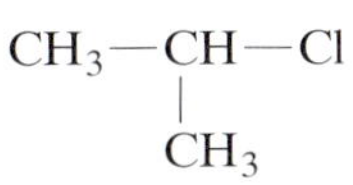

b. $CH_3—CH_2—CH_2—OH$

38. What would be one danger of using pure methane as a household fuel for heating and cooking?

39. Name each of the following compounds:

a. 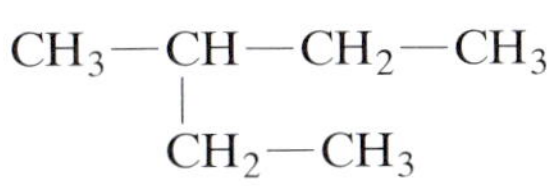

b. 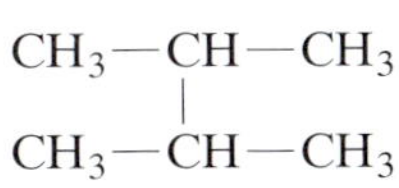

c. 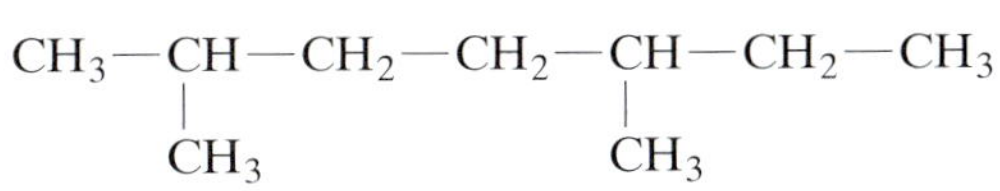

Compare compound (a) with compound (b) of Exercise 8. Are these two compounds isomeric alkanes? Are they mutually identical? Are they identical compounds with different names? Are they different compounds with the same names? Answer these questions after comparing compound (b) with compound (e) of Exercise 8. Repeat for a comparison of compound (c) with compound (d) of Exercise 8.

40. Can two different compounds have the same name? Can two compounds identical in *every* other way have different names? Explain.

41. Because alkanes burn readily in air they are useful as components of commercial fuels. What is another property or characteristic of alkanes, especially the higher alkanes, that makes them useful in consumer products, even in products that are not used as fuels?

42. Both the methyl free radical and the chlorine atom contain an unpaired electron. When two methyl free radicals combine they form ethane, $CH_3—CH_3$. When two chlorine atoms combine, they form a diatomic chlorine molecule, Cl_2. Write the condensed structure of the molecule formed when a methyl free radical combines with a chlorine atom.

43. Write the molecular formula of heptadecane.

44. How would using a warm beaker rather than a cool one in the demonstration that opened this chapter have affected our observations of a candle burning in an inverted beaker?

45. How would you distinguish between food that is "organic" and food that is not "organic"? Apply the chemist's definition of "organic" to food and describe how this affects your interpretation of the term *organic food*.

46. Can any *inorganic* substance be isolated from living or once-living material? Explain.

47. Give the molecular structures and the IUPAC names of all five isomeric hexanes.

48. Draw molecular structures for all the isomers of (a) C_2H_6 (one isomer), (b) C_2H_5Cl (one isomer), (c) C_3H_7Cl (two isomers), (d) C_4H_{10} (two isomers), and (e) C_5H_{12} (three isomers).

49. Should we classify benzene as a saturated hydrocarbon or as an unsaturated hydrocarbon? Explain.

50. What was the significance of Wöhler's laboratory synthesis of urea?

51. Suppose a sealed vial of a liquid is recovered from an ancient tomb in Egypt and writing on a clay tablet accompanying the vial is translated to indicate that the liquid was made from the juice of berries, but there is no indication of the age of the liquid. How might you go about determining how long ago the liquid was prepared from the berries? What technique might you use?

52. We can speculate that life exists in some other part of the universe, perhaps in a form far different from what we find here on Earth. On Earth, life is based on compounds of the element carbon. If life were to occur in another part of the universe, and if that life were based on an element other than carbon, suggest what element that might be. Justify your answer with reference to the periodic table and the relationships among elements within it.

53. Suppose that the carbon atoms of alkanes weren't tetrahedral but were *square planar* instead. That is, suppose the carbon atom were at the center of a simple square and the hydrogens or other carbons bonded to it were at the corners of the square and that they all lay in the same plane, as they do on this sheet of paper. Knowing that atoms and groups bonded to each other by covalent bonds can rotate freely around those bonds, predict how many isomeric methanes could exist if all carbons of alkanes were square planar. How many ethanes? Propanes? Butanes?

Web Questions

Use the Web to answer or accomplish each of the following. Report your Web source for each response.

1. Section 6.1 describes the red-orange dye *henna* and the antimalarial agent *quinine* as organic compounds that can be extracted directly from plants. Identify two other useful organic compounds that can be extracted directly from plants.

2. Many Web sites offer "organic" foods for sale. Describe some common characteristics of these foods described as "organic."

3. Find commercial sources for samples of each of the first 10 alkanes, methane through decane.

4. Where are the official headquarters of the IUPAC? Where are its administrative headquarters? Where and when was the latest meeting of its general membership held? Where and when is the next meeting scheduled?

5. What is the IUPAC name for the unbranched hydrocarbon $C_{46}H_{94}$?

6. Write a brief history and description of *benzene*. Discuss the contributions of Michel Faraday and Friedrich August Kekulé to our knowledge and understanding of benzene. Describe where benzene was discovered, where or how it is currently obtained, what it looks like and how it behaves under ordinary conditions of temperature and pressure, and what it is used for. Discuss the hazards of handling the compound and of absorbing it into the body.

7. In addition to those presented in this chapter, identify aromatic compounds: a) isolated from plant material, b) not found in nature but manufactured commercially, used c) in prescription medicines, d) as food additives, e) in the manufacture of plastics, f) in dyes.

Petroleum

THE DRIVING FORCE OF ENERGY

Pumping oil up, out of the earth.

Petroleum and Strong Tea

Fossil fuels are fuels formed from the partially decayed animal and vegetable matter of living things that inhabited the earth in eras long past.

Most of the world's energy comes from the burning of organic compounds, whether they represent the organic matter of wood and other plant-based material or the hydrocarbons of natural gas, coal, and petroleum. These hydrocarbon-rich **fossil fuels** were most likely formed from the partially decayed animal and vegetable matter of living things that inhabited the earth in eras long past. The combustion of fossil fuels accounted for about 85% of all the energy consumed in the United States during the decade of the 1990s. Not only do the fossil fuels (especially petroleum and natural gas) provide us with most of our energy, but they also serve as the source of virtually all our hydrocarbons and the organic compounds derived from them.

The origin of the word "petroleum" reflects the origin of the viscous fluid itself. It's a combination of two Latin terms, *petra*, meaning rock, and *oleum*, oil; petroleum is an oil extracted from rocks that lie just below the surface of the earth. In the decade of the 1990s this oil—known as "crude oil" or simply "crude" as it comes directly from the ground—provided about 40% of all the commercial energy consumed in the United States. In this chapter we'll focus on the hydrocarbons of petroleum and one of the most important sets of products containing these hydrocarbons: gasoline, kerosene, heating oil, and jet fuel.

To help us understand just how we obtain these fuels and other valuable products from crude petroleum, we'll begin by looking at some of the science behind a common kitchen phenomenon. If you've ever boiled anything in a covered pot of water, you've probably already seen something like this happen.

Make a glass of strong, deeply colored tea. Put it into a small pot with a lid, making sure you don't fill the pot more than about a quarter full. Put the pot's lid into the freezer. Heat the tea until it's boiling gently in the open pot. As the tea is heating, wrap several ice cubes in a towel. When the tea comes to a boil, cover the pot with the cold lid, allow the tea to continue to boil for a few seconds, then remove it from the heat. Now cool the top of the lid by rubbing it with the wrapped ice cubes, being careful to keep the towel far from the heating element so as not to ignite it. Rub the top of the lid with the wrapped ice cubes for about 45 seconds, or until it's cool, then quickly wipe the top of the cool lid clean of any water from melted ice cubes. Now remove the cover and tilt it over a glass so the water that's condensed on the inside of the lid drains out into the glass. The few drops of water that run into the glass from the bottom of the cover are clear and clean.

Distilling pure water from strong tea. Boil the tea, cool the lid and the vapors that come in contact with it, and collect the condensed water.

Put the lid back on the pot and repeat the procedure several times. (You don't need to chill the lid in the freezer again.) You'll find that you are collecting water that's about as pure as you can make it there in your own kitchen. If the tea is boiling briskly, there might be a little color because of tea splattering up from the pot onto the bottom of the cover. Otherwise, the water you collect is quite pure, containing perhaps only a little of the more fragrant, low-boiling organic compounds of the tea leaves.

In doing all this you have separated water from the mixture we call "tea" by the process of **distillation**. The water you collected is *distilled water*. To summarize what happened, you boiled the water, converting it from liquid water into steam, or water vapor. The hot water vapor condensed to liquid water when it came into contact with the cold lid. You collected this condensed and purified water by removing the lid from the pot and allowing the water to drain into another container, the clean glass. The importance of this process of distillation is that it allows us to separate liquids of different boiling points and to separate liquids from solids dissolved in them. By carrying out a distillation carefully, with sophisticated equipment, we can separate a pure liquid substance completely from any other liquids or any solids mixed with it. In this way we can purify a liquid to a very high degree. The principles you used in distilling water from the strong tea are the same as those used in the separation of the hydrocarbons contained in petroleum. ■

Distillation is the process of purifying a liquid by boiling it in one container, condensing its vapors to a liquid, and collecting the separated, condensed liquid in another container.

■ QUESTION

Is the wood obtained from trees cut down either by a commercial logging company or by a private individual considered to be a fossil fuel? Explain.

7.1 Our Thirst for Oil

Distilling petroleum yields a variety of **petroleum fractions**, mixtures of petroleum hydrocarbons with similar boiling points. These hydrocarbons can make up as much as 98% of petroleum, as they do in the highest grade of crude oil from Pennsylvania, or as little as 50%, as in the oil from some regions of California. The hydrocarbons vary over a wide range of **volatility**, or ease and speed of evaporation from a liquid to a vapor. They range from methane, a low-boiling, gaseous component of the natural gas often associated with crude oil (Sec. 6.2), to very high boiling, waxy, solid alkanes, such as hexacontane, $C_{60}H_{122}$, and its isomers.

Petroleum fractions are mixtures of components of petroleum with similar boiling points that are obtained through the distillation of petroleum. The **volatility** of a substance is a measure of the ease and speed of its transformation from a liquid to a vapor.

In addition, petroleum can also contain organic compounds of nitrogen, oxygen, and sulfur, as well as traces of metals that are incorporated into the organic molecules. Nonetheless it's the relatively volatile hydrocarbons ranging from about 5 to 20 carbons arranged in chains of various degrees of branching that interest us here. These are the hydrocarbons of the fuels that run our cars, trucks, buses, ships, airplanes, and other forms of transportation.

In this chapter we'll learn how crude petroleum is converted into refined gasoline, how the internal combustion engine converts the chemical energy of gasoline's hydrocarbons into the mechanical energy that runs our cars, how cars are modified to minimize gasoline's environmental effects, and how gasoline is modified to make our cars run more efficiently.

As we proceed, keep in mind the societal impact of the chemistry we examine. In 2000, the United States consumed a little more than 7 billion barrels of oil. Of this, 3 billion barrels (roughly 125 billion gallons) was in the form of gasoline. The United States consumed more than a quarter of all the petroleum used by all the nations of the world, more than the next four largest users—Japan, China, Germany, and Russia—combined.

In the United States alone, more than 200 million motor vehicles of all kinds, including cars, vans, trucks, buses, and motorcycles, travel over 2,400,000,000,000 (2.4×10^{12}) miles each year. With figures like these it's clear that every day enormous numbers of people in the United States and throughout the rest of the world initiate countless chemical reactions as they slide into the driver's seat and turn the ignition key. It's also evident that the chemicals of gasoline and the chemical reactions that take place in a car's engine produce an enormous economic and environmental impact on our world.

■ QUESTION

What property of a liquid is characterized by its *volatility*?

7.2 The Four-Stroke Cycle

When you turn the ignition key, you begin a sequence of chemical reactions. First you send a stream of electrons from the battery (Chapter 11) through a starting motor, which is connected to the engine itself. The electric current from the battery turns the starting motor, the starting motor turns the engine, and the engine begins consuming a mixture of gasoline and air. As soon as the spark plugs begin firing, the gasoline hydrocarbons react with the oxygen of the air, and the chemical energy released through this reaction keeps the engine running.

Almost all cars that run on gasoline have engines with four, six, or eight cylinders (Fig. 7.1). Each cylinder is a hollow tube, only a few centimeters in diameter, bored into the engine block. Each holds a close-fitting piston that rides smoothly up, nearly to the top of the cylinder, and down to the bottom. Each full passage of the piston from one end of its travel to the other, upward or downward, is a *stroke*. A cleverly designed arrangement of rods, gears, and various linkages converts the linear motion of the piston's strokes into the rotary motion of the wheels. In the simplest engines, two valves at the top of the cylinder open and close in rhythm with the movements of the piston, and a spark plug ignites the gasoline–air mixture at precisely the right moment. When all this operates just as it should, the engine converts the energy released by the burning gasoline into the energy of motion, and the car runs smoothly. (Some engines are more complex, with more than two valves in the cylinder and with other refinements of the basic design. But since the process

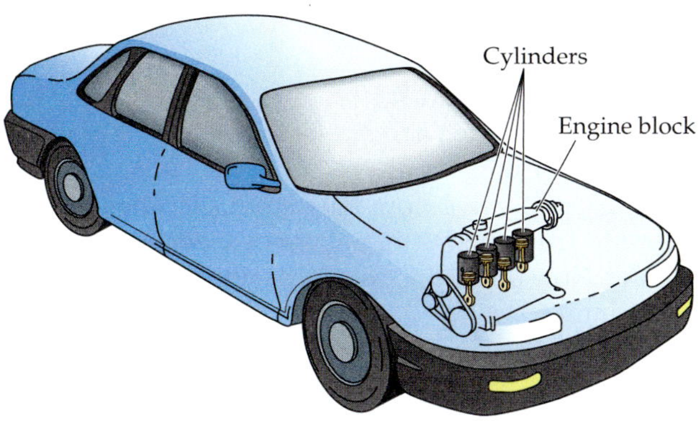

Figure 7.1 Schematic of a four-cylinder engine.

of converting the chemical energy of gasoline into the energy of motion is fundamentally the same in all gasoline engines, we'll stick to the simplest case.)

The entire process takes four strokes of the piston to accomplish, hence the name "four-stroke internal combustion engine." ("Internal combustion" refers to engines designed to burn their fuel inside the engine. Engines that run from fuel burned outside the engine—steam engines, for example—could be called "external combustion" engines.) As shown in Figure 7.2, the four strokes are:

- The *intake stroke*: With the intake valve open (and the exhaust valve closed), the piston moves downward from the top of the cylinder, drawing a mixture of gasoline and air into the cylinder.

Figure 7.2 (a) The beginning of the intake stroke. (b) The middle of the intake stroke. (c) The beginning of the compression stroke. (d) The beginning of the power stroke. (e) The beginning of the exhaust stroke.

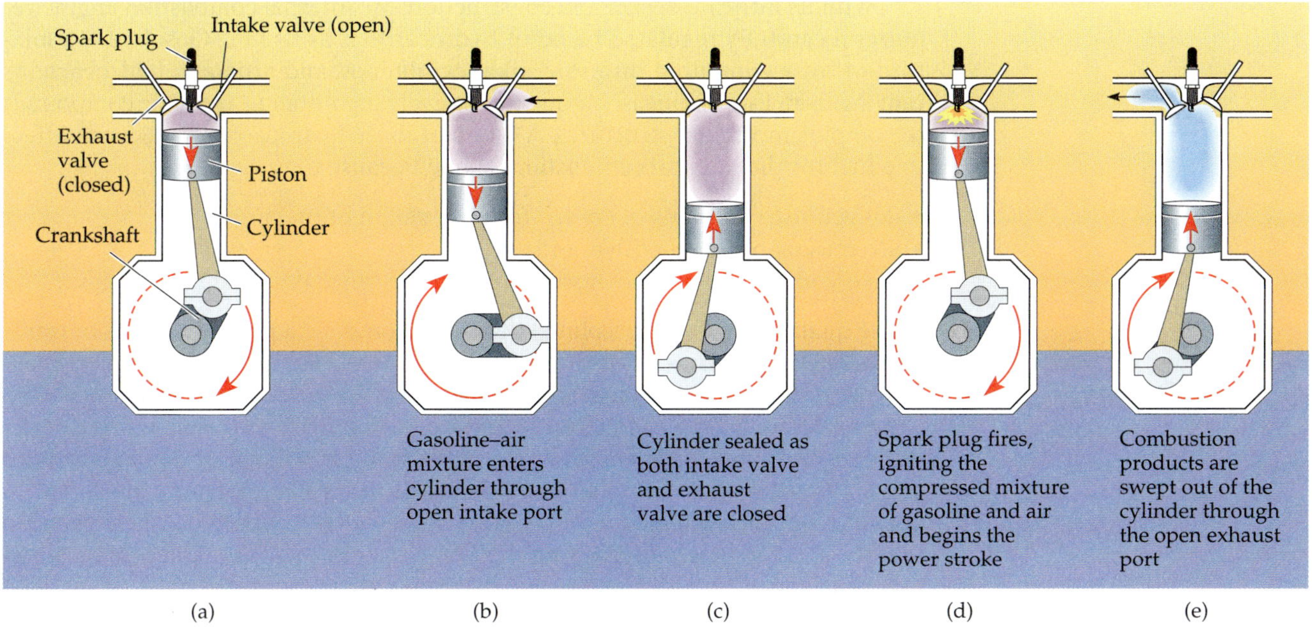

- The *compression stroke*: With both valves closed, the piston moves upward, compressing the gasoline–air mixture.
- The *power stroke*: With both valves still closed, the spark plug fires, igniting the compressed mixture of gasoline and air. The energy released by the burning hydrocarbons forces the piston downward.
- The *exhaust stroke*: With the exhaust valve now open (and the intake valve closed), the piston moves upward from the bottom of the cylinder, pushing the exhaust gases out of the cylinder. Now the entire process begins again as the piston moves downward in a new intake stroke.

The exhaust gases, which enter the atmosphere as they leave the car's tailpipe, consist of the expected carbon dioxide and water produced when hydrocarbons burn (Chapter 6) as well as a variety of other gases. Inefficient combustion adds unburned hydrocarbons and carbon monoxide to the exhaust gases, and small quantities of nitrogen- and sulfur-containing impurities (that carry over from the crude petroleum to the commercial gasoline) contribute oxides of nitrogen and sulfur, which add to the problem of acid rain and other forms of pollution (Chapter 14).

■ QUESTION

During the operation of the four-stroke, internal combustion engine, which strokes (if any) are in operation under the following sets of conditions? (a) Both the intake valve and the exhaust valve are closed. (b) Only one of these valves is open. (c) Both valves are open at the same time.

7.3 The Volatility of Hydrocarbons

As we've seen, turning the ignition key starts a complex process in which the hydrocarbons of gasoline react with the oxygen of air, forming a mixture of exhaust gases and releasing the energy that moves the car. We'll examine the effects of some of the exhaust gases in Chapter 14 and we'll look more closely at energy itself in Chapter 8. Here we'll consider the characteristics of the gasoline that moves the car and how it is obtained from the raw petroleum.

With its intricate design and construction, the internal combustion engine requires a carefully regulated blend of hydrocarbons as its fuel. Gasoline is a mixture of over a hundred different alkanes, alkenes, and aromatic hydrocarbons ranging from low-boiling isobutane to relatively high-boiling aromatic compounds of complex molecular structures. A blend of these hydrocarbons makes an effective fuel for the internal combustion engine because they

- deliver large amounts of energy for each gallon of fuel used,
- are reasonably inexpensive, and
- have volatilities (ease of vaporization) in the range required by the engine.

These qualities make the gasoline-powered automobile a practical means of transportation. We wouldn't find cars very useful if the fuel delivered so little energy and carried us such a short distance that we had to fill the tank twice a day or if gasoline cost, say, $50 or $100 a gallon. Nor would we be happy with a gasoline that doesn't vaporize and ignite in the engine or one that's so volatile that it simply doesn't burn efficiently when the spark plug fires. This third factor, the gasoline's ability to vaporize to just the right extent, is critical to the effective operation of the engine. So that the spark ignites a gasoline–air mixture containing just the right proportion of hydrocarbons for smooth and complete oxidation, the gasoline must be blended from hydrocarbons that lie within a specific range of volatilities.

Generally, the lower the boiling point of a liquid, the more volatile it is at any particular temperature, which is another way of saying that the lower the liquid's boiling point, the more readily it evaporates. In any given family of compounds, such as the hydrocarbons, boiling points *increase* (and volatilities decrease) with *increasing* molecular weight as long as we're dealing with molecules of the same general shape, such as straight-chain alkanes. Boiling points *decrease* with *increasing* branching of the carbon chain among isomeric hydrocarbons (those of the same molecular formula and therefore the same molecular weight; Sec. 6.6). We can see these effects in Tables 7.1 and 7.2.

Table 7.1 The Effect of Chain Length on Boiling Point

Straight-Chain Alkane	Structure	Molecular Weight	Boiling Point (°C)
Methane	CH_4	16	−164
Ethane	$CH_3—CH_3$	30	−89
Propane	$CH_3—CH_2—CH_3$	44	−42
Butane	$CH_3—CH_2—CH_2—CH_3$	58	−0.5
Pentane	$CH_3—CH_2—CH_2—CH_2—CH_3$	72	36
Hexane	$CH_3—CH_2—CH_2—CH_2—CH_2—CH_3$	86	69
Heptane	$CH_3—CH_2—CH_2—CH_2—CH_2—CH_2—CH_3$	100	98
Octane	$CH_3—CH_2—CH_2—CH_2—CH_2—CH_2—CH_2—CH_3$	114	126
Nonane	$CH_3—CH_2—CH_2—CH_2—CH_2—CH_2—CH_2—CH_2—CH_3$	128	151
Decane	$CH_3—CH_2—CH_2—CH_2—CH_2—CH_2—CH_2—CH_2—CH_2—CH_3$	142	174

Table 7.2 The Effect of Chain Branching on Boiling Point

Name	Structure	Branching	Boiling Point (°C)
Isomeric C_4H_{10} Alkanes			
Butane	$CH_3—CH_2—CH_2—CH_3$	Four-carbon chain, no branching	−0.5
Isobutane (2-methylpropane)	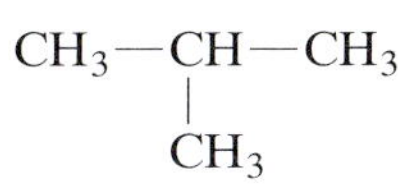	Three-carbon chain, one methyl branch	−11.6
Isomeric C_5H_{12} Alkanes			
Pentane	$CH_3—CH_2—CH_2—CH_2—CH_3$	Five-carbon chain, no branching	36
Isopentane (2-methylbutane)	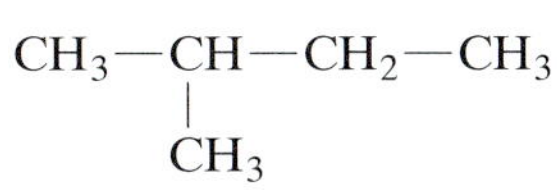	Four-carbon chain, one methyl branch	28
Neopentane (2,2-dimethylpropane)	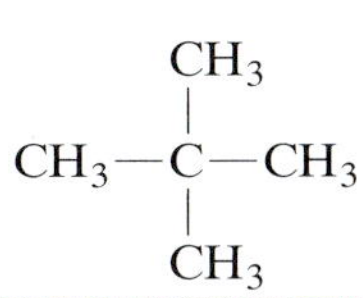	Three-carbon chain, two methyl branches	9.5

Table 7.1 lists the boiling points of some of the unbranched alkane hydrocarbons. For hydrocarbons of the same molecular shape—straight chains of carbons, for example—the boiling point increases along with molecular weight. For isomeric hydrocarbons Table 7.2 shows that the greater the degree of branching, the lower the boiling point. Isobutane and neopentane (2-methylpropane and 2,2-dimethylpropane), with their combination of relatively low molecular weights and relatively high degrees of branching, have much lower boiling points and are much more volatile than heptane, for example, which is a higher-molecular-weight, unbranched alkane. They also boil at lower temperatures than their straight-chain isomers, butane and pentane.

On the one hand, gasoline must be rich in volatile hydrocarbons to ensure that enough hydrocarbon vapors mix with the air in the cylinders to start a cold engine on a cold morning. If the gasoline–air mixture contains too little of the organic compounds, nothing burns when the spark plug fires. On the other hand, gasoline must not be too rich in volatile hydrocarbons. If it were, the ignited gases would be largely hydrocarbons; there wouldn't be enough oxygen in the mixture to burn the vaporized gasoline efficiently. The results would be poor ignition, roughness, poor fuel economy, and even vapor lock, a condition in which a bubble of gasoline vapor blocks the flow of fuel to the engine. With vapor lock a hot engine dies and won't start again until it cools off. Since gasoline must be equally efficient in Miami in July and in Chicago in January, manufacturers blend their gasolines to match the season and the locale of use.

■ QUESTION

Which one of the following alkanes has the highest boiling point? Which one has the lowest? (a) 2,3-dimethylbutane, (b) octane, (c) 3-methylpentane, (d) 3-methylheptane. Explain.

7.4 Knocking

An engine's **compression ratio** is the ratio of the maximum volume of the gasoline–air mixture, at the beginning of the compression stroke, to the volume of the compressed mixture, as the spark plug fires.

To squeeze as much energy out of the burning gasoline as possible, an engine must be designed with a high compression ratio. The **compression ratio** is simply the ratio of the maximum volume of the gasoline–air mixture at the beginning of the compression stroke to the volume of the fully compressed mixture as the spark plug fires at the end of the compression stroke (the beginning of the power stroke; Fig. 7.3). The higher the compression ratio, the greater the compression of the gasoline–air mixture when it is ignited, and the more powerful the thrust it delivers to the piston on its way down in the power stroke. In effect, the higher the compression ratio, the more energy we can squeeze out of the hydrocarbons, up to the limit of the chemical energy they contain. Imagine a wet sponge, for example. The more you squeeze it, the more water you get out of it, up to the limit of the water it holds. The more energy we can squeeze out of the hydrocarbons of gasoline, the more power the engine can provide.

The average automobile engine of 1925 had a compression ratio of 4.4. That's low by today's standards. In modern engines compression ratios run as high as 10. In 1925 the gasoline–air mixture was compressed to between one-fourth and one-fifth its original volume before it was ignited; today the mixture can be compressed to as little as one-tenth its original volume. As a result, today's engines can get much more energy out of hydrocarbon combustion than the engines of 1925.

Knocking is a rapid pinging sound produced by preignition or irregular combustion of the gasoline–air mixture in the cylinders of an internal combustion engine.

But there's a limit to all this compression: *knocking*. **Knocking** is a rapid pinging or knocking sound that comes from an engine when it is pushed to produce a lot of power quickly. Knocking usually occurs when a car is accelerating, especial-

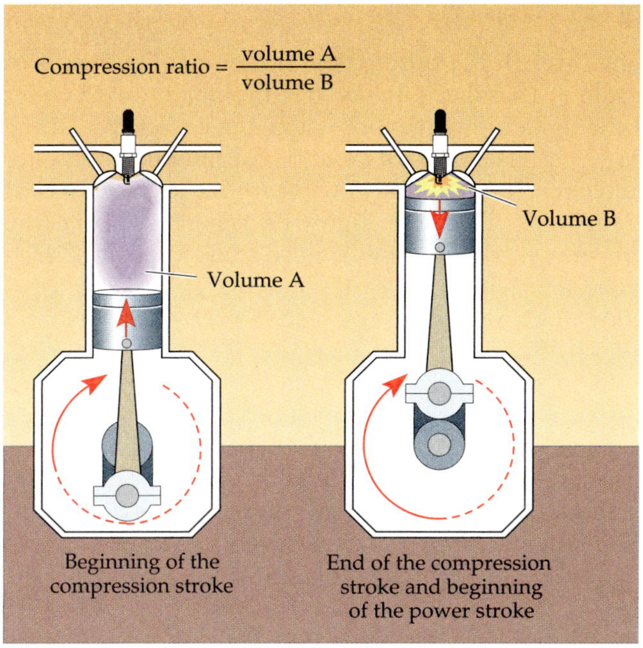

Figure 7.3 The compression ratio is volume A divided by volume B.

ly while going uphill or when pulling a trailer. In severe cases an engine can knock while the car is simply cruising along a level highway.

Knocking is the sound of actual explosions in the cylinders. Normally the gasoline–air mixture burns smoothly when the spark plug fires. The combustion of the compressed hydrocarbons begins at the spark plug and proceeds outward smoothly and evenly, like ripples in a pond. Sometimes knocking results from *preignition*, an ignition of the mixture before the piston has risen to the point where the spark plug normally fires. Knocking can also result from a combustion that begins spontaneously at one or more spots in the cylinder, either before or just as the plug fires, as though several stones had dropped into a pond, producing a set of irregular, overlapping ripples (Fig. 7.4). Or it can be a combination of the two. In any case, the resulting irregular, uncontrolled combustion produces a series of small explosions that we hear as knocking. It can cause loss of power, inefficient and uneconomical use of fuel, and, in severe cases, it can produce pits or fractures in the top surface of the piston. With any particular grade of gasoline, the higher the compression ratio, the greater the likelihood of knocking.

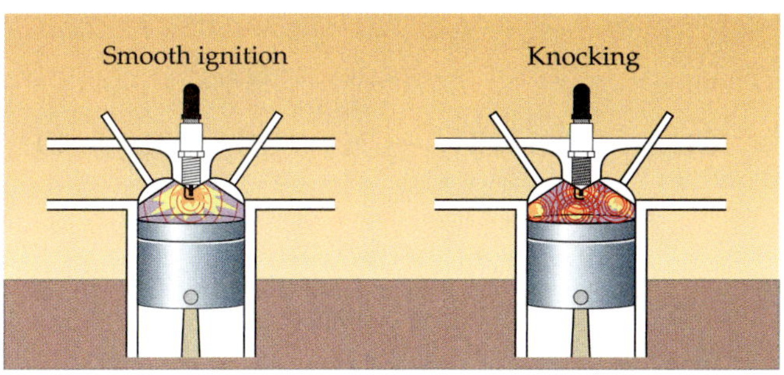

Figure 7.4 Smooth ignition and knocking.

■ QUESTION

Explain what engine *knocking* is and identify its source or origin.

7.5 Octane Rating

The elimination of knocking through the production of high-quality blends of gasoline requires some measure of a gasoline's ability to burn smoothly even under the rigorous conditions of the modern high-compression engine. To look at the problem from another angle, we need a means for describing a gasoline's *resistance* to knocking. Studies of the tendencies of various hydrocarbons to knock in test engines reveal one consistent trend: The more highly branched an alkane, the greater its tendency to burn smoothly and evenly and to resist knocking. 2,2,4-Trimethylpentane, for example, is a highly branched isomer of C_8H_{18} that consists of a five-carbon chain bearing three methyl groups; it shows very little tendency to knock. In connection with gasoline this isomer is customarily (and mistakenly) called "isooctane" and even simply "octane." Heptane, on the other hand, is completely unbranched and knocks readily, even under mild conditions (Fig. 7.5).

The **octane rating** or **octane number** of a gasoline is a measure of the gasoline's resistance to knocking.

Mixtures of these two alkanes are used in assigning **octane ratings** or **octane numbers** to commercial gasolines. Because of its considerable ability to burn smoothly and to resist knocking, 2,2,4-trimethylpentane is assigned an octane rating of 100; heptane, with its great tendency to knock, receives an octane rating of zero. Mixtures of the two are given octane ratings equal to the percentage of the octane they contain.

To determine the octane rating of any particular blend of gasoline we simply compare the knocking tendencies of the particular blend itself with those of mixtures of "octane" and heptane. If, for example, a blend of gasoline has knocking tendencies identical to those of a mixture of 85% "octane" and 15% heptane, under standard test conditions, we assign the blend an octane rating of 85. In summary, straight-chain, unbranched alkanes have a great tendency to knock and are assigned low octane numbers, while both highly branched alkanes and aromatic hydrocarbons such as benzene have little tendency to knock and receive high octane numbers. Octane numbers of representative hydrocarbons and other compounds appear in Table 7.3.

It's worth noting that there are several slightly different ways to measure the antiknock quality of a gasoline. Two stand out. One provides a "research" octane rating, R, that emphasizes the ability of a gasoline to burn smoothly when you start out with the gas pedal to the floor. The other yields a "motor" octane rating, M, more appropriate to cruising along an expressway. Since neither one is useful under all driving conditions, the number actually posted at the pump is an average of these two, R and M. For a gasoline with an average rating of 87, for example, you might see the following on the pump:

$$\frac{R + M}{2} = 87$$

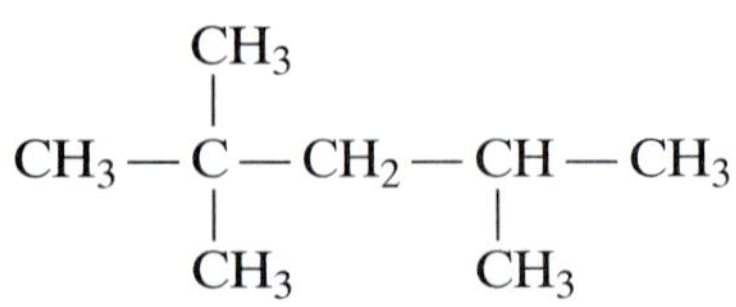

2,2,4- Trimethylpentane, known as "octane" or "isooctane", a highly branched alkane, shows little tendency to knock. Octane rating = 100.

$CH_3-CH_2-CH_2-CH_2-CH_2-CH_2-CH_3$

Heptane, a straight-chain, unbranched alkane, knocks readily, even under moderate conditions. Octane rating = 0.

Figure 7.5 A combination of 2,2,4-trimethylpentane and heptane is used to evaluate octane ratings.

Three grades of gasoline and their octane ratings.

Table 7.3 Octane Ratings

	Approximate Octane Rating
Octane	−20
Heptane	0
Pentane	60
Regular gasoline	87
Premium gasoline	93
2,2,4-Trimethylpentane	100
Ethanol	105
Methanol	105
Benzene	105
Methyl *tert*-butyl ether	115
Ethyl *tert*-butyl ether	118

■ QUESTION

What's the difference between the "octane" that is given an octane rating of 100 and the *octane* of Table 7.3, with its octane rating of −20?

7.6 A Problem of the 1920s and Its Solution: Leaded Gasoline

The problem back in the 1920s was finding an economical way of increasing octane ratings to match the increased compression ratios of the powerful engines that the public wanted. One solution was to modify the molecular structure of hydrocarbons obtained from petroleum. Increasing their branching, for example, would raise the octane ratings of the gasoline. But this is an expensive process, one we'll examine later when we discuss the chemistry of petroleum refining (Sec. 7.12, Fig. 7.10).

Tetraethyllead, or more simply "lead", is a chemical that was added to gasoline to inhibit knocking.

In 1922 Thomas Midgley, working at General Motors, discovered another way to raise octane ratings that's both simple and inexpensive. He found that adding less than 0.1% of **tetraethyllead** to a gallon of gasoline, roughly one teaspoon per gallon, increases the octane number of the gasoline by 10 to 15 points. Note in the following structure that *Pb* is the chemical symbol for lead and $CH_3—CH_2—$ is the *ethyl* group. The molecule is made up of four ethyl groups covalently bonded to one lead atom:

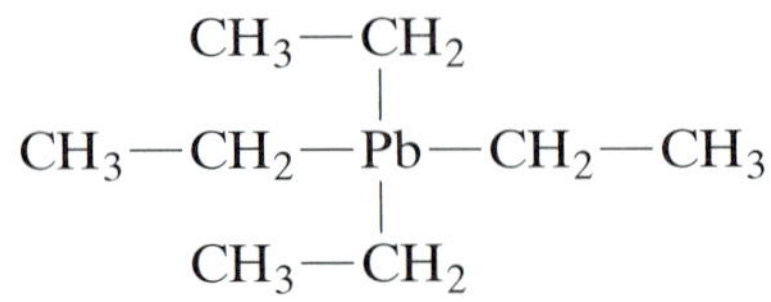

or $(CH_3—CH_2)_4Pb$

tetraethyllead
(also known as *lead tetraethyl*, or simply "lead")

The high octane ratings achieved with inexpensive leaded gasoline permitted the production of more powerful engines that would be free of knocking.

■ QUESTION

What is the molecular formula of tetraethyllead?

7.7 A Problem of the 1970s: The Case of the Poisoned Catalyst

Ideally, the hydrocarbons of gasoline burn completely to yield only carbon dioxide, water, and energy. But an internal combustion engine is not an ideal device for the combustion of hydrocarbons. Under the real operating conditions of a real engine oxidations can range through the entire spectrum of chemical reactivity. The chemical mixture discharged by the engine can contain carbon dioxide and water (from complete oxidation), hydrocarbons (from failure of any oxidation to occur), carbon monoxide (from partial oxidation; Sec. 6.14), and other gases resulting from oxidation of impurities within the gasoline.

The gasoline that comes out of a service station pump isn't always an ideal mixture of pure hydrocarbons; it often contains traces of impurities that incorporate elements other than carbon and hydrogen. Sulfur, a contaminant that comes to gasoline from poorer grades of petroleum, burns to form sulfur dioxide, SO_2; nitrogen, another contaminant, undergoes oxidation to form various nitrogen oxides. Both sulfur dioxide and the oxides of nitrogen are major atmospheric pollutants, each contributing to urban smog and to the worldwide problem of acid rain (Chapter 14). Any oxidation of sulfur- or nitrogen-containing impurities produces atmospheric pollutants.

By the middle of the 20th century increasing numbers of automobiles were spewing these gases—unburned hydrocarbons, carbon monoxide, sulfur dioxide, and various oxides of nitrogen—out of their tailpipes and into the air we breathe. In 1970, spurred by the threat of steadily increasing atmospheric pollution from industrial and commercial sources as well as from automobiles, the U.S. Congress acted. In the Clean Air Act of 1970 Congress specifically required that 1975 model cars emit no more than 10% of the carbon monoxide and hydrocarbons that came from their 1970 counterparts and that emissions of other pollutants be reduced as well.

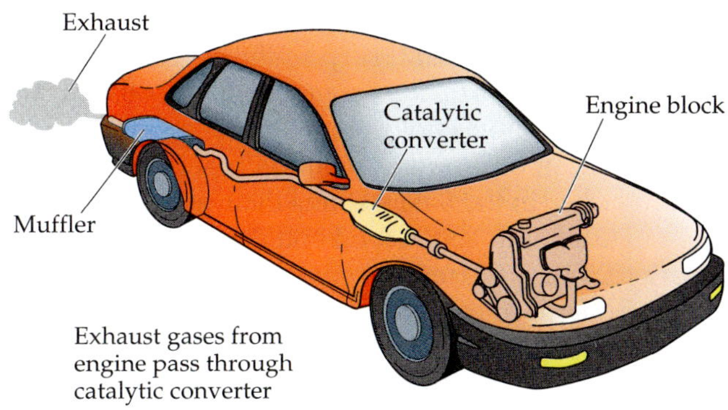

Figure 7.6 Catalytic converter.

For a chemical solution to a chemical problem, automobile manufacturers turned to the *catalytic converter*. Since 1975, virtually all automobiles built or imported into the United States have been equipped with this device. A **catalyst** is any substance that speeds up a reaction but is not itself a reactant. The catalyst simply allows the reactants to come together to produce the very same products they would have generated in its absence, but much faster or under much milder conditions. In another context, our bodily enzymes are also catalysts. They allow us to digest and metabolize food under the mild conditions of temperature, acidity, and alkalinity that our bodies tolerate easily. In the absence of enzyme catalysts, digestion and metabolism would take place far too slowly to maintain life.

A **catalyst** is a substance that speeds up a reaction but is not itself a reactant.

The automobile's catalytic converter consists of a large canister mounted between the engine's exhaust system and the tailpipe, containing finely divided platinum, palladium, and sometimes other substances as well (Fig. 7.6). As the exhaust gases from the engine pass through the converter, the finely divided particles of platinum and palladium catalyze the complete oxidation of unburned hydrocarbons and carbon monoxide into water and carbon dioxide, thereby decreasing the amounts of these atmospheric pollutants leaving the exhaust pipe.

The catalysts themselves are sensitive to a form of deterioration called *catalytic poisoning*, in which another chemical coats their surfaces and renders them ineffective. Lead, in particular, can poison a platinum or a palladium catalyst. It's been estimated that using as little as two tanks of leaded gasoline can completely destroy the activity of a catalytic converter. For this reason Congress prohibited the use of leaded gasoline in cars equipped with catalytic converters (unless, in an emergency, no unleaded gasoline is available).

As older cars not equipped with the converters left active use and as the demand for leaded gasoline decreased, high-octane unleaded gasoline steadily replaced leaded gasoline as the commercially available fuel for automobiles. Today leaded gasoline is no longer available at service stations.

■ QUESTION

What automotive air pollutants are reduced by the use of catalytic converters? What pollutants are unaffected or perhaps increased?

7.8 Beyond Lead

Even if Congress hadn't acted to protect catalytic converters, it's likely that lead would no longer be used as a gasoline additive. Lead and its compounds are poisonous not only to catalysts but to humans as well. Lead poisoning, which affects

Removing old paint containing lead from the interior of a building.

the nervous system, the organs, the reproductive system, and fetuses, was once so common a hazard that it's known by the medical term *plumbism*, from the Latin word for the metal, *plumbum*. Years ago colorful lead compounds were used widely in consumer products, as pigments in household paint, and in ceramic glazes. Poisoning of infants and young children occurred as paint on baseboards and indoor walls flaked off and children picked at the flakes, eating them as attractive candies. Lead pigments in glazes can leach into acidic food and drink that are stored in ceramic utensils for long periods. Today the federal government rigorously limits the amount of lead that may be used in commercial paints or that may leach from a glaze. Yet, paint in very old buildings and glazes on ceramic souvenirs brought back from other countries may still contain hazardous levels of lead. Because of this threat, several commercial home-testing kits have become available to consumers. Many of those familiar with lead and its effects believe that lead entering the atmosphere from cars using leaded gasoline has been equally hazardous to living things. The removal of lead from gasoline protects not only automobile catalysts and the environment but consumers as well.

An **octane enhancer** is a gasoline additive used to increase the octane rating.

With tetraethyllead no longer acceptable as a gasoline additive, other approaches must be used to produce high-octane gasoline. Among these are (1) the use of octane-boosting additives other than tetraethyllead, which are known as **octane enhancers**, and (2) the more extensive and more intricate processing of petroleum as it is converted to gasoline. One of the first and most widely used of the lead-free octane enhancers is methyl *tert*-butyl ether, also known as MTBE:

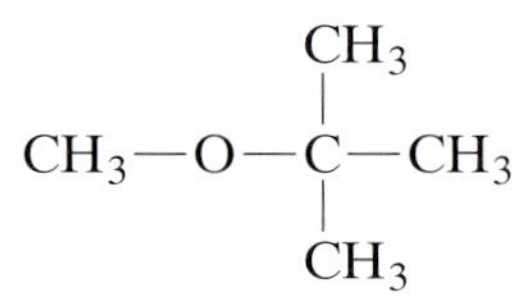

methyl *tert*-butyl ether

(The *tert*-butyl group forms when the 3°—*tertiary*—hydrogen of isobutane is removed.)

■ QUESTION

Even if the internal combustion engine were so efficient in burning gasoline that no catalytic converters were needed to protect the environment, it's likely that "leaded" gasoline would no longer be available. Why?

7.9 The Triple Life of MTBE: Octane Enhancer, Oxygenate, Water Pollutant

An **ether** is a compound in which an oxygen atom is bonded to the carbons of two organic groups, which may be identical or different in structure.

Methyl *tert*-butyl ether (MTBE) illustrates several different aspects of the extraordinary chemistry of ordinary things. First, it provides us with an example of a class of organic compounds known as ethers. An **ether** is a compound in which an oxygen atom is bonded to the carbons of two organic groups, of the same or different structures. In MTBE the two groups are different: one is a methyl group, the other a *tert*-butyl group. In diethyl ether the two ethyl groups bonded to the oxygen have the same structure (Sec. 6.7). Like alkanes, the lower-molecular-weight ethers are flammable, but they are otherwise relatively unreactive. MTBE itself has been used as an octane enhancer in gasoline since the late 1970s, but aside from MTBE and a few others, lower-molecular-weight ethers have few uses in consumer products.

In addition to acting as an octane enhancer, MTBE also serves as an **oxygenate**, a gasoline additive that contains oxygen within its own molecules and that improves the efficiency of hydrocarbon combustion. As we've already seen for charcoal (Sec. 6.14) and for the hydrocarbons of gasoline (Secs. 7.2 and 7.7), combustion in the presence of insufficient oxygen leads to the formation of carbon monoxide rather than carbon dioxide. Inefficient combustion in an engine can contribute carbon monoxide and unburned hydrocarbons to the exhaust gases. To decrease emissions of these pollutants, the U.S. Congress amended the Clean Air Act (Sec. 7.7) in 1990 to require the addition of oxygenates to the gasoline sold in certain metropolitan areas that fall below federal air quality standards. In effect, the addition of oxygenates causes this *reformulated gasoline* (*RFG*) to carry some of the needed oxygen along with its own hydrocarbons. The law specifies that the RFG must contain a minimum of 2% oxygen by weight. For MTBE, this translates into 11% by volume. MTBE is the most widely used oxygenate in RFG; in 1999 it was present in 87% of all RFG produced.

An **oxygenate** is a gasoline additive that contains oxygen and that improves the efficiency of hydrocarbon combustion.

An MTBE warning label on a gasoline pump in California: "The state of California has determined that the use of this chemical presents a significant risk to the environment."

Although MTBE effectively serves a double function as octane enhancer (raising the octane rating of a gasoline) and oxygenate (reducing exhaust gas pollutants), the ether appears to pose a threat of contamination to drinking water. The tendency for underground gasoline storage tanks to leak occasionally, combined with MTBE's solubility in water, its ability to move rapidly through porous earth, and its resistance to natural forces of degradation, has led to the discovery of MTBE in small amounts in ground water and drinking water at various locations.

Because of its potential as a drinking water contaminant, the use of MTBE as a gasoline additive is ending or being phased out in several regions of the United States. One of the leading candidates for its replacement is ethanol, $CH_3{-}CH_2{-}OH$. Although ethyl alcohol appears to be environmentally benign, its relatively high cost, short supply, and technical difficulties in distributing ethanol-enriched gasoline work against its use. (About 6–8% of the U.S. corn crop goes to the production of ethanol, through fermentation. Widespread adoption of ethanol as a gasoline additive could increase the demand for corn considerably.)

■ QUESTION

Which is more effective as an octane enhancer, MTBE or ethanol? (You may wish to refer to Table 7.3.)

7.10 Petroleum Refining: Distillation Revisited

We can now look back at our opening demonstration for a clearer understanding of how the hydrocarbons of raw petroleum are converted into useful consumer products such as gasoline, kerosene, and other fuels. Distillation, as we noted, allows us to separate liquids of different boiling points and to separate liquids from dissolved solids. For their simplest distillations, chemists use an apparatus only a little more intricate than the pot of boiling water of our demonstration (Fig. 7.7). As the liquid contained in a flask boils, its vapors rise above the bulb of a thermometer, passing into a condenser (usually cooled by a stream of running water). There the hot vapors cool and condense back to a liquid, which drains into a collection flask. The thermometer registers the boiling point of the condensed liquid.

Distillations are ideal procedures for separating the various fractions of hydrocarbons that make up crude oil. At a typical petroleum refinery the crude oil is heated in a vessel connected to the base of a tall *fractionating tower* (Fig. 7.8).

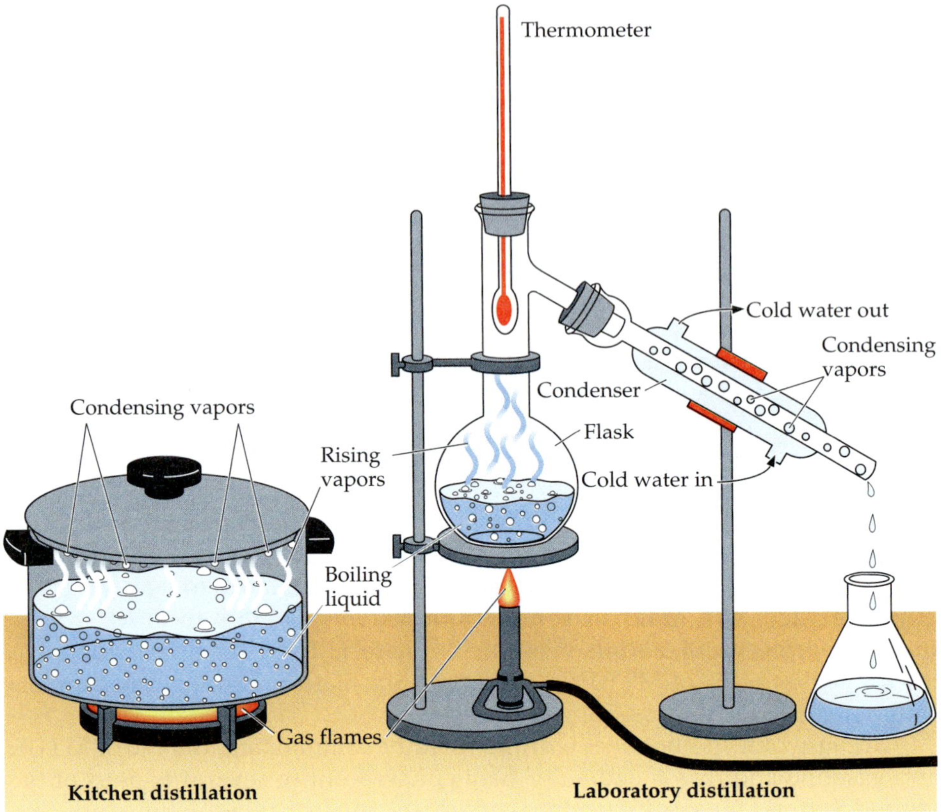

Figure 7.7 Distillations.

A fractionating tower. Distillation of crude oil in towers like this separates the hydrocarbons of the petroleum into fractions useful as gasoline, kerosene, heating oil, jet fuel, and so on.

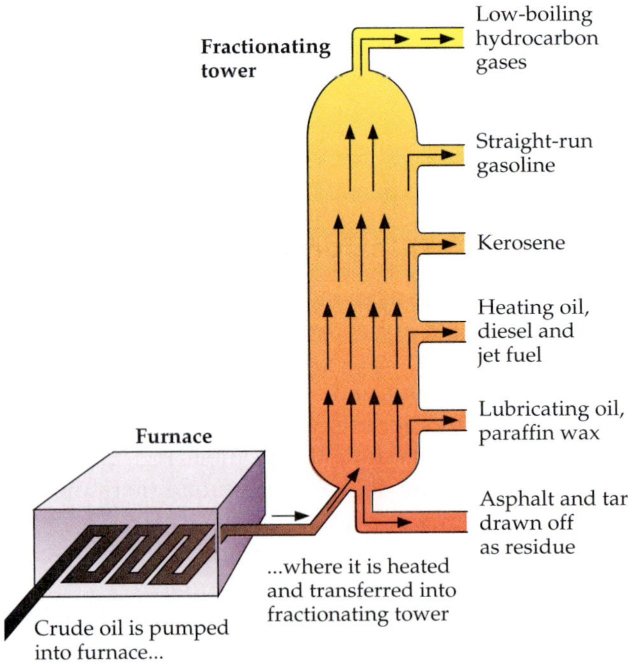

Figure 7.8 Schematic diagram of fractional distillation of crude oil.

Tank truck delivering gasoline to a gas station.

The hydrocarbon vapors pass up, into the fractionating tower and separate into groups of compounds according to their boiling points. The vapors of the most volatile hydrocarbons—those with the lowest boiling points—rise to the top of the tower, where they condense to liquids and are drawn off. Hydrocarbon mixtures with higher boiling points, the less volatile fractions, condense and are drawn off from the lower regions of the tower.

In this way the crude petroleum is separated by *fractional distillation* into the mixtures of hydrocarbons represented in Table 7.4. Since the boiling point of a hydrocarbon is related to its molecular weight and, therefore, to the number of carbon atoms in its molecules, fractional distillation produces a separation of the hydrocarbon constituents of petroleum according to their carbon content. Gasoline produced directly from the hydrocarbons of crude oil, separated from other fractions by distillation, is known as *natural gasoline* or *straight-run gasoline*. Distilling at higher temperatures than straight-run gasoline are the hydrocarbons that make up such products as kerosene, heating oil, and jet fuel (Table 7.4).

Table 7.4 Hydrocarbon Fractions Obtained on the Distillation of Petroleum

Approximate Boiling Range (°C)	Carbon Atoms per Molecule	Fraction
Below 200	4–12	Straight-run gasoline
150–275	10–14	Kerosene
175–350	12–20	Heating oil, diesel, and jet fuel
350–550	20–36	Lubricating oil, paraffin wax
Residue	Over 36	Asphalt, tar

■ QUESTION

What facet of molecular structure other than the molecular weight or carbon content of its molecules affects the boiling point of an alkane?

7.11 Petroleum Refining: Catalytic Cracking

Distillation alone isn't sufficient to transform petroleum into gasoline for today's engines. One reason is that there just aren't enough of the right kind of hydrocarbons in petroleum to provide the modern world with all the gasoline it demands. Although the composition of crude oil varies considerably, depending on the region of the world it comes from, generally not much more than about 20% of any barrel of crude oil consists of hydrocarbons that can be blended into a gasoline mixture as they come from the fractionating tower.

In the early part of this century, with relatively few cars on the road, kerosene, which was used for heating, was the most valuable product of petroleum refining. The 20% or so of straight-run gasoline obtained from distillation easily satisfied the relatively small demand for fuel for the internal combustion engine. As the automobile displaced the horse and buggy, the demand for gasoline grew until natural limitations in the supply of straight-run gasoline threatened to strangle the relatively young automobile industry. Just before the outbreak of World War I a solution to the problem appeared in the form of a chemical process known as *cracking*. As the term implies, large covalent molecules are cracked, or broken apart into smaller ones. The process yields several smaller molecules from one larger parent.

Cracking a C_{12} hydrocarbon molecule of the kerosene fraction, for example, could yield two C_6 hydrocarbon molecules (Fig. 7.9). A C_6 hydrocarbon is more useful than a C_{12} hydrocarbon in blending a gasoline mixture because of its greater volatility. Cracking a C_{30} hydrocarbon could increase the amount of heating oil, or kerosene, or gasoline from a barrel of petroleum. In any case, cracking can increase the amount of low-molecular-weight hydrocarbons obtained through petroleum refining, but only at the expense of the higher-molecular-weight hydrocarbons. By using a **catalytic cracking** process, the yield of gasoline from a barrel of petroleum

Catalytic cracking is a process by which large hydrocarbon molecules are converted into two or more smaller hydrocarbon molecules.

Figure 7.9 Catalytic cracking.

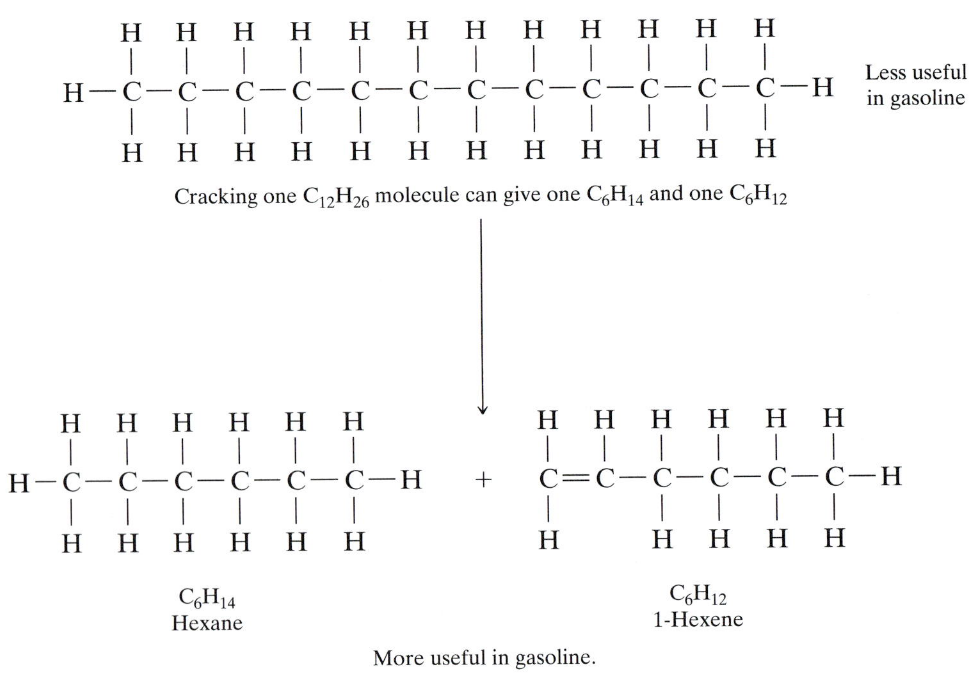

can be increased from 20% to about 50%, *but at the expense of the higher-boiling hydrocarbons, including kerosene and home heating oil.*

As long as winters are warm and supplies of crude oil are ample, there's plenty of kerosene, heating oil, and gasoline to satisfy everyone. But cold winters and petroleum shortages can drain our commercial reserves of heating oil, forcing petroleum refiners to perform a careful balancing act to maintain ample supplies of both gasoline and heating oil through catalytic cracking and other processes. Coupled with severe restrictions in our imports of petroleum, as a result of political actions, wars, or economic conditions, cold winters can force on our society the uneasy choice between producing enough gasoline to satisfy everyone (but at the expense of cold homes), and producing enough kerosene and heating oil to generate winter warmth for all (but at the expense of gasoline shortages).

The technology we would use in putting the choice into effect requires only a knowledge of the chemistry of petroleum refining and the existence of the refineries, catalysts, and equipment to do the job. But the choice itself—deciding how we use this technology in a moment of crisis, for mobility or for warmth—is societal and political in character. It requires wisdom in addition to knowledge.

■ QUESTION

Suppose that one product of cracking a C_9 alkane is

$$CH_3—CH_2—CH_2—CH_2—CH{=}CH_2$$

What is the other hydrocarbon product?

7.12 Petroleum Refining: Reforming

Petroleum refineries not only separate the various petroleum fractions through distillation and convert large hydrocarbon molecules into several smaller ones through catalytic cracking, they also reorganize the molecular shapes of molecules through **catalytic reforming**. In this refining process the carbon skeletons of unbranched or only slightly branched hydrocarbons are reorganized into much more highly branched molecules and into cyclic hydrocarbons (Sec. 6.10), usually with an increase in octane numbers.

When the carbon skeleton of an alkane is rearranged to form a more highly branched isomer, the reforming process is **isomerization** (Fig. 7.10). In another reforming step, **cyclization**, chains of five or six carbons are converted to cycloalkanes with an accompanying loss of hydrogen (Fig. 7.11). Loss of several hydrogens from a cyclohexane ring can produce an aromatic hydrocarbon through **aromatization** (Fig. 7.12). Aromatization of petroleum hydrocarbons through catalytic reforming serves as a major commercial source of aromatic hydrocarbons, which are used as industrial solvents and in the manufacture of pharmaceuticals, plastics, synthetic rubber, and other consumer products.

In **catalytic reforming** hydrocarbon molecules are reorganized into more useful structures of the same carbon content.

In **isomerization** the reforming results in the formation of a more highly branched isomer.

Cyclization converts noncyclic structures into cyclic molecules.

Aromatization converts cyclohexane rings into aromatic rings.

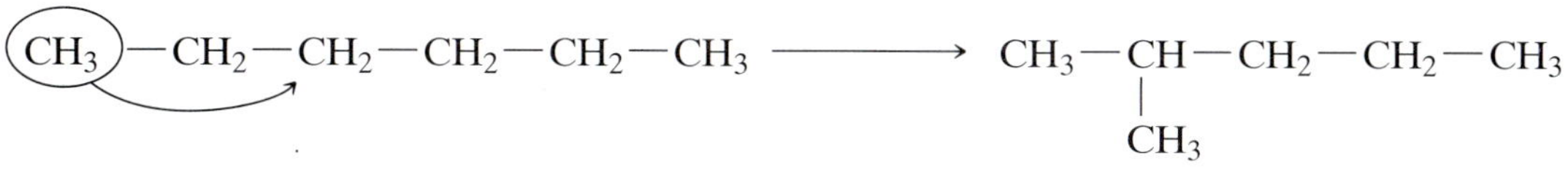

Lower octane rating　　　Higher octane rating

Figure 7.10 Isomerization of hexane to isohexane (2-methylpentane).

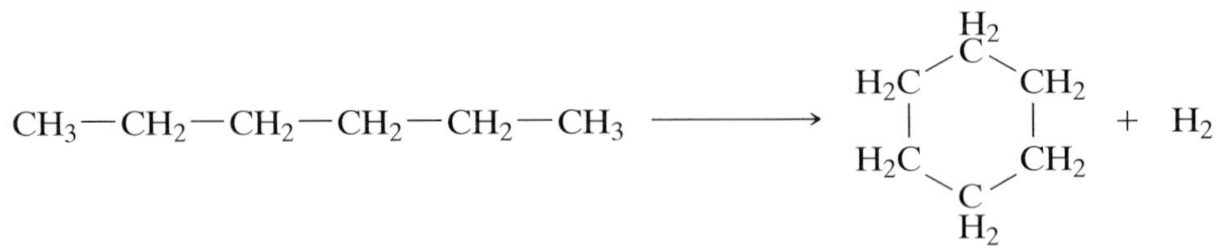

Figure 7.11 Cyclization of hexane to cyclohexane.

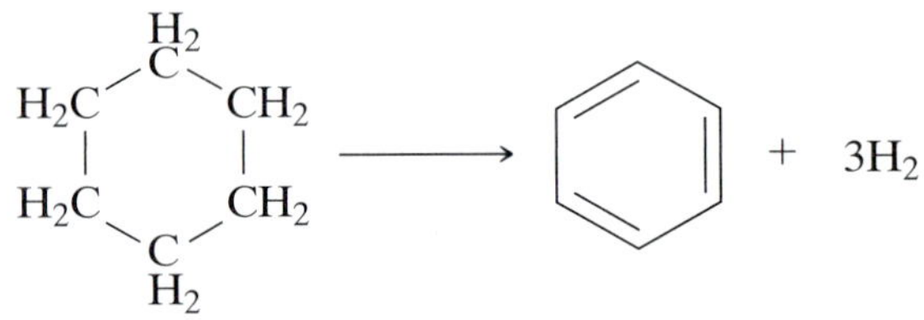

Figure 7.12 Aromatization of cyclohexane to benzene.

■ QUESTION

Name two products produced by the isomerization of pentane. What is produced through the aromatization of cyclohexane?

PERSPECTIVE

Energy for the Cars of Tomorrow

In this chapter we have examined hydrocarbons as a source of energy, with particular emphasis on the hydrocarbons of gasoline and their use in the internal combustion engine. It's an appropriate emphasis since more than 40% of all the petroleum used in the United States now ends up as gasoline for our cars. Overall, two-thirds of all the petroleum used goes, in one form or another, into transportation of all kinds.

Driving is a dirty business. We add more pollutants to our environment by running our internal combustion engines than we do with any other single activity. Auto exhaust, even exhaust presumed to be clean, contributes about a quarter of all the carbon dioxide discharged into the air above the United States, as well as almost all of the carbon monoxide in and above our cities. (Although carbon dioxide isn't a health hazard, we'll examine its role as an atmospheric pollutant in Sec. 8.14.) Moreover benzene, one of the hydrocarbons of gasoline, is a known *carcinogen*, a cancer-causing agent. Next time you fill your tank, look for a note on the gas dispenser telling you that "Long-term exposure to vapors has caused cancer in some laboratory animals."

It seems clear that improving the health of our environment (and ourselves) requires, among other things, limiting or eliminating tailpipe emissions. This can take several forms. Automobile manufacturers can improve the fuel efficiency of new cars so that they consume less gasoline and produce less exhaust gas. Gasoline manufacturers can formulate cleaner burning gasolines with improved methods of refining and improved gasoline formulations. Still another route would be to abandon hydrocarbon-rich gasoline and turn instead to other sources of energy, the *alternative fuels*.

Liquified Petroleum Gas (LPG) Liquified petroleum gas (LPG), often referred to simply (and erroneously) as propane, is a liquified mixture of hydrocarbons that normally exist as gases. Its composition varies, but it's usually largely propane with the isomeric butanes and smaller amounts of methane, ethane, and other low-molecular-weight hydrocarbons also present.

Liquified petroleum gas has been used extensively as a fuel, in general applications for more than 90 years and to propel vehicles for more than 60 years. Today it's the most widely used alternative fuel. Only gasoline and diesel fuel surpass LPG as a fuel for transportation. It's a clean-burning fuel, producing little carbon monoxide and fewer oxides of nitrogen than gasoline. But it is still a fossil fuel, with all the disadvantages that this implies (Chapter 8). Leaks of LPG result in a dense, flammable, low-lying, in-

Installing a tank of natural gas.

visible cloud of gas that ignites easily. In some localities cars powered by LPG are prohibited from entering tunnels or parking garages.

Compressed Natural Gas Compressed natural gas is another fossil fuel that's used as an alternative fuel. Composed principally of methane, with small amounts of other alkanes, natural gas is used widely for cooking and heating. Plentiful, clean burning, and often at a cost not far from gasoline's, it ranks with LPG as a major competitor to gasoline. For use in vehicles, natural gas must be compressed and stored in heavy, sturdy tanks. Its disadvantages are the short driving range it provides—about 100 miles—the heavy and awkward fuel tanks it requires, and the complexity of refueling. It is more suitable to fleet vehicles than to private automobiles.

Ethanol Ethanol (ethyl alcohol) belongs to a class of fuels known as *renewable* fuels. Unlike fuels obtained from petroleum, natural gas, or coal, ethanol comes principally from the fermentation of grain and other crops, including potatoes and corn, and is thus a renewable resource. (*Grain alcohol* is another name for ethanol.) The fossil fuels, on the other hand, are nonrenewable; none are being formed today to regenerate what we draw from the earth's current reserves. Add to this ethanol's octane rating of 105, its low impact on the environment, and its relatively low toxicity, and ethanol becomes an attractive replacement for hydrocarbon fuels.

Among its disadvantages are its relatively high cost (compared to gasoline), low energy content (which would require large fuel tanks), and scarcity. Converting all the grain grown in the United States each year into ethanol would supply only about a quarter of the fuel needed. One practical solution to the problem of scarcity is the use of mixtures of ethanol and gasoline, such as 85% ethanol–15% gasoline, a blend known commercially as E85. Major automobile manufacturers currently produce a limited number of cars designed to run on either gasoline or gasoline–ethanol blends.

Methanol Methanol (methyl alcohol) can be produced from a variety of sources, including wood, the fossil fuels coal and natural gas, and even garbage. (Originally generated by heating wood to a high temperature, methanol is sometimes known as *wood alcohol*.) Methanol's octane rating is high, close to ethanol's, and it generally burns more cleanly than gasoline. A mixture of 85% methanol–15% gasoline is available commercially as M85.

But methanol also comes with a few disadvantages. It corrodes common varieties of steel, has a relatively low energy content, and is toxic. As a result, fuel tanks would have to be large, made of expensive, corrosion-resistant stainless steel, and well sealed. Moreover, one of the few pollutants methanol does produce is especially nasty. Incomplete combustion of methanol generates formaldehyde, which is a carcinogen, an irritant, and an air pollutant. Exhaust systems would have to be redesigned to keep it out of the environment.

A bus powered by methanol.

An electric car.

Electricity No matter how efficient the engine, a car that get its power from the combustion of a hydrocarbon or an alcohol or some other kind of organic compound must produce exhaust gas emissions of some sort, even if they're no more than water and carbon dioxide. Cars that meet the ideal of no emissions whatever, that produce no exhaust gases, that add nothing at all to the atmosphere, are known as *zero-emission vehicles*, or ZEVs. One source of power that meets this test is electricity, released by storage batteries mounted in the car. Electric cars are classified as ZEVs. Because of concerns about air pollution, the state of California requires that by 2003 at least 10% of all cars sold within the state must be ZEVs.

Despite the absence of emissions, electric cars come with problems of their own. The major disadvantages of electric cars are their short driving ranges, the long times required for recharging their batteries, and the uncertainty of where all the electric power needed to recharge large numbers of batteries will come from. Part of the solution to the last problem comes from the nature of energy itself. Moving vehicles powered by other sources of energy are slowed and stopped by the force of the brakes. These convert the energy of the motion to heat as they press against the rotating wheels. Electric cars use a process known as *regenerative braking*, in which an electric generator slows the moving car by converting its energy of motion into electricity rather than heat. This regenerated electricity is fed to the battery to help maintain its charge. We'll have more to say about the various forms of energy in the next chapter.

Despite energy economies produced by regenerative braking and similar techniques, the wholesale conversion of our transportation system from fossil fuels (or other fuels that rely on combustion) to electric power could produce the most serious problem of all. Electricity itself is a secondary source of energy in the sense that we generate it from some other, primary source such as fossil fuels, nuclear reactions, falling water, sunlight, the wind, or chemical reactions. Switching from burning hydrocarbons to electric batteries won't, in itself, decrease the total amount of energy we consume; it will simply shift the burden of producing that energy from the internal combustion engine to some other power source. Unless we use wisdom in addition to technical knowledge, we might find that in replacing gasoline with electric batteries we have simply moved our problems of energy production from one location to another.

Fuel Cells The fuel cell can be described as a "gaseous battery," a term applied to it by its English inventor, William Grove, in 1839. It's designed not to accept, store, and release electricity, as a common battery does, but rather to induce chemical reactions to release energy in the form of electricity rather than heat. In a common form of the cell, hydrogen combines with oxygen on the surface of the cell's catalysts (Sec. 7.7) to form water and produce electricity (Fig. 7.13):

$$2H_2 + O_2 \longrightarrow 2H_2O + \text{energy} \quad \text{(electricity)}$$

The oxygen for the cell is obtained from the atmosphere; the hydrogen can either be stored on the vehicle or produced catalytically from a hydrocarbon or other hydrogen-con-

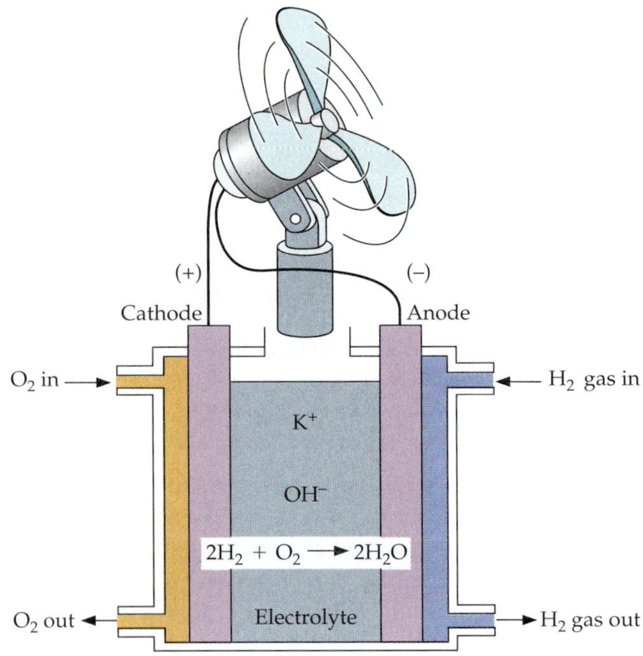

Figure 7.13 The essentials of a typical fuel cell.

taining fuel carried on the vehicle. Although not truly a zero-emission device—it produces water as a product—the cell is essentially pollution free. As with other forms of alternative power, fuel cells come with their own disadvantages. They are currently large, heavy, and much more expensive than other sources of power.

Combinations, Cost, Convenience, and Cleanliness Still another possibility is the *hybrid electric vehicle* (*HEV*), which carries both gasoline and electric engines. In a *parallel HEV* both engines are connected to the wheels; driving conditions determine which one is running. In a *series HEV* the gasoline engine serves principally to recharge the batteries. In the end a balance of cost, convenience, and cleanliness will probably determine our choice of fuel for tomorrow's car. "Cost" will include the cost of the fuel and the engine, and the expense of maintaining a clean environment. "Convenience" will reflect both the car's driving range and the simplicity of charging or exchanging a battery or filling a tank with fuel. "Cleanliness" will be a matter of the kind of environment in which we choose to live. In any case, we will have choices before us, both as individuals and as a society. Whatever fuel or combination of fuels we do choose, it will be chemistry, in one form or another, that delivers the energy. Making the choices intelligently will require an understanding of the chemistry that lies behind them. ■

EXERCISES

For Review

1. Following are a statement containing a number of blanks, and a list of words and phrases. The number of words equals the number of blanks within the statement, and all but two of the words fit correctly into these blanks. Fill in the blanks of the statement with those words that do fit, then complete the statement by filling in the two remaining blanks with correct words (not in the list) in place of the two words that don't fit.

__________ is a mixture of __________ that provide power to a car as they react with oxygen within the cylinders of the __________ to produce __________ and __________ and to release energy. In a four-stroke engine, the __________ ignites the gasoline–air mixture at the end of the __________, sending the piston downward. The __________, which follows, translates the energy of the gasoline into the power that moves the vehicle.

As the __________ (the ratio of the maximum volume of the gasoline–air ratio at the beginning of the compression stroke to that of the fully compressed mixture at the end of the compression stroke) of an engines increases, the tendency of a gasoline to __________ also increases. The ability of a particular gasoline blend to resist this tendency is reflected by its __________, which is measured by comparing the gasoline blend to a mixture of __________________ and ______________, which is also known as "octane" and "isooctane."

Gasoline itself is derived from a naturally occurring mixture of hydrocarbons known as ____________________. Although a certain amount of gasoline hydrocarbons (known as natural or straight-run gasoline) can be obtained directly from this mixture by ____________________, additional gasoline can be produced by __________________, a process in which the molecules of higher boiling fractions, such as __________, are broken down into smaller molecules that have properties more suitable to gasoline.

benzene	kerosene
catalytic cracking	knock
compression ratio	octane rating
compression stroke	methyl *tert*-butyl ether
distillation	petroleum
gasoline	power stroke
hydrocarbons	spark plug
internal combustion engine	2,2,4-trimethylpentane water

2. Define or identify each of the following:

a. catalyst	b. catalytic converter
c. crude petroleum	d. fossil fuel
e. intake stroke	f. isomerization
g. oxygenate	h. petroleum refining
i. reforming	j. tetraethyllead
k. ZEV	

3. Identify the four strokes of the internal combustion engine and describe what happens during each.

4. What are the *three* products of the *complete* oxidation of a hydrocarbon? What lethal substance is produced on *incomplete* oxidation of a hydrocarbon?

5. What two characteristics of a hydrocarbon's molecular structure affect its volatility?

6. What is a major advantage of increasing the compression ratio of an engine? What characteristic of a gasoline must be changed as the compression ratio increases?

7. What purpose does the catalytic converter of an automobile serve?

8. Into which chemical is the carbon monoxide (CO) of exhaust gases converted as the gases pass through the catalytic converter?

9. (a) Why was tetraethyllead added to gasoline? (b) What was the original reason for discontinuing the addition of tetraethyllead? (c) What additional benefit comes from removing this additive? (d) Name another chemical additive that can serve the same function as tetraethyllead. (e) What refining process produces the same characteristic in gasoline as that obtained by adding tetraethyllead?

10. What is the function of distillation in the refining of petroleum?

11. In what way do compounds that occur naturally in petroleum but that are *not* hydrocarbons contribute to environmental pollution?

12. Petroleum, which does not contain large quantities of aromatic hydrocarbons, is a major source of aromatic hydrocarbons. Explain why and describe the chemistry involved.

13. Does the catalytic cracking of petroleum produce straight-run gasoline? Explain.

14. What *two* refining steps are needed to convert hexane into benzene?

15. What is the function of: (a) an oxygenate, (b) an octane enhancer?

16. Identify an oxygenate that is not an ether. What functional group does it possess?

17. What is one disadvantage to the use of ethanol as an oxygenate?

A Little Arithmetic and Other Quantitative Puzzles

18. On average, how many cylinders are in the power stroke at any given moment in a four-cylinder engine? In an eight-cylinder engine? In a six-cylinder engine?

19. Using the data presented in Section 7.1, calculate the number of miles per gallon provided by the average motor vehicle in the United States. Identify at least one source of error in this calculation.

20. Draw the condensed structure of the isomer of C_8H_{18} that has the *greatest number* of methyl groups. Write the IUPAC name of this isomer. What prediction can you make about the probable numerical value of its octane number?

21. Repeat Exercise 20 for the isomer of C_8H_{18} that has the *smallest number* of methyl groups.

22. What correlation do you think may exist between the ratio of primary and secondary carbons on a pair of isomeric alkanes and the relative magnitude of their octane numbers? Explain.

23. Isopentane (2-methylbutane) is a major component of gasoline. Give its molecular structure and write a balanced equation for its combustion (see Sec. 6.8).

24. (*Note:* This exercise deals with data presented in Section 7.1.) Assuming that the number of vehicles on the road and the number of miles we drive remain constant, how many gallons of gasoline would we save each year for each mile-per-gallon increase in the energy efficiency of our cars?

Think, Speculate, Reflect, and Ponder

25. The distillation of water from the tea in the pot of the opening demonstration allows you to collect water that is quite pure, except for possible contamination by the more volatile components of the tea. What is left behind in the pot as you distill the water from the pot and collect it in the glass?

26. Arrange the following hydrocarbons in order of increasing boiling point, with the lowest boiling hydrocarbon first and the highest boiling last: (a) nonane, (b) ethane, (c) butane, (d) heptane, (e) methane.

27. Arrange the following hydrocarbons in order of increasing boiling point, with the lowest boiling hydrocarbon first and the highest boiling last: (a) 2-methylbutane, (b) butane, (c) 2,2-dimethylpropane, (d) 2-methylpropane, (e) pentane.

28. Arrange the following hydrocarbons in order of increasing boiling point, with the lowest boiling hydrocarbon first and the highest boiling last: (a) octane, (b) 2-methylbutane, (c) hexane, (d) 2,2-dimethylpropane, (e) heptane.

29. Arrange the following hydrocarbons in order of increasing boiling point, with the lowest boiling hydrocarbon first and the highest boiling last: (a) 2,2-dimethylpentane, (b) 2-methylhexane, (c) 2,2,3-trimethylbutane, (d) heptane.

30. Based on what you know of the effect of the molecular structure of alkanes on their boiling points, why is it difficult to predict whether 2,2,3-trimethylheptane or octane will have the lower boiling point?

31. Aromatization of a cyclic hydrocarbon, A, molecular formula C_7H_{14}, produces *toluene*, C_7H_8. Write the name of hydrocarbon A and give its molecular structure.

32. In Section 6.10 we saw that adding water to an alkene produces an alcohol. Methyl *tert*-butyl ether is manufactured by adding a certain organic compound to 2-methyl-1-propene:

$$\begin{array}{c} CH_2{=}\underset{\displaystyle CH_3}{\underset{|}{C}}{-}CH_3 \end{array}$$

What compound is added to the double bond of this alkene to produce methyl *tert*-butyl ether?

33. You have a mixture of two different liquid compounds and you attempt to separate them from each other by distillation. You find, however, that no matter how carefully you distill them and no matter how complex and intricate a distillation apparatus you use, you cannot separate the two components. What can you conclude about these two liquids?

34. Which of the following components of auto exhaust emissions would you expect to be *decreased* by the use of a catalytic converter: H_2O, isopentane, CO_2, CO?

35. Which of the components of exhaust emissions described in Exercise 34 would you expect to be *increased* by the use of catalytic converters?

36. Would you expect the chemical composition—the specific hydrocarbons used to blend the gasoline and their proportions in the mixture—of a specific brand of gasoline to be the same in Miami in July as it is for the same brand sold in Chicago in January? Explain.

37. Suppose you were given samples of two different liquids. Describe how you could determine which one is the more volatile.

38. Engines are more likely to stall as the result of vapor lock after the car has been traveling for a while than when it is just started. Why?

39. You are in charge of designing a new form of the internal combustion engine. One of your workers has suggested moving the valves and the spark plug from the top of the cylinder to the side of the cylinder, midway between the top and the bottom of the piston's travel. Are there any advantages to this idea? Are there any disadvantages? Explain.

40. Describe one way our society can produce a significant decrease in our consumption of petroleum. Describe your reasoning.

41. In the event of a severe and prolonged petroleum shortage, we may have to choose between producing enough heating oil to keep us warm during a particularly cold winter and producing enough gasoline to keep our cars running. Explain why we might have to make this choice and describe the chemistry involved.

42. What would be a major advantage of using the combustion of hydrogen (H_2) as a source of energy in a car's engine? What might be a major disadvantage?

43. Which of the compounds in the question at the end of Section 7.3 would you expect to have the highest octane number? The lowest? Describe your reasoning.

44. Can a compound have an octane rating higher than 100? Lower than zero? If your answer to either of these is *no*, explain why not. If your answer to either of these is *yes*, explain their significance.

45. (a) If the combination of a new engine design and the development of the ideal hydrocarbon fuel produced *only* CO_2 and H_2O as combustion products, would you agree to the removal of catalytic converters from automobiles or to the sale of new cars without them? Explain. (b) If catalytic converters were removed, would you agree to the return of leaded gasoline to raise octane ratings?

46. Suppose that you now own a car and could replace it with a new one that is equivalent or better in all respects, is far less expensive to own and operate, and doesn't pollute the atmosphere in any way. Suppose, though, that this new car has to be refueled (or recharged if it is an electric car) every 10 miles you drive. Would you replace your current car with this one? Would you replace it if the new car required refueling or recharging every 20 miles? 50 miles? 100 miles? 250 miles? 1000 miles? How does the *range* of a car affect your attitude toward it?

47. Prepare a table listing sources of energy that may replace today's gasoline for our cars of tomorrow. List several of the advantages and disadvantages of each.

48. Suppose that all motor vehicles that now run on fuels derived from petroleum were replaced by electric, battery-powered cars. How would you suggest we produce the electricity needed to recharge the battery so as to minimize pollution of all kinds? What source of energy would you use for the production of the large quantities of electricity that would be needed?

Web Questions

Use the Web to answer or accomplish each of the following. Report your Web source for each response.

1. When and where was the first oil well drilled within the continental United States? Is it still in operation?

2. What's the average, typical, or most common compression ratio of the cars that drive on our highways? What's the range of compression ratios of racing cars that participate in events like the Indy 500? What fuels do the racing cars use? What are their octane ratings?

3. Describe the concerns about air quality that led to the passage of the Clean Air Act of 1970. What amendments, if any, have been added to the act since 1970?

4. What octane enhancers have been used or proposed for gasoline, in addition to methyl *tert*-butyl ether? What are the advantages and disadvantages of each?

5. What is the location of the nearest petroleum refinery? What is its annual output of refined petroleum products?

6. Prepare a graph showing the total amount of gasoline used by cars in the United States each year since 1920.

7. Describe the current prospects for the use of solar power in automobiles.

CHAPTER 8

Working with Chemistry

FOOD, FUEL, AND ENERGY

Energy for living.

Warming Up with Work

You can duplicate the results of several of the most important experiments ever carried out in the history of science by just rubbing your hands together (Fig. 8.1). Rub them together lightly, without putting much effort into it. Not much happens. Now bear down. Press them together hard and rub vigorously. Put some energy into it and they get hot, instantly. The greater your effort, the hotter they get. Be careful! With enough work you could produce a blister.

What's producing the heat? It's not warm blood rushing to the surface. You can demonstrate that by folding up two newspapers, placing one on a sturdy surface, and rubbing the other firmly against it. The harder you rub one paper against the other, the warmer they get. You can start a fire by rubbing two sticks together next to dry leaves. The more you work at it, the sooner the fire starts. All these activities have one thing in common: Your work generates heat. The more energy you use in rubbing your hands together, the more work you do; the more work you do, the hotter they get. The same holds true for newspapers, sticks, or anything else. Energy, work, heat, and their connections to each other and to chemistry are all topics of this chapter.

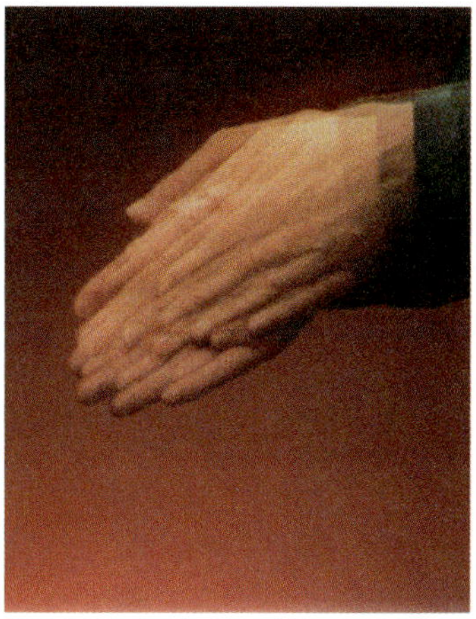

Figure 8.1 Doing work generates heat. As you rub your hands together, they become warm.

8.1 Fuel, Chemicals, Engines, and Energy

Like the mechanical, gasoline engine of earlier chapters, your body is also an engine, converting the chemical energy of fuel into work. But unlike the car's mechanical engine your body serves as a complex biological engine. It's fuel consists of the nutrients of your food. You use the chemical energy of these nutrients in doing your work.

In earlier chapters we saw how a car's engine works. Gasoline enters the cylinders of the internal combustion engine, burns, releases its energy, and pushes the pistons downward. As a result, the car moves. When we run out of gasoline the engine stops and the car stops. Chemistry comes into play when we view the gasoline as an assortment of hydrocarbon molecules that react with molecules of oxygen under the influence of a spark. All the atoms and bonding electrons of both the hydrocarbon molecules and the oxygen molecules reorganize, transforming these reactants completely (ideally) into water and carbon dioxide molecules and, most important to us, releasing energy in the process. We saw the same sort of reaction when we lit the candle for the opening demonstration of Chapter 6. This same kind of chemical process—different in many important ways, similar in others—also takes place as the chemicals of food release their energy within your body.

Different as they obviously are, your human body and a car's engine do have this in common: Both use energy stored in chemical structures to do work. We've already examined the chemical components of petroleum and gasoline. Later, in Chapters 15 to 17, we'll look more closely at the chemistry of the nutrients of our food, the chemicals that provide energy and substance to our bodies. But now, in this chapter, we'll focus on the principal benefit that petroleum, food, and all other forms of fuel provide: the energy to do our work. All engines, in short, run on fuel, and fuel delivers its energy through chemistry.

■ QUESTION

What fuel does a candle use to produce its energy? When a car runs out of fuel, the car stops. What happens when a lit candle runs out of fuel? What happens when an animal runs out of its fuel?

8.2 Potential Energy, Kinetic Energy, and Heat

We've seen molecular energy converted into heat by an internal combustion engine, by a burning candle, and by two hands rubbing against each other. We can also find energy produced through the effects of motion itself, even motion accompanied by insignificant amounts of friction. Plenty of energy is released and plenty of work can be done by the very act of motion, as in the expansion of combustion gases during a cylinder's power stroke, the forward movement of a car, the swing of a baseball bat, or by a body diving into a swimming pool.

Potential energy is energy stored in an object or a substance, often as a consequence of its location or composition. **Kinetic energy** is the energy of motion.

Energy is the capacity to do work.

There are other kinds of energy as well. The electrical energy of a battery can light a flashlight or start a car's engine moving even before the first drop of gasoline ignites. The thermal energy of a glowing wire or a flame can produce light and heat.

While energy comes in many forms, it's convenient to highlight two categories: **potential energy**, which can be thought of as stored energy, and **kinetic energy**, the energy of motion. **Energy** itself is simply the capacity to do work. When you do work, you use energy; when you use energy, you do work. The more energy you have, the more work you can do.

Potential energy.

Kinetic energy.

Work.

Potential energy can be stored in two different ways:

- through a body's location or position, and
- through a substance's composition.

If you climb up onto a diving board over a swimming pool, for example, your body acquires a certain amount of potential energy. The higher you climb—the higher your location above the water—the more potential energy you have. As you stand at the end of the board, motionless, your energy is *potential* since nothing is happening. You have the potential for doing work, but you aren't doing any as long as you stand motionless. When you jump off the board you convert your potential energy into *kinetic energy*, the energy of motion. As you land in the water the kinetic energy of your falling body does work in moving the water, creating a splash and generating noise.

Before you jump, your potential energy lies in your position, your location some distance above the surface of the water. But potential energy can come from the composition of a substance as well as from its position. The hydrocarbons of gasoline and paraffin wax, for example, contain great amounts of chemical potential energy, regardless of their location. Igniting a mixture of these hydrocarbons and oxygen releases this potential energy, which can be converted (for gasoline) into the kinetic energy of a moving car or (for paraffin wax) into the heat and light of a candle's flame. Similarly, food contains chemical potential energy that's converted into a body's heat, the actions of the muscles, the functioning of the organs, and so on—all involving chemical processes. This potential energy of gasoline, paraffin wax, and food derives from their chemical composition.

While energy presents itself in many forms—biological, chemical, electrical, thermal, and more—energy itself is usually described most simply and most conveniently in terms of its equivalent in heat (Fig. 8.2). We'll look now at the origins of our modern view of heat and we'll see how heat is related to work and to chemical energy.

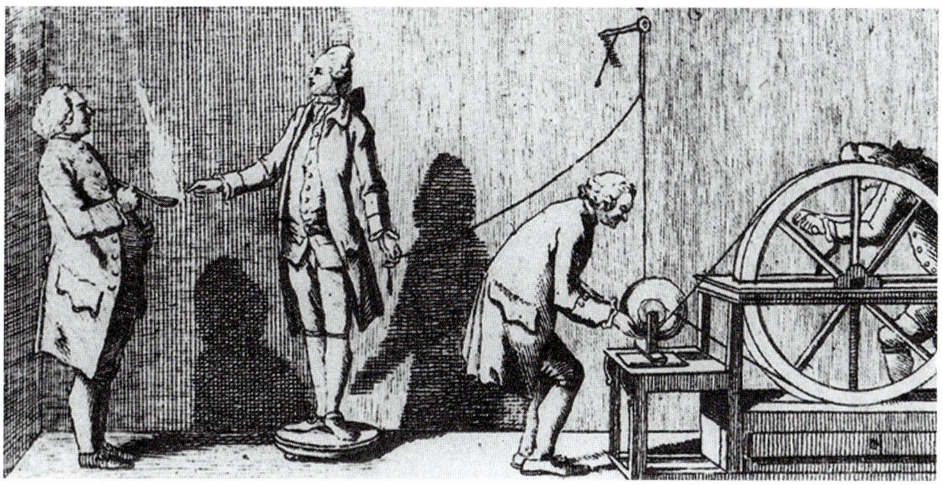

Figure 8.2 A 1774 engraving showing what we now recognize as transformations of energy from one form into another. From right to left we see transformations of (1) biological energy into the kinetic energy of a spinning wheel; (2) the wheel's kinetic energy into the potential energy of static electricity; (3) the electrical potential energy into the heat of an electrical spark. Finally, the spark's heat ignites the flammable material held in the spoon, releasing its chemical energy. Overall, after several transformations the biological energy does work through the chemical energy of the material in the spoon.

■ **QUESTION**

Imagine a child swinging on a playground swing. At what point(s) does the child have the greatest amount of kinetic energy? The greatest amount of potential energy?

8.3 How Francis Bacon Died Attempting to Discover the Benefits of Refrigeration

Francis Bacon, who wrote that heat results from motion.

In the history of science, ideas sometimes get lost for a while. Even an especially shrewd and perceptive idea, one that could lead to great advances in our understanding of the world around us, can vanish for a period of time. It's ignored and it goes underground while other, less useful and sometimes more bizarre theories blossom, attract the best minds, and then wither away as the earlier, more productive idea comes forth once again to revitalize the exploration of science. That's exactly what happened to Francis Bacon's theory of heat.

Francis Bacon, born in 1561 to a family of considerable influence and power in the court of Queen Elizabeth I of England, was a statesman, an essayist, and the father of inductive logic and modern experimental science. Back in 1620, with considerable insight, Bacon wrote about his idea that heat results quite simply from *motion*: "[The] very essence of heat," he proposed, "is motion, and nothing else." Heat, Francis Bacon concluded, is neither more nor less than a "brisk agitation" of the very particles that compose matter (Fig. 8.3).

Bacon died in the practice of those very same investigative experimental methods that he championed so vigorously. On a snowy March day in 1626, while riding in his coach somewhat north of London, he was seized by the idea that cold—simple cold—might prevent the decay of meat. He stopped his carriage and rushed out into the snow to buy a freshly butchered piece of poultry. To test his theory, Bacon gathered up some snow with his bare hands and stuffed it into the animal. He never knew the results. Instead of discovering the modern technique of preserving food by refrigeration, Francis Bacon caught cold during the experiment and soon died of pneumonia.

The modern refrigerator validates Bacon's idea that keeping food cold might help preserve it

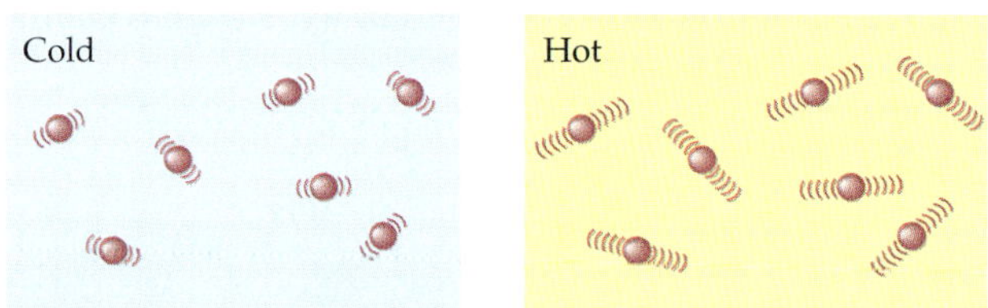

Figure 8.3 Heat results from the "brisk agitation" of the particles of matter.

■ QUESTION

Applying Bacon's theories of heat in a modern context, suppose you have a cold piece of copper and a hot piece of copper. Using the terminology of modern chemistry, what would you say is moving more "briskly" in the hot piece of copper than in the cold one?

8.4 What Happened When Count Rumford Bored Cannons

Bacon's theory—that the basis of what we call heat lies in the motion of particles of matter—was overwhelmed for over 150 years by the caloric theory of heat. (*Calorie* and *caloric* come from the Latin *calor*, meaning heat.) Until nearly the very last year of the eighteenth century most scholars believed that heat was a mysterious sort of fluid, which they called "caloric," that flowed spontaneously from a hot body to a cooler one. This "caloric" was a very strange substance. It had no weight (substances neither gain weight as they grow warm nor lose weight as they cool) but nonetheless occupied space (things tend to expand as they heat up and shrink as they cool) and permeated all things.

The scientific reign of this mysterious caloric fluid was ended partly through the work of Benjamin Thompson, Count Rumford. The story is a classic example of how a simple but elegant experiment can affect the course of science. Benjamin Thompson, born in 1753 in Woburn, Massachusetts, moved to England during the American Revolution. Later he traveled to continental Europe and, after several years of service to the Duke of Bavaria, he was named Count Rumford of the Holy Roman Empire.

While supervising the manufacture of cannons in the Bavarian city of Munich, the Count became intrigued by the amount of heat generated in the process of hollowing out their solid cores. It was well known at the time that boring out cannons produced both a tremendous amount of heat and lots of very fine metal shavings. According to the prevailing caloric theory of heat, the solid metal of the (unbored) cannons supposedly had a greater capacity for holding the mysterious caloric fluid than did the finely ground shavings. As the metal of the cannons' cores was ground out into the very small particles of the shavings, it lost its capacity to hold the caloric. This lost caloric was supposedly perceived by the workers as heat.

Count Rumford chose to test this theory with an ingenious experiment. With a well-dulled cannon borer (to produce a considerable amount of friction), he attacked a solid cannon blank immersed in 12 kg of cold water (26.5 pounds of water according to his report; the metric system was not in use at the time). Two and a half hours of steady grinding brought the water to a boil and generated 269 g of finely powdered metal (or 4145 grains as he measured it in the units of his day). Count Rumford then demonstrated that exactly the same amount of heat was required to raise the temperature, by 1°C, of both the 269 g of finely powdered metal and an equivalent 269 g of solid, unbored cannon metal. With this demonstration he showed that the metal had lost none of its capacity for holding heat as it was ground down. The great quantity of heat that raised the mass of water to boiling, then, did not come from a mysterious caloric fluid drained out of the cannon blank. All this heat came, instead, from the two and a half hours of hard work in grinding out the core of the cannon.

In 1798 Count Rumford described to the British Royal Society the results of this investigation into the nature of heat as "An Enquiry concerning the Source of Heat which is excited by Friction." Heat, the Count showed, isn't a substance at all. Heat comes from the vigorous movements of the grinding. Heat is the physical equivalent of work. Shades of Francis Bacon!

The caloric theory of heat has long since passed into the history of quaint scientific ideas, but it left us with the *calorie* as a unit of energy. A **calorie** (*cal*) is the amount of heat (or energy) needed to raise the temperature of one gram of water by one degree Celsius. One thousand calories, the amount of heat needed to raise the temperature of one *kilogram* of water by one degree Celsius, make up one **kilocalorie** (kcal) (Fig. 8.4).

A **calorie** is the amount of heat (or energy) needed to raise the temperature of one gram of water by one degree Celsius. A **kilocalorie** is 1000 calories.

The technique of measuring the amount of heat that is equivalent to a specific amount of energy or work is called *calorimetry*, and the device used in carrying out the measurement is a *calorimeter*. You can carry out a very simple bit of calorimetry with a kilogram of water, a 1-liter laboratory flask, a thermometer, and a candle. Measure the temperature of the kilogram of water in the liter flask, then heat the water for a short time and measure its temperature again. The increase in temperature, in degrees Celsius, gives you the number of kilocalories absorbed by the kilogram of water in the flask. Very roughly, heating for 1 minute with

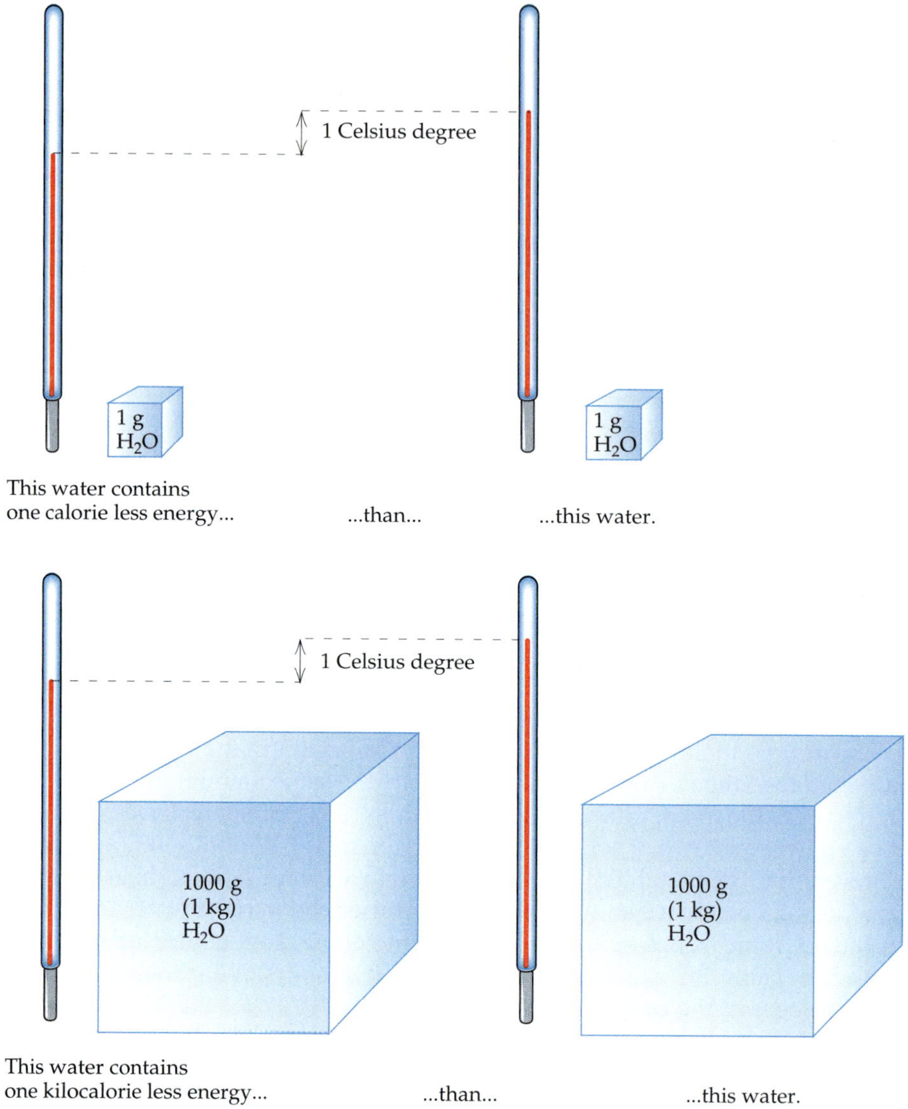

Figure 8.4 The calorie and the kilocalorie.

Figure 8.5 Preparing to use a calorimeter to measure the amount of heat released during a chemical reaction.

a common household candle adds about 1 kilocalorie of energy to the water in the flask. The exact value depends, of course, on the specific conditions and techniques you use. A more sophisticated version of a calorimeter appears in Figure 8.5.

E X A M P L E

Warm Water

Suppose you heat 0.5 kg of water for a short time and find that the temperature of the water has risen by 3°C. How many calories did you add to the water?

■ We can approach problems of this kind either intuitively—by thinking them through—or through the technique of units cancellation, a more formal method discussed in detail in Appendix C. We'll approach this example intuitively.

Since 1 calorie raises the temperature of 1 gram of water by 1°C and you started with 0.5 kg, or 500 g, it would take 500 calories to raise the temperature of the entire 500 g by 1°C, and three times that, or 1500 calories, to raise the temperature by 3°C.

Therefore you added 1500 calories, or 1.5 kilocalorie to the water in the flask.

■ QUESTION

How many kilocalories of work did Count Rumford do simply in heating the 12 kg of water from (let's assume) 20°C to 100°C in his two and a half hours of boring work?

8.5 A Legacy of Joules

Despite the elegance of Rumford's work with the cannons, wrong ideas die hard. It took more than Bacon's theory and Rumford's experiment to bring us to where we are today. One of those who helped refine the connection between work and heat was the English physicist James Prescott Joule, who lived from 1818 to 1889. He established that work of *any* kind has its exact equivalent in heat. Joule's legacy is twofold: a legacy of demonstration and a legacy of measurement. He *demonstrated* that mechanical work, electrical work, and chemical work all produce heat, and that the amount of heat produced is directly proportional to the amount of work done and is absolutely independent of any other factors. In Joule's own words, "The amount of heat produced by friction is proportional to the work done and independent of the nature of the rubbing surfaces." He also *measured*, quite accurately, the numerical ratio of the amount of work done to the amount of heat produced.

James Prescott Joule proved that all work produces heat and determined the relationship between the amount of work done and the quantity of heat produced.

A **joule** is the work done by one watt of electricity in one second. A joule is equivalent to 0.24 calorie.

In his honor we have, in addition to the calorie, the *joule* as a unit of work or energy. One **joule** is the work done by one watt of electricity in one second. A joule is equivalent to just under 0.24 calorie. For our purposes we can summarize this as:

$$1 \text{ joule} = 1 \text{ watt} \times 1 \text{ second} \qquad 1 \text{ joule} = 0.24 \text{ calories} \qquad 1 \text{ calorie} = 4.2 \text{ joules}$$

EXAMPLE

Bright Light

How many kilocalories of energy does it take to keep a 100-watt light bulb lit for 1 minute?

■ Since the number of joules of work is determined by both the number of watts used and the time span during which the work is being done, measured in seconds, we can write

$$\text{joules} = \text{watts} \times \text{seconds}$$

$$\frac{\text{joules}}{\text{watts} \times \text{seconds}} = 1$$

Knowing also that 1 calorie = 4.2 joules, and using units cancellation (Appendix C), we write

$$100 \text{ watts} \times 1 \text{ minute} \times \frac{60 \text{ seconds}}{\text{minute}} \times \frac{\text{joules}}{\text{watts} \times \text{seconds}} \times \frac{\text{cal}}{4.2 \text{ joules}} \times \frac{\text{kcal}}{1000 \text{ cal}} = 1.4 \text{ kcal}$$

It takes about 1.4 kcal of energy to keep a 100-watt light burning for 1 minute.

While both the joule and the calorie provide us with convenient units for measuring work and energy, the joule is now favored for scientific measurements and calculations. Nonetheless, we'll stick to the calorie in our discussions of energy since it's still widely used in nutritional comparisons. We can always convert from one to the other by using the relationship described above.

Before leaving the giants of science and the work they did to provide us with our modern view of energy, let's look at the work you did in our opening demonstration. By rubbing your palms together you produced Bacon's "brisk agitation," the very friction that Rumford recognized as the source of heat generated in grinding out cannons. And you generated heat equally well by rubbing together the surfaces of your palms or the surfaces of two newspapers. Joule would have applauded. Moreover, the amount of heat you generated was directly proportional to how hard you rubbed—to the amount of work you did. Joule would have applauded even harder, doing more work himself and generating some heat. Once again, what we know of the universe comes directly from the questions we ask, the experiments we do, and our interpretation of the results. It's the scientific method in action.

■ QUESTION

How many *joules* of work did Count Rumford do in raising the temperature of 12 kg of water from 20°C to 100°C? For how many hours would this much energy keep a 40-watt light bulb glowing at full brightness?

8.6 The Storehouse of Chemical Energy

We've seen that burning hydrocarbons consume oxygen, produce carbon dioxide and water, and release energy in the form of heat and light. Following are balanced chemical equations (Sec. 6.14) for the combustion of two specific hydrocarbons, the pentane of gasoline and a 30-carbon alkane of paraffin wax:

$$C_5H_{12} + 8\,O_2 \longrightarrow 5\,CO_2 + 6\,H_2O + \text{energy}$$

$$2\,C_{30}H_{62} + 91\,O_2 \longrightarrow 60\,CO_2 + 62\,H_2O + \text{energy}$$

The actual amounts of energy released by the combustion of some of the lower molecular weight alkanes appear in Table 8.1. The values shown in the table are determined by a much more accurate form of calorimetry than the one of Section 8.4 and are obtained from equal numbers of molecules of each of the alkanes represented: 6.02×10^{23} molecules, which is known as a *mole* of molecules, much as 12 molecules are a dozen molecules (Sec. 9.3). In effect, this permits us to compare the energy contents of various alkanes at their molecular levels.

Even at first glance we can see a direct connection between the molecular weight of the alkane and the amount of heat it releases on combustion. Clearly, the larger the alkane molecule—the greater its carbon content—the more energy it releases. But the basis for this relationship isn't obvious. The combustion of a molecule of pentane releases very nearly four times as much energy as the combustion of a molecule of methane. Yet the pentane molecule contains 5 times as many carbons, 3 times as many hydrogens, and 3.4 times as many atoms (of carbon and hydrogen combined) as the methane molecule. Superficially, there doesn't seem to

Table 8.1 Heats of Combustion of Alkanes (as Gases)

Alkane	Molecular Formula	Approximate Number of Kilocalories Released[a]
Methane	CH_4	213
Ethane	C_2H_6	373
Propane	C_3H_8	531
Butane	C_4H_{10}	687
Pentane	C_5H_{12}	845

[a] All alkanes are represented by equal numbers of molecules, the same number as are present in 16 g of methane (1 *mol* of methane; Sec. 9.3)

be anything about the pentane and methane molecules that accounts for the ratio of 4:1 in their release of energy on combustion. Similarly, there's no obvious, quantitative correlation between the way any of the other alkane molecules are put together and the amount of energy they release. Yet a very real, quantitative correlation with molecular structure does exist. We can see what it is by recognizing an important generalization: *Almost all the chemical energy contained in a molecule is stored in its chemical bonds.*

While all chemical bonds are storehouses of energy, some hold more energy than others. If, in any specific reaction, the total amount of energy contained in all the bonds of all the *products* is *less* than the total amount of energy in all the bonds of all the *reactants*, the difference in energy between the product bonds and the reactant bonds is liberated as heat. Heat is released during the reaction because some of the energy stored in the bonds of the reactant molecules is liberated to the surroundings. To sum up, the products contain less energy than the reactants; the difference is liberated to the environment.

When an alkane burns, for example, the total energy of all the O=C=O bonds and all the H—O—H bonds among the products is less than the total energy of all the C—H and C—C bonds of the alkane and all the O=O bonds of the oxygen that's consumed in the reaction. As a result the reaction releases the energy difference as heat. A reaction of this sort, one that releases energy as it proceeds, is called **exothermic**.

An **exothermic** reaction is a reaction that releases heat as it proceeds.

Conversely, if the total energy of all the bonds of all the products is *greater* than that of all the reactants, the reaction must absorb energy from its surroundings. In this case the chemicals absorb energy from their surroundings and store it as the energies of their bonds. This kind of reaction is **endothermic**. Almost all the chemical reactions that occur spontaneously in our everyday world are exothermic. We've seen this in the burning of gasoline and candle wax.

An **endothermic** reaction is a reaction that absorbs heat as it proceeds.

Notice that since it's the *difference* in bond energies between the reactants and the products of a reaction that determines whether a reaction will be endo- or exothermic, the reverse of any exothermic reaction must be an endothermic reaction. The reactants of one direction become the products of the other. The reverse of alkane combustion, for example, must be endothermic:

$$\text{exothermic:} \quad CH_4 + 2O_2 \longrightarrow CO_2 + 2H_2O$$

$$\text{endothermic:} \quad CO_2 + 2H_2O \longrightarrow CH_4 + 2O_2$$

[You can get a rough idea of the importance of covalent bonds in determining the energy of a chemical reaction by calculating the energy released for each covalent bond of the alkanes of Table 8.1. Methane, for example, has four covalent bonds per molecule, all of them C—H bonds. Burning 16 g of methane—which happens to have the same numerical value as methane's molecular weight but in units of grams rather than amu (Sec. 3.13) and contains 6.02×10^{23} methane molecules, or 1 mole of molecules (Sec. 9.3)—generates 213 kcal of energy. That amounts to about 53.2 kcal per covalent bond. The very same number of ethane molecules, with a total of seven bonds per molecule (six C—H and one C—C), generates 373 kcal. That's about 53.3 kcal/bond. The remaining alkanes also produce close to 53 kcal/bond when they burn. This constancy of 53 kcal/bond points to the importance of covalent bonds in determining the energy of a chemical reaction.

It's important to recognize that this is not a rigorous analysis, especially since it doesn't take into account the covalent bonds of the oxygen consumed or the carbon dioxide and water that are formed. This constancy of 53 kcal/bond comes to us as something of a lucky accident, one that we can't expect to be repeated in extended examinations of other reactions.]

■ QUESTION

How much energy would you expect to be released by the combustion of hexane, using the same number of molecules of hexane as were used for each of the alkanes of Table 8.1?

8.7 Energy for the Human Engine

We can return now to the comparison of the human body and a car's gasoline engine (Sec. 8.1). We can view the human body as an incredibly complex engine that operates in the physical world like any other engine, using energy to perform work and functioning always under the constant supervision of every known law of the physical universe. Just as hydrocarbons provide the fuel for mechanical engines, the chemicals of food furnish the fuel for the biological engines of our bodies. In this important sense, then, our bodies are like the internal combustion engines of cars. They use fuel to produce energy and to do work. Gasoline fuels our cars; food fuels our bodies. As we proceed through our examination of food as fuel, we'll use the term *metabolism* in discussing the conversion of food into energy and the physical substance of our bodies. **Metabolism** encompasses all the chemical reactions that take place in any living organism, including those that release the chemical potential energy of our foods. With this in mind we can draw a more distinct parallel between the *combustion* of gasoline as its hydrocarbons provide energy to a car through a rapid reaction with oxygen, and the *metabolism* of food as its components provide energy to a living body through a slower reaction with oxygen.

Metabolism is the combination of all the chemical reactions that take place in any living organism.

We'll view the human body, then, as if it were an ordinary engine, one that acquires, stores, and uses energy, always in complete accordance with the universal laws of the physical sciences. In brief, we'll look at this biological machine as if it were no more than an engine that uses food as fuel, expends energy in doing work in various forms, and stores its excess energy as body fat. And with all this we'll think of energy in terms of heat.

■ QUESTION

To continue the analogy between the human engine and a car's engine, what's the human equivalent to filling a car's tank with gasoline?

8.8 The Energy Equation

One of the physical laws that controls the operation of all machines is the Law of Conservation of Energy: Energy can neither be created nor destroyed in a chemical reaction but can only be converted from one form into another (Sec. 4.11). (We've seen in Chapters 4 and 5 that matter and energy can be interconverted. For this discussion, though, we'll stick to the larger region of the physical world in which we feel more at home and in which the Law of Conservation of Energy holds well.) With energy being conserved, *all* the energy that comes into our bodies as food must be accounted for; it must be either used or stored. We can state this requirement through an *Energy Equation* that relates energy consumption (from food), energy expenditure (in several ways we'll soon examine), and energy storage (as body fat):

$$\text{Energy In} = \text{Energy Out} + \text{Energy Stored}$$

Eating: energy in.

To put the whole, sometimes complex relation between body energy and human nutrition into simple terms, either we burn up the energy of the food we eat or we store it as fat. The Energy Equation tells us that if we want to lose weight we have to make sure that "Energy Out" is larger than "Energy In." In more common terms this means exercising more and/or eating less. (Of course to gain weight we eat more and/or exercise less so that "Energy In" becomes larger than "Energy Out.")

One more matter needs to be dealt with. As we saw in Section 8.4, physical scientists define the *calorie* as the energy (or amount of heat) needed to raise 1 gram of water by 1°C. For many years, and largely in fields outside the physical sciences, this calorie has been known occasionally as the "small calorie." In a parallel fashion the kilocalorie (1000 cal) is sometimes called the "large calorie." This same kilocalorie, or "large calorie," is also a common unit in the field of nutrition, where it's always written as a capitalized *Calorie* and abbreviated as a capitalized *Cal*. Each of these Calories of nutritional energy, then, represents the same amount of energy as 1 kcal, or 1000 of the "small" calories.

In actual practice, there's seldom any confusion between the Calorie (1000 cal) and the calorie. To avoid mistaking one for the other, we'll take care never to begin a sentence with the word "calories" unless the meaning is quite clear. When the word is capitalized in these discussions we'll know that we're referring to the nutritionist's Calorie.

Exercise, one form of energy expenditure.

Energy stored as fat can be recovered to be spent in various ways, including Sumo wrestling.

■ QUESTION

How many calories are there in 0.1 Calorie? How many Calories are there in 0.1 calorie?

8.9 Energy Out: Exercise, Specific Dynamic Action, and Basal Metabolism

The human body uses energy in three different ways, through

- exercise,
- specific dynamic action, and
- basal metabolism.

Exercise, the path we know best, includes more than the standard pushups, jogging, swimming, tennis, and the like. Exercise, in the sense we're using here, is the composite of *all* the physical work we do with our bodies. The rate of energy expenditure through exercise ranges from the enormous kinetic output demanded by the most vigorous physical activities, such as a strenuous game of tennis or competitive running and swimming, to just sitting quietly, letting time pass. (Even while sitting we're exercising in this particular sense. We're burning up a minute amount of energy as we use muscles to keep our balance and to keep our head up and eyes open.) We use our muscles, more or less strenuously, in an enormous range of physical activities, and we do an equally great range of work in all these activities. Table 8.2 presents some estimates of the rates at which we use energy (Calories) in various activities. Table 8.3 presents the numbers of Calories of typical foods and their equivalents in minutes of exercise.

Specific dynamic action, (*SDA*), accounts for the energy consumed in digesting and metabolizing food and converting its energy into our own, for use through exercise and basal metabolism. It's the price we pay, in energy, for extracting energy from food. In this sense it's analogous to the energy cost of refining petroleum into gasoline, heating oil, diesel fuel, and other products through distillation, isomerization, and similar refinery processes (Chapter 7).

We can sometimes sense SDA in operation as we grow a little warmer and find our heart beating a bit faster after a meal. The warmth and the higher pulse are both signs that our body is hard at work digesting food and using up energy through SDA. (Drowsiness after a large meal comes from the diversion of blood from the brain to the digestive system. The body is working hard to process all the fuel that's just been brought in and—first things first—digestion takes momentary precedence over mental alertness.)

The actual amount of energy lost through SDA depends on what we're eating. For fat it's about 4% of the energy content of the food; for carbohydrates, rough-

Table 8.2 Approximate Rates of Energy Expenditure for a 70-kg (154-lb) Person

		Calories Expended	
Level	Examples	Per Minute	Per Hour
Very light	Sitting, reading, watching television, writing, driving	1.0–2.5	60–150
Light	Slow walking, washing, shopping, light sports such as golf	2.5–5.0	150–300
Moderate	Fast walking, heavy gardening, moderate sports such as bicycling, tennis, dancing	5.0–7.5	300–450
Heavy	Vigorous work, sports such as swimming, running, basketball, and football	7.5–12.0	450–720

Table 8.3 Calorie Content of Typical Foods and Their Exercise Equivalents for a 70-kg (154-lb) Person

		Time Spent (min)				
Food (Portion)	Calories	Lying Down, Resting	Walking	Bicycle Riding	Swimming	Running
Apple (large)	101	78	19	12	9	5
Bacon (2 strips)	96	74	18	12	9	5
Carrot, raw	42	32	8	5	4	2
Chicken, fried ($\frac{1}{2}$ breast)	232	178	45	28	21	12
Egg, boiled	77	59	15	9	7	4
Egg, fried	110	85	21	13	10	6
Halibut steak ($\frac{1}{4}$ lb)	205	158	39	25	18	11
Hamburger sandwich	350	269	67	43	31	18
Malted milk shake	502	386	97	61	45	26
Orange juice (1 glass)	120	92	23	15	11	6
Pizza, cheese ($\frac{1}{8}$)	180	138	35	22	16	9
Potato chips (1 serving)	108	83	21	13	10	6
Soda (1 glass)	106	82	20	13	9	5
Tuna fish salad sandwich	278	214	53	34	25	14
Shrimp, French fried (1 piece)	180	138	35	22	16	9

ly 6%; for protein, around 30%. But measuring SDA is not a simple matter. A complex interaction among our foods causes the digestion of one foodstuff to affect the SDA demanded by another. As a result, the total SDA resulting from any particular meal isn't necessarily the sum of the individual SDAs we might assign to each of the macronutrients of the meal. As a very rough estimate, about 10% of the energy we get from food is spent in SDA. That is, something like 10% of food energy goes toward the very act of extracting and using the remaining 90%.

Basal metabolism, the third of the three routes, accounts for all the work that goes on inside our bodies just to keep us alive. It's the energy we use to keep our heart beating, our lungs expanding and contracting, and our liver, kidneys, and all other organs working to maintain life.

The standard technique for measuring basal metabolism determines our total energy output in the absence of both exercise and specific dynamic action. In this measurement a person lies at rest and awake after 12 hours of fasting. Lying at rest eliminates virtually all energy expenditure by exercise. The added 12-hour fast precludes energy expenditure through specific dynamic action. In a healthy adult the energy spent in this condition—through basal metabolism—amounts to roughly 1 Calorie per hour per kilogram of body weight. Basal metabolism increases with any bodily stress, desirable (such as pregnancy or lactation) or not (illness, for example).

Another useful measure of energy expenditure is *resting metabolism*, which represents the combination of basal metabolism and SDA. In effect, resting metabolism measures our entire energy output excluding the various forms of exercise we do throughout the day.

All of the energy produced by animal bodies comes from the metabolism of the nutrient chemicals of food. The body's metabolic chemistry releases the potential energy of food by processes similar to the way combustion releases the energy of hydrocarbons: through the formation of products—largely carbon dioxide and water—that contain less total energy than the reactants. In a typical example a series of metabolic reactions converts carbohydrates and oxygen into carbon dioxide, water, and biological energy. The following equation sums up the metabolic reactions for the carbohydrate *sucrose*, the table sugar of the opening demonstration of Chapter 1:

$$\underset{\text{sucrose}}{C_{12}H_{22}O_{11}} + 12\,O_2 \longrightarrow 12\,CO_2 + 11\,H_2O + \text{energy}$$

Although similar to combustion in its results, metabolism differs from combustion in two important ways:

- Metabolism releases the energy of chemical bonds more slowly than combustion, without the accompanying light and high temperatures.
- The chemical reactions of metabolism take place under much milder conditions than combustion, conditions that don't endanger body tissue.

■ QUESTION

The population of the entire world is estimated at about 6 billion people. (a) Assuming that the average person spends 200 Calories an hour in the activities of exercise throughout the day (Table 8.2), calculate the *annual* energy output, through exercise alone, by all the humans in the entire world. (b) Calculate the *annual* average daily use of energy through basal metabolism alone, again by all the humans in the entire world. Assume an average body weight of 50 kg (110 lb). (c) Neglecting the effects of specific dynamic action, what is the total annual energy output of all humans on earth? Reviewing the contents of the Appendices may help in these calculations.

8.10 Energy In: The Macronutrients

And where does all this biological energy come from? From the macronutrients of our food. The food components we call macronutrients are the three classes of chemicals that make up the great bulk of our food supply, the chemicals that serve as the only source of energy for the human machine. Specifically, they are the chemicals we classify as:

- fats and oils,
- carbohydrates, and
- proteins.

Through the elegant, intertwined complex of chemical reactions that constitute human metabolism, the chemical energy of these nutrients becomes transformed into the human energy of our lives (Fig. 8.6). We'll examine in detail the structure, properties, and chemistry of these three macronutrients in Chapters 15, 16, and 17. Here, in this Section, we'll look more closely at the energy they furnish.

Figure 8.6 Energy and the human machine.

The actual quantity of energy provided by each of these classes of macronutrients is well known. Fats and oils yield 9 Cal per gram, carbohydrates 4 Cal per gram, and proteins 4 Cal per gram. The amount of energy each of these chemicals delivers to our bodies is independent of both the kind of food in which it occurs and the presence or absence of other macronutrients. One gram of fat or oil provides 9 Cal whether it comes from, say, butterfat, peanuts, beef, or corn. A meal containing 10 g of fat, 15 g of carbohydrate, and 20 g of protein, for example, would deliver to our bodies

$$10 \text{ g fat} \times \frac{9 \text{ Cal}}{\text{g fat}} = 90 \text{ Cal} \quad \text{(from fat)}$$

$$15 \text{ g carbohydrate} \times \frac{4 \text{ Cal}}{\text{g carbohydrate}} = 60 \text{ Cal} \quad \text{(from carbohydrate)}$$

$$20 \text{ g protein} \times \frac{4 \text{ Cal}}{\text{g protein}} = 80 \text{ Cal} \quad \text{(from protein)}$$

$$\text{total} = 230 \text{ Cal}$$

EXAMPLE

More Fat

Suppose we change our illustration by increasing the amount of fat in the meal by 5 g while keeping the weights of the carbohydrate and protein the same. That is, we go from 10 to 15 g fat. What's the new energy content?

■ At 9 Cal/g of fat, we've added

$$5 \text{ g fat} \times \frac{9 \text{ Cal}}{\text{g fat}} = 45 \text{ Cal}$$

The meal now delivers 230 + 45 = 275 Cal.

EXAMPLE

Comparing Calories

Now suppose that instead of adding 5 of fat to the meal, we add 5 g of carbohydrate. What's the new energy content?

■ At 4 Cal/g of carbohydrate, we've added

$$5 \text{ carbohydrate} \times \frac{4 \text{ Cal}}{\text{g carbohydrate}} = 20 \text{ Cal}$$

The meal now delivers 230 + 20 = 250 Cal.

Adding fat to our diet increases the number of calories we consume far more quickly than adding carbohydrate or protein.

EXAMPLE

Percentage of Fat

You may have seen packages of prepared foods, such as lunchmeats, stating that they contain only, let's say, 5% fat. Assuming that this is a percentage of the total weight of the macronutrients, what percentage of the total calories does this amount of fat provide?

■ To make our calculations easy, we'll assume that we have 100 g of a food that's 5% fat by weight. Thus it contains 5 g of fat and 95 g of a combination of carbohydrates and proteins. Since both carbohydrates and proteins deliver 4 Cal/g, we can calculate the total number of Calories in the 100 g portion as

fat:	5 g × 9 Cal/g =	45 Cal
carbohydrates and proteins:	95 g × 4 Cal/g =	380 Cal
	total =	425 Cal

The percentage of total calories provided by the fat is

$$\frac{45 \text{ Cal}}{425 \text{ Cal}} \times 100 = 10.6\%$$

In this case, fat provides 10.6% of the total number of calories in the food, even though it makes up only 5% of the weight.

As macronutrients, the fats and oils provide us with the most concentrated form of food energy available, 9 Cal/g, which is 2.25 times as much as the energy content of carbohydrates and proteins. This high energy density of fats and oils makes fat an ideal storage medium for reserve supplies of energy in animal bodies. When the *Energy In* portion of the energy equation exceeds the *Energy Out* factor, our bodies store the surplus of energy as fat.

For comparison, alcohol, or more correctly ethyl alcohol (Sec. 6.7), another substance that some of us consume along with food, provides 7 Cal/g. Thus alcohol stands a bit closer to fats and oils than to carbohydrates and proteins in the number of Calories per gram it can add to our total intake.

■ QUESTION

The nutrition information panel on a can of a typical commercial chicken noodle soup reveals that one serving contains 13 g of protein, 15 g of carbohydrates, and 5 g of fat. (a) How many Calories does one serving of this soup provide? (b) What percentage of these Calories comes from fat? (c) How many hours of basal metabolic activity would one serving provide to the average person of the Question at the end of Section 8.9?

8.11 The Source of (Almost) All Our Energy

All the commercial and industrial energy producers on earth generate about 10^{17} kcal each year, roughly 10 times our estimated energy output of the world's entire human population (See the Question at the end of Sec. 8.9). This energy comes from fossil fuels of all kinds, from water moving through dams and waterfalls, from nontraditional sources such as solar panels and windmills, and from nuclear fission. Worldwide, about 85% of the total output of commercial and industrial energy comes from fossil fuels, 7% from water power, 7% from nuclear power, and the rest from solar panels, windmills, and other sources. Almost all this energy—all the energy produced by all the plants and animals of the earth and well over 90% of all the energy produced by commercial and industrial sources—originates in reactions that take place over 90 million miles away, at the surface of the sun.

(The exceptions—nonsolar sources of energy, those not related to the sun's output—include nuclear energy, which we examined in Chapter 5, and geothermal energy. We see the release of geothermal energy in the action of volcanoes, geysers, and naturally warm springs. In what follows we will consider only energy derived directly or indirectly from the sun.)

As we saw in Section 4.14, the sun's energy originates in countless nuclear fusion reactions that continuously convert the sun's protium atoms into helium and energy. A small fraction of this energy travels to the earth in the form of light and both ultraviolet and infrared radiation. The light provides us with our daylight; the

Banks of wind turbines convert the energy of the wind into electricity.

ultraviolet radiation produces, among other things, suntans and sunburns; the infrared radiation comes to us as heat.

But solar radiation furnishes energy in other ways as well. The sun's heat provides energy for the evaporation of the earth's surface waters, its oceans, rivers, lakes, and streams. Once in the atmosphere as water vapor and clouds, the water raised from the surface of the earth condenses and returns as rain and snow. Back on the earth's surface, the water returns to its origins, traveling downward from higher elevations toward sea level. Where the water moves quickly and forcefully in its downward journey, hydroelectric plants can convert its kinetic energy into electric power.

The atmospheric effects of the sun's radiation produce surface winds that can power electric generators connected to windmills. In addition, solar panels can be used to convert solar radiation directly into electricity. Wind power and solar panels represent small but growing sources of energy.

■ QUESTION

(a) Identify one means of generating electricity that's driven directly or indirectly by the energy of solar radiation. (b) Identify one that is not.

8.12 Why We Breathe

A larger and more important source of energy originates in the transformation of the sun's radiant energy into the biological and chemical energy of plants. Through the process of *photosynthesis* plants use the energy of solar radiation to convert water, atmospheric carbon dioxide, and various elements of the soil into the structural, nutrient, and other chemicals of the plants themselves. Through photosynthesis plants convert the radiant energy of the sun into the chemical energy of plant molecules.

In its overall chemistry, plant photosynthesis can be thought of as the reverse of animal metabolism. In the photosynthesis of sucrose, for example, the net chemistry of the process is the reverse of the net chemistry of sucrose metabolism (Sec. 8.9):

$$12\,CO_2 + 11\,H_2O + \text{solar energy} \longrightarrow \underset{\text{sucrose}}{C_{12}H_{22}O_{11}} + 12\,O_2$$

Notice that in producing metabolic energy, animals *consume* oxygen and *release* carbon dioxide; in harnessing solar energy through photosynthesis, plants *release* oxygen and *consume* carbon dioxide. This interchange of oxygen and carbon dioxide takes place at the cellular level. In higher animals, with lungs and circulatory systems, the blood carries inhaled oxygen from the lungs to the cells and the waste carbon dioxide from the cells back to the lungs.

We breathe in order to maintain our supplies of continuously consumed O_2 and to expel our continuously generated CO_2. Our sense of needing to breathe is generated by the accumulation of CO_2 in our lungs. This accumulation generates a gasping response when we hold our breath. A deficiency of O_2, with no buildup of excess of CO_2 in our lungs, as when we might breathe normally in an atmosphere deficient in oxygen, can produce a loss of consciousness accompanied (for a while at least) by normal breathing. We breathe, then, as much to exhale accumulated carbon dioxide as to inhale fresh supplies of oxygen.

■ QUESTION

What is the source of the energy that drives the process of photosynthesis? Into what form of energy is this transformed within plants?

8.13 Fossil Fuels and the Carbon Cycle

The **biomass** is the total mass of all life, including its waste products, residues, and remains.

In the **carbon cycle** carbon shuttles among the earth's various natural storage facilities for the element and its compounds. The carbon moves from one region to another largely as carbon dioxide.

In their ability to remove carbon dioxide from the atmosphere and incorporate the carbon into their own structures, plants play an important role in the cycling of carbon among the earth's atmosphere, land, water, and **biomass**, the total mass of all life, including its waste products, residues, and remains. This plant-mediated process forms part of an extensive system of carbon transfer known as the *carbon cycle*.

The **carbon cycle** shuttles carbon, principally as carbon dioxide, from one of earth's storage facilities to another. The providing region is the *source*; the recipient is the *sink*. Depending on which way the carbon moves, any given storage region may be a source at one time, a sink at another. Because of the importance of atmospheric carbon dioxide in regulating surface temperatures (Secs. 8.14 and 8.15), the earth's atmosphere plays an important role as both a sink and a source. Photosynthesis acts to move carbon, as carbon dioxide, from the atmosphere (the source in this case) into plants (the sink). The dissolving of atmospheric carbon dioxide into the world's surface waters (here, another sink) acts in the same way, moving CO_2 from the atmosphere into the oceans and other bodies of water.

Activities that release carbon into the atmosphere (now a sink), include animal metabolism, the burning of fossil fuels and wood, and the release of dissolved CO_2 from surface waters. If the rate of addition of carbon dioxide into the atmosphere from its sources equalled the rate of transfer of carbon dioxide out of the atmosphere into its sinks, the carbon cycle would be a stable process and the concentration of atmospheric CO_2 would be constant over long periods of time. But this doesn't appear to be the case. There is evidence that the rates of transfer are not equal and that the concentration of atmospheric carbon dioxide is slowly rising. The destabilizing activity appears to be the use of fossil fuels as energy sources.

Unlike the generation of nuclear, hydroelectric, wind, solar, or geothermal power, liberating energy by burning fossil fuels releases CO_2 into the atmosphere

Burning gasoline and other fossil fuels adds carbon dioxide to the atmosphere.

and serves as a carbon source. And unlike fuels such as wood and other plant material, which are regenerated relatively rapidly, the formation of fossil fuels doesn't serve as a carbon sink, at least not within any imaginable time frame. Fossil fuels can add CO_2 to the atmosphere, but they can't remove it in any practical sense.

The ultimate origin of fossil fuels is uncertain. They appear to be the remnants of living things that once thrived on the surface of the earth and its oceans, perhaps tens of millions of years ago. That's the significance of the "fossil" in *fossil fuels*. The most widely accepted of several current explanations holds that petroleum originated as the organic matter of *plankton*, microscopic plants that float on the surface of bodies of water. As these plankton died and settled to the bottom long ago, they mixed with fine marine sediments and were buried, undecayed or only partially decayed, and only partially consumed by still other organisms. Over tens of millions of years, subjected to heat, pressure, and chemical activity, their residues lost the oxygen of their molecules. The remaining carbon and hydrogen formed the hydrocarbons of what we now know as petroleum. Coal is believed to have been formed in a similar manner, but from larger, multicellular plants, including trees and similar plants, growing in prehistoric swampland. Death, burial, incomplete decay, heat, pressure, and other forces converted all this prehistoric plant material to today's coal, oil, and natural gas.

These, then, are our fossil fuels—nonrenewable resources that contribute an added stress to our contemporary carbon cycle.

■ QUESTION

Identify two sources and two sinks for atmospheric carbon.

8.14 Carbon Dioxide: A Greenhouse Gas

An accumulating body of evidence suggests that this additional carbon dioxide could change the world's climate radically through the **greenhouse effect**. This is the same effect that allows us to grow temperature–sensitive plants in a cold climate. The glass walls and roof of a greenhouse allow the sun's warming radiation

The **greenhouse effect** is the warming of the earth by solar heat trapped through the insulating effect of atmospheric gases.

A greenhouse. The sun's radiation passes through the glass panels and is transformed to heat, which is trapped inside the building.

to penetrate into the enclosed space. There, inside the greenhouse, the radiation is absorbed and its energy is transformed into heat, which doesn't leave the glass enclosure nearly as easily as the solar radiation enters. The trapped heat keeps the temperature within the greenhouse high enough for the plants to grow and thrive.

Earth's atmosphere behaves similarly. Solar radiation penetrates through the atmospheric gases easily and its energy is converted to heat at and near the earth's surface. The carbon dioxide and other atmospheric **greenhouse gases** (Table 8.4) trap this newly generated heat in much the same way as the glass traps the heat within the greenhouse (Fig. 8.7). And it's a good thing they do. Life as we know it

A **greenhouse gas** is an atmospheric gas that contributes to the greenhouse effect.

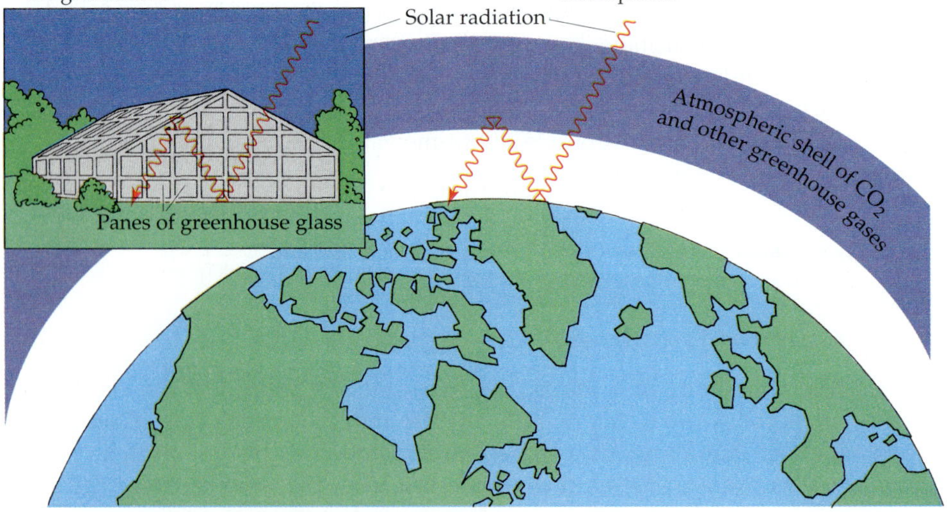

Figure 8.7 The greenhouse effect.

Table 8.4 The Most Important Greenhouse Gases Currently Produced by Human Activity

Gas		
Name	Formula	Source
Carbon dioxide	CO_2	Burning fossil fuels, wood, and other plant materials
Methane	CH_4	Biomass activity in solid-waste landfills, cattle farming, rice growing, coal mining, petroleum refining, use of natural gas
Ozone	O_3	Burning plant materials, smog, engine exhausts
Nitrous oxide	N_2O	Use of nitrogen-containing fertilizers

Note: Another class of greenhouse gases are the chlorofluorocarbons, such as CCl_2F_2. Once produced for use as refrigerants and in the preparation of plastic foam, they are no longer being manufactured commercially. Nonetheless they are still present in the atmosphere because of their resistance to degradation.

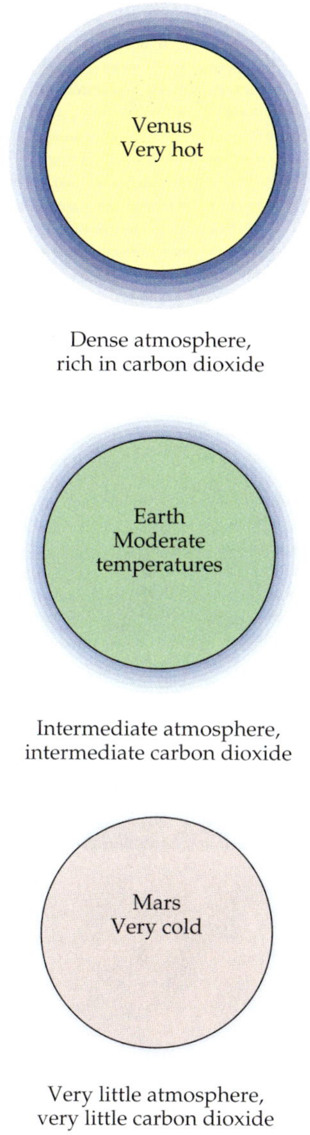

Figure 8.8 Planetary greenhouse effects.

couldn't be sustained on earth if it weren't for the presence of just the right amount of carbon dioxide and other greenhouse gases in our atmosphere. To see why, we'll compare our own planet with its two nearest neighbors, Venus and Mars.

Venus is surrounded by an atmosphere that's about 95% carbon dioxide and is compressed to about 95 times the density of earth's atmosphere. The surface of Venus roasts at close to 450°C, well above the melting point of lead. The surface of Mars, on the other hand, is covered by an atmosphere that's also about 95% carbon dioxide but is less than 1% as dense as ours. Temperatures on the Martian surface range from a high of a bit below 40°C at noon on a summer day down to a nighttime low near −80°C. Neither one of those two planets provides a surface temperature that's hospitable to the kinds of life we find on Earth. Venus is too hot and Mars is too cold.

While Venus, Earth, and Mars receive progressively smaller amounts of solar radiation because of their progressively greater distances from the sun, the dramatic differences in their surface temperatures almost certainly result from differences in the atmospheric greenhouse effect on the three planets (Fig. 8.8). The dense, insulating layer of carbon dioxide around Venus efficiently traps the solar warmth that penetrates through to the planet's surface. Like a greenhouse, Venus holds the heat that reaches it from the sun and keeps it from radiating back into space. As a result, Venus is a hot planet. Mars, with its thin, almost nonexistent atmosphere, returns its heat to space easily. Mars is cold.

Earth's atmosphere, which contains only about 0.03% carbon dioxide and has a density between the two extremes of Venus and Mars, maintains just the right temperature range for the forms of life we know so well.

■ QUESTION

What two conditions or effects are responsible for a much higher surface temperature on Venus than on earth? What two conditions or effects are responsible for a much lower surface temperature on Mars?

8.15 Global Warming: Will Palm Trees Soon Grow in Alaska?

Several observations indicate that earth's temperature is increasing and that this warming trend is due at least partly to the increasing use of fossil fuels.

Landscapes of Venus, Earth, and Mars. Greenhouse effects on these three planets produce striking differences in their surface environments. Earth supports a variety of life, while the surfaces of Venus and Mars are barren.

Throughout history humans have harnessed various sources of energy to do our work. In earlier times we used animals to pull plows, wind to move ships, falling water to grind grain, wood to feed fires and provide heat. But the middle of the 18th Century introduced a new era in energy usage. What came to be known as the *Industrial Revolution* generated a demand for energy that far exceeded what animals, wind, water, and burning wood could provide. In Great Britain, the development of an efficient steam engine and the expanded mining of coal led to the transformation of a primarily agricultural society into a newly industrial society. With coal now readily available and a new steam engine able to convert coal's energy into industrial work, the era of fossil fuels had begun.

As the Industrial Revolution spread from Great Britain to what we now call the industrialized world, the use of fossil fuels expanded and the CO_2 content of the atmosphere began to creep upward. In 1958 atmospheric concentrations of carbon dioxide were estimated to have risen by about an eighth from their levels of the 1750s. Today they are roughly a third higher and are growing at an estimated 0.4% each year. Virtually the entire increase is attributed to human activities, largely through the burning of fossil fuels for transportation, heating, and industrial production. Worldwide, our current use of fossil fuels releases an estimated 20 billion tons of carbon dioxide annually. Roughly half of this remains in the atmosphere, with the remainder transferred to various carbon sinks.

Along with an increasing atmospheric CO_2 concentration, the average temperature of the earth's surface appears to be rising as well. Figure 8.9 shows changes in the average temperature of the earth's surface, in °F, annually since 1880.

An irregular but generally upward trend is clear, with 1998 the warmest year on record. Moreover, the temperature increase seems to be accelerating. During the 20th century as a whole, the temperature rose at a rate of about 0.6°C/century; during its last quarter the rate accelerated to nearly 2°C/century. Evidence of a general warming trend lies in the melting of the Arctic ice cap and a decrease in the area and depth of mountain glaciers throughout the world, a rise in sea level, earlier and longer growing seasons in many regions, and a spread in the territorial range of several species of plants and animals to higher elevations and higher latitudes, which they once shunned because of the cold.

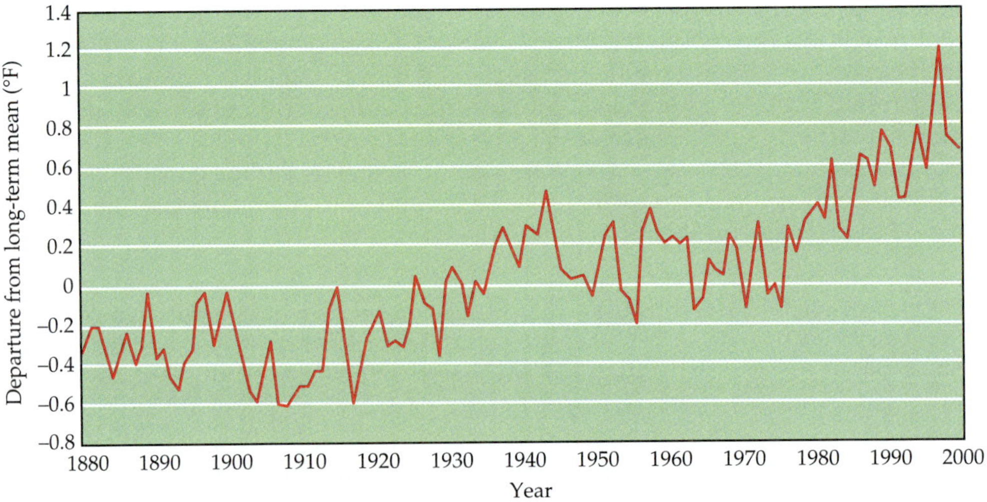

Figure 8.9 Global temperature changes, 1880–2000. *Source:* U.S. National Climatic Data Center, 2001. Available on-line: http://www.epa.gov/globalwarming/climate/index.html.

The greenhouse effect itself (Sec. 8.14) is real. Carbon dioxide and other greenhouse gases do act to retain earth's surface heat. But while a large portion of the increase in earth's surface temperature results directly from human activities, part of the warming trend may also reflect normal variations in the planet, the kinds of variations that brought the latest ice age, some 20,000 years ago, and the subsequent warming to today's moderate climate. Both these climate changes—both the cooling and the warming—took place long before humans dominated the planet.

A report, "Climate Change Science," issued in 2001 by the National Academy of Sciences, a private, nonprofit group of distinguished scientists and engineers, analyzes the contributions of both human activity and nonhuman influences on earth's climate. It concludes that the current warming of the earth's surface probably does result largely from human activity but that climate variations unconnected with human behavior may also play a significant role in global warming.

As for the future, an increase of a degree or two in earth's average surface temperature, spread over a century, may not seem like much. But during that century what might seem to be an insignificant change in average temperature could produce summers in New York much like those in Miami today, and palm trees on Alaskan beaches. Global warming could induce sufficient melting of polar ice caps to raise sea levels, with serious results for beaches, seaside settlements, and water supplies. However gradual it might be, a persistent global warming could also generate public health concerns through an increase in the level of tropical diseases and heat-related illnesses.

Will Alaska look like this as a result of the greenhouse effect?

■ QUESTION

If earth's average surface temperature keeps increasing at the same rate as for the last quarter of the 20th century, what level will it reach in the year 2100?

PERSPECTIVE

Energy in the Balance

We began this chapter with a simple demonstration in which you generated heat by rubbing your hands together; we concluded by asking whether the large-scale use of fossil fuels might produce dramatic changes in our global climate. Between these two lies a great deal of both well-established science and tentative analysis.

We can now see that the opening demonstration carried us stepwise along a well-established path of energy creation, transfer, and use, originating in nuclear fusion at the sun's surface and ending in our own daily activities. The path of energy, from its origin to biological use by us and other living things, follows a sequence of:

1. fusion of the sun's protium to form helium and release energy,
2. radiation of energy from the sun to the earth,
3. conversion into the chemical energy of plant materials by photosynthesis,
4. transfer of plant nutrients to animals that eat plant products, and finally
5. metabolic conversion of the chemical energy of the plant nutrients (and, for animals that feed on the products of other animals, animal nutrients derived from them) into the energy of animal bodies.

As humans and other forms of animal life use this energy, carbon cycles repeatedly between plants and animals, transferred back and forth between animal metabolism and plant photosynthesis by way of atmospheric CO_2. With plant photosynthesis balanced against animal metabolism, this continuous cycling produces little net effect on atmospheric concentrations of CO_2. As a result, the combined, energy-consuming activities of all the animal life on earth by the path outlined above has little impact on the planet's climate.

But our global society also relies heavily on consuming the sun's energy delivered by a different path, one that follows a fork after step 3. By this alternative path we use energy stored not by contemporary, readily regenerated plant materials, but captured instead by organisms living millions of years ago, stored under the earth's surface, and delivered to us now by what we call fossil fuels. With today's combustion of fossil fuels we add a new, long-lasting burden of carbon to the worldwide cycle, increasing the CO_2 content of the atmospheric sink for centuries to come as this newly added carbon is transferred slowly to other sinks. The analysis by the National Academy of Sciences described in Section 8.15 estimates that the transfer of carbon dioxide from the earth's atmosphere to other sinks is a slow process, with more than a third of the added carbon dioxide still retained in our atmosphere after 100 years, and more than 5% even after 300 years.

How, then, do we achieve a balance of energy production and climate stability? The answers are neither clear nor simple. They carry us from the well-understood chemistry of work, energy, and fuels into the ambiguities of economics, public policy, and societal concerns, on both national and global scales. Among a host of possibilities are the following.

Converting our vehicles from gasoline to electricity: The Perspective of Chapter 7 examines this possibility. We might expect that converting our cars, trucks and busses to electric power would cut sharply into our use of gasoline and other fossil fuels, at least for transportation. Yet the energy that these fuels now deliver would have to be replaced, kilocalorie for kilocalorie, by its electrical equivalent.

Obtaining the needed electrical energy from oil-, coal-, or gas-powered generating plants would simply transfer the source of the added carbon from the automobile's exhaust to the generator's smokestack. Although energy might be generated more efficiently, with less total CO_2 emission, at a few, central generating plants than in a large number of individual gasoline-powered vehicles, a transition from the internal combustion engine to an electric motor would not solve the greenhouse problem unless the source of the electric power is something other than fossil fuels.

Expanding our use of nuclear power: We might turn to nuclear power to generate the electricity needed for transportation and for home and industrial use as well. But Chapter 5 reveals the problems in this approach, including the construction of additional nuclear plants, the accumulation of additional nuclear wastes, and a public reluctant to see nuclear plants built in or near their own localities.

Using ethanol and other alternative fuels: The advantages and disadvantages of using ethanol as an alternative to gasoline were discussed in the Perspective of Chapter 7. Although burning ethanol adds carbon dioxide to the atmosphere, ethanol itself is a renewable resource. The carbon released on combustion of ethanol cycles back into plants, which provide additional ethanol through fermentation. Replacing fossil fuels with ethanol would require a large increase in the proportion of agriculture devoted to growing corn and other grains that ferment easily. Other possibilities for the direct release of energy, or for the generation of electricity, include wind-powered generators and solar panels (Sec. 8.11).

Lowering the levels of greenhouse gases other than carbon dioxide: The greenhouse gases of Table 8.4 are far from equal in their ability to trap heat. Methane, for example, is about 30 times as effective as carbon dioxide. Even a small reduction in methane emissions—through improvements in petroleum refining, cattle farming, rice growing, and even through more careful handling and transfer of methane itself—can help minimize the general trend toward global warming.

Energy conservation: The less energy we use, even with the continued use of fossil fuels, the smaller its impact on our climate. Energy conservation can take many forms: increasing the fuel efficiency of our vehicles and the energy efficiency of our appliances, improving the insulation of our buildings, making short trips by foot, bicycle, or public transportation rather than by car—in general using energy more effectively, more efficiently, and with less waste.

In summary, each of these possibilities may have its appeal. Some might be less expensive than others; some might produce a smaller impact on our standard of living; some might be more appropriate to a short-term approach, others to the longer term. But none is perfect; none offers an ideal, prompt, inexpensive solution to the overall balance between energy use and climate stability.

One important factor lies in the convenience of fossil fuels themselves. They are currently plentiful and cheap, at least in comparison to other sources. And the infrastructure for their delivery and use—the combination of tankers, tank trucks, pipelines, gasoline stations, industrial generators, and internal combustion engines—is already in place.

In the final analysis, we might find it necessary to balance the size of the polar ice caps, the depth of mountain glaciers, the level of the oceans, the length of the growing season, and the range of migration patterns against our own choice of energy usage, whether we walk or drive to the local drugstore. Certainly, though, whatever wisdom we may use either in counteracting a greenhouse effect or in adapting our society to its effects, our actions will have among their foundations an understanding of the chemistry of energy and the fuels we use to produce it. ■

EXERCISES

For Review

1. Following are a statement containing a number of blanks, and a list of words and phrases. The number of words equals the number of blanks within the statement, and all but two of the words fit correctly into these blanks. Fill in the blanks of the statement with those words that do fit, then complete the statement by filling in the two remaining blanks with correct words (not in the list) in place of the two words that don't fit.

__________ is a form of __________, which is simply the capacity to do __________. The __________, which is the amount of energy needed to raise the temperature of __________ by __________, is a common unit of nutritional energy. Another unit, the __________, is more widely used in the physical sciences. It is the amount of work done by 1 watt of __________ in 1 second. Chemical energy is stored as the energy of __________. If the total amount of energy stored among all the __________ of a chemical reaction is less than the total amount of energy stored among the __________, energy is released as the reaction occurs. The actual measurement of energy and work is accomplished through the technique of __________.

Carbohydrates, proteins, and __________ are the macronutrients that provide us with all of our bodily energy. We use the energy that they and our other macronutrients provide to do physical work, for the __________ necessary to the digestion and metabolism of our food and for the __________ that keeps our organs running and our bodies alive.

1 g of water	heat
1° Celsius	joule
basal metabolism	products
Calorie	reactants
calorimetry	specific dynamic action
chemical bonds	watt
electricity	work
energy	

2. Identify the contribution made by each to our understanding of the physical nature of heat: (a) Francis Bacon; (b) Benjamin Thompson, Count Rumford; (c) James Prescott Joule.

3. How might Francis Bacon have explained the results of Count Rumford's experiment?

4. Which of the following are examples primarily of kinetic energy and which are examples primarily of potential energy:

a. a waterfall
b. a person standing on a diving board, about to jump into a swimming pool
c. a diver who is in the process of entering the water at the end of the dive
d. an electric battery standing alone on the floor of a room
e. a stream of α particles
f. a coiled, compressed spring
g. the earth as it orbits the sun
h. a car moving at 0.5 mile per hour

5. Write the Energy Equation for human nutrition and describe each of its terms.

6. Using the terms of the Energy Equation, describe the conditions under which a person's weight (a) increases; (b) decreases; (c) remains constant.

7. What are the three processes by which we use or expend the energy provided by food?

8. What is resting metabolism?

9. What are the three macronutrients of our foods?

10. Which macronutrient contributes the greatest amount of energy per gram to the body?

11. How do our bodies store unused energy?

12. Label each of the following as exothermic or endothermic:

a. burning hydrocarbons of a candle
b. nuclear fission
c. steam condensing to water
d. ice melting to water
e. electricity passing through the filament of a light bulb
f. water boiling to steam
g. nuclear fusion
h. chlorine gas reacting with sodium metal
i. electrolysis of water to hydrogen and oxygen

13. Give an example of a substance produced by nature, other than food or petroleum products such as gasoline, that we value for its chemical potential energy.

14. Describe the connection between the Industrial Revolution and the average temperature of the earth's surface.

15. Describe the atmospheric greenhouse effect and explain why carbon dioxide is classified as a greenhouse gas.

16. Name three other gases, in addition to carbon dioxide, that are classified as greenhouse gases and describe their principal sources.

17. In what way does the use of gasoline as a fuel contribute to the greenhouse effect? Why doesn't the use of corn-derived ethanol as a fuel contribute to the greenhouse effect?

18. Suppose we discovered a way to modify the combustion products of fossil fuels by converting CO_2 emissions into CH_4 emissions through the chemical reaction

$$CO_2 + 2H_2O \xrightarrow{\text{catalyst}} CH_4 + 2O_2$$

What effect, if any, would this have on the greenhouse effect?

A Little Arithmetic and Other Quantitative Puzzles

19. A person in normal health and weighing 165 lb (75 kg) fasts for 36 hours while lying still in bed. What is the total number of Calories lost during the last 24 hours of the fast?

20. Suppose this same 165-lb person chooses to lie in bed for a month, without engaging in any of the normal, daily physical activities that constitute exercise in its broadest sense. Taking into account all the factors involved in resting metabolism, calculate the average daily intake of food Calories needed by this individual to maintain body weight at a constant level.

21. The following are the amounts of each of the macronutrients provided by one standard serving of several commonly available packaged foods. How many Calories does one serving of each of these foods provide?

a. Bumble Bee chunk light tuna in water

carbohydrate	0 g
protein	12 g
fat	2 g

b. V8 vegetable juice

carbohydrate	9 g
protein	1 g
fat	0 g

c. Campbell's chicken broth

carbohydrate	2 g
protein	3 g
fat	2 g

d. Farm Best low-fat milk

carbohydrate	11 g
protein	8 g
fat	5 g

e. Breyers mint chocolate chip ice cream

carbohydrate	17 g
protein	3 g
fat	9 g

f. Planter's Sweet-N-Crunchy peanuts

carbohydrate	15 g
protein	4 g
fat	8 g

g. Charles Chips potato chips

carbohydrate	15 g
protein	1 g
fat	10 g

22. How long would it take a 70-kg person to burn up the energy provided by one standard serving of each of the foods (a-g) in Exercise 21 through the operation of basal metabolism alone?

23. How long would it take this same 70–kg person to burn up the energy provided by one standard serving of each of the foods (a-g), in Exercise 21 through the energy expenditure of 5 Cal per minute during a moderate exercise such as brisk walking?

24. Suppose you weigh 154 lb and have just eaten a lunch consisting of a glass of orange juice, a hamburger, a malted milk shake, and a serving of potato chips. How long would you have to run to use up the calories you absorbed from the lunch? (Refer to Table 8.3.)

25. Repeat Exercise 24, but this time for a lunch consisting of a glass of orange juice, a tuna fish sandwich, a raw carrot, and a glass of soda.

26. How many calories does it take to heat 25 g of water from 40 to 50°C?

27. How many calories are released as 25 g of water cools from 60 to 50°C?

28. What is the final temperature of the 50 g of water formed when 25 g of water at 60°C is mixed with 25 g of water at 40°C? Assume that all of the heat released by the warmer water is absorbed by the cooler water.

29. What is the final temperature of the 50 g of water formed when 10 g of water at 60°C is mixed with 40 g of water at 40°C?

30. A 100-watt electric light bulb was placed in a calorimeter and turned on. After a period the temperature of 1000 g of water in the calorimeter rose by exactly 10°C. Using the information in Section 8.5, calculate the length of time the light bulb was on.

31. How long would the nutritional energy provided by one serving of the potato chips described in Exercise 21 g keep a 100-watt light bulb burning?

32. Which is larger, the total annual production of energy by all the commercial and industrial plants on earth or the total annual production of energy by all the people on earth, through the combination of exercise and basal metabolism. The Question at the end of Section 8.9 may be useful here.

Think, Speculate, Reflect, and Ponder

33. A person who is ill with a fever is releasing energy more rapidly than normal. This energy raises the body temperature, which we detect as the fever. Which one of the three energy-expending processes increases as a result of the illness?

34. You have an electric hot plate that transfers heat directly to whatever is on it with negligible losses. How could you determine the wattage rating of the hot plate experimentally?

35. It takes energy to accelerate a car from zero to 50 miles an hour. Through the Law of Conservation of Energy we know that the kinetic energy of the car traveling at 50 miles an hour has to come from the transformation of some other form of energy into the car's kinetic energy. Where does the car's energy come from? It also takes energy to bring a car traveling at 50 miles an hour to a complete stop. The kinetic energy of the car has to be transformed into some other form of energy. Into what kind of energy is the car's kinetic energy transformed as it slows to a stop?

36. Table 8.3 shows that a fried egg provides 110 Cal, while a boiled egg gives us 77 Cal. Since eggs are often fried in butter, margarine, or vegetable oil, what do you suppose is the source of the additional calories in the fried egg? What do you think would be the calorie content of an egg fried without the use of butter, margarine, or oil (in a nonstick frying pan)?

37. Describe a device or a process by which the following transformations occur:

a. chemical energy is transformed into electrical energy
b. chemical energy is transformed into heat
c. chemical energy is transformed into kinetic energy
d. electrical energy is transformed into chemical energy
e. electrical energy is transformed into heat
f. electrical energy is transformed into kinetic energy
g. kinetic energy is transformed into electrical energy
h. kinetic energy is transformed into heat

38. Francis Bacon and Count Rumford used the scientific method in the investigations described in this chapter. State the question each asked and describe the experiment each devised to answer the question, the results each obtained (or would have obtained, had Bacon lived), and the conclusions drawn.

39. Using commonly available materials, how would you demonstrate that (a) work produces heat and (b) heat can be used to do work?

40. Planting 100 million trees has been suggested as a means for decreasing the greenhouse effect of our atmosphere. Describe the reasoning behind this proposal.

41. What would have been the fate of the carbon of plankton if the plankton had been completely metabolized by other organisms instead of being trapped, only partially decomposed, at the bottom of bodies of water and eventually converted into fossil fuels?

42. If you wanted to use sucrose containing C-11 in a PET investigation (Sec. 5.9), how might you obtain a quantity? Assume that you have a supply of C-11 on hand and describe each step you would take in the conversion of the C-11 into sucrose containing this isotope.

43. Suggest some *benefits* that might come from a gradual warming of the earth's surface through the greenhouse effect.

Web Questions

Use the Web to answer or accomplish each of the following. Report your Web source for each response.

1. In addition to the calorie and the joule, another unit of heat or energy in widespread use, especially outside the sciences, is the *british thermal unit* or *btu*. Write a brief description of the btu, including its definition, its origin, and the areas in which it is principally used.

2. Describe how a person's basal metabolism is measured in a medical setting.

3. Summarize briefly federal regulations for the listing of macronutrients and their associated Calories on the labels of packages of processed foods.

4. What is the estimated mass, in kilograms, of the earth's entire biomass?

5. What is the estimated mass, in kilograms, of all the carbon dioxide currently present in the earth's atmosphere?

6. What is the temperature of the moon's surface when it is lit by the sun, during the lunar day? What is the temperature when the surface is dark, during the lunar night? How do these temperatures differ from comparable temperatures on earth? What is the source or cause of this difference between the moon's surface temperatures and the earth's?

7. Describe additional evidence for the greenhouse effect, beyond the phenomena discussed in Section 8.15.

8. Describe any current arguments or evidence that the greenhouse effect is not real, or that the burning of fossil fuels is not a significant contributor to the greenhouse effect.

9. Describe the current status of each of the proposals, listed in the Perspective, for achieving a balance of energy production and climate stability.

Arithmetic of Chemistry

CONCENTRATING ON POLLUTION

Would you drink this water?

The Glass Where Pollution Begins

Figure 9.1 1. Put a teaspoon of salt into the standard measuring glass and fill it to the mark with water. Allow the salt to dissolve. *(continues on next page)*

According to one of the less formal principles of chemistry, "There's a little bit of everything in anything." This implies that if we could detect and measure the most exceedingly small quantities, down to the very molecules and atoms that make up any particular substance, we'd find a bit of whatever we might look for in anything we choose to examine. As a result we can say quite accurately that nothing is pure; everything is contaminated with something else, to one degree or another. If we consider pollution to be the contamination of any substance with another, undesirable material, then the question we ought to ask isn't *whether* any particular substance is polluted but, rather, "*What's the extent* of the pollution?" To put it a bit differently, our own concerns in our everyday world of ordinary things aren't so much with the very *fact* of contamination (the unavoidable presence of undesirable materials in our air, water, food, consumer goods, and other aspects of our environment) but rather with the actual *level* of contamination that exists, the concentrations of these disagreeable materials. Coupled with this, we must ask what levels of contamination we as individuals and as a society are willing to tolerate and what levels we will not accept.

Let's take a concrete example, one we'll return to repeatedly in this discussion. Start with eight drinking glasses, a ruler, a marking pen, and some table salt and table sugar. The glasses need not all be the same size and shape, but it's easier if they are. Put a mark near the top of one of the glasses (the smallest, if they aren't all the same size) so that you can fill it with the same volume of water repeatedly. That's your standard glass; you'll use it to fill the others. Mark the remaining seven from 1 to 7, or line them up in a row so you can always tell which is which.

Put a teaspoon of salt into the standard glass, fill it to the mark with warm tap water, and stir until the salt dissolves. Now pour this salt water into glass 1. Using the ruler and the marking pen, put a mark on the side of glass 1, 1/10th of the way down from the water's surface. With this mark as a guide, transfer 1/10th of the salt water in glass 1 (and therefore 1/10th of the salt) back to the standard glass. (The measurement will be most accurate if all the glasses have straight, vertical sides, but even if they don't, the mark will still do nicely for this demonstration.)

Now once again fill the standard glass to its own mark with warm tap water and stir the solution a bit so that the salt is evenly distributed. You have just diluted the salt water from glass 1 by a factor of 10. Empty this new, diluted salt water mixture into glass 2.

Repeat the procedure until each of the seven glasses holds a standard glass of salt water. The first glass contains salt water at a concentration of one teaspoon of salt per glass of water. Each of the remaining six contains 10% of the salt in the one before it. Figure 9.1 shows the preparation of the first two.

Figure 9.1 *(continued)* 2. Pour the standard glass of saltwater into glass 1 and place a mark one-tenth of the way down from the top of the saltwater solution. 3. Using the mark on glass 1 as a guide, pour one-tenth of the solution from glass 1 back into the empty standard measuring glass. 4. Dilute the saltwater solution in the standard measuring glass by adding enough fresh water to bring the level of the solution up to the mark. Now you can pour this solution from the standard measuring glass into glass 2. Repeat the process so that the concentration of the salt in each glass of the set is one-tenth the concentration of the salt in the preceding glass.

Taste the water in each glass, starting with the most dilute. A good way to do this is to use a fresh set of three glasses. Have a friend pour a little water from glass 7 into one of them and put tap water into the other two so that you don't know which is which. Now taste the water in the three glasses. Can you tell which has the salt in it? Repeat this procedure, going successively to the lower numbered glasses (with higher concentrations of salt) until you're sure you can taste the salt. At what dilution can you first taste the salt: 1/1000 teaspoon per glass? 1/100 teaspoon? 1/10 teaspoon? 1 teaspoon? You probably find that the water in glass 1 is too salty to drink. Would you call the water in glass 1 polluted with salt? Do you think you can drink the water in glass 7? Would you call the water in that glass polluted? What fraction of a teaspoon of salt is there in glass 7? Now repeat this entire procedure with solutions of sugar instead of salt.

As we proceed through this chapter, this demonstration will help us understand how we count chemical particles, what concentrations are and how we describe them, and the importance of measuring levels of pollution and dealing with pollution in a quantitative fashion. ■

9.1 A Pair, a Six-Pack, a Dozen, and More Than All the Stars in the Sky

We buy, sell, and count out things in units. They can be units of weight, as in a pound of hamburger; units of volume, as in five gallons of gas; units of length, as in two yards of fabric; or units of area, as in three acres of timberland. Sometimes the units reflect a natural or a common way of using things. We buy socks in pairs, for example. Since we have two feet, the pair is a good unit for socks. Lose one sock and you'll have to buy two of them to replace it. We usually buy soft drinks in six-packs and eggs in dozens. Sometimes we buy and sell very small, inexpensive things in packages of 50 or 100 or 250 simply for convenience. We're not likely to find run-of-the-mill paper clips or rubber bands for sale in ones and twos. More likely we'd buy a box of 50 or 100 or so (Fig. 9.2).

Atoms, molecules, and ions give us the problem of units on a scale unlike any other. In even the smallest quantity of material we might handle, the enormous numbers of chemical particles present simply overwhelm our ability to count. Terms like pair and dozen are useless; counting into the thousands or millions or billions is hopeless. There are more atoms of carbon in a single lump of charcoal than there are stars in the sky. Our common language fails us. We have to use new ideas and new words to count out the particles of chemistry.

We also have to use arithmetic. Like all sciences, chemistry functions through experimental observations of the world around us and through mathematical manipulations of the results they produce. Whether we're counting socks, cans of soda, sheets of paper, or atoms and molecules, we have to use arithmetic to manipulate the numbers. Understanding the extraordinary chemistry of even the simplest things often requires familiarity with exponential or scientific notation, facility in the use of the metric system, and the ability to perform simple calculations involving values and measurements that carry a variety of units.

Appendices A, B, C, and D at the back of this book provide a brief introduction to the mathematical tools of the chemist. A review of them may be helpful before continuing with this chapter. Appendix E contains a detailed solution to Exercise 26 at the end of this chapter.

Figure 9.2 A pair of socks, a six-pack of soda, a dozen eggs, and a box of 100 paper clips.

■ QUESTION

What number of individual items or smaller units make up (a) an ordinary deck of cards, without the jokers; (b) a baseball team (on the playing field); (c) a piano duet; (d) a kilogram; (e) a decade?

9.2 Counting Atoms and Molecules: Part I

Our examination of how we count molecules returns us to the charcoal grill of Section 6.14. Back there we used the example of a charcoal briquette burning in a backyard grill to begin an examination of balanced equations. We defined a balanced equation as one in which each atom of each element of the products, to the right of the arrow, exactly balances an atom of the same element among the reactants, to the left of the arrow. We saw, for example, that when a charcoal briquette—composed almost entirely of elemental carbon—burns in the open air, the carbon combines with oxygen to form carbon dioxide according to the balanced equation

$$C + O_2 \longrightarrow CO_2$$

Here, one atom of carbon and two atoms of oxygen (bonded to each other in a diatomic oxygen molecule) serve as the reactants. The same single carbon atom and two oxygen atoms form the product molecule, carbon dioxide.

The following two Examples review the process of balancing an equation and extend the chemistry to the combustion of some of the alternative fuels discussed in the Perspective of Chapter 7.

EXAMPLE

Starting with Ethanol

As we saw in the Perspective of Chapter 7, ethanol (ethyl alcohol, $CH_3—CH_2—OH$) shows promise as an alternative fuel for automobiles. Write and balance the chemical equation for the combustion of ethanol to carbon dioxide and water.

■ As in the Examples of Section 6.14, we start with the unbalanced equation

$$\textit{unbalanced:}\quad C_2H_6O + O_2 \longrightarrow CO_2 + H_2O$$

Initially we can focus on the two carbon atoms and six hydrogen atoms of the ethanol. To balance ethanol's two carbons we'll include *two* carbon dioxide molecules in the product. For the six hydrogens we'll use *three* water molecules:

$$C_2H_6O + O_2 \longrightarrow 2CO_2 + 3H_2O$$

As written above we now have seven oxygen atoms among all the molecules in the products but only three among the reactants. If we place three diatomic oxygen molecules in the reactant side of the equation we now have the full complement of seven oxygen atoms, and the equation is balanced:

$$\textit{balanced:}\quad C_2H_6O + 3O_2 \longrightarrow 2CO_2 + 3H_2O$$

Each side of the equation now contains two carbon atoms, six hydrogens, and seven oxygens.

EXAMPLE

Problems with Methanol

Methanol ($CH_3{-}OH$, methyl alcohol) gives us additional problems, both when used as a fuel—as described in the Perspective of Chapter 7—and when balancing the equation for its combustion:

$$\textit{unbalanced:}\quad CH_4O + O_2 \longrightarrow CO_2 + H_2O$$

■ Proceeding as we did with ethanol, we balance the single carbon atom and four hydrogen atoms in the molecular formula of methanol with one carbon dioxide molecule and *two* water molecules in the products:

$$CH_4O + O_2 \longrightarrow CO_2 + 2H_2O$$

This gives us four oxygen atoms among the products but only three in the reactants. We now face a two-part problem:

- The products contain an even number of oxygen atoms while the reactants contain an odd number.
- Each diatomic oxygen molecule we might add to the reactant side of the equation introduces two oxygen atoms and thereby maintains an odd number of oxygen atoms among the reactants.

Since we can't balance the equation by adding oxygen molecules to the reactants, we'll add another methanol molecule. With this additional methanol molecule we now have to specify, among the products, two carbon dioxide molecules (for two carbon atoms now present among the reactants) and four water molecules (for the eight hydrogens):

$$2CH_4O + O_2 \longrightarrow 2CO_2 + 4H_2O$$

This gives us eight oxygens (an even number) among the products and four oxygens (another even number) among the reactants. With the two methanol molecules contributing two oxygen atoms, we can obtain the needed total of eight oxygen atoms by using three oxygen molecules:

$$\textit{balanced:}\quad 2CH_4O + 3O_2 \longrightarrow 2CO_2 + 4H_2O$$

The combustion equation for methanol is now balanced.

With an understanding of what constitutes a balanced equation we can now return to the backyard grill and move further into its chemistry by asking: How much oxygen does the carbon in the sack of charcoal consume as it burns?

A single lump of commercial charcoal, the kind that's made for use in outdoor grills, weighs about 36 g. If we make the reasonably accurate assumption that the charcoal is pure carbon, it's easy enough to calculate exactly how much oxygen it takes to convert the entire 36 g of charcoal into carbon dioxide.

A charcoal briquette that weighs 36 grams.

Here's how. We start with the balanced equation

$$C + O_2 \longrightarrow CO_2$$

which tells us that one atom of carbon combines with one diatomic molecule of oxygen to form one molecule of CO_2. We also know, from the periodic table, that the mass of a carbon atom is 12 amu (Sec. 2.4) and that the mass of one atom of oxygen is 16 amu. Naturally, the mass of a *diatomic molecule* of oxygen is twice this, or 32 amu. If we had a balance that could give us the mass of single atoms and molecules, we'd see something like Figure 9.3. One atom of carbon has a mass of 12 amu; one molecule of oxygen, 32 amu. Now we can answer a question that helps us along: "What mass of oxygen will react completely with 12 amu of carbon to give carbon dioxide as a product?" The answer must be "32 amu of oxygen" since we have to work with whole atoms and entire molecules. One whole carbon atom has a mass of 12 amu and one entire oxygen molecule (two oxygen atoms) has a mass of 32 amu.

The balances in Figure 9.3 are completely imaginary; now we'll turn to real scales and actual weights. Figure 9.4 shows two scales calibrated in grams. One of the scales holds a piece of charcoal made up of 12 *grams* of carbon. Now we ask the question, "What weight of oxygen reacts with 12 *grams* of carbon to produce carbon dioxide?" To answer this question we don't have to know how many carbon atoms are on the scale in Figure 9.4. We need only weigh out 32 *grams* of oxygen to have the same number of oxygen molecules on hand as we have carbon atoms, whatever that number may be. We know that's so because the ratio of the mass of a molecule of oxygen to the mass of an atom of carbon is 32 : 12. The same ratio holds true for their weights.

Figure 9.3 The mass of one carbon atom and of one diatomic oxygen molecule, expressed in atomic mass units. The balances are completely imaginary.

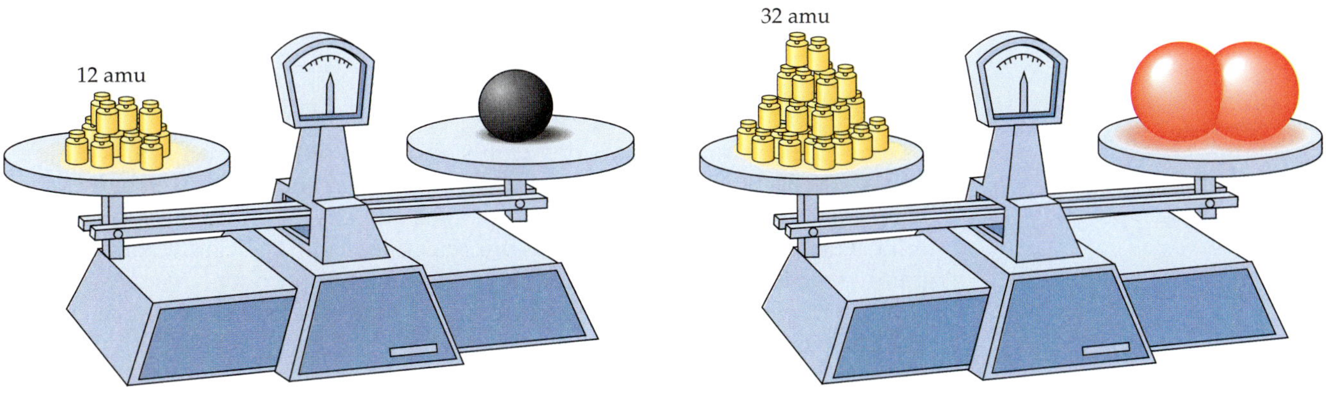

One atom of the most common isotope of carbon has a mass of 12 atomic mass units

One diatomic oxygen molecule, consisting of two atoms of the most common isotope of oxygen, has a mass of 32 atomic mass units

Figure 9.4 Equal numbers of carbon atoms and oxygen molecules.

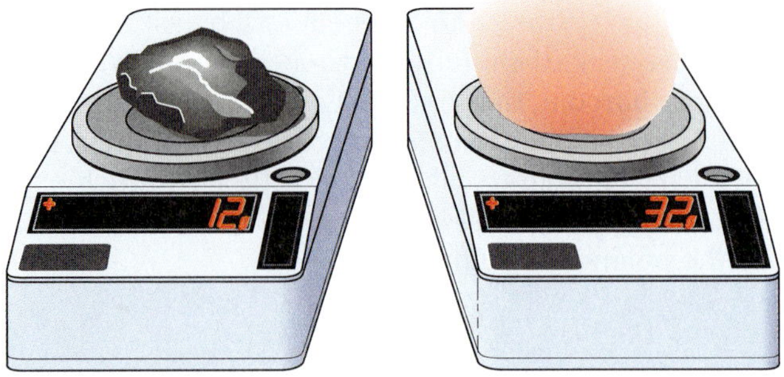

There are as many diatomic oxygen molecules in 32 g of oxygen as there are carbon atoms in 12 g of carbon

What's important about all this is that *we have counted out atoms by weighing them*. In actual practice we could weigh out 12 g of carbon as charcoal. For 32 g of oxygen we could place a cylinder containing oxygen on a scale and open its valve until the weight of the cylinder and the oxygen it contains changes by 32 g.

One mole of carbon, of lead, of water, and of sulfur.

We don't know (just yet) how many carbon atoms and how many oxygen molecules we are dealing with in Figure 9.4, but we do know that the two scales hold *equal numbers of the two chemical species*. And since the balanced equation tells us that carbon atoms and oxygen molecules react in a 1 : 1 ratio to produce carbon dioxide, all the oxygen atoms in 32 g of oxygen will react with all the carbon atoms in 12 g of carbon, with neither carbon nor oxygen left over.

Now we can answer our original question, "How much oxygen does the carbon in the sack of charcoal consume as it burns?" We now know that one molecule of diatomic oxygen reacts with one atom of carbon, and we know that the ratio of the mass of a molecule of oxygen to the mass of an atom of carbon is 32 : 12, or 8 : 3. The mass (or weight) of oxygen we need is $\frac{8}{3}$ of whatever weight of carbon we have. If we assume that each lump of charcoal weighs 36 g, we need

$$36 \text{ g of carbon} \times \tfrac{8}{3} = 96 \text{ g of oxygen}$$

for each lump of charcoal in the bag.

■ **QUESTION**

What weight of CO_2 results when 36 g of charcoal burns completely in air?

9.3 Avogadro and the Mole

As long as the ratio of masses or weights of oxygen molecules and carbon atoms is 32 : 12 (8 : 3)—the ratio of their respective molecular and atomic weights—we are dealing with equal numbers of carbon atoms and oxygen molecules. With this in mind, it doesn't matter in what units our scales are calibrated: grams, ounces, kilograms, pounds, tons, or some other units. All will do. Nor do we have to work with whole multiples of 12. Simple algebra tells us that if we have x grams of carbon, then it takes $\frac{8}{3}$ times x grams of oxygen to convert all the C to CO_2.

As simple as the algebra is, it's far easier and much more convenient to handle the calculations of chemistry by defining a specific number of chemical particles in the same sort of way we define specific numbers of socks (a pair), soft drink cans (a six-pack), eggs (a dozen), or sheets of printer paper (a ream). The set of chemical particles chemists have worked out, the *mole*, has the same relation to atoms, molecules, and ions that a pair has to socks, a six-pack to cans of soda, and a dozen to eggs. That is, just as a pair, a six-pack, or a dozen represent specific numbers of items, a mole also represents a specific number of items, which happen to be chemical particles. *A mole is a specific number of chemical particles.*

A **mole** is the number of atoms of carbon-12 in exactly 12 g of carbon-12. In practice, a *mole* of a substance is the number of chemical particles contained in its atomic, molecular, or ionic weight, expressed in grams.

By its precise, chemical definition, a **mole** (abbreviated *mol*) is the amount of any substance that contains the same number of chemical particles as there are atoms in exactly 12 g of carbon-12. Looked at from a different angle, the number of atoms of carbon-12 in exactly 12 g of this particular isotope of carbon defines the number of chemical particles in a mole. (Carbon-12 is the isotope of carbon with a mass of 12 amu; Sec. 2.6.) In practice, if you weigh out a quantity of an element or a compound equal to its atomic or molecular weight expressed in *grams*, you have 1 mole of atoms or molecules of the substance. (Be careful: The amount of a chemical that makes up a mole is its *mass* expressed in grams. Normally, at the surface of the earth the weight of a substance is the same as its mass, as we noted in Sec. 2.3.)

To use the mole successfully, we need only recognize the simplifying fact that *one mole of any chemical substance contains the same number of chemical particles as one mole of any other chemical substance*. One mole of carbon atoms (12 g of C) reacts completely with 1 mole of oxygen molecules (32 g of O_2) because 1 mole of carbon and 1 mole of molecular oxygen contain exactly the same numbers of carbon atoms and oxygen molecules.

Amadeo Avogadro, for whom the number of chemical particles in a mole is named.

Although knowing the exact number of particles in a mole isn't critical to using the whole idea successfully, chemists are never satisfied with ignorance. We know now that 1 mole of anything consists of 6.02×10^{23} chemical particles. This number is called **Avogadro's number** in honor of the Italian physicist Amedeo Avogadro, who was born in the year of the Declaration of Independence and whose pioneering work led to the way we now count chemical particles.

Avogadro's number, 6.02×10^{23}, is the experimentally determined number of chemical particles in a mole.

Avogadro's number is staggeringly large. In terms of time, Avogadro's number of seconds, 6.02×10^{23} seconds, stretches out to about 2×10^{14} centuries, which is roughly a million times the best current estimates of the age of the entire universe. In money, just 1 mole of pennies would fill all the space within the moon's orbit around the earth in stacks of about 400 pennies each (Fig. 9.5). Distributed

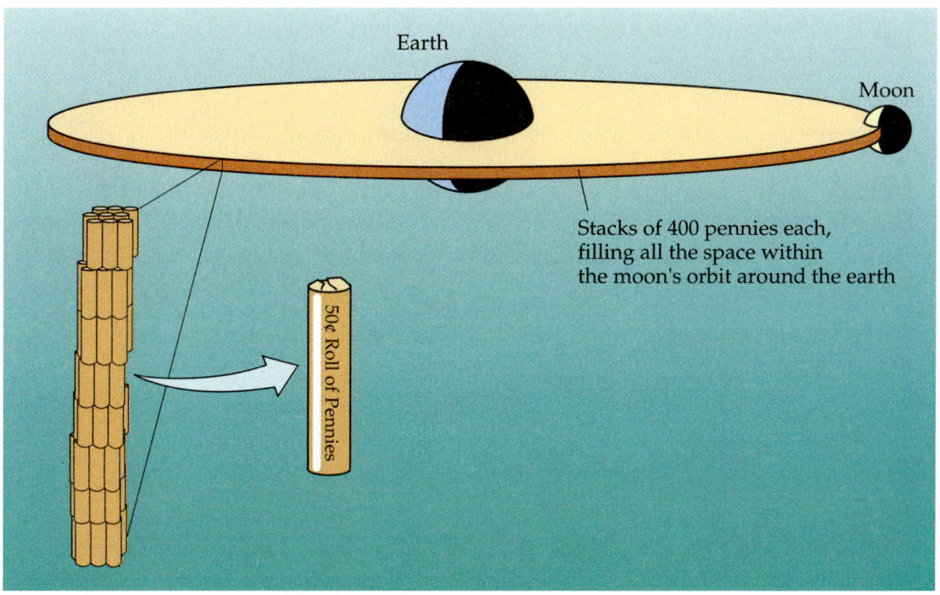

Figure 9.5 A mole of pennies.

Figure 9.6 There are 3.68×10^{23} atoms of aluminum in the can, 1.58×10^{23} atoms of silver in the coin, and 79×10^{23} molecules of water in the glass.

equally to everyone in the earth's population of about 6 billion people, a mole of pennies would give each one of us a trillion dollars.

In terms of chemical particles, an aluminum soft drink can, which weighs about 16.5 g, contains about

$$16.5 \text{ g Al} \times \frac{\text{mol Al}}{27.0 \text{ g Al}} \times \frac{6.02 \times 10^{23} \text{ atoms Al}}{\text{mol Al}} = 3.68 \times 10^{23} \text{ atoms of aluminum}$$

A 1-oz coin of pure silver contains

$$1 \text{ oz Ag} \times \frac{28.3 \text{ g}}{\text{oz}} \times \frac{\text{mol Ag}}{108 \text{ g Ag}} \times \frac{6.02 \times 10^{23} \text{ atoms Ag}}{\text{mol Ag}}$$

$$= 1.58 \times 10^{23} \text{ atoms of silver}$$

As another illustration, an ordinary 8-oz glass of water contains about 13.1 mol of water, which amounts to 13.1 times Avogadro's number, or just under 79×10^{23} molecules of water (Fig. 9.6). By any measure that's a lot of molecules.

Getting a good value for the number hasn't been simple. Our best estimate comes from X-ray measurements of various crystalline substances, including diamonds and a form of calcium carbonate ($CaCO_3$) known as calcite.

Before continuing, it might be good to distinguish clearly between the *number* of a group of objects or items and their *mass* or *weight*. Suppose, for example, you had a dozen feathers and a dozen elephants. You would have the same *number* of each (12, a dozen), but would the two groups have the same mass? Then again, suppose you have 3 tons of feathers and 3 tons of elephants. You would have the same mass of each, 3 tons. But would you expect to have the same number of elephants as feathers?

We'll examine the mole and Avogadro's number further in the following examples.

E X A M P L E

Golden Moles

Which has more atoms, 1 mole of gold or 1 mole of lead?

■ A mole is a specific number of chemical particles. One mole of gold contains just as many atoms as 1 mole of lead (just as there are as many socks in a pair of socks as there are shoes in a pair of shoes).

E X A M P L E

Aluminating Helium

Compare the number of atoms in 27 g of aluminum and 4 g of helium.

■ Here we have different masses of two elements, aluminum and helium. Yet the two samples contain the same number of atoms. Since 27 g of aluminum represents the atomic weight of aluminum, in grams, and 4 g of helium represents the atomic weight of helium, again in grams, the two contain the same number of atoms, 1 mole of the atoms of each element.

E X A M P L E

Hydrogen or Oxygen?

Which represents more atoms, 1 gram of hydrogen atoms or 1 gram of oxygen atoms?

■ In this example we have the same masses of hydrogen atoms and oxygen atoms. Since the atomic weight of oxygen is 16, 1 gram of oxygen represents

1/16 of a mole of oxygen atoms. On the other hand, 1 gram of hydrogen, atomic weight 1.008, represents very nearly 1 mole of hydrogen atoms. Since 1 mole of any chemical substance must contain exactly the same number of chemical particles as 1 mole of any other chemical substance, there must be almost 16 times as many atoms of hydrogen in 1 gram of hydrogen as there are oxygen atoms in 1 gram of oxygen.

E X A M P L E

One Gold Atom

What's the mass of one gold atom?

■ We know that the atomic weight of an element, expressed in grams, contains 1 mole of that element. We also know that 1 mole of any chemical substance contains 6.02×10^{23} atoms, molecules, or ions of that substance.

Since the atomic weight of gold is 197 amu (a value we can find in the periodic table or in any table of atomic weights), 197 *grams* of gold must contain 1 *mole* of gold, or 6.02×10^{23} *atoms* of gold.

The product of the two fractions,

$$\frac{197 \text{ grams of gold}}{\text{mole of gold}} \times \frac{\text{mole of gold}}{6.02 \times 10^{23} \text{ gold atoms}}$$

gives the result

$$32.7 \times 10^{-23} \frac{\text{grams of gold}}{\text{gold atom}} \quad \text{or} \quad 3.27 \times 10^{-22} \text{ grams per gold atom}$$

One gold atom has a mass of 3.27×10^{-22} grams. Compare this Example with the Question at the end of Section 2.2

E X A M P L E

Hydrogen Atoms

How many hydrogen atoms are there in 18.02 g of water?

■ The molecular weight of water is

$$\begin{array}{lrl} \text{H:} & 2 \times 1.008 \text{ amu} = & 2.016 \text{ amu} \\ \text{O:} & 1 \times 16.00 \text{ amu} = & \underline{16.00 \text{ amu}} \\ & & 18.02 \text{ amu} \end{array}$$

This mass of water (18.02 g) represents 1 mole of water. Since there are two hydrogen atoms for each water molecule, there must be twice as many hydrogen atoms as there are water molecules in *any* quantity of water molecules. In 18.02 g of water (1 mole of H_2O) there must be *two* moles of hydrogen atoms, or $2 \times 6.02 \times 10^{23}$, or 12.04×10^{23} hydrogen atoms.

■ Q U E S T I O N

Which contains the greater number of carbon atoms: 1 gram of CO_2 (atomic weight 44 amu) or 1 gram of CO (atomic weight 28 amu)? Explain your reasoning.

9.4 Counting Atoms and Molecules: Part II

Now, by using the mole and understanding its connection to Avogadro's number, we can count out the chemical particles that take part in any reaction.

One of the most important of all the reactions our society uses is the burning of the hydrocarbons of gasoline, natural gas, kerosene, jet fuel, diesel fuel, and other products derived from petroleum (Chapter 7). In Chapter 8 we saw evidence that the carbon dioxide produced by this combustion contributes to the gradual warming of the earth's surface. We'll look now at the quantitative connection between the amount of the fuel that's used and the amount of carbon dioxide generated.

We'll look specifically at the combustion of *methane*, a relatively simple compound with the molecular formula CH_4. Methane is the principal component of natural gas, a fuel widely used for cooking and heating and in many industrial processes, including the manufacture of steel. Methane burns according to the balanced equation

$$CH_4 + 2O_2 \longrightarrow CO_2 + 2H_2O$$

This equation tells us that *one* molecule (or 1 mole) of methane reacts with *two* molecules (or 2 moles) of diatomic oxygen to produce *one* molecule (or 1 mole) of carbon dioxide and *two* molecules (or 2 moles) of water (Fig. 9.7). (Naturally, other quantities of these chemicals can react with each other as well, but always in that same ratio.) This must be the case since in this balanced equation every atom of every element that appears among the products also appears among the reactants.

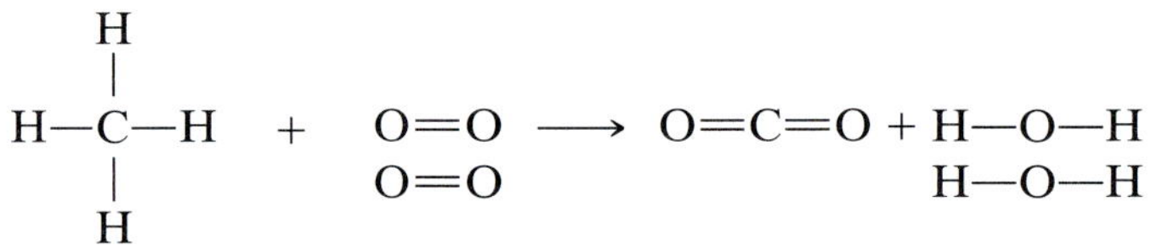

1 carbon atom, 4 hydrogen atoms, and 4 oxygen atoms   1 carbon atom, 4 hydrogen atoms, and 4 oxygen atoms

$$CH_4 \quad + \quad 2O_2 \longrightarrow CO_2 \quad + \quad 2H_2O$$

Figure 9.7 The reaction of one molecule of methane with two molecules of oxygen to form one molecule of carbon dioxide and two molecules of water.

We can summarize the logic of all this as follows: Since matter may be neither created nor destroyed in a chemical reaction (Sec. 4.11), every atom appearing on one side of the arrow must appear on the other side as well. To obtain this condition, we balance equations. A balanced equation, in turn, tells us the ratio of the moles (or atoms or molecules or ions) of all the substances that make up its reactants and its products. From this information we can calculate the mass of each chemical involved in the reaction, given the mass of any one of them. The following example shows how.

E X A M P L E

Five Easy Steps

What weight of carbon dioxide is generated for every 100 g of methane that we use as fuel?

■ We can solve this problem in five simple steps, represented graphically as follows:

	Step 2 Weight of methane		*Step 5* Weight of carbon dioxide
Step 1 Balanced equation:	$CH_4 + 2O_2$	$\longrightarrow$	$CO_2 + 2H_2O$
	Step 3 Moles of methane		*Step 4* Moles of carbon dioxide

Step 1: Write the balanced equation for the reaction. This gives us the relationships among the numbers of particles of each chemical of the reactants and each chemical of the products:

$$CH_4 + 2O_2 \longrightarrow CO_2 + 2H_2O$$

Step 2: Identify the *known weight* of any of the chemicals taking part in the reaction. In this case we are told that 100 g of methane is used. Since we want to know how much carbon dioxide appears for every 100 g of methane that disappears, we'll start with 100 g of methane:

$$\underset{}{\overset{\textit{100 g}}{CH_4}} + 2O_2 \longrightarrow CO_2 + 2H_2O$$

Step 3: Determine how many moles this corresponds to. For the molecular weight of methane, we find

C:	1 × 12.01 amu =	12.01 amu
H:	4 × 1.008 amu =	4.032 amu
		16.04 amu, which we round off to 16.0 amu to maintain three significant figures (Appendix D)

Thus, there are 16.0 g of methane (and 6.02×10^{23} molecules of CH_4) in every mole of methane. Using units cancellation (Appendix C), we convert 100 g of methane to

$$100 \text{ g } CH_4 \times \frac{\text{mol } CH_4}{16 \text{ g } CH_4} = 6.25 \text{ mol } CH_4$$

$$\underset{\textit{6.25 mol}}{\overset{\text{100 g}}{CH_4}} + 2O_2 \longrightarrow CO_2 + 2H_2O$$

Step 4: Convert this to the number of moles of the chemical whose weight we want. The balanced equation tells us that every molecule (or mol) of methane that reacts produces one molecule (or mol) of carbon dioxide. Therefore the disappearance of 6.25 mol of CH_4 must be accompanied by the appearance of 6.25 mol of CO_2:

$$\underset{\text{6.25 mol}}{\overset{\text{100 g}}{CH_4}} + 2O_2 \longrightarrow \underset{\textit{6.25 mol}}{CO_2} + 2H_2O$$

Step 5: Calculate the weight of the substance we're interested in. For the molecular weight of CO_2 we have

C:	1 × 12.01 =	12.01 amu
O:	2 × 16.00 =	32.00 amu
		44.01 amu, which we round off to 44.0 amu to maintain three significant figures

This means that 6.25 mol CO_2 weighs

$$6.25 \text{ mol } CO_2 \times \frac{44.0 \text{ g } CO_2}{\text{mol } CO_2} = 275 \text{ g } CO_2$$

$$\underset{6.25 \text{ mol}}{\overset{100 \text{ g}}{CH_4}} + 2O_2 \longrightarrow \underset{6.25 \text{ mol}}{\overset{\mathit{275\ g}}{CO_2}} + 2H_2O$$

We have now answered the original question. Every 100 g of CH_4 that disappears during combustion generates 275 g of CO_2.

EXAMPLE

Water, Too and Two

Water, too, is formed as methane burns. How much water is generated by the 100 g of methane?

■ Here, we can repeat the first three steps of the preceding example. But for the fourth step we must recognize that *two* molecules (or moles) of water form for every one molecule (mole) of methane that reacts:

$$\underset{\mathit{6.25\ mol}}{\overset{100 \text{ g}}{CH_4}} + 2O_2 \longrightarrow CO_2 + 2H_2O$$

Step 4: Since 6.25 moles of methane exist among the reactants, twice this amount of water—12.5 moles of H_2O—must be formed:

$$\underset{6.25 \text{ mol}}{\overset{100 \text{ g}}{CH_4}} + 2O_2 \longrightarrow CO_2 + \underset{\mathit{12.5\ mol}}{2H_2O}$$

Step 5: Since the molecular weight of water is

H: 2×1.008 amu = 2.016 amu

O: 1×16.0 amu = 16.0 amu

18.016 amu, or 18.0 amu to three significant figures

a mass of

$$12.5 \text{ mol } H_2O \times \frac{18.0 \text{ g } H_2O}{\text{mol}} = 225 \text{ g } H_2O$$

results from the reaction:

$$\underset{6.25 \text{ mol}}{\overset{100 \text{ g}}{CH_4}} + 2O_2 \longrightarrow CO_2 + \underset{12.5 \text{ mol}}{\overset{\mathit{225\ g}}{2H_2O}}$$

■ QUESTION

Using this five-step approach, determine how many grams of oxygen it takes to react completely with 100 g of methane.

9.5 Getting to the Right Solution Takes Concentration

Although weighing quantities of chemicals provides a fine method for counting out atoms, molecules, and ions, we can often count chemical particles far more conveniently and easily by measuring volumes of liquid solutions. You make up a **solution** by dissolving one substance, the **solute**, in another substance, the **solvent**. The result is a solution of the solute in the solvent. Solutions are completely homogeneous mixtures of solutes and solvents. That is, the composition and properties of any solution are perfectly uniform throughout the entire solution. Sometimes it takes a bit of stirring or shaking to achieve this homogeneity, as when you dissolved the salt and sugar in water in the opening demonstration. With enough stirring, the sodium and chloride ions of the salt and the sucrose molecules of the sugar were distributed evenly throughout the water. The sodium chloride and the sucrose were the solutes; the water was the solvent; and the salt water and the sugar water were the solutions.

A **solution** is a homogeneous mixture of one substance, the **solute**, dissolved in another, the **solvent**. The solute is present in a smaller proportion than the solvent.

If we know how much of the solute is dissolved in a specific volume of the solution, we know the **concentration** of the solute. Knowing this value allows us to measure out an exact quantity of solute by measuring a volume of the solution itself. In effect, we can count out chemical particles by measuring volumes of solutions.

The **concentration** of a solution expresses the quantity of solute dissolved in a specific quantity solution. The quantity of solution is usually stated as a volume.

In the demonstration that opened this chapter we saw examples of various concentrations of a solute (salt) in a solution (salt water). As we dissolved one teaspoon of sodium chloride in a glass of water, we prepared a solution whose

The five steps in preparing a solution of a specific concentration. (a) Adding a known weight of solute to a flask etched with a mark to allow filling it to an accurately measured, predetermined volume. (b) Adding a portion of the solvent. (c) Swirling the mixture to dissolve the solute in the solvent. (d) Adding enough solvent to bring the surface of the solution exactly to the etched mark. This produces an accurately measured volume of solution, usually a multiple or simple fraction of a liter. (e) Shaking the stoppered flask to distribute the solute uniformly throughout the solution.

(a) (b) (c) (d) (e)

concentration was *one teaspoon per glass*. While we don't normally use units of this sort, our measurements with the standard glass of water in that demonstration make this a perfectly valid, if nonstandard and not very precise, description of the concentration of salt water in glass 1. Recall that we transferred 1/10 of the salt from the first glass into glass 2 by pouring out 1/10 of the solution. The concentration of the sodium chloride in glass 2 thus became 0.1 teaspoon per glass. In the next glass the concentration was 0.01 teaspoon per glass. The same principles and definitions of that demonstration apply equally to the much more accurate measurements carried out in scientific laboratories.

QUESTION

What's the concentration of the sodium chloride in glass 7?

9.6 Molarity

The terms and units available for expressing concentrations are almost as varied as the kinds of solutions we can prepare. They range from the not-very-scientific and imprecise "teaspoon per glass" of our demonstration, through some of the terms that appear on the labels of our consumer products (which we'll examine in the next section), to well-defined units based on the concept of the mole. This last unit—the mole—makes it particularly easy to get a quick count of chemical particles of a solute by measuring the volume of a solution in an inexpensive, readily available, calibrated cup, beaker, flask, or cylinder. That beats carrying around a heavy, cumbersome, and expensive balance or scale.

The **molarity** of a solution refers to the number of moles of solute per liter of solution.

For counting chemical particles by measuring volumes of solutions we commonly use concentrations expressed in *molarity* (M). The **molarity** of a solution refers simply to *the number of moles of solute per liter of solution*. A 1 M (one molar) solution contains one mole of solute in each liter of solution; a 2 M solution contains two moles of solute per liter of solution; and so on.

Molarity provides an especially easy way to count chemical particles. Simply multiply the volume of a solution (in liters) by its concentration (in molarity) and you have the number of moles of solute in the volume of solution at hand. Two liters of a 1 M solution contain a total of 2 mol of the solute, but then so does 1 liter of a 2 M solution, or 0.5 liter of a 4 M solution:

Volume of Solution		Concentration of Solution		Number of Moles of Solute
2 liters	×	1 mol/liter	=	2 mol of solute
1 liter	×	2 mol/liter	=	2 mol of solute
0.5 liter	×	4 mol/liter	=	2 mol of solute

EXAMPLE

Cleaning Up

Solutions of ammonia are sometimes used as household cleaners. How many moles of ammonia, NH_3, are there in 1.2 liters of a solution that is 0.50 M in ammonia?

■ Using the definition of molarity as the number of moles of solute per liter of solution, we have

$$1.2 \text{ liters} \times \frac{0.50 \text{ mol } NH_3}{\text{liter}} = 0.60 \text{ mol } NH_3$$

EXAMPLE

More Ammonia

We are still using this same 0.50 *M* solution of ammonia, but this time we need 1.8 mol of the solute. What volume of the solution do we use to get 1.8 mol of ammonia?

■ This time we start with the number of moles we need and we divide by the value of the molarity to give us the answer in liters:

$$1.8 \text{ mol } NH_3 \times \frac{\text{liter}}{0.50 \text{ mol } NH_3} = 3.6 \text{ liters}$$

EXAMPLE

Mixing Up More

We have just run out of the 0.50 *M* solution of ammonia and must make up more. We prepare the solution 5 liters at a time. What weight of ammonia do we need to prepare 5.0 liters of 0.50 *M* ammonia? To convert between moles of ammonia and grams of ammonia, we use the molecular weight of ammonia:

$$\begin{array}{lll} \text{N:} & 1 \times 14.01 \text{ amu} = & 14.01 \text{ amu} \\ \text{H:} & 3 \times 1.008 \text{ amu} = & \underline{3.024 \text{ amu}} \\ & & 17.03 \text{ amu,} \end{array}$$

which we round off to 17 amu since we're using only two significant figures for volume and concentration

■ In this case we know that we need 5.0 liters of solution, we know that the solution must contain 0.50 mol of ammonia per liter, and we know that there are 17 g of ammonia per mole. This knowledge gives us

$$5.0 \text{ liters} \times \frac{0.50 \text{ mol}}{\text{liter}} \times \frac{17 \text{ g } NH_3}{\text{mol}} = 42 \text{ g } NH_3$$

Our desired solution contains 42 g of ammonia in 5.0 liters of solution.

In a final illustration we stretch our definition of molarity just a bit to find the molarity of water molecules themselves in any ordinary sample of water. It's only a small stretch since molarity normally deals with the concentration of solute molecules in a solvent. In this case we're considering water to be both the solute and the solvent.

With a density of 1000 g per liter for water and a molecular weight of 18.0 amu for the water molecule, we have

$$\frac{1000 \text{ g water}}{\text{liter of water}} \times \frac{\text{mol of water}}{18.0 \text{ g water}} = 55.6 \text{ mol of water per liter}$$

With 55.6 mol of water molecules per liter of water, the molarity of water is 55.6 *M*. This gives water, of all our common liquids, the highest concentration of molecules in a given volume.

■ QUESTION

(a) Calculate the molarity of the salt water and the sugar water of glass 1 of the opening demonstration. Assume that the standard glass holds 400 mL of each solution and that the rounded teaspoons of sodium chloride and sucrose you used in preparing the solutions weighed 23.4 g and 17.1 g, respectively. Use 23.0 for the atomic weight of Na, 35.5 for Cl, and 342 for the molecular weight of sucrose. (b) Calculate the molarity of the solution of glass 7.

9.7 Percentage Concentrations

Although molarity is a common measure of concentration in scientific studies, the solutions we prepare in our daily lives employ more common units. We use a teaspoon or two of sugar in our coffee or tea and a couple shakes of salt (if any at all) in our soup. A quarter-cup of detergent does the wash, and a squeeze of the plastic bottle does the dishes. We take medicines a tablet or two or a teaspoon at a time. In each of these cases we're dealing with the concentration of a solute in a solution. A teaspoon of sugar in coffee or tea produces a concentration of one teaspoon per cup. The single teaspoon of sugar is the solute and the single cup of sweetened beverage is the solution. A quarter-cup of detergent gives us a concentration of 0.25 cup per laundry tub. A squeeze of dishwashing detergent into a sinkful of dishes gives us an imprecise concentration of one squeeze per sink. Even two aspirin tablets and a glass of water produce a certain concentration of *acetylsalicylic acid*, the active component of the tablet, per human body (Sec. 23.1). The unit of concentration of the salt water we prepared in our opening demonstration was "teaspoon per glass."

Vinegar contains acetic acid at a concentration of 5% by weight.

When we look at the labels on our commercial goods, we often find concentrations of solutes expressed as a percentage (%) of weight or volume. The acidity of vinegar, for example, results from a combination of simple organic acids, principally *acetic acid*, dissolved in water. The label of a typical bottle of ordinary vinegar reveals that the product is "diluted with water to 5% acidity." This is a form of commercial shorthand (or maybe simply jargon) that's meant to indicate the concentration of acids in vinegar. Commercial vinegar normally contains 5 g of acetic acid (Fig. 9.8) in every 100 g of vinegar. A percentage concentration expressing the weight of solute for every 100 units of weight of the solution is a weight/weight percentage, or w/w %.

A 3% solution of H_2O_2 in water forms the common antiseptic *hydrogen peroxide*, while rubbing alcohol is a solution of 70% *isopropyl alcohol* and 30% water.

Figure 9.8 Commercial solutes, solvents, and solutions.

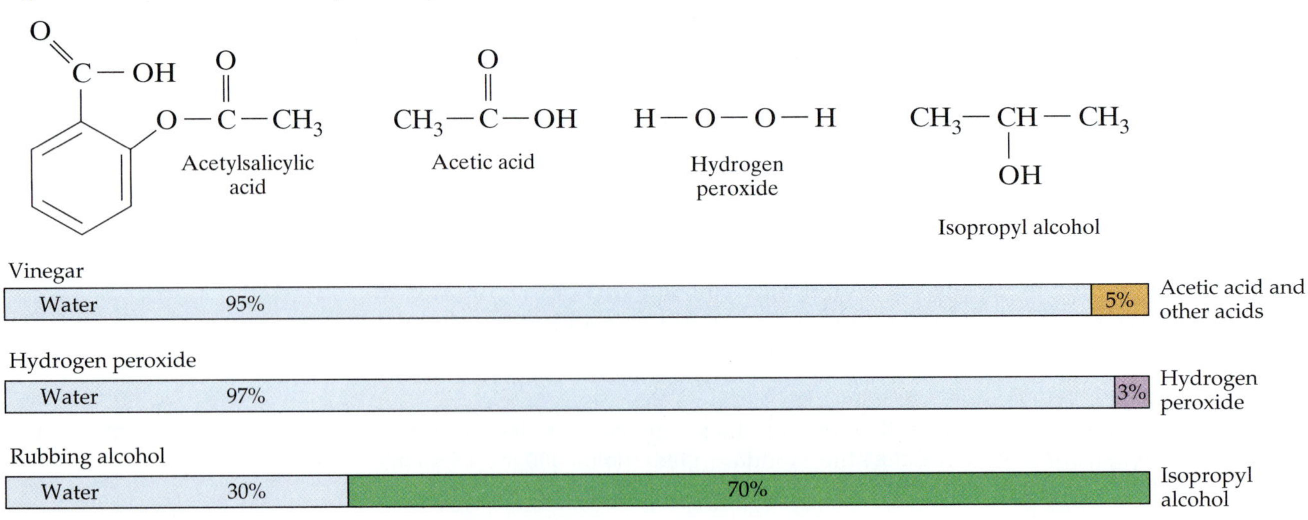

As far as the language of chemistry is concerned, water is the solute in rubbing alcohol since water is the minor component of the mixture. Isopropyl alcohol is the solvent in this case. In chemical usage the only difference between "solute" and "solvent" lies in the relative proportions of the two.

Figure 9.8 shows the percentage compositions (by weight) of vinegar, commercial hydrogen peroxide, and rubbing alcohol.

■ QUESTION

A typical 8-fluid-ounce cup of coffee weighs about 240 g, while a teaspoon of sugar weighs about 5 g. What is the weight-percentage concentration of sugar in the coffee when you sweeten a typical cup of coffee with two teaspoons of sugar? (Don't forget to add the weight of the added sugar to the total weight of the solution.)

9.8 Molarities of Common Solutions: Vinegar, Hydrogen Peroxide, and Sweet Coffee

To get a sense of the range of molarities of some common solutions, we'll calculate the molarities of

- acetic acid in vinegar,
- hydrogen peroxide in the commercial solution we might find in a drugstore, and
- the sucrose in a cup of sweet coffee.

To simplify all our calculations, we'll assume that each of these has a density of exactly 1 (Sec. 13.1). That is, we'll assume that 1 liter of each has a mass of exactly 1000 grams. While this isn't quite true, it's close enough for a calculation of approximate molarities. Since these are only very rough calculations, we'll find the molarities to only one significant figure (Appendix D).

For vinegar, we'll also assume that all of the acidity of vinegar is due to acetic acid, which has the molecular formula $C_2H_4O_2$, and that the acetic acid is present as a 5% (w/w) solution: 5 g of acetic acid per 100 g of vinegar.

EXAMPLE

Acetic Acid in Vinegar

With the assumptions made above, what is the molarity of acetic acid in vinegar?

■ The molecular weight of acetic acid is

$$\begin{aligned}
\text{C:}\quad & 2 \times 12.01\ \text{amu} = 24.02\ \text{amu} \\
\text{H:}\quad & 4 \times 1.008\ \text{amu} = 4.032\ \text{amu} \\
\text{O:}\quad & 2 \times 16.00\ \text{amu} = \underline{32.00\ \text{amu}} \\
& \qquad\qquad\qquad 60.05\ \text{amu, which we round off to 60 amu}
\end{aligned}$$

At a concentration of 5% (w/w) and with the assumptions we've made, there are 5 g of acetic acid per 100 g of vinegar. Starting with this concentration we get

$$\frac{5\text{ g } C_2H_4O_2}{100\text{ g vinegar}} \times \frac{1\text{ mol } C_2H_4O_2}{60\text{ g } C_2H_4O_2} \times \frac{1000\text{ g vinegar}}{\text{liter vinegar}} = \frac{0.8\text{ mol } C_2H_4O_2}{\text{liter vinegar}} = 0.8\ M$$

The concentration of acetic acid in vinegar is 0.8 M.

EXAMPLE

Drugstore Hydrogen Peroxide

Now for the hydrogen peroxide, H_2O_2.

■ The molecular weight of H_2O_2 is:

H: 2×1.008 amu = 2.016 amu

O: 2×16.00 amu = 32.00 amu

34.02 amu, which we'll round off to 34 amu

At a concentration of 3% (w/w) and with the assumptions we've made, there are 3 g of hydrogen peroxide per 100 g of the commercial solution. Starting with this concentration, we get

$$\frac{3\text{ g } H_2O_2}{100\text{ g solution}} \times \frac{1\text{ mol } H_2O_2}{34\text{ g } H_2O_2} \times \frac{1000\text{ g solution}}{\text{liter solution}}$$

$$= \frac{0.9\text{ moles } H_2O_2}{\text{liter hydrogen peroxide solution}} = 0.9\ M$$

The concentration of hydrogen peroxide in the commercial solution is 0.9 *M*.

EXAMPLE

Sweet Coffee

Using the information given in the question at the end of Section 9.7, a molecular formula of $C_{12}H_{22}O_{11}$ for sucrose (table sugar), and units conversion data taken from Appendix B, we can calculate the approximate molarity of the sugar in the cup of coffee.

■ For the molecular weight of the sucrose we have

C: 12×12.01 amu = 144.12 amu

H: 22×1.008 amu = 22.176 amu

O: 11×16.00 amu = 176.00 amu

342.296 amu, which we'll round off to 342 amu

Recalling that we use two teaspoons of sugar, at 5 g each, we get

$$\frac{10\text{ g sucrose}}{250\text{ g coffee}} \times \frac{1\text{ mol sucrose}}{342\text{ g sucrose}} \times \frac{1000\text{ g coffee}}{\text{liter of coffee}} = 0.1\ M$$

Two teaspoons of table sugar in an 8-oz cup of coffee produces coffee that's about 0.1 *M* in sucrose.

■ QUESTION

Which represents the greatest number of molecules of solute per liter of solution: (a) the acetic acid in vinegar, (b) the hydrogen peroxide in the commercial solution, or (c) the sucrose in the coffee of the preceding example?

9.9 You're 167 in a Trillion: Expressing Exceedingly Small Concentrations

Even though molarity is a useful concentration term for chemists, and percentages are informative for many consumer products, the very small concentrations of the pollutants that affect our environment require a different approach.

It's useful to express exceedingly small concentrations, such as those of food contaminants and environmental pollutants, in terms of parts per thousand, parts per million, parts per billion, and so on. These terms are closely related to the percentages we've already examined. The term *percent* literally means "per hundred." A one percent solution contains one unit of solute in each 100 units of solution, whatever the units happen to be. We can also speak of concentrations in terms of parts per thousand, which are the same as tenths of a percent. The coffee in the question at the end of Section 9.7, for example, contains 10 g of sugar in 250 g of the entire mixture. That amounts to 4%, or 4 parts per hundred, which is the same as 40 parts per thousand.

One *part per million* (one **ppm**) represents a particularly convenient concentration unit since it's the concentration of one milligram of one substance distributed throughout one kilogram of another. A concentration of 1 ppm, then, is the same as a concentration of one milligram per kilogram, 1 mg/kg. One milligram is a thousandth part of a gram and a gram is a thousandth part of a kilogram. One-thousandth of one-thousandth is one-millionth:

The concentration term **ppm** refers to parts per million, or milligrams of solute per kilogram or liter of solution.

$$\underset{\text{one-thousandth}}{\frac{1}{1000}} \times \underset{\text{one-thousandth}}{\frac{1}{1000}} = \underset{\text{one-millionth}}{\frac{1}{1{,}000{,}000}}$$

As an illustration, 35 mg of one substance in every kilogram of another would amount to a concentration of 35 mg/kg, or 35 parts per million, or 35 ppm.

Consumer products provide plenty of examples of concentration terms. Labels of canned fruit drinks, for example, often reveal the presence of 0.1% sodium benzoate, the sodium salt of benzoic acid (Sec. 10.8), as an added preservative. A tenth of one percent amounts to a concentration of one part per thousand.

The concentration of potassium iodide, KI, in iodized table salt is even smaller. Traces of iodide ion in the diet help prevent the enlargement of the thyroid gland, a condition known as *goiter*. To provide this dietary iodide, KI is added to commercial table salt, NaCl, to the extent of about 7.6×10^{-5} g of KI per gram of NaCl. The following example shows how we can convert this concentration into ppm.

EXAMPLE

Concentrate on Salt

Translate the concentration of potassium iodide in table salt into ppm.

■ A million is 1,000,000, or 10^6. We know that the concentration of KI in table salt is

$$\frac{7.6 \times 10^{-5}\ \text{g KI}}{1\ \text{g table salt}}$$

and we want to know how many grams of KI there are in 10^6 g of table salt.

The quickest way to get the answer is to multiply both the numerator and the denominator by 10^6.

$$\frac{7.6 \times 10^{-5}\ \text{g KI}}{1\ \text{g table salt}} \times \frac{10^6}{10^6} = \frac{7.6 \times 10\ \text{g KI}}{10^6\ \text{g table salt}}$$

$$= 7.6 \times 10\ \text{ppm KI} = 76\ \text{ppm KI}$$

The concentration of KI in table salt is 76 ppm.

Botulinum toxin, the virulent poison of spoiled food, provides us with another illustration of the importance of even very low concentrations of substances. As we'll see in Section 20.4, botulinum toxin is the most powerful biologically produced toxin known. A dose of as little as 1×10^{-9} g of this substance can kill a 20-g mouse. At that level the concentration of the poison in the mouse's body amounts to 1 g of toxin per 20×10^{9} g of mouse. To state this more directly, we can multiply both the weight of the toxin and the weight of the mouse by 50, which reveals that the toxin is lethal to the mouse at a concentration of only 50 parts per trillion!

For comparison, consider yourself as one individual person among all the people on earth. With the world's total population of about 6 billion humans, you yourself make up roughly one person in 6 billion, or 167 parts per trillion of all of those alive today on our planet. Remarkably, the lethal concentration of botulinum toxin in a mouse's body is even less than your own "concentration" among all those now living.

■ QUESTION

Zinc is important to human health. A deficiency of zinc can lead to stunted growth, incomplete development of the sexual organs, and poor healing of wounds. Among foods richest in this element are liver, eggs, and shellfish, which contain zinc at levels ranging from about 2 to 6 mg per 100 g. Express this range of zinc concentrations in parts per million. (Note that an excess of zinc can be just as bad as a deficiency. Dietary excesses of this element can cause anemia, which is a deficiency of red blood cells, kidney failure, joint pain, and other medical problems. A balanced diet provides most of us with all the zinc we need.)

9.10 The Very Purest Water on Earth

The presence of dissolved minerals and organic compounds in the water we drink provides a fine example of the significance of the concentrations of solutes. When we think of pure water we may imagine fresh rainwater falling from a crisp sky or clear water bubbling up from a mountain spring. The very purest water on earth isn't either of these. The very purest of all water comes from a chemist's laboratory. It's water that has passed through columns of specially prepared cleansing resins or water that has been distilled repeatedly under carefully controlled laboratory conditions to remove all traces of contaminants, as demonstrated in the opening demonstration of Chapter 7. All other water, including pristine rainwater and the very purest drinking water, carries a variety of chemical impurities in a range of concentrations.

Even the very purest rain contains dissolved carbonic acid, H_2CO_3.

To clarify the nature and origin of these impurities, we'll follow the rain as it drops to earth and becomes part of our water supply. As rain falls it absorbs, to varying degrees, the gases that make up our atmosphere (Section 14.1). The nitrogen, oxygen, and carbon dioxide of air all dissolve to a very small extent in water. Near room temperature a maximum of about 0.01 g of nitrogen, 0.05 g of oxygen, and 3.4 g of carbon dioxide can dissolve in 1 liter of water. (The small bubbles that rise from heated tap water just before it starts to boil are bubbles of these very same dissolved gases. Almost all gases are much less soluble in hot water than in cold water. They tend to escape from their water solutions as the water warms up.)

Part of the carbon dioxide that enters rainwater reacts with the water itself to form carbonic acid (Chapter 10):

$$\underset{\text{carbon dioxide}}{CO_2} + \underset{\text{water}}{H_2O} \longrightarrow \underset{\text{carbonic acid}}{H_2CO_3}$$

As a result all rainwater is very slightly acidic. The absorption of still other atmospheric gases, especially industrial pollutants and automobile exhaust gases, can increase the acidity of rainwater sharply, well beyond its normal acidity, to produce what is known as *acid rain*. We'll have more to say about acids and about how acid rain affects our environment in Chapter 14.

Rain that falls onto land soaks into the earth, runs off to our rivers, lakes, and other bodies of water, or simply evaporates and reenters the atmosphere. Even some of the rain that seeps into the ground finds its way, underground, to lakes and rivers. Much of it, though, enters layers of porous rock and earth lying just below the surface of the land, where it accumulates as large reservoirs of fresh water known as *groundwater*.

■ **QUESTION**

Assuming that drinking water has a density of 1000 g per liter, what is the maximum concentration, in ppm, of N_2 and of O_2 in the water we drink?

9.11 The Water We Drink

Despite the abundance of rivers, streams, and lakes throughout most of the populated land area, chances are that the water coming from your kitchen faucet or from a public water fountain was drawn up from the reservoir of groundwater. Reaching the surface through wells or springs, this source provides the drinking water for about half the people in the United States and about three-quarters of those of us who live in large cities. More water is present in this subsurface layer than in all of our rivers, streams, and lakes combined.

Pollution in a rural stream.

In chemical terms, groundwater and the water we draw from our rivers and other bodies of water are solutions of solutes in a solvent. Water is the solvent; the substances the water picks up in its travels from the clouds, through the earth, and into our faucets are the solutes. It's a combination of the chemistry of the solutes and their concentrations that determines whether the water is polluted.

Some of the solutes have been in earth's water since rain began to fall on the newly formed planet. Partly because of rainwater's normal acidity (resulting from the carbonic acid it contains) and partly because water itself is a very good solvent for many substances, the rainfall that passes through the soil picks up a variety of minerals from the earth itself. As a result, all the waters of the earth, including those that feed our public and private water supplies and those that furnish commercially bottled "natural" or "mineral" drinking water, contain a variety of minerals in a range of concentrations. Among these are calcium, iron, magnesium, potassium, and sodium cations and fluoride anions. The concentrations of representative minerals found in natural spring water of the French Alps appear in Table 9.1. Minerals such as these are usually harmless or even beneficial to most

Table 9.1 Concentration of Several Minerals in Natural Spring Water of the French Alps

Mineral	Concentration, mg/liter (ppm)
Calcium	78
Magnesium	24
Sodium	5
Potassium	1

Table 9.2 Maximum Contamination Levels of Minerals in Community Water Systems Permitted by the Safe Drinking Water Act of 1974

Contaminant	Maximum Permitted Level, mg/liter (ppm)
Arsenic	.05
Barium	1.
Cadmium	0.010
Chromium	0.05
Lead	0.05
Mercury	0.002
Selenium	0.01
Silver	0.05
Sodium	160.

Commercially bottled water.

of us at their typical levels in natural and in commercially bottled drinking water, and at even higher levels as well. Some of us, for example, take mineral supplements that provide calcium, iron, or magnesium at levels far higher than those found in drinking water.

There are hazards, though. A few of the chemicals that can enter our water supplies are particularly toxic even at what may seem to be very low concentrations. Contamination by these substances comes as rainwater picks up residues of agricultural fertilizers and pesticides (known as *agricultural runoff*) and from industrial, urban, and household wastes dumped onto the surface or injected just below it. Any of these can be carried by the natural flow of rainwater into groundwater.

To ensure the safety and high quality of public drinking water, the U.S. Congress passed the Safe Drinking Water Act of 1974, which establishes, among other things, maximum drinking water levels for specific, potentially hazardous chemical contaminants. Some of the minerals controlled by the Act appear in Table 9.2, while representative organic compounds appear in Table 9.3. When present in concentrations

Table 9.3 Maximum Contamination Levels of Organic Compounds in Community Water Systems Permitted by the Safe Drinking Water Act of 1974

Compound and Use	Maximum Permitted Level (mg/liter)
Endrin (an insecticide no longer manufactured or used in the United States)	0.0002
Lindane (an insecticide)	0.004
Methoxychlor (an insecticide)	0.1
2,4-D (an herbicide)	0.1
Silvex (an herbicide)	0.01
Trihalomethanes (solvents and other uses): bromodichloromethane, dibromochloromethane, bromoform, chloroform	0.10

An abandoned gasoline station. A potential source of groundwater pollution.

greater than those shown in the tables, the minerals and the organic substances exceed the safety standards set by the Safe Drinking Water Act. As with all legislation, levels of contamination permitted by the Safe Drinking Water Act of 1974 can be revised when conditions change and knowledge improves. Because of increased concern about the hazards of arsenic in drinking water, the Environmental Protection Agency (Sec. 20.8) is considering lowering the permitted concentration of arsenic from the current 0.05 ppm to 0.01 ppm.

Curiously, the Act also applies to some kinds of commercially bottled water but not to others. The federal standards for tap water apply as well to "bulk" or "commodity" water, the kind that comes in large jugs and can be dispensed through water coolers. "Mineral water," which we'd expect to be high in minerals anyway, is exempt from the standards of the Safe Drinking Water Act, as are both seltzer and club soda. These last two are considered soft drinks and are covered by regulations of the Food and Drug Administration (Sec. 20.8).

Of course, not all water pollution comes from accidents or the careless or deliberate dumping of waste chemicals or from agricultural runoff. Some results simply from the way our society operates. Just as the high concentration of cars in and near our cities, for example, contributes to air pollution (Chapter 14), the large number of gasoline stations that serve these cars are potential sources of pollution. Old and rusting underground tanks in these stations (some of which have been left abandoned in the ground as the stations closed) may leak gasoline, which eventually travels into the groundwater. Whether this presents a major threat to our drinking water is still uncertain, but it does illustrate one of the many less visible sources of contamination.

■ QUESTION

Which of the following, if any, exceed federally permitted levels when they are present in our public water supplies at a level of 0.5 mg/liter: (a) lead, (b) sodium, (c) barium, (d) mercury, (e) silver?

9.12 *That's* the Glass Where Pollution Begins

We can return now to the seven glasses of salt water we prepared in our opening demonstration and point directly to the glass where pollution begins. Here's how we can identify it.

First, since we can expect variations in our sense of taste and our sensitivity to any contaminant, we have to find a reference or a criterion for pollution that's common to all of us. Perhaps the most useful is the legal standard of Table 9.2, 160 ppm sodium. Having chosen 160 ppm as the standard for pollution, we can now apply a little simple arithmetic.

Since the mass of a sodium ion is about 23.0 amu and the mass of a chloride ion is about 35.5 amu, the sodium ion makes up about

$$\frac{23.0 \text{ amu}}{(23.0 + 35.5) \text{ amu}}$$

or roughly 39% of the mass of any quantity of NaCl.

Like the teaspoon of sugar in the question that follows Section 9.7, a teaspoon of table salt has a mass of about 5 g. This means that the sodium in the teaspoon of table salt we added to the first glass has a mass of about 39% of 5 g, or roughly 2 g. Although glasses vary quite a bit in size, we wouldn't be far off if we estimated that the glass holds about 250 mL of water. Estimating that the density of the solution is about 1 g per mL, we can conclude that our solution contains about 2 g of sodium in about 250 g of solution. Even though these are rough approximations, we can see now that the concentration of the sodium in the first glass is close to 8 g per kilogram of solution, or (since there are 1000 milligrams per gram) 8000 ppm. That's real pollution by any standard.

In each succeeding glass the concentration of sodium chloride, and of sodium, drops by a factor of 10. The sodium concentration in the second glass, then, is 800 ppm, which is still above the legal limit. For the third glass the sodium concentration is 80 ppm, which is now below the standard for sodium in drinking water. Working our way backward from glass 7, which holds the most dilute solution, we can now point to glass 2 and say "*That's* the glass where pollution begins." This entire approach illustrates the use of a combination of fundamental chemistry and simple arithmetic to help us work through environmental problems.

■ QUESTION

Suppose we were off by 50% in any of our estimates. Suppose, for example, that the teaspoon of table salt has a mass of 7.5 g, or the glasses hold only 125 g of water. How would this affect our determination that glass 2 is the glass where pollution begins?

PERSPECTIVE

The Arithmetic of Pollution

Given the variety of sources of water pollution and the variety of the pollutants themselves, it's unrealistic to expect that our water can ever be made *completely* free of every foreign substance imaginable. Moreover, we'd have to understand what it would mean for water to be *completely* free of every other substance. As our means of chemical detection and measurement become increasingly sensitive and sophisticated, we become able to detect contaminants at concentrations far below those we could smell or taste or see or that we might reasonably expect to be hazardous. We could imagine, for example, a method of detection that would reveal the presence of a *single molecule* of, say, a pesticide, or a component of gasoline, or a dry cleaning fluid, or a noxious industrial waste in an 8-oz glass of water. At that concentration, one molecule of contaminant per 79 $\times$ 10^{23} molecules of water (Fig. 9.6), our language provides no terms of reference equivalent to parts per million, parts per billion, or the like. Would we realistically call the water "polluted"? Surely not.

Nor would we consider a glass of water to be polluted if it contained two, three, or ten molecules of another substance. Yet at some higher level, some threshold concentration, we would just as surely call the water polluted and unfit to drink. We do, indeed, expect the concentrations of any contaminants in our drinking water to remain well below the levels at which they might be hazardous or might give the water an offensive taste, odor, or appearance, but we cannot realistically expect the total and complete absence of anything but molecules of H_2O in our drinking water.

When we consider the arithmetic of pollution we can better understand the questions of the opening demonstration. What must be asked is not "Is there *any* contaminant in our drinking water?" but rather "What are the *limits* of contamination we can reasonably accept?" Understanding the answer to this more critical question requires an understanding of how we count chemical particles and what their concentration terms mean. ■

EXERCISES

For Review

1. Following are a statement containing a number of blanks, and a list of words and phrases. The number of words equals the number of blanks within the statement, and all but two of the words fit correctly in these blanks. Fill in the blanks of the statement with those words that do fit, then complete the statement by filling in the two remaining blanks with correct words (not in the list) in place of the two words that don't fit.

One ________ of a substance is equal to its atomic or molecular weight expressed in ________ and contains 6.02 $\times$ 10^{23} chemical particles, which is also known as ________ of particles. Since 1 mole of any chemical substance contains the same number of ________ as 1 mole of any other chemical substance, we can calculate the weight of one chemical that will react completely with a given weight of another by using their ________ in our calculations. We can also use ________ of solutions rather than weights in these calculations as long as we know the ____________ of the solution, which is a measure of the amount of ___________ dissolved in a specific volume of the ___________. A solution that contains 1 mole of solute per ____________ of solution has a concentration of 1 ________.

atomic or molecular weights	mole
Avogadro's number	ppm
concentration	solute
grams	solution
kilogram	volumes
liter	

2. Explain, define, or describe the significance of each of the following:

a. agricultural runoff
b. groundwater
c. ppm
d. The Safe Drinking Water Act of 1974
e. weight/weight percentage (w/w %) concentration

3. Write the chemical equation for the reaction that occurs as charcoal burns (a) in the open air; (b) in an enclosed space, with a limited supply of air.

4. (a) What is the danger in heating a closed room by burning charcoal briquettes in a grill, placed in the room? (b) Why isn't this a hazard when the grill is used outdoors?

5. Is it necessary to know the numerical value of Avogadro's number to carry out calculations using quantities of moles? Explain.

6. Write the balanced reaction for the combustion of methane to produce carbon dioxide and water.

7. Why are concentrations expressed in *molarity* particularly useful to chemists?

8. What unit of concentration is used most frequently in consumer products?

9. What unit of concentration is equivalent to 1 milligram of solute per kilogram of solution?

10. Name a unit of concentration useful in discussions of extremely small concentrations, as in pollution studies.

11. Describe three processes that remove freshly fallen rainwater from the ground.

A Little Arithmetic and Other Quantitative Puzzles

12. How many (a) *left* shoes are there in 15 *pairs* of shoes? (b) dozen yolks would you get from half a dozen eggs? (c) moles of sodium cations are there in half a mole of sodium chloride?

13. How many moles of hydrogen atoms does it take to produce 1 mole of diatomic hydrogen molecules?

14. What is the weight of:

a. 1 mole of neon
b. 2 moles of zinc
c. 1 mole of diatomic nitrogen
d. 3 moles of diatomic hydrogen gas
e. 1/2 mole of chloride ions
f. 1 mole of chlorine atoms
g. 1 mole of diatomic chlorine molecules
h. 2 moles of neutrons

15. What weight of (a) sodium contains 1 mole of sodium atoms? (b) chlorine atoms contains 1 mole of chlorine atoms? (c) sodium chloride contains 1 mole of sodium ions? (d) sodium chloride contains 1 mole of chloride ions? (e) CO_2 is produced when 1 mole of carbon combines with 1 mole of diatomic oxygen molecules?

16. Assuming that rubbing alcohol is 70% (w/w) isopropyl alcohol, and that its density is 1 kg/liter:

a. How many grams of water and how many grams of isopropyl alcohol are there in a liter of rubbing alcohol?

b. How many moles of water and how many moles of isopropyl alcohol (molecular weight 60) are there in a liter of rubbing alcohol?

c. If we decide which is the solute on the basis of *weight*, is isopropyl alcohol the solute or the solvent?

d. If we decide on the basis of the *number of moles*, is isopropyl alcohol the solute or the solvent?

e. Considering water as the solute and the isopropyl alcohol as the solvent, what is the molarity of the water in the solution?

f. Considering isopropyl alcohol as the solute and water as the solvent, what is the molarity of the isopropyl alcohol in the solution?

17. Which one of the following contains the greater number of moles of sucrose, (a) or (b), or do they contain the same number? (a) 4 liters of a 0.125 *M* sucrose solution; (b) 0.375 liter of a 1.5 *M* sucrose solution?

18. A typical aspirin tablet contains 0.325 g of the active ingredient, acetylsalicylic acid. What is the average concentration, expressed as a percentage by weight (w/w %), of the acetylsalicylic acid in the body of a 165-lb (75-kg) person who has just taken two aspirin tablets?

19. In determining the molarity of the sucrose in the coffee of the example of Section 9.8, we assumed that the coffee has a density of 1, or 1000 g per liter. With all the solutes in the coffee, including the 10 g of sugar, our assumption of 1000 g per liter is probably too low. If we assumed a density of, let's say, 1100 g per liter, what molarity (to one significant figure) would we now get?

20. Chlorine gas consists of diatomic molecules, Cl_2. The balanced chemical equation for the reaction of sodium metal with chlorine gas to produce sodium chloride is $2Na + Cl_2 \longrightarrow 2NaCl$.

a. What weight of chlorine gas does it take to react completely with 4.6 g of sodium?

b. What weight of sodium does it take to react completely with 1.42 g of chlorine gas?

c. If we started with exactly 10 g of sodium and 10 g of diatomic chlorine gas, would either of these elements be left over at the end of the reaction? If so, which one?

21. The use of fuels with small amounts of sulfur-containing impurities results in the formation of sulfur trioxide, SO_3, an atmospheric pollutant. The SO_3 dissolves in rain and other forms of atmospheric water to form sulfuric acid, H_2SO_4, according to the chemical equation $SO_3 + H_2O \longrightarrow H_2SO_4$. Through the generation of sulfuric acid, SO_3 contributes to the formation of acid rain (Sec. 14.5)

a. How many moles of SO_3 are there in 8 g of sulfur trioxide?

b. How many moles of sulfuric acid form when 8 g of sulfur trioxide dissolve in water?

c. What weight of sulfuric acid forms when 8 g of sulfur trioxide dissolve in rainwater?

22. You have 1 liter of a solution that contains selenium at a level of 15 parts per billion and 1 liter of another solution that contains barium at a level of 3 parts per million. You now mix the two solutions together to produce 2 liters of a single solution. Does the new solution produced by mixing the two original liters together meet federal drinking water standards for selenium? For barium?

23. The Safe Drinking Water Act of 1974 permits a maximum of 160 mg of sodium per liter of drinking water. What is the maximum *molarity* of sodium chloride permitted in community drinking water by the Safe Drinking Water Act? (You may find Section 9.12 useful in working this problem.)

24. How many sodium cations are there in all of the water in glass 7 of the demonstration at the beginning of the chapter?

25. Prepare a table showing the molarity of the sugar water in each of the seven glasses of the second part of the demonstration that opened this chapter. Which glass comes closest to the molarity of two teaspoons of sugar in a cup of coffee? Refer to the examples in Section 9.8.

26. If you count all of the protons and all of the neutrons in the body of a person weighing 165 lb and add the sum of all the protons you have counted to the sum of all the neutrons you have counted, what is the combined total? This is not a trick question and you don't have to perform tiresome or complex calculations. All that's required is a good understanding of the material in this chapter and a little thought. The solution appears in Appendix E, along with a discussion of the chemical principles involved.

27. As we saw in Chapter 7, carbon monoxide is also produced by the incomplete combustion of gasoline in a car's engine. To help reduce air pollution *catalytic converters* (Sec. 7.7) are used on cars to promote the combination of carbon monoxide with oxygen, thus transforming the CO into CO_2. Write the balanced equation for the reaction of CO with O_2 to form CO_2.

28. Write the balanced equation for the combustion of methane in an enclosed space, with a limited amount of oxygen, to produce carbon monoxide and water instead of carbon dioxide and water.

29. Write balanced equations for the complete combustion of methanol ($CH_3{-}OH$), ethanol ($CH_3{-}CH_2{-}OH$), and pentane (C_5H_{12}, a major hydrocarbon of gasoline). Calculate the weight of carbon dioxide produced by the complete combustion of 100 g of each of these compounds. Assuming that all three of these compounds are equally efficient when used as a fuel for an internal combustion engine, rank them in order of the amount of carbon dioxide released to the atmosphere per mile of automobile travel. Based on the content of Section 8.14, which would contribute most to an increase in the greenhouse effect if used as an automotive fuel? Which would contribute least?

30. The sun is 150,000,000 km from the earth and the diameter of a penny is 1.9 cm. If Avogadro's number of pennies were used to build a road from the earth to the sun and the road were just one layer of pennies deep, how many pennies wide would the road be? (This problem is worked in some detail in Appendix C.)

Think, Speculate, Reflect, and Ponder

31. Why is it not possible, either in theory or in practice, to calibrate a scale in units of moles?

32. Suppose you have a scale you know to be accurate, but you don't know what units of weight it measures. You have a quantity of acetic acid and you want to measure a quantity of water that contains the same number of water molecules as there are acetic acid molecules. Could you use the scale to do that? Explain. Suppose you want to prepare a 1 *M* solution of acetic acid in water. Could you use the scale to do that? Explain.

33. Suppose you have an analytical instrument that could tell you quickly exactly how many molecules of an agricultural insecticide there are in a glass of water. What standard(s)—the number of insecticide molecules, the odor, color, taste of the water, or any other criterion—would you use to determine whether the water is fit to drink? (No governmental drinking-water standards have yet been set for this insecticide.)

34. Regardless of any legal definition, which glasses in the opening demonstration would you consider to be polluted with salt? Which, if any, would you consider safe to drink?

35. Define "pollution" in your own words.

36. (a) Under what conditions would you consider water to be "polluted"? (b) Would you say that all "polluted" water is unfit to drink? (c) Can water that is unfit to drink not be "polluted"? (d) Can water that is "polluted" be safe to drink?

37. Describe three specific actions you would recommend to improve the purity of your drinking water. Which one would you want carried out first?

38. We normally think of community water purification processes as designed to *remove* impurities from water. Yet they can also *add* chemicals to water to benefit the public health and welfare. Many communities, for example, add chlorine to water to kill disease-causing microorganisms. Some add fluoride salts to retard tooth decay. Do you favor the addition of chemicals to public drinking water to promote the public health and welfare? Do you think that small quantities of essential nutrients or vitamins ought to be added to drinking water? Do you think that nothing ought to be added to the water since everyone in the community must drink the same water and some may not wish to have anything added to their drinking water? Do you think the public welfare ought to be more important than individual wishes in making these decisions? Discuss your answers.

39. Here's breaking news, just in! It's just been discovered that the value long accepted for Avogadro's number is wrong! The value used for so long has just been shown to be off by at least 25%, although no one knows just yet exactly how much or in which direction. Which answers in these end-of-chapter exercises will have to be changed as soon as we learn the correct value of Avogadro's number, and which can we let stand as they are?

Web Questions

Use the Web to answer or accomplish each of the following. Report your Web source for each response.

1. Summarize the life of Amedeo Avogadro with a brief biography that includes the dates of his birth and death, his nationality, background, education, and a summary of his scientific contributions.

2. Among the concentration terms used in chemistry is *molality*. Define molality. Describe how it differs from molarity and how and why it is used in chemistry.

3. As described in Section 9.9, both sodium benzoate and potassium iodide are added to processed foods in concentrations that can be reported in parts per thousand. Find another food additive included in processed foods in concentrations easily reported as parts per thousand.

4. Describe a recent incidence of botulinum poisoning (sometimes referred to as botulism poisoning). Include the circumstances under which it occurred and its effects or results.

5. Identify the source of the municipal drinking water distributed in your area. Does it come from surface water, groundwater, or some other source?

6. Find the maximum permitted level, in community drinking water, of an element, ion, or compound other than those listed in Tables 9.2 and 9.3.

7. Give the details of a major accidental spill, commercial or industrial discharge, or similar event that caused or threatened significant environmental pollution within your state. Include any action taken by state or local authorities, and the outcome.

8. Describe national, state or local requirements or restrictions that apply in your region for the disposal of containers of unused paint; for the disposal of used motor oil; for exhaust emissions of gasoline-powered lawn mowers or other household machinery.

Acids and Bases

IF IT TASTES SOUR IT MUST BE AN ACID

If it tastes sour, it's an acid.

Breath with the Strength of Red Cabbage

(a)

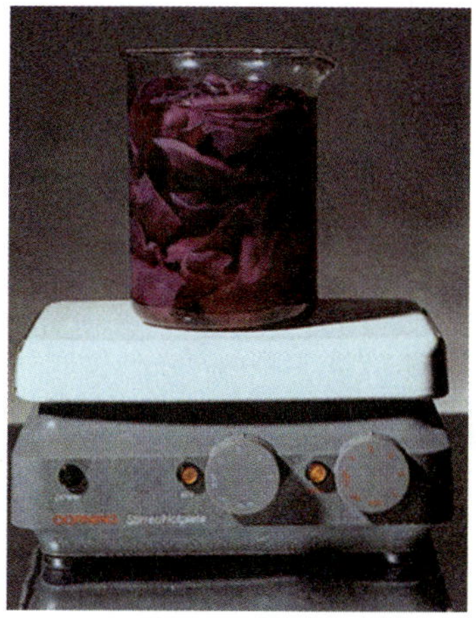

Figure 10.1 The red cabbage breath test. (a) Heating leaves of red cabbage in water to extract the acid–base indicator that gives the cabbage its color.
(continues on next page)

Here's a bit of chemical magic you can use to entertain, impress, or embarrass your friends. You'll need a leaf of red cabbage, a little household ammonia, a little vinegar, some water, two glasses, and a drinking straw. Announce that you're about to perform a breath test. (Make it a breath test for anything you'd like: alcohol, bad breath, sweet breath, the residual odor of cabbage, etc.)

It works like this (Fig. 10.1). First, put a few drops of ordinary household ammonia into about half a glass of water. The exact amounts of ammonia and water don't matter much; what you need is a very dilute solution of ammonia in water. Identify the solution in some way so that no one mistakes it for plain water and then put it aside for a moment.

AMMONIA IS A POISON!...Be sure to mark clearly anything containing ammonia and place the marked container out of reach. Don't drink or taste the ammonia and don't let any ammonia solutions get on your skin or clothes. Don't let anyone else come into contact with the ammonia.

Next, break about half of a large leaf of red cabbage into small pieces. A leaf with a deep purple color is best. Boil the pieces in just enough water to cover them. Keep the water boiling for about 2 minutes to extract the dye that gives the cabbage its color; then remove the mixture from the heat and set it aside to cool. Pour the cool, deep blue water extract of the cabbage into a clear, colorless glass and discard the boiled cabbage pieces.

What you have now is a blue solution of cabbage leaf extract. Add *just* enough of the dilute ammonia solution you prepared earlier to this cabbage extract to turn it from deep blue to an emerald green. A few drops of the solution usually work nicely. Don't add more ammonia solution than just enough to produce the green color. This is now your breath test solution. Pour a bit of the test solution into another clear, colorless glass, enough to cover the bottom to a depth of a few centimeters (about an inch) and put the straw into the glass. Finally, have the person whose breath is to be "tested" blow a steady stream of bubbles through the straw into the green test solution and watch for a color change. (It's a good idea to try out this stunt yourself before going public with your claims.)

The total volume of breath that has to be blown into the solution depends on just how it was prepared. Several lungfuls might be necessary. In any case, with continuous blowing into the solution the color soon turns from emerald green back to a distinct blue, close to the shade of the cabbage dye originally extracted by the boiling water. Having more of the green test solution nearby for comparison makes it easier to see the color change. So does looking at the solution against a white background, with good lighting.

You can make up whatever stories you wish about what this color change says of the state of your friend's breath, but you ought to point out that things could be much worse. Pour a little vinegar into the (now blue) solution and its color changes to a

(b) (c) (d)

(e)

(f)

Figure 10.1 *(continued)* (b) Pouring out the solution of the indicator. (c) Adding just enough dilute ammonia to turn the indicator solution green. (d, e) Blowing into the slightly basic solution to make it slightly acidic and turn its color from green to blue. (f) Adding vinegar turns the blue solution pink.

An **acid–base indicator** is a dye that changes, loses, or acquires color as a solution containing the dye changes in acidity or basicity.

brilliant pink! The fact is that the color change from green to blue doesn't reflect sweet breath or bad breath or the odor of cabbage or any other food or drink. It reveals only one thing characteristic of all human breath and of the breath of all animals: Exhaled breath is acidic, and that's something all of us can be thankful for.

Actually, what appears to be a clever stunt is no more than a chemical demonstration that you exhale plenty of carbon dioxide as you breathe and that the carbon dioxide dissolves in water to form an acid. This, in turn, takes us into the chemistry of acids and bases and, later, to an examination of how our bodies use the food we eat. [Incidentally, the color of the solution changes from blue to pink as you add the vinegar because the acetic acid in the vinegar (Sec. 9.7) is even more acidic than the solution of the carbon dioxide from your breath, so the acetic acid causes an additional color change.]

Very briefly, here's how the "breath test" works. It uses two acids, one base, and an *acid–base indicator*. The acids are the solution of carbon dioxide formed as your friend blows into the "test" solution and the acetic acid of the vinegar; the base is the ammonia of the solution you prepared; and the acid–base indicator is the dye you extracted from the cabbage leaf. **Acid–base indicators** are dyes that change color (or become either colored or colorless) as solutions containing them change in acidity or basicity.

When you add the dilute ammonia solution to the extracted cabbage dye, which is a member of a class of compounds known as *anthocyanins*, you make the test solution slightly basic and the molecular structure of the dye changes, turning the color of the entire solution to an emerald green. By blowing your exhaled breath into the test solution you add carbon dioxide, which forms carbonic acid, neutralizes the basic ammonia, and eventually turns the solution slightly acidic. With the change from basicity to acidity, the molecular structure of the dye changes to one that appears blue. Adding the acetic acid makes the solution even more acidic and produces further molecular changes in the dye, changing its color to red. We'll examine some of these transformations and the chemistry behind them in more detail as this chapter unfolds. ■

10.1 The Real Litmus Test

Before entering the world of acids and bases we'll spend a few moments with an indicator, *litmus*, that has added color to the English language as well as to chemistry. This oldest and best known of the acid–base indicators is a pink mixture of compounds extracted from certain small plants found principally in the Netherlands. Included in this mixture is a dye, *erythrolitmin*, that turns red in acid, blue in base. Small strips of paper impregnated with litmus serve as a highly convenient diagnostic tool, *litmus paper*, which is used to test liquids and moist solids for acidity and basicity.

Wetting a strip of litmus paper with a water solution of an acid produces a distinctly red spot; water solutions of bases turn the paper blue. This is the litmus test for acids and bases. To accentuate the color changes, commercial litmus paper is pretreated to produce blue litmus paper for use in testing acids, which give a red spot that shows up clearly against the blue background, and red litmus paper for testing bases, which give a blue spot against a red background (Fig. 10.2).

The term "litmus test," at first narrowly applied to the simple and definitive test that establishes beyond question the acidic or basic properties of substances, has expanded to describe simple and definitive tests of political candidates, social issues, economic policies, and related matters. The "litmus test" of social issues is often a simple question whose answer determines the fate of an entire social or political program, just as the chemical litmus test determines simply and quickly whether we are dealing with an acid or a base.

Figure 10.2 Acids turn blue litmus red. Bases turn red litmus blue.

■ **QUESTION**

What is the color of litmus paper when it's moistened (a) with ammonia? (b) with vinegar?

10.2 What Are Acids and Bases? Phenomenological Definitions

In reply to the question "What are acids and bases?" there is only one honest and truly realistic answer: *It depends*. If you mean "How can I tell if a certain substance that I have here in a bottle is an acid or a base?" you'll need a phenomenological definition. That is, you'll need a definition that depends on phenomena, on activities or properties that you can sense. It's a definition that hinges on what you can see, taste, feel, or touch.

In this phenomenological sense, water solutions of all acids

- taste sour,
- turn litmus red, and
- react with certain metals, such as iron and zinc, to liberate hydrogen gas.

A sour taste may be the world's oldest indication of acidity. Our word *acid* comes from the Latin *acidus*, an adjective meaning sour or having a sharp taste.

Water solutions of all bases, on the other hand,

- taste bitter,
- turn litmus blue, and
- feel slippery.

Acids and bases in consumer products. Some are hazardous and must be used with care.

WARNING...Of all these diagnostic tests, only the litmus test is safe enough for general use with a substance of uncertain identity. *Tasting any questionable*

Pieces of zinc metal react with hydrochloric acid to produce hydrogen gas.

In **neutralization** an acid and a base react to produce a *salt*.

A **salt** is a compound (other than water) produced by the reaction of an acid with a base.

material to learn whether it's sour or bitter or rubbing it between your fingers to determine whether it feels slippery is a dangerous activity. **Although many of our foods—lemons and vinegar, for example—contain acids and taste distinctly acidic, other substances may be composed of acids (or bases) powerful enough to destroy skin and mucous membranes and cause great harm. Some substances we may come in contact with are highly poisonous in ways that have nothing to do with acidity or basicity. Even many consumer products are toxic and/or corrosive and can cause much damage if used carelessly or in ways other than those specified by their labels. Never taste or touch any material you aren't sure of. It could be quite dangerous. We'll examine some of our ideas about safety in more detail in Chapter 20. The remaining phenomenon, the reaction of acids with metals like iron and zinc, can also be hazardous. The product is hydrogen gas, which is dangerous in itself because of its flammability.**

In addition to the characteristics we've just described, another useful property that distinguishes acids from all other kinds of chemical substances is their ability to neutralize bases. Similarly, bases can neutralize acids. In this acid–base **neutralization** an acid and a base react chemically with each other to produce a *salt*, a compound that has no (or much weaker) acidic or basic properties. A **salt** is a compound (other than water) produced by the reaction of an acid and a base.

A particularly simple acid–base neutralization occurs, for example, when hydrochloric acid (HCl, hydrogen chloride) reacts with sodium hydroxide (NaOH). Hydrochloric acid, usually available in hardware stores by its commercial name, *muriatic acid*, is a strong acid useful for cleaning metals, masonry, cement, and stucco. Sodium hydroxide, better known as household *lye*, is a powerful base often effective in clearing clogged drains. Both sodium hydroxide and hydrochloric acid are dangerous, corrosive chemicals that must be handled with great care.

Hydrochloric acid and sodium hydroxide react with each other to form sodium chloride (a salt) and water:

HCl	+	$NaOH$	$\longrightarrow$	$NaCl$	+	H_2O
hydrogen chloride	and	sodium hydroxide	react to produce	sodium chloride, a salt	and	water

Notice that the hydrogen of the HCl and the OH of the NaOH combine to form water, while a combination of the Cl of the HCl and the Na of the NaOH produces NaCl. This resulting sodium chloride is a salt in two closely related senses. It's not only the common table salt that we use on our food and that began our study of chemistry in Chapter 1, but it's also a salt in the generic sense. That is, sodium chloride is produced, along with water, by the reaction of HCl (an acid) with NaOH (a base). Sodium chloride is a perfectly neutral salt, neither acidic nor basic in any sense at all.

Other salts include potassium iodide, KI, which provides the iodide of iodized salt (Sec. 9.9); magnesium sulfate, $MgSO_4$, better known as *epsom salts* and sometimes used as a laxative; monosodium glutamate (MSG), an additive that enhances or intensifies the flavor of food and is often used in Chinese food; and calcium carbonate, $CaCO_3$, common chalk.

Acid		**Base**		**Salt**		
HI	+	KOH	$\longrightarrow$	KI	+	H_2O
hydrogen iodide		potassium hydroxide		potassium iodide		

$$H_2SO_4 + Mg(OH)_2 \longrightarrow MgSO_4 + 2H_2O$$

sulfuric acid + magnesium hydroxide → magnesium sulfate

$$H_2CO_3 + Ca(OH)_2 \longrightarrow CaCO_3 + 2H_2O$$

carbonic acid + calcium hydroxide → calcium carbonate

■ QUESTION

Write balanced equations to show how each of the following salts can be produced by the reaction of an acid with a base: (a) K_2SO_4, (b) CaI_2, and $MgCO_3$.

10.3 What Are Acids and Bases? Answers at the Molecular Level

We don't have to know anything at all about the molecular structure of a substance, then, to decide whether it's an acid. All we need to do is dissolve a bit of it in water and test it with litmus, taste it (but only if we are absolutely certain of its safety), observe whether it reacts with zinc or similar metals to produce hydrogen gas, and determine whether it neutralizes bases. If it does all of these, it's an acid. Similarly, we can determine whether something is a base by observing a few of its properties.

But chemistry is interested in more fundamental phenomena than color changes, tastes, and the like. In chemistry the question "What are acids and bases?" has a deeper meaning, more like: "What is the single characteristic of a molecule that gives it all the properties of an acid or of a base?"

The answer isn't simple. In fact, there is no single, unequivocal answer. Attempts at defining an acid at a fundamental level started long ago, early in the history of chemistry. In 1778 Antoine Lavoisier, a French chemist, proposed (incorrectly) that all acids were formed by the combination of oxygen—an element that had only recently been discovered—with other elements. The name given to this new element, oxygen, is formed by combining the Greek prefix *oxy-* (sharp or pungent) with the Greek suffix *-gen* (forming or producing). It reflected Lavoisier's belief that the element generates acids. The connection between oxygen and acidity was reasonable, given the state of knowledge of that era: All acids whose compositions were known at the time of the American Revolution did, indeed, contain oxygen. (Because of his connection to the French aristocracy, Lavoisier was executed at the guillotine in 1794 by agents of the French Revolution. He was a brilliant scientist who contributed much to our knowledge of chemistry despite his mistaken theory of acidity.) Several examples of oxygen-containing acids (some of which were not known to Lavoisier) appear in Table 10.1.

Table 10.1 Typical Oxygen-Containing Acids

Formula	Name	Occurrence or Application
H_3BO_3	Boric acid	Mild antiseptic; component of eye drops and rectal suppositories
H_2CO_3	Carbonic acid	Forms whenever carbon dioxide dissolves in water
HIO_3	Iodic acid	Disinfectant
HNO_3	Nitric acid	Manufacture of fertilizers, explosives, dyes, pharmaceuticals
$HClO_4$	Perchloric acid	Manufacture of explosives; plating of metals
H_3PO_4	Phosphoric acid	Manufacture of fertilizers, detergents; a food additive, especially in soft drinks
H_2SO_4	Sulfuric acid	Manufacture of fertilizers, explosives, dyes, paper, and various consumer products

In the first decade of the next century, the discovery that the elements hydrogen and chlorine (without the participation of oxygen) formed hydrochloric acid (HCl) shifted attention from oxygen to hydrogen as the acid-forming element. Partly to accommodate the role of hydrogen, the Swedish chemist and physicist Svante August Arrhenius proposed in 1887 that an acid is anything that produces hydrogen ions, H^+, in water. (He also proposed that a base is anything that produces hydroxide ions, OH^-, in water.) The proton that forms the basis of Arrhenius's definition is the very same particle that forms the nucleus of the hydrogen atom.

EXAMPLE

HCl In Water

Show that HCl is an acid according to Arrhenius's definition.

■ In Section 3.14 we saw a hydrogen atom and a chlorine atom combine to form the covalent molecule, HCl:

$$\mathrm{H}\cdot + \cdot\ddot{\underset{..}{\mathrm{Cl}}}: \longrightarrow \mathrm{H}:\ddot{\underset{..}{\mathrm{Cl}}}:$$

When this covalent HCl dissolves in water, it ionizes (Sec. 3.15) to a proton and a chloride anion:

$$\mathrm{H}:\ddot{\underset{..}{\mathrm{Cl}}}: \longrightarrow \mathrm{H}^+ + :\ddot{\underset{..}{\mathrm{Cl}}}:^-$$

Since HCl ionizes in water to produce a proton, HCl is an acid.

Johannes Brønsted. He and Thomas Lowry extended Arrhenius's definition of an acid to anything that can transfer a proton to another chemical species. They defined a base as anything that can accept a proton.

Unlike Lavoisier's proposal, Arrhenius's view is perfectly valid. When dissolved in water, an acid *does* generate protons (and a base *does* generate hydroxide ions). But since much important chemistry takes place in the complete absence of water, Arrhenius's definition is too limited for widespread use. To expand our chemical horizons beyond aqueous solutions, and even beyond solutions of any kind, two chemists, the Dane Johannes Brønsted and the Englishman Thomas M. Lowry, in 1923 independently defined an **acid** as anything that can transfer a proton to another chemical species and a **base** as anything that can accept a proton. Water may be a solvent, but it needn't be.

By the Brønsted–Lowry definition, an **acid** is any substance that can transfer a proton to another substance. A **base** is a substance that can accept a proton.

You can literally see the importance of the step that Brønsted and Lowry took if you place an open bottle of ammonia next to a bottle of hydrochloric acid (Fig. 10.3). The liquid hydrochloric acid that's available commercially is actually a solution of hydrogen chloride in water; similarly, the ammonia is a solution of gaseous ammonia in water. As the open bottles stand next to each other, molecular HCl and molecular NH_3 both vaporize from their respective solutions and escape into the atmosphere as gases.

The fog that forms above the two open bottles results from a reaction of gaseous HCl with gaseous NH_3 to form the solid, crystalline salt *ammonium chloride*, NH_4Cl. Neither water nor any other solvent is present as the acid and the base combine in the open atmosphere (Fig. 10.4):

$$\mathrm{HCl} + \mathrm{NH_3} \longrightarrow \underset{\text{ammonium chloride, a salt}}{\mathrm{NH_4Cl}}$$

Figure 10.3 Vapors of ammonia and hydrogen chloride combine above the bottles to form ammonium chloride, which appears as a fog. Since this reaction takes place in the absence of water, the Brønsted–Lowry definition of acids and bases is superior to the Arrhenius definition in this case. The reaction that takes place is $NH_3 + HCl \longrightarrow NH_4Cl$.

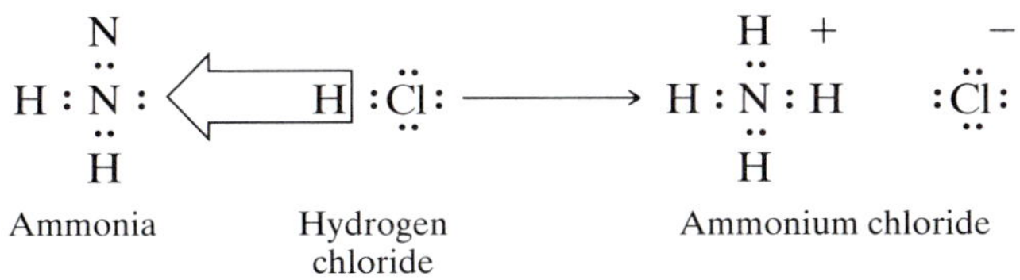

Figure 10.4 Gaseous ammonia reacts with gaseous covalent HCl to form ionic ammonium chloride.

Here, in this gaseous neutralization reaction, Brønsted and Lowry give us our definition of choice. The HCl is an acid because it donates a proton to the ammonia molecule, and NH_3 is a base since it accepts a proton from the HCl. We can see what's happening more clearly if we write the ammonium chloride as an ionic compound, $NH_4^+Cl^-$. In this reaction a proton, H^+, is transferred from the HCl to the NH_3, forming NH_4^+, the *ammonium ion*.

Notice that acids and bases are intimately connected by the Brønsted–Lowry definition. For anything to act as an acid, a base *must* be present to accept a proton; for anything to act as a base, an acid *must* be present to provide the proton. In essence, the Brønsted–Lowry definition views an acid–base reaction as no more than the simple transfer of a proton (H^+) from one substance (the acid) to another (the base).

Svante August Arrhenius defined an acid as anything that produces hydrogen ions in water.

But the story of acids and bases doesn't stop here. To remove even the proton as a necessary part of the definition, Gilbert N. Lewis, the chemist who gave us the electron-dot structures for the valence shells of atoms (Sec. 3.4), proposed an even more general definition that focuses on electron pairs rather than protons. Still other definitions are even more general, but none of them need concern us here. It's enough to understand that no single definition of acids and bases fits all situations. For us, the Brønsted–Lowry definition offers a nice combination of simplicity and generality.

E X A M P L E

Which Is Which?

In the reaction

LiOH	+	HI	$\longrightarrow$	LiI	+	H_2O
lithium hydroxide		hydrogen iodide		lithium iodide		water

which compound is the acid, which is the base, and which is the salt?

■ Since there's a proton present here, we can use it as a guide along with the Brønsted–Lowry definition that an acid releases a proton to a base and a base accepts a proton from an acid. The proton of the HI clearly leaves the HI and moves to the LiOH, forming H_2O. The HI, then, is the acid and the LiOH is the base. As the proton departs from the HI it leaves the iodide anion, I^-, behind, and as the OH of LiOH becomes part of the water molecule it leaves the lithium cation, Li^+, behind. The combination of the lithium cation and the iodide anion produces the salt lithium iodide, LiI.

■ Q U E S T I O N

Is it possible for a base to exist in the absence of an acid (a) in terms of the Arrhenius definition? (b) in terms of the Brønsted–Lowry definition? Explain.

10.4 Amphoteric Water

Having examined some of the phenomena that identify acids and bases experimentally, and some of the definitions that focus on molecular structures, we'll summarize what we now know of acids and bases in Table 10.2 and turn to the

Table 10.2 Definitions of Acids and Bases

Method	Acids	Bases
By Physical Phenomena		
Taste	Sour	Bitter
Feel	Not applicable	Slippery
Effect on metals	Liberates H_2 on reaction with iron, zinc, and tin	Not applicable
Effect on litmus	Red	Blue
By Chemical Structure		
Arrhenius	Generates protons in water	Generates hydroxide ions in water
Brønsted–Lowry	Transfers a proton to a base	Accepts a proton from an acid

unusual acid–base properties of one of our most ordinary substances, water. By far, water is the one compound we consume in the greatest quantities. Depending on our age and the amount of fat in our bodies, water makes up half to three–quarters of our weight. About 2.5 liters of water enter our bodies directly each day through the food and drink we consume. Another quarter of a liter comes from the chemical oxidation of food inside our bodies. By one route or another, about a thousand liters of water pass through our bodies each year.

Question: Is all this water we consume an acid, a base, neither, or both? **Answer:** It's both an acid and a base, in equal measure and at the same time. In Figure 3.15, we saw that water ionizes reversibly to provide both protons and hydroxide ions. To show that ionization and recombination are going on at the same time, we write the overall reaction with two arrows, showing reactions proceeding simultaneously forward and backward:

$$H_2O \rightleftharpoons H^+ + OH^-$$

Water, then, is both an acid and a base by the Arrhenius definition since it provides both hydrogen ions and hydroxide ions (in water, of course).

Water is both an acid and a base by the Brønsted–Lowry definition as well since one water molecule can transfer a proton from itself, acting as an acid, to another water molecule, which acts as a base. As we saw in Figure 3.14 the oxygen atom of water holds eight electrons in its valence shell. Two pairs of these electrons form the two covalent bonds to the hydrogens of the molecule, while the remaining two pairs serve as nonbonding electrons. Each pair of these nonbonding electrons is available for covalent bond formation to a proton. The resulting species, H_3O^+ (Fig. 10.5), is the **hydronium ion**:

The **hydronium ion**, H_3O^+, forms when a proton is bonded to a water molecule.

$$\underset{}{2H_2O} \rightleftharpoons \underset{\text{hydronium ion}}{H_3O^+} + OH^-$$

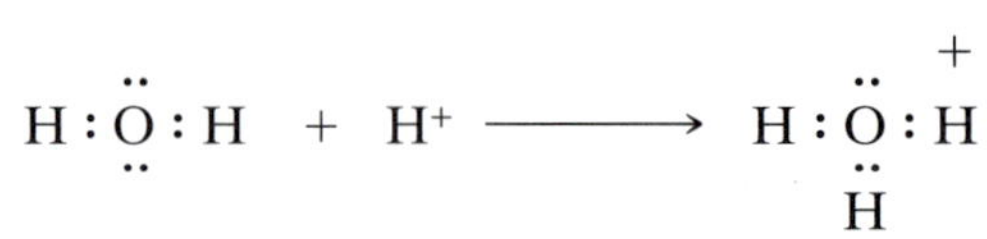

The hydronium ion

Figure 10.5 The hydronium ion. One of the oxygen's two pairs of nonbonding electrons forms a covalent bond with a proton.

Substances that can behave as either acids or bases are called *amphoteric* from the Greek *amphoteros*, meaning "either of two." Water, then, is amphoteric.

Actually, protons themselves don't exist as identifiable species in water. Partly because of the proton's enormous affinity for any base and partly because of the very high concentration of water molecules in the body of the water itself (Sec. 9.6), any and all protons that might form in water bond firmly to a free electron pair of a water molecule to produce the hydronium ion.

$$H^+ + H_2O \longrightarrow H_3O^+$$

In a more modern statement of the Arrhenius definition, then, an acid is anything that generates hydronium ions in water.

■ QUESTION

Recognizing that free hydrogen ions do not exist in water, complete the chemical equation for the ionization of HCl in water:

$$HCl + H_2O \longrightarrow Cl^- + ?$$

10.5 Dynamic Equilibrium: Hurly-Burly in Pure Water

Pure water is perfectly neutral, neither the slightest bit acidic nor the slightest bit basic. Yet a measurable concentration of hydronium ions does exist even in the very purest water. Although this may seem paradoxical, we've already seen that water molecules ionize to produce hydrogen ions, which proceed immediately to form hydronium ions as they combine with water molecules that haven't ionized:

$$H_2O \longrightarrow H^+ + OH^-$$

$$H^+ + H_2O \longrightarrow H_3O^+$$

or, as we saw in the preceding section,

$$2H_2O \rightleftharpoons H_3O^+ + OH^-$$

As the two arrows indicate—one showing a reaction in one direction, the other showing a reaction in the opposite direction—water's ionization is reversible. Even as some water molecules are ionizing, previously formed hydrogen ions leave their hydronium ions and recombine with hydroxide ions to form new, covalent water molecules.

This simultaneous ionization and recombination is a dynamic process. At any instant countless numbers of covalent water molecules are ionizing to hydronium and hydroxide ions, here and there throughout any particular sample of water. At the same instant equally countless hydronium and hydroxide ions are recombining (again, here and there throughout the sample) to regenerate covalent water molecules. When the rate of ionization equals the rate of recombination—when the rates of the forward reaction and the reverse reaction are equal—there is a *dynamic equilibrium*: Throughout the sample, water molecules forever ionize and ions forever recombine to regenerate water molecules. Yet despite all the hurly-burly of random ionizations and recombinations at equilibrium, the actual concentrations of both the hydronium and hydroxide ions in pure water are fixed and measurable values and are constants at any given temperature.

The molar concentrations (Sec. 9.6) of all the transient hydronium and hydroxide ions that exist at equilibrium are fixed at any given temperature and, in chemically pure water, are always equal to each other. They must equal each other in pure water since the ionization of each water molecule produces an equal number of the two, one hydronium ion and one hydroxide ion. Experimental measurements show that each of these ions is present in pure, neutral water at a concentration of almost exactly 0.0000001 mole per liter at 25°C. For brevity, we write $[H_3O^+]$ for "the molar concentration of H_3O^+" and we express the value of the molar concentration in exponential notation (Appendix A), with the capital M as the symbol for moles/liter.

$$[H_3O^+] \quad = \quad 1 \times 10^{-7}\,M$$

The molar concentration of the hydronium ion is 0.0000001 M

Equally, $[OH^-] = 10^{-7}\,M$. Thus at 25°C, pure water, which is perfectly neutral and neither the slightest bit acidic nor basic, contains both OH^- anions and H_3O^+ cations, each at a concentration of $10^{-7}\,M$. We should note that this value of 10^{-7} holds only at a temperature of 25°C. At higher temperatures the value of $[H_3O^+]$ is somewhat higher; at lower temperatures it's a bit lower.

■ QUESTION

Suppose you have a solution of HCl in water that has a hydronium ion concentration of $[H_3O^+] = 0.001\,M$. Write this hydronium ion concentration in exponential notation. Now suppose you have a solution in which $[H_3O^+] = 10^{-5}$. Write the hydronium ion concentration of this solution in conventional, decimal notation.

10.6 Strong Acids and Weak Acids

In parallel to what we just saw for the ionization of water, the value of $[H_3O^+]$ produced by any particular concentration of an acid depends on the extent to which the acid ionizes. We've already seen that the very slight ionization of water generates a very small concentration of hydronium ions in pure water. Similarly, the ionization of an acid in water produces hydronium ions, as we saw in the question at the end of Section 10.4 for the ionization of HCl:

$$HCl + H_2O \longrightarrow Cl^- + H_3O^+$$

As we might expect, the higher the fraction of the molecules that ionize, the greater the concentration of the hydronium ions it produces (for any particular concentration of the acid itself). Since hydrogen chloride ionizes completely and irreversibly in water, every HCl molecule that enters water produces one Cl^- ion and one H_3O^+ ion. Since *all* of the HCl molecules ionize instantaneously and irreversibly to produce these ions, once the HCl molecules enter the water, the hydronium ion concentration of the HCl solution exactly equals the concentration of the HCl. (More accurately, the hydronium ion concentration is the same as what the concentration of the HCl molecules *would have been* if they had remained intact. There are no HCl molecules left since all of them have ionized to H^+ and Cl^-.)

EXAMPLE

Splitting Up

What is the hydronium ion concentration of a solution prepared by dissolving 0.01 mole of pure HCl in 1 liter of water?

■ Here we have a solution that would be 0.01 molar in HCl if the molecule didn't ionize. But each HCl molecule ionizes irreversibly to produce one cation and one anion:

HCl	$\longrightarrow$	H^+	+	Cl^-
0.01 mol		0.01 mol (which reacts with the water present to form 0.01 mol of H_3O^+)		0.01 mol

Since the 0.01 mole of HCl ionizes to form 0.01 mol of hydronium ions in water, the value of $[H_3O^+]$ is 0.01 *M*.

Like water molecules (and unlike HCl molecules), acetic acid molecules ionize reversibly:

$$CH_3-\overset{\overset{\displaystyle O}{\|}}{C}-O-H \rightleftharpoons CH_3-\overset{\overset{\displaystyle O}{\|}}{C}-O^- + H^+$$

acetic acid, unionized	the anion produced by the ionization of acetic acid, the *acetate* anion

As a result, not all the acetic acid molecules in solution are ionized at the same time. Since only a fraction of these molecules are producing H^+ at any given moment, the concentration of H^+ at any instant is less than the concentration of acetic acid molecules introduced into the solution. Naturally, the greater the fraction of the acetic acid molecules ionized at any moment, the greater the concentration of the hydronium ions and the more acidic the solution. We saw in the previous example that $[H_3O^+]$ for a 0.01 *M* solution of HCl in water is 0.01 *M*. For the same molar concentration of acetic acid in water—0.01 *M* acetic acid—$[H_3O^+]$ is 0.0004. Thus the 0.01 *M* acetic acid solution is much less acidic than the 0.01 *M* HCl solution.

Since a particular concentration of hydrogen chloride renders a solution more acidic than does the same concentration of acetic acid, we consider hydrogen chloride to be a stronger acid than acetic acid. In fact, any acid that ionizes completely and irreversibly in water (like hydrogen chloride) is considered to be a *strong acid*, while an acid that ionizes reversibly and therefore only partially (like acetic acid) is a *weak acid*. Typical strong acids are hydrochloric acid and the nitric and sulfuric acids of Table 10.1. The remaining acids of Table 10.1 serve as examples of weak acids.

Because the term "weak acid" simply means that a particular acid doesn't ionize completely in water, calling any particular acid a "weak acid" doesn't tell us much about the actual extent of ionization, or the concentration of the hydronium ion in a solution of that acid. Furthermore, since the acidity of a solution—the concentration of the hydronium ion—depends on both the concentration of the acid and on the extent to which the acid ionizes, some weak acids (that ionize to a great extent, but not completely) can be considered as stronger acids than other weak acids (that ionize to a much smaller extent). We'll examine the relative acidities of some common weak acids further in Section 10.8.

■ QUESTION

Which member (if either) of each of the following pairs would you expect to show a higher hydronium ion concentration: (a) 0.01 *M* HCl or 0.0001 *M* HCl; (b) 0.01 *M* acetic acid or 0.0001 *M* acetic acid; (c) 0.01 *M* HCl or 0.0001 *M* acetic acid; (d) 0.01 *M* HCl or 0.01 *M* acetic acid?

10.7 pH: The Measure of Acidity

In Section 10.5 and in the questions at the end of Sections 10.5 and 10.6, we catch brief glimpses of a technical problem in writing numerical values of hydronium ion concentrations: long strings of zeroes between the decimal and the significant digits. We can make things simpler for ourselves by writing the value of $[H_3O^+]$ as an exponent of 10. In Section 10.5, for example, we saw that we can express $[H_3O^+]$ for pure water at 25°C—0.0000001 *M*—as 10^{-7}; in the same way we can write the values of $[H_3O^+]$ for 0.01 *M* HCl and 0.0001 *M* HCl in the question at the end of Section 10.6 as 10^{-2} *M* and 10^{-4} *M*, respectively.

We can carry this a step further by dispensing with both the *10* and the negative sign. This is exactly what the Danish biochemist S. P. L. Sørensen did in 1909 when he proposed that concentrations of H^+ (or, as we now know, H_3O^+) be treated as exponential values. Following Sørensen's recommendation, we now consider $[H_3O^+]$ in terms of **pH**, which is defined as *the negative logarithm of the hydronium ion concentration.* The pH of a solution is a measure of the solution's acidity. To find the pH we

The **pH** of a solution is a measure of the solution's acidity. The pH is the negative logarithm of its hydronium ion concentration.

- write the concentration of $[H_3O^+]$ as a power of 10 and
- use the exponent of 10, but
- reverse its sign.

The letters *pH* represent the *p*ower of the *H*ydrogen (or *H*ydronium) ion. As a symbol for acidity, pH reflects nicely the international character of chemistry. The letter *p* begins the English word *power* as well as its French and German equivalents, *puissance* and *Potenz*. At the time of Sørensen's suggestion, English, French, and German were the world's dominant scientific languages.

The following mathematical expression of pH uses exponential notation. You may want to review the topic, which is discussed in Appendix A:

$$pH = -\log[H_3O^+]$$

For neutral water,

$$\begin{aligned} [H_3O^+] &= 10^{-7} \\ pH &= -\log[H_3O^+] \\ &= -\log 10^{-7} \\ &= -(-7) \\ pH &= +7 \end{aligned}$$

The pH of neutral water, then, is 7.

EXAMPLE

The Basic Egg

A hydronium ion concentration of 0.00000001 *M* is common for fresh eggs. What is the pH of a fresh egg?

■ Here, the hydronium ion concentration translates into a value of 10^{-8} *M*. All we need to do now is write the exponent of 10 (which is −8), but with its sign reversed: +8. A common pH for a fresh egg is 8.

E X A M P L E

How Acidic is a SopHt Drink?

A typical pH of a soft drink is 3.0. What's the value of $[H_3O^+]$ for a soft drink?

■ In this case we reverse the process. Since the pH is 3, the exponent of 10 must be the negative of +3, which is −3. Using this value as the power of 10 that corresponds to the $[H_3O^+]$, we get $[H_3O^+] = 10^{-3}\ M = 0.001\ M$.

In the first example, we see that the hydronium ion concentration of an egg is *lower* than the hydronium ion concentration of pure water. To understand the significance of this, we'll take a closer look at the hydronium concentration of solutions of acids and bases in water. While we can increase the H_3O^+ concentration of a solution by adding acid, such as HCl, or increase the OH^- concentration by adding base, such as NaOH, the *numerical product* obtained by multiplying $[H_3O^+]$ by $[OH^-]$ is always constant, regardless of the addition of acid or base to the water. Since the value of this product remains fixed at any specific temperature, adding acid not only increases $[H_3O^+]$ but lowers $[OH^-]$ as well. Similarly, adding a base to water increases the hydroxide concentration and lowers the hydronium ion concentration. In water at 25°C this *ion product constant* of water, K_w, equals the product of the two molar concentrations:

K_w	$=$	$[H_3O^+] \times [OH^-]$	$=$	$10^{-7} \times 10^{-7} = 10^{-14}$
The ion product constant of water	is	the product of the hydronium ion concentration times the hydroxide ion concentration	and,	at 25°C, equals 0.00000000000001, which is a constant at this temperature.

Thus at 25°C the product of the hydronium ion concentration times the hydroxide ion concentration, $[H_3O^+] \times [OH^-]$, is always 10^{-14}.

Only when $[H_3O^+]$ equals 10^{-7} (and $[OH^-]$ also equals 10^{-7}) is water perfectly neutral. Any increase in the hydronium ion concentration above 10^{-7} (with an associated decrease in hydroxide ion concentration) produces an acidic solution; any decrease in $[H_3O^+]$ (and increase in $[OH^-]$) renders the solution basic.

This means that in acidic solutions the H_3O^+ concentration is greater than 10^{-7} *M*, the exponent of 10 is *less negative* than −7, and the pH is smaller than 7. In basic solutions the reverse is true; with the hydronium ion concentration less than 10^{-7}, the exponent of 10 is *more negative* than −7 and the pH is greater than 7. A fresh egg, then, is slightly basic.

To summarize, at 25°C:

- The pH of an acidic solution is less than 7.
- The pH of a neutral solution equals 7.
- The pH of a basic solution is greater than 7.

What's more, the definition of pH requires that for every *tenfold* increase in the $[H_3O^+]$ concentration there is a decrease of *one unit* in the pH; for every *tenfold* decrease in the acidity there is an increase of *one unit* in the pH. Table 10.3 summarizes the relation between acidity and pH.

Table 10.3 Acidity and pH

	pH		$[H_3O^+]$
Strongly acidic	1	10^{-1}	0.1
	2	10^{-2}	0.01
	3	10^{-3}	0.001
	4	10^{-4}	0.0001
	5	10^{-5}	0.00001
Weakly acidic	6	10^{-6}	0.000001
Neutral	7	10^{-7}	0.0000001 (at 25°C)
Weakly basic	8	10^{-8}	0.00000001
	9	10^{-9}	0.000000001
	10	10^{-10}	0.0000000001
	11	10^{-11}	0.00000000001
	12	10^{-12}	0.000000000001
	13	10^{-13}	0.0000000000001
Strongly basic	14	10^{-14}	0.00000000000001

E X A M P L E

Blood and Seawater

Which is more acidic, human blood, with a pH of 7.3–7.5, or seawater, with a pH in the range 7.8–8.3?

■ In Table 10.3 we see that the *lower* or less positive the value of pH, the *higher* the value of $[H_3O^+]$ and the *higher* the acidity of the solution. Since human blood has a lower pH than seawater, the hydronium concentration in blood must be higher than it is in seawater.

E X A M P L E

Sour Milk

The pH of fresh cow's milk is about 6.5. When milk spoils we sometimes say it's gone *sour* because of the taste it develops. From what you now know of chemistry, would you say the pH of spoiled milk is higher, lower, or the same as that of fresh milk?

■ We now know that one of characteristics of acids is that they taste sour. We can infer, then, that as milk spoils it becomes more acidic. With an increase in acidity, the pH of a substance drops. Therefore it's reasonable to assume that the pH of spoiled milk is lower than the pH of fresh milk.

In Section 10.8 we'll find that the acidity of sour milk comes from the formation of an acid known as *lactic acid*.

■ **Q U E S T I O N**

(a) A typical pH for household ammonia is 11. To what value of $[H_3O^+]$ does this correspond? (b) What is the value of $[OH^-]$ in household ammonia? (c) The hydronium ion concentration of an average tomato is about 0.001 *M*. What's the pH of the average tomato?

10.8 The Acids of Everyday Life

Usually pH is measured with either a strip of test paper or an instrument called a *pH meter* (Fig. 10.6). While the red and blue colors of litmus paper show the presence of acidity and basicity very clearly, they give no indication of the strength of the acid or base. Other, more sensitive types of test strips are impregnated with combinations of acid–base indicators that turn various colors as the pH changes. While these "universal" test strips offer fast, convenient, and cheap indications of pH, they give only approximate values. For more precise measurements the pH meter is the instrument of choice. It's essentially a specially designed voltmeter connected to a pair of electrodes that are dipped into the solution being examined. Most common pH meters provide values accurate to about 0.01 pH unit.

Table 10.4 presents the pH of some common substances. Notice that our stomach's digestive juices (sometimes called *gastric juices*) are quite acidic and that our food and drink range from the very high acidity of citrus fruits, such as lemons, to neutral or weakly basic substances like drinking water, fresh eggs, and crackers. The stomach's acidity, which comes from HCl secreted by the cells of its lining, promotes the digestion of food proteins by the enzymes of the gastric juices. *Pepsin*, for example, is a stomach enzyme that cleaves the very large molecules of proteins into smaller, more easily handled fragments. Pepsin does its chemical work best at a pH of 1.5–2.5 and quits functioning when the pH rises to between 4.0 and 5.0. Both good health and the digestion of proteins require an acidic stomach.

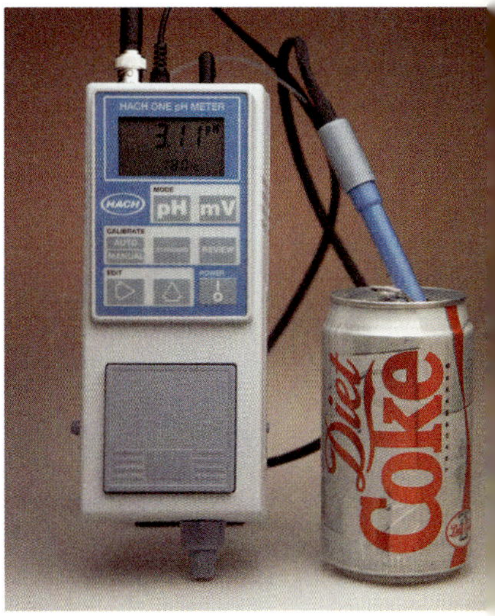

Figure 10.6 A pH meter, an instrument used for the very accurate measurement of hydrogen ion concentrations.

Table 10.4 pH of Some Common Substances

	pH	Material
Acidic	1.0–3.0	Gastric juices
	2.2–2.4	Lemons
	2.4–3.4	Vinegar
	2.5–3.5	Soft drinks
	3.0–3.4	Sour pickles
	3.0–3.8	Wine
	3.0–4.0	Oranges
	4.0–4.4	Tomatoes
	4.8–7.5	Human urine (usually 6.0)
	5.6	Carbonated water, rainwater
	6.3–6.6	Cow's milk
	6.4–6.9	Human saliva (during rest)
	6.6–7.6	Human milk
	6.5–8.0	Drinking water
Neutral	7.0	Pure water
Basic	7.0–7.3	Human saliva (while eating)
	7.3–7.5	Human blood
	7.6–8.0	Fresh eggs
	7.8–8.3	Seawater
	8.4	Sodium bicarbonate, saturated
	9.4	Calcium carbonate, saturated
	10.5	Milk of magnesia
	10.5–11.9	Household ammonia

The acids and bases we encounter most frequently are the acids and bases of our foods, our consumer products, and our environment. Except for the HCl of our gastric fluids, the most common of the acids we live with are organic acids, those containing carbon. Of these, a major group consists of the *carboxylic acids*, which are acids characterized by the *carboxyl* functional group (Sec. 6.12). The carboxyl group itself is a combination of a *carbonyl group* (a carbon doubly bonded to an oxygen) and a *hydroxyl group* (an —OH group). The carboxyl group is often written as $—CO_2H$ for brevity:

carbonyl group | hydroxyl group | carboxyl group, the structural unit of a carboxylic acid

The simplest of all carboxylic acids, with the lowest molecular weight, is formic acid, the acid that causes the sting of red ant bites. Its name comes from the Latin word for ant, *formica*. Acetic acid is the major organic acid of vinegar, while carbonic acid forms whenever carbon dioxide dissolves in water. Citric acid is responsible for the acidity of lemons, limes, grapefruit, and all other citrus fruit:

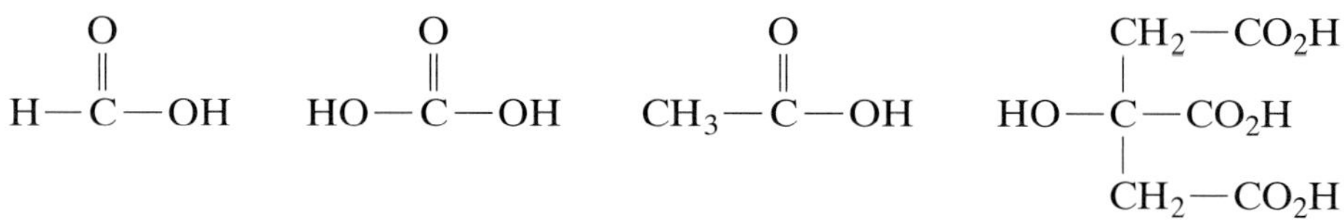

formic acid | carbonic acid | acetic acid | citric acid

Other important carboxylic acids include oxalic acid, propionic acid, butyric acid, benzoic acid, and lactic acid. Oxalic acid is a toxic *di*carboxylic acid (two carboxyl groups per molecule) that occurs widely in plants, including rhubarb, spinach, and sorrel:

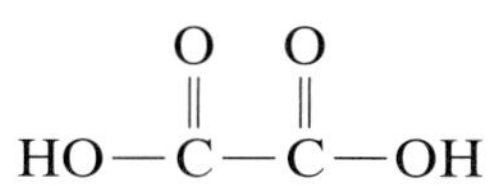

oxalic acid

Although eating pure oxalic acid can be fatal, its concentration in most edible plants is probably too low to present a serious hazard. It's been estimated that an average person would have to eat about 4 kg of spinach at one meal, almost 9 lb of the vegetable, to consume the lowest dose of oxalic acid known to be fatal to a human. Not all parts of all food plants, though, are as innocuous as spinach. For example, only the cooked stalks of rhubarb are ordinarily eaten. The plant's leaves contain such high concentrations of the acid and its salts that even small portions of them can be poisonous, especially to children. Oxalic acid itself is a good rust and stain remover and is used in various commercial cleaning preparations.

Propionic acid, $CH_3—CH_2—CO_2H$, and butyric acid, $CH_3—CH_2—CH_2—CO_2H$, are important partly for their odors. The characteristic odor of propionic acid contributes to the flavor of Swiss cheese; butyric acid, whose name comes from the Latin *butyrum*, for butter, gives rancid butter its peculiar odor. Salts of propionic acid, especially the sodium and calcium salts, are added to cheeses and to bread and other baked goods to retard the growth of mold and help preserve freshness.

Benzoic acid consists of a carboxyl group bonded directly to a benzene ring:

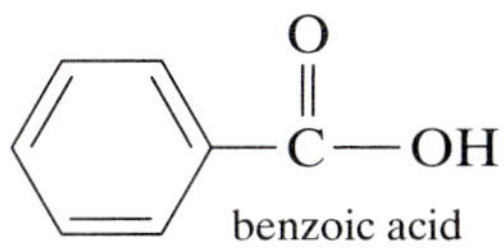

benzoic acid

To consumers, benzoic acid is less important in itself than as its sodium salt, sodium benzoate (also known as benzoate of soda). This salt is used as a preservative in canned and bottled fruit drinks and also in baked goods.

Lactic acid is the acid that gives sour milk its sharp taste. Bacterial fermentation of *lactose*, the principal sugar of milk, produces the acid:

$$\text{lactose (milk sugar)} \xrightarrow{\text{bacterial fermentation}} CH_3-\underset{\displaystyle OH}{\underset{|}{CH}}-\overset{\displaystyle O}{\overset{||}{C}}-OH$$

lactic acid,
the acid of sour milk

Lactic acid is also important to us in another, more immediate way: in our own bodies. As we've seen in Chapter 8, our biological energy comes from chemical oxidation of the foods we eat. We've seen that we obtain energy from our nutrients by oxidizing them to carbon dioxide and water, much as a car's engine derives energy from the hydrocarbons of gasoline. But our bodies are far more complex than the internal combustion engine. Converting the chemicals of our foods to carbon dioxide and water requires an enormous variety of steps involving a huge assortment of intermediate compounds. For example, as we metabolize glucose (a sugar) through muscular activity, lactic acid is generated in our muscle tissue and is then decomposed into CO_2 and H_2O. With heavy exercise, our metabolic chemistry forms lactic acid more rapidly than it decomposes. It's this transient accumulation of lactic acid that causes an aching, tired sensation in our muscles when we use them vigorously. As our body disposes of the accumulated lactic acid, the discomfort disappears and we sense a recovery from the muscular ache and fatigue.

Figure 10.7 summarizes the occurrences of some representative carboxylic acids.

Figure 10.7 Common acids include the citric acid of limes, lemons, oranges, grapefruit, and citrus juices, the propionic acid of Swiss cheese, and the oxalic acid of rhubarb.

In terms of commercial production, sulfuric acid (H_2SO_4) is by far our leading industrial chemical. In raw tonnage produced, about 44 million tons in the United States in 2000, it swamps any of its rivals. Most of it ends up as a reactant or a catalyst (Sec. 7.7) in the manufacture of various consumer products, including pharmaceuticals, plastics, synthetic fibers, and synthetic detergents. More calcium oxide (CaO, or lime; roughly 22 million tons in 2000) is produced than any other base. Much of it goes into the production of building materials, the generation of still other industrial chemicals, and water and sewage treatment.

■ QUESTION

Name or identify each of the following carboxylic acids: (a) gives rancid butter its foul odor; (b) toxic, it is found in small concentrations in raw spinach and rhubarb; (c) forms in our muscles as the result of metabolic activity; (d) produces the flavor of Swiss cheese; (e) used as its sodium salt to preserve canned and bottled fruit drinks; (f) responsible for the taste of sour milk; (g) produces the acidity of citrus fruit; (h) generates the sting of red ant bites; (i) the major organic acid of vinegar.

10.9 Antacids: Bases That Fight Cannibalism

Carbon dioxide is released when an Alka-Seltzer tablet is dropped into water. As the sodium bicarbonate and citric acid of the tablet react in water, they form carbonic acid. With increasing concentration the carbonic acid decomposes to carbon dioxide and water.

Although an acidic stomach is necessary for good health, excessive stomach acidity can produce the discomforts commonly called "acid indigestion" and "heartburn" as well as contribute to the pain of gastric ulcers. Like much of our food, our body tissues and organs, including the walls of the stomach itself, are made of protein. The enzymes of the stomach fluids that help digest protein can't discriminate between the protein of food and the protein of the stomach walls. They would digest the stomach itself as easily as a steak if it weren't for special protective devices, such as an alkaline mucous lining that resists the action of the stomach's acid and its enzymes. This barrier keeps the stomach from being eaten away by its own juices. Sometimes, though, things get out of hand. When the body's defenses fail, a combination of stomach acid and pepsin (perhaps accompanied by certain bacteria) can attack the stomach wall in an act of chemical cannibalism. The result is a gastric ulcer.

Relief from the discomforts of excess stomach acid, which can include the pain of ulcers, can often be obtained from antacids such as Alka-Seltzer, Milk of Magnesia, Rolaids, Titralac, and Tums, or simply some *baking soda* in water. Each year Americans buy a quarter of a billion dollars worth of these and other antacid products simply for the relief of pain brought on by maverick hydronium ions of the stomach.

(*Baking soda* is the commercial term for *sodium hydrogen carbonate*, $NaHCO_3$, which is also known by an older name, *sodium bicarbonate*. The anion HCO_3^- can be called the *hydrogen carbonate* or the *bicarbonate* ion. For simplicity, and to be consistent with the term most often used commercially, we'll use the older *bicarbonate* here.)

The antacid's function isn't to bring the stomach fluids to a complete acid–base neutrality of pH 7. Such an action would shut down digestion completely and could shock the walls into flooding the stomach with fresh acid in what's called "acid rebound." Instead, a good antacid neutralizes enough of the HCl in the gastric juices to alleviate the pain and discomfort yet allows normal stomach action to proceed. Even if as much as 90% of the stomach's HCl were to be neutralized, the pH would still be a healthy 2.3.

The bases most widely used in antacids, providing an inexpensive combination of safety, effectiveness, and efficiency, include:

Aluminum hydroxide, $Al(OH)_3$
Calcium carbonate, $CaCO_3$
Magnesium carbonate, $MgCO_3$
Magnesium hydroxide, $Mg(OH)_2$
Sodium bicarbonate, $NaHCO_3$
Potassium bicarbonate, $KHCO_3$

Magnesium oxide, MgO, is also used as an antacid ingredient since it reacts with water to form magnesium hydroxide:

$$MgO + H_2O \longrightarrow Mg(OH)_2$$

The active ingredients of several of the more popular antacid preparations appear in Table 10.5. On reaction with HCl, the sodium bicarbonate of Alka-Seltzer and other commercial antacids forms carbonic acid, which is a much weaker acid than HCl and which decomposes readily (and reversibly) to carbon dioxide and water:

$$NaHCO_3 + HCl \longrightarrow NaCl + \underset{\text{carbonic acid}}{H_2CO_3}$$

$$H_2CO_3 \rightleftharpoons CO_2 + H_2O$$

Each Alka-Seltzer tablet contains 1.0 g of sodium bicarbonate, 0.3 g of potassium bicarbonate, and 0.8 g of citric acid. When you drop the tablet into water, the citric acid and the sodium and potassium bicarbonates dissolve. They react with each other to produce carbonic acid, which decomposes to carbon dioxide by the very same sort of reactions shown above for the neutralization of HCl. The fizz of the Alka-Seltzer tablet comes from this chemical release of the carbon dioxide. You can produce the very same effervescence by squeezing a little lemon juice onto some household baking soda, which is a commercial form of sodium bicarbonate. The citric acid of the lemon juice reacts with the sodium bicarbonate to produce carbon dioxide just as it does when you drop an Alka-Seltzer tablet into water.

There's considerable choice among the various brands and formulations of antacid tablets. The selection could be made, for example, on the basis of the particular base used. People with hypertension (high blood pressure) who have been advised to avoid sodium might wish to choose an antacid that doesn't contain sodium bicarbonate ($NaHCO_3$). On the other hand, those who are concerned about the possible bone degeneration resulting from a loss of calcium, a condition known as *osteoporosis*, might choose an antacid formulated with calcium carbonate ($CaCO_3$).

Table 10.5 Common Antiacids

Principal Active Ingredient	Representative Antiacid	Neutralization Reaction
Sodium bicarbonate Potassium bicarbonate	Alka-Seltzer	$NaHCO_3 + HCl \longrightarrow NaCl + H_2CO_3$ $(H_2CO_3 \longrightarrow H_2O + CO_2)$
Magnesium hydroxide	Phillips' Milk of Magnesia	$Mg(OH)_2 + 2HCl \longrightarrow MgCl_2 + 2H_2O$
Dihydroxyaluminum sodium carbonate	Rolaids	$Al(OH)_2NaCO_3 + 4HCl \longrightarrow AlCl_3 + NaCl + 2H_2O + H_2CO_3$
Calcium carbonate	Titralac Tums	$CaCO_3 + 2HCl \longrightarrow CaCl_2 + H_2CO_3$
Aluminum hydroxide Magnesium carbonate Magnesium hydroxide	Di-Gel	$Al(OH)_3 + 3HCl \longrightarrow AlCl_3 + 3H_2O$ $MgCO_3 + 2HCl \longrightarrow MgCl_2 + H_2CO_3$ $Mg(OH)_2 + 2HCl \longrightarrow MgCl_2 + 2H_2O$

■ QUESTION

Some commercial antacids claim only to neutralize "excess" stomach acidity. Why is it undesirable to neutralize completely all stomach acidity? What would result from bringing the pH of stomach fluids all the way up to 7?

10.10 Red Cabbage, Antacid Fizz, and Le Châtelier's Principle

We're now in a position to understand why the cabbage breath test works and how the color change of the cabbage extract is connected to the fizz that an Alka-Seltzer tablet produces. In addition we'll also see how breathing helps maintain a stable pH for your blood.

In the opening demonstration, the ammonia that's added to the deep blue solution of cabbage extract makes the solution slightly basic and turns the color of the indicator solution to an emerald green. As someone blows into the basic test solution of the opening demonstration, the carbon dioxide of the exhaled breath dissolves in the water to form carbonic acid, which immediately reacts with the dissolved ammonia, neutralizing it.

Formation of carbonic acid as carbon dioxide dissolves in water:

$$\underset{\text{carbon dioxide}}{CO_2} + H_2O \rightleftharpoons \underset{\text{carbonic acid}}{HO{-}\overset{\overset{\displaystyle O}{\|}}{C}{-}OH}$$

Neutralization of ammonia by carbonic acid to form a salt, ammonium carbonate:

$$\underset{\text{carbonic acid}}{HO{-}\overset{\overset{\displaystyle O}{\|}}{C}{-}OH} + 2NH_3 \longrightarrow \underset{\text{ammonium carbonate}}{NH_4^{+}\ {}^{-}O{-}\overset{\overset{\displaystyle O}{\|}}{C}{-}O^{-}\ {}^{+}NH_4}$$

Then, after all the ammonia has been converted to ammonium carbonate, the excess carbonic acid ionizes and produces hydronium cations and bicarbonate anions, and the indicator turns to a light blue color as the test solution becomes slightly acidic.

Ionization of excess carbonic acid, with water acting as a base as it accepts the proton released by the carbonic acid and forms the hydronium ion:

$$\underset{\text{carbonic acid}}{HO{-}\overset{\overset{\displaystyle O}{\|}}{C}{-}OH} + H_2O \rightleftharpoons \underset{\text{hydronium ion}}{H_3O^{+}} + \underset{\text{bicarbonate anion}}{{}^{-}O{-}\overset{\overset{\displaystyle O}{\|}}{C}{-}OH}$$

Adding the vinegar at the end of the demonstration introduces acetic acid into the solution. The acetic acid is a much stronger acid than carbonic acid; it makes the test solution far more acidic than the carbon dioxide of breath does. With the marked increase in acidity, the indicator turns a vivid pink or red.

Ionization of acetic acid to form a hydronium ion and an acetate anion:

$$\underset{\text{acetic acid}}{CH_3{-}\overset{\overset{\displaystyle O}{\|}}{C}{-}OH} + H_2O \rightleftharpoons \underset{\text{acetate anion}}{CH_3{-}\overset{\overset{\displaystyle O}{\|}}{C}{-}O^{-}} + H_3O^{+}$$

One of the key factors in this demonstration is the *formation* of the carbonic acid. If it weren't for the reaction of carbon dioxide with the water to form carbonic acid, the demonstration wouldn't work. We've got to use the opposite view to explain the production of fizz when we drop an Alka-Seltzer tablet into water. Here the key is the *decomposition* of the carbonic acid. In this case, if it weren't for the decomposition of carbonic acid to carbon dioxide and water, there would be no fizz.

Both these phenomena depend on the operation of the following equilibrium reaction:

$$CO_2 + H_2O \rightleftharpoons H_2CO_3$$

Viewed in the forward direction, from left to right, it's the reaction that makes the cabbage breath test work; in the reverse direction, from right to left, it's responsible for the fizz.

What determines the direction of the equilibrium reaction? What determines whether carbon dioxide and water are converted to carbonic acid, or carbonic acid is converted to carbon dioxide and water? The answer depends on the rules of chemistry, on the conditions at hand, and on a chemical principle known as **Le Châtelier's principle**, which states that *placing a stress on an equilibrium causes the equilibrium to shift so as to relieve the stress*. It's a principle that operates in all chemical equilibria. Henri Louis Le Châtelier, the French chemist who first proposed this principle, lived from 1850 to 1936 and worked primarily on the chemistry of industrial processes and products.

Le Châtelier's principle states that when a stress is placed on a system in equilibrium, the system tends to change in a way that relieves the stress.

Le Châtelier's principle explains nicely both the formation and the decomposition of carbonic acid. The key lies in a simple bit of chemistry: In any equilibrium reaction, the ratio of the concentrations of the products to the concentrations of the reactants tends to remain constant. In the equilibrium of carbonic acid, carbon dioxide, and water, for example,

$$H_2CO_3 \rightleftharpoons CO_2 + H_2O$$

the ratio

$$\frac{[CO_2] \times [H_2O]}{[H_2CO_3]}$$

tends to remain at a constant value. Thus anything that might act to increase the concentration of carbon dioxide would also produce an increase in the concentration of the carbonic acid so as to keep the numerical value of the ratio constant. Similarly, anything acting to increase the concentration of carbonic acid would also act to increase the concentration of carbon dioxide.

As you blow your exhaled breath through the straw, the CO_2 reacts with the water to form H_2CO_3. The continuing absorption of carbon dioxide as you continue to blow into the straw keeps increasing the concentration of the CO_2 and places a stress on the equilibrium. To relieve this stress and to maintain a constant value of the ratio described above, the equilibrium shifts to convert the additional carbon dioxide into a higher concentration of carbonic acid (Fig. 10.8). The process continues until the water is saturated with carbon dioxide.

The reverse of this process occurs when you dissolve an Alka-Seltzer tablet in water or add the citric acid of lemon juice (or the acetic acid of vinegar) to a sodium bicarbonate solution. The acid reacts with the bicarbonate ion to produce carbonic acid:

$$H^+ + HCO_3^- \rightleftharpoons H_2CO_3$$

As the concentration of carbonic acid grows, the equilibrium shifts to convert carbonic acid to water and carbon dioxide, which escapes as a gas. Once again the

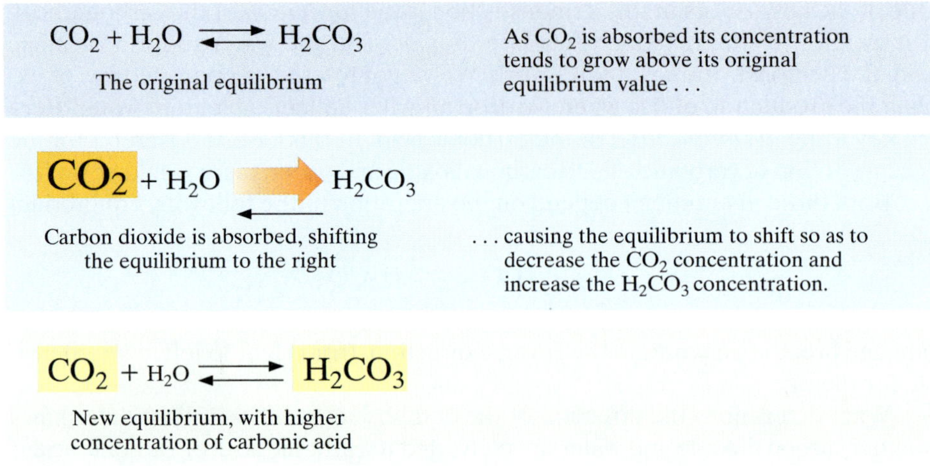

Figure 10.8 Le Châtelier's Principle in operation.

numerical value of the ratio is held constant. In this case the very large quantity of carbonic acid that is produced puts a stress on the system, which is relieved as carbonic acid decomposes into carbon dioxide and water (Fig. 10.9).

Le Châtelier's Principle also describes how your blood's pH is maintained within a narrow range. Blood is slightly basic; its pH, normally 7.4, doesn't vary by more than about one tenth of a pH unit in a healthy person. In maintaining this narrow range of pH through all the stresses of a normal life, the body uses an acid–base equilibrium, with carbonic acid as the acid and water as the base. (Recall that water is amphoteric; Sec. 10.4.):

$$H_2CO_3 + H_2O \rightleftharpoons H_3O^+ + HCO_3^-$$

Figure 10.9 The chemistry of the fizz.

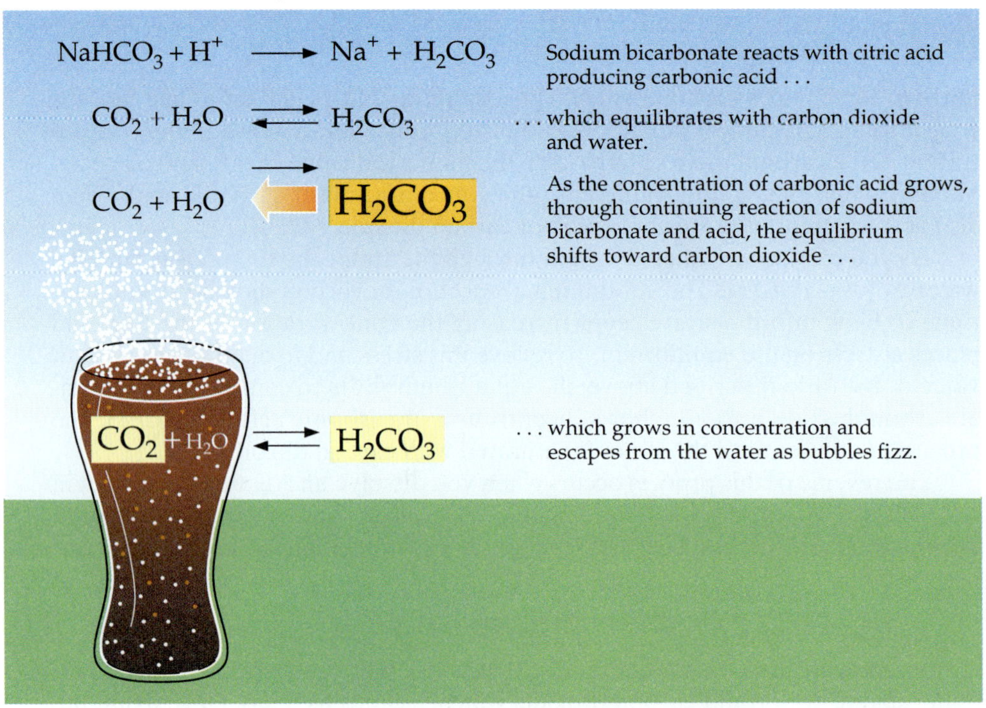

It works like this. If the pH of the blood begins to drop, because of disease or some other factor, the concentration of the hydronium ion rises. This places a stress on the equilibrium and brings Le Châtelier's principle into action. In response to the stress, the bicarbonate anion combines with the acid and shifts the equilibrium to the left, thereby decreasing the concentration of the hydronium ion and maintaining a pH of 7.4.

If the reverse occurs and the pH begins to rise, the concentration of the hydronium ion tends to drop and the equilibrium shifts to the right. Carbonic acid ionizes to restore the pH to its desirable level of 7.4.

All this, of course, depends on the body's having some means for regulating the blood's supply of carbonic acid. With the blood's carbonic acid concentration rising as the equilibrium shifts to the left and dropping as it shifts to the right, quantities of carbonic acid must be moved into and out of the body. This particular piece of work—the regulation of carbonic acid concentrations—takes place as the blood passes through the lungs, and it operates through an equilibrium we've seen before:

$$H_2CO_3 \rightleftharpoons CO_2 + H_2O$$

With an excess of carbonic acid, carbon dioxide passes from the blood into the lungs; with a deficiency the reverse occurs. Once again Le Châtelier's principle comes into play.

■ QUESTION

Hyperventilation, which occurs with very rapid and very deep breathing, results in a rapid loss of carbon dioxide from the body through the lungs. What happens to the pH of the blood as the direct result of hyperventilation? Does the pH immediately rise or drop or does it remain unchanged? Explain.

PERSPECTIVE

The Lesson of Le Châtelier

We began our examination of acids and bases with a demonstration of the effect of the acid we exhale, CO_2, on the dye of red cabbage. Proceeding from there we have expanded our scope to include other acids, bases, their applications in the world around us, and their interactions with each other. Among the topics we have covered are some that have applications well beyond the limits of acid–base chemistry, applications that at times seem to grow in importance as their scope broadens. One of these is Le Châtelier's Principle. In its narrowest form Le Châtelier's Principle allows us to predict the results that follow from changing the conditions of a chemical equilibrium. Taken at its most literal, it explains why carbonic acid can either come into existence or decompose, depending on the prevailing conditions. More broadly, we can see its operation in the more figurative equilibria of our society and our environment.

Placing a stress on *any* equilibrium—chemical, biological, societal, environmental, or personal—must cause the equilibrium to change. We have already seen an illustration in the greenhouse effect (Sec. 8.14). As we burn fossil fuels we increase the carbon dioxide content of our atmosphere. It's a stress of sorts on the equilibration of carbon dioxide between its free existence in the atmosphere and its incorporation into the material of plants. This stress may lead to a slow warming of the earth.

In the Perspectives of Chapters 7 and 8 we touched on the possibility that by replacing our gasoline-powered cars with electric vehicles we might relieve one environmental stress, only to replace it by another as we search for power sources to meet the expanded demands for electricity. At present we can only speculate what equilibria will be affected by the new stress and how they will change.

As we continue in our study of chemistry and its impact on our daily lives, we'll see additional examples of Le Châtelier's Principle, described in still other terms. In any case, we can be certain that any change in our actions as a society will produce a stress of some kind on our environment, and that our environment, obeying the Principle described by Le Châtelier, will respond in one way or another. ■

EXERCISES

For Review

1. Following are a statement containing a number of blanks, and a list of words and phrases. The number of words equals the number of blanks within the statement, and all but two of the words fit correctly into these blanks. Fill in the blanks of the statement with those words that do fit, then complete the statement by filling in the two remaining blanks with correct words (not in the list) in place of the two words that don't fit.

All acids have certain properties in common. When dissolved in water they produce a _________ taste, they turn _________ from blue to red, and they react with metals such as iron and ___________ to liberate ___________. Water solutions of _____________, on the other hand, taste _________, turn litmus from _________ to _________, and produce a _________ sensation when rubbed between the fingers. At the molecular level, the _________ definition describes an acid as anything that can transfer a _________ to a base.

Even though pure water is perfectly neutral, its covalent molecules ionize in an _________ reaction to form protons and _________ anions. The protons form bonds to other water molecules, converting them to _________ ions. A common measure of acidity, _________ is the negative of the logarithm of the hydronium ion concentration. Among the most common acids are _________ acid, which is the principal acid of vinegar; _________ acid, which occurs in citrus fruit; and _________ acid, which is found in milk. Commercially, the acid manufactured in the largest quantity is _________ acid.

acetic	hydronium
bases	hydroxide
bitter	litmus
blue	pH
Brønsted–Lowry	proton
carbon dioxide	red
citric	slippery
equilibrium	sour
formic	zinc
hydrogen	

2. Define or identify each of the following:

a. amphoteric
b. antacid
c. anthocyanin
d. carboxylic acid
e. Le Châtelier's principle
f. neutralization
g. pepsin
h. a salt

3. What are the acids of the cabbage breath test demonstration that opened this chapter? What base is used in that demonstration?

4. What color—red or blue—would you expect each of the following to produce when added to litmus: (a) milk of magnesia, (b) wine, (c) seawater, (d) a soft drink, (e) tomato juice?

5. Match the following:

______ a. acetic acid	1. amphoteric
______ b. benzoic acid	2. formed as milk turns sour
______ c. butyric acid	3. found in rancid butter
______ d. carbonic acid	4. ionizes completely in water
______ e. citric acid	5. gives lemons a sour taste
______ f. hydrochloric acid	6. occurs in spinach and rhubarb
______ g. lactic acid	7. the major acid of vinegar
______ h. oxalic acid	8. contains a carboxyl group bonded directly to a benzene ring
______ i. propionic acid	9. decomposes to carbon dioxide and water
______ j. sodium bicarbonate	10. baking soda
______ k. water	11. forms in cheese; its calcium salt is used as a preservative in baked goods

6. Identify each of the following as an acid, a base, or a salt; give its chemical name (if it is shown as a chemical formula or structure); name a consumer product in which it occurs; and, if possible, describe its function in that product:

a. $Al(OH)_3$
b. $CaCO_3$
c. CH_3CO_2H
d. H_2SO_4
e. HCl
f. KI
g. $Mg(OH)_2$
h. $MgSO_4$
i. NaF
j. $NaHCO_3$
k. NaOH
l. NH_3
m.

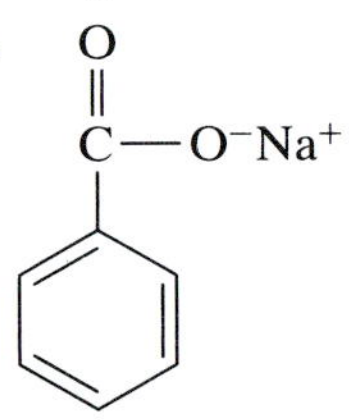

7. Define a salt in terms of acids and bases.

8. Suppose the nucleus of a hydrogen atom enters a quantity of water. What is produced by the interaction of the hydrogen nucleus with a water molecule?

9. What do we mean when we say that water is amphoteric?

10. What is the difference, if there is any difference, between a proton and a hydrogen ion?

11. What gas is formed when you drop an Alka-Seltzer tablet into water? Write the chemical reaction that produces this gas.

12. Hydrogen fluoride (HF), an acid, reacts with sodium hydroxide to produce a salt that's used in toothpaste to help prevent tooth decay. Write the chemical equation for the neutralization of hydrogen fluoride with sodium hydroxide. What is the name of the salt that forms in this reaction?

13. Name six acids that occur, as the acids themselves or as their salts, in substances we normally eat or drink and name the food or drink that contains each.

14. What connection is there between sour milk and tired muscles?

15. What characteristic of molecular structure is common to all carboxylic acids?

16. Give the names of two carboxylic acids that can be found in the human body.

17. Give a specific example of a tricarboxylic acid.

18. Would the data in Table 10.4 lead you to expect (a) lemon juice, (b) milk, or (c) seawater to have the same effect as vinegar on the test solution of the opening demonstration? Explain.

19. What's the difference between a weak acid and a strong acid? Give an example of each.

20. Identify each ion or compound in the following reaction (other than water) as an acid, a base, or a salt:

$$CH_3{-}CO_2H + NaOH \longrightarrow CH_3{-}CO_2^-Na^+ + H_2O$$

21. In Section 10.2 we reviewed three tests that can be used to determine whether a substance is an acid. A fourth test comes from a reaction of the substance in question with ordinary chalk, which is mostly calcium carbonate, $CaCO_3$. Complete and balance the following equation for the reaction of HCl with $CaCO_3$ and explain how the test could be used to determine whether the substance is an acid:

$$HCl + CaCO_3 \longrightarrow CaCl_2 + ?$$

22. Write the chemical reaction for the generation of fizz or foam when you squeeze lemon juice into water containing dissolved baking soda.

A Little Arithmetic and Other Quantitative Puzzles

23. Suppose you prepare a solution by dissolving 1 mole of pure HCl gas in 100 liters of water. (Recall that HCl ionizes completely in water.) What is (a) the molarity of H_3O^+ in this solution? (b) the pH of this solution?

24. Suppose you prepare a solution by dissolving 1 mole of pure NaOH in 100 liters of water. (NaOH ionizes completely in water.) What is (a) the molarity of OH^- in this solution? (b) The molarity of H_3O^+ in this solution? (c) The pH of this solution?

25. What is the concentration of Cl^- in a solution of HCl that has a pH of 2?

26. Hydrogen fluoride is a weak acid; it does not ionize completely. What can you say about the pH of a solution prepared by adding 1 mole of HF to a quantity of water and then adding enough pure water to dilute the mixture to 100 liters? What can you say about the molarity of the fluoride anion in this solution?

27. What is (a) the pH of a solution in which $[H_3O^+]$ is 0.0001 *M*? (b) the pH of a solution in which $[H_3O^+]$ is 0.0000000001 *M*? (c) the pH of neutral water? (d) the molarity of $[H_3O^+]$ in a solution whose pH is 3?

28. Can the pH of water be: (a) greater than 14? (b) negative? Explain.

29. How many liters of 0.1 *M* NaOH solution does it take to neutralize (a) 1 liter of 0.1 *M* HCl? (b) 0.5 liter of 0.2 *M* HCl? (c) 3 liters of 0.01 *M* HCl?

30. How many liters of 0.5 *M* HCl solution does it take to neutralize (a) 0.5 liter of 0.1 *M* NaOH? (b) 1 liter of 0.5 *M* NaOH? (c) 0.1 liter of 2 *M* NaOH?

31. By analogy to pH, how would you define pOH?

32. Based on your definition of pOH, what would always be true of the sum

$$pH + pOH$$

in any acidic or basic solution in water at 25°C?

Think, Speculate, Reflect, and Ponder

33. In the opening demonstration, what would you observe if you added the dilute ammonia solution, drop by drop, to the pink mixture you obtained after adding the vinegar?

34. Acids and bases react with each other to produce neutral solutions. Could you obtain a neutral solution by mixing, in the proper proportions: (a) vinegar and sodium bicarbonate; (b) the juice of sour pickles and carbonated water; (c) seawater and milk of magnesia.

35. Name a consumer product or a household substance other than vinegar or a fruit juice that would produce a fizz when added to baking soda.

36. Why is it so much more convenient to express acidities in terms of pH rather than in terms of the molarity of the hydrogen ion?

37. Sodium cyanide, NaCN, is a highly toxic salt that's handled in the form of pellets. It is sometimes used to generate the lethal gas hydrogen cyanide, HCN, in exterminating rats and insects in large buildings. Write a reaction for the production of HCN by dropping pellets of NaCN into a solution of hydrochloric acid.

38. For each of the following sets, pick the single compound that doesn't belong with the others. Explain your choice in each case.

1. (a) HCl, (b) $CH_3{-}CO_2H$, (c) H_2CO_3, (d) NH_3, (e) H_2SO_4
2. (a) acetic acid, (b) benzoic acid, (c) nitric acid, (d) lactic acid, (e) propionic acid
3. (a) LiOH, (b) CO_2, (c) MgO, (d) NH_3, (e) $Al(OH)_3$
4. (a) $CaCO_3$, (b) $NaHCO_3$, (c) $Mg(OH)_2$, (d) $Al(OH)_3$, (e) NaOH

39. As we'll see in Chapter 11, galvanized nails are coated with zinc to protect them against corrosion and rust. If you cover several galvanized nails with vinegar you will soon see small bubbles of gas forming on the nails and rising to the surface of the vinegar. Why? Describe the chemical reactions taking place. Why could this be a hazardous thing to do? Would you expect the same result if you used iron nails that are not galvanized? Would you expect the same result if you used lemon juice instead of vinegar? Explain your answers.

40. We've seen in Section 10.5 that there *are* ions in pure water, after all. In the demonstration that opened Chapter 1, why didn't the light bulb with the wires dipping into the pure water (which contained hydronium cations and hydroxide anions) glow as brightly as the one with leads dipping into the sodium chloride solution? (*Hint:* Assume that the salt water of that demonstration contained sodium chloride at the same molar concentration as glass 1 of the opening demonstration to Chapter 9.)

41. Describe how you might prepare strips of acid–base indicator paper (which could be used in much the same way as litmus paper) from a common material available to you in supermarkets.

42. Many acid–base indicators change color as a result of changes in their molecular structure as the pH of a solution changes. Some change color as a result of structural changes brought on as the molecule acquires a proton, H^+, with a drop in pH. To what general class of compounds do such indicators belong?

43. Acid deposits around a car's battery terminals can be cleaned with a solution of sodium bicarbonate. Assuming that the acid is sulfuric acid, complete and balance the equation for the neutralization reaction

$$NaHCO_3 + H_2SO_4 \longrightarrow ?$$

44. Consider a solution of carbonic acid in water, with the equilibrium mixture consisting of carbonic acid, the bicarbonate anion, and the proton:

$$H_2CO_3 \rightleftharpoons HCO_3^- + H^+$$

In which direction, forward or reverse, would this equilibrium shift if a small amount of acid (which would act to increase $[H^+]$) were added to the solution? In which direction would the equilibrium shift if a small amount of a strong base such as NaOH (which would act to decrease $[H^+]$ through the reaction $OH^- + H^+ \longrightarrow H_2O$) were added?

45. Write equilibrium reactions similar to those of Section 10.10 to show how the pH of the blood is maintained near 7.4 when (a) an excess of base enters the blood and (b) an excess of acid enters the blood.

46. In addition to H_2CO_3 and $NaHCO_3$, another combination of an acid and a base—NaH_2PO_4 and Na_2HPO_4—also acts to maintain blood at a stable pH of 7.4. Recognizing that NaH_2PO_4 is the more acidic of these two components, and that Na_2HPO_4 is the more basic component, write a chemical reaction that shows how this system changes as the blood becomes more acidic than pH 7.4. Write a reaction that shows how it changes as the blood becomes more basic.

47. *Malonic acid*, with a molecular formula $C_3H_4O_4$, is a dicarboxylic acid containing three carbons. Draw its molecular structure.

48. Some chemical principles have their counterparts in social activities as well as in the chemical laboratory. Describe the operation of Le Châtelier's Principle (or something closely resembling it) in a social situation.

49. "An acid cannot exist in the absence of a base." Is this statement true or false? Explain.

50. We've seen in Section 10.4 that water's amphoterism requires a restatement of the Arrhenius definition of an acid. Does this same amphoterism require a restatement of the Arrhenius definition of a base? If it does, restate Arrhenius's definition to accommodate water's amphoterism. If not, explain why Arrhenius's definition of a base is compatible with water's amphoterism.

51. Does water's amphoterism also require a restatement of the Brønsted–Lowry definition of an acid? If it does, restate the Brønsted–Lowry definition to accommodate water amphoterism. If not, explain why not.

52. The term *litmus test* is sometimes used in describing a political issue that, in itself alone, can determine a voter's support for a candidate. Give several examples of "litmus test" issues that can, in themselves, determine a voter's choice of candidates.

53. Give a phenomenological definition of something you are familiar with outside the realm of chemistry.

Web Questions

Use the Web to answer or accomplish each of the following. Report your Web source for each response.

1. Anthocyanins are a class of compounds found in plants. Give the name and a reference to the structure of a specific anthocyanin. Name several plants other than red cabbage that contain anthocyanins.

2. Write a brief history of the use of litmus as an acid–base indicator. Describe how litmus paper is currently made.

3. *Phenolphthalein* is another useful acid–base indicator. This chemical was once used as the active ingredient in a common consumer product. What was that product, what was phenolphthalein's function in the product, and why is phenolphthalein no longer used for that purpose?

4. Name two other chemicals, other than anthocyanins, litmus, and phenolphthalein, that can be used as acid–base indicators.

5. Summarize the lives of Johannes Brønsted and Thomas M. Lowry. and describe briefly how they came, independently, to the same definition of an acid and a base.

6. Identify the chemist Felix Hofmann and describe his connection to an aromatic carboxylic acid that can be found in almost every medicine cabinet.

7. Not all biologically important acids are carboxylic acids. Give the molecular structure of *uric acid*, explain why it is not classified as a carboxylic acid, and summarize its effects on human health and well being.

Oxidation and Reduction

THE ELECTRICITY OF CHEMISTRY

Electricity in the form of lightning.

Galvanized Tacks, Drugstore Iodine, and Household Bleach

In this chapter we examine the behavior of the electrons in the valence shells of atoms and we learn how we can use them both to store and generate electricity. A simple but intriguing demonstration starts us off. Cover the bottom of a glass with a layer of small, galvanized tacks, the kind you can get in any hardware store. (*Galvanized* tacks, garbage cans, and other metal products are covered by a thin layer of zinc to protect them from corrosion.) Now add some tincture of iodine to the glass, the kind drugstores sell as antiseptic solutions. Use just enough of the iodine solution so that it doesn't quite cover every bit of the topmost tacks. It's better to use too little of the tincture of iodine than too much.

The iodine solution covering the tacks now has the very dark, purple–violet color of elemental iodine itself. But if you let the glass stand for about half an hour you'll see the liquid fade slowly to a very pale yellow or lose its color completely. The actual time it takes for the color to fade depends on several factors. Swirling the glass gently, for example, can speed up the decolorization.

So that you can see clearly what happens next, carefully pour the solution itself into another glass, leaving the tacks behind. Now add a few drops of liquid household bleach to the pale solution. Immediately the very dark purple violet of the iodine returns. If the color appears to form clumps, producing a lumpy appearance within the liquid, add a few drops of vinegar. The vinegar causes the clumps to dissolve, producing a dark solution that looks very much like the iodine you used originally. Figure 11.1 sums up this entire procedure.

What you've just seen results from the sort of chemistry we'll cover in this chapter. Both the loss of color from the iodine solution and the regeneration of the original color take place as electrons are transferred from atoms of one element to atoms of another. Specifically, the purple–violet elemental iodine (I_2) of the tincture of iodine reacts with the zinc coating on the tacks and is converted to the colorless iodide anion (I^-) as the zinc atoms (Zn) are transformed into zinc cations (Zn^{2+}).

With the conversion of the purple–violet I_2 to the colorless iodide ion, the intense color of the solution fades. The liquid bleach that you add to restore the color contains the hypochlorite anion (ClO^-), which reacts with the I^-, reconverting it to the original purple-violet I_2. The color returns.

The clumps that form appear as hydroxide ions of the basic liquid bleach react with zinc ions to produce zinc hydroxide, which isn't soluble in water:

$$\underset{\text{zinc cations}}{Zn^{2+}} + \underset{\text{hydroxide anions}}{2OH^-} \longrightarrow \underset{\text{zinc hydroxide}}{Zn(OH)_2}$$

The acetic acid of the added vinegar reverses the process:

$$Zn(OH)_2 + 2CH_3\text{–}CO_2H \longrightarrow Zn^{2+} + 2H_2O + 2CH_3\text{–}CO_2^-$$

1. The equipment: household bleach, galvanized tacks, tincture of iodine, and vinegar.

2. Add the iodine solution to the tacks, nearly covering them.

3. The color of the iodine fades as time passes.

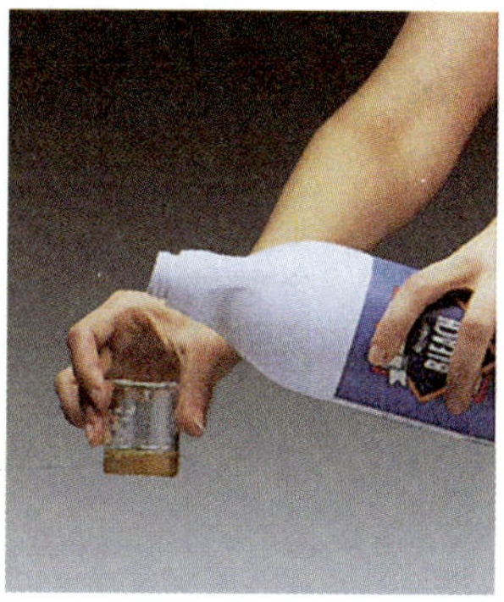

4. Adding a few drops of bleach to the faded solution . . .

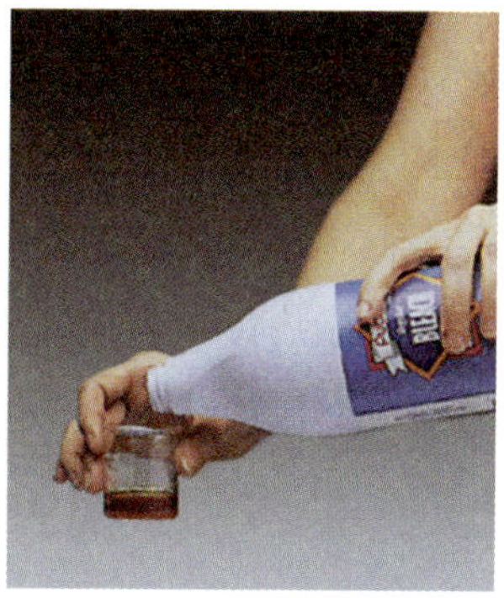

5. . . . restores the original color.

6. Adding a little vinegar produces a clear, intensely colored solution.

Figure 11.1 Color changes with galvanized tacks, iodine, and household bleach.

We'll examine in more detail *how* and *why* these changes take place—why the zinc atoms on the galvanized tacks cause the deeply colored iodine molecules to be transformed into colorless iodide ions and why the chlorine of the household bleach reconverts the iodide to iodine and restores the color—after we've examined some of the chemistry that takes place when electrons move from one chemical particle to another. It's this same kind of chemistry that runs one of our most common, everyday consumer products, the electric battery. ■

11.1 Inside the Flashlight Battery

Almost anywhere we go we can find batteries to provide light from our flashlights, to operate our clocks, watches, cameras, cellular phones, portable radios, CD players, and computers, and to start our cars. These small, transportable storehouses of electricity are available in a variety of forms for a variety of uses. They can be found as the relatively inexpensive, common flashlight battery, known variously as the standard, classic, or heavy-duty battery (or, more technically, as the carbon-zinc dry cell), or as the more expensive and longer lasting or more specialized alkaline, lithium, mercury, or silver oxide batteries, or as rechargeable batteries, such as the lead–acid batteries that start our cars and the nickel–cadmium batteries that run our rechargeable power tools.

Figure 11.2 A flashlight battery.

An **electric current** is a flow of electrons. A **circuit** is the path the electrons follow.

The inside of an ordinary flashlight battery—a standard, classic, or heavy-duty carbon-zinc dry cell—looks much like what's shown in Figure 11.2. The outside of the battery itself is a cylindrical zinc cup, sheathed in a protective paper cover. This cup is filled with a moist, black paste of ammonium chloride (NH_4Cl), zinc chloride ($ZnCl_2$), and manganese dioxide (MnO_2). Sticking down into the center of this mass is a porous rod of carbon. A metal cap covers the top of the rod and forms one of the battery's contacts; the bottom of the zinc shell is the other contact. Plenty of insulation protects the carbon rod and its metal cap from direct mechanical and electrical contact with the shell.

Although the carbon-zinc battery is one of our simpler consumer products, the chemistry that goes on inside that wet, black paste is much too complex to be described in detail here. We can say something, though, about how these batteries provide us with electrical power when we put them into a flashlight, for example. When we turn the flashlight on we complete a **circuit**, which is a path followed by electrons. The circuit of a flashlight allows the electrons to leave the zinc casing of the battery, travel to the bulb over a portion of the circuit that's built into the flashlight itself, move through the bulb (causing it to light up), and then back along the remainder of the circuit to the batteries, which they enter through the small, round contact on the top of the metal cap. From that contact the electrons pass through the carbon rod attached to it and into the body of the battery. There they take part in the chemical reactions that occur inside the black paste. What we've just described is an **electric current**, a flow of electrons along a particular path or circuit.

As the electrons leave the zinc casing, the zinc atoms that lose the electrons become transformed into zinc cations, Zn^{2+}. It's the flow of these electrons—from the metal casing of the battery through the electrical circuits of our appliances and back into the core of the battery—that provides the power to our radios, flashlights, clocks, and the like. (The metal casings of alkaline, mercury, lithium, and other kinds of batteries also provide the electrons these batteries release. Just as with the carbon-zinc battery, the chemistry that goes on inside these batteries is too complex to be covered here.)

■ QUESTION

What element forms the metal casing of a common flashlight battery? Name three compounds (other than water) that form the black paste held within this metal casing. What element forms the porous rod embedded in this paste?

11.2 Two Potatoes, a Clock, and a Cell Named Daniell

The release of electrons and the conversion of metals into their cations occur in other electrical devices as well as in batteries. One of these creations is a cleverly designed digital clock that runs from the power provided by two potatoes, or so it seems. The clock itself is attached to two wires, each ending in a different metal plate: one copper, the other zinc. Pushing each of these plates into its own potato and joining the two potatoes to each other by a third wire (also with ends of copper and zinc) starts the clock running as though it were connected to a commercial battery (Fig. 11.3).

Using potatoes isn't critical here. Moist soil will do, as will an ordinary soft drink, or even salt water. Actually, anything that acts as an electrolyte (Sec. 1.1) will run the clock. The secret lies not in the potatoes or in any specific electrolyte, but in the chemical identity of the embedded metals themselves. Together with the

Figure 11.3 A clock run by a chemical reaction.

electrolyte, these zinc and copper strips make up an **electrochemical cell** (a cell or battery that uses chemical reactions to produce electricity) similar to the one invented in 1836 by John Frederic Daniell, an English physical scientist who had been appointed the first professor of chemistry at King's College, London, in 1831.

An **electrochemical cell** is a cell or battery that produces electricity from chemical reactions.

We can build one form of the *Daniell cell* easily with two beakers, one containing a solution of zinc sulfate, $ZnSO_4$, and a strip of zinc that dips into the liquid; the other holding a solution of copper sulfate, $CuSO_4$, and a copper strip.

Copper sulfate and zinc sulfate are electrolytes that ionize when they dissolve in water. Copper sulfate produces copper cations, Cu^{2+}, and sulfate anions, SO_4^{2-}; zinc sulfate produces zinc cations, Zn^{2+}, and sulfate anions. The sulfate anion is an example of a polyatomic ion, with four oxygen atoms covalently bonded to a single sulfur atom and bearing, as an entire group, a charge of 2−. In $ZnSO_4$ the two negative charges of the SO_4^{2-} anion balance the two positive charges of the Zn^{2+} cation; in $CuSO_4$ they balance the two positive charges of the Cu^{2+} cation.

The two solutions are joined by a *salt bridge* that allows ions to move physically from one beaker to the other. This bridge need be no more than a strip of cloth soaked in a solution of sodium chloride, with its ends dipping into the two solutions (Fig. 11.4). The sodium chloride serves as an electrolyte, allowing an electric current to flow between the beakers just as the electrolyte allowed current to flow between the two wires of the electric light demonstration in Section 1.1. The combination of the two solutions, the two strips of metal, and the salt bridge makes up the Daniell cell.

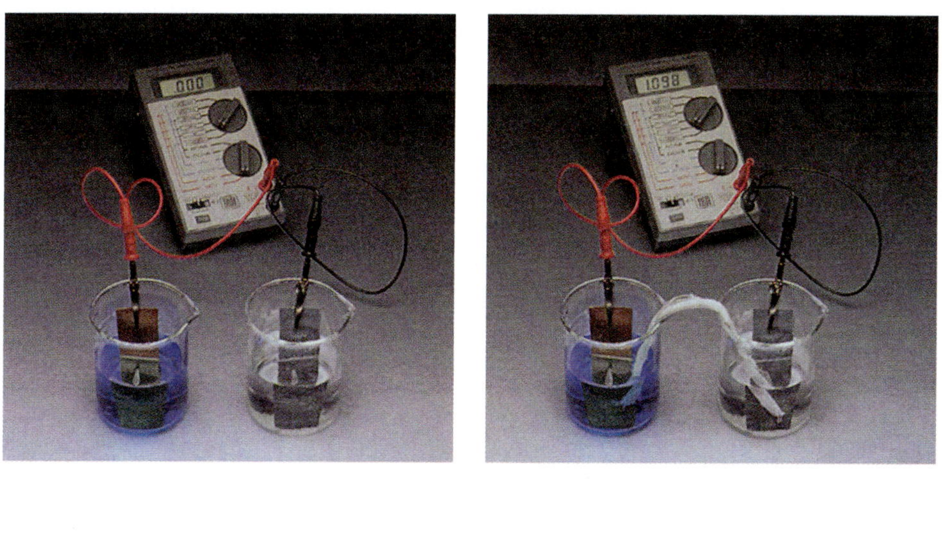

Figure 11.4 A form of the Daniell cell. The sodium chloride of the salt bridge acts as an electrolyte and allows ions to flow between the two beakers, generating almost 1.1 volts.

Connecting a voltmeter to the two metal electrodes of the cell completes the electrical circuit and gives a reading of very nearly 1.1 volts. The response of a voltmeter gives us proof that the cell is producing an electric current. (We'll have more to say about just what this reading indicates in later sections.) Similarly, the combination of the zinc strips and the copper strips of the two–potato clock, connected by the electrolyte consisting of the water and the ions within the potatoes, generates a current that runs the clock (Fig. 11.3).

Although it's more cumbersome than our modern batteries, the Daniell cell is just as much a battery as any of the small sources of electrical power we put into our flashlights, radios, and cellular phones. Understanding how it works sheds light on how our more modern batteries work.

■ QUESTION

Would a Daniell cell work equally well if the salt bridge were replaced by a length of copper wire? Explain.

11.3 The Blue Disappears, and So Does the Zinc

The simplest and most direct way of learning how the Daniell cell operates is by using the scientific method (Sec. 1.4) and asking a simple, straightforward question: What happens to the cell as it produces the current? To answer the question we simply build a Daniell cell, connect the zinc and copper strips to each other, and watch the cell over time.

At first the zinc and the copper remain clean and bright and the copper sulfate solution shows the intense blue color of its Cu^{2+} ions. Slowly, over hours or days, depending on how we construct the cell and the conditions of its operation, the deep blue color of the copper sulfate solution fades and the copper plate darkens. In the other part of the cell the zinc strip develops pits, erodes, and eventually disappears, especially if it's very thin.

These same changes occur much more quickly and dramatically if we simply place a strip of clean zinc directly into a blue copper sulfate solution. Almost immediately the strip darkens. With time it crumbles into the solution as the blue color of the copper sulfate fades. While all this is going on, the dark coating on the zinc strip thickens and flakes off as fine rust–colored or black granules of copper metal (Fig. 11.5).

Figure 11.5 Zinc metal and copper sulfate. As time passes, the zinc strip erodes, dark granules of copper appear, and the blue color of the copper solution fades. Zinc atoms of the metal strip are oxidized to colorless zinc ions as blue copper ions in the solution are reduced to copper atoms. Meanwhile, copper metal accumulates at the bottom of the beaker. No change occurs in the beaker holding the copper strip immersed in a solution of zinc sulfate.

Continuing with our use of the scientific method, we can ask what happens if we place a strip of clean *copper* directly into a *zinc sulfate* solution. The answer, which we can obtain by actually trying this experimentally (Fig. 11.5), is that nothing at all happens.

■ QUESTION

We can infer that copper ions in water produce a blue color by noting that a solution of copper sulfate in water, which contains copper ions and sulfate ions, is blue and that a solution of zinc sulfate in water, which contains zinc ions and sulfate ions, is colorless. Why does this observation lead us to conclude that the copper ions of the copper sulfate solution produce the blue color?

11.4 Our Investigation Bears Fruit: Redox

The simplest explanation for all these observations is that the zinc atoms of the strip of zinc metal, which certainly isn't soluble in water, lose two electrons each and become zinc ions (Zn^{2+}), which *are* soluble in water. That's why the zinc strip erodes: It's converted into zinc ions, which dissolve in the water. In addition, each of the blue copper ions (Cu^{2+}), which are also soluble in water, gains the two electrons lost by each zinc atom. In this way the copper ions become transformed into particles of copper metal, which is *not* soluble in water (Fig. 11.6). The copper metal forms the rust–colored or black solid granules that appear as the blue copper ions of the solution are converted into copper atoms. The blue color disappears because the copper ions—the source of the color—disappear. Our observation that nothing happens when we place a copper strip into a zinc sulfate solution adds weight to our conclusion. Electrons flow from the zinc atoms to the copper ions, not from the copper atoms to the zinc ions.

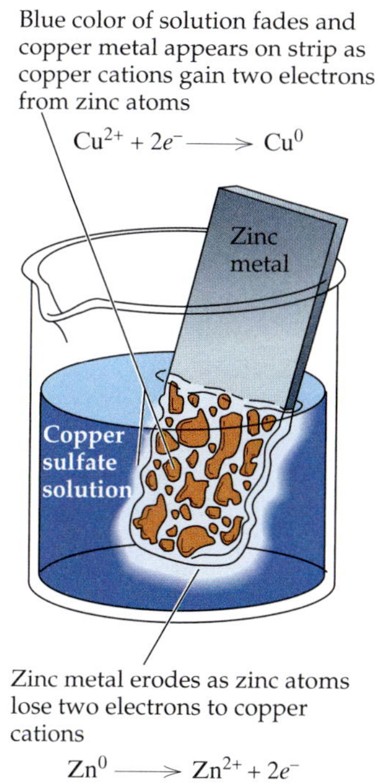

Figure 11.6 Zinc metal in copper sulfate solution.

The entire process, then, occurs as each zinc *atom* loses two electrons to a copper *ion*. As a result, the zinc atom, which we now write as Zn^0 to emphasize that it bears no net electrical charge, is transformed into a zinc cation. The copper cation is transformed into a copper atom, which we now write as Cu^0 for the same reason:

Zn^0	+	Cu^{2+}	$\longrightarrow$	Zn^{2+}	+	Cu^0
atoms that form the zinc strip; each of these loses two electrons to a copper ion		blue ions in solution; each of these gains two electrons from a zinc atom		colorless ions in solution		atoms that form the rust-colored or black granules

The overall reaction, the transfer of one or more electrons from one atom or molecule to another, is known as a **redox reaction**, which we examine in the next section. The redox reaction of the Daniell cell is an example of an **electrochemical reaction**, a chemical reaction that can produce a flow of electrons from one location to another, or a reaction that's caused by such a flow. In this particular redox reaction the sulfate anions remain unchanged; they don't take part in any way.

A **redox reaction** takes place with a transfer of one or more electrons from one chemical species to another. An **electrochemical reaction** can produce a flow of electrons from one location to another or is caused by this flow.

■ QUESTION

Suppose that as the zinc metal is converted into zinc ions (which are soluble in water), *all* the copper ions in solution are converted to copper metal. What new compound, a combination of a cation and a balancing anion, would then exist in the colorless solution?

11.5 Redox, a Combination of Reduction and Oxidation

Reduction is the gain of electrons.

The term *redox* itself comes from a combination of the two words "*red*uction" and "*ox*idation." The word *reduction* has many meanings in both the English language generally and in the science of chemistry in particular. But when we're dealing specifically with reactions involving the transfer of electrons, **reduction** refers to the *gain* of electrons by a chemical particle. In this sense the blue Cu^{2+} ions of the solution are being *reduced* as they gain the two electrons lost from the zinc metal:

$$\text{reduction of the } Cu^{2+} \text{ ion:} \quad Cu^{2+} + 2e^- \longrightarrow Cu^0$$

Oxidation is the loss of electrons.

Oxidation has several meanings, too. In a broad and general sense, the word *oxidation* indicates the gain of oxygen by any substance. In electrochemical reactions, though, **oxidation** refers specifically to the *loss* of electrons by a chemical particle. In the reaction of copper ions with zinc atoms, for example, the zinc atoms lose two electrons to the copper ions and are oxidized:

$$\text{oxidation of the zinc atom:} \quad Zn^0 \longrightarrow Zn^{2+} + 2e^-$$

Oxidation, then, is the loss of electrons; reduction is the gain of electrons. As a result, whenever any chemical species loses an electron and undergoes oxidation, the electrical charge it bears becomes more positive (or less negative). Conversely, as any chemical species gains an electron and undergoes reduction, its electrical charge becomes more negative (or less positive). For an easy way to remember what happens to electrons in oxidations and reductions, just think of an OIL RIG. You'll know instantly that *O*xidation *I*s *L*oss, *R*eduction *I*s *G*ain (of electrons).

Electrons, of course, can't be taken from or lost into thin air. Each electron acquired by any ion, atom, or molecule *must* be lost by something else. The converse is also true. Electrons lost from anything *must* be gained by something else. What this amounts to is that nothing can be reduced unless something else is oxidized and nothing can be oxidized unless something else is reduced. The electrons gained by the copper ions as they are reduced come from the zinc atoms as they are oxidized. Since the copper ions are the agents that remove the electrons from the Zn^0 and cause it to be oxidized, the copper ions constitute the **oxidizing agent** in this reaction. Similarly the zinc atoms, the agents that cause the copper ions to acquire electrons and become reduced to Cu^0, act as the **reducing agent**. It comes down to this: The substance being reduced is the oxidizing agent, and the substance being oxidized is the reducing agent.

An **oxidizing agent** aquires electrons from something else and causes it to be oxidized; a **reducing agent** releases electrons to something else and causes it to be reduced.

To emphasize that reductions and oxidations must occur together, hand-in-hand, the overall reaction

$$Cu^{2+} + Zn^0 \longrightarrow Cu^0 + Zn^{2+}$$

and others like it are called *redox reactions*. Note that we can consider a redox reaction to be the sum of a reduction reaction and an oxidation reaction that combine to produce the overall redox reaction. The two electrons that appear on both sides of the arrows, among both the reactants and the products, cancel each other and can be eliminated in writing the redox reaction itself:

$$\begin{array}{rr} \text{reduction:} & Cu^{2+}\cancel{2e^-} \longrightarrow Cu^0 \\ \text{oxidation:} & Zn^0 \longrightarrow Zn^{2+} + \cancel{2e^-} \\ \hline \text{SUM (redox):} & Cu^{2+} + Zn^0 \longrightarrow Cu^0 + Zn^{2+} \end{array}$$

In this sense a redox reaction is simply a combination of a reduction reaction and an oxidation reaction that occur simultaneously.

Since each of the individual oxidation and reduction reactions makes up *half* of the overall redox reaction of an electrochemical cell, each one is known as a **half-cell reaction**. The redox reaction, then, is a combination of two half-cell reactions, a *reduction half-cell* and an *oxidation half-cell*.

Each of the oxidation and reduction reactions that combine to form a redox reaction is a **half-cell reaction**.

■ QUESTION

The reaction of calcium and chlorine described in Section 3.12 can be rewritten as $Ca + Cl_2 \longrightarrow CaCl_2$. In this reaction, what is being oxidized? What is being reduced? Write reactions for the two half-cells of this redox reaction.

11.6 The Daniell Cell Explained

We've chosen the Daniell cell as a specific example of an electrolytic cell. It generates electricity as the zinc atoms of the zinc strip lose their electrons and become oxidized to zinc ions. These ions enter the solution and, over time, the zinc strip erodes to nothing. As the metal strip is eroding, the solution it stands in becomes enriched with a surplus of zinc cations.

The electrons released by the zinc atoms travel up, along the wire of the electrical circuit, to the voltmeter, where they register 1.1 volts; then they move along the other wire, down to the copper strip. Once they are in the portion of the copper strip that's in contact with the solution, the electrons combine with the copper ions of the solution and reduce them to copper metal, which deposits on the strip as rusty or black solids. With this loss of copper cations, the copper sulfate solution develops a deficiency of cations.

As a surplus of cations develops in the zinc sulfate solution and a deficit of cations develops in the copper sulfate solution, there has to be a way to reestablish an electrical balance. That's where the salt bridge comes in. It allows cations to travel over it, away from the surplus of positive charge in the zinc sulfate solution and toward the deficit of positive charge in the copper sulfate solution. This charge imbalance is also corrected by the travel of anions through the salt bridge in the opposite direction. As these migrations occur, the movement of ions through the salt bridge maintains electrical neutrality in both solutions. The cell continues to produce an electrical current as long as there are any zinc atoms left to be oxidized and copper ions left to be reduced (Fig. 11.7).

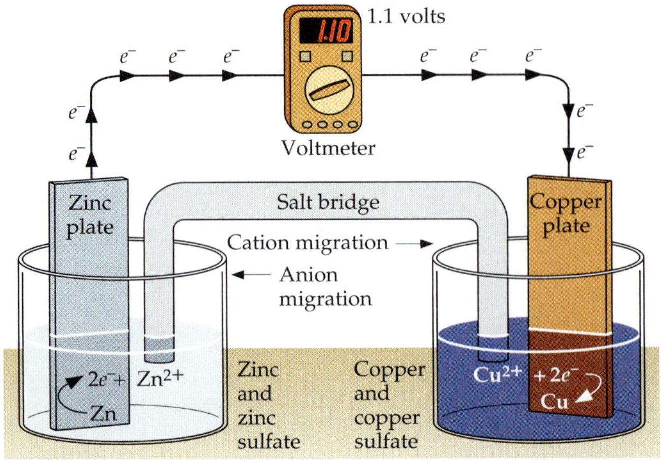

Figure 11.7 Chemical and electrical changes in the Daniell cell.

■ **QUESTION**

We prepared the salt bridge of our Daniell cell by soaking a strip of cloth in a sodium chloride solution. As the cell begins to operate, would you expect the sodium and chloride ions of the bridge to move? If your answer is *yes*, in which direction would the sodium ions move and in which direction would the chloride ions move? If your answer is *no*, explain why not.

11.7 Electrical Voltage: Putting Pressure on an Electron

The **volt** is a unit of electrical potential, the pressure that moves electrons from one point to another.

As long as we stick to strips of zinc and copper, regardless of their size or shape, we'll get the same reading from the meter, a little under 1.1 volts. Understanding just what this voltage reading represents is a key to understanding how electrolytic cells and other kinds of batteries operate.

The **volt** is a unit of electrical potential, which represents the tendency of electrons to move from one point on the circuit to another. The greater the voltage, the greater the pressure that moves the electron through the circuit. It's a bit like water pressure. The greater the water pressure in a pipe, the greater the force that moves the water along from point to point inside the pipe. The greater the voltage, the greater the force that moves an electron along from point to point within the circuit.

Alessandro Volta demonstrating his discovery to the French Academy in 1801.

The word *volt* honors Alessandro Volta, an Italian physicist who published a description of the world's first electrical battery, the *Volta pile*, in 1800. Volta's pile consisted of a series of disks made of silver, paper moistened with a salt solution, and zinc. This trio was repeated over and over to form a tall pile. (In later versions, copper successfully replaced the silver.) As with the Daniell cell, the voltage delivered by the Volta pile does not depend on the size or shape of the disks.

The **amp** or **ampere** is a unit of electrical current, or the rate of flow of electrons.

While the volt is a unit of electrical potential, the **ampere** (which is often shortened simply to **amp**) measures the rate of flow of the current in much the same way that a unit like gallons per minute measures the rate of flow of water. As amperage increases, there's an increase in the number of electrons traveling through a circuit during any particular period. André Marie Ampère, the French physicist for whom the unit is named, was a contemporary of Volta and, like the Italian physicist, is remembered for his pioneering work in electricity and magnetism.

Under normal conditions it takes a combination of a high voltage and a high amperage, such as we might find in a lightning bolt or inside a television set, to pose a hazard to humans. (This is the equivalent of a lot of water flowing with a high pressure, as we might find in a flooding river.) The voltage of ordinary consumer batteries, though, is too low to do us any harm. An automobile battery, for example, can deliver a current measuring in hundreds of amperes, but at such a low potential, 12 volts, that it simply can't cause us serious injury under ordinary conditions. (Here we can draw a parallel to a large body of water, a wide river for example, flowing slowly.)

A high voltage that delivers an insignificant number of amperes is equally harmless to people. The spark you produce as you walk across a carpet on a dry day and touch a light switch or another person carries thousands of volts, but its infinitesimal current can't hurt us. (This would be like a small but sharp jet of water, the kind useful in cleaning teeth and massaging gums.) Nonetheless, the spark's high voltage can do considerable damage to sensitive electrical and magnetic equipment. It can, for example, change or destroy data stored in computers and on magnetic disks and tape.

■ QUESTION

By increasing the areas of the zinc and copper plates in the Daniell cell, we can increase the amperage of the current flowing through the circuit without affecting the voltage we measure with the voltmeter. How does increasing the size of the zinc and copper plates affect the electrical pressure that moves the electrons along? How does increasing the size of the plates affect the rate of flow of the electrons through the circuit?

11.8 Standard Reduction Potentials

Since it's the zinc strip that disappears and the dark, finely divided copper metal that appears as the color of the copper ions fades, we have concluded that the zinc atoms are being oxidized and the copper cations are being reduced. As a result, the electrons that pass through the meter with a force or electrical potential of 1.1 volts must be moving *from* the zinc strip *to* the copper strip rather than in the reverse direction.

To take our conclusion one step further, there must be a greater potential for electrons to leave zinc atoms than there is for electrons to leave copper atoms. Similarly, the Cu^{2+} cations must have a greater potential for acquiring two electrons than the Zn^{2+} cations have. We've observed indirectly, then, that the copper cations, which acquire electrons spontaneously in the cell, have a greater electrical potential for reduction, or a greater **reduction potential**, than the zinc cations. Similarly, the zinc atoms, which lose electrons spontaneously to the copper cations, have a greater electrical potential for oxidation, a greater **oxidation potential**, than the copper atoms. We measure these potentials for the gain and loss of electrons as electrical voltages, much as we measure the potential for the movement of water from one point to another in terms of water pressures.

A **reduction potential** is the voltage produced by or required for the addition of electrons to an atom or ion. An **oxidation potential** is the voltage produced by or required for the removal of electrons from an atom or ion.

Although we can't isolate an individual oxidation or reduction reaction from the redox combination, we can nevertheless measure the electrochemical potential of any one of these individual half-cell reactions relative to a universally accepted standard reaction—the reduction of two hydrogen ions to a hydrogen molecule:

$$\text{reduction of hydrogen ions:}\quad 2H^+ + 2e^- \longrightarrow H_2$$

standard reduction reaction

In this reduction of hydrogen ions, the addition of an electron to each of two protons produces two hydrogen atoms, which combine through the sharing of their two electrons to form a diatomic hydrogen molecule, H_2 (Fig. 11.8 and Sec. 3.13).

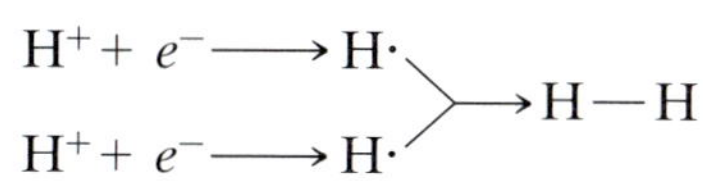

Figure 11.8 Reduction of two hydrogen ions to a hydrogen molecule.

Table 11.1 Standard Reduction Potentials (25°C)

Half-Cell Reaction	Potential (volts)
$Li^+ + e^- \longrightarrow Li$	−3.04
$K^+ + e^- \longrightarrow K$	−2.93
$Cs^+ + e^- \longrightarrow Cs$	−2.92
$Ca^{2+} + 2e^- \longrightarrow Ca$	−2.87
$Na^+ + e^- \longrightarrow Na$	−2.71
$Mg^{2+} + 2e^- \longrightarrow Mg$	−2.38
$Al^{3+} + 3e^- \longrightarrow Al$	−1.66
$2H_2O + 2e^- \longrightarrow H_2 + 2OH^-$	−0.83
$Zn^{2+} + 2e^- \longrightarrow Zn$	−0.76
$Fe^{2+} + 2e^- \longrightarrow Fe$	−0.44
$Cd^{2+} + 2e^- \longrightarrow Cd$	−0.40
$PbSO_4 + 2H^+ + 2e^- \longrightarrow Pb + H_2SO_4$	−0.36
$Ni^{2+} + 2e^- \longrightarrow Ni$	−0.26
$2H^+ + 2e^- \longrightarrow H_2$	0
$Cu^{2+} + 2e^- \longrightarrow Cu$	+0.34
$I_2 + 2e^- \longrightarrow 2I^-$	+0.54
$Ag^+ + e^- \longrightarrow Ag$	+0.80
$ClO^- + H_2O + 2e^- \longrightarrow Cl^- + 2OH^-$	+0.84
$Br_2 + 2e^- \longrightarrow 2Br^-$	+1.07
$O_2 + 4H^+ + 4e^- \longrightarrow 2H_2O$	+1.23
$Cl_2 + 2e^- \longrightarrow 2Cl^-$	+1.36
$Au^+ + e^- \longrightarrow Au$	+1.69
$PbO_2 + H_2SO_4 + 2H^+ + 2e^- \longrightarrow PbSO_4 + 2H_2O$	+1.69
$F_2 + 2e^- \longrightarrow 2F^-$	+2.87

Increasing Strength of Reducing Agent (arrow pointing up)

Increasing Strength of Oxidizing Agent (arrow pointing down)

The **standard reduction potential** of a substance is the value of its reduction potential as compared with the reduction of the hydrogen ion, which is defined as zero volts.

If we now define the electrochemical potential of this particular reduction (arbitrarily) as exactly zero volts, we can measure all other half-cell potentials with respect to it. These measurements, standardized at 25°C and at specific concentrations of the ions, produce a series of **standard reduction potentials** (Table 11.1) that are universally applicable. The numerical values of Table 11.1 express the electrical pressure for the gain or loss of electrons from the chemical particles shown in the table, all in units of volts. Thc potential for the standard reduction reaction—the reduction of hydrogen ions to a hydrogen molecule—is simply defined as zero volts:

$$2H^+ + 2e^- \longrightarrow H_2 \quad \text{(defined as zero volts)}$$

As long as reaction temperatures and the concentrations of ions in solution are equivalent in all cases, reductions with positive standard reduction potentials occur more readily than the reduction of two protons to H_2; those with negative potentials occur less readily than the standard reduction. To put this a bit differently, we can say that the more positive the standard reduction potential, the greater the driving force for the ion, atom, or molecule to acquire electrons and for the reduction to occur. Similarly, the less positive (or more negative) the standard reduction potential, the smaller the driving force for the reactant to acquire an electron and to be reduced.

In view of this, any chemical with a large, positive standard reduction potential has a strong tendency to acquire electrons from some other substance (that's capable of releasing electrons) and to oxidize it. As a result, a chemical with the large, positive standard reduction potential is itself a strong oxidizing agent, readily capable of oxidizing (removing electrons from) many other substances. To summarize, *the more positive the standard reduction potential of a chemical, the greater its tendency to behave as an oxidizing agent* (Sec. 11.5).

■ QUESTION

Which element, atom, molecule, or ion in Table 11.1 is the strongest oxidizing agent? Which is the weakest?

11.9 Oxidation Potentials

The reverse of each of the reductions in Table 11.1 is an oxidation. Reversing any of the reduction half-cells, then, gives us an oxidation half-cell. In reversing the direction of the chemical equation we must also reverse the sign of the voltage associated with it. (This is roughly equivalent to switching the high- and low-pressure ends of a water pipe to reverse the flow of water.) To find an oxidation potential, then, we simply reverse the direction of the corresponding reduction reaction—the one that gives us the oxidation half-cell we're looking for—and we reverse the algebraic sign of its standard reduction potential. For example, the +2.87 volts for the reduction of molecular fluorine becomes −2.87 volts for the oxidation potential of the fluoride anion:

$$\text{oxidation of fluoride ion:}\quad 2\,F^- \longrightarrow F_2 + 2e^- \quad (-2.87\text{ volts})$$

In parallel to standard reduction potentials, this relatively large, negative number indicates that there's very little tendency for a fluoride anion to release its acquired electron and to be oxidized to F_2. The fluoride anion is such an extremely weak reducing agent that it's not normally considered to be a reducing agent at all. After all, once a powerful oxidizing agent acquires an electron from another substance, it's hardly likely to give up that acquired electron easily, so it can't be much of a reducing agent once it has secured the electron.

Notice that the standard reduction or oxidation potential doesn't depend on the number of atoms, molecules, or electrons involved in the reaction. The potential is the same for the oxidation of a single fluoride ion to a fluorine atom as it is for the oxidation of two fluoride ions to a diatomic fluorine molecule:

$$\text{oxidation of fluoride ion:}\quad F^- \longrightarrow F^0 + e^- \quad (-2.87\text{ volts})$$

■ QUESTION

What is the most powerful *reducing agent* in Table 11.1? Explain.

11.10 Why Batteries Work

We can now use our understanding of redox reactions to see just how and why batteries work. Batteries work because electrons flow spontaneously from reducing agents to oxidizing agents. When we plug a battery into a flashlight or other portable electrical device, we provide a circuit for the electrons to follow as they move from the battery's reducing agent to its oxidizing agent. In flowing through that circuit the electrons provide the electrical current that runs the device. What's required is a redox reaction that takes place spontaneously within the battery and causes it to send its electrons out from the oxidation half-cell, through the electrical circuit, and into the reduction half-cell. The data of the table of standard reduction potentials (coupled with the oxidation potentials we can derive from them):

- tell us whether any particular redox reaction occurs spontaneously, and
- allow us to calculate the resulting voltage of the electrochemical cell.

To learn whether any particular redox reaction will occur spontaneously, simply add the voltages of its two component half-cells, the reduction half-cell and the oxidation half-cell. If the resulting redox voltage is *positive*, the redox reaction that is the sum of the two half-cell reactions *can* occur spontaneously with the release of energy. (Remember that we obtain the sign of the oxidation half-cell by reversing the sign of its corresponding reduction half-cell.) The numerical value we calculate for the redox voltage is the voltage we can expect to obtain from the electrochemical cell or the battery.

If the sum is *negative*, the redox reaction *cannot* occur spontaneously and no current will flow. Nevertheless, we can cause a reaction with a negative redox potential to take place by adding energy to the cell from an external source, as we'll see in Section 11.14. As we do, though, the electrochemical cell no longer *produces* energy; rather, it *consumes* energy.

E X A M P L E

Daniell's Potential

Show that the Daniell cell generates electricity spontaneously and calculate the voltage it produces.

■ For the Daniell cell, we add the electrochemical potentials for the reduction of Cu^{2+} and the oxidation of Zn^0 to get

$$\begin{array}{lrcll} \text{reduction:} & Cu^{2+} + 2e^- & \longrightarrow & Cu^0 & (+0.34\ \text{volt}) \\ \text{oxidation:} & Zn^0 & \longrightarrow & Zn^{2+} + 2e^- & (+0.76\ \text{volt}) \\ \hline \text{SUM (redox):} & Cu^{2+} + Zn^0 & \longrightarrow & Cu^0 + Zn^{2+} & (1.10\ \text{volts}) \end{array}$$

Since the sum is positive we can expect the Daniell cell to produce a current (as we know it does). The electrical potential the cell will deliver if we build it to exact specifications of ion concentrations, temperature, and so forth, is 1.10 volts, which is just about the reading we got from the voltmeter.

E X A M P L E

Galvanized Tacks and Iodine

In the opening demonstration we saw that galvanized tacks (tacks coated with zinc) will decolorize iodine. Now let's see if we could have predicted the same result by looking at the redox reaction. Knowing that I_2 is purple-red in color, and I^- is colorless, we can now ask: Will galvanized tacks cause the purple-red color of iodine to vanish? This now amounts to asking: Will the following reaction take place spontaneously?

$$Zn^0 + I_2 \longrightarrow Zn^{2+} + 2I^-$$

■ Here we add the half-cells for the oxidation of Zn^0 and the reduction of I_2 to get the redox reaction and the redox potential:

$$\begin{array}{lrcll} \text{oxidation:} & Zn^0 & \longrightarrow & Zn^{2+} + 2e^- & (+0.76\ \text{volt}) \\ \text{reduction:} & I_2 + 2e^- & \longrightarrow & 2I^- & (+0.54\ \text{volt}) \\ \hline \text{SUM (redox):} & Zn^0 + I_2 & \longrightarrow & Zn^{2+} + 2I^- & (1.30\ \text{volts}) \end{array}$$

Since the redox voltage is positive, we can conclude that this reaction will, indeed, take place spontaneously and, as we have already observed, will cause the color of the iodine to vanish.

EXAMPLE

The Color Returns

Again in the opening demonstration we saw that adding a chlorine bleach to the nearly colorless solution causes the color of the iodine to return. To see why—and knowing that the oxidizing agent in the bleach is ClO^-, as we saw in the discussion of the opening demonstration—we'll now ask: Will the ClO^- oxidize the iodide ion to molecular iodine and therefore restore the color?

■ For the oxidation of iodide ion to molecular iodine, the oxidation half-cell is:

$$\text{oxidation:}\quad 2I^- \longrightarrow I_2 + 2e^- \quad (-0.54\ \text{volt})$$

With ClO^- as the oxidizing agent, the reduction half-cell is:

$$\text{reduction:}\quad ClO^- + H_2O + 2e^- \longrightarrow Cl^- + 2OH^- \quad (+0.84\ \text{volt})$$

Thus the redox reaction is:

$$\text{redox:}\quad 2I^- + ClO^- + H_2O \longrightarrow I_2 + Cl^- + 2OH^- \quad (+0.30\ \text{volt})$$

With a positive redox potential, the ClO^- of the bleach oxidizes the iodide ion spontaneously, regenerating the I_2 as the deep color of the molecular iodine returns.

EXAMPLE

Stable Salt

Now for another question. Why don't the sodium cations and the chloride anions of table salt react with each other to form sodium metal and chlorine gas?

■ In this case, we're asking why the following redox reaction doesn't occur spontaneously in a salt-shaker:

$$2Na^+ + 2Cl^- \longrightarrow 2Na^0 + Cl_2$$

Once again we add an oxidation potential and a reduction potential to get a redox potential. (Notice that although it's necessary to double all the chemical species in the reduction reaction so that the number of electrons matches the number used in the oxidation, the reduction voltage remains unchanged.)

$$\text{oxidation:}\quad 2Cl^- \longrightarrow Cl_2 + 2e^- \quad (-1.36\ \text{volts})$$

$$\text{reduction:}\quad 2Na^+ + 2e^- \longrightarrow 2Na^0 \quad (-2.71\ \text{volts})$$

$$\text{SUM (redox):}\quad 2Na^+ + 2Cl^- \longrightarrow 2Na^0 + Cl_2 \quad (-4.07\ \text{volts})$$

The negative redox potential explains why the sodium cations and the chloride anions don't react with each other spontaneously. And it's a good thing they don't. If they did, all the table salt in our kitchens and restaurants would be churning away right now, with sodium cations and chloride anions interacting with each other, producing very dangerous sodium metal and poisonous chlorine gas (Sec. 3.2).

■ QUESTION

In Section 10.2 we saw that acids liberate hydrogen gas when in contact with certain metals, such as iron and zinc. Using the data of Table 11.1, show that the reactions of iron and zinc with an acid spontaneously generate hydrogen gas by redox reactions. Identify three other metals that will spontaneously liberate hydrogen gas from an acid. Identify three metals that will not.

11.11 Galvanized Tacks, Drugstore Iodine, and Household Bleach Revisited

The middle two examples of the previous section answer the questions of *how* and *why* galvanized tacks cause the disappearance of the color of tincture of iodine and liquid household bleach brings it back, questions that were asked in the chapter-opening demonstration. As for *how*, the examples show us that the decolorization of iodine by galvanized nails and the regeneration of the purple-violet color of the iodine by household bleach take place by redox reactions. In reducing the iodine molecules to iodide ions, the zinc metal acts as a reducing agent and transfers its electrons to iodine molecules. The purple-violet iodine molecules are reduced to colorless iodide ions. In restoring the color, the solution of bleach, which contains a reserve of ClO^-, oxidizes the colorless iodide ions back to purple-violet iodine molecules. As for *why*, the answer is that the redox potential of each of these reactions is positive. Each one proceeds spontaneously by the transfer of electrons from a reducing agent to an oxidizing agent.

■ QUESTION

Which, if any, of the following metals will decolorize tincture of iodine: (a) nickel, (b) copper, (c) silver, (d) magnesium, (e) calcium? Explain your answer for each.

11.12 Energy versus Rate

We've seen that the sum of the two half-cell voltages—the redox potential—represents the measured voltage of the entire cell. Standard reduction potentials not only provide the voltages of batteries, whether real or proposed, but they also tell us something about the energies of chemical reactions. Sodium and chlorine, for example, react explosively when they come into contact with each other under ordinary conditions. The sum of the oxidation potential of sodium and the reduction potential of chlorine is over 4 volts, a sizable value compared with the voltages generated by the Daniell cell (1.10 volts), by a common flashlight battery (1.5 volts), or by one of the cells of the lead–acid storage battery (about 2 volts, as we'll see shortly). We should note, though, that the vigor of the reaction between sodium and chlorine, or between any other substances for that matter, depends not only on the *amount* of energy released, which is indicated by the electrochemical potential, but also on the *rate* at which the energy is liberated. Some reactions that release plenty of energy take place so slowly that not only are they *not* explosive, they're too slow even to be observed under ordinary conditions. That isn't the case with sodium and chlorine.

■ QUESTION

Using the half-cell reactions of Table 11.1, write the redox reaction that would release the greatest amount of energy. That is, write the redox reaction that produces the greatest redox voltage.

11.13 Ideal Batteries and Real Batteries

At the other end of the table from fluorine lies lithium. With its reduction potential of −3.04 volts, the lithium cation shows very little tendency to pick up an electron and be reduced. On the other hand, reversing the half-cell reaction and changing the algebraic sign of the potential to +3.04 volts reveals that lithium metal has a very strong tendency to lose an electron and to be oxidized to the lithium cation:

$$Li^0 \longrightarrow Li^+ + e^- \quad (+3.04 \text{ volts})$$

Lithium metal, then, is a powerful reducing agent. In its redox reaction with fluorine, lithium produces a potential of +5.91 volts, as you can see in the following example.

EXAMPLE

High Voltage

Show that the redox reaction of lithium and fluorine produces a potential of +5.91 volts.

■ For the redox reaction we have

oxidation:	$2Li^0 \longrightarrow 2Li^+ + 2e^-$	(+3.04 volts)
reduction:	$F_2 + 2e^- \longrightarrow 2F^-$	(+2.89 volts)
SUM (redox):	$2Li + F_2 \longrightarrow 2Li^+ + F^-$	(+5.91 volts)

Although the cell's high voltage might make it an ideal battery for many uses, fluorine gas is far too reactive and far too hazardous to be used in any consumer product (Chapter 1, Perspective). Of course we're not restricted to half-cell combinations of the highest possible potentials in our search for practical, commercial batteries. We can get potentials far greater than 6 volts from batteries made of reasonably safe, convenient, economical materials simply by connecting several batteries in series, with the positive terminal of one connected to the negative terminal of another, as we do when we place two common 1.5-volt flashlight batteries head-to-tail in flashlights. For a two-battery flashlight this alignment gives a total, combined voltage of about 3.0 volts, the sum of the voltages of each individual battery.

Another of the more common, popular, practical, commercial batteries is the alkaline battery, built much like the carbon-zinc battery but containing potassium hydroxide in addition to the other substances described in Section 11.1. Like the carbon-zinc battery, the alkaline battery delivers about 1.5 volts, but over a much longer lifetime. Other more specialized batteries include the lightweight lithium battery, in which lithium rather than zinc is oxidized and which provides over 3 volts; the mercury battery, which delivers a very constant 1.3 volts from a redox reaction of zinc and an oxide of mercury, HgO; and the small and long-lived silver oxide battery, which provides about 1.5 volts from a reaction of zinc and silver oxide, Ag_2O. The rechargeable nickel–cadmium battery, consisting principally of cadmium and an oxide of nickel, and the rechargeable lead–acid battery, which we'll examine in the next section, round out the list of the most widely used commercial batteries.

■ QUESTION

The Daniell cell itself isn't used as a commercial battery. Suggest at least one reason why it isn't.

11.14 Chemistry That Starts Cars

We turn now to a battery that's more practical than the Daniell cell and that produces electricity by a simpler set of chemical reactions than the batteries of the preceding section. It's the lead–acid battery that starts our cars (Fig. 11.9). The

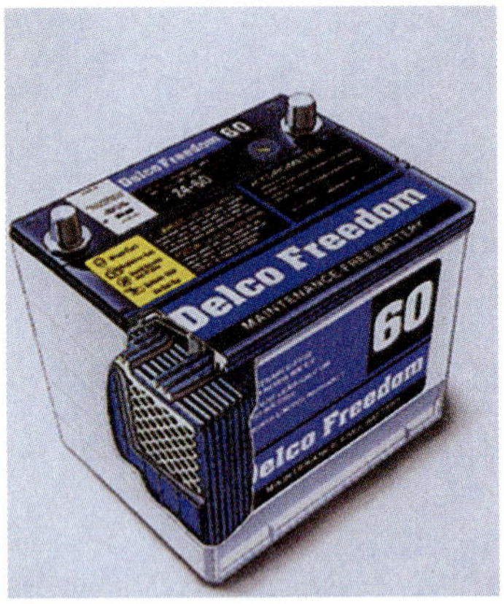

Figure 11.9 Construction of the lead-acid storage battery.

The **anode** of a battery is the part that provides the electrons flowing through the external circuit. The **cathode** is the part of the battery that receives the returning electrons.

battery is relatively inexpensive, can be discharged and recharged repeatedly, lasts from about three to five years (depending on its construction and use), and is small enough and light enough to be installed in cars and trucks.

Inside this battery two sets of plates stand immersed in a solution of sulfuric acid and water. One set, made of a spongy form of metallic lead, serves as the *anode*; the *cathode* consists of plates of lead dioxide, PbO_2. (The **anode** is the part of the cell from which electrons leave as they travel through the circuit. Since whatever furnishes these electrons loses them through an oxidation, the oxidation half-cell reaction always takes place at the anode. The **cathode** is the part of the battery that receives the returning electrons. They are the electrons that are consumed in the reduction half-cell reaction that takes place at the cathode.) When you start a car, the battery jolts the engine into action by a combination of oxidation of the lead metal of the anode and reduction of the lead dioxide of the cathode.

In the oxidation half-cell the spongy lead reacts with sulfuric acid, producing lead sulfate and releasing two protons and two electrons in the process. Neither the sulfate anion (SO_4^{2-}) nor the protons (H^+) undergo any change in their oxidation state throughout this half-cell reaction. Only the lead is oxidized, from an oxidation state of zero (Pb^0) to an oxidation state of 2+ (Pb^{2+}). It's the lead atoms that lose the two electrons:

$$\text{oxidation:} \quad Pb^0 + H_2SO_4 \longrightarrow PbSO_4 + 2H^+ + 2e^- \quad (+0.36 \text{ volt})$$

In the reduction half-cell the lead dioxide reacts with sulfuric acid and two protons to form lead sulfate and water. Once again it's the lead, this time the Pb^{4+} ions of the lead dioxide, that undergoes a change in oxidation state. These lead ions pick up two electrons and are reduced from an oxidation state of 4+ (in the PbO_2) to 2+ (in the $PbSO_4$):

$$\text{reduction:} \quad PbO_2 + H_2SO_4 + 2H^+ + 2e^- \longrightarrow PbSO_4 + 2H_2O \quad (+1.68 \text{ volts})$$

Adding both half-cells produces the redox reaction that takes place when you turn the ignition key; adding the two half-cell voltages gives the electrical potential of the cell as it discharges:

$$\text{discharge:} \quad Pb + PbO_2 + 2H_2SO_4 \longrightarrow 2PbSO_4 + 2H_2O$$

(+2.04 volts, which we can round off to +2 volts for convenience)

The lead–acid battery discharges and provides energy whenever there is a demand that's not met by the car's alternator or generator, such as when you start the car or run the radio or CD player with the engine off, or if you park and forget to turn the lights off. That bit of forgetfulness is often enough to drain all the electrochemical energy out of the battery, leaving it dead, unable to provide enough current to start the car again.

(Although both an alternator and a generator produce electricity, the alternator produces an alternating current, much like house current but at a considerably lower voltage; the generator produces a direct current, the kind we get from batteries. One or the other is attached to the engine to furnish electricity while the engine is running.)

The lead–acid battery offers one great advantage over the common flashlight battery: It can be recharged. As long as the engine is running, whether you're driving the car or simply letting it idle at a stoplight or in neutral, the electric current produced by the generator or the alternator restores the electrochemical energy of the battery by reversing the redox reaction. The current enters the battery and reconverts the lead sulfate and water into sulfuric acid, spongy lead metal, and lead dioxide:

$$\text{charge:} \quad 2\,PbSO_4 + 2\,H_2O \longrightarrow Pb + PbO_2 + 2\,H_2SO_4$$

$$(-2.04 \text{ volts, which we again round off to } -2 \text{ volts})$$

The negative potential of the redox reaction tells us that a battery won't recharge itself spontaneously. (Common sense tells us the same thing.) To carry out this redox reaction electrical energy must be put into the battery, as we saw in Section 11.10. Notice that the electrical potential produced by the discharge redox reaction is 2 volts. The case of the standard 12-volt automobile battery actually encloses six of these 2-volt cells in series, head-to-tail as in a flashlight.

■ QUESTION

What change takes place in the composition of the battery fluid—the liquid that bathes the battery's plates—as the battery is recharged by the generator or alternator?

11.15 Niagara Falls Fights Typhoid through Electrolysis

With one exception, all the reactions we've seen in this chapter produce an electric current or a voltage by means of a chemical change occurring in a battery. In that single exception, the recharging of a lead–acid storage battery, electrical energy *supplied* to the battery *produces* a chemical change within it. Lead sulfate and water are transformed into lead, lead dioxide, and sulfuric acid.

Electrical energy can also be used to generate a variety of other chemical changes that don't occur spontaneously, including the *electrolysis* of water into hydrogen and oxygen. The suffix *-lysis* implies a cleavage, rupture, or decomposition; **electrolysis** produces the decomposition of a substance into its component parts by means of electricity.

Electrolysis is the decomposition of a substance by means of electricity.

In 1800 William Nicholson, an English writer, lecturer, and experimenter in what was then known as "natural philosophy," read Volta's newly published description of his Volta pile (Sec. 11.7) and built one of his own. Nicholson attached two platinum wires and used them to pass an electric current from the pile through a container of water. The resulting electrolysis of the water produced bubbles of what proved to be hydrogen gas at one wire and oxygen at the other, always in a constant ratio of two volumes of hydrogen to one of oxygen (Fig. 11.10). The consistency of this ratio provided early evidence that a water molecule contains twice as many hydrogen atoms as oxygen atoms. (The reasoning behind this conclusion will become evident in Sec. 12.10.)

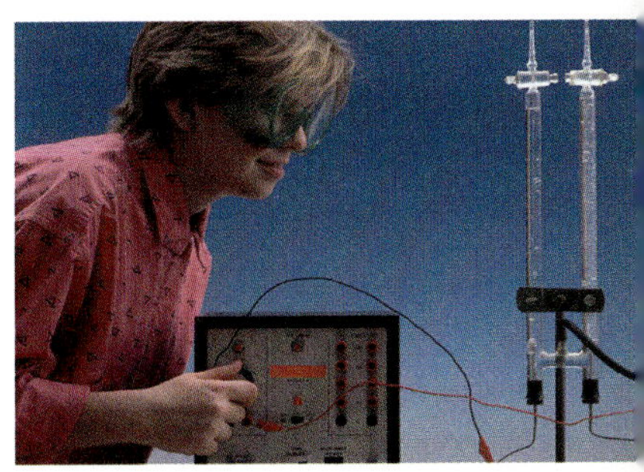

Figure 11.10 The electrolysis of water. Twice as much gas (hydrogen) is being produced in the tube on the left as in the one on the right (oxygen).

$$2\,H_2O \xrightarrow{\text{electrolysis}} 2\,H_2 + O_2$$

$2\,H_2O$	$2\,H_2$	O_2
two molecules of water	two molecules of hydrogen, each containing two hydrogen atoms	one molecule of oxygen containing two oxygen atoms

Today electrolysis represents a valuable process in the chemical industry, useful for generating a variety of commercially important chemicals. The electrolysis of a solution of sodium chloride in water, for example, produces chlorine gas at the anode (from oxidation of the chloride ions in the solution) and hydrogen at the cathode. At the cathode a net reduction of water's protons to H_2 leaves the hydroxide anion, OH^-, in the remaining solution. The combination of this anion with Na^+ left over from the oxidation of the chloride anions results in the formation of NaOH, *sodium hydroxide*, within the remaining solution:

$$Na^+ + OH^- \longrightarrow \underset{\text{sodium hydroxide}}{NaOH}$$

Sodium hydroxide is an important industrial chemical, useful in the production of a variety of commercial products. As a direct consumer chemical itself, though, its application is limited to use as highly caustic *lye* for opening clogged sinks and drains. It is an extremely corrosive and hazardous substance that must be used with great care.

Electrolysis on a commercial scale consumes enormous quantities of electricity and can be carried out economically only in regions where electric power is cheap and plentiful. In one of these locations, Niagara Falls, the power of falling water is harnessed to provide the needed energy. The chlorine produced in the electrolysis reaction is a valuable industrial raw material for manufacturing a large number of consumer products ranging from simple, rugged plastic plumbing fixtures to pharmaceuticals with intricate and complex molecular structures.

As a strong oxidizing agent, chlorine also provides an important tool for safeguarding the public health. Its oxidizing power makes it deadly to bacteria, including those that cause typhoid fever. The large–scale chlorination of public water supplies, swimming pools, and sewage, made possible by the commercial production of inexpensive chlorine, has played a major role in eliminating epidemics of typhoid fever and other public health threats in developed nations. (Chlorine gas can be deadly to humans, too. It was one of the poison gases used in the trench warfare of World War I, as we saw in Section 3.2.)

■ QUESTION

What three chemicals are produced through the electrolysis of sodium chloride solutions? What three chemicals would you expect to be produced by the electrolysis of a solution of sodium bromide?

11.16 Redox in Everyday Life

Our contact with redox and electrolysis in our everyday lives isn't limited to batteries, or the benefits of disinfection through chlorination. Many of our metals, for example, come to us through one of these processes. Commercial quantities of aluminum are produced through the electrolysis of hot solutions of aluminum oxide, Al_2O_3, dissolved in cryolite, Na_3AlF_6:

$$2\,Al_2O_3 \xrightarrow{\text{electrolysis}} 4\,Al + 3\,O_2$$

The aluminum oxide is obtained through the processing of bauxite, an aluminum-containing ore.

The highly purified form of silicon needed for the production of silicon chips used in computers and calculators comes from a redox reaction of highly purified silicon chloride, $SiCl_4$, with hydrogen:

$$SiCl_4 + 2\,H_2 \longrightarrow Si + 4\,HCl$$

Even the copper plating of baby shoes and similar objects depends on the kinds of electron transfers we've examined in this chapter. Copper plating—the transfer of copper atoms from a copper strip to another object—takes place when metallic copper and the object to be plated are placed in a solution of an electrolyte and are connected to the poles of a battery. The copper is connected to the anode; the plating takes place at the cathode. To facilitate the plating, objects that aren't good conductors of electricity are often coated with a thin layer of graphite, a form of carbon much like the central core of a flashlight battery.

Up to this point we've examined some of the more useful aspects of oxidation and reduction: storing and obtaining electrical energy from batteries, and useful aspects of electrolysis. There's another, undesirable side of redox though: **corrosion**. Although we normally think of corrosion as affecting metals, it's defined more generally as the erosion and disintegration of any material as the result of chemical reactions. The term **rust** applies specifically to the corrosion of iron that produces various scaly, reddish–brown oxides of iron.

Corrosion is the erosion and disintegration of a material resulting from chemical reactions. **Rust** is the corrosion of iron to form various reddish-brown oxides of iron.

Corrosion of all kinds does billions of dollars of damage each year in the United States alone. The redox reaction of metals, of which rusting is only one example, is accelerated by moisture, the presence of acids or bases, and various pollutants of the air and water. Because of the variety and complexity of the chemical reactions that lead to corrosion, we'll consider here only a few that affect us most directly. One results from the direct oxidation of metals that are exposed to air and water. (Here we're using *oxidation* in a more conventional sense, as the combination of something with oxygen.)

The direct oxidation of metals in the presence of air or water results from a series of redox reactions involving a combination of the metal itself, oxygen, and water. Iron, for example, rusts more rapidly in humid air than in dry air. These reactions result in the formation of *metal oxides*, chemical combinations of the metal and oxygen. Perhaps surprisingly, the formation of these oxides can either accelerate the corrosion or retard it, depending on the nature of the metal and the oxide it forms. Iron rusts rapidly because its oxides are granular and flaky. Once formed, the oxides fall or rub off the metal easily and expose fresh surfaces to additional corrosion. Aluminum, too, reacts quickly with oxygen to form *aluminum oxide*, Al_2O_3. This aluminum oxide, unlike the oxides of iron, forms a thin, tough, protective coating on aluminum that retards further oxidation. Thus, although aluminum metal oxidizes easily, it is highly resistant to *corrosion*.

The largest commercial use for metallic zinc isn't in manufacturing batteries, but in **galvanizing** other metals, especially iron and steel products, to provide them with a thin surface film of zinc metal that protects them from corrosion (chapter-opening demonstration). Galvanizing is usually carried out in any of three ways:

Galvanizing is a process that provides a protective zinc coat to metals.

- by dipping the metal into hot, molten zinc;
- by spraying molten zinc onto the metal's surface; or
- by reducing zinc cations directly onto the metal surface through an electrochemical redox reaction.

The zinc protects iron products by reacting with components of the atmosphere to form a protective film, much as aluminum does, and also by oxidizing more readily than the iron it's protecting. If the coating of zinc should crack in spots so that the iron's surface becomes exposed to the atmosphere, the zinc metal corrodes before the iron does. In essence, the zinc is used as a sacrificial metal to protect the more valuable iron. Since galvanizing a large metal object presents practical difficulties, metal tanks used for underground storage of gasoline and other hazardous liquids are often protected from oxidation by attaching a sacrificial block of aluminum, magnesium, or zinc. Because these three metals are more readily oxidized than the iron of the tanks, they protect it from corrosion.

Of course, some metals are simply inherently resistant to any form of corrosion. Gold's high resistance to oxidation (and other forms of corrosion as well), coupled with its ability to conduct electricity well, makes it useful as a coating for electrical connections. Since the ability of the gold coating to conduct a current is high to begin with and doesn't deteriorate through corrosion, many of the more expensive units of stereo and other electronic equipment, for example, use gold-coated parts and connectors.

Galvanic or **bimetallic** corrosion result from the contact of two different metals separated by a thin layer of electrolyte.

Another form of corrosion results from the contact of two different metals separated by a thin layer of electrolyte. Because it results from the generation of an electric current between the two metal parts, it's sometimes called **galvanic** or **bimetallic** corrosion. This type of corrosion is produced by the same sort of redox reaction we saw in the Daniell cell. Anyone who has teeth containing metallic fillings and has inadvertently bitten a piece of metal foil has felt the jolt of the electric current generated in galvanic corrosion. This kind of corrosion is most severe in metallic objects, such as boating equipment, that are in constant contact with salt water.

■ QUESTION

What is *galvanized* metal? Describe two ways galvanizing helps keep iron from rusting.

11.17 Vitamin C, Drugstore Iodine, and Household Bleach ... A Final Visit

With our knowledge of redox we can now understand why the opening demonstration works as it does. The zinc reduces the deeply colored I_2 to colorless I^-, and the Cl_2 of the bleach oxidizes the I^- back to I_2.

Although these redox reactions illustrate a nice piece of chemistry, they probably don't do much for us in our daily lives. Here's one that does. Place a drop or two of drugstore iodine on a piece of cotton cloth. Use a worthless scrap that won't be missed if something goes wrong. If this were an article of clothing or a valuable piece of fabric, you might consider the fabric ruined by the iodine stain. But you can remove the stain by rubbing it with a moist tablet of vitamin C. Dampen the tablet and rub it over the spot. You'll find that the color of the iodine vanishes.

It might seem to be magic, but it's simply redox in operation once again. Vitamin C is an organic compound we'll examine in more detail in Chapter 19. It's not only needed in our diets for good health, but since it happens to be oxidized very easily it's also a good reducing agent. The structure of vitamin C and the chemistry of its oxidation are too complex to be examined here, but the redox reaction converts the I_2 into colorless I^- just as in the opening demonstration.

This example of everyday chemistry works well for removing iodine stains from clothing and similar articles. But be careful not to bleach the fabric until you rinse out the residual I^-. The bleach can reoxidize any remaining I^- to I_2 and thereby cause the stain to reappear, just as in the opening demonstration. It's best to rinse the area thoroughly, immediately after you remove the stain, to wash out the residual I^- and the vitamin C.

(In a related bit of redox chemistry, the reaction of Cl_2 and I^- has been used to reveal the illegal trapping of egg-bearing lobsters. To escape detection, lobster harvesters sometimes use a Cl_2 solution to help remove the eggs from the lobster's shell. The illegal harvesting comes to light when parts of the lobsters' shells are dipped into a solution of KI and the residual Cl_2 on the shell oxidizes the I^- to the telltale color of I_2.)

PERSPECTIVE

Electrochemical Reactions for the Cars of Yesterday ... and Tomorrow?

One of our themes in this survey of the extraordinary chemistry of ordinary things concerns our use of energy. In this chapter we've investigated how we use the transfer of electrons to produce energy from one of the most ordinary of our everyday things, the common battery. Another theme we touch on is the application of chemistry to the needs of society. These two themes come together as we ask whether we can use the energy delivered by an electrochemical cell as an alternative to the petroleum-based fuels that run our cars, trucks, and buses.

Can we run our cars and vans on electric batteries? Years ago the answer was yes. Today, despite all the complexities of our more modern society, we may be forced to answer yes once again. To understand why, let's first look at the electric battery's principal competition, the gasoline engine. The original four-stroke internal combustion engine, the kind that runs on gasoline, was built in 1876. With its many improvements over the years, we now use this same engine in today's cars. (We examined the operation of the internal combustion engine and the chemistry of gasoline in Chapter 7.) In 1900, though, this gasoline-powered internal combustion engine ran only a very small minority of the automobiles in existence. At the beginning of the 20th century steam-powered cars and cars that ran on electric batteries dominated the marketplace, sharing the road almost equally. Combined, they outsold cars with gasoline engines by better than three to one. Yet by 1917, only 17 years later, more than 98% of all cars were running on the same sort of gasoline engines we use today. Steam-powered cars had all but disappeared and the electric car was dying rapidly. The battery-operated electric cars in existence today are no more than experimental cars or have only specialty uses, such as golf course carts.

The causes of the death of the electric car were varied and complex. They included limited driving ranges and low speeds, and the great weight and long recharge times of the batteries. Continuous improvements to the internal combustion engine and its associated transmission, as well as subtleties of consumer and producer psychology, also contributed to the extinction of the electric car.

Yet the internal combustion engine brought its own problems, principally air pollution (Chapters 8 and 14). Exhaust emissions from the internal combustion engines of highway vehicles—cars, vans, trucks, buses, and so on—contribute most of the carbon monoxide and substantial amounts of all the other pollutants discharged into the air over the United States (Sec. 14.4). In some regions the pollution produced by the internal combustion engine has led to an intolerable degradation of air quality. To reduce the engine's contribution to air pollution, California, Maine, Massachusetts, New York, and Vermont have moved to require the gradual introduction of zero-emission (Perspective, Chapter 7) or low-emission cars into the new-car market. We can expect cleaner air as these newer cars displace older, polluting vehicles from our roads and highways.

With current technology, the only hope for achieving a zero-emission goal appears to be through the reintroduction of electric cars, which produce no exhaust gases at all. Can today's technology revitalize the electric car? Perhaps. And if it does, what will supply the electricity that moves these cars?

While the lead–acid battery is practical for starting engines, a combination of its weight, cost, and limited capacity—which results in a severely limited driving range—makes other kinds of batteries seem more attractive. Batteries made of nickel and cadmium, iron, or zinc; or of lithium and iron, sulfur, or aluminum; or of sodium and sulfur; or still other combinations of redox systems have all seemed promising, but testing has revealed some form of defect or disadvantage in each. Perhaps, ultimately, cars, trucks, buses, and vans will run on the energy supplied by *fuel cells* (Fig. 7.13, Perspective, Chapter 7). Since water is the only chemical product of the redox reaction that takes place within the fuel cell, the cell is essentially pollution free. Fuel cells have already provided the electrical power (and water) for the space shuttle; one day they may also move our cars.

Certainly much work must still be done if the cars of the future are to be powered by batteries or fuel cells. Experimental cars that run on batteries and on yet other sources of energy, even sunlight, are currently being built and tested. Considering the history of the automobile, if we could learn to use electrochemical reactions of one kind or another to power our vehicles efficiently, the car of the future could well turn out to be the car of the distant past. ■

EXERCISES

For Review

1. Following are a statement containing a number of blanks, and a list of words and phrases. The number of words equals the number of blanks within the statement, and all but two of the words fit correctly into these blanks. Fill in the blanks of the statement with those words that do fit; then complete the statement by filling in the two remaining blanks with correct words (not in the list) in place of the two words that don't fit.

A __________ reaction is one involving two __________ reactions, one of them an oxidation, the other a reduction. When a substance is oxidized, it __________; when a substance is reduced, it __________. The energy of the gain or loss of electrons is described in terms of an electrical unit, the volt, and is shown in a table of experimentally measured __________. If the sum of the potential of the oxidation half-cell and the reduction half-cell is __________, the redox reaction they form will occur spontaneously, with the release of energy. If the sum of the voltages is __________, the redox reaction will not occur spontaneously, so energy must be added to the mixture (or to the substance) if a reaction is to occur.

An early version of an __________ was the __________, which used the oxidation of metallic __________ and the reduction of __________ ions to generate an electrical voltage. A more modern version is the __________, which uses a redox reaction consisting of the oxidation of __________ and the reduction of __________ to generate the electrical current needed to start an automobile engine. Electrical energy is returned to the battery as an electrical current produced by the car's generator or alternator converts __________, the product of the energy-liberating redox reaction, back into lead and lead dioxide.

Daniell cell	loses electrons
electrochemical cell	negative
gains electrons	positive
half-cell	redox
hydrogen	standard reduction potentials
lead	
lead–acid storage battery	water
lead dioxide	zinc

2. Define, describe, or explain each of the following:

a. anode	f. electrolysis
b. cathode	g. fuel cell
c. electrical circuit	h. oxidizing agent
d. electric current	i. reducing agent
e. electrochemical reaction	j. salt bridge

3. What element is used to coat galvanized metal products?

4. What happens when the anode and cathode of an electrochemical cell are connected to each other?

5. What's the common term we use for the commercial electrolytic cells that provide power to our flashlights and other portable electrical devices?

6. (a) What metal makes up the outer casing of a common or heavy-duty flashlight battery? (b) What is the function of that metal? (c) What element makes up the porous rod that forms the inner core of the battery? (d) What is that rod's function? (e) What three compounds, other than water, make up the bulk of the battery's interior?

7. What substance does an "alkaline" battery contain in addition to the three compounds referred to in Exercise 6(e)?

8. Do electrons move from oxidizing agents to reducing agents or from reducing agents to oxidizing agents?

9. As electrons depart from the zinc casing of a flashlight battery, they leave behind positively charged zinc cations that cause the zinc casing to acquire an overall positive electrical charge. Is the zinc casing the battery's anode or cathode? (For help, see Sec. 1.3.)

10. What's the significance of a positive potential for a redox reaction? A negative potential?

11. How can you cause a reaction with a negative redox potential to occur?

12. You are given a table of standard reduction potentials. (a) How do you determine the *reduction* potential of a reduction half-cell? (b) How do you determine the *oxidation* potential of an oxidation half-cell? (c) How do you determine the *redox* potential of a redox reaction?

13. As we saw in Chapter 3, elemental sodium and elemental chlorine react with each other to form sodium chloride. Although chlorine normally exists as diatomic molecules (Cl_2), we can write the reaction of sodium atoms with chlorine atoms as $Na + Cl \longrightarrow Na^+Cl^-$. In this reaction as written, what is being oxidized and what is being reduced? What is the oxidizing agent and what is the reducing agent? Write the individual half-cell reactions that combine to form the redox reaction. You may want to review the discussion of the reaction in Chapter 3.

14. Which one of the following has the greatest tendency to acquire an electron? Which has the greatest tendency to lose an electron? (a) Zn^0, (b) Cl^-, (c) Br_2, (d) Cs^+, (e) a mixture of $PbSO_4$ and H_2O, (f) H^+, (g) H_2O, (h) Cl_2.

15. Name and write the chemical formulas of all the products of the electrolysis of (a) pure water, (b) a solution of sodium chloride in water, and (c) a solution of potassium chloride in water.

16. Potassium is a hazardous metal that reacts rapidly and explosively with water. In contact with water, potassium produces potassium hydroxide (KOH) and hydrogen gas, which ignites as a result of the heat generated in the reaction. (a) In view of this (and referring to the table of standard reduction potentials), can water act as an oxidizing agent? (b) Can water also act as a reducing agent? (c) Name two halogens that water is capable of reducing.

17. What determines whether a substance that *can* act as either an oxidizing agent or a reducing agent (like water) *does*, in fact, act as an oxidizing agent or a reducing agent?

18. (a) What is the difference between the *oxidation* of a metal that's caused by components of the atmosphere and the *corrosion* of a metal that's caused by components of the atmosphere? (b) Give an example of oxidation of this kind that *accelerates* corrosion. (c) Give an example of oxidation of this kind that *protects against* corrosion.

19. Large metal storage tanks used for holding gasoline underground are too large to be coated with zinc by any practical galvanizing process. How are they protected from corrosion?

20. What's the principal advantage of an electric car over the kinds of cars we use today? What were some of the factors that led to the displacement of electric cars, early in this century, by cars with internal combustion engines?

A Little Arithmetic and Other Quantitative Puzzles

21. What voltage would you obtain from a Daniell cell made of copper and silver sheets and solutions of their cations?

22. Can steel wool, which contains a large percentage of iron, be used in place of the galvanized tacks of the opening demonstration? Justify your answer by calculating the redox potential for the reduction of iodine by iron to produce FeI_2.

23. What, if anything, would you expect to observe if you put

a. a strip of zinc metal into a solution of copper sulfate
b. a strip of magnesium metal into a solution of copper sulfate
c. a strip of copper metal into a solution of zinc sulfate
d. a strip of copper metal into a solution of magnesium sulfate
e. a strip of silver metal into a solution of copper sulfate
f. a strip of copper metal into a solution of silver sulfate

24. The lead–acid automobile battery consists of plates of spongy lead and plates of lead dioxide. As the battery discharges, the lead and the lead dioxide are converted to lead sulfate. Would a similar battery consisting of plates of spongy lead and plates of *lead sulfate* provide any current? If your answer is yes, what voltage would you expect? If your answer is no, explain why it would not work. Repeat this question for a battery consisting of plates of lead dioxide and lead sulfate.

25. How do we manage to get 12 volts from a car battery when the redox voltage of the electrochemical reaction taking place within it is only 2 volts?

26. Suppose we modify the table of standard reduction potentials and define the reduction of F_2 to fluoride ions as our standard, replacing the reduction of hydrogen ions to H_2, which is the current standard. If we now define the reduction potential for fluorine as zero (arbitrarily, just as the reduction potential of hydrogen ions to H_2 was originally defined arbitrarily as zero):

a. How would this change all the other values in the table?
b. What would be the new reduction potential for hydrogen ions?
c. How would this change affect the *measured* value for the voltage of the Daniell cell?
d. How would this change affect the *calculated* value for the voltage of the Daniell cell?
e. Would we still be able to use the new table to determine whether a redox reaction can occur spontaneously?
f. If your answer to part (e) was yes, explain how we would now use our modified table.
g. If your answer to part (e) was no, explain why we would not be able to use our modified table for this purpose.
h. Does it matter which reduction reaction we choose as our standard? Explain.
i. Why do you think the reduction of hydrogen ions was chosen as the standard?

Think, Speculate, Reflect, and Ponder

27. The more expensive, "heavy duty," "longer lasting," or "longer life" carbon-zinc batteries have cups made of thicker zinc than the less expensive, "standard" carbon-zinc batteries. Why? How does a thicker shell of zinc help extend the operating life of a battery?

28. If you remove an iodine stain from a piece of cotton by rubbing the stain with a moist vitamin C tablet and neglect to rinse the area afterward, a dark color can reappear even before the fabric is washed. Suggest a reason why.

29. Household bleach is normally used to *remove* colored stains from fabrics. Yet when we decolorize a solution of tincture of iodine with galvanized tacks, as we did at the opening of this chapter, adding household bleach *produced* a color in the nearly colorless solution. Explain why this happened.

30. Liquid household bleaches are good oxidizing agents. If you add a few drops of a liquid bleach to a clear, colorless solution of KBr in water, the solution turns orange–brown. What do you conclude from this observation? (*Note:* The products of the *reduction* of the liquid household bleach are colorless.)

31. Buffalo, New York, is a major center for the production of chlorine. Suggest a reason for this.

32. If we carry out the electrolysis of pure water, the reaction proceeds very slowly and produces oxygen at the positive terminal. If we add table salt to the water, the electrolysis proceeds much faster and produces chlorine at the positive terminal. Explain both of these phenomena and show that they are related to each other.

33. Suppose there were one sodium *cation* and one chlorine atom in a (very) small box and you added one electron to this (very) small box. Both the sodium cation and the chlorine atom need only one more electron to achieve an electron octet in their valence shells. Where would the electron go: to the sodium cation or to the chlorine atom? Explain.

34. What does a table of standard reduction potentials tell us about the likelihood of an explosion when we mix two substances together? Explain.

35. A freshly constructed Daniell cell made of copper metal, copper sulfate, zinc metal, and zinc sulfate generates a voltage of about 1.1 volts. What voltage is produced by the same cell after it has been operating long enough to discharge completely the blue color of the copper sulfate solution and turn it completely colorless? Describe your reasoning.

36. Would a Daniell cell work equally well if the salt bridge were replaced by a strip of cloth wet with pure water? Explain.

37. Could you operate a two-potato clock from zinc and copper strips that dip into a solution of table salt in water? Could you run one from zinc and copper strips in a solution of table sugar in water? Explain.

38. Which of the following factors affect the life of the two-potato clock? Explain your answer for each.

a. the thickness of the zinc plates
b. the thickness of the copper plates
c. the size of the potato
d. the humidity of the atmosphere

39. The Daniell cell uses metal plates of copper and zinc, and operates through a redox reaction involving the oxidation of zinc atoms to zinc cations and the reduction of copper cations to copper atoms. The two-potato clock also uses metal plates of copper and zinc. The redox reaction that runs the clock uses an oxidation half-cell in which zinc atoms are oxidized to zinc cations. But there are virtually no copper ions present in a potato to take part in a reduction half-cell. Considering what you know about raw potatoes as well as the table of standard reduction potentials, what do you think is being reduced as the two-potato clock operates?

40. Over a period of time the zinc plates of the two-potato clock will erode and must be replaced. Even though (ungalvanized) iron nails will also corrode, they are inexpensive and readily available, so we might consider them a convenient replacement for the zinc. Do you think iron nails would make a satisfactory substitute for the zinc plates? Explain.

41. Summarize your response to someone who tells you that the successful operation of a two–potato clock from wires plugged into two potatoes proves that potatoes provide you with healthful energy, since they are obviously capable of providing the clock with power.

42. Give an example of corrosion taken from a demonstration described in this chapter.

43. Food that stands in contact with air for a long time sometimes deteriorates as some of the components of the food react with the oxygen of the air. Vitamin C is added to some foods not only as a supplemental nutrient but also to protect against this kind of deterioration. How does vitamin C provide this protection?

44. The use of natural gas as a fuel for an internal combustion engine produces very little of the kinds of pollution generated by gasoline. Natural gas, which consists largely of methane, burns according to the balanced equation

$$CH_4 + 2O_2 \longrightarrow CO_2 + 2H_2O$$

With this in mind, would an electric car powered by a fuel cell have any advantage over the use of natural gas in a car with an internal combustion engine? Refer to the concluding sections of Chapter 8 in your answer.

45. A fuel cell using the chemical combination of hydrogen and oxygen to generate electricity might use the oxygen of the air. How might the hydrogen be obtained?

46. Give two reasons why electric cars eventually gave way to cars powered by the internal combustion engine.

47. What do you believe is the greatest superiority of an electric car compared with a gasoline-powered car? What do you believe is the greatest superiority of a gasoline-powered car compared with an electric car?

48. Would you be in favor of a federal regulation requiring that all cars now on the road be replaced by electric cars, over a reasonable period? Explain.

49. Submarines run on battery-powered engines when they are submerged and on diesel engines (a variation of an internal combustion engine) when they are on the surface. Why don't they run on battery-powered engines all the time?

50. A battery using a redox reaction that converts lithium metal to lithium cations and fluorine gas to fluoride anions would provide about 5.9 volts. Why is a battery of this composition unlikely to become a widely used consumer product?

51. You are the owner of a small firm that produces all kinds of electric batteries. You wish to develop a new battery that uses a redox reaction (with a positive redox potential) never before used in a commercial battery. What are some of the factors you would consider in deciding whether to manufacture this new battery?

52. Some chemists claim that *all* chemical reactions are redox reactions. In the simple *ionization* of water, what is being oxidized according to this view? What is being reduced? Refer to Section 3.15.

Web Questions

Use the Web to answer or accomplish each of the following. Report your Web source for each response.

1. Some brands of toothpaste and mouthwash claim to be particularly effective in brightening your teeth, eliminating mouth odors, killing oral bacteria, etc., because of a special oxidizing ingredient they contain. What is this oxidizing chemical added to these oral cleansing agents? How does it work?

2. Write a short biography of André Marie Ampère, John Frederick Daniell, or Alessandro Volta. Describe when the scientist lived, where he received his education, what his contributions to the science of electricity were, and what honors he received.

3. Describe the commercial process for galvanizing iron and steel products to protect them from corrosion.

4. What is the current status of the use of fuel cells to provide power to automobiles?

5. Who was Charles Martin Hall? Describe the contribution he made to one of the topics of this chapter.

6. Summarize the history of the lead–acid storage battery, from its invention in 1859 to the present. Identify its inventor and describe its original form or construction. What were some of the major advances in its development and when was each made?

Solids, Liquids, and Especially Gases

THE STATES OF MATTER

Splashing water: droplets of a liquid scattering through a gas.

Squeezing Air Out of a Bottle

Here's another parlor stunt that illustrates an important scientific principle. Announce that you are going to squeeze air out of a glass bottle with your bare hands. Before making this announcement, put an empty glass soda bottle in a freezer for about half an hour. (You can leave the paper label on the bottle, but for best results remove any insulation that might be wrapped around the glass.) Remove the bottle from the freezer (you may want to do this out of viewing range so that no one else knows you are using a well-chilled bottle), invert it, wrap your warm hands around it, and place its mouth well below the surface of a saucer or glass of water. Act as if you are squeezing the bottle, *but don't actually put any force on it.* Squeezing a glass bottle can be hazardous. If the bottle is flawed it could break in your hands, producing a nasty cut. As the air inside the bottle warms up, bubbles start coming out of the mouth, as though you were, indeed, squeezing air out of it.

In a variation of this stunt, you can set the cold bottle upright on a table, wet the rim, and place a coin, such as a quarter, over the bottle's mouth. (Wetting the rim produces a weak but effective seal between the coin and the glass.) Wrap your hands around the bottle, as before. Soon, as the air inside the bottle warms, the coin starts making a pinging noise as the air in the bottle expands and forces its way out, lifting the coin periodically.

Figure 12.1 "Squeezing" air out of a bottle. Warming the air in a cold bottle causes the air to expand and escape, making a coin pop (as shown in top photo) and producing bubbles (as shown on bottom photo).

The secret is in the science. At constant pressure, any given quantity of air, such as the air inside the cold bottle, takes up more room when it's warm than when it's cold. As the heat of your hands warms the air inside the bottle, the gas lightly trapped inside expands and leaves the bottle either as bubbles, in the first form of this stunt, or in short bursts that lift the coin, in the second form (Fig. 12.1).

In this chapter we'll learn more about gases, as well as solids and liquids. Together, solids, liquids, and gases form the three *states of matter* we encounter in our daily lives. We can find water, for example, as a solid (ice), a liquid (which we refer to simply as "water"), and a gas (steam). After examining these three states of matter, we'll look more closely at gases and how they behave. We'll describe the relationships among the volume, pressure, and temperature of a quantity of gas, and we'll see how these relationships affect us in our daily lives.

12.1 The States of Matter

We define anything we contact in our daily lives as a solid, liquid, or gas by the way it retains its shape and volume. **Solids** have distinct, fixed volumes and well-defined shapes. A piece of ice is a solid. It occupies a specific amount of space and holds its own shape. It might have the form of an ice cube we get from a freezer tray, or it might be one of the irregular chunks of ice we find in bags of ice from supermarkets or convenience stores.

Figure 12.2 Solid, liquid, gas.

Liquids, too, occupy fixed volumes, but liquids have no shapes of their own. Except for small droplets that tend to become spheres, liquids always take the shape of whatever container they're held in. When a piece of ice melts, the liquid water that forms takes the shape of the container. **Gases**, on the other hand, always acquire both the shape and the volume of their container (Fig. 12.2). When we boil water the liquid water becomes a gas, which we call steam if it's freshly formed from the boiling water. More generally it's known as water vapor, especially if it is cooler than steam and part of a humid atmosphere. Steam or water vapor has no shape or size of its own; instead it takes both the shape and size of whatever container holds it.

Solid, liquids and **gases** are three states or phases of matter. **Solids** maintain their own volumes and shapes. **Liquids** maintain their own volumes but take the shapes of their containers. **Gases** take both the volumes and shapes of their containers.

We can change a substance from a solid to a liquid and then to a gas simply by heating it. Heat an ice cube and it melts. Heat water and it boils. In the reverse direction, cooling converts a gas to a liquid and then to a solid. Steam condenses to liquid water as it cools, as we saw in the demonstration that opened Chapter 7. Place trays filled with liquid water in a freezer and you get ice cubes. Whether an element or a compound exists as a solid, a liquid, or a gas depends principally on its temperature.

Often substances become transformed from one state to another when we heat them because they decompose or take part in chemical reactions with other elements or compounds, rather than because of melting or boiling. Hold a match to the bottom of a wax candle and you'll find that the solid hydrocarbons of the paraffin wax melt to a liquid, then harden again when you remove the match and allow them to cool. The hydrocarbons have melted and then resolidified, remaining the same compounds throughout. But hold the match to the wick, as we did in the opening demonstration for Chapter 6, and the hydrocarbons vaporize, mix with the atmosphere, and react chemically with oxygen to produce carbon dioxide and water. When we refer to the effects of heat in this chapter we're assuming that no chemical reaction takes place and that subsequent cooling gives us the same substance we started with.

Extremely high temperatures, beyond those we normally find here on earth, can even produce a fourth state of matter, a **plasma**. Temperatures near those of the surface and interior of the sun not only cause molecules to disintegrate into their component atoms but also cause the electrons of atomic shells to be stripped away from the nuclei, producing a form of matter in which positively charged nuclei and negatively charged electrons move about randomly. Since the properties of a plasma are far different from those of solids, liquids, and gases, plasmas are regarded as a fourth state of matter. We won't consider plasmas further.

A **plasma** is a state of matter in which electrons have been stripped out of their atomic shells to produce positively charged nuclei and negatively charged electrons, but without the existence of atomic structure.

■ QUESTION

Which of the three common states of matter—solids, liquids, or gases—(a) maintain their own volumes, no matter what container holds them; (b) maintain their own shapes, no matter what container holds them?

12.2 What Happens When Solids Melt

We can melt ice and boil water by heating them. We'll now see why. Since heat is a form of energy, heating anything increases the energy of its chemical particles. One result is that, with an increase in kinetic energy, the particles move faster (Sec. 8.2). This follows from one of the more fundamental rules of chemistry and physics: *The average kinetic energy of the chemical particles of any substance depends directly on its temperature*. The higher the temperature of anything, the greater the average energy of its atoms, ions, or molecules. Heating an ice cube or a pot of water or steam increases the average energy of all of the H_2O molecules within it. We'll examine now why this increased energy causes solids (like ice) to melt and liquids (like water) to boil.

Molecules and ions made up of two or more atoms can increase their kinetic energy by

- rotating faster around their center of mass (equivalent to a body's center of gravity)—much like the blades of a fan that's switched from low to high—with an energy called *rotational energy*;
- vibrating faster with no change in their center of mass—much like a vibrating spring—with an energy called *vibrational energy*; and
- moving from one location to another, with an energy called *translational energy*.

In what follows we're interested only in changes in the translational energy.

Solids melt to liquids when they are heated because of an increase in the translational energies of their atoms, ions, or molecules. Solids keep their own shapes because the translational energies of their particles are especially small, too small to tear them out of a rigid order fixed by the cohesive forces—the forces of attraction that hold the particles close to one another and that keep them in the solid state. In crystalline solids, such as sodium chloride, sucrose, and ice, the particles remain firmly fixed next to each other in the orderly arrangement of the crystalline lattice, unable to move about within the bulk of the material. In these materials the movements of the particles are limited to small vibrations about their fixed positions in the crystal lattice (Sec. 3.10).

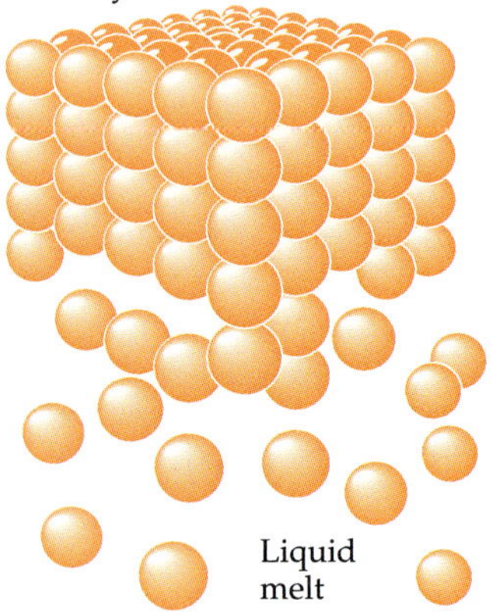

Figure 12.3 At its melting point, the chemical particles of a crystalline solid leave the orderly lattice and achieve a greater freedom of movement in the liquid melt.

A **melting point** is the temperature at which a solid becomes a liquid. The liquid returns to the solid state at this same temperature.

When you heat a solid, though, it absorbs energy. Heat a solid to a temperature that's high enough, and the movements of its particles become sufficiently vigorous to tear them away from their neighbors and out of their lattice positions. As they move out of the lattice and begin to move about freely within the bulk of the material, we observe that the solid melts to a liquid.

As a pure, crystalline solid melts, all the added energy goes toward giving the particles greater freedom of movement rather than toward increasing the temperature. This specific temperature at which a pure solid melts to a liquid, or a liquid becomes a solid as it cools, is the substance's **melting point** (Fig. 12.3). Substances like fats and some plastics aren't crystalline materials. Instead of melting at sharply defined temperatures, they soften gradually to liquids. Without the rigor of fixed positions in a crystal lattice, the molecules of these *amorphous* solids (meaning "without form or shape") can move about more and more freely as the temperature increases.

■ QUESTION

Name or describe (a) a moving object (other than a fan blade) that possesses rotational energy, (b) a moving object (other than a vibrating spring) that possesses vibrational energy, and (c) a moving object that possesses translational energy. Now name or describe a moving object that possesses simultaneously any two of these three forms of kinetic energy.

12.3 What Happens When Liquids Boil and Evaporate and When Solids Sublime

Even in the liquid state the average translational energy of the chemical particles is still relatively small. Although the particles are now moving about freely, with enough energy to keep them out of the fixed shape of a crystalline or amorphous solid, the remaining cohesive forces are still strong enough to keep them trapped, for the most part, in the form of a liquid. We say "for the most part" because while the *average* translational energy is too low to convert the entire bulk of the liquid into a gas, a few of the particles have energies far lower than average, and a few have energies far higher than average. These few particles with the exceptionally high translational energies overcome the remaining cohesive forces and escape into the vapor state in the process we call *evaporation* (Fig. 12.4).

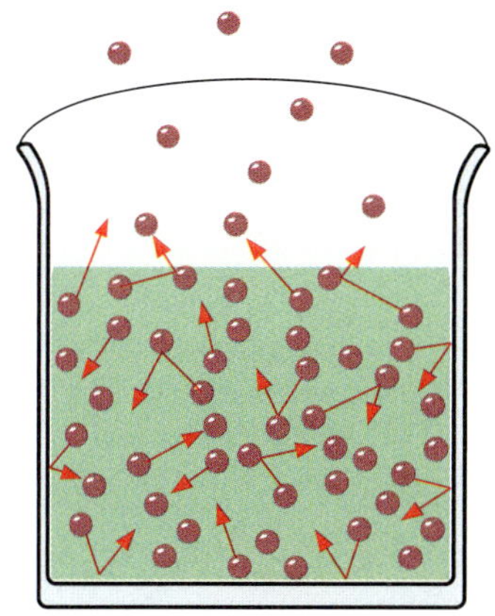

Figure 12.4 Liquids evaporate as molecules with high translational energies escape from the liquid into the vapor.

When the temperature of the entire liquid reaches the boiling point, *all* the chemical species have translational energies high enough to escape the pull of their neighbors, so all the particles escape into the vapor phase. As this happens simultaneously throughout the entire bulk of the liquid, we see it boil. Chemical particles in the gaseous state have translational energies far higher than the forces that would bind them to their neighbors so they move about freely, virtually independent of each other.

The temperature at which a liquid is transformed into a gas, or a gas into a liquid, is the material's **boiling point**. Since boiling points are particularly sensitive to changes in the external pressure, they're usually reported as temperatures at normal atmospheric pressure. Table 12.1 presents the melting points and boiling points of some common chemicals.

A **boiling point** is the temperature at which a liquid becomes a gas and the gas becomes a liquid, usually at normal atmospheric pressure.

(The particles of some substances move directly from the solid state to the gaseous state without passing through a liquid state in a process called **sublimation**. At normal atmospheric pressure solid carbon dioxide, also known as Dry Ice, sublimes directly to gaseous carbon dioxide.)

Sublimation occurs when a solid becomes a gas without first passing through the liquid state.

Table 12.1 Melting Points and Boiling Points of Typical Substances

Substance	Common Source or Use	Melting Point (°C)	Boiling Point (°C, atmospheric pressure)
Acetic acid	Vinegar	16.6	118
Ammonia	Household cleaner	−78	−33
Benzene	Gasoline hydrocarbon	5.5	80
Citric acid	Citrus fruit	153	(Decomposes)
Ethyl acetate	Food flavoring, solvent	−83.6	77
Ethyl alcohol	Beer, wine, liquors	−117	78.5
Gold	Coins, jewelry	1064	3080
Hydrogen chloride	Muriatic acid	−115	−85
Oxygen	Atmosphere	−218	−183
Potassium iodide	Iodized salt	681	1330
Propane	Fuel for grills	−190	−42
Sodium chloride	Table salt	801	1413
Sodium hydroxide	Lye	318	1390
Sucrose	Table sugar	185	(Decomposes)
Toluene	Paint remover	−95	110.6
Water	Water	0	100

■ QUESTION

As shown in Table 12.1, sodium chloride and potassium iodide boil at much higher temperatures than do water, ethyl alcohol, and propane. What kind of bonding—covalent or ionic—occurs in each of these compounds? What do these boiling points indicate about the relative strengths of the forces of attraction between ions of ionic compounds and between molecules of covalent compounds?

12.4 The Gas We Live In

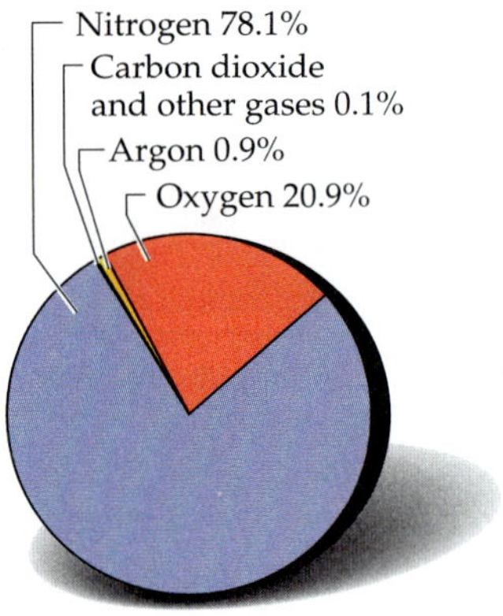

Figure 12.5 Composition of dry air.

The **troposphere** is the region of the earth's atmosphere that rises from the earth's surface and that produces the phenomena of our weather. The **stratosphere** lies above the troposphere and holds the ozone layer. The **mesosphere** lies above the stratosphere. The **ionosphere** is a higher region of the atmosphere that contains ionized gases and serves to reflect shortwave radio transmissions.

Now, with a knowledge of some of the forces at work in solids, liquids, and gases, we can turn to the gas we live in, our *atmosphere*. This gas that sustains our lives, moment by moment, enters our lungs as a complex mixture of simple substances. Nitrogen makes up 78.1% of dry air, measured by volume and at sea level; oxygen, 20.9%; argon, 0.9%; and carbon dioxide, 0.03%. The infinitesimal remainder, in order of decreasing proportions, consists of a combination of neon, helium, methane, krypton, hydrogen, oxides of nitrogen, and xenon (Fig. 12.5). Humid or wet air also contains water vapor, which lowers the percentages of the other gases in the total combination; polluted air carries with it noxious fumes, or suspensions of solids or liquids in the form of smoke or haze.

As we travel upward through the atmosphere we find that its composition and physical characteristics change with altitude, creating the regions of Figure 12.6. Immediately above the surface of the earth lies the region known as the **troposphere**, which gives us our clouds, wind, rain, and snow. The temperature of the air drops steadily as we ascend, until we reach the **stratosphere**. There the air is dry, clear, and cold, with a relatively uniform temperature of about −55°C at the earth's middle latitudes. The stratosphere, as we see in Figure 12.6, is the home of the *ozone layer*, which we'll examine in more detail in Chapter 14. Above the stratosphere the temperature now rises with increasing altitude, then drops in the **mesosphere**, and then rises again as we enter the **ionosphere**, a region in which the gases that make up the air are ionized by the sun's radiation.

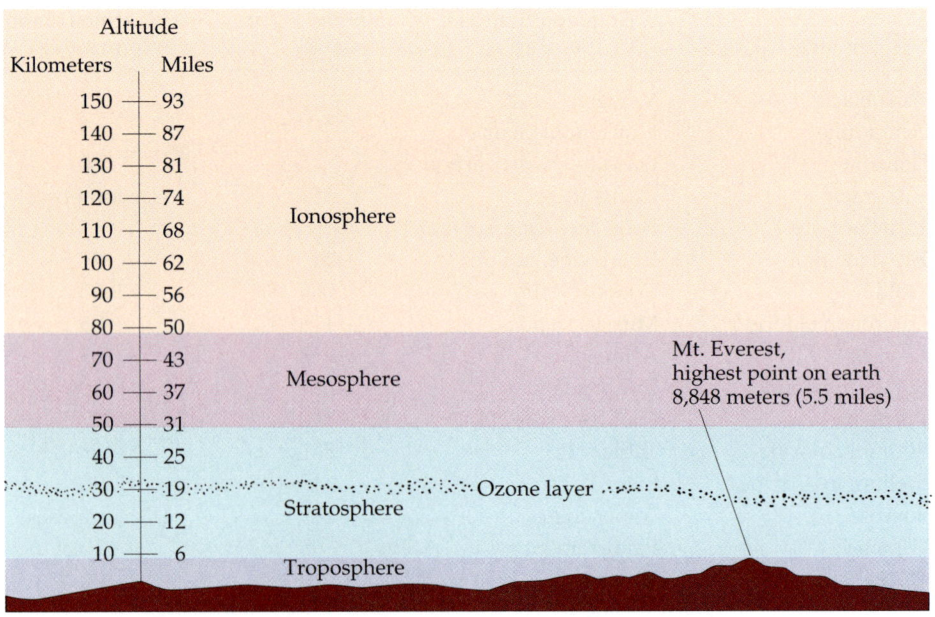

Figure 12.6 The regions of the atmosphere.

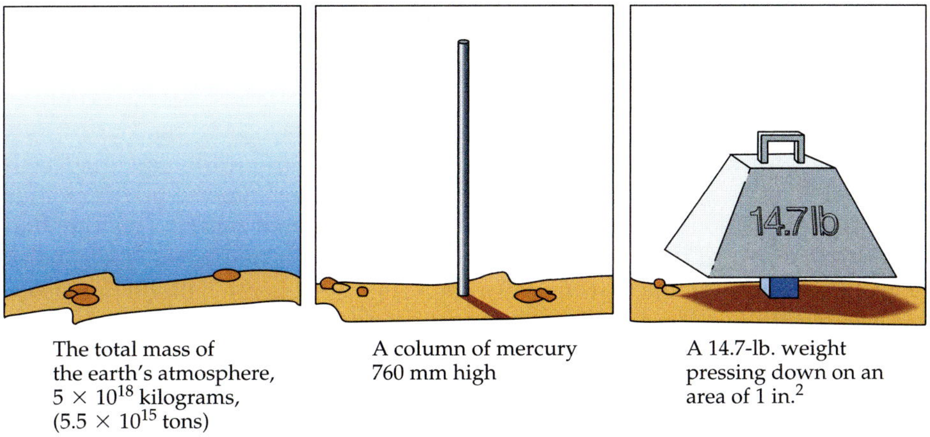

Figure 12.7 Atmospheric pressure at sea level, 760 mm-Hg, 14.7 lb/in.^{2}

This layer of ionized gases reflects radio waves effectively and makes shortwave radio transmissions possible.

The combined weight of all the gases above any particular point, either on the earth's surface or anywhere within the atmosphere itself, generates the **atmospheric pressure** at that spot. At sea level the total mass of the entire atmosphere, estimated at about 5×10^{18} kg (about 5.5×10^{15} tons), produces the same pressure, on average, as a column of mercury (chemical symbol, Hg) 760 mm high. Average sea-level pressure, then, is 760 mm-Hg, which is equivalent to about 14.7 lb/in^2. (Fig. 12.7). Each of these values represents a pressure of *one atmosphere*. For comparison, one atmosphere of pressure is the same as the pressure of a layer of water that's about 10.3 m deep.

The **atmospheric pressure** at any point on the earth's surface or above it is the pressure generated by the combined weight of all the atmospheric gases above that point.

We'll use the hyphenated *mm-Hg* as a unit of *pressure*, distinct from the unhyphenated *mm Hg*, which would represent the actual metric height of a physical column of mercury. An alternative unit for the *mm-Hg* is the *torr*, named for the Italian physicist and mathematician Evangelista Torricelli (1608–1647). Torricelli, who served as Galileo's secretary during the last few months of that great physicist's life, was the first to suggest that the atmosphere has mass and that it exerts a pressure on everything within it, just as the water of the sea presses on everything submerged. In 1643, the year following Galileo's death, Torricelli confirmed his ideas with his invention of the *barometer*, an instrument for measuring atmospheric pressure (Fig. 12.8). *Barometer* itself is a combination of two Greek words, *baros* for weight or pressure, and *metros*, meaning measure. Measurements of atmospheric pressure are often referred to as *barometric pressure*.

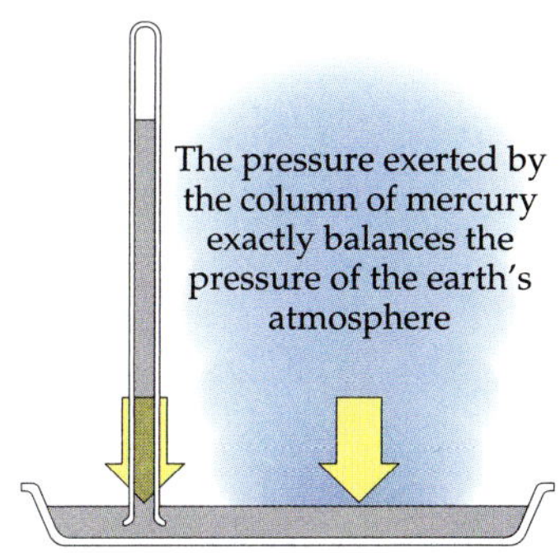

Figure 12.8 The Torricelli barometer.

Atmospheric pressure drops rapidly with altitude, decreasing to about 90% of its sea-level value at a height of just under 900 m (a little under 3000 ft) and to half the sea-level value at about 5500 m (18,000 ft). Changes in the density of the atmospheric gases above the earth's surface produce regions of high atmospheric pressure and low atmospheric pressure, the "highs" and "lows" of our weather forecasts. A rising atmospheric pressure usually forecasts a change to bright, sunny skies, while the reverse indicates cloudy, rainy weather ahead.

■ QUESTION

What would you expect the average barometric pressure to be at an altitude of 5500 m (18,000 ft)?

12.5 The Kinetic-Molecular Theory: From Ricocheting Billiard Balls to Ideal Gases

Pumping air into a tire increases its pressure.

The **kinetic-molecular theory of gases** explains the behavior of gases by assuming that they are made up of point-sized, perfectly resilient, constantly moving chemical particles.

If you've ever used a hand pump to inflate a tire, you know that the pumping gets harder as the tire becomes fully inflated. Maybe you've checked a tire's pressure when the tire is cold, before starting on a trip, and then again after it has warmed up from some travel. If you have, you know that the pressure rises as the tire and the air inside it heat up. Perhaps you've watched a hot-air balloon being prepared for flight. As a flame warms the air inside, the heat makes the air less dense than the surrounding atmosphere, giving the balloon buoyancy, and finally lifting it upward. If you've emptied an aerosol can, down to the last bit of its contents, you've noticed a considerable drop in the pressure as the very last bit of propellent leaves the nearly empty canister. Maybe you've also read the label warning of a possible explosion if you toss the empty can into an incinerator.

Each of these gaseous phenomena we observe in today's world parallels in some way the observations made by a handful of 17th- and 18th-century scientists who were the first to study gases in a systematic way. Today we sum up the knowledge they gathered about the behavior of gases and the theories they formulated to explain that behavior into what we now call the **kinetic-molecular theory of gases**. According to this model, gases act—ideally, at least—as though their individual, component molecules or atoms were perfectly resilient, infinitesimal billiard balls, taking up no space whatever. They behave like completely resilient, elastic spheres that have no diameters, circumferences, or volumes, and that move about continuously, bouncing merrily off each other and off the walls of their containers, losing no energy in the process and exerting absolutely no attraction for one another. Any gas made up of chemical particles that behave in exactly this way is an *ideal gas*.

According to this view, the physical properties of any gas—or at least any gas that behaves as if it were an ideal gas—depend *only* on the *kinetic* energy (the energy of motion) of its *molecules* (or its individual atoms if it's an inert gas or another substance that contains only unbonded atoms). That's why it's called the *kinetic-molecular* theory. These physical properties remain completely independent of the space actually filled by the molecules, and completely independent of any real forces of attraction that draw them together. What's more, the kinetic energy of the molecules depends entirely on the temperature of the gas. The higher the temperature, the greater its molecular energy (Fig. 12.9). The pressure of the gas in an enclosed container results from repeated collisions of the particles of the gas with the walls of the container.

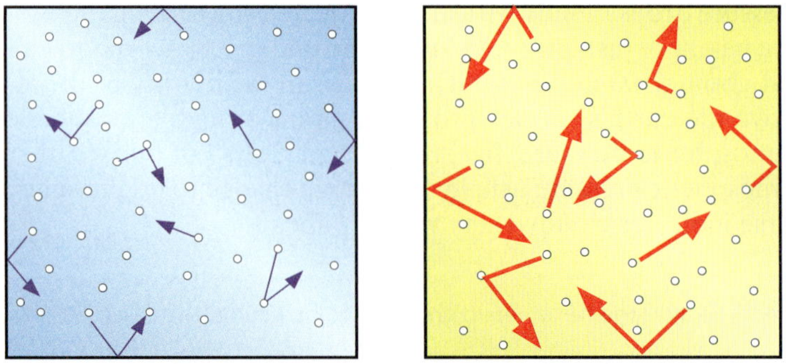

Figure 12.9 Molecular motion in an ideal gas. The higher the temperature, the greater the energy of molecular motions.

Of course atoms and molecules of real gases do occupy an extremely small but finite amount of space, and they do exert small but measurable attractions for each other. Otherwise no gas would ever condense into a liquid; there would be, for example, no such substance as liquid water. Nevertheless, this idea of point-sized billiard balls careening about with an energy that depends entirely on their temperature does give us a very good description of the ideal behavior of a gas, particularly at high temperatures and low pressures. Under these conditions all gases follow a pattern of relationships easily described by simple mathematical equations known as the *ideal gas laws*. The higher the temperature and the lower the pressure of the gas, the more nearly the gas approaches the ideal and the more closely it follows these laws.

Robert Boyle, discoverer of the relationship between the pressure and volume of a gas.

The gas laws come to us from the observations and the insights of the early scientists who studied the ways gases respond to changes in pressure, temperature, and volume, and who formulated the mathematical equations that describe their behavior. In the next few sections we'll examine the properties of gases through the eyes of the most important of these scientists:

Robert Boyle (Irish–English, 1627–1691)
Jacques Alexandre César Charles (French, 1746–1823)
Joseph Louis Gay-Lussac (French, 1778–1850)
William Thomson, Lord Kelvin (British, 1824–1907)
Amedeo Avogadro (Italian, 1776–1856)
John Dalton (English, 1766–1844)
William Henry (English, 1775–1836)

■ **QUESTION**

What characteristics would the atoms or molecules of a real gas have to have in order for the gas to behave *exactly* like an ideal gas?

12.6 Robert Boyle: Pressure and Volume

Robert Boyle, the fourteenth child of the first Earl of Cork, Ireland, became the first of the great gas law investigators shortly after the middle of the 17th century. In 1657 Boyle read of a newly devised air pump. With some ingenuity he improved its original design, built a model, and began studying the effects of pressure on the volume occupied by a quantity of air. Soon he described what we now know as **Boyle's Law**: *With the temperature and the number of moles of a quantity of gas held constant, the volume of the gas varies inversely with its pressure.* That is, with a fixed temperature and a fixed number of moles, as the pressure rises, the volume of the gas shrinks; as the pressure drops, the volume grows.

Boyle's Law states that with the temperature and the number of moles of a quantity of gas held constant, the volume of the gas varies inversely with its pressure.

Take a specific, fixed weight (a fixed number of moles) of a gas and keep it at a constant temperature. As you increase the pressure on the gas you squeeze it into a smaller volume. Lower the pressure and the gas expands into a larger volume. More quantitatively, double the pressure and the volume of the gas drops to half (Fig. 12.10). Decrease the pressure to half, and the volume doubles. To whatever extent the pressure changes, the volume change is exactly the inverse. We can express this mathematically as

$$\text{pressure} \times \text{volume} = \text{a constant}$$

or

$$P \times V = k$$

Figure 12.10 The effect of pressure on a gas. Double the pressure on the gas, at constant temperature, and its volume decreases to half.

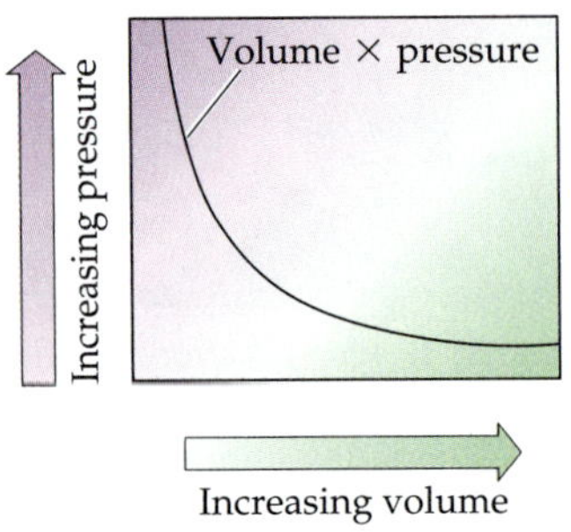

Figure 12.11 Boyle's Law: the pressure–volume relationship for a fixed quantity of gas maintained at constant temperature.

This equation tells us that with the temperature and the total number of molecules held constant, the product of the pressure and the volume of any fixed weight of a gas is always constant. Plotting an infinite number of pressures and the infinite number of corresponding volumes gives the curve shown in Figure 12.11.

The kinetic-molecular theory of ideal gas behavior explains Boyle's Law nicely. The pressure the gas exerts on the walls of its container results from the force of its molecular collisions with the walls. The greater the number of collisions at any instant, the greater the pressure on the container. Squeeze the gas into half its original volume and the rate at which the molecules bounce off the walls doubles. So does the pressure. Double the volume of the gas and the rate of collisions drops to half. So does the pressure. (This works both ways. Double the pressure exerted *on* the gas and the pressure exerted *by* the gas doubles in response. The only way this can occur, with the temperature and the total number of molecules kept constant, is for the gas's volume to decrease to half.) Figure 12.12 illustrates the connection between the kinetic-molecular theory and Boyle's Law.

E X A M P L E

A Little Less Pressure

Suppose that at sea level you fill a balloon, made of a perfectly elastic substance, to a volume of exactly 1 L. Now, keeping its temperature constant, you carry it up the side of a mountain to an altitude of some 900 m, where the atmospheric pressure is exactly 90% of its value at sea level (Sec. 12.4). What would the new volume of the balloon be?

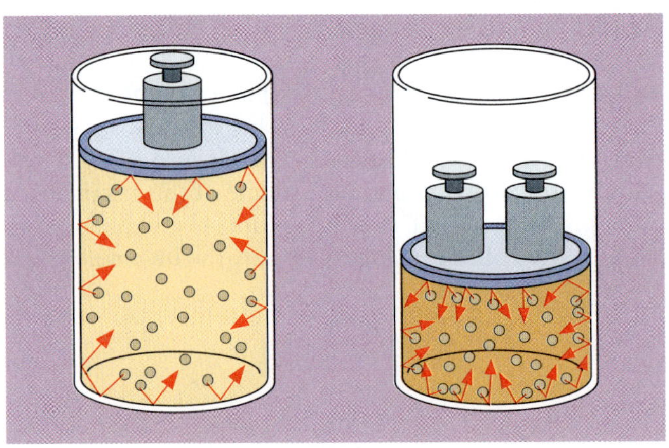

Figure 12.12 Boyle's Law and the kinetic-molecular theory of gases.

■ Although in solving this problem we use the equation for Boyle's Law, $P \times V = k$, we certainly don't have to know the actual numerical values of all the measurements involved, nor do we have to calculate the value of k. Since we're working with two different pressures (which we can call P_1 and P_2 to represent the first and second pressures, respectively) and also with two different volumes (V_1 and V_2, the first and second volumes), we can set up a ratio that eliminates k entirely. Since

$$P_1 \times V_1 = k$$

and

$$P_2 \times V_2 = k$$

and since the two ks are identical, then

$$P_1 \times V_1 = P_2 \times V_2$$

and

$$\frac{P_1}{P_2} = \frac{V_2}{V_1}$$

Since P_2 is 90% of P_1, we can set P_1 at 100 and P_2 at 90 without being concerned about the units, which cancel out in the fraction anyway. With just a little rearrangement this gives

$$\frac{V_2}{1.0\ \text{L}} = \frac{100}{90}$$

or

$$V_2 = 1.0\ \text{L} \times \frac{100}{90} = 1.1\ \text{L}$$

The volume of the balloon at the 900-m level is 1.1 L.

We'll use this sort of approach whenever possible as we consider the effects of conditions on the characteristics of a gas.

■ QUESTION

The volume of a quantity of a gas held at constant temperature and 760 mm-Hg pressure is 100 mL. What pressure does it take to reduce the volume to 95 mL?

12.7 Can a Gas Shrink to Nothing at All?

Born 89 years after Boyle first read of the newly devised air pump, J. A. C. Charles turned to science from a career in the French Ministry of Finance. In 1787, four years after he became the first person to inflate a balloon with hydrogen gas (which is less dense than air) and the first to use a balloon to ascend some 3 km into the atmosphere, Charles discovered that a quantity of gas kept at a constant pressure expands as it warms and contracts as it cools.

In 1802, Joseph Louis Gay-Lussac, lecture demonstrator at the French Ecole Polytechnique (and soon to become professor of physics at the Sorbonne), independently discovered what Charles had learned earlier about the effect of temperature on gases. Gay-Lussac went beyond Charles' works, though, as he discovered that for every one-degree (Celsius) change in temperature, this change in volume amounts to 1/273 of the volume the gas occupies at 0°C.

Charles' Law states that with the pressure and the number of moles of a quantity of gas held constant, the volume of the gas varies directly with its temperature.

The combination of Charles' and Gay-Lussac's studies gives us the gas law now known generally as **Charles' Law**, in recognition of his earlier description of the principle: *With the pressure and the number of moles of a quantity of gas held constant, the volume of the gas varies directly with its temperature*. Mathematically, this amounts to

$$V = k \times T$$

with T representing the temperature.

William Thomson, Lord Kelvin, devised the Kelvin temperature scale, in which temperature measurement begins at absolute zero and moves upward in Celsius degrees.

The numerical value of the temperature used in this equation comes from a scale designed by the British physicist William Thomson, who was given the title of Lord Kelvin by Queen Victoria, partly in recognition of his contributions to the design, manufacture, and placement of the first trans-Atlantic telegraph cable. The *Kelvin* or *absolute* temperature scale begins 273 degrees below the Celsius zero (−273°C), at **absolute zero**, and moves upward with degrees the same size as those of the Celsius scale. The zero of the Celsius thermometer becomes +273 on the Kelvin or absolute scale (Fig. 12.13). (The degree symbol, °, is not used in expressing temperatures on the Kelvin scale.) As the term *absolute* suggests, Lord Kelvin's zero represents the lowest possible temperature. Nothing can become colder than absolute zero, 0 K or −273°C. To convert from a Celsius temperature to Kelvin, add 273 to the Celsius reading; from Kelvin to Celsius, subtract 273 from the Kelvin value.

Absolute zero is the lowest possible temperature. It is 0 K, or −273°C.

The physical relationship that Charles and Gay-Lussac independently discovered implies that an ideal gas, with its volumeless chemical particles, would vanish completely at absolute zero. Mathematically,

$$V = k \times T,$$

If $T = 0$, then

$$V = k \times 0 = 0$$

But all gases show ideal behavior only at *high* temperatures (and low pressures), not at temperatures anywhere near absolute zero. The real volumes of gas particles and the real attractive forces they exert on each other cause all real gases to liquify as temperatures drop and pressures rise. The ideal world of bouncing, point-sized billiard balls fades into the real world of substance, of actual atoms and molecules. At a pressure of 1 atmosphere (1 atm), for example, helium liquifies at −268.9°C, or 4.1 K. The boiling points of some representative gases appear in Table 12.2. These are the temperatures at which the liquids boil to become real gases and at which the real gases condense to liquids.

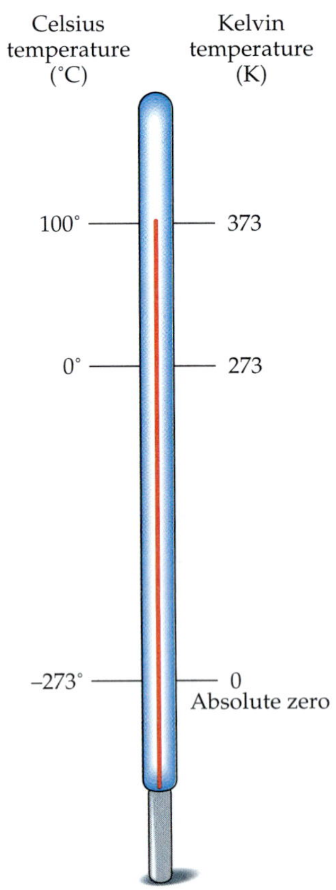

Figure 12.13 Celsius and Kelvin (absolute) temperature scales.

Table 12.2 Boiling Points of Representative Gases

Gas	Formula	Approximate Boiling Point (°C, 1 atm)
Water	H_2O	+100
Chlorine	Cl_2	−35
Hydrogen chloride	HCl	−85
Methane	CH_4	−164
Oxygen	O_2	−183
Fluorine	F_2	−188
Nitrogen	N_2	−196
Hydrogen	H_2	−259
Helium	He	−269

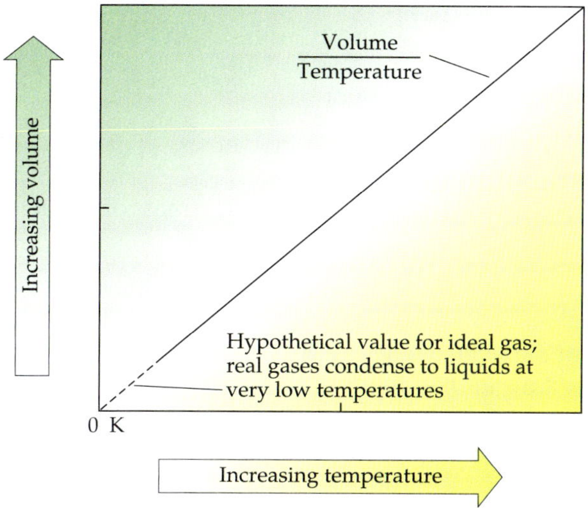

Figure 12.14 Charles' Law: the temperature–volume relationship for a fixed quantity of gas maintained at a constant pressure.

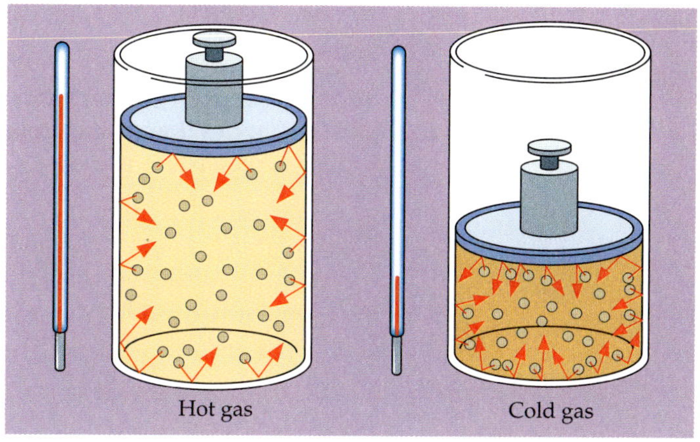

Figure 12.15 Charles' Law and the kinetic-molecular theory of gases: At constant pressure, as the temperature increases so does the volume.

The relationship between the volume of an ideal gas and its temperature appears graphically in Figure 12.14. As with Boyle's Law, the kinetic-molecular theory works nicely here too. As the temperature of a gas drops, so does the kinetic energy of its molecules. At the lower temperature molecules don't bounce off one another or off the walls of their container with quite as much energy or as often as they do at the higher temperature. To maintain the constant pressure on the walls of the container, the volume of the gas must shrink. On the other hand, as the temperature rises the energy and frequency of collisions also increase, and so does the volume if the pressure is to remain constant (Fig. 12.15).

E X A M P L E

The Shrinking Balloon

To illustrate Charles' Law, suppose we inflate a balloon to a volume of 1 L at 27°C, then place it in a refrigerator kept at 7°C. Now we'll ask: What is the volume of the balloon when the gas cools to the refrigerator's temperature?

■ We can solve the problem with the same kind of ratio we used for Boyle's Law. Since $V_1 = k \times T_1$ and $V_2 = k \times T_2$, the ratio becomes

$$\frac{V_2}{V_1} = \frac{T_2}{T_1}$$

To find V_2 we'll use Kelvin temperatures of $(27 + 273) = 300$ K for T_1 and $(7 + 273) = 280$ K for T_2. This gives us

$$\frac{V_2}{1\text{ L}} = \frac{280}{300}$$

$$V_2 = 1\text{ L} \times \frac{280}{300} = 0.933\text{ L} = 933\text{ mL}$$

■ **Q U E S T I O N**

A gas kept at constant pressure has a volume of 10 L at 25°C. At what Celsius temperature would the gas have a volume of 20 L?

12.8 Putting Them All Together: The Combined Gas Law

So far we've seen two laws that describe the behavior of gases:

- Boyle's Law, which relates pressure and volume for a fixed mass of gas held at a constant temperature, and
- Charles' Law, which relates volume and temperature for a fixed mass of gas held at a constant pressure.

Instead of treating these as two distinct laws, independent of each other, we can combine them into a single law that describes the interrelationship of the pressure, volume, and temperature of a fixed mass of gas. We can write this combination of Boyle's Law and Charles' Law in the following convenient form:

$$\frac{P_1V_1}{T_1} = \frac{P_2V_2}{T_2}$$

EXAMPLE

Counteracting Forces

To illustrate the use of the combined gas law, we'll examine a balloon voyage, Gay-Lussac made in 1804, which took him to an altitude of 7000 m. At that height the atmospheric pressure drops to about 300 mm-Hg and the temperature could easily cool to a chilly −33°C. Suppose Gay-Lussac carried with him a small balloon filled to 1 L at sea level at a pressure of 760 mm-Hg and a temperature of 27°C. What would its volume ultimately be when he reached a height of 7000 m?

■ This 1-L balloon would be subjected to two contrasting effects during the ascent. The pressure drop as the balloon rose would tend to cause it to expand, while the decrease in temperature would tend to make it contract.

The answer to the question requires using the combined equation. With

$$P_1 = 760 \text{ mm-Hg} \quad V_1 = 1 \text{ L} \quad T_1 = (27 + 273) = 300 \text{ K}$$
$$P_2 = 300 \text{ mm-Hg} \quad T_2 = (-33 + 273) = 240 \text{ K}$$

we can rewrite the equation

$$\frac{P_1V_1}{T_1} = \frac{P_2V_2}{T_2}$$

as

$$\frac{760 \text{ mm-Hg} \times 1 \text{ L}}{300 \text{ K}} = \frac{300 \text{ mm-Hg} \times V_2}{240 \text{ K}}$$

Solving for V_2,

$$\frac{1 \text{ L} \times 760 \text{ mm-Hg} \times 240 \text{ K}}{300 \text{ mm-Hg} \times 300 \text{ K}} = 2.03 \text{ L}$$

We find, then, that the decrease in pressure would produce a much greater effect than the decrease in temperature. The balloon's volume would almost exactly double.

EXAMPLE

Warning!

Labels on spray cans warn not to dispose of the empty can in an incinerator since it might explode. Why would the can explode?

■ The answer comes from the combined gas law. Even when the can appears to be empty, a small amount of gas remains inside at a pressure equal to atmospheric pressure. Since the can is sealed, the volume of this remaining gas remains constant, $V_1 = V_2$, and the combined gas law becomes

$$\frac{P_1}{T_1} = \frac{P_2}{T_2}$$

This tells us that as the temperature of the gas in the sealed can rises, so does the pressure. At some high temperature the pressure of the sealed gas would become great enough to rupture the can and produce an explosion.

■ QUESTION

Suppose we start again with a balloon filled to a volume of 1 L at a pressure of 760 mm-Hg and a temperature of 27°C—the same initial conditions as in the first example of this section. Now suppose we again reduce the external (or atmospheric) pressure to 300 mm-Hg. To what Celsius temperature would we have to cool the balloon to keep its volume at 1 L? (Assume that the gas inside the balloon behaves like an ideal gas.)

12.9 Gay-Lussac's Law

Despite the importance of his work in relating volume to temperature, Gay-Lussac is probably better known for yet another discovery, one that now bears his name. **Gay-Lussac's Law** states that *when gases react with one another at constant temperature and pressure, they combine in volumes that are related to each other as ratios of small, whole numbers*. If the product is also a gas, its volume is also related to the volumes of the reacting gases as a small, whole number.

Gay-Lussac's Law states that when gases react with one another at constant temperature and pressure, they combine in volumes that are related to each other as ratios of small, whole numbers.

For example, if we set up conditions so that hydrogen gas and nitrogen gas react to produce ammonia gas, we'd find that the volume of the hydrogen that enters into the reaction is exactly three times the volume of the reacting nitrogen. We'd also find that the volume of the gaseous ammonia produced is exactly twice the volume of the nitrogen used. Notice that these proportions are not only small, whole numbers, but they're in exactly the same ratio as the numbers of the molecules present in the balanced equation for the reaction:

$$3H_2 + N_2 \longrightarrow 2NH_3$$

EXAMPLE

HCl in a Balloon

Suppose we mix 1 L of H_2 and 1 L of Cl_2 in a balloon and allow the elements to react to produce HCl. (Assume that the material that forms the balloon is resistant to each of these chemicals and to the conditions of the reaction.) If we keep the balloon and its contents at constant pressure and temperature, what is the volume of the balloon at the end of the reaction?

■ First we'll write the balanced equation for the reaction. (To review writing and balancing equations, refer to Sections 6.14 and 9.2.) Since both hydrogen and chlorine are diatomic, we have

$$H_2 + Cl_2 \longrightarrow 2HCl$$

With an implied 1 in front of both the H_2 and the Cl_2, the gaseous hydrogen and chlorine must react with each other in equal volumes. That is, 1 L of hydrogen and 1 L of chlorine react completely with each other, with none of either element remaining after the HCl is formed. The 2 before the HCl indicates that the volume of the HCl formed is twice the volume of the H_2 (or the Cl_2).

Thus a total of 2 L of the mixture of H_2 and Cl_2 reacts, and a total of 2 L of HCl is formed. We conclude that the volume of the balloon doesn't change.

■ QUESTION

When gaseous oxygen and gaseous hydrogen react to form water, in what volume ratio do the hydrogen and the oxygen react? (As in the Example, writing a balanced equation for the reaction helps.)

12.10 Avogadro Carries Us Further Along

Avogadro's Law states that equal volumes of different gases (at the same temperature and pressure) contain equal numbers of atoms or molecules.

Every time you blow up a balloon or pump up a flat tire, you see still another gas law in action, one that's closely related to Gay-Lussac's. Amedeo Avogadro (Sec. 9.3) carried Gay-Lussac's reasoning one step further. Avogadro recognized that since the volume ratios of reacting gases are identical with the molecular ratios of balanced equations, the volume of a gas must reflect the number of atoms or molecules within it. This gives us **Avogadro's Law**: *Equal volumes of different gases (at the same temperature and pressure) contain equal numbers of atoms or molecules*. Put another way, at constant temperature and pressure the volume of a gas is directly proportional to the number of moles present, or

$$V = k \times n$$

where k is a constant and n is the number of moles. The more moles there are, the larger the volume. The more air we put into a deflated balloon or tire, the bigger it gets.

If we're using a hand pump to put air into a tire, the volume hardly changes after we've inflated it a bit. Under this condition of nearly constant volume, the pressure we're working against increases with each additional stroke, so the pumping becomes harder as we progress.

■ QUESTION

Suppose you had 10 L of a gas composed of individual hydrogen atoms. Then, at constant temperature and pressure, these individual hydrogen atoms combined to form diatomic hydrogen molecules, H_2. What would be the volume of this gaseous H_2?

12.11 John Dalton and William Henry Explain Why Soft Drinks Go Flat

Dalton's Law states that the total pressure of a mixture of gases equals the sum of the partial pressures of each of the gases in the mixture.

The **partial pressure** of each gas is the pressure each gas would exert, at the same temperature and in the same volume, in the absence of all the other gases.

John Dalton and William Henry round out our early explorers of the properties of gases. Dalton, who was born in northern England in 1766 and spent his life teaching mathematics and science in Manchester, stated what we now know as **Dalton's Law**: *The total pressure of a mixture of gases equals the sum of the partial pressures of each of the gases in the mixture*. The term **partial pressure** refers to the pressure each gas would exert, at the same temperature and in the same volume, in the absence of all the other gases.

The total pressure of any mixture of gases, then, is

$$P_{(\text{total})} = P_A + P_B + P_C + \cdots$$

Table 12.3 Partial Pressures of the Major Atmospheric Gases

Gas	Percentage of Atmosphere (by Volume)	Mole fraction	Approximate Partial Pressure (mm-Hg)
Nitrogen	78.1	0.781	594
Oxygen	20.9	0.209	159
Argon	0.9	0.009	7
Carbon dioxide	0.03	0.0003	0.2

where P_A represents the partial pressure of gas A, P_B represents the partial pressure of gas B, and so on. For air, the total pressure represents the sum of the partial pressures of each of its component gases—N_2, O_2, and so on. Since the partial pressure of each component gas reflects the fraction of its molecules in the total number of molecules of the entire mixture, we can use the *mole fraction* of each component gas to calculate its partial pressure. (The *mole fraction* represents the ratio of the number of moles of each component to the total number of moles present.) Partial pressures for the gases that make up the air we breathe appear in Table 12.3.

Here again we see that still another one of the gas laws, Dalton's Law, supports the kinetic-molecular theory of the ideal gas. As the particles of any particular gas bounce around within a container, unaffected by each other, they're equally unaffected by the chemical particles of any other gas that may be present. Each gas, acting independently of the others, exerts its own pressure, with the total pressure equal to the sum of each of the individual, partial pressures.

William Henry, our final gas law scientist, was born in 1775 in Manchester, the same city where John Dalton would later find his career as a teacher. Henry's discovery that, at a fixed temperature, *the quantity of a gas that dissolves in a liquid depends directly on the pressure of that gas above the liquid*, has become **Henry's Law**. That is, the higher the pressure of a gas above a liquid, the greater the amount of dissolved gas in the liquid. On the other hand, as the pressure of a gas above a liquid drops, so does the amount of the gas that's dissolved in the liquid.

Henry's Law states that at a fixed temperature the quantity of a gas that dissolves in a liquid depends directly on the pressure of that gas above the liquid.

Now we can explain nicely why there's a rush of gas when we open a bottle of soda, why effervescence begins after the container is opened, and why a bottle of soda eventually goes flat if it stands open to the atmosphere. Knowing that the sparkle of a carbonated drink comes from its dissolved carbon dioxide, we can put the gas laws into operation:

1. *Henry's Law*: To put plenty of CO_2 into the drink, the container is sealed under a high pressure of the gas.
2. *Boyle's Law*: When we open the can or bottle, the pressure above the liquid drops quickly to atmospheric pressure and the pressurized gas in the space immediately above the drink expands and escapes into the room with the familiar hissing sound.
3. *Dalton's Law*: With the decreased total pressure of all the gases above the drink, the partial pressure of the gaseous CO_2, which is part of that gaseous mixture, drops as well.
4. *Henry's Law, again*: The concentration of the CO_2 dissolved in the drink shrinks proportionately.

Figure 12.16 The gas laws and a carbonated drink.

As the drink stands unstoppered, the very low partial pressure of the atmospheric CO_2 (0.2 mm-Hg, Table 12.3) allows almost all the CO_2 to escape from its liquid solution, causing the soda to go flat (Fig. 12.16).

■ **QUESTION**

Fish, which consume oxygen just as all other animals do, survive on atmospheric oxygen that dissolves in water. Name a gas present in water in a greater concentration than oxygen. Explain.

12.12 Don't Shake a Warm Bottle of Soda; Also Some Advice About an Aquarium

A combination of the laws of gases, including their low solubility in a warm liquid, causes this to happen when you shake a bottle of warm soda before you open it.

Henry's Law tells us all we need to know about the actual solubility of a gas in a liquid, but it says nothing about how fast the gas dissolves or how fast it comes out of solution. It doesn't tell us anything about the rates of the two processes or how to change them.

We know from experience that a carbonated beverage loses its CO_2 slowly. We can watch the bubbles form on the sides and bottom of a glass and rise to the top slowly, over a reasonably long period, before the drink goes flat and the entire process stops. But shaking a bottle or a can of soda, especially a warm one in which the solubility of the CO_2 is low, often causes the drink to foam up and spill out of the container when we open it.

This foaming results from a process called *nucleation*. Shaking the soda causes microscopic bubbles of the gas in the space above the drink to become trapped inside the liquid. These serve as nuclei around which dissolved CO_2 can come out of solution to form gas bubbles quickly as we open the can or bottle and the external pressure drops sharply. With the sudden drop in the external pressure and the rapid formation of a large number of CO_2 bubbles around the gaseous nuclei, the drink foams up and gushes out.

Nucleation can also occur on the surface of a solid. Try adding a few crystals of table sugar to a freshly poured soft drink. As soon as the sugar granules enter the liquid, they provide a surface for nucleation and result in a rush of gas bubbles.

The reverse process, dissolving a gas, requires a high pressure or a large surface area, preferably both. At atmospheric pressure a large area of contact between the gas and the liquid helps speed the flow of the gas into solution. Like other animals, fish need oxygen for life; they depend on the atmospheric oxygen that dissolves at a partial pressure of 159 mm-Hg (for dry air; Table 12.3) through the surface of the waters of lakes, streams, oceans, and home aquaria. To provide plenty of oxygen for an aquarium containing several fish, it's usually necessary to pump air into the aquarium water as a stream of small bubbles. The smaller the bubbles the better, because as a sphere shrinks its volume decreases more rapidly than its surface area. Plenty of small air bubbles present a larger surface area, per volume of gas, than a few large ones. This high ratio of surface area to the volume of gas permits an effective transfer of the oxygen into the water.

Because of their higher ratio of surface area to volume, small bubbles of gas transfer their oxygen to water more effectively than large bubbles.

■ QUESTION

Aside from temperature, what is one factor that determines *how much* N_2 can dissolve in a given quantity of water? What is one factor that determines *how fast* the N_2 dissolves?

12.13 The Art and Science of Breathing

The simple act of breathing provides us with the most common and the most important application of the gas laws to our everyday lives. We breathe simply to exchange the oxygen of the air for the carbon dioxide produced by the body's cells as they metabolize macronutrients (Sec. 8.12). As an illustration, the cellular oxidation of glucose produces carbon dioxide, water, and energy:

$$\underset{\text{glucose}}{C_6H_{12}O_6} + \underset{\text{oxygen}}{6\,O_2} \longrightarrow \underset{\text{water}}{6\,H_2O} + \underset{\text{carbon dioxide}}{6\,CO_2} + \text{energy}$$

Our cells use or store the energy that's generated, and the water becomes part of our general physical inventory. But the carbon dioxide, a waste product, has to be transported by the blood from the cells to the lungs so that it can be eliminated in exhaled breath.

In a normal, healthy adult, the blood reaches the lungs carrying its cargo of CO_2 at a concentration equivalent to what we'd find in a solution that's under a partial pressure of 45 mm-Hg of carbon dioxide at normal body temperature. Henry's Law tells us that at any specific temperature the concentration of a gas in a solution is directly related to its partial pressure above the fluid. Knowing this we can use P_{CO_2} = 45 mm-Hg (or simply "45 mm-Hg") to define the concentration of carbon dioxide in the blood reaching the lungs. (Here, P_{CO_2} represents the partial pressure of CO_2; Sec. 12.11.)

This blood coming into the lungs through the blood vessels also carries residual, unused oxygen at a concentration of 40 to 45 mm-Hg. Within the lungs, the blood—carrying its dissolved gases—passes through the capillaries of the *alveoli*, the very small sacs that terminate the bronchial network. We'll leave the blood here in the alveolar capillaries for a moment as we examine the gases that fill the alveolar sacs themselves.

A tank of compressed gas furnishes oxygen at a partial pressure high enough to allow normal breathing underwater.

As we inhale, the pressure inside our expanding lungs drops slightly and the outside air, traveling from a region of higher external atmospheric pressure to a slightly lower internal pressure, enters the lungs. Once inside, the air's oxygen moves continuously to regions of ever lower partial pressure until it comes, finally, to the interior of the alveoli. Here the gaseous oxygen's partial pressure is at its lowest, about 100 mm-Hg. Within the sac itself, and still moving from a region of higher partial pressure to one of lower pressure, the oxygen passes through the thin membrane of the capillary wall and enters the fluid (blood) that's moving within the blood vessel (Fig. 12.17). With the passage of oxygen across the wall and into the capillary, the concentration of oxygen within the blood increases from the 40 to 45 mm-Hg of venous blood to arterial blood's 100 mm-Hg.

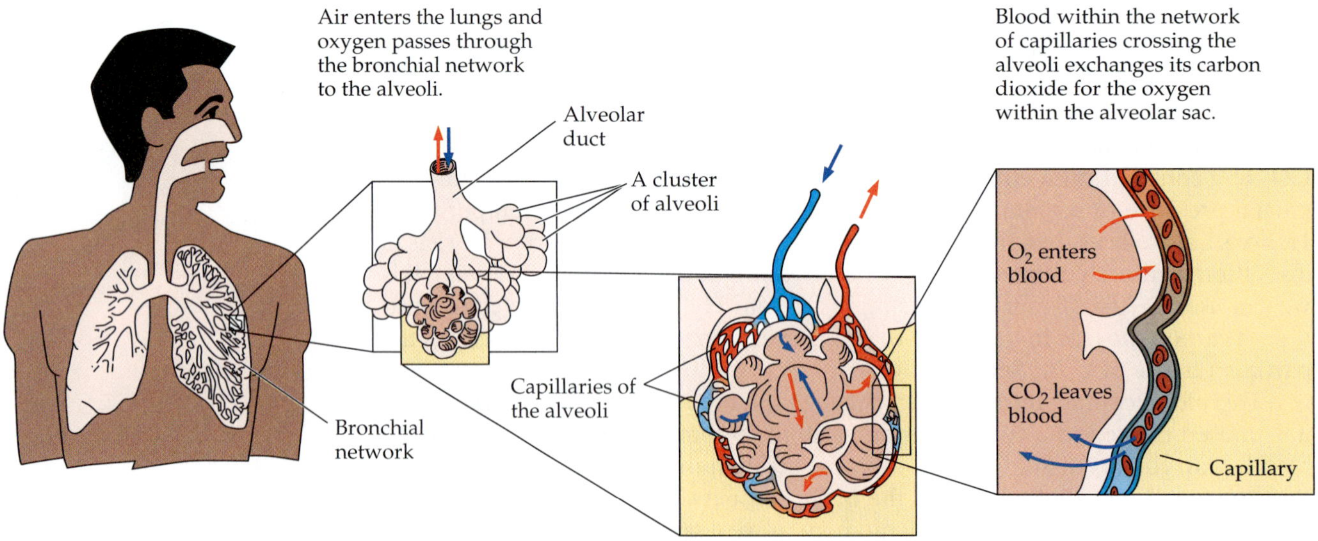

Figure 12.17 The lungs and the alveoli.

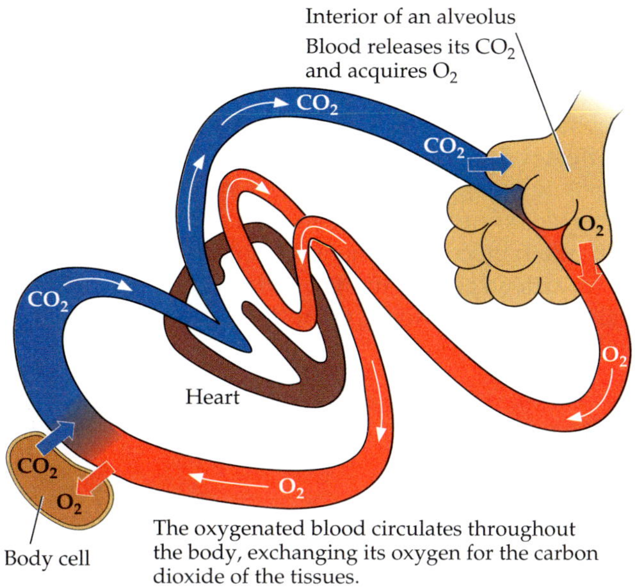

Figure 12.18 The movement of oxygen and carbon dioxide within the body.

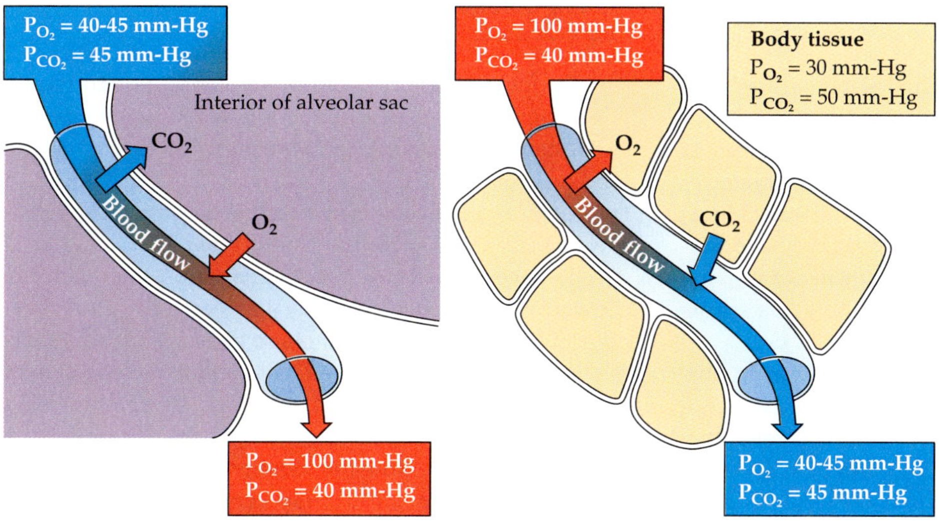

Figure 12.19 Partial pressures and gas flows to and from the blood. Both oxygen and carbon dioxide flow from regions of high partial pressure to regions of low partial pressure.

Meanwhile, the dissolved carbon dioxide travels in the opposite direction. With partial pressures of 45 mm-Hg in the blood and only a few tenths of a mm-Hg within the alveolar sac, the carbon dioxide passes through the membranes and into the alveoli to become part of the exhaled breath. The blood itself, now enriched with oxygen and diminished in carbon dioxide, leaves the lungs for its journey through the arteries to the cells, where it finds partial pressures of 30 mm-Hg or less for O_2 and at least 50 mm-Hg for CO_2. The process is reversed as the gases flow again from higher pressures to lower pressures, and the cycle is renewed (Figs. 12.18 and 12.19).

■ **QUESTION**

Two gases constitute 99% of the (dry) air we inhale: nitrogen (78.1%) and oxygen (20.9%). Four gases make up virtually all of the air we exhale: nitrogen (74.9%), oxygen (15.3%), carbon dioxide (3.7%), and another gas that accounts for just over 6% of our exhaled breath. What is this fourth gas that makes up an even larger fraction of our breath than does carbon dioxide, and where does it come from?

12.14 How You Squeezed Air Out of the Bottle

In the opening demonstration we saw that by wrapping your hands around a cold, rigid bottle you can make a coin (resting on the bottle's mouth) pop periodically. Similarly you can produce bubbles from a cold bottle that's inverted in water. Now we can understand that the effect is an application of Charles' Law, which relates the volume of a gas to its absolute temperature (Sec. 12.7). As your hands warm the bottle and the air that's trapped inside, the kinetic-molecular theory takes over. As the temperature increases, the molecules of the gas trapped inside the bottle begin moving about faster and, with no change in the applied (atmospheric) pressure, the gas expands and escapes. This expanding, escaping gas produces the popping of the coin and the bubbling in the water.

■ **QUESTION**

What do you think would happen if you inverted a *warm* bottle in the water and placed a towel soaked in *cold* water around it? Explain.

PERSPECTIVE

Pure Science and Applied Science

One of the key words in the preceding section may have passed by unnoticed. It's the word *application*, in the sentence that starts "Now we can understand that the effect is an application of Charles' Law, . . ." It's important because it points out the difference between two forms of science: *pure* and *applied*. Each is important. Without pure science, there would be no applied science; without applied science, pure science could become no more than a sterile exercise in the scientific method.

We can think of pure science as science for its own sake, pursued for no other reason than to improve our understanding of the universe. Most of the science described in this chapter is pure science. Boyle, for example, used the newly invented air pump simply to study the effect of pressure on the volume of a gas; he didn't put it to any particular, practical use. Charles, too, was pursuing knowledge for its own sake in his development of Charles' Law. In their studies, Boyle, Charles, and the others were studying some part of the universe without regard to the practical applications of their studies.

Torricelli's invention of the barometer illustrates applied science. As we saw in Section 12.4, Torricelli used his understanding of atmospheric pressure to devise and build an instrument to measure it. Here we see applied science at work as Torricelli used his knowledge of the atmosphere to produce a practical scientific instrument.

In a even broader sense, the gas laws—derived from the operation of pure science—have given us the knowledge of how the gases of our blood behave (Sec. 12.13) and the intellectual tools for the development of medical techniques for the use of gaseous anesthetics and for the diagnosis and treatment of diseases of the lungs and the circulatory system. Here the most practical aspects of applied medicine depend for their very existence on the investigations of Boyle, Charles, and the others, all in pursuit of pure science.

In medicine and many other areas, without the knowledge and understanding that only pure science can provide, the products of applied science could not exist. ■

EXERCISES

For Review

1. Following are a statement containing a number of blanks, and a list of words and phrases. The number of words equals the number of blanks within the statement, and all but two of the words fit correctly into these blanks. Fill in the blanks of the statement with those words that do fit, then complete the statement by filling in the two remaining blanks with correct words (not in the list) in place of the two words that don't fit.

A __________ is a substance that holds its shape and volume, regardless of the shape and volume of its container; a __________ has its own volume but takes the shape of its container; a __________ takes both the volume and shape of its container. A solid melts when the __________ of its chemical particles increases beyond the cohesive forces that hold them firmly next to one another, in a specific shape. The temperature at which a solid becomes a liquid is known as its __________. As the temperature of the liquid continues to increase and its particles acquire even more energy, they are able to break away entirely from each other and enter the gas phase. The temperature at which this conversion from a liquid to a gas occurs throughout the entire liquid is known as the liquid's __________. Even below this temperature, a certain fraction of the chemical particles have enough energy to escape from the liquid to the gas phase, at the liquid's surface, producing __________.

boiling point	melting point
crystal lattice	solid
gas	sublimation
liquid	

2. Complete the following statement using the same conditions as described for Exercise 1.

According to the __________, an __________ consists of point-sized particles, with no __________ whatsoever, that bounce off each other in perfectly __________ collisions, without losing any __________ in the process. The __________ of a gas enclosed in a container results from the collisions of the particles with the walls of the container. If we increase the temperature of an ideal gas held in a container of constant volume, the pressure __________ and always remains directly proportional to the __________. If we increase the pressure on a gas held at __________, the volume of the gas __________ proportionately.

absolute temperature	increases
absolute zero	kinetic-molecular theory of gases
constant temperature	pressure
energy	specific quantity of a gas
ideal gas	volume

3. Explain, describe, or define each of the following:

absolute temperature	barometer
alveoli	mm-Hg
atmospheric pressure	P_{CO_2}

4. Match each of the names with the contributions described.

______ a. Amedeo Avogadro
______ b. Robert Boyle
______ c. J. A. C. Charles
______ d. John Dalton
______ e. Joseph L. Gay-Lussac
______ f. William Henry
______ g. William Thomson, Lord Kelvin
______ h. Evangelista Torricelli

1. With the temperature and the number of moles of a quantity of gas held constant, the volume of the gas varies inversely with its pressure, or $P \times V = k$.
2. The total pressure of a mixture of gases equals the sum of the partial pressures of each of the gases in the mixture.
3. Equal volumes of gases at the same temperature and pressure contain equal numbers of atoms or molecules.
4. Invented the barometer.
5. The quantity of a gas that dissolves in a liquid depends directly on the pressure of that gas above the liquid.
6. Devised the absolute temperature scale.
7. When gases react with each other at constant temperature and pressure, they combine in volumes that are related to each other as ratios of small, whole numbers.
8. With the pressure and the number of moles of a quantity of gas held constant, the volume of the gas varies directly with its temperature, or $V = k \times T$.

5. Describe how:

a. The volume of a gas varies with changes in pressure if the temperature is kept constant.
b. The volume of a gas varies with changes in temperature if the pressure is kept constant.
c. The pressure of a gas varies with changes in temperature if the volume is kept constant.

6. What determines the energy of the molecules of an ideal gas?

7. Describe two ways in which the chemical particles that make up a real gas differ from those of an ideal gas.

8. What gas makes up the largest percentage of the air we inhale?

9. Explain why shaking a bottle of soda causes it to fizz excessively when you open it.

10. What are the three kinds of kinetic energy a molecule can exhibit? Which one of these three resembles a ballet dancer twirling about in one spot on the stage? Which one resembles the energy of a sprinting runner? Which one is analogous to the energy you expend while your chest expands and contracts as you breathe, yet you remain standing in one place?

11. Rank the molecules of water, ice, and steam in order of the amount of kinetic energy each possesses. Start with the substance whose molecules have the highest kinetic energy.

12. Describe, in terms of temperature, kinetic energy, and cohesive forces, what happens as steam is cooled to water and the water is cooled to ice.

13. Explain the difference between pure science and applied science and give an example of each.

A Little Arithmetic and Other Quantitative Puzzles

14. You have a certain volume of gas kept at atmospheric pressure. You increase the temperature of the gas by 100°C and find that its volume doubles. What was the original temperature of the gas?

15. At a certain temperature and pressure, 1 L of diatomic hydrogen gas, H_2, weighs 0.1 g. What would 1 L of helium weigh under those same conditions?

16. If, at a certain temperature and pressure, one mole of diatomic hydrogen molecules, H_2, occupies a volume of 20 L, what would be the volume of one mole of hydrogen *atoms* under those same conditions?

17. Suppose you fill a cold tire to 32 psi (lb/in.^{2}) 17°C and run it until it reaches a temperature of 37°C. What's the new pressure? (Assume that the volume of the tire doesn't change.)

18. You have a balloon that contains 2 L of hydrogen and 1 L of oxygen for a total of 3 L of gas. If the hydrogen and oxygen combine within the balloon to produce gaseous water (without bursting the balloon), will the balloon grow larger or smaller as a result? Remember, when the water forms there's no hydrogen or oxygen left.

19. Suppose that the smallest decrease in the volume of a balloon that you could see easily is 20%. (That is, the balloon would have to shrink to 80% of its original volume for the decrease in volume to be visible.) To what *Celsius* temperature would you have to cool a balloon that was filled at 27°C, so that you could actually see it contract? Assume that atmospheric pressure remains constant.

20. A mountain climber fills three 1-L balloons with air at 760 mm-Hg and 27°C and carries each one to the top of Mt. Everest, the highest mountain in the world, Mt. McKinley, the highest mountain in the United States, and Ben Nevis, the highest mountain in the British Isles. Given the data below, calculate the volume of the balloon at the top of each of these mountains.

	Mt. Everest	Mt. McKinley	Ben Nevis
Height			
m	8848	6194	1343
ft	29,028	20,320	4,406
Average atmospheric pressure, mm-Hg	221	345	647
Average temperature, °C	−42	−25	+7

For each peak, is the drop in pressure or the drop in temperature the dominant factor in determining the final volume of the balloon?

21. Now let's take a 1-L balloon down to the bottom of Death Valley, California. It's the lowest point in the United States, 86 m below sea level, with a mean atmospheric pressure close to 770 mm-Hg. The hottest temperature ever recorded in the United States was in Death Valley: 57°C (135°F) on July 10, 1913. Suppose we take our balloon, filled at sea level on a pleasant day (760 mm-Hg and 27°C), and carry it to the depths of Death Valley on a day that equals the record for temperature (770 mm-Hg and 57°C). Will it expand or will it contract, and what will the new volume be?

Think, Speculate, Reflect, and Ponder

22. Avogadro's Law tells us that the more air we put into a balloon, the greater the number of molecules of gas there are within the balloon. This leads to an increase in the balloon's size and we might expect it to lead to an increase in the pressure of the gas as well. Yet it usually takes much more effort to start inflating a small party balloon than to continue once it's been blown up a bit. Why does the needed pressure *decrease*, at first, as more air enters the balloon? (*Note:* The answer to this question has nothing to do with the gas laws. This question is included to illustrate an important point: Our everyday world displays many different, interwoven phenomena. The gas laws we have examined describe only one set of these phenomena. All the rules of the physical world, including the laws governing the behavior of stretching rubber and those of the gases, operate together to affect the way ordinary things act.)

23. If a helium balloon breaks loose, it rises into the atmosphere and at some point it bursts. Explain why.

24. Every known element and compound can be obtained in a solid or a liquid form. Explain why this demonstrates that no real gas behaves exactly like an ideal gas.

25. If a gas behaved *exactly* like an ideal gas, at what temperature would it liquefy? Explain.

26. Explain why there are no *ideal liquid laws* or *ideal solid laws*.

27. Which one of the gases in Table 12.2 behaves most like an ideal gas? Which one behaves least like an ideal gas? What can you conclude about the forces of attraction between the molecules of each of these two gases?

28. Explain why the partial pressure of a gas is an effective measure of the concentration of a gas in a solution.

29. It's sometimes necessary to carry out laboratory experiments in liquids that are completely free of dissolved gases. What simple technique should ordinarily free a liquid of all dissolved gases?

30. Henry's Law applies at constant temperature. If we vary temperature but keep pressure constant, we find that the solubility of any gas in a liquid drops with increasing temperature. What is the chemical composition of the bubbles that form in a heated pot of tap water long before it begins to boil?

31. Gases move from regions of high pressure to regions of low pressure. What force keeps the gases of the earth's atmosphere from diffusing into interplanetary space, where the pressure is essentially zero?

32. As we inflate a bicycle tire, what shape and volume does the air take as it enters the tire? When a bicycle tire goes flat and the air that was inside it escapes into the open, what shape and volume does the air take as it leaves the tire?

33. Why is the air in jet aircraft flying at high altitudes pressurized?

34. Why are many small bubbles more effective than a few very large bubbles in maintaining a supply of dissolved air in a fish tank?

35. Which would you expect to produce more bubbles faster in a carbonated drink, a little granulated sugar or the same weight of powdered sugar? Why?

36. Why is it necessary for mountain climbers to carry a supply of oxygen and an oxygen mask with them on their ascent to a high peak?

37. Given the following statements, use Charles' Law to explain the origin of thunder:

a. Lightning produces extremely high temperatures and has been responsible for starting fires during thunderstorms.

b. Explosions result from the extremely rapid expansion of gases.

Web Questions

Use the Web to answer or accomplish each of the following. Report your Web source for each response.

1. Summarize the life of Evangelista Torricelli, the inventor of the barometer, with emphasis on his contributions to science.

2. What are the normal temperature ranges of the four regions of the earth's atmosphere? How are these temperature variations explained?

3. How does the total pressure of the gas or vapor above the surface of a liquid affect the liquid's boiling point? Use this relationship between pressure and boiling point to explain the operation of the kitchen utensil known as a *pressure cooker*.

4. Who was James Dewar and what contributions did he make to the topics of this chapter?

5. Summarize the history of the condensation of helium to a liquid. Who was the first person to accomplish this? When was it done?

6. What causes the physical condition known as *the bends*? How do divers avoid this condition? How is it treated when it does occur?

7. Explain what the *critical temperature* of a substance is. What effect does the critical temperature have: (a) on the conversion of a gas to a liquid? (b) on the conversion of a liquid to a gas?

Surfactants

SOAPS AND DETERGENTS: CLEANING UP WITH CHEMISTRY

Surface tension supports an insect on water.

With Nerves as Steady as a Chemical Bond

Here's another piece of chemical magic. Unlike the cabbage breath test of Chapter 10, this demonstration requires a little advance preparation. You begin by "floating" a thumbtack on the surface of a glass of cold water. We know, of course, that thumbtacks don't actually float on water, but there's an old parlor stunt in which you place a thumbtack, or some other small metal object like a needle or a paper clip, *very gently* on the surface of a glass of water. Instead of sinking, the metal seems to float. Actually it isn't floating at all; it's resting on the water's surface, supported by a phenomenon known as *surface tension*. We'll have more to say about surface tension later, but for now the key to this feat lies in placing the thumbtack *gently* onto the surface. If you have trouble with this step you can use a loop of fine wire shaped as shown in Figure 13.1.

Now challenge someone to duplicate what you're about to do next. Announce that you are about to poke the end of a toothpick into the water so very carefully—with nerves as steady as a chemical bond—that the thumbtack will remain undisturbed on the surface. The challenge is to push the toothpick into the water repeatedly, but so gently that the tack doesn't plummet to the bottom.

Push the toothpick ever so gently through the surface about midway between the tack and the edge of the glass. With a little practice you'll be able to do this easily without sinking the tack. A cylindrical toothpick, tapered to a point at each end, is better for this stunt than one of those flat toothpicks, broad at one end and narrow at the other.

Now give the toothpick to your friend. With the first stab into the water, no matter how gentle, the tack drops to the bottom of the glass. The secret of the stunt lies in a twist of chemistry. Better yet, it lies in a twist of the toothpick.

To make the magic work, secretly coat one end of the toothpick with a drop of a colorless liquid detergent just before you begin the proceedings. *You* poke the *dry* end of the toothpick into the water, but as you hand it to your challenger you invert the toothpick (unobserved) so that your friend sticks the detergent-coated end into the water.

Rotating the toothpick as you pass it over takes a little sleight of hand and a bit of practice, but it's really the chemistry that does the job. Just as soon as the detergent touches the water, the tack drops sharply to the bottom of the glass. Figure 13.2 describes it all. (Using a toothpick whose ends have identical shapes decreases very slightly the likelihood of discovery. That's why the symmetrical toothpicks are a bit better for this stunt.)

In this chapter we'll examine surface tension and its origin. We'll see how detergents affect surface tension and how they help us clean clothes and other household goods. We'll look at some of the similarities and differences between soaps and detergents, and we'll see why synthetic detergents are more efficient than soaps in regions with hard water. In doing all this, we'll use the examination of soaps and detergents to reveal some of the extraordinary chemistry brought to us by a box of ordinary detergent.

Figure 13.1 "Floating" a tack on water.

Figure 13.2 Place a tack on the surface of a glass of water. The tack remains on the water's surface as you gently push a toothpick into the water. The tack drops as soon as someone else pushes the toothpick through the water's surface, no matter how gently. The secret lies in the chemistry of liquids, surfaces, and detergents.

13.1 Density, and How Insects Walk on Water

The real magic of the opening demonstration lies in the chemistry of surface tension and of soaps and detergents. If you've ever seen an insect skimming across the surface of a puddle or pond, you've seen surface tension in action. To understand the effects of surface tension—how it allows insects to walk over water and tacks to rest on its surface—we'll look at one of the fundamental properties of matter: *density*.

The **density** of a substance is its mass per unit volume. Density is stated in units of grams per milliliter or grams per cubic centimeter.

The **density** of any substance is simply its mass per unit of volume. In scientific work, the accepted units of density are grams per milliliter (g/mL) and grams per cubic centimeter (g/cm^3). Since a milliliter is the same as a cubic centimeter, the two measures of density are equivalent and interchangeable. We calculate densities by the formula

$$\text{density} = \frac{\text{mass}}{\text{volume}}$$

(An easy way to determine the volume of an object is to immerse it in water and determine the volume of water it displaces. Naturally, if the object can be damaged by water some other technique must be used.)

EXAMPLE

Statuesque Density

A small statue has a mass of 99 g and a volume of 45 cm^3. What is the density of the statue?

■ To find density, we divide mass by volume. In this case

$$\text{density} = \frac{\text{mass}}{\text{volume}}$$

$$\text{density} = \frac{99\text{ g}}{45\text{ cm}^3} = 2.2\text{ g/cm}^3$$

The density of the statue is 2.2 g/cm^3.

Be sure that you recognize the difference between the *weight* of an object and its *density*. We sometimes think of lead, for example, as a heavy metal. But what we really mean is that it's dense. A single lead fishing sinker weighs only a few grams and is small enough to be held in your hand. A huge tree, though, can be heavy enough to crush a house if the tree falls in a storm. While the single tree is certainly much heavier than the sinker, the tree is nonetheless far less dense than the sinker. The key here is that 1 cm^3 of the sinker (lead) is heavier than the same volume—1 cm^3—of the tree (wood). Lead is more dense than wood, but small things made of lead can weigh less than large things made of wood. Densities of some typical solid substances appear in Table 13.1.

Table 13.1 Average Densities of Typical Solids

Substance	Density (g/mL or g/cm^3)
Aluminum	2.7
Cork	0.2
Diamond	3.5
Glass (common)	2.6
Gold	19.3
Ice	0.9
Lead	11.4
Pyrite (a worthless mineral called "fool's gold" and easily mistaken for gold)	5.0
Quartz	2.6
Sodium chloride (as rock salt)	2.2
Sucrose	1.6
Wax (candles, paraffin)	0.9
Wood (average values)	
Balsa	0.1
Birch	0.6
Maple	0.7
Oak	0.8
Teak	0.9

Our standard for density is water, which has a density of 1.0 g/mL. Anything that's less dense than water—wood for example—floats; anything that's more dense—lead—sinks. That's why trees float in water but lead weights sink. Drop a nickel into a glass of water or a concrete block into Lake Erie and it sinks to the bottom. Metals and concrete are more dense than water—they have a greater mass than the volume of water they displace—so they sink. Anything with a smaller mass than its equivalent volume of water floats. Fats and oils float because they're intrinsically less dense than water. Boats float because the air trapped inside their hulls gives them an average density less than water's. Empty buckets float for the same reason. Some brands of bar soap, Ivory for example, float because of an enormous number of microscopic bubbles of air whipped into them as they're manufactured. Ducks float because a film of oil on the surface of their feathers traps plenty of air next to their bodies (Fig. 13.3).

Figure 13.3 Ducks float on water, supported by the buoyancy of air trapped by their feathers. Water's surface tension prevents it from penetrating the oily layer of their feathers and displacing the air.

But the tack you've been using *doesn't* float. The metal tack is denser than water and there's no trapped air to give it buoyancy. The tack is supported in a small dent on the water's surface, held up by the same phenomenon that supports an insect's feet as it walks across a pond: *surface tension*.

■ **QUESTION**

Which is heavier, a block of aluminum with a volume of 10 cm^3 or a block of lead with a volume of 2 cm^3? Which is more dense, the block of aluminum or the block of lead?

13.2 Surface Tension

Surface tension supports both the insect and the tack, and keeps them from dropping into the water even though they're both more dense than water, because water's surface—the surface of any liquid for that matter—behaves a little differently from the bulk of its interior. As we saw in Sections 12.2 and 12.3, virtually all the water molecules of any sample of water that's well below its boiling point exert relatively large forces of attraction on all their neighboring molecules—those to their sides, in front and behind, and above and below. Throughout the bulk of the liquid the total force of attraction exerted by any one water molecule on all its neighbors is dispersed spherically, in all directions.

At the water's surface, though, things are different. Above the surface there exist only occasional molecules of the gases of the atmosphere and the very few water molecules that have escaped by evaporation. The only nearby water molecules are below and to the sides of those at the surface. At the surface, then, the forces of attraction become focused toward the sides and downward. This particularly strong attraction of the surface molecules for each other and for the molecules immediately below them results in the cohesion of the surface that we call surface tension and that keeps bugs and tacks and other dense but lightweight things from sinking (Fig. 13.4). It's the same phenomenon that allows you to fill a glass to its brim and then add a little more water so that the surface of the water bulges up very slightly above the rim of the glass.

As your friend pokes the toothpick into the water, the small bit of detergent on its tip lowers the surface tension sharply and the tack drops to the bottom. Now we'll examine the connection between the detergent's molecular structure and its effect on surface tension.

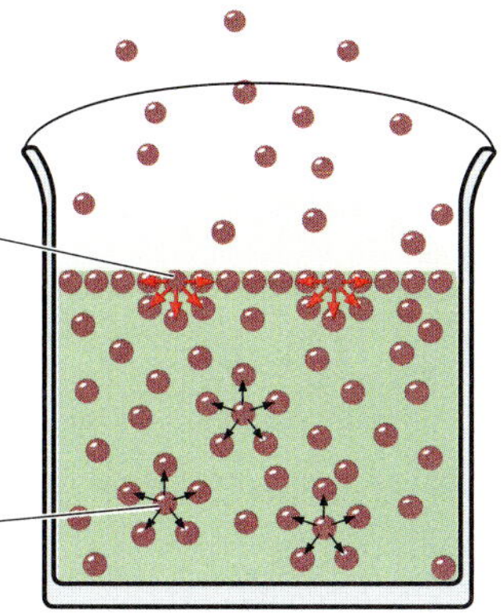

Figure 13.4 The origin of surface tension.

■ QUESTION

Describe a simple way to lower water's surface tension.

13.3 Soap, a Detergent

Detergent comes from the Latin *dētergēre*, meaning to wipe off or to clean. A detergent is anything that cleans, especially if it removes oily or greasy dirt. To see how detergents work, we'll look first at one particular kind of detergent, a soap.

Each time you use a bar of soap you go back a long way, chemically, into an uncertain history. There's reason to believe the Babylonians knew how to make soap almost 5000 years ago. While the Phoenicians and ancient Egyptians may have manufactured it, too, most historians give the Romans the real credit for discovering soap, or at least for writing down the secrets of its preparation and passing them along. The Romans knew that heating goat fat with extracts of wood ashes, which contain bases, produces soap. They also used lye (sodium hydroxide, NaOH), a stronger base than the ash extracts and more effective in converting fat into soap. The word *lye* itself is connected to soap and soapmaking through an intricate linguistic path that includes a long list of words from Latin, Greek, Old English, Old Irish, and other languages, with meanings such as lather, wash, bathe, and even ashes.

Knowledge of soapmaking dimmed for a while after Rome fell, then traveled from Italy to Germany, France, Spain, and then, in the 14th century, to England. The manufacture of soap came to America in 1608 with the arrival of Polish and German soapmakers in Jamestown, Virginia. Until the rise of commercial soapmaking in the 19th century, the process remained a household art, practiced much as the Romans must have made soap some 20 centuries earlier, or even as the Babylonians may have produced it long before the Romans.

Soaps are detergents in the sense that they help clean oily and greasy dirt from fabrics, metals, our skin and hair, and the like. But soaps make up a very narrow class of detergents. We restrict the term *soap* to the *sodium* (or, less often, *potassium*) *salts of long-chain carboxylic acids*. As we saw in Section 10.8, a carboxylic acid is marked by the presence of a carboxyl group, $-CO_2H$. With the anion of the carboxyl group balanced by a sodium cation and tied by a covalent bond to a long chain of $-CH_2-$ groups that terminate in a CH_3- group, we have a soap molecule. Sodium palmitate, the sodium salt of palmitic acid, is a typical soap. (It's a salt in the same sense that NaCl, KI, $MgSO_4$, and $CaCO_3$ are salts; see Sec. 10.2.)

$$CH_3{-}CH_2{-}CH_2{-}CH_2{-}CH_2{-}CH_2{-}CH_2{-}CH_2{-}CH_2{-}CH_2{-}CH_2{-}CH_2{-}CH_2{-}CH_2{-}CH_2{-}CO_2H$$

or

$$CH_3{-}(CH_2)_{14}{-}CO_2H$$

palmitic acid

$$CH_3{-}(CH_2)_{14}{-}CO_2^{-}Na^{+}$$

sodium palmitate, a soap

With sodium palmitate serving as a specific illustration, we can generalize the molecular structure of a soap molecule as

$$CH_3{-}(CH_2)_n{-}CO_2^{-}Na^{+}$$

where the subscript *n* represents an even number, usually ranging from 8 to 16. (The entire carbon chain, including both the methyl carbon and the carboxylate carbon, runs from 10 to 18 carbons.) With this structure a soap molecule possesses two opposing chemical tendencies.

On the one hand, the long chain of methylene groups that ends in the methyl group

$$CH_3{-}(CH_2)_n{-}$$

resembles quite closely the long chains of the hydrocarbon molecules of Chapter 6. Like the molecules of gasoline and mineral oil, this part of the soap molecule tends to dissolve readily in materials that are or that resemble hydrocarbons, but not in water. All these long chains of $-CH_2-$ groups of soaps and of hydrocarbons and hydrocarbon-like materials intermingle easily, but they don't mix readily with the H_2O molecules of water.

The other end of the molecule, though, is ionic:

$$-\overset{\displaystyle O}{\overset{\|}{C}}-O^{-}Na^{+}$$

Like sodium chloride and other ionic compounds, that ionic end tends to dissolve in water, but not in hydrocarbon solvents. As a result we have here one molecule with two opposite and contradictory tendencies. Part of the molecule contains a *hydrophilic* structure; the ionic end is attracted toward water molecules but shuns hydrocarbons and other oily and greasy substances. The other part, the long hydrocarbon chain, is a *hydrophobic* structure; it shuns water but mixes easily with those very oily, greasy substances that repel the hydrophilic part (Fig. 13.5). With its long chain of $-CH_2-$ groups, this hydrophobic portion of the soap molecule follows the old adage that oil and water don't mix.

(Both *hydrophilic* and *hydrophobic* start with *hydro-*, which is taken from a Greek word for water. The endings *-philic* and *-phobic* come from Greek words meaning loving and fearing, respectively. In this sense, the hydrophilic end of the soap molecules loves or seeks out water; the hydrophobic end fears or shuns water.)

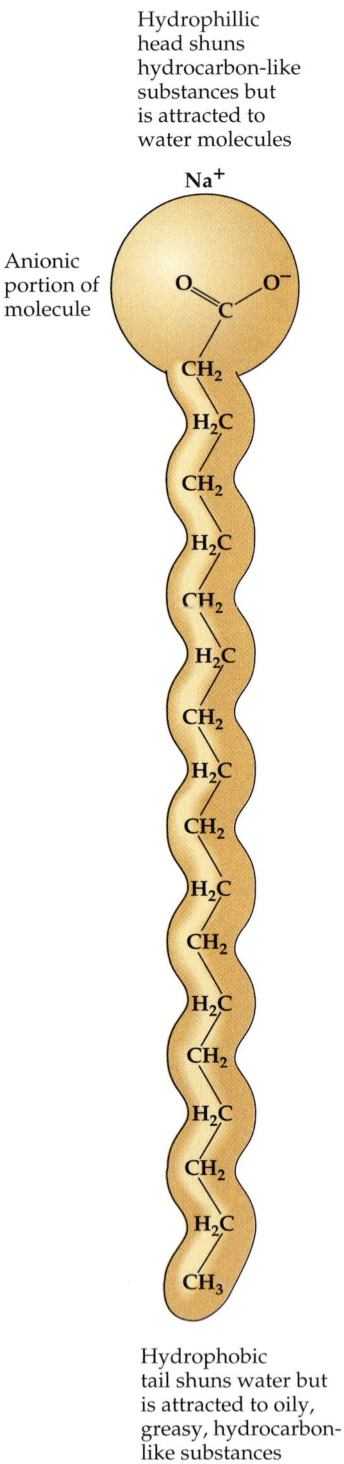

Figure 13.5 A typical soap molecule.

■ **QUESTION**

What structure forms the hydrophobic portion of a soap molecule? What structure forms the hydrophilic portion?

13.4 Surfactants

All detergent molecules, like those of soaps, consist of a hydrophilic portion and a hydrophobic portion. When they enter water, detergent molecules head for the single location where both tendencies can be accommodated: the surface. There the hydrophilic end of the molecule becomes comfortably embedded among the water molecules that make up the surface while the hydrophobic tail sticks up, away from the water molecules. As shown in Figure 13.6, the detergent molecules become interspersed among the molecules at the water's surface. In disrupting this tightly knit layer of molecules, the detergent interferes with the strong attractive forces that the surface water molecules normally exert on each other and thereby lowers the surface tension. This is why the tack drops when the toothpick introduces the detergent into the water.

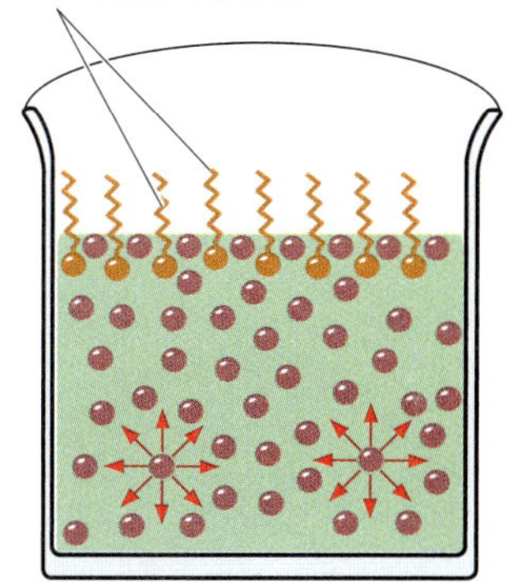

Figure 13.6 Surface effect of detergent molecules.

Soap, detergents, and any other substances that accumulate at surfaces and change their properties sharply, especially by lowering the surface tension, are *surface-active agents* or, more briefly, *surfactants*. Substances other than surfactants can also change surface tension, but not nearly as effectively as surfactants. Even a small amount of a surface-active agent can produce dramatic effects. At a concentration of 0.1%, for example, soap lowers water's surface tension by almost 70%. In contrast, sodium hydroxide, which is not a surfactant and which diffuses homogeneously throughout its water solutions rather than concentrating at the surface, actually raises water's surface tension very slightly. At a 5% concentration, for example, sodium hydroxide increases water's surface tension by about 6%.

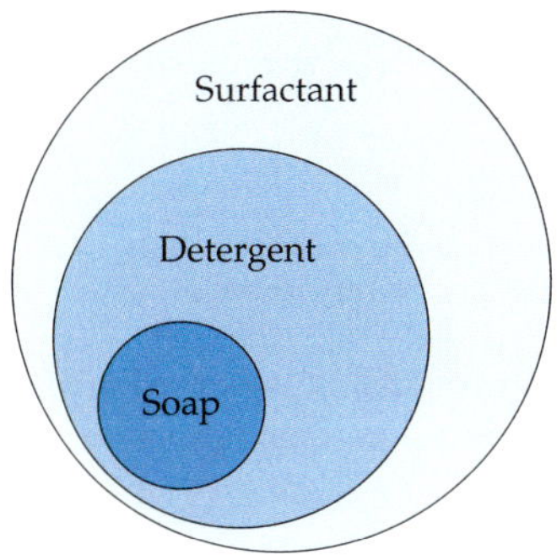

Figure 13.7 All soaps are detergents; all detergents are surfactants.

Figure 13.7 illustrates the relationship among the three terms: *surfactant*, *detergent*, and *soap*. All surfactants act at surfaces. Some surfactants, the detergents, are also good cleaning agents. Some detergents, the soaps, are sodium or potassium salts of long-chain carboxylic acids. Other detergent molecules have different structures, yet all detergents have a typical molecular structure in common: a long, hydrophobic carbon chain that resembles a molecular "tail," which is connected to a hydrophilic "head." In Section 13.11 we'll examine some detergents that aren't soaps.

■ **QUESTION**

What do all surfactants, detergents, and soaps have in common? What do all detergents have in common? Identify a substance that is a surfactant but not a detergent.

13.5 Micelles, Colloids, and John Tyndall

As the water's surface becomes filled with surfactant molecules and even more detergent is added, the additional detergent molecules soon become crowded out of the surface. They begin shielding their hydrophobic tails from water molecules in a different way. They begin clustering into *micelles* within the bulk of the water.

Micelles are submicroscopic globules or spheres of one substance distributed throughout another, usually a liquid. As detergent molecules accumulate into micelles in the interior bulk of the water, their hydrophilic heads form the spheres' surfaces and their hydrophobic tails point inward, shielded from the water molecules in an accommodating environment made up of other, similar, hydrophobic hydrocarbon chains (Fig. 13.8).

Micelles are submicroscopic globules or spheres of one substance distributed throughout another, usually a liquid.

Although these detergent micelles are well dispersed in the water, they aren't actually dissolved. They're present as a *colloid* rather than as the solute of a solution (Sec. 9.5). A colloidal dispersion differs from a true solution largely in the size of the

A **colloid** is made up of particles of one substance dispersed throughout another. The particles of the dispersed substance range in diameter from about 10^{-7} to 10^{-4} cm.

Figure 13.8 Cross section of a spherical detergent micelle.

dispersed particles. In a **colloid** the particles run from about 10^{-7} to 10^{-4} cm in diameter. In contrast, the diameter of a chloride ion is 3.6×10^{-8} cm; for a sodium ion it's barely 2×10^{-8} cm. As we've seen in earlier chapters, sodium chloride dissolves in water to form a true solution of Na^+ cations and Cl^- anions.

There's no sharp, well-defined division between the size of a particle dispersed as a colloid and one that's dissolved in a true solution. As with acids and bases, we have to rely on observable phenomena to determine which we're dealing with. The easiest way to see the difference between a solution and a colloidal dispersion is to shine a light through each of them. Shining a flashlight beam through a solution of sodium chloride in water doesn't show much. The beam passes through the clear solution without producing any observable effect. But shine the beam through a mixture of soap and water, even one containing very little soap, and you can see the beam's path clearly illuminated, especially if you view the mixture against a dark background. This *Tyndall effect*, produced as the colloidal particles scatter the light to all sides, was discovered by John Tyndall, a British physicist born in Ireland in 1820. It's one of the more easily observed properties of a colloidal dispersion, one that readily distinguishes it from a true solution (Fig. 13.9).

The Tyndall effect isn't limited to detergents. A few drops of milk in a glass of water also show the path of the light. Milk is essentially an aqueous mixture of colloidal fats and proteins along with dissolved lactose (milk sugar) and minerals. You can even see the Tyndall effect outdoors on a foggy night as the fog—a colloidal dispersion of water in air—scatters the headlight beams of automobiles.

With one exception, colloids can form when each of the three states of matter—solids, liquids, and gases (Chapter 12)—disperse in each of the others. Since all gases are infinitely soluble in each other, no gas forms a colloidal dispersion in any other gas. Examples of the various kinds of dispersions appear in Table 13.2.

John Tyndall, discoverer of the Tyndall effect, delivering a popular lecture on science.

Table 13.2 Colloidal Dispersions

	Common Name	Example
Solid dispersed in		
Solid		Ruby glass—glass colored red by a colloidal dispersion of solid gold particles
Liquid		Clay, paint, putty, toothpaste
Gas	Aerosol	Smoke
Liquid dispersed in		
Solid	Gel	Gelatin, jelly
Liquid	Emulsion	Milk, salad dressing
Gas	Aerosol	Fog
Gas dispersed in		
Solid		Popcorn
Liquid	Foam	Whipped cream
Gas		(Does not exist)

Figure 13.9 The Tyndall effect is evident as a beam of light passes through a colloidal dispersion of water droplets in the atmosphere.

■ QUESTION

What's the simplest way to determine whether a liquid mixture represents a solution or a colloidal dispersion?

13.6 How Soap Cleans

Now we can begin to understand why soap acts as a detergent. Soap cleans by:

1. decreasing water's surface tension, making it a better wetting agent;
2. converting greasy and oily dirt into micelles that become dispersed in the soapy water, and
3. keeping the grease micelles in suspension, thereby preventing them from coalescing back to large globules of grease that could be redeposited on a clean surface.

Each of these functions has its basis in the structure of the soap molecule.

Our common sense tells us that water wets whatever it touches, but if we look closely we find that water isn't a particularly effective wetting agent after all. Examine the waxy surface of a well-polished car just after a rain and you'll see the water forming small beads rather than spreading evenly over the surface. Water doesn't wet the waxed surface very well. Look at your clothing or at the top of an umbrella after you've been out in a light drizzle and you'll see the rain forming small beads on the fabric before it penetrates into the cloth. Generally, water simply doesn't do a very good job of wetting most ordinary substances.

Surface tension holds water in the form of small beads on the leaves of the Japanese Root Iris.

In washing, though, water must penetrate well and deeply into the substance we want to clean. As we saw in the opening demonstration, a detergent lowers water's surface tension. In its action as a detergent, then, soap first lowers surface tension so that the water carrying the micelles can get to the dirt (Fig. 13.10).

When the soap micelles reach the embedded dirt, the soap molecules that form these micelles once again find themselves at a surface. This time it's the surface between the water and the grease that makes up (or carries with it) most of the dirt. Now, as the hydrophilic heads of the soap molecules remain surrounded by water molecules, the soap micelles break up and the hydrophobic hydrocarbon tails, which had remained in the interior of the spherical micelles, become embedded in the grease. With this greasy dirt providing as compatible a chemical

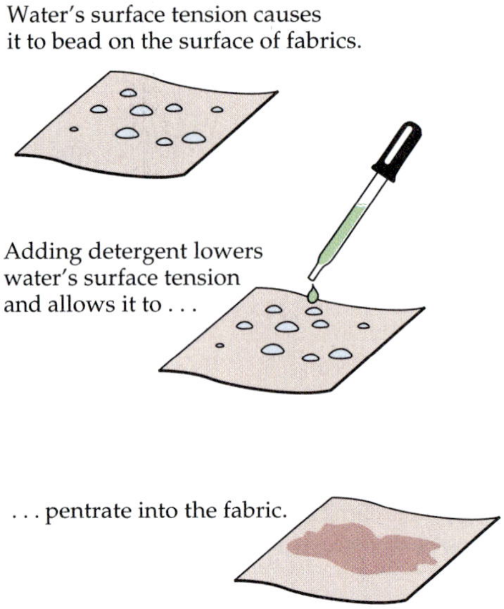

Figure 13.10 Detergents enable water to penetrate rapidly into fabrics.

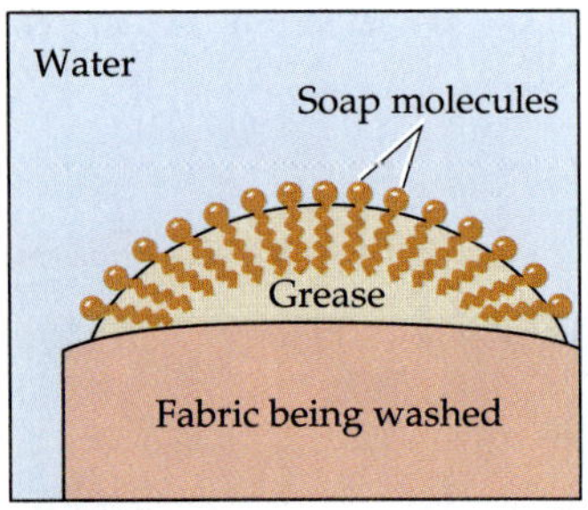

Figure 13.11 Water, grease, and soap.

environment for the hydrophobic tails as the water provides for the hydrophilic heads, the tails are just as much at home in the grease as the heads are in the water. With their heads embedded in the water and their tails in the grease, the soap molecules effectively nail these two phases together (Fig. 13.11).

Agitation now breaks the grease into micelles whose surfaces are covered by the negatively charged carboxylate groups, the hydrophilic $—CO_2^-$ groups of the embedded soap molecules (Fig. 13.12). With a coating of negative electrical charges enveloping the entire surface of each micelle, the grease droplets repel each other and remain suspended in the wash water instead of coalescing and redepositing on the material being cleaned. In the end, the suspended droplets go down the drain with the wash water. (While all this is going on, the sodium ions move about freely and independently in the wash water, just as they did in the electrolyte solution of the light bulb demonstration that began Chapter 1.)

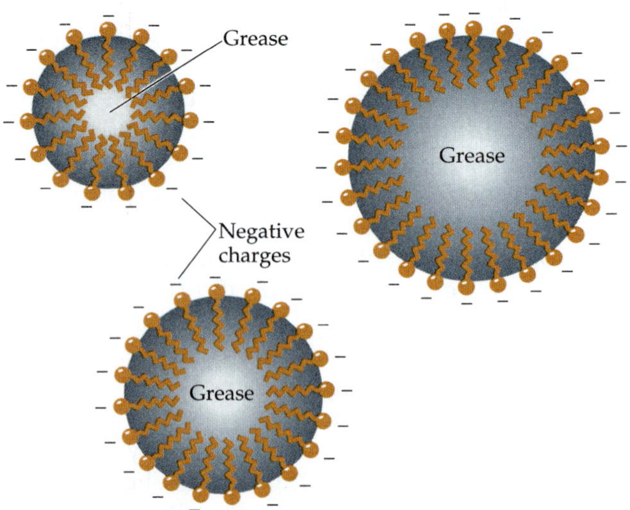

Figure 13.12 Grease micelles with embedded soap molecules.

Since the anionic carboxylate groups of the soap molecules place a cover of negative electrical charge on the micelles' surfaces, soap falls into the class of *anionic detergents*.

■ QUESTION

All surfactants lower surface tension, but only some surfactants are detergents. What functions does a chemical perform in addition to lowering surface tension that allows us to classify it as a detergent?

13.7 Esters and the Manufacture of Soap

However soap is made, whether by boiling goat fat and wood ashes in ancient kettles (Sec. 13.3) or through the modern methods of 20th-century factories, it's all the same chemically. Soap comes from the hydrolysis of naturally occurring fats and oils, which are themselves members of a class of organic compounds known as *esters*. [An ester is a compound whose molecules contain the functional group (Sec. 6.12)

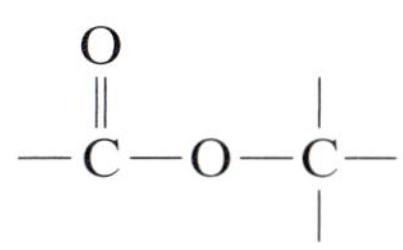

while **hydrolysis** is the decomposition of a substance, or its conversion to other substances, through the action of water.]

Hydrolysis is the decomposition of a substance, or its conversion to other substances, through the action of water.

All naturally occurring fats and oils are esters of fairly complex molecular structures. We'll start instead with a simpler ester, *ethyl acetate*, which is used as a food flavoring and as a common, commercial solvent found in many brands of nail polish and paint remover. (Ethyl acetate doesn't occur in fats, though.)

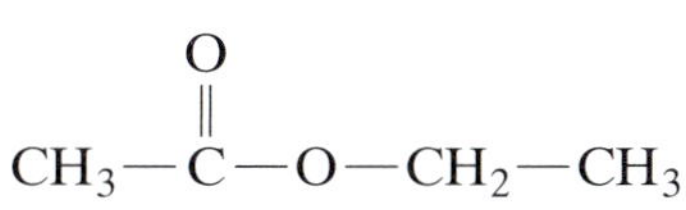

ethyl acetate

Ethyl acetate results from the reaction of acetic acid and ethyl alcohol in the presence of a small, catalytic amount of a strong acid. As the acetic acid and the ethyl alcohol react, the oxygen atom of the alcohol becomes bonded to the carbonyl carbon of the acid (Sec. 10.8), while the H of the alcohol's —OH group and the entire —OH of the acid leave as a molecule of water, a by-product of this *esterification* reaction (Fig. 13.13). Although the carboxylic acid that takes part in this esterification is itself an acid, it's a relatively weak one: The reaction requires the presence of a much stronger acid, such as HCl or H_2SO_4, as a catalyst. Catalysts, which affect the rate of a reaction but are not themselves reagents (Sec. 7.7), are usually written above the reaction arrow along with required conditions, such as heat and pressure.

Figure 13.13 An acid-catalyzed esterification.

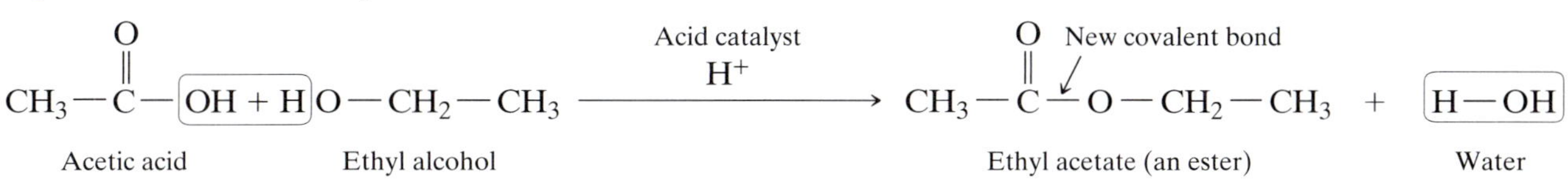

The name *ethyl acetate* reveals that the compound has structural connections to both *ethyl* alcohol and *acetic* acid. The suffix *-ate* tells us we're dealing either with an ester or with a salt of a carboxylic acid, such as *sodium acetate*:

$$CH_3-\overset{\overset{\large O}{\|}}{C}-O^-Na^+$$

sodium acetate

In naming esters we give the name of the alcohol first, then the acid, with the suffix *-ate* replacing the acid's *-ic*. For example, the ester formed by the reaction of isopropyl alcohol (Sec. 6.8) with benzoic acid (Sec. 10.8), in the presence of a strong acid, is *isopropyl benzoate* (Fig. 13.14).

Heating an ester with water in the presence of either an acid or a base produces a hydrolysis, with water adding to the ester in a reversal of the esterification reaction. As shown in Figure 13.15, when the hydrolysis is carried out in the presence of an acid, the reaction becomes precisely the reverse of the esterification. Here an ester reacts with water, forming an alcohol and a carboxylic acid. Carrying out the hydrolysis in the presence of a base, such as sodium hydroxide, produces the salt of the carboxylic acid rather than the acid itself. In this case the carboxylic acid and the base react with each other in a neutralization reaction (Sec. 10.2). In the presence of a base, and with the formation of the salt of the acid rather than the acid itself, the hydrolysis is called a **saponification** (Fig. 13.16), a term that brings us to the chemistry of soapmaking.

A **saponification** is a hydrolysis of an ester carried out in the presence of a base.

Figure 13.14 Isopropyl benzoate.

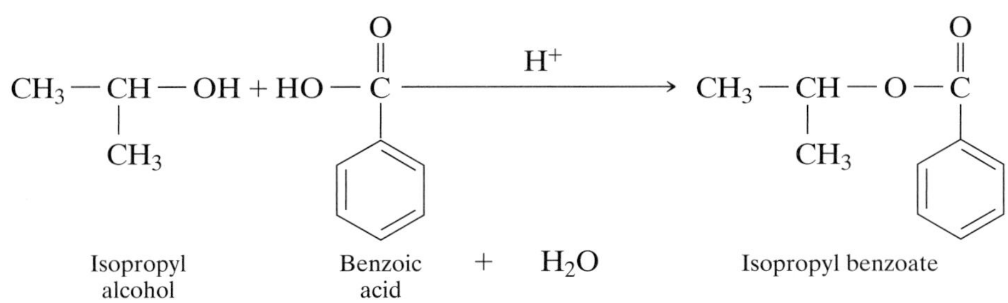

Figure 13.15 An acid-catalyzed hydrolysis. An ester reacts with water, producing an acid and an alcohol.

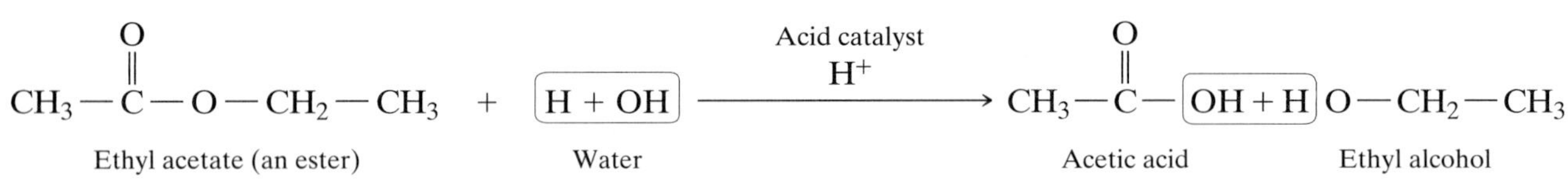

Figure 13.16 A basic hydrolysis; a saponification.

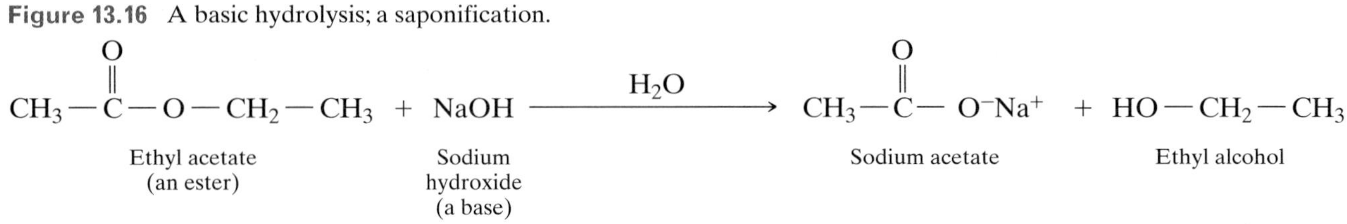

■ **QUESTION**

What is the name of the molecule formed by reversing the orientation of the *ester* functional group of ethyl acetate (switching the positions of the —CO— group and the —O— group)? Section 10.8 contains information that might prove useful.

13.8 From Fats to Soap

Ethyl acetate, as we've seen, is a simple ester formed from an alcohol (ethyl alcohol) that bears a single —OH group. The fats and oils are more complex esters, formed from *glycerol*, a *triol* (or *tri*hydroxy alcoh*ol*) bearing three —OH groups. Glycerol is also known more commonly as *glycerin*, sometimes spelled with an added *e*, *glycerine*. Glycerol is used in many hand and body lotions to soften the skin.

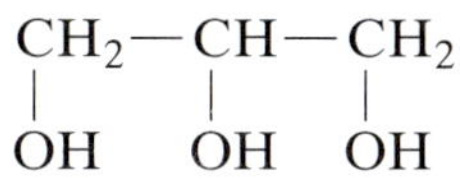

glycerol, also known as glycerin or glycerine

Each of glycerol's three —OH groups constitutes the hydroxyl functional group of an alcohol, so each one can enter into ester formation with a carboxylic acid. The result is a *triester* (Fig. 13.17) in which the three organic groups of the acids (R, R′, and R″ of Fig. 13.17) can be identical or can differ from each other. The generalized triester of Figure 13.17 bears a variety of generic names. Most commonly, it's described as a *glyceride* or a **triglyceride**. While the structures of the three acids that combine with glycerin can vary, almost all of the acids formed by saponifying fats and oils show certain structural similarities. Virtually all of them consist of straight, unbranched chains containing even numbers of carbons. Sodium salts of the **fatty acids** containing from 10 to 18 carbons make the best soaps (Table 13.3). Triglycerides are the principal organic compounds of animal fats and vegetable oils, and fatty acids are the acids they generate through hydrolysis.

Triglycerides are the principal organic compounds of animal fats and vegetable oils. **Fatty acids** are the acids they generate through hydrolysis.

To sum up, then, we can produce a soap by heating a triglyceride—an animal fat or a vegetable oil—in an aqueous solution of sodium hydroxide (Fig. 13.18). Hydrolysis of the triglyceride generates both glycerol, which remains dissolved in the water, and the sodium salts of the various acids that make up the triglyceride. These salts of the fatty acids congeal at the surface as the mixture cools and we scoop them up as soap.

Figure 13.17 A triester of glycerol.

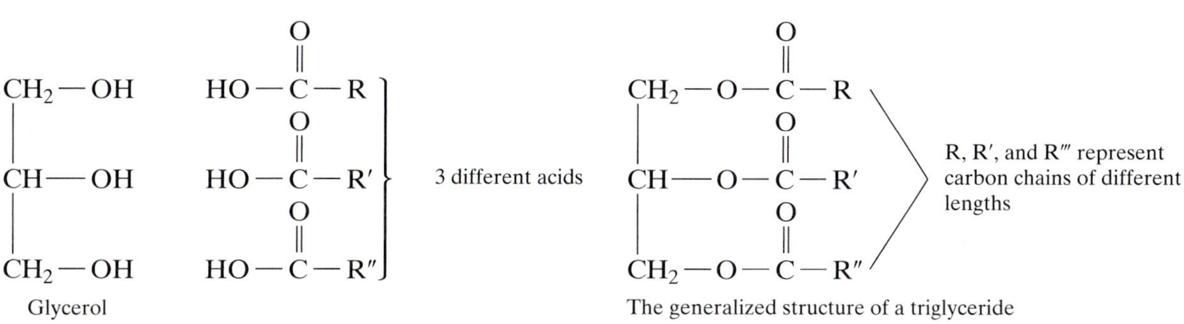

Table 13.3 Common Fatty Acids of Soap

Structure and Name	*n* of $CH_3—(CH_2)_n—CO_2H$
$CH_3—CH_2—CH_2—CH_2—CH_2—CH_2—CH_2—CO_2H$ Caprylic acid	6
$CH_3—CH_2—CH_2—CH_2—CH_2—CH_2—CH_2—CH_2—CH_2—CO_2H$ Capric acid	8
$CH_3—CH_2—CH_2—CH_2—CH_2—CH_2—CH_2—CH_2—CH_2—CH_2—CH_2—CO_2H$ Lauric acid	10
$CH_3—CH_2—CH_2—CH_2—CH_2—CH_2—CH_2—CH_2—CH_2—CH_2—CH_2—CH_2—CH_2—CO_2H$ Myristic acid	12
$CH_3—CH_2—CH_2—CH_2—CH_2—CH_2—CH_2—CH_2—CH_2—CH_2—CH_2—CH_2—CH_2—CH_2—CH_2—CO_2H$ Palmitic acid	14
$CH_3—CH_2—CH_2—CH_2—CH_2—CH_2—CH_2—CH_2—CH_2—CH_2—CH_2—CH_2—CH_2—CH_2—CH_2—CH_2—CH_2—CO_2H$ Stearic acid	16
$CH_3—(CH_2)_7—CH{=}CH—(CH_2)_7—CO_2H$ Oleic acid	—
$CH_3—(CH_2)_4—CH{=}CH—CH_2—CH{=}CH—(CH_2)_7—CO_2H$ Linoleic acid	—

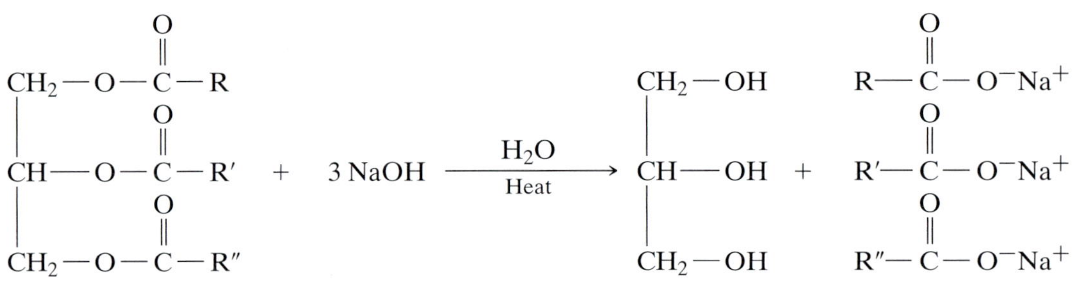

A triglyceride — Glycerol — Soaps

Figure 13.18 The saponification of a triglyceride.

■ QUESTION

What part of their chemical structures is common to all *triglycerides*? How do the chemical structures of the various triglycerides differ from each other?

13.9 A Problem: Hard Water

As we saw in Section 9.11, the drinking water of many regions contains various minerals, dissolved in slightly acidic rainwater as it filters through the soil. Water that's rich in the salts of calcium, magnesium, and/or iron is called *hard water*. (*Soft water*, on the other hand, is virtually free of these minerals.) The mineral cations of hard water combine with fatty acid anions and remove them from water as waxy, insoluble salts. In regions where the water is particularly hard, you can actually see these precipitates deposited as gray rings, known as *curd*, or *soap curd*, around bathtubs and sinks after washing with soap. This curd is made up of the calcium, magnesium, and/or iron salts of the fatty acids of soap:

$$2\,CH_3—(CH_2)_n—CO_2^{-}\,{}^{+}Na + Ca^{2+} \longrightarrow$$

$$CH_3—(CH_2)_n—CO_2^{-}\,Ca^{2+}\,{}^{-}O_2C—(CH_2)_n—CH_3 + 2\,Na^{+}$$

curd, or *calcium soap*
(the ring that forms around the bathtub in regions with hard water)

Hard water wastes soap because much of the soap that would otherwise be used in cleaning is consumed in curd formation as it reacts with the mineral ions of hard water. In hard water, a certain amount of soap has to be used up initially in combining with the mineral cations and removing all of them from solution before additional soap can act effectively as a detergent. What's more, this curd deposits on the surface of laundered clothes, dulling the cloth and giving washed goods a slightly gray appearance.

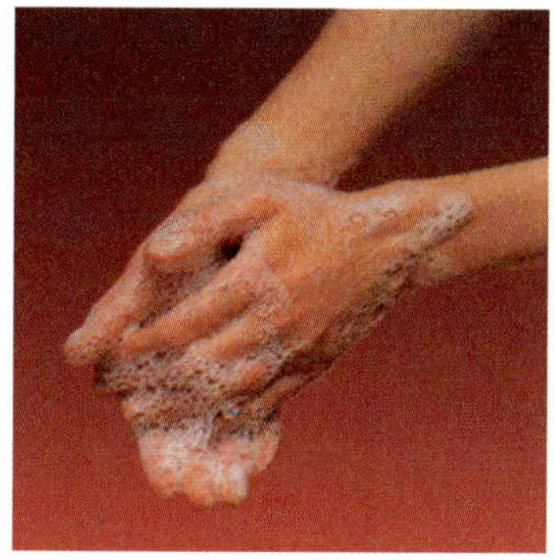

Soap forms rich suds in soft water. In hard water soap is less efficient. The mineral cations of hard water combine with the soap, forming curds and reducing the soap's effectiveness.

You can see the effect of hard water on soap by first shaking up some soap in distilled water. A piece of common bar soap will do nicely for this demonstration, but laundry detergent won't work for reasons we'll come to soon. Distilled water, which you can get in a supermarket or convenience store, is free of the mineral ions of hard water and generates a good deal of suds with the soap. Now add some hard tapwater to the sudsy mixture or, if you live in an area of soft water, add about a tablespoon of milk, which contains considerable calcium. (Skim milk also works well, as does a solution of an antacid that contains calcium.) The suds vanish as you stir the hard water or the milk into the soapy water because the fatty acid anions are converted into their insoluble calcium salts. You'll probably be able to see the curd swirling in the water.

■ QUESTION

The demonstration described in this Section asks you to shake a piece of common bar soap in distilled water, then add some hard tapwater to the mixture. What would you expect to see if you reversed the procedure by shaking the soap first in some hard tapwater, then adding some distilled water to the mixture?

13.10 One Solution: Water Softeners

One solution to the problem caused by hard water is to remove the cations that produce the hardness. Commercial water softeners do this by trading sodium ions for the water-hardening cations—the calcium, magnesium, and iron ions—and thereby converting hard water to soft water.

The typical water softener installed in a home consists of a tank containing a specially prepared polymer, a storage bin for sodium chloride, and regulators to control the flow of water (Fig. 13.19). (Polymers are materials made of extraordinarily large molecules; Chapter 21). The polymers used in these water softeners have large numbers of anionic functional groups, which allow the polymeric molecules to hold large numbers of cations. You activate or *charge* the water softener by running concentrated salt water through it. The sodium ions of the salt water displace whatever cations might be on the polymer and stick to the polymer in the tank. Then, when hard water passes through the tank, the cations of the hard water take their turn at displacing the sodium cations from the polymer.

With the exchange of calcium, magnesium, and iron ions of the hard water for the sodium ions held by the polymer, the water passing through becomes softened and the polymer eventually becomes saturated with the ions of the hard water.

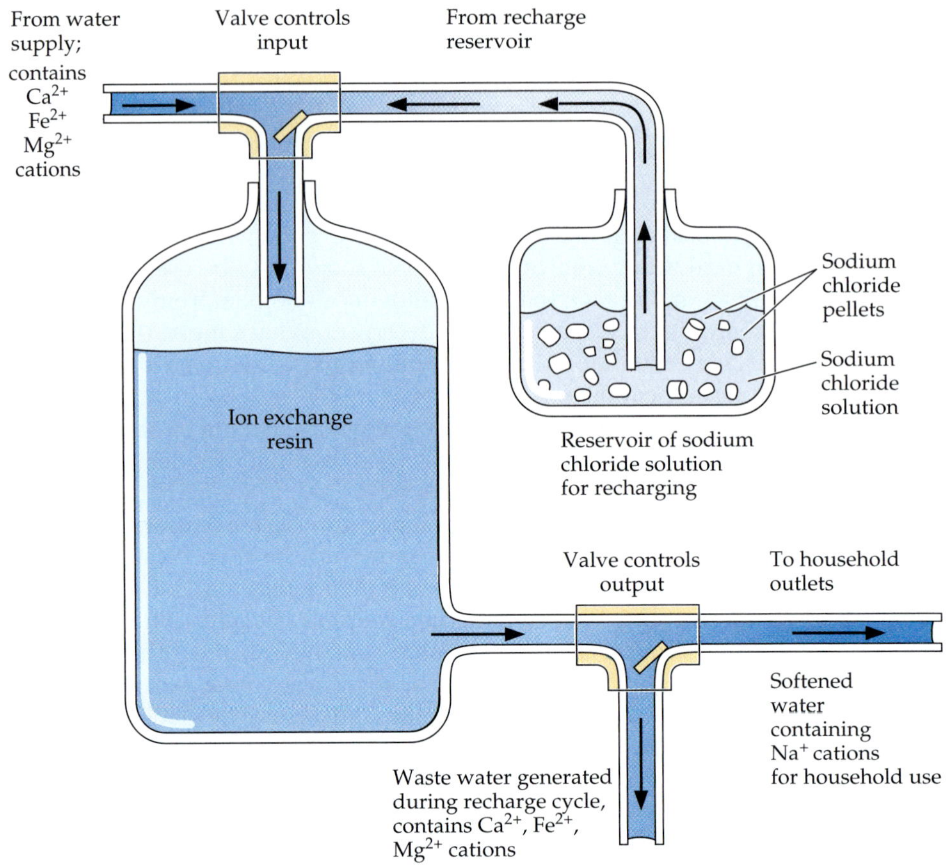

Figure 13.19 A home water softener.

Because of this gradual accumulation of the ions of the hard water, the water softener must be recharged with sodium ions periodically by passing salt water (formed from the sodium chloride pellets in the storage bin) through it. With the offending ions removed from the water, the sodium carboxylates—the soaps—now remain dispersed in the water and can act as effective detergents. (Resinous polymers of the kind used to replace one ion for another as a fluid passes through them are known as *ion exchange resins*.)

QUESTION

What happens to the calcium, iron, and other ions of hard water that have accumulated on the ion exchange resin as they are replaced by sodium ions during the recharging process? Where do these cation of hard water go?

13.11 Another Solution: Synthetic Detergents

As an alternative to softening hard water, we can use a detergent that remains dispersed even in the presence of cations that would inactivate a soap. In the 1940s chemical science and technology solved the problem of using hard water for washing by devising economical, commercially useful *synthetic detergents*. In the most successful of these, the *alkylbenzenesulfonates*, a sulfonate functional group, $—SO_3^-$, rather than a carboxylate group, $—CO_2^-$, acts as the hydrophilic structure:

$—CO_2^-$	$—SO_3^-$
a carboxylate group	a sulfonate group

The advantage of the sulfonate anion is the great water solubility of virtually every one of its mineral salts. Unlike soaps, alkylbenzenesulfonates remain dispersed and effective in hard water. [The term *alkyl* refers to the group produced by removal of a hydrogen atom from an alkane (Sec. 6.3). Alkyl groups we've already seen include the methyl group, $CH_3—$, the ethyl group, $CH_3—CH_2—$, the propyl group, $CH_3—CH_2—CH_2—$, and the isopropyl group.]

$$\begin{array}{l} CH_3—CH— \\ \quad\quad\;\; | \\ \quad\quad\; CH_3 \end{array}$$

A variety of synthetic laundry detergents.

Since there's no simple, economical way to place a sulfonate group directly onto a long hydrocarbon chain in large-scale production, commercially successful synthetic detergents don't mimic exactly the structure of a soap molecule. Some commercially feasible reactions do exist, though, that allow chemists to place both a hydrocarbon chain and a sulfonate group onto the same benzene ring (Sec. 6.11). Combining these reactions in successive steps, which lie beyond the scope of our examination, produces the synthetic alkylbenzenesulfonate detergents.

With a variety of synthetic reactions at hand, the chemist can tailor the structure of a detergent molecule to specific needs. *Sodium lauryl sulfate*, for example, is an anionic detergent particularly suitable to toothpastes and other toiletries (Chapter 22).

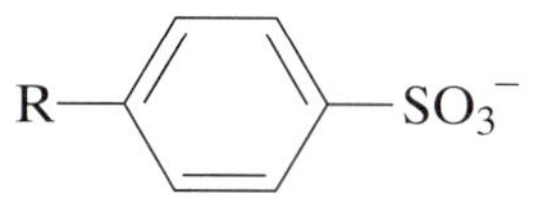

an alkylbenzenesulfonate anion
(R represents an alkyl group in the form of a long hydrocarbon chain.)

(In a *sulfate* an oxygen atom lies between the sulfur of the hydrophilic group and the carbon of the hydrophobic group; in a sulf*on*ate the sulfur is bonded directly to the carbon; Fig. 13.20).

Figure 13.20 Typical detergents.

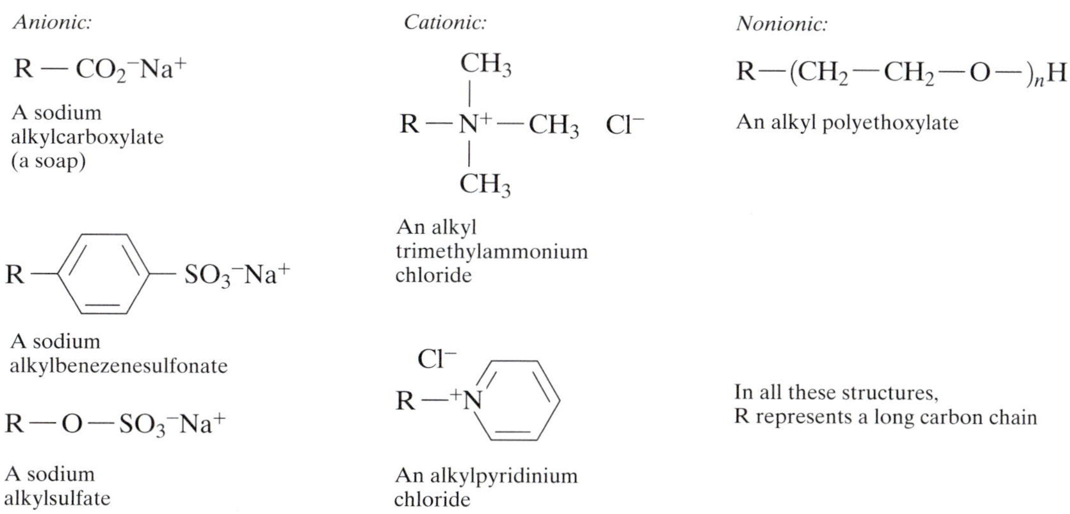

Anionic detergents, which make up the great bulk of all synthetic detergents, are particularly effective at cleaning fabrics that absorb water readily, such as those made of natural fibers of cotton, silk, and wool. *Nonionic* detergents, many of which have large numbers of covalently bonded oxygens in their hydrophilic structures, are especially useful in cleaning synthetic fabrics, such as polyesters (Chapter 21). Most nonionic detergents are liquids that produce little foam. They are used, along with anionic detergents, in formulating dishwashing liquids and liquid laundry detergents.

Most *cationic* detergents are ammonium salts (Sec. 10.3) that also happen to be effective germicides. They're used in antiseptic soaps and mouthwashes, and also in fabric softeners since their positive charges adhere to many fabrics that normally carry negative electrical charges. Molecular structures of some typical nonionic and cationic detergents appear in Figure 13.20.

■ QUESTION

What would you expect to see if you repeated the demonstration described at the end of Section 13.9, but used a synthetic detergent rather than a piece of bar soap? Explain.

13.12 What's in a Box of Detergent?

Now, with our understanding of what detergents are and how they work, we can ask: What's actually in a box of laundry detergent, the kind we get off a store shelf? The answer is: A mixture of chemistry and consumerism, and of fact and fantasy. We'll observe this intermingling as we take a detailed look at its contents.

Naturally, the most important ingredient of a detergent formulation is the surfactant itself (Table 13.4). In the decades since synthetic detergents were introduced, the anionic alkylbenzenesulfonates have been the principal surfactants in detergent formulations. Today, though, this dominance may be shifting toward non-

The ingredients in a box of detergent.

Table 13.4 Components of a Typical Detergent Formulation

Component	Example	Function
Surfactants	Sodium alkylbenzenesulfonates	Detergency
Builders	Phosphates, zeolites	Soften water and increase surfactant's efficiency
Fillers	Sodium sulfate: Na_2SO_4	Add to bulk of detergent and keep detergent pouring freely
Corrosion inhibitors	Sodium silicates: Na_2SiO_3, $Na_2Si_2O_5$, Na_4SiO_4	Coat washer parts to inhibit rust
Suspension agents	Carboxymethylcellulose (CMC)	Help keep dirt from redepositing on fabric
Enzymes	—	Remove protein stains, such as grass and blood
Bleaches	Perborates	Remove stains
Optical whiteners	Fluorescent dyes	Add brightness to white fabrics
Fragrances	—	Add fragrance to both the detergent and fabrics
Coloring agents	—	Add blueing effect

ionic surfactants, such as the alkyl polyethoxylates of Figure 13.20, as the result of a growing concern for the environment. Among the environmental advantages of the nonionic detergents is their smaller need for *builders*, which are essentially water–softening agents. (The term *builders* reflects the ability of these added agents to "build up" the detergent power of the surfactant.) Although all classes of synthetic detergents are superior to soaps in hard water, the nonionic detergents are even more effective than the anionic detergents. With less builder needed for the nonionic detergents, their formulations can be more concentrated and their packages smaller. Compact packaging decreases the amount of discarded packing material in our trash and garbage; concentration results in smaller amounts of detergents passed into the environment with the waste wash water.

For many years phosphates, such as sodium tripolyphosphate, $Na_5P_3O_{10}$, seemed to be the perfect builders. Their low cost, very low toxicity, and general absence of hazard, coupled with their ability to bind firmly to the ions of hard water and thereby reduce their chemical activity, made them the leading builders. Yet phosphates suffer from one overwhelming defect: They are superb nutrients for the algae and other small plants that grow on the surfaces of lakes and streams. Algae, nourished by a steady supply of phosphates, can cover the surface of a body of water and prevent atmospheric oxygen from reaching the marine life below. The resulting death of fish and other aquatic animals, sometimes occurring on a large scale in lakes and rivers covered by algae, led many states to ban the use of phosphates as detergent builders. The most promising substitute is a class of compounds of aluminum, silicon, and oxygen, known as *zeolites*. When added to hard water as part of a commercial detergent, the sodium salts of zeolites act like ion exchange resins, exchanging their sodium ions for the ions that produce the hardness and thereby softening the water.

One of the major functions of *fillers*, primarily sodium sulfate, Na_2SO_4, was to give consumers a sense that they were getting their money's worth. At one time a bulky box of detergent was seen as giving the consumer a lot of detergent for the cost. The bulk was provided principally by the added sodium sulfate (which also made the granular detergent a bit easier to pour). With increasing consumer awareness and a growing concern for the environment, the amount of filler used in a box of detergent has dropped from about 35% to 10% or, in some cases, none at all.

Other ingredients include corrosion inhibitors to protect the machinery that agitates the wash; suspension agents to help keep the grease, dirt, and grime in suspension so they don't redeposit on the fabrics; enzymes to help decompose proteins of blood, grass, and other difficult stains; and oxygen-containing bleaches to help remove stains. All of these contribute to the surfactant's ability to produce a clean product.

Some of the remaining ingredients may add more to the appearance of a clean wash than to its reality. Optical whiteners are organic compounds that coat the fabrics and convert the invisible ultraviolet component of sunlight (Secs. 14.8 and 22.13) into an almost imperceptible blue tint. The effect of this blue tint, which can also be produced by blue coloring agents added to the detergent, is to add a bit of brilliance to white fabrics and give them the appearance of extra cleanliness. Finally, added perfumes and fragrances produce what many regard as a pleasant odor in the finished wash.

■ QUESTION

Why were phosphates originally added to detergent formulations? Why were they subsequently removed?

13.13 A Return to the Dropping Tack

We can see now that the opening demonstration summarizes much of the chemistry of this chapter. Even though the tack is more dense than water, it remains on the surface, held up by the surface tension that results from the strong attractive forces among the water molecules at the surface. When you add the detergent, its molecules gather at the surface, with their hydrophilic heads among the water molecules and their hydrophobic tails pointed upward, away from the water. As the detergent disperses among the water molecules at the surface, its molecules disrupt the attractive forces among them, producing a sharp drop in the surface tension. With decreasing surface tension, the density of the tack becomes the controlling factor and the tack drops. We've also seen that this same surfactant effect enhances the cleaning action of soaps and synthetic detergents.

PERSPECTIVE

Where Do Detergents Belong?

Although this chapter focuses on soaps and detergents and the chemistry that makes them work, you can find another, more subtle theme as well. As you progress through the discussion of surface tension, surfactants, micelles, and the like, and into their applications in commercial detergents, you may find that the borders between pure chemistry and applied chemistry (Chapter 12, Perspective) occasionally grow dim and perhaps vanish in places. The same holds true for the sometimes arbitrary divisions among the various branches of chemistry, such as organic and inorganic chemistry, and even between chemistry itself and other sciences, including biology, geology, and physics.

For example, it's usually reasonable to define chemistry as the science that examines the composition and properties of matter and the changes it can undergo, as we did in Section 1.2. In a similar vein, we might also hold that the science of physics deals (in part) with studies of the forces of attraction and repulsion between bodies. According to this view, then, a discussion of the forces that hold molecules close to each other in a liquid, and that cause them to interact more strongly at a liquid's surface than in its interior—and that produce the phenomenon of surface tension—belongs more properly to physics than to chemistry. Yet without an appreciation of how these forces act, the chemistry of a detergent doesn't make much sense. Thus the border between these two areas of science grows less distinct in the study of detergents.

Again, it's possible to maintain that since detergent molecules contain such large numbers of carbon atoms, their study rests more appropriately within an examination of organic compounds. Yet without understanding the effects of such inorganic cations as Ca^{2+}, Mg^{2+}, Fe^{2+}, and Na^{+} on both soaps and anionic detergents, it's difficult to understand the effects of hard water on soaps and detergents. Moreover, organic detergents act within the confines of water, a fully inorganic solvent. Here again, the border drawn between two regions of chemistry seems to vanish.

Finally, it might seem that a study of commercial soaps and synthetic detergents, a set of widely used consumer products, certainly involves an examination of applied chemistry. Yet without an understanding of the fundamental science behind the properties of surfaces and micelles, the reactions of esters, the process of ion exchange, and a host of other structures, reactions, and phenomena, the box of detergent you find on your grocer's shelf might not even exist.

One of the themes of this chapter, then, is that while it's often convenient and sometimes necessary to divide science into a variety of arbitrary categories, our comprehension of the universe we live in cannot be limited by the boundaries we draw. They are borders and divisions that we ourselves construct and that we ourselves can and must remove when they obstruct our larger view of our universe. ■

EXERCISES

For Review

1. Following are a statement containing a number of blanks, and a list of words and phrases. The number of words equals the number of blanks within the statement, and all but two of the words fit correctly into these blanks. Fill in the blanks of the statement with those words that do fit; then complete the statement by filling in the two remaining blanks with correct words (not in the list) in place of the two words that don't fit.

Both soaps and synthetic detergents consist of molecules made up of a __________ portion, which is usually a long, hydrocarbon chain, and a ___________ portion. A __________ functional group forms the hydrophilic end of the __________ molecule, while the class of synthetic detergents known as __________ carry a __________ functional group at their hydrophilic end. The combined effects of the two classes of functional groups cause both soaps and synthetic detergents to accumulate at the surface of water and to lower the water's __________. As a result, both of these classes of detergents belong to the larger category of __________.

__________ contains __________ cations. Because these ions form water-insoluble salts with long-chain __________, soaps are inefficient detergents in hard water and form precipitates known as __________. The sulfonate salts of these ions are generally soluble and thus __________ are much more useful in regions where the water is hard.

alkylbenzenesulfonates	hydrophobic
calcium, iron, and magnesium	micelles
	nonionic detergent
carboxylate	soap
carboxylic acids	sulfonate
hard water	surfactants
hydrophilic	synthetic detergents

2. Define, identify, or explain each of the following:

a. anionic detergent	i. hydrolysis
b. calcium soap	j. lye
c. colloidal dispersion	k. micelle
d. density	l. saponification
e. detergent builder	m. soft water
f. ester	n. triglyceride
g. fatty acid	o. Tyndall effect
h. glycerol	p. zeolite

3. Suppose you find each of the following on the surface of a lake: (a) an empty metal bucket, (b) a metal paper clip, (c) a bar of soap, (d) an aluminum rowboat, (e) an aluminum nail, (f) a cork stopper, (g) a model airplane made of balsa wood, (h) a piece of ice, (i) a wax candle. Which is floating and which is supported by surface tension?

4. Explain what we mean when we say that a detergent molecule (a) makes water wetter; (b) acts as a molecular nail.

5. The structure

$$CH_3-(CH_2)_n-\overset{\overset{CH_3}{|}}{\underset{\underset{CH_3}{|}}{N^+}}-CH_3$$

represents the cation of a class of cationic detergents. What part of this molecule makes up (a) its hydrophobic tail? (b) its hydrophilic head?

6. What two functional groups are generated when an ester reacts with water by hydrolysis in the presence of a strong acid?

7. What trihydroxy alcohol is produced by the hydrolysis of *any* triglyceride?

8. Using the information in Table 13.3, write the structures of (a) glyceryl triacetate, (b) glyceryl tripalmitate, and (c) ethyl palmitate. Which if any of these would undergo hydrolysis in the presence of NaOH? Which, if any, would produce a soap on hydrolysis with NaOH?

9. Write the structure of *methyl benzoate*. Write complete chemical reactions for the hydrolysis of methyl benzoate in the presence of (a) HCl; (b) KOH. Name each of the organic products produced in these reactions.

10. When the ancient Romans made soap, what did they use as the source of (a) triglycerides? (b) bases?

11. Can there be a surfactant that is *not* a soap? Can a soap exist that is *not* a surfactant? Explain.

12. Name or give the molecular structure of all the products of each of the following reactions:

a. heating ethyl acetate with a solution of sodium hydroxide in water

b. heating methyl acetate with a solution of HCl in water

c. saponification of methyl palmitate

d. heating a mixture of CH_3-CH_2-OH, $CH_3-CH_2-CO_2H$, and a catalytic quantity of HCl

13. What would you expect to happen if you add a teaspoon of each of the following to the sudsy mixture produced by shaking soap in distilled water: (a) table salt; (b) $MgSO_4$ (epsom salts, Sec. 10.2); (c) ground chalk ($CaCO_3$, Sec. 10.2)?

14. What property of cationic detergents that are ammonium salts makes them useful in mouthwashes?

15. List five ingredients of a commercial laundry detergent, other than the surfactant, and describe the function of each.

16. Which of the ingredients of a typical detergent formulation, listed among those of Table 13.4, play no part in the removal of dirt, grease, stains, or other objectionable material from fabrics?

17. Why have many communities banned the use of phosphates as detergent builders?

A Little Arithmetic and Other Quantitative Puzzles

18. You have three sheets of plastic, each 1 cm thick: *A* is a square that's 2 cm on each side and weighs 2 g; *B* is a rectangle that's 1 cm on one side and 3 cm on the other, and weighs 4 g; *C* is a rectangle 2 × 2.5 cm and weighs 3 g. Which, if any, of the plastic sheets will float on water?

19. You immerse a small statue made of aluminum into a measuring cup containing water and find that the statue has a volume of 25 mL. What does the statue weigh?

20. Another small statue has a mass of 1 kg. Immersing the statue in water reveals that it has a volume of 52 cm^3. Which of the materials of Table 13.1 is the statue probably made of?

21. What's the volume of a 1-oz coin made of pure gold (1 oz = 28.3 g)?

22. While exploring a remote area, you find a large, yellow nugget lying, on the ground. The nugget looks like gold, yet you suspect it may be *pyrite*, a virtually worthless yellow mineral sometimes called "fool's gold" because it's so often mistaken for the real thing. You find that the nugget weighs 50 g and displaces 2.6 mL of water. Is it more likely to be fool's gold or the real thing?

23. How large does a piece of aluminum have to be to weigh the same as 10 cm^3 of lead?

24. We've seen in Section 13.7 that colloidal particles range from about 10^{-7} to 10^{-4} cm in diameter. A simple calculation shows that spheres with this range of diameters run from about 5×10^{-22} to 5×10^{-11} cm^3 in volume. We also know that there are 6.02×10^{23} water molecules in 18 g of water and that 1 g of water occupies very nearly 1 cm^3. Given these values, calculate the diameter a water molecule would have if it were a perfect sphere. Using this approximate volume of a water molecule, predict whether a mixture of 1 mL of water in 100 mL of ethyl alcohol would show a Tyndall effect. Describe your reasoning.

Think, Speculate, Reflect, and Ponder

25. You have a cube, 1 cm on a side, of a substance that has *exactly* the same density as water. Describe what would happen if you place the cube in a body of water so that its top surface is 1 cm below the surface of the water.

26. In the opening demonstration we made a tack drop to the bottom of a glass of water by adding a small amount of detergent to the water. Would an aluminum rowboat sink in a pond if enough detergent were added to the pond? Explain.

27. Would you expect the surface tension of cold water to be greater than, less than, or exactly the same as the surface tension of water that's near its boiling point? Explain.

28. A felt eraser removes chalk marks made on a chalkboard. Is the eraser acting as a detergent? Is it acting as a surfactant? Describe your reasoning.

29. What is the source of the mineral ions present in hard water?

30. Would you expect a soap to act as an effective detergent in seawater? Would you expect a synthetic detergent (an alkylbenzenesulfonate) to act as an effective detergent in seawater? Explain.

31. What determines whether the mixture of one substance within another produces a solution or a colloidal dispersion?

32. Which of the following would you expect to show a Tyndall effect: (a) smoke from a backyard grill, (b) a mixture of helium and argon, (c) sugar water, (d) a well shaken vinegar–and–oil salad dressing, (e) the interior of a cloud?

33. Why doesn't the identity or chemical behavior of *anions* present in water affect the water's hardness?

34. Could you use a solution of alkylbenzenesulfonate detergent in distilled water instead of soap in distilled water to distinguish between the sample of hard water and the sample of soft water? Explain.

35. If we used distilled water to wash our clothes, would the wash water act as an electrolyte? Explain.

36. Again you have two samples of water, but this time each is hard water. One is very hard, with a high concentration of calcium ions; the other is not as hard, with a lower concentration of calcium ions. Show how you could use a solution of soap in distilled water and a dropper or other measuring device to determine which is the harder water.

37. The sodium chloride of a water softener's storage bin must be replaced periodically as the NaCl is consumed in the formation of salt water. What would result if calcium chloride rather than sodium chloride were mistakenly added to an empty storage bin?

38. Why is it reasonable to conclude from the content of this chapter that goat fat consists of molecules that contain long hydrocarbon chains?

39. Suppose that you live in an area with very hard water and have just found that you are allergic to all available synthetic detergents. You find that you must use soap for all your household cleaning. What would you do to use soap most effectively?

40. No colloidal dispersion of one gas in another is known to exist. Yet when gaseous HCl and gaseous NH_3 are mixed, a fog results that clearly produces a Tyndall effect. Explain why.

41. How could you use a bar of soap to convert hard water to soft water?

42. Considering how soaps and synthetic detergents remove greasy dirt from fabrics that are washed in water, do you think a synthetic detergent might be useful for removing small quantities of water from gasoline? Explain.

Web Questions

Use the Web to answer or accomplish each of the following. Report your Web source for each response.

1. Summarize the early history of soap, from ancient civilizations to the fall of the Roman Empire. Include the legend of the role Mt. Sapo played in the origin of the word *soap*.

2. Write a brief biography of John Tyndall with emphasis on his contributions to science and to science education, especially the explanation of scientific concepts to the general, nonscientific public.

3. What are the principal applications or uses of (a) anionic detergents; (b) cationic detergents; (c) nonionic detergents. In what kind or class of product might each of these be found?

4. Describe a medical, commercial, or industrial use for surfactants that is not related to their ability to act as detergents.

5. What is a *meniscus* and how is it related to surface tension?

6. Describe how the degree of hardness of water is determined. What units or scales of measurement are used to express the hardness of water? What range is considered soft water? Moderately hard water? Hard water?

Chemicals, Pollution, and the Environment

THE MEANING OF POLLUTION

A major source of urban air pollution.

How to Generate Air Pollution in Your Own Kitchen

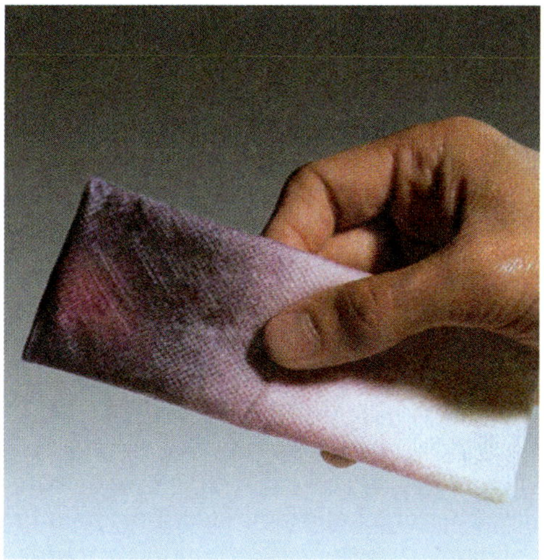

You can see how acid rain forms by using some red cabbage leaves, a pot of water, a few matches, a candle, and a piece of paper towel. A toothpick and some vinegar are also helpful. Start by tearing up a few red cabbage leaves and dropping them into some boiling water to extract the anthocyanin acid–base indicator, as you did for the opening demonstration of Chapter 10. While the anthocyanin dye is being extracted, light a candle and set it aside in a safe place. Also tear a match out of a book of matches and set it aside as well.

Now fold a piece of paper towel into a long, slender shape. It ought to look a bit like a narrow, flat ruler, perhaps 8 to 10 inches long. Hold one end and dip several centimeters of the other end into the cooled anthocyanin solution. The wet end should appear distinctly blue. As you'll recall from Chapter 10, the color of the anthocyanin solution is a good indication of its pH. A blue solution is nearly neutral. As it becomes slightly basic, it turns green. As it becomes acidic, it turns pink or red.

With one hand hold the dry end of the folded towel so that the wet, blue end is just above and a little to the side of the candle flame. With your other hand hold the match in the flame until it ignites (Fig. 14.1a). Quickly move the newly ignited match so that it flares up just under the wet part of the paper towel. You can allow the flames to reach the wet paper, but be careful to keep the match (and the candle flame) away from the dry part of the paper towel so that the towel itself doesn't ignite (Fig. 14.1b). Keep the match just under the wet part of the towel until the initial flare has subsided, then blow the match out. Place it on a surface that won't burn or scorch. You'll need it again in a moment.

You will now find that the burning match has produced a distinct red spot within the blue part of the zone that was just above the flare (Fig. 14.1c). The color of the red spot is similar to the color generated by the vinegar you added to the blue anthocyanin solution in Chapter 10. Like that color change, this one also indicates that something acidic dissolved in the water of the wet paper towel. You can demonstrate this by

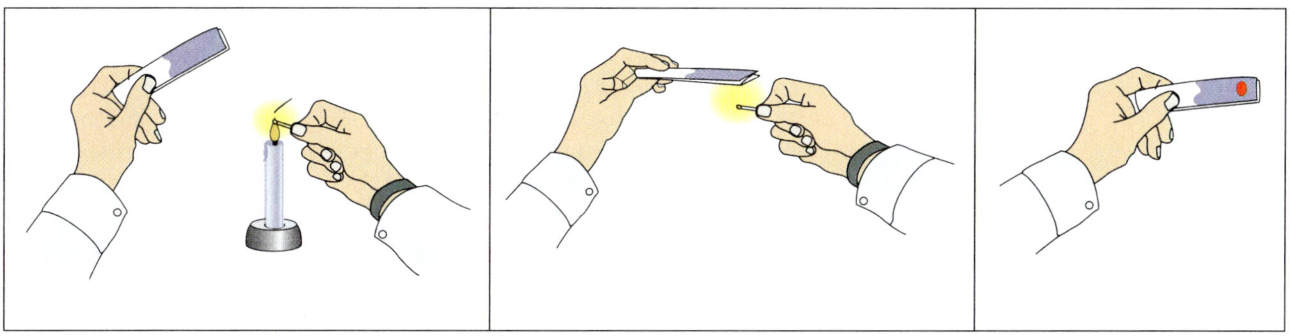

Figure 14.1a
Ignite the match in the candle flame and . . .

Figure 14.1b
. . . as it continues to flare move it quickly under the wet, blue portion of the paper towel.

Figure 14.1c
The match flare produces a red spot on the blue paper.

transferring a drop of vinegar to another portion of the blue region of the towel. That, too, produces a red spot. (A toothpick makes a convenient carrier for the drop.) The similarity of the colors of the two red spots indicates that the flaring match made the water about as acidic as the vinegar. We can now ask two questions:

1. What acid caused the appearance of the red spot above the newly lit match?
2. Where did this acid come from?

In considering the first question we might initially label carbonic acid, H_2CO_3, as the culprit, since we know that carbon dioxide, CO_2, is a common product of combustion and that carbon dioxide forms carbonic acid as it dissolves in water (Sec. 10.10). Yet carbonic acid can't be the source of the red spot. We saw in the opening demonstration of Chapter 10, for example, that blowing exhaled breath (with its content of CO_2) into a slightly basic anthocyanin solution turns the solution from green to a shade of blue. We didn't get a red or pink color until we added vinegar. That color change came from acidification of the solution by the acetic acid of the vinegar. Moreover, in Table 10.4 we saw that the pH of vinegar (2.4–3.4) is much lower than that of a solution of carbonic acid in water (carbonated water, 5.6). We have to conclude that the ignited match introduced something much more acidic than CO_2 into the atmosphere and into the water on the paper towel. We'll learn what it was after we examine the second question.

As for the origin of the acidic substance, we can use the scientific method to demonstrate that it came from the initial flare of the match rather than from combustion of the paper of its stem. For this, simply relight the extinguished match in the candle flame and hold the burning stem under a fresh portion of the anthocyanin solution that remains on the wet towel. No additional reddening occurs, indicating that the acid didn't come from combustion of anything contained in the paper of the match stem. The acid must have come from something in the flammable head of the match.

For effective ignition, the head of a match contains several flammable substances, all of which ignite easily and give off considerable heat when they burn. That's why the match flares up when you strike it. Among these is sulfur or one of its compounds that burns in air to form the gas sulfur dioxide, SO_2. This sulfur dioxide is one of the substances responsible for the acrid odor you sometimes smell just after you strike a match. The sulfur dioxide produced by the flaring match dissolves in the water of the wet paper towel and forms sulfurous acid, H_2SO_3:

$$\underset{\text{sulfur dioxide}}{SO_2} + \underset{\text{water}}{H_2O} \longrightarrow \underset{\text{sulfurous acid}}{H_2SO_3}$$

Sulfurous acid is far more acidic than carbonic acid and even a bit more acidic than acetic acid. It's the source of the red spot. Sulfur dioxide formed by the combustion of traces of sulfur compounds in petroleum and its products dissolves in the water droplets of clouds and falls to the earth as acid rain. Even more acidic and corrosive is sulfuric acid, H_2SO_4, which forms as SO_2 reacts with the oxygen of the atmosphere and slowly oxidizes to sulfur trioxide, SO_3. Sulfur trioxide generates sulfuric acid as it dissolves in water:

$$\underset{\text{sulfur trioxide}}{SO_3} + \underset{\text{water}}{H_2O} \longrightarrow \underset{\text{sulfuric acid}}{H_2SO_4}$$

Sulfuric acid dissolved in rainwater is responsible for much of the environmental damage produced by *acid rain*. We'll have more to say about acid rain and other forms of pollution later in this chapter.

While earlier chapters dealt largely with the chemistry of substances we choose to use or to consume, or with substances whose benefits to society generally outweigh their risks, this chapter has a different focus. Here we'll examine substances, like the sulfur dioxide of the match, that pollute, contaminate, or otherwise degrade our environment, substances that we'd prefer not to consume and that we'd rather be without, substances whose risks certainly outweigh any benefits they might offer.

In the process we'll examine what we mean by *pollution* and how we might recognize and define it. We'll also examine answers to the question that's implicit in the title of this opening demonstration: When we light a match in the kitchen, do we actually pollute the air? As we proceed, we'll examine some of the chemicals that contaminate our air, water, and land. We'll conclude by looking at what happens when we throw something "away," and we'll examine the time scales of pollution.

14.1 The Elements of Our Environment

Just as we can divide the familiar substances of our physical universe into solids, liquids, and gases (Sec. 12.1), it's convenient to separate our physical environment into the solid ground beneath our feet, the water we drink and use for other purposes, and the air we breathe. In more scientific terms the ground is part of the **lithosphere**, which is the hard, rigid shell of the earth, about 100 km (roughly 60 miles) thick. (The word *lithosphere* comes from the Greek *lithos* for stone.)

The **lithosphere** is the hard, rigid shell of our planet. The surface layer of the lithosphere forms the earth's crust.

A relatively thin, surface layer of the lithosphere forms the earth's crust, the familiar combination of rocks, sand, gravel, limestone, humus, and all the other solid substances we walk on, pave over, and build our buildings on, and that plants grow from. The crust holds the oceans, rivers, lakes, and groundwater (Sec. 9.10) of our environment. Combined, all these waters make up the **hydrosphere**. Above the crust and hydrosphere lies the **atmosphere**, the mixture of gases that surrounds the earth. Together, the earth's crust, the hydrosphere, and the atmosphere constitute the physical environment we live in.

All the waters of the earth's crust make up the **hydrosphere**.

The **atmosphere** is the body of gases that surrounds the earth.

A portion of the earth's lithosphere, hydrosphere, and atmosphere: Ecola State Park, Oregon.

Table 14.1 The 10 Most Abundant Elements of the Earth's Crust

Element	Average Percentage (by weight)
Oxygen	46
Silicon	28
Aluminum	8
Iron	6
Calcium	4
Sodium	3
Magnesium	2
Potassium	2
Titanium	0.6
Hydrogen	0.1

Table 14.2 The 10 Most Abundant Elements of the Earth's Oceans

Element	Average Percentage[a] (by weight)
Oxygen	86
Hydrogen	11
Chlorine	2
Sodium	1
Magnesium	0.1
Sulfur	0.09
Calcium	0.04
Potassium	0.04
Bromine	0.006
Carbon	0.003

[a]The total percentage is more than 100 because of the effects of rounding.

Table 14.3 The 10 Most Abundant Gases of the Earth's Atmosphere

Gas	Average Percentage (by volume in dry air)
N_2	78
O_2	21
Ar	0.9
CO_2	0.03
Ne	0.002
He	0.0005
CH_4	0.0002
Kr	0.0001
H_2	0.00005
N_2O	0.00005

The crust, which makes up only about 0.3% of the mass of the entire planet, varies quite a bit in thickness. It averages about 8 km (5 miles) under the oceans and roughly 45 km (almost 30 miles) beneath the continents. The crust is composed principally of silicates, compounds of silicon and oxygen that make up about three–quarters of its mass. Table 14.1 shows the 10 principal elements of the crust.

As in the crust, oxygen is the dominant element of the hydrosphere. With its 16 amu of mass, oxygen makes up $\frac{16}{18}$ or 88.9% of the mass of a water molecule. Yet because of the presence of soluble salts and organic matter in the hydrosphere, the average percentage of oxygen is actually a little less than this. The actual percentages of the various elements of the hydrosphere vary from one body of water to another and from freshwater lakes and rivers to the world's oceans. Table 14.2 lists the 10 principal elements of the oceans, and Table 14.3 shows the 10 leading gases of the atmosphere.

■ **QUESTION**

Silicates, compounds of silicon and oxygen, are the most abundant compounds of the earth's crust. What is the most abundant compound of the earth's (a) atmosphere (b) hydrosphere?

14.2 Pollution: A Case of Misplaced Matter

Now, with our view of the physical environment as a combination of the solid land beneath us, the waters held by the land, and the gaseous atmosphere above, we can begin our examination of the effects of chemicals on the environment. Often, when we bring the topic of chemicals into a discussion of the environment, the word *pollution* comes to mind. It's a word with many meanings, some emotional, some rigorous, and some simply hard to pin down with any precision.

At one end of this spectrum are precise numerical values of concentrations, such as those that specify the legal limits of contaminants permitted in community water systems. Set by the Safe Drinking Water Act of 1974 and subsequent amendments, these describe pollution in quantitative terms (Tables 9.2 and 9.3). If any of the specified contaminants lies above its legal limit, we can consider the water to be polluted (Sec. 9.12). In a sense, these are the limits set by society and made explicit in our written laws.

At the other extreme are more personal or more subjective limits set by each of us, as individuals. For an illustration we can return to the set of saltwater solutions prepared in the opening demonstration of Chapter 9 and ask again, "Which is the glass where pollution begins?" Various answers are possible, each reflecting our own individual sense of what constitutes a polluted environment. Some of us might choose the glass that first gives a salty taste. In that case we define pollution subjectively, with limits depending on our individual perceptions of the world about us. (Here, for example, the sensitivity of our own taste buds plays a large role in how we define pollution.) Some might choose the glass in which federal limits for sodium are exceeded. Others might want to learn at what concentration the sodium in our drinking water might eventually produce harmful effects to our health, by increasing our risk of hypertension for example (Sec. 3.3). Still others might maintain that any amount of sodium chloride in drinking water, no matter how little, constitutes pollution.

Clearly, we have to agree on a suitable, working definition of pollution. One that can serve us nicely in this chapter is that **pollution** occurs when an excess of a substance generated by human activity is present in the wrong environmental location. It's not precise and it won't stand up to a rigorous scientific analysis, but we'll find that it works well. Let's look at some of its implications.

Pollution occurs when an excess of a substance generated by human activity is present in the wrong environmental location.

As one of its strengths, this kind of definition helps us understand what we mean by the "purity" of our environment. We sometimes speak about having "pure" water to drink and "pure" air to breathe. Yet if by "pure" we mean "uncontaminated by any other substance," there's surely not a single drop of truly pure water in the entire lithosphere, hydrosphere, or atmosphere. Even the "purest" rainwater, for example, isn't pure at all. It contains the dissolved atmospheric gases of Table 14.3, absorbed as the raindrops fall through the atmosphere, and carbonic acid, formed by the reaction of absorbed carbon dioxide with the water of the raindrop (Sec. 9.10). Nor can we draw pure water from wells or springs. We're likely to find salts of calcium, magnesium, sodium, potassium, and other minerals dissolved in water drawn from the lithosphere. The only truly pure water, uncontaminated by any other substance, comes from the chemist's laboratory (Sec. 9.10).

Nor is our air "pure." Not only is the atmosphere a mixture of several gases (Table 14.3), including varying amounts of water vapor, but it's often contaminated by ozone generated by lightning and other natural phenomena (Secs. 1.2 and 12.4) and by volatile organic compounds produced by animals and vegetation. When we speak of environmental purity, then, we often refer to air and water that's wholesome and pleasant to consume as well as free of disease-causing chemicals and biological organisms. In this sense, pollution becomes an esthetic as well as a scientific topic.

Lightning, a generator of ozone and of nitrogen oxides.

Our working definition of pollution contains two more, important implications. First, we can't label any chemical or any mixture of chemicals as a pollutant unless we specify where it is. What counts in environmental pollution is the combination of substance and location. Ozone provides a good illustration. As we'll see in more detail in Section 14.7, ozone is an irritating, toxic gas. In the troposphere, the region of the atmosphere that touches the surface of the earth and holds the air we breathe, ozone is clearly a pollutant. Many kilometers higher, in the stratosphere, ozone becomes a lifesaving gas as it protects us from devastating effects of the sun's intense ultraviolet radiation (Sec. 14.8). Simply moving the ozone of the troposphere into the stratosphere would convert an atmospheric pollutant into a lifesaving shield.

Decaying limb of a white birch tree releases chemicals to the atmosphere and to the soil.

Our definition also suggests that we use the term *pollution* for the degradation of our environment that results specifically from human activities, excluding natural events beyond our control. This usage reflects not only the relatively short span of human activities in our planet's history but also their scale. Isotopic dating techniques show that the earth's surface is probably 3.5 to 4 billion years old (Sec. 5.10). Fossils indicate that life has existed on earth for several billions of these years. And while plants and animals resembling those we see on earth today have been around for several million years, the discharge of chemicals into the environment by humans, on an industrial scale at least, has been going on for only a few hundred years.

Having survived periodic catastrophes such as volcanic eruptions, floods, and forest fires ignited by lightning, it might seem reasonable that the forms of plant and animal life now existing on earth can continue to survive repetitions of these events. Yet we can only speculate whether they can tolerate the continued discharge of masses of chemicals created by humans on a scale never seen before. In this sense pollution is something that comes with human activity.

To illustrate, we can look at the contrast between the discharge of sulfur dioxide and sulfur trioxide resulting from volcanic eruptions and that created by human activities. (We saw one example of the human generation of sulfur dioxide in the opening demonstration.) In June 1991, Mt. Pinatubo, in the Philippines, erupted in what was probably the greatest volcanic blast of the 20th century. In a series of explosions the volcano blew an estimated 15 to 20 million tons of sulfur dioxide into the stratosphere.

Although dramatic, eruptions of this size are rare. A 1982 eruption of Mexico's El Chicon volcano spewed about half as much sulfur dioxide into the atmosphere; continual eruptions of Mt. Kilauea in Hawaii since 1986 have released roughly 350,000 tons of sulfur dioxide each year, for a total of about 3.5 million tons per decade. Yet industrial and other human activity in the United States alone has been putting about 20 million tons of sulfur oxides into the air *every year* since

Eruption of Mt. Pinatubo, Philippines, June 1991.

A forest fire contaminates the atmosphere.

about 1950. By one estimate, about 90% of all the sulfur that enters the global atmosphere comes from human activities.

■ QUESTION

What is your own definition of *pollution*? How would you use your definition to locate the glass of salt water that marks the beginning of pollution?

14.3 Air Pollution and Energy

Industrial air pollution.

Each year the peoples of the world consume roughly 9×10^{16} kcal of energy, in all its forms. That's enough energy to heat 9×10^{14} kg—about 9×10^{11} (900,000,000,000) tons—of water from its freezing point to its boiling point. A major portion of all this energy comes from burning coal, natural gas, and fuels derived from petroleum: gasoline, kerosene, fuel oil, and jet fuel (Chapters 7 and 8).

Burning these fossil fuels produces plenty of water and the greenhouse gas carbon dioxide, which contributes to global warming as we saw in Section 8.14. But these fossil fuels rarely consist of hydrocarbons alone. Petroleum usually contains impurities, including small amounts of sulfur compounds and sometimes even traces of compounds of metals such as nickel and vanadium. The amounts of these nonhydrocarbon components vary with the origin of the petroleum. The sulfur content, for example, ranges from far less than 1% in some grades from wells in Sumatra and in Mississippi and Louisiana, to several percent in supplies from the Middle East, California, and Wyoming. A similar range of sulfur impurities occurs in coal. In petroleum, most of the sulfur is incorporated into the molecular structures of organosulfur compounds (organic compounds containing sulfur). Most of the sulfur of coal is combined with iron in the mineral *pyrite* or *iron pyrite*, FeS_2.

Although the processing of raw petroleum and coal removes many of these impurities from the finished products, eliminating every trace of all impurities becomes extremely expensive. Removing the last bit of each impurity would raise the cost of fuel (and thus the cost of energy) prohibitively. We'll have more to say later about the choices we must make between the cleanliness of our environment and the level at which our society operates. Now, though, we'll examine some of the pollutants generated by burning coal and fuels derived from petroleum; then we'll look at some of the ways we can keep these pollutants from invading our environment.

■ QUESTION

What is the valence of iron in *pyrite*? (You might find both Section 3.12 and the periodic table useful here.)

14.4 Air Pollution: The Major Pollutants

Here we'll look at some of the air pollutants—especially oxides of sulfur, oxides of nitrogen, and carbon monoxide—that result from the use of combustion as a source of energy, and some of the volatile, or readily vaporized compounds contained in the materials that we use in our daily lives.

Since some of the impurities of the crude fossil fuels remain in the finished products, when we burn coal, gasoline, and similar fuels we oxidize not only their hydrocarbons but their impurities as well. The sulfur of the pyrite that remains in coal, for example, oxidizes to sulfur dioxide, an irritating gas with a harsh, acrid odor described in this chapter's opening demonstration. The reaction for the oxidation is

$$\underset{\text{pyrite or iron pyrite}}{4\,FeS_2} + \underset{\text{oxygen}}{11\,O_2} \longrightarrow \underset{\text{iron oxide}}{2\,Fe_2O_3} + \underset{\text{sulfur dioxide}}{8\,SO_2}$$

A **primary air pollutant** is a pollutant that enters the air as the direct result of a specific activity.

Sulfur dioxide forms as coal and petroleum products burn; it enters the air as a **primary air pollutant**, an air pollutant generated as the direct result of a specific activity. While sulfur oxidizes readily to sulfur dioxide during the combustion, further oxidation isn't quite as easy. Yet once sulfur dioxide enters the atmosphere and remains in contact with oxygen of the air, it can oxidize further to sulfur trioxide, SO_3. This reaction can occur by several different paths, including a direct reaction with an oxygen molecule:

$$\underset{\text{sulfur dioxide}}{2\,SO_2} + \underset{\text{oxygen}}{O_2} \longrightarrow \underset{\text{sulfur trioxide}}{2\,SO_3}$$

A **secondary air pollutant** is a pollutant formed by the further reaction of a primary air pollutant.

Like sulfur dioxide, sulfur trioxide is a highly irritating gas. It can produce choking, irritating sensations at concentrations as low as 1 ppm (Sec. 9.9). Sulfur trioxide dissolves in atmospheric water droplets to produce sulfuric acid, a major component of acid rain (Sec. 14.5) and a **secondary air pollutant**—a pollutant formed by the further reaction of a primary air pollutant:

$$\underset{\text{sulfur trioxide}}{SO_3} + \underset{\text{water}}{H_2O} \longrightarrow \underset{\text{sulfuric acid}}{H_2SO_4}$$

For simplicity, sulfur dioxide and sulfur trioxide are sometimes combined into the category of *sulfur oxides* and represented by the generalized formula SO_x.

Nitrogen oxides, NO_x, form another group of air pollutants. Unlike the sulfur of SO_x, the nitrogen of these pollutants comes principally from the air itself rather than from impurities in fuels. Molecules of the nitrogen and oxygen of the air combine at very high temperatures to form *nitric oxide*, NO, a colorless gas:

$$\underset{\text{nitrogen}}{N_2} + \underset{\text{oxygen}}{O_2} \longrightarrow \underset{\text{nitric oxide}}{2NO}$$

Both the high temperatures of lightning and those of combustion chambers of the internal combustion engine are effective in converting nitrogen and oxygen to NO. The nitric oxide formed in engines is discharged along with all the other combustion products and enters the atmosphere as a primary air pollutant. Once in the atmosphere, nitric oxide reacts with additional oxygen to form *nitrogen dioxide*, NO_2, a red-brown, toxic gas that causes irritation to the eyes and the respiratory system:

$$\underset{\text{nitric oxide}}{2NO} + \underset{\text{oxygen}}{O_2} \longrightarrow \underset{\text{nitrogen dioxide}}{2NO_2}$$

At levels of 100 to 200 ppm, nitrogen dioxide can produce an inflammation of the lungs that often appears minor at first but that can become fatal in a few days. Fortunately, nitrogen dioxide is not a major air pollutant nationally. Monitoring stations throughout the United States report an average level of less than 0.1 ppm. Nonetheless, much higher levels can accumulate in the air above urban areas, especially during periods of heavy traffic.

A layer of photochemical smog over New York City.

Further reaction of nitrogen dioxide with atmospheric oxygen and water produces still other oxides of nitrogen, as well as nitric acid, HNO_3, another component of acid rain. In addition, the energetic ultraviolet radiation of sunlight (Sec. 14.8) can dislodge one of nitrogen dioxide's oxygen atoms. Reaction of this oxygen atom with a diatomic oxygen molecule produces ozone. The newly generated oxygen atom can also react with unburned hydrocarbons of auto exhausts to produce a variety of new pollutants.

To make matters worse, these new pollutants can react further with oxygen molecules and with nitrogen oxides to produce additional pollutants. The complex combination of all these products of the initial interaction of sunlight and nitrogen dioxide is known as **photochemical smog**. We see it as the brown haze that sometimes forms over our cities. The initial reaction and the formation of ozone appear below.

Photochemical smog is the complex combination of all the products resulting from the initial interaction of sunlight and nitrogen dioxide and subsequent reactions involving the bimolecular oxygen of the atmosphere as well as hydrocarbon pollutants.

Initial reaction of nitrogen dioxide with sunlight:

$$\underset{\text{nitrogen dioxide}}{NO_2} \xrightarrow{\text{sunlight}} \underset{\text{nitric oxide}}{NO} + \underset{\text{oxygen atom}}{O}$$

Formation of ozone:

$$\underset{\text{oxygen atom}}{O} + \underset{\text{oxygen molecule}}{O_2} \longrightarrow \underset{\text{ozone}}{O_3}$$

The brown haze of photochemical smog is especially noticeable during periods of **thermal inversion**, which occur during abnormal weather. Normally, air temperature decreases with altitude, with warmer layers of air lying near the ground and cooler air above. Occasionally a layer of warm air moves over a cooler layer below. This inversion of the normal temperature profile traps the cooler air and all the gases in it, including pollutants, until more normal atmospheric conditions return.

A **thermal inversion** occurs when a layer of warm air lies above a layer of cooler air, trapping it and any pollutants within it.

Still another pollutant, one that we can't see and that produces no sense of irritation, is carbon monoxide, CO (Sec. 6.14). Known as the silent killer because it is odorless, tasteless, and invisible, its major symptom is a drowsiness sometimes accompanied by headache, dizziness, and nausea. A product of the incomplete combustion of carbon or organic compounds, such as the hydrocarbons of gasoline, CO is primarily a pollutant of cities and usually fluctuates with the flow of traffic. Like using charcoal grills to warm closed, unventilated rooms (Sec. 6.14), running a car's engine in an enclosed garage can generate lethal concentrations of carbon monoxide.

Naturally, these pollutants aren't the only ones affecting the quality of the air we breathe. Atmospheric pollution by metals and inorganic compounds (compounds not containing carbon, Sec. 6.1) can come from a variety of sources. Incinerators, for example, can discharge traces of metals used in the printing dyes of product labels and packages. There's even concern about pollution resulting from cremation. An estimated 360 kg (about 800 lb) of mercury used in dental fillings enters the atmosphere each year as a result of cremations. Other pollutants released through cremation include the compounds of metals used in medical procedures and treatments.

VOC represents volatile organic compounds that can produce air pollution; **VOS** represents volatile organic solvents with the same potential.

Organic compounds often become air pollutants simply by evaporating. Those that evaporate easily are called **volatile organic compounds (VOC)** or **volatile organic solvents (VOS)**. (A volatile substance is one that evaporates readily.) These include

- volatile hydrocarbon components of consumer products such as paint thinners, roof tar, and glazing compounds, which are often listed on ingredients labels as "petroleum distillates";
- evaporating solvents and propellants of personal care and household products such as nail polish, deodorants, after-shave lotions, hair sprays, and insecticides; and
- ethyl acetate (Sec. 13.7) and more exotic additives that you exhale when you chew gum and breath fresheners for what advertisers call a "fresh, clean breath."

Yet, even with all these additional sources, the generation of energy through the combustion of fossil fuels remains by far the major contributor to air pollution. The United Nations Environment Program and the World Health Organization estimate that potentially dangerous levels of air pollutants of every category threaten the health of over 1 billion people throughout the world. Of these, well over half face hazardous levels of sulfur dioxide. Table 14.4 lists major sources of air pollutants in the United States.

Table 14.4 Major Sources of Air Pollution in the United States, 1998

	Millions of Tons				
	Carbon Monoxide	Nitrogen Oxides	Sulfur Dioxide	Volatile Organic Compounds	Total
Fuel combustion					
Highway vehicles	50.4	7.8	0.3	5.3	63.8
Electric utilities	0.4	6.1	13.2	0.1	19.8
All other nonvehicle sources	5.0	4.1	3.5	0.8	13.4
Industrial emissions (other than fuel combustion)	3.6	0.8	1.5	1.4	7.3
Waste disposal and recycling	1.2	0.1	0	0.4	1.7

SOURCE: *U.S. Statistical Abstracts, 2000*, Table 394.

■ QUESTION

(a) Write a single, balanced equation for the conversion of atmospheric nitrogen and atmospheric oxygen into nitrogen dioxide under the influence of the high temperatures of lightning or of combustion within the internal combustion engine. (b) Write a single, balanced equation for the formation of ozone from a mixture of nitrogen dioxide and atmospheric oxygen. (c) Now write a single, balanced equation for the formation of ozone from the high-temperature reaction of atmospheric nitrogen and atmospheric oxygen.

14.5 Air Pollution: Acid Rain

The formation of sulfuric acid and nitric acid as secondary air pollutants results in a form of pollution known as *acid rain*. (The term was first used in 1852 to describe the results of atmospheric pollution produced by the beginnings of the Industrial Revolution in the midlands of England. Industry in that region and at that time derived its energy from nearby coal deposits.) By now "acid rain" has come to describe all forms of rain and snow whose excessive acidity causes environmental mischief, including the destruction of vegetation and marine life and the etching and corrosion of buildings and works of art that are exposed to the weather.

In a real sense all rain is acidic, with or without air pollution, simply because of the natural presence of carbon dioxide in the earth's atmosphere. Like the carbon dioxide of breath that turns the cabbage water acidic in the opening demonstration of Chapter 10, the carbon dioxide of the atmosphere dissolves in raindrops and forms carbonic acid as the drops fall to the ground (Fig. 14.2). The chemical reaction is the same as the one that produces the color change in the cabbage water of Chapter 10:

$$CO_2 + H_2O \longrightarrow H_2CO_3$$

Through this reaction atmospheric carbon dioxide can dissolve in rain drops to produce an acidic solution with a pH as low as 5.6 (Table 10.4). As a result, the purest rain, falling through an unpolluted sky, can reach earth as an acidic solution of pH 5.6. Rainfall would still be acidic, then, even if all factories were shut down and all cars remained parked. (At this pH, the acidity of rainfall stands somewhere between that of milk and fresh tomatoes; Table 10.4.)

Since the pH of rainwater containing carbonic acid can be as low as 5.6, we might consider rain whose pH is *lower* than 5.6 to be acid rain. Indeed many scientists did just that until about 1980. We know now, though, that other phenomena, not at all related to human activities, also work to decrease the pH of rainwater. Volcanic eruptions, for example, release gaseous sulfur dioxide and hydrogen sulfide, both of which are converted eventually to atmospheric sulfuric acid. Even the action of the bacteria of the soil contributes to the presence of sulfuric and nitric acids. As a result, even rain that falls far from inhabited or industrialized areas, through apparently uncontaminated air, shows a pH lower than 5.6—generally about 5.0.

Yet it's clear that the great majority of the air pollutants that eventually form acids—as much as 90% of all the atmospheric sulfur, for example—result from human activity, primarily the burning of fossil fuels. It's also clear that this activity is the source of the rainfall and atmospheric moisture that show remarkably low pH values. Rain that falls in the western regions of New York State, for example, can carry a pH as low as 4.1; during the summer the bottoms of clouds above the eastern United States often show a pH of about 3.6 (and occasionally as low as 2.6); a storm in Scotland in 1974 produced rainfall of pH 2.4; several years ago a fog in Los Angeles generated a pH of 2.0. For comparison, the pH of vinegar falls in the range 2.4–3.4 (Table 10.4). Much of this acidity appears to trace back to the sulfur-containing impurities of fossil fuels. As winds carry atmospheric contaminants across national borders, acid rain becomes an

Figure 14.2 The natural acidity of rain.

Graphic damage done to outdoor artwork by the action of acid rain and air pollution over a period of 60 years. The sandstone statue is located at the Herten Castle in Westphalia, Germany.

international issue. Air pollutants produced in one country end up in the rainfall of another. Canada, in particular, appears to suffer from acid rain coming from pollutants released in the United States.

The effects of acid rain on living things seem to result from a subtle complex of factors rather than from a direct, corrosive destruction of vegetation and marine life by the acids themselves. Changes in the pH of fresh water and resulting changes in the balance of its minerals seem to affect the reproduction and survival of many species of fish and other marine life; pH variations in soils can place a biological stress on trees, decrease the quantity and quality of soil nutrients available to them, and make the trees more susceptible to lethal diseases.

Acid rain and perhaps other forms of corrosive air pollution appear to be responsible for still other forms of environmental decay. In several European cities, statues and other works of art exposed to the weather show signs of severe erosion. While this form of destruction can't be repaired easily, at least some of the biological results of acid rain do appear to be reversible. With tighter regulation of emissions from automobiles and factories, the acidities of some lakes in the northeastern United States and in eastern Canada seem to be dropping, with an increase in the variety of their marine life.

■ QUESTION

Name one natural phenomenon, completely unrelated to any human activity, that can produce a pH lower than 5.6 in rainfall.

14.6 Air Pollution: Solving the Problem

What can be done about acid rain and other forms of air pollution? Several options are available, including

- the use of alternative energy sources,
- removal of pollutants from the products of combustion,
- improvement in the efficiency of the combustion process itself, and
- energy conservation.

We will look briefly at the first two, which make use of chemicals and chemical processes.

The Perspective of Chapter 7 presents a discussion of several alternative energy sources for motor vehicles. Of these, natural gas is particularly appealing because of its small potential for air pollution. This fuel contains few sulfur impurities; it burns cleanly with little formation of sulfur or nitrogen oxides. While electric cars powered by batteries might seem to be a completely pollution–free alternative, their use might only shift environmental concerns from one source or location to another. After all, if we switched from the internal combustion engine to electric batteries as a source of power, the total quantity of energy now provided by gasoline, diesel fuel, and similar refinery products would have to be replaced by an equivalent quantity of electricity needed to recharge the newly introduced transportation batteries. The burden of producing this additional electricity would have to be taken up by existing or new electric power generators, which would in itself introduce new environmental concerns.

In any case, cars and other vehicles account for a small fraction of the total emissions of SO_x and NO_x (but most of the CO) in the United States. As shown in Table 14.4, electric utilities and other stationary facilities produce the majority of these pollutants. While these plants might be designed to use alternative energy sources, more practical or more immediate approaches include improving the efficiency of the combustion process itself and removing pollutants from exhaust gases before they reach the atmosphere. We've seen an illustration of this last technique in the catalytic converters used in cars (Sec. 7.7).

Several different, practical approaches can reduce pollutants in industrial exhausts, including *electrostatic precipitation*, *filtration*, and *scrubbing*. Electrostatic precipitation removes *particulates* and *aerosols*, which are small particles of liquids or solids dispersed in smoke (Table 13.2). (Smaller particles, less than 10^{-3} mm in diameter, make up the aerosols; the larger particles are the particulates.) As shown in Figure 14.3, exhaust gases pass between two charged vertical plates or electrodes. The particles pick up electrons supplied by the negative electrode and move to the more positive one. There the liquid particles accumulate and flow to the bottom of the collector; with agitation the solid particles drop off.

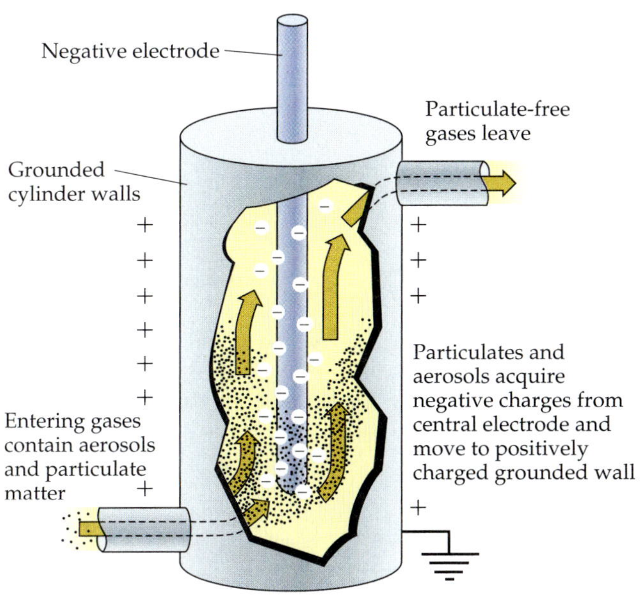

Figure 14.3 Schematic cutaway diagram of electrostatic precipitator.

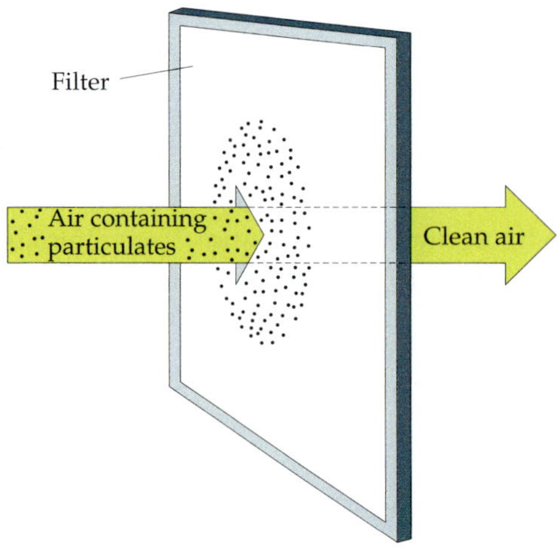

Figure 14.4 Particulate removal by filtration.

Industrial filters work mechanically, much like air-conditioner or vacuum cleaner filters. Exhaust gases pass through, leaving particulates and aerosols behind on the filter (Fig. 14.4). Scrubbers operate by passing exhaust gases through water, often present as a fine spray (Fig. 14.5). Scrubbers that force the gases through a slurry of calcium carbonate, $CaCO_3$, or magnesium hydroxide, $Mg(OH)_2$, are especially useful for removing SO_2. The sulfur dioxide dissolves in the water and reacts with the calcium carbonate or the magnesium hydroxide to form a salt. If magnesium hydroxide is used, the magnesium sulfite ($MgSO_3$) that's formed can be isolated and heated to regenerate SO_2. This recovered sulfur dioxide can be collected and used as a raw material in other commercial processes.

Removal of SO_2 by scrubbing with a slurry of $Mg(OH)_2$:

$$\underset{\text{sulfur dioxide}}{SO_2} + \underset{\text{magnesium hydroxide}}{Mg(OH)_2} \longrightarrow \underset{\text{magnesium sulfite}}{MgSO_3} + \underset{\text{water}}{H_2O}$$

Regeneration of SO_2:

$$\underset{\text{magnesium sulfite}}{MgSO_3} \longrightarrow \underset{\text{sulfur dioxide}}{SO_2} + \underset{\text{magnesium oxide}}{MgO}$$

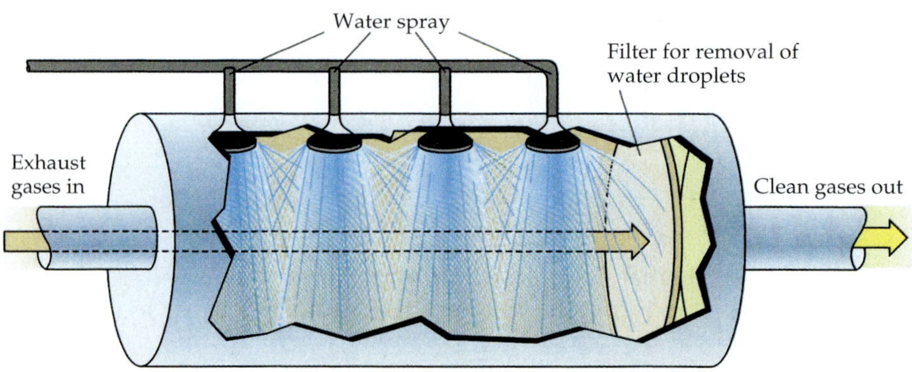

Figure 14.5 Schematic cutaway diagram of a scrubber.

Which of the three methods of removing air pollutants from exhaust gases—electrostatic precipitation, filtration, and/or scrubbing—operate(s) by producing chemical changes in the pollutants? That is, which operate(s) by producing changes in the chemical composition of the pollutants as they are removed?

■ QUESTION

14.7 Ozone: A Gas That Can Kill

Ozone, O_3, a triatomic form of elemental oxygen, illustrates nicely the problem of trying to pin the label "pollutant" on any particular substance without taking into consideration its location and its concentration (Sec. 14.2). Like so many other chemicals, ozone can be beneficial or harmful, depending on the circumstances. In the atmosphere near the surface of the earth—in the air we breathe—ozone is an irritating, toxic gas; high above us, in the ozone layer of the stratosphere, this same gas forms a lifesaving shield that protects us from the catastrophic effects of overwhelming ultraviolet solar radiation. As we examine ozone, we'll look at it first as a pollutant of the troposphere, then, in Section 14.8, as a shield in the stratosphere.

As indicated in Figure 14.6, O_3 molecules contain more energy and are less stable than the more common, diatomic form of oxygen, O_2. Ozone forms from oxygen molecules in a variety of ways, each of which results in the conversion of oxygen to ozone through the addition of energy to the O_2. As electrical currents pass through the air, from lightning or from a sparking motor or electrical appliance, for example, these high-energy currents transform diatomic oxygen molecules into triatomic ozone molecules, as shown in the balanced equation of Figure 14.6. Ozone also forms as an indirect result of the operation of the internal combustion engine, as we saw in Section 14.4.

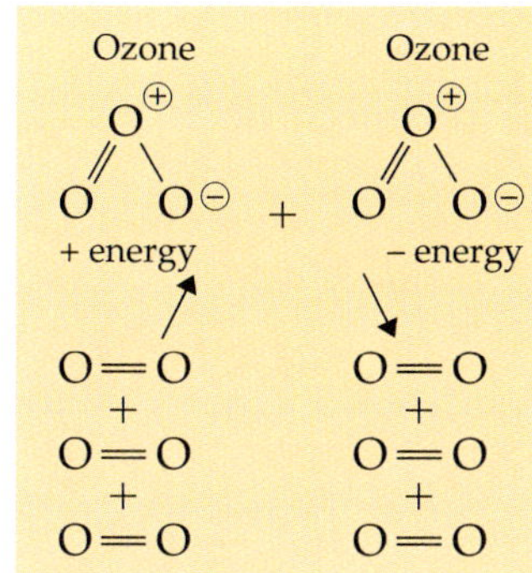

$$3\,O_2 \underset{-\text{ energy}}{\overset{+\text{ energy}}{\rightleftharpoons}} 2\,O_3$$

Figure 14.6 Oxygen and ozone. The addition of energy to oxygen molecules converts them to ozone molecules. The loss of energy from ozone molecules converts them to oxygen molecules.

However they may have been formed, triatomic ozone molecules eventually lose their added energy and decay back to the less energetic, more stable diatomic oxygen molecule (Fig. 14.6). Because of this, concentrations of ozone in the atmosphere of urban areas usually reach their maximum during the day, with both intense automobile traffic and intense sunlight contributing to their formation, then drop at night as traffic decreases and sunlight vanishes.

Transient, relatively low atmospheric concentrations of ozone formed during electrical storms can add a pleasant, fresh quality to the air. But breathing ozone over a longer term, ranging from a few minutes to several hours, and at concentrations as low as 1 ppm in air (such as we might find consistently near sparking machinery, electrical generators, and some types of photocopiers) can lead to sore throats, general bronchial irritation, coughing, and fatigue. Much higher concentrations can kill. Ozone is also lethal to lower forms of life, including bacteria. In Europe ozone is widely used as a disinfectant for community water supplies, much as chlorine is used in the United States.

Would you expect higher concentrations of ozone in the air of cities in the Northern Hemisphere during the winter or the summer? Explain.

■ QUESTION

14.8 Ozone: A Gas That Protects Life

In the stratosphere ozone shows us a different face. There, because of its ability to absorb the sun's ultraviolet radiation, ozone plays the role of a lifesaving gas. The sun's ultraviolet rays contain far more energy than the infrared radiation (heat) and light that accompany them. Like many chemicals, this ultraviolet radiation presents us with both benefits and risks. For example, in what may be one of their few beneficial effects ultraviolet rays promote the formation of vitamin D in our skin (Sec. 19.5). Yet this same radiation can also destroy living cells and tissue. As anyone who has suffered a sunburn knows, acute exposure to ultraviolet radiation can produce severe and painful burns. A tan, sometimes considered a sign of good health or of easy living, is actually the skin's response to this damaging radiation. The darkening, which we interpret as a tan, acts to protect the skin from radiation damage. Still, long-term exposure to ultraviolet radiation, even at levels far below those that burn, can cause cataracts and premature aging of the skin and other skin disorders, including cancer.

Sunburns, cataracts, and prematurely aged and cancerous skin are among the effects of long exposure to the relatively weak ultraviolet radiation that reaches the earth's surface. Exposure to *intense* solar radiation, unfiltered by the ozone of the stratosphere, could present a lethal hazard to all forms of unprotected life, ranging from the lowly microorganisms that thrive on and just beneath the surface of the oceans, to the exposed, higher land animals, including humans. What protects us from these potentially lethal levels is the layer of ozone—the *ozone layer*—that surrounds the earth within the stratosphere, centered some 30 to 35 km (about 20 miles) above the earth's surface. Its average altitude, shape, and dimensions vary over a large range with changes in latitude, the seasons, and the intensity of solar radiation. It forms as oxygen atoms, torn out of their diatomic molecules by the intense solar radiation at higher levels of the atmosphere, work their way downward into the stratosphere and combine with diatomic oxygen to form triatomic ozone (Fig. 14.7).

Absorption of ultraviolet radiation by an ozone molecule within the ozone layer regenerates an oxygen atom and an oxygen molecule. Left undisturbed over the millennia, this cycling between the formation and destruction of ozone molecules has produced a reasonably steady concentration of ozone that has allowed surface life to evolve and thrive under its invisible shield (Fig. 14.8). Should the ozone layer vanish now, life as we know it might not easily survive on the surface of the earth.

Figure 14.7 Formation of the ozone layer.

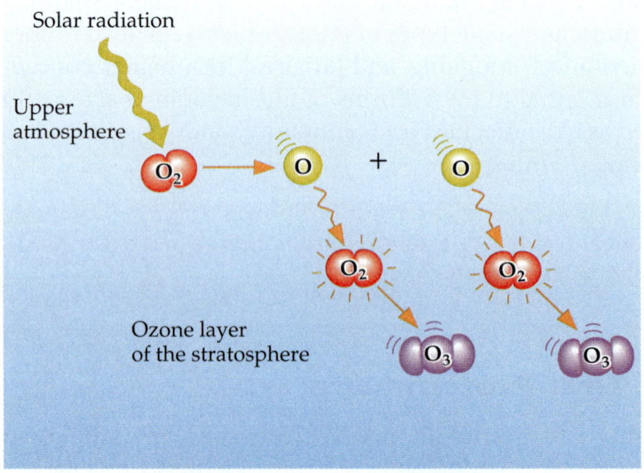

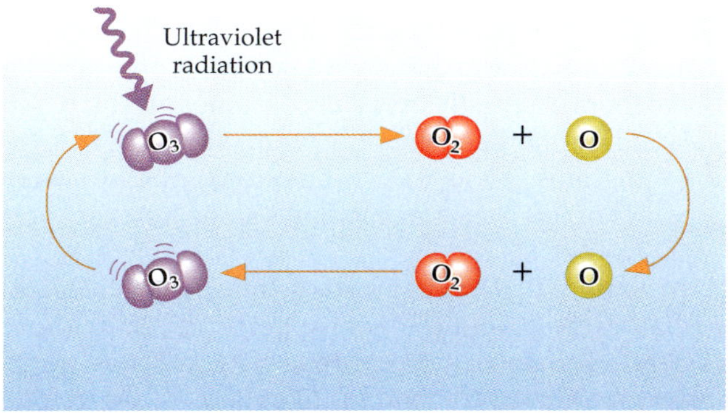

Figure 14.8 Ozone–oxygen cycling in the ozone layer.

■ QUESTION

The ozone layer absorbs much of the sun's ultraviolet radiation, permitting only a fraction of the radiation to reach the earth's surface. Describe one way in which ultraviolet radiation harms humans and one way in which it provides a benefit.

14.9 Ozone: Preserving the Shield

A reasonably stable ozone layer has existed in the upper atmosphere for countless centuries as a result of a steady balance between the formation of ozone in the stratosphere and its natural decay to oxygen molecules and atoms. Today, though, something new and unexpected threatens the integrity and perhaps the very existence of this shield. The threat comes from a set of highly volatile, relatively stable organic compounds containing bromine and/or chlorine substituents. Virtually the only source of these compounds is through human activity, especially industrial production. They are either not at all flammable or burn only under unusual conditions.

Because of their low human toxicity and very low or nonexistent flammability they have been considered relatively safe compounds, some of which have been used extensively in consumer products or services. Many are no longer found in commercial application because of their threat to the ozone layer. The following are among the most important:

- Carbon tetrachloride, CCl_4: cleaning fluid and spot remover, both commercially and in the home, and fire extinguisher
- Methyl bromide, CH_3Br: pesticide especially useful for fumigating soil and killing insects on crops
- Methyl chloroform, $CH_3—CCl_3$: industrial solvent and raw material in the commercial production of other compounds
- Chlorofluorocarbons (CFCs), such as trichlorofluoromethane, CCl_3F, and hydrochlorofluorocarbons (HCFCs), such as 1,1-dichloro-1-fluoroethane, $CCl_2F—CH_3$: aerosol propellants for the canned sprays and foams of consumer products; refrigerants in mechanical air conditioners and refrigerators; foam-producing agents in the manufacture of highly porous, lightweight plastics, plastic foams, and insulation; solvents for cleaning circuit boards and similar electronic parts
- Halons, a group of polyhalogenated alkanes, such as bromochlorodifluoromethane, $CBrClF_2$: fire extinguishers

Because of their threat to the ozone layer, both the manufacture and use of these halogenated alkanes (also known as *halocarbons*) have been discontinued or severely restricted. Their menace originates, paradoxically, in their low chemical reactivity, low or nonexistent flammability, low toxicity, and low cost. Properties like these might be expected to make them especially useful in consumer products and applications, and indeed they were. In the peak year for production of CFCs for example (1986), over a billion kilograms (about 1.25 million tons) of CFCs were produced. That's about half a pound of this class of compounds alone for every person on earth.

F. Sherwood Rowland (top), Mario J. Molina (middle), and Paul Crutzen (bottom). Work of Rowland and Molina, based on Crutzen's earlier discoveries, led to the recognition that atmospheric chlorofluorocarbons could destroy the ozone layer.

But partly because of their volatility, their widespread use resulted in an equally widespread entry of these halocarbons into the atmosphere, especially those used as aerosol propellants. And their low chemical reactivity and low water solubility—normally desirable properties in compounds like these—lead to long lifetimes within the atmosphere. They had plenty of time to drift upward, in a journey that might take 7 to 10 years, into the stratosphere where the ozone layer lay. There they might remain for decades, even (for CCl_2F_2) more than a century, to work their mischief.

In the upper levels of the atmosphere, the CFCs and other halocarbons absorb the sun's ultraviolet radiation and decompose into various chemical products, including chlorine and bromine atoms. In this respect halocarbons mimic the ozone molecule as it decays to an oxygen molecule and an oxygen atom. But there's an important difference here. The *oxygen atom* that's produced through the interaction of ultraviolet radiation with ozone can recombine with a diatomic oxygen molecule to regenerate the protective ozone. The *chlorine* and *bromine* atoms of the halocarbons, though, react with ozone by a complex sequence of reactions to generate *oxygen molecules*, which do not regenerate ozone as quickly as do the *oxygen atoms* produced from ozone by ultraviolet radiation.

Formation of chlorine atoms:

$$\text{chlorofluorocarbons} \xrightarrow{\text{ultraviolet radiation}} \underset{\text{chlorine atoms}}{\text{Cl}} + \text{other products}$$

Reactions leading to the conversion of ozone to oxygen molecules:

$$2\,O_3 \xrightarrow{\text{catalytic action of chlorine atoms}} 3\,O_2$$

$$O_3 + O \xrightarrow{\text{catalytic action of chlorine atoms}} 2\,O_2$$

The result is a relatively rapid depletion of ozone. With the rate of depletion overwhelming the natural restorative process that put the ozone there in the first place, the result is an erosion or destruction of the ozone layer itself. Through the chemical reactions that operate in this conversion of ozone to molecular oxygen, a single chlorine atom or a single bromine atom that's split off a halocarbon molecule could, *in theory* (but hardly in practice, under the conditions of the stratosphere), destroy the entire ozone layer.

This ozone-destructive potential of atmospheric CFCs was recognized as early as 1974 by F. Sherwood Rowland, professor of chemistry at the University of California, Irvine, and his co-worker, Mario J. Molina.

[For demonstrating that compounds produced by human activity might cause erosion of the ozone layer, both Rowland and Molina shared the 1995 Nobel Prize for Chemistry with Paul Crutzen of the Max-Planck-Institute for Chemistry in Germany. Professor Crutzen laid the foundation for Rowland and Molina's discovery by demonstrating, in 1970, that NO and NO_2 (Sec. 14.4) could decompose ozone catalytically.]

In 1978, in response to warnings of possible damage to the ozone layer and a potential increase in the intensity of ultraviolet radiation that reaches the surface of the earth, the United States banned the use of CFCs as aerosol spray propellants in consumer products. A decade later an international panel of scientists reviewing fluctuations in the ozone layer found evidence that it is indeed slowly shrinking, presumably through the action of halocarbons that have already entered the stratosphere.

The strongest evidence for this erosion lies in the discovery of a sharp drop every fall in the density of the part of the ozone layer that lies over the Antarctic. In 1993 this annual drop in ozone density, known as an *ozone hole*, reached a record size, covering an area almost equal to all of North America. Although ozone depletion over northern regions doesn't approach the proportions observed in Antarctica, the loss of stratospheric ozone over the Arctic grew to a record level of its own briefly in 1996. To protect the ozone layer, our vital shield against the sun's ultraviolet radiation, more than 140 nations agreed in 1987, through a treaty known as the *Montreal Protocol* (and its later revisions), to end or severely restrict the production of CFCs and related halocarbons. Despite the provisions of the Montreal Protocol, ozone depletion will be a continuing concern because of the long lifetimes of halocarbons already released into the atmosphere.

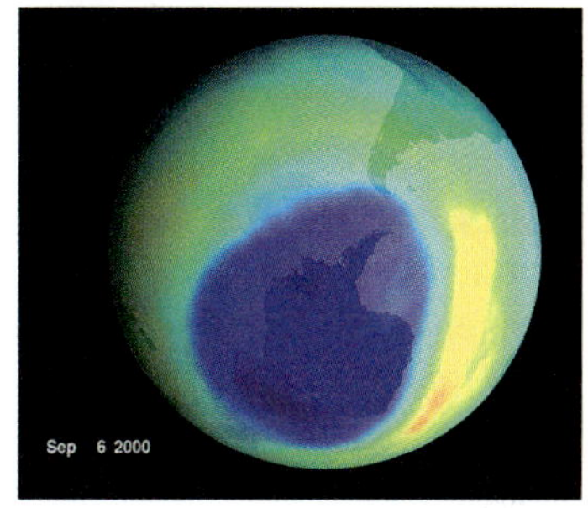

Computer-enhanced image of the Antarctic ozone hole, September 2000. Its record size is about three times that of the entire United States.

■ QUESTION

What aspect of molecular structure do the compounds discussed in this Section all have in common? What aspect of environmental effect do these compound all have in common?

14.10 Water and Water Pollution

When we speak of the water that's used in human activities, what often comes to mind is a glass of pure, clear, cold water or a simple beverage that quenches thirst. We might also include the water we use for washing ourselves and our clothes, dishes, and cars, as well as the water of recreational activities such as fishing, swimming, and boating. Excluding the salt water that we use for recreation, the water available for the ordinary uses of everyday life is often scarce. We might even say that the amount of fresh, nonsaline (not salty) water available to us for our common, daily activities is only a drop in the bucket of all the waters of the earth's hydrosphere and crust.

The earth's crust holds about 2×10^{21} kg (2×10^{18} tons) of water spread out over more than 70% of its surface. Of all this, we're able to reach easily only about 0.3% for use as fresh, drinkable water (Table 14.5). But it's a vital 0.3%. Currently all the peoples of the earth remove—and largely return—about 4300 km^3 (roughly 1000 cubic miles) of fresh water each year from our lakes, rivers, and streams and from the **groundwater**, the water in the rocks and soil of the crust, lying in a range of a few to a few thousand feet below the surface.

Groundwater is water within the earth's crust, lying just below the surface.

With all the water we use for drinking (as water itself and as the principal component of our beverages), for washing, and for preparing food, we might think that more water is used in the direct support of our lives and cleanliness than for

Worldwide, more fresh water is used for agriculture and irrigation than for anything else.

Table 14.5 Distribution of Water in the Earth's Crust and Hydrosphere

	Volume (cubic miles)	Percentage[a]
Salt water		97.5
Oceans	317,000,000	
Inland seas and saltwater lakes	25,000	
Fresh water, easily available		0.3
Groundwater, within half-mile of surface	1,000,000	
Lakes	30,000	
Rivers	300	
Fresh water, not easily available		2.5
Antarctic icecap	6,300,000	
Groundwater, more than half-mile from surface	1,000,000	
Arctic icecap and glaciers	680,000	
Total, rounded	325,000,000	

[a] Percentages add to more than 100% because of rounding.

any other single purpose. Far from it. Globally, about 67% of all the fresh water drawn from the earth goes toward irrigation and other agricultural needs. The needs of individual humans for drinking, washing, and other domestic activities account for only about 10%.

Biological contamination of water results from the presence of disease-causing microorganisms.

Several forms of pollution affect our water supplies, including *biological*, *thermal*, *sedimentary*, and *chemical*. **Biological contamination** results from the presence of disease-causing and life-threatening microorganisms, especially in drinking water. To remove these microorganisms, community water systems of the United States, which provide water to more than 90% of the population, generally use disinfection with chlorine gas (Sec. 11.15) or ozone. (Chlorination is more common in the United States; ozone is more often used in Europe.) Further purification takes place through filtration, which removes microorganisms resistant to these disinfectants.

Thermal pollution occurs when environmental harm comes from the warming of a body of water by the discharge of waste coolant.

Thermal pollution often occurs through the warming of a body of water by the discharge of waste coolant. Gases, including the oxygen needed by both land and aquatic animals, are less soluble in warm than in cold water (Sec. 9.10). Thus a rise in the temperature of a body of water can deprive fish and other aquatic animals of needed oxygen. A rise in temperature can also, in itself, shorten the lives and affect the reproductive cycles of some species.

Sedimentary pollution results from the accumulation of suspended particles.

Sedimentary pollution results from the accumulation of suspended particles. These can be particles of soil washed into a body of water by the runoff of rainwater from the land or particles of insoluble organic and inorganic chemicals carried along by runoff or by waste water. These sediments pollute in many ways, including the simple blocking of sunlight. A drop in the amount of light that penetrates into a body of water interferes with photosynthesis by aquatic plants and decreases the ability of aquatic animals to see and to find food. Sediments also carry absorbed chemical and biological pollutants with them. Worldwide, sediments account for the greatest mass of pollutants and generate the greatest amount of water pollution. Sediments can be removed from water by a variety of techniques, including

- filtering, a procedure similar to the one used for removing particles from polluted air (Sec. 14.6);

- settling, in which particles move to the bottom of standing water under the influence of gravity; and
- coagulation, in which an agent such as aluminum sulfate causes small particles of sediment to clump together, forming larger, more easily removed particles.

Perhaps the most troubling of all is **chemical pollution**, caused by the presence of harmful or undesirable chemicals. Chemical pollution is a bit different and a bit more subtle than other forms of pollution. Thermal pollution has little effect on the potability of water. Sedimentary pollution is usually highly visible and easily removed, at least from public water supplies. Even biological pollution seems in some ways less hazardous than chemical pollution since microorganisms can usually be destroyed simply by boiling the water they infect. Furthermore, they make their presence known almost immediately through testing or through the outbreak of an epidemic. The illness of the unfortunate first victims usually triggers a "boil order" from local health authorities, which reduces further harm.

Chemical pollution results from the presence of harmful or undesirable chemicals.

But chemical pollution isn't as simple a matter. The harmful effects of chemical pollution, like the effects of hazardous substances in general, can be subtle and can take time to make themselves felt. Chemical pollutants that find their way into the waters of the earth are a varied group of substances, including

- the common salt that's used to melt the winter ice of northern roads and that seeps into the ground with spring thaws and rains;
- agricultural fertilizers that increase crop yields and that are carried into our groundwater and freshwater lakes and rivers by the runoff of rainwater;
- leftover household paints, used motor oil, and similar consumer products that are thoughtlessly dumped onto the ground and into sewers; and
- the notorious toxic dumps and major oil spills that make headline news.

We'll examine chemical pollution of our freshwater supplies briefly, first through the role that chemicals play in biological contamination, then through the chemicals that act as pollutants in their own right.

■ QUESTION

In what ways do *thermal pollution* and *sedimentary pollution* affect the presence or absence of chemicals in a body of water?

14.11 The Role of Chemicals in Biological Contamination

Not all forms of pollution are necessarily harmful to all living things. Some are actually highly beneficial, but only to certain classes of life. Some chemicals, for example, can act to the advantage of one class of living things and at the same time to the detriment of another. Nitrates (salts containing the nitrate anion, NO_3^-) and phosphates (salts with a phosphate anion, PO_4^{3-}), for example, are major components of agricultural fertilizers, the wastes of farm animals, and sewage. As we might expect from their presence in commercial fertilizers, these compounds are effective plant nutrients. When they and similar plant nutrients enter the waters of still lakes and slow-moving streams, they produce a rapid growth of surface plant life, especially algae. As these minuscule plants grow, they form a mat that can cover the surface, sealing off the rest of the water from the oxygen of the air. Deprived of dissolved oxygen, fish and other aquatic

Eutrophication produces a mat of algae on the surface of a body of water.

animals virtually disappear from these waters. Plant life then thrives at the expense of animal life. This form of water pollution, which operates through selective stimulation of plant life at the expense of animals, is called **eutrophication**, from Greek words meaning well nourished.

Eutrophication is a form of water pollution in which plants, well nourished by pollutants, thrive at the expense of aquatic animals.

Dissolved oxygen plays another role as well: in the quantitative measurement of pollution. Many of the wastes dumped into the hydrosphere consist of organic compounds that serve as nourishment to microorganisms or that are degraded by them to simpler products. In either case, much of the biological activity generated by organic wastes takes place under *aerobic* conditions, requiring the use of oxygen. Like us, the more active these microorganisms are, the more oxygen they consume. By measuring the **biochemical oxygen demand (BOD)** in a body of water, we can estimate the total of all the biochemical processes that consume oxygen—including the aerobic activity of the microorganisms—and thus indirectly determine the amount of organic material present. Other measures of organic pollution include determination of the amount of **dissolved oxygen (DO)** in the water and observation of any decrease in dissolved oxygen content, which can be viewed as the **dissolved oxygen deficit (DOD)**. The smaller the amount of dissolved oxygen, or the greater the deficit, the greater the microbiological activity and, presumably, the greater the amount of organic material in the water.

Biochemical oxygen demand (BOD), **dissolved oxygen (DO)**, and **dissolved oxygen deficit (DOD)** are all measures of the aerobic activity of aqueous microorganisms and, thus, indirect measures of organic pollutants.

■ **QUESTION**

Identify each of the following quantitative measures of water pollution and explain whether a high value or a low value indicates the presence of organic material in the water: (a) BOD; (b) DO; (c) DOD.

14.12 Groundwater, Aquifers, and Chemical Pollution

We have about three times as much groundwater (fresh water lying in the earth's crust, just beneath the surface) available to us as there is in all the lakes and rivers of the world (Table 14.5). As we saw in Section 9.11, about half the people of the

United States, including three–quarters of those who live in large cities, get their drinking water from wells or springs that tap into this groundwater.

Much of this groundwater resides in **aquifers**, large underground layers of porous rock, often limestone, from which fresh water can be drawn through wells. The groundwater of these porous rock formations makes up about 97% of the earth's freshwater supplies. Some of these aquifers, the *fossil aquifers* of deserts and other regions, contain water that seeped into them perhaps thousands of years ago. This "fossil water" remains trapped in the aquifers as a nonrenewable resource. Once the fossil aquifer is dry, the particular society that depends on it must find another source of water. Among regions that rely on fossil aquifers are northwestern Texas, the African nation of Libya, and the Arabian peninsula. The people of Saudi Arabia use fossil water for about three-quarters of their freshwater supplies.

An **aquifer** is a large layer of porous rock that holds fresh water and from which water can be drawn by wells.

Rechargeable aquifers of other, less arid and more fertile regions stand in contrast to the fossil aquifers. The water within rechargeable aquifers isn't static, but flows into, through, and out of their porous rock, at rates ranging from a few centimeters to a meter or so each day. The water flows into the aquifer from the ground and can move both to and from lakes and rivers. In dry seasons, the groundwater flowing from these aquifers helps keep streams flowing; in wet seasons and after heavy rains, water flows into the aquifers, replenishing the supply of groundwater.

One of the largest rechargeable aquifers is the Biscayne Aquifer that lies under the southeastern tip of Florida. This aquifer, which resembles a tilted triple-layer cake (Fig. 14.9), covers an area of many thousands of square miles and supplies more than 300 million gallons of fresh water each day to the residents of Miami-Dade

Figure 14.9 Cross section of the world's most productive aquifer—the Biscayne Aquifer.

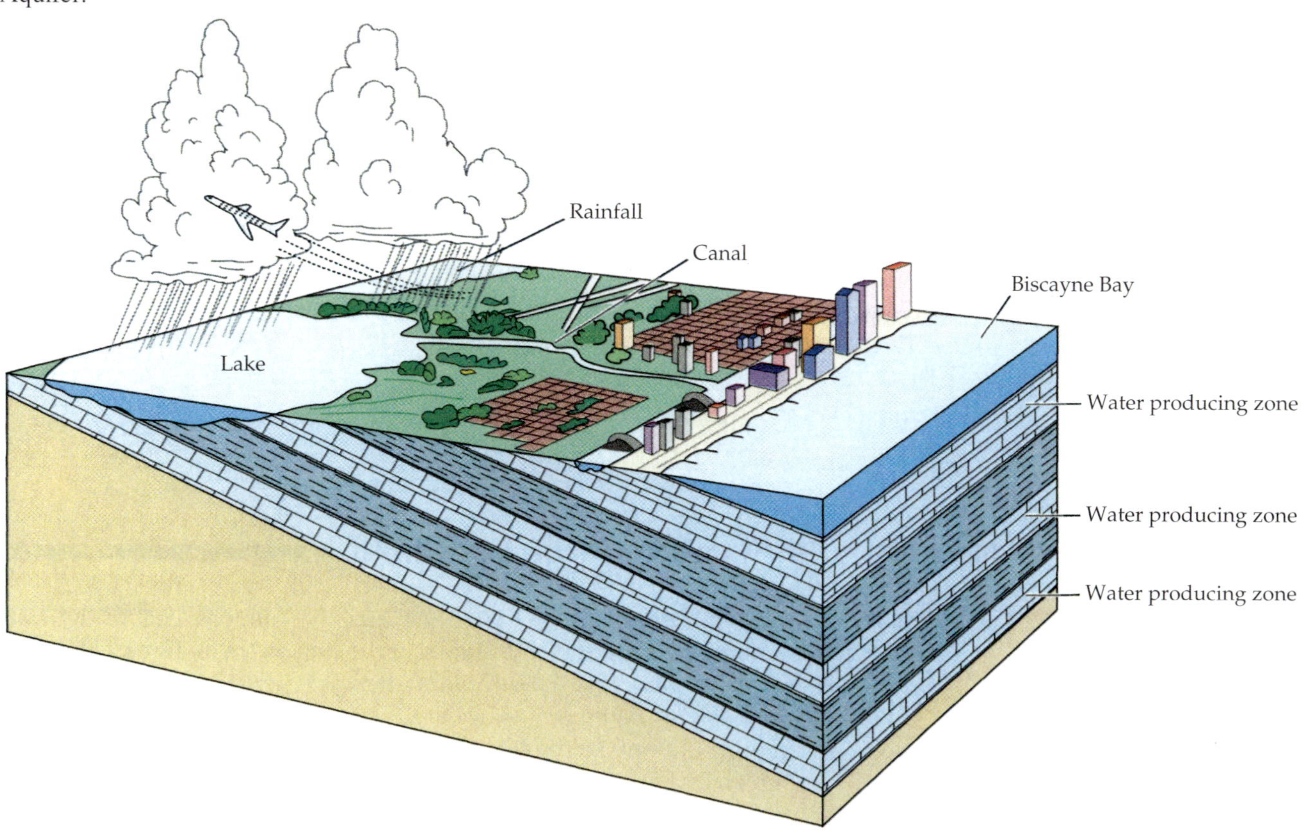

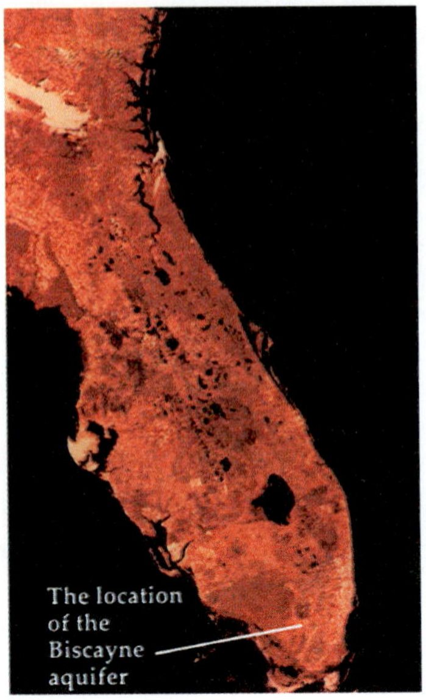

The Florida peninsula. The Biscayne Aquifer lies at the state's southeastern tip.

County and nearby areas, including the cities of Miami and Miami Beach. The aquifer runs from about 25 to 45 m (80 to 150 ft) deep along the southeast coast of the peninsula, along the Atlantic Ocean, to less than 3 m (10 ft) deep well inland, about halfway to the western shore and the Gulf of Mexico. It consists of three thick, sloping layers of porous, water-containing limestone separated by two thinner layers of clay and sand. The water flows eastward in the limestone layers, toward the ocean.

While the porosity of the rock, the periodic rains, and the ease with which rain falling on the surface of the land percolates down into the aquifer all contribute to the continuous recharging of the aquifer itself, they also contribute to pollution of the water it supplies. Carried by the rain or seeping into the ground on their own, pollutants that lie on the surface of the land have easy access to much of the groundwater.

To protect its drinking water, Miami-Dade County, the metropolitan area served by the Biscayne Aquifer, has established concentration limits or other guidelines for a large number of chemicals, substances, and conditions, some of which appear in Table 14.6. These either add to or narrow the federal standards of the Safe Drinking Water Act (Sec. 9.11). The standards of Table 14.6 apply not only to drinking water but also to substances that can become part of the drinking water by seeping into the ground from the surface of the land. One of the surest safeguards for the purity of our drinking water is to keep pollutants off the surface of the land. We'll examine this topic in the next section.

Table 14.6 Pollution Limits for the Water of Miami-Dade County, Florida (Selected Pollutants and Conditions)

	Maximum Concentration Allowed (ppm; mg/L)		
Substance or Condition	Sewage and Industrial Waste Water Discharges	Discharges into Sanitary Sewers	Drinking Water
Arsenic	0.05	0.1	0.01
Chromium	1.0	1.0	
Copper	0.5	0.5	1.0
Lead	0.05	0.3	
Mercury	None	0.01	
Dissolved oxygen	2.0 ppm[a]		
Sulfides	0.2		
Zinc	1.0	1.0	5.0
Biochemical oxygen demand	30.	200.	
Oil and grease	30.	100.	
pH	6.0–8.5[b]		
Suspended solids	40.	200.	
Total metals		2.0	
Total organic compounds		2.0	

[a] Minimum allowed.
[b] Range permitted.

■ **QUESTION**

What is an aquifer? How does water enter an aquifer and how do we obtain water from one?

14.13 Chemicals on the Surface of the Land and in the Air and Water

Since many substances can move easily and quickly from one part of the environment to another, sorting out pollutants according to the region of the environment that they impact isn't a simple matter. What we recognize as a land pollutant one day can be an air or a water pollutant the next. We've already seen how an air pollutant can be transformed into a water pollutant as the sulfur oxides of air pollution come back to damage the earth and its waters through the medium of acid rain.

Examples of land contaminants that can become water pollutants include the volatile organic compounds vinyl chloride (also known as 1-chloroethene) and vinylidene chloride (1,1-dichloroethene). Among other uses, these two compounds serve as raw material in the manufacture of plastics. Vinyl chloride, which provides the plastic known as PVC (Sec. 21.12), is listed by the government as a carcinogen, while vinylidene chloride, used to produce plastics such as the food-wrapper *Saran* (Sec. 21.12), is an irritant that can cause damage to internal organs:

$CH_2{=}CHCl$	$CH_2{=}CCl_2$
vinyl chloride 1-chloroethene	vinylidene chloride 1,1-dichloroethene

If they are spilled onto the land, perhaps as part of accidental waste discharges from manufacturing operations, these compounds can work their way into our water supply and become water pollutants. Several years ago these two compounds were found in municipal well water drawn from some sections of the Biscayne Aquifer. To remove them, the water was cleaned by *air stripping* (Fig. 14.10), a process that's essentially the reverse of the scrubbing of Figure 14.5.

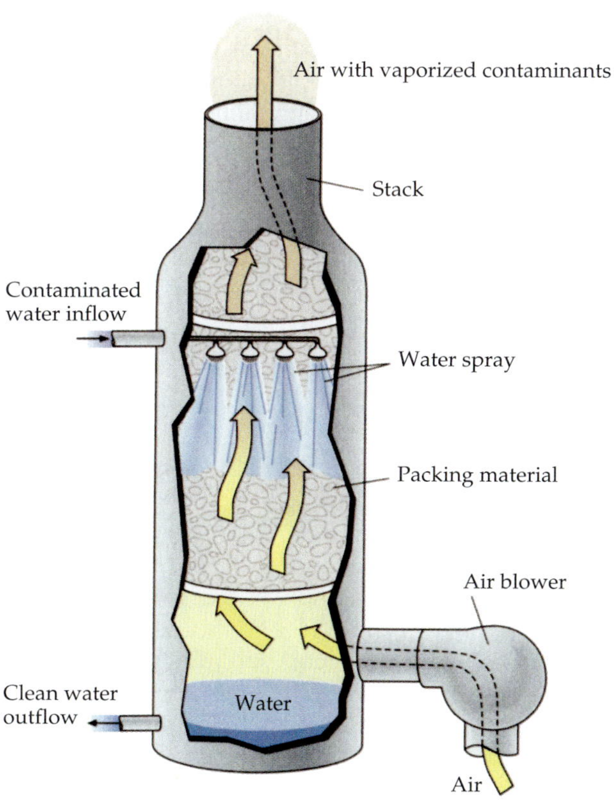

Figure 14.10 Schematic diagram of an air stripper used for removal of VOC.

In air stripping, volatile solutes are removed from water by spraying the water into a moving stream of air. The volatile pollutants evaporate from the small droplets of the fine spray of contaminated water and are carried into the atmosphere by the stream of air. The process can be used when the concentrations of the pollutants carried into the atmosphere by the stripper's exhaust are within acceptable limits. Stripping reduced the concentration of vinyl chloride in the aquifer water from an unacceptable 0.000021 g/liter of water (21 parts per billion, or ppb, Sec. 9.9) to an acceptable 0.00000003 g/liter (0.03 ppb); levels of vinylidene chloride dropped from 5 ppb to 0.02 ppb.

■ QUESTION

In what sense is air stripping the reverse of scrubbing?

14.14 Controlling Pests of the Land: Insecticides and Herbicides

Chemicals that can create hazards on the land usually don't announce themselves, unless they're among the pesticides sprayed on lawns and gardens. It's common practice for commercial sprayers to post signs on newly treated lawns warning people to keep off the grass. Among the chemicals used to control pests of the land are the *insecticides* used to eliminate flying and crawling insect pests; *herbicides*, which destroy weeds and other unwanted plant growth; *fungicides* for the control of plant-damaging fungi; and various chemicals used for the destruction of rats, mice, nematodes, and other animal pests. We'll look now at some useful insecticides and herbicides.

Insecticides, which are (ideally) poisons far more lethal to insects than to other forms of life (opening demonstration, Chapter 20), have been used for hundreds of years to protect agricultural plants from hungry insects. Early insecticides were mostly inorganic compounds containing arsenic, lead, and/or mercury. Unfortunately, while effective at killing insects they were also poisonous to humans. Once applied, they persisted in the soil, were taken up by growing plants, and became part of the food supply.

Pesticides used on lawns and gardens can pose health hazards to humans and pets.

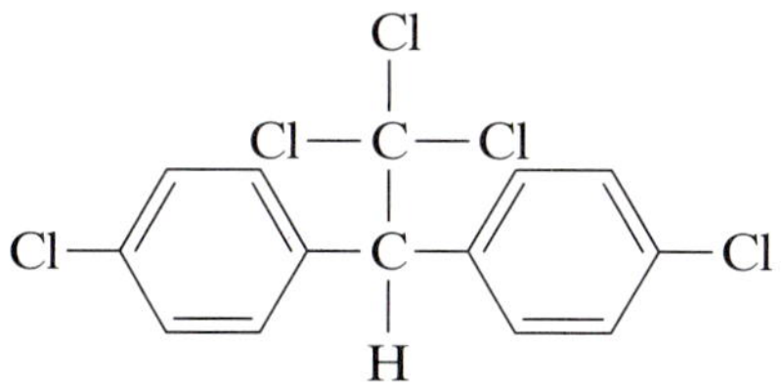

Figure 14.11 The insecticide DDT. Also known as dichlorodiphenyltrichloroethane and 1,1,1-trichloro-2,2-bis (*p*-chlorophenyl) ethane.

The search for greater selectivity led to use of *nicotine* extracted from tobacco plants (Sec. 23.10) and the *pyrethrins*, a group of compounds isolated from chrysanthemums. The turn to organic compounds led to the examination of *DDT*, a shortened form of a common (but ambiguous) chemical name, *dichlorodiphenyltrichloroethane*, for the structure shown in Figure 14.11. A more precise name, 1,1,1-trichloro-2,2-bis(*p*-chlorophenyl)ethane is rarely found outside chemical reference works. First prepared in 1874 by a German chemist, the compound was demonstrated to be a potent insecticide in 1939 by Paul Müller of Switzerland. In 1948 Müller received the Nobel Prize in medicine and physiology for his work with DDT.

Widely used during and after World War II to remove bodily pests from soldiers and others caught up in the turmoil, DDT produced no readily apparent harm to humans. An apparently safe substance, DDT became the insecticide of choice after the war. Soon, though, it became apparent that a combination of the following could eventually lead to an environmental disaster:

- its resistance to degradation, reflected by its persistence in the environment (DDT is an example of a *chlorinated insecticide*; organic compounds containing chlorine atoms bonded to benzene rings are particularly resistant to degradation);
- its great solubility in fat, including the fat of milk and the body fat of animals, with the potential for damage to internal organs;
- its harmful effects on reproduction of birds, fish, and other animals, especially by weakening the shells of eggs; and
- its decreasing effectiveness as many species of insects developed resistance.

The effects of DDT and its potential for harm were described clearly by the American biologist Rachel Carson in her 1962 book *Silent Spring*. Partly as a consequence of this warning, the use of DDT as an insecticide was severely restricted in the United States in 1973. Nonetheless, DDT is still manufactured in the United States for export to countries where its use is freely permitted.

Spraying crops with pesticides.

As in so many other areas of the interaction between science and society, the use of pesticides balances risks against benefits, or one kind of risk against another kind. Many insects, for example, present serious hazards to society. Locusts and grasshoppers destroy crops, mosquitos spread malaria, and other insects do their own deadly work. Yet chemicals used to eliminate these hazards present hazards of their own, as we've just seen for DDT. In balancing one risk against another, chemists and biologists have developed a broad range of strategies for insect control.

Figure 14.12 The insecticides malathion and diazinon.

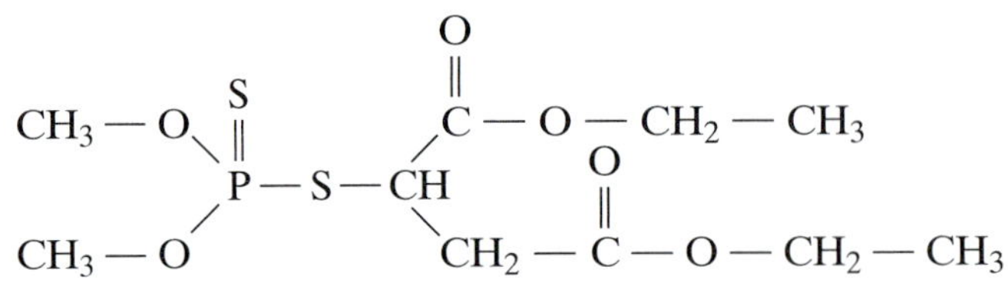

Malathion

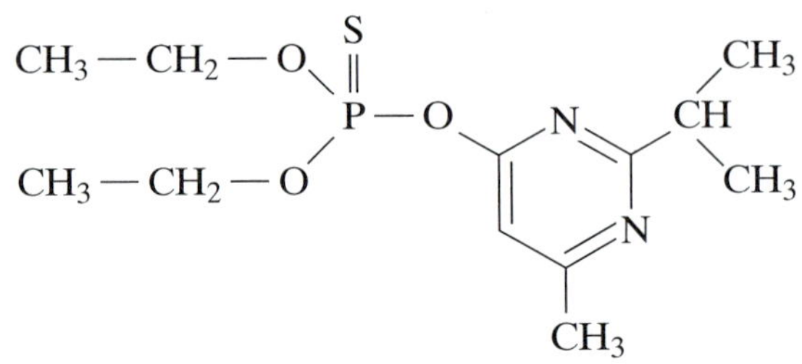

Diazinon

Newer insecticides include some that are substantially more toxic than DDT yet decay rapidly when exposed to environmental conditions. Several insecticides containing phosphorus, such as *parathion* (Sec. 20.5) and the *malathion* and *diazinon* of Figure 14.12, are quite deadly, yet in the presence of water they decompose quickly to relatively harmless fragments. Because of this relatively rapid decomposition, their environmental lifetimes are brief, measured in days rather than the years or even decades of DDT. Despite their relatively short environmental lifetimes, the hazards they pose to humans and wildlife have placed their manufacture and use under close review and increasing restriction.

Among a host of other chemical approaches are (relatively) nonlethal compounds that remove insect pests without killing them. An example is a set of *juvenile hormones* (Fig. 14.13), which control portions of the life cycle of insects. The larvae, which have not yet developed into mature insects, synthesize these compounds themselves as growth regulators. Their production stops when the larvae are about to mature into the adult form. The application of juvenile hormones to larvae at just the right stage produces insects that are unable to reproduce.

With all our concern about the environmental effects of insecticides, it might seem surprising that the use of herbicides in the United States surpasses that of pesticides. In fact, the total mass of all herbicides spread on the land exceeds the total combined mass of all insecticides, fungicides, and every other kind of pesticide. On a weight basis, herbicide use runs about 3.5 times insecticide use.

Herbicides are vital to modern commercial farming for the removal of weeds and other undesirable plant growth. Weeds compete with valuable plants, including commercial crops, for water and for soil nutrients, slowing their growth. On

Figure 14.13 One of several juvenile hormones.

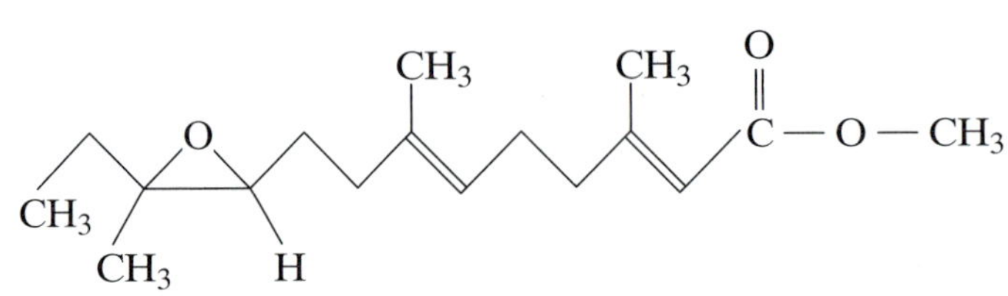

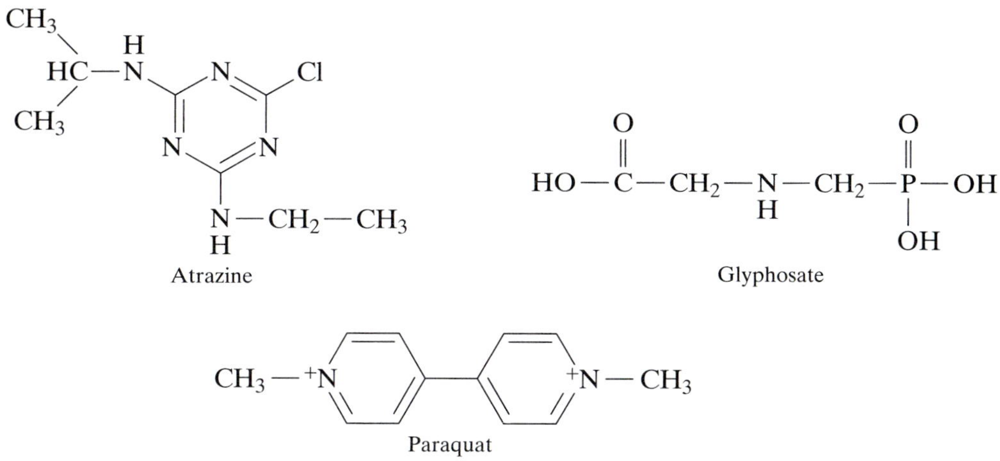

Figure 14.14 The herbicides atrazine, glyphosate, and paraquat.

farms this competition decreases the yield of commercial crops per acre planted. Without the use of herbicides, weeds would have to be removed by machinery (or by hand), increasing the time, labor, and expense of producing food for humans and farm animals.

Commonly used herbicides include *atrazine*, *glyphosate*, and *paraquat* (Fig. 14.14). Atrazine is effective for weed removal on farmland. Glyphosate, which is structurally related to the amino acid glycine (Sec. 17.2), is a broad-spectrum herbicide used in both agricultural and consumer-oriented herbicide preparations. Purchase and application of paraquat, the most toxic of the three, is restricted to professionals. Aerial spraying of paraquat has been used to destroy marijuana fields in both the United States and Latin America.

■ **QUESTION**

What characteristic of parathion, malathion, and diazinon makes them superior to DDT as insecticides? What characteristic might favor DDT (rather than parathion, malathion, or diazinon) as an insecticide?

14.15 Waste Disposal: Throwing Things Away

Where is "away" when you throw something away? Unless you place it on a rocket to the moon, put it into orbit in outer space, or bury it deep in the lithosphere, below the level of groundwater, it lands somewhere in our environment. Actually, we never really throw anything away. We either move it from where it is to somewhere else, out of sight, or we destroy it. Even then we have to put the pieces somewhere, whether they're large enough to hold in our hands or a few wisps of assorted atoms and molecules.

The problems of throwing things away—waste disposal—range from the commonplace to the truly catastrophic. Much of what we throw away each day goes into the trash can and ends at trash dumps and landfills as municipal solid wastes. In 1996 we in the United States discarded just over 200 million tons of these solid wastes—paper, metals, glass, plastics, cans, jars, boxes, packing materials, yard cuttings, and similar materials—at a rate of 4.3 pounds per person per day. With a little over a quarter of all this waste recycled or composted, some 150 million tons

Cleaning up household wastes before they become environmental pollutants.

of unrecoverable municipal solid wastes went into our dumps and landfills. These municipal dumps and landfills may or may not be suitable resting places for all of this waste material, but they are certainly *not* suitable depositories for the hazardous chemicals in these wastes, those that can find their way as pollutants into our groundwater.

Our daily concerns with potentially hazardous chemical wastes arise from ordinary things like exhausted but not quite empty pesticide spray cans, small amounts of paint left over from redecorating, used motor oil collected during a driveway oil change, small amounts of leftover garden and pool chemicals, and even the unknown contents of old bottles with no labels. Although it may be tempting to pour these wastes down the kitchen drain, into a toilet, down a storm drain, or even onto the ground itself, that's often a certain route into the groundwater.

Used engine oil is a particular problem, if only because of its volume. Each year engine maintenance in the United States produces over 1.3 billion gallons of used oil. Of this, an estimated 200 million gallons is disposed of carelessly, in ways that could take it to the groundwater that gives us our drinking water. (About 800 million gallons is recycled.) Used engine oil is a particularly nasty pollutant not only because the oil itself can contaminate groundwater, but also because of the traces of various metals and additives it contains, as well as various decomposition products.

We might take comfort in knowing that many gasoline stations and garages throughout the country serve as collection centers for used engine oil, especially the oil drained in home driveways. To a consumer, throwing this used oil "away" means no more than carrying it to one of these collection centers. We might expect this oil to be recycled, both conserving petroleum and protecting the environment, but it doesn't quite work that way. Although a small amount of this reclaimed oil is, indeed, recycled into useful lubricants, over 90% is added to fuel mixtures and burned, quite legally, by a variety of users. There's little governmental regulation of the burning of reclaimed oil because it isn't defined legally as a hazardous waste, which would make its disposal subject to strict regulations. Much of the used oil generated in the United States, then, becomes a real or a potential air pollutant rather than a real or potential water pollutant. Even with this possibility, taking drained oil to a collection center is by far the best way to throw it "away."

■ QUESTION

Disposing of used paint thinner by pouring it onto the ground poses the risk of contaminated groundwater. Instead of risking contamination of groundwater, would pouring the paint thinner into a pan and letting it evaporate outdoors be a better method of disposal? Explain.

14.16 The Scales of Environmental Harm

Environmental damage operates on many scales of both time and quantity. The cumulative environmental harm resulting from the daily disposal of small amounts of wastes by large numbers of individual consumers rarely reaches catastrophic proportions. It seldom provokes newspaper headlines or television reports. Industrial accidents and the discharge or dumping of industrial wastes on a massive scale are different matters.

One of the world's worst environmental catastrophes occurred in Japan in the middle of the past century. Industrial wastes containing tons of mercury were poured into the bay at Minamata, near the southernmost point on the islands that make up the country. Biological processes in the ocean converted the mercury into highly toxic organomercury compounds, which entered the food chain and worked their way up the chain and into the tissues of fish that made up the local diet. The mercury poisons in the fish caused the deaths of hundreds of people and severe mental and physical disorders in thousands of others.

Sealing industrial wastes in drums and burying them in shallow, mass graves represents another hazardous way of throwing things "away." In the 1940s and

Cleaning up massive toxic waste pollution at Love Canal, New York.

early 1950s, some 20,000 tons of toxic chemicals were buried in a 10-square-block area near Niagara Falls, New York, known as Love Canal. The chemicals, most of them in sealed drums, were covered over with soil and forgotten. Years later a small community and a school were built on the site. By 1976 the chemicals had begun leaking out of the corroded drums; eventually they contaminated the entire Love Canal area. By 1980 the resulting harm to the local environment and to the people living there resulted in the relocation of more than 230 families, isolation of a large part of the area, and declaration of Love Canal as a federal disaster area.

An industrial accident at Bhopal, India, generated a more dramatic catastrophe. Early in December, 1984, a combination of equipment failures and/or human errors caused an explosion at a plant that manufactured methyl isocyanate, $CH_3—N=C=O$, a deadly compound used in the manufacture of pesticides. The explosion released a cloud containing an estimated 23,000 kg (50,000 lb) of methyl isocyanate. As the cloud drifted over Bhopal it killed 2000 people and injured another 200,000, all within a few hours.

Dramatic events and tragic accidents like these can divert our attention from more subtle yet even more damaging environmental hazards, those that work over longer time scales than abrupt explosions and instantaneous devastation. Some activities can be especially insidious since their benefits, initially at least, seem to outweigh their (known) risks. Cigarette smoking, for example, once accepted by society as a harmless pleasure, is now connected with heart and lung damage and cancer. And it appears that smokers not only injure themselves, but harm others as well with the environmental pollutants they exhale, now classified as **environmental tobacco smoke (ETS)**. As nonsmokers inhale this ETS they involuntarily undergo *passive smoking*, a phenomenon that's believed to produce, over a period of time, many of the same effects as active smoking. According to the Environmental Protection Agency, ETS in the United States annually produces

ETS represents **environmental tobacco smoke**, an environmental hazard to smokers and nonsmokers alike.

Environmental tobacco smoke can be a major indoor air pollutant.

bronchitis, pneumonia, and other respiratory problems in 150,000 to 300,000 infants and very young children, aggravates existing cases of asthma in 200,000 to 1,000,000 children, and causes 3000 lung cancer deaths in adult *non-smokers*.

If this much harm were caused by a single, catastrophic environmental event, it would surely be placed in the same class as the disasters at Bhopal and even Chernobyl (Sec. 5.2). Unlike explosions, with their instantaneous consequences, some forms of environmental damage take years, even decades, to make themselves known. In our search for environmental safety, then, we find that the hazards we are willing to accept now are necessarily those we know of now; other hazards, unknown and unpredictable today, can materialize far in the future. The search for safety has a time scale of its own.

■ QUESTION

Should environmental tobacco smoke be classified as an atmospheric pollutant? Justify your answer.

14.17 Polluting the Environment, One Match at a Time

We can now return to our opening demonstration and view it in a new perspective. The sulfur of the match we lit becomes oxidized to sulfur dioxide, a primary air pollutant. Once in the atmosphere the sulfur dioxide oxidizes slowly to sulfur trioxide, a secondary air pollutant. Left free to disperse in the atmosphere, the sulfur trioxide can enter a droplet of water somewhere above the earth's surface and return as acid rain.

Does lighting a match pollute the environment?

Did we contribute to environmental pollution by lighting the match? There's no easy answer, partly because there's no simple, universal definition of pollution. The quantity of chemicals introduced into the environment by that single match can hardly have any detectable effect on the earth's atmosphere, hydrosphere, or crust. But light another match, then another, and another; light millions upon millions of matches until the amount of sulfur released approaches that of a volcanic eruption, or all the factories on earth, and yes, there will be a noticeable, undesirable change in the environment. There will be pollution.

Can we produce a pollution-free world, then, by banning all matches, and all cars, factories, plastics, personal care products, and everything else that might release or end up as unwanted substances in the environment? And would we want to? We'll examine these questions further in the following Perspective.

PERSPECTIVE

The Myth of a Pollution-Free World

Pollution will always be with us, no matter how we define it. If we define a pollutant as *anything* that damages the environment, even if it comes only from forces of nature, then volcanoes and lightning strikes, with their contributions of sulfur oxides, nitrogen oxides, and ozone, will always be sources of environmental pollution. If, as we have done in this chapter, we add the restriction that pollution comes only from human activity, then as long as humans exist on earth there will be environmental pollution. Even before recorded history, when small bands of humans or prehumans slept in caves and roamed and hunted in an otherwise pristine environment, the smoke of protective fires they lit to guard against wild animals surely polluted the night air.

Even in a modern society that acts with care and wisdom, what the majority considers to be a desirable addition to the environment may appear to be a benefit to some but a pollutant to others. To many, the chlorination of municipal water supplies to prevent epidemics represents a lifesaving activity of government. To others, who drink only bottled water or who boil their tap water to purify it, chlorination of public drinking water is an act of pollution.

Fluoridation of water supplies represents another, perhaps more controversial matter. Strong evidence exists that the intake of very small concentrations of fluorides by children strengthens their teeth and bones and inhibits tooth decay. At the level of only 1 ppm in drinking water, fluorides produce a sharp reduction in rates of tooth decay. Such low concentrations of fluorides have shown no evidence of harm to humans.

To reduce cavities among children, many communities have added small amounts of fluoride salts to their municipal drinking water. Many approve of this as a public health measure. Others point out that even though the presence of fluorides does nothing to improve an adult's dental health, adults are required to face, involuntarily, whatever risks there may be in fluoridation of public water supplies, however small they may be. As a result, they maintain, adults derive no benefits to counterbalance the risks. Some suggest that fluoridation of drinking water in schools may be the best way to protect those whose teeth are vulnerable while avoiding any risk for those whose teeth are not. To some, then, the fluoridation of public water supplies represents responsible action by local governments. To others it is an act of pollution.

As with many of the other topics covered in this chapter, there are no easy answers. Only an understanding of the extraordinary chemistry of our everyday world—even of so ordinary an act as striking a match—will enable us even to begin considering the questions. Starting with the salt water that introduces Chapter 9 and continuing through this chapter, we've seen that pollution has several definitions and wide limits. For many of us, the consequence is an environment that will always present to us the challenge of contaminants. ■

EXERCISES

For Review

1. Following are a statement containing a number of blanks, and a list of words and phrases. The number of words equals the number of blanks within the statement, and all but two of the words fit correctly into these blanks. Fill in the blanks of the statement with those words that do fit, then complete the statement by filling in the two remaining blanks with correct words (not in the list) in place of the two words that don't fit.

Three regions of the earth make up the physical environment in which we live: the __________, which includes all the gases we breathe; the __________, which includes all the water of the earth's surface; and the crust, which is the hard, uppermost portion of the __________. Chemically, __________ is the dominant element of the gases surrounding the earth's surface, __________ is the major element of the water of our environment, and __________ are the dominant compounds of the solid earth we live on.

Among the major pollutants of our atmosphere are the __________ that are components of volcanic gases and that are formed through oxidation of the __________ of fossil fuels, and the __________ formed by the combination of nitrogen and oxygen molecules at the high temperatures generated in combustion chambers and by lightning. Worldwide, __________ are the principal pollutants of the oceans and other bodies of water.

atmosphere	oxygen
hydrosphere	ozone
lithosphere	sulfur
nitrogen	sulfur oxides
nitrogen oxides	water

2. Following are a statement containing a number of blanks, and a list of words and phrases. The number of words equals the number of blanks within the statement, and all but two of the words fit correctly into these blanks. Fill in the blanks of the statement with those words that do fit, then complete the statement by filling in the two remaining blanks with correct words (not in the list) in place of the two words that don't fit.

Nitric oxide and sulfur dioxide are among the __________ produced directly by the burning of fossil fuels. Released into the atmosphere, these are converted to the __________ sulfur trioxide and nitrogen dioxide. In further reaction, sunlight strips an oxygen off a molecule of nitrogen dioxide and initiates a series of reactions that produces a complex mixture of air pollutants known as __________. Air pollutants that result from the evaporation of organic solvents from our consumer products are classified as __________. Similarly, other organic compounds that pollute by evaporation are known as __________. One of the newly recognized air pollutants is the tobacco smoke exhaled by smokers, __________, which can harm nonsmokers who inhale it.

Pollution of the hydrosphere can occur as __________, which produces environmental harm through the heating of bodies of water, usually when they are used to cool commercial processes; __________, which results from the accumulation of suspended particles in water; __________, produced by the presence of microorganisms in water; and __________, from the presence of undesirable chemicals in water. Sometimes, the presence of certain chemicals in bodies of water encourages the growth of some forms of plant life, to the detriment of aquatic animal life. This form of pollution, in which some forms of life are well nourished at the expense of others, is called __________. Chemical pollution can be measured through the aerobic activity of microorganisms that metabolize the chemicals and thus generate a high __________. As the microbiological activity increases, the __________ decreases, resulting in a __________.

biochemical oxygen demand (BOD)	environmental tobacco smoke (ETS)
biological contamination	passive smoking
chemical pollution	primary air pollutants
chlorofluorocarbons (CFCs)	secondary air pollutants
	sedimentary pollution
dissolved oxygen (DO)	thermal pollution
dissolved oxygen deficit (DOD)	VOCs
	VOSs

3. What names do we give to the portions of the earth that form (a) our gaseous environment, (b) our liquid environment, and (c) our solid environment? (d) Another segment of our environment is known as the *biosphere*. What does the biosphere consist of?

4. Name or otherwise identify two primary air pollutants and two secondary air pollutants. Describe their origins or sources.

5. What element found among the impurities of both coal and fuels derived from petroleum is a major contributor to air pollution?

6. The sulfur of sulfur oxides originates in impurities of the fossil fuels we burn to produce energy. What is the principal source of the nitrogen of the nitric oxides that pollute the air?

7. Write the molecular formulas of (a) iron pyrite, (b) nitrogen dioxide, (c) nitric oxide, (d) magnesium hydroxide, and (e) magnesium sulfite.

8. Describe how each of the following removes pollutants from exhaust gases: (a) filtration, (b) scrubbing, (c) electrostatic precipitation.

9. Write balanced reactions for:

a. formation of nitric oxide from nitrogen and oxygen

b. conversion of sulfur dioxide to sulfur trioxide

c. reaction of sulfur dioxide with magnesium hydroxide

d. formation of nitrogen dioxide from nitric oxide and oxygen

e. formation of ozone from a sequence of reactions initiated by the action of sunlight on nitrogen dioxide

10. Name another air pollutant, in addition to sulfur dioxide, that can be transformed into a water pollutant through the medium of acid rain.

11. The oxidation of sulfur, carried out through several chemical steps, eventually produces sulfur trioxide, SO_3. Write a balanced chemical equation showing the reaction of SO_3 with water to produce sulfuric acid, H_2SO_4.

12. As shown in Table 14.4, what is the major source of each of the following air pollutants? (a) CO, (b) NO_x, (c) SO_2, (d) VOC.

13. High levels of nitric oxide can be found in the atmosphere after thunderstorms. Why? What is its origin?

14. The burning of coal containing pyrite impurities contributes to the presence of sulfuric acid in the atmosphere. Write a single balanced equation for the conversion of the pyrite impurity of coal into atmospheric sulfuric acid. Assume that water vapor is present in the atmosphere.

15. Worldwide, what is the most widespread form of water pollution?

16. Identify two chemical pollutants responsible for eutrophication.

17. What *two* factors or conditions limit the use of air stripping to remove pollutants from water?

18. Worldwide, what is the major use of fresh water drawn from the earth?

19. Worldwide, what is the major form of water pollution?

20. Name and describe four types of water pollution. Explain how each damages aquatic life.

21. Give the names and molecular formulas of three acids found in acid rain.

22. How can a substance that nourishes and sustains certain forms of aquatic life act as a pollutant?

23. Describe the similarities and differences between the use of a scrubber to remove pollutants from industrial exhaust gases and the use of an air stripper to remove pollutants from water.

24. "Minamata disease" is a name given to symptoms that develop from eating fish contaminated with organomercury compounds. What's the origin of this term?

25. For each of the following, identify or describe the hazardous substance involved and describe how it entered the environment: (a) Bhopal, (b) Love Canal, (c) Minamata.

A Little Arithmetic and Other Quantitative Puzzles

26. What societal function or human activity contributes the greatest combined weight of (a) sulfur and nitrogen oxides, (b) carbon monoxide, and (c) VOCs to the pollution of the atmosphere? (Refer to Table 14.4.)

27. We might define "acid rain" as any rain with a pH lower than 5.6. Why would we use 5.6 as the upper limit for our definition? What is one weakness of such a definition?

28. According to the pollution limits shown in Table 14.6, could 10 liters of solution that contains a total of 0.6 milligram of arsenic be discharged legally as industrial waste water? Could this solution be discharged legally into sanitary sewers? Does this solution meet the drinking water standard of Table 14.6?

29. How much water (if any at all) that's uncontaminated by arsenic would you have to add to the 10 liters of solution of Exercise 28 in order to meet the arsenic standard for each of the categories of Table 14.6?

Think, Speculate, Reflect, and Ponder

30. Section 14.3 tells us that FeS_2, also known as *pyrite* or *iron pyrite*, is a sulfur-containing impurity of coal. What is the color of FeS_2. Refer to Table 13.1 for help.

31. Exercise 29 assumes that—to use a phrase sometimes applied to pollution problems—dilution is the solution to pollution. Do you think that adding fresh water to polluted water to reduce the concentration of contaminants is an effective method for reducing pollution? Explain.

32. The photograph in Section 14.1 shows a beach, where we can see portions of the lithosphere, the hydrosphere, and the atmosphere. Name or describe a location on earth where we might be able to see only (a) the lithosphere and the atmosphere, (b) the atmosphere and the hydrosphere, and (c) the lithosphere and the hydrosphere.

33. Seawater is too salty to drink. Is seawater polluted by salt?

34. How could energy conservation help reduce air pollution?

35. Assuming that there's no decrease in our use of energy, what would you recommend as the best way to reduce air pollution due to carbon monoxide?

36. Under what circumstance is ozone beneficial to life? In what circumstances is it harmful?

37. We've defined pollution as an excess of a substance that's generated by human activity and that's present in the wrong environmental location. What is the effect of including the phrase "that's generated by human activity" in the definition? Does including this phrase strengthen or weaken the definition?

38. Halocarbons generally have low chemical reactivities, low or nonexistent flammability, and low toxicity. Using the definition of *pollution* presented in Section 14.2, explain why halocarbons might be considered to be pollutants.

39. How would you respond to the suggestion that we spray solutions of concentrated ammonia into the atmosphere to counteract the effects of acid rain?

40. How would you respond to the claim that since all rain is acidic, the use of the term acid rain is misleading?

41. Describe the connection between air pollution and the generation of energy.

42. Natural gas consists almost entirely of pure hydrocarbons. Explain why the use of natural gas as a fuel is less likely to lead to acid rain than the use of other fossil fuels, such as petroleum.

43. Discuss the environmental impact of replacing electric utility generating plants that burn coal or petroleum with generating plants that use nuclear power. Refer to Chapter 5.

44. Regions of the world dependent on the water of fossil aquifers can expect to face shortages of fresh water at some time in the future. Yet even some nations that depend on rechargeable aquifers are now finding that their freshwater resources are not sufficient. This has nothing to do with pollution or with changing patterns of rainfall. Suggest a reason why, in the absence of either increased pollution or decreased rainfall, some nations face shortages of fresh water.

45. What does the term "environmental persistence" mean when it is applied to insecticides and herbicides? Under what conditions might environmental persistence be undesirable? Under what conditions might it be desirable?

46. Give an example, other than water or gasoline, of a substance that is useful or beneficial to us, in one place but that acts as a pollutant in another.

47. You are a saltwater fish, one that lives its entire life in the ocean but cannot survive in fresh water. Would you use the term "polluted" to describe an environment of fresh water, suitable for humans to drink?

Web Questions

Use the Web to answer or accomplish each of the following. Report your Web source for each response.

1. Find several definitions of pollution that are different from the one presented in this chapter.

2. Prepare a list of air, water, and ground pollutants other than those discussed in this chapter.

3. What is the lowest pH reported for acid rain falling on the United States in the past five years?

4. Describe the most recent observations of the ozone hole over both the Arctic and the Antarctic.

5. Write a brief summary of the scientific work done by F. Sherwood Rowland, Mario J. Molina, and Paul Crutzen that earned the Nobel Prize in 1995.

6. One of the principal uses of CFCs was as refrigerants in air conditioners and refrigerators. Now that their use has been sharply curtailed in order to protect the ozone layer, what chemicals have replaced them as refrigerants?

7. Identify the chemical known as TCDD and describe how it is related to the topics of this chapter.

8. Describe the major incident of land pollution that occurred in Times Beach, Missouri, in 1971. Discuss its cause, its consequences, and the current situation at Times Beach.

9. Describe the 1976 industrial accident at Seveso, Italy. Discuss its cause, its environmental consequences, and the current situation at Seveso.

10. Summarize current medical evidence for the effects of passive smoking on human health.

CHAPTER 15

Fats and Oils

THE TRIGLYCERIDES WE EAT

Fats and oils in the foods we eat.

Vegetable Oil Chemistry

Fundamentally, the food we eat is neither more nor less than a huge assortment of chemicals. We view them in forms ranging from essentially pure, individual compounds, such as the sucrose and sodium chloride of table sugar and table salt, to the complex combinations that form our fruits, vegetables, grains, meats, dairy products, etc. But in its essence, food is simply a combination of chemical substances, each with its own chemical formula, its own chemical properties, its own characteristic chemical reactions, and its own chemical, health-related effects.

You can watch a bit of food chemistry in action with a little drugstore iodine and some peanut oil and sunflower oil. Put equal amounts of the two oils into two heat-resistant containers and add a few drops of drugstore iodine solution to each. (Use the same kind of tincture of iodine you used in the opening demonstration of Chapter 11.) It's important to add the same amount of iodine to each of the oils. You can use 10 to 20 mL (a little less than a fluid ounce) of each of the oils, along with a few drops of the iodine solution. The iodine solution won't dissolve in the oils, so swirl each carefully and gently to disperse the iodine as small droplets.

Now let each mixture of oil and iodine stand for a while. How long depends on the temperature. After several minutes you'll see the sunflower oil lose its darker tint. (If the tincture of iodine is still present as dispersed droplets, you might see the color of the droplets fade away.)

If you don't see any change after 10 or 15 minutes at room temperature, heat the containers gently in some water, so the oils don't splatter. Heating them simultaneously ensures that they are both heated to the same extent for the same amount of time. You'll find that the peanut oil retains the red-violet color of the iodine somewhat longer, or it may not lose the excess color at all (Fig. 15.1). (If the peanut oil loses its color at the same time as the sunflower oil you may have to add a little more iodine to each to see the difference between the two oils.)

All this comes about because peanut oil and sunflower oil are composed of chemicals that react with iodine, which is also a chemical. The color change results from these chemical reactions, and the difference we observe in the way the two oils react with iodine reflects differences in their molecular structures. The components of food, in short, are chemicals. Their structures are chemical, their reactions are chemical, and they serve our nutritional needs as chemicals.

In this and the next few chapters we'll explore some of these chemicals we call food. First we'll examine the chemistry of the *macronutrients*—the fats and oils, the carbohydrates, and the proteins that make up the bulk of our foods (Sec. 8.10). We'll view them as the chemical fuel that provides the energy for our human engine (Sec. 8.7) and as the raw material that provides the physical substance of our human bodies.

After we examine the role proteins play in the chemistry of heredity, we'll explore the *micronutrients*. These are the minor components of our foods, the minerals and vitamins that enable our bodies to use the macronutrients efficiently. We'll also explore some of the additives in our foods and the risks and benefits they carry.

Figure 15.1 The effect of polyunsaturation. Adding a few drops of iodine to peanut oil and sunflower oil produces a red color in each. The color in the sunflower oil disappears more quickly than the color in the peanut oil because the sunflower oil is more highly polyunsaturated.

15.1 Fats and Oils: The Triglycerides We Eat

We've already encountered fats and oils in Section 8.10, where we found that at 9 Cal/g they provide us with our densest form of nutritional energy. We saw them again, this time as triglycerides, in our examination of soaps and soap making (Sec. 13.8). We found that triglycerides are esters of the triol *glycerol* with three fatty acids (Figs. 15.2 and 13.17) and that we can split the triglyceride into glycerol and its component fatty acids by *hydrolysis*, the cleavage of the molecule through reaction with water (Sec. 13.7). Soap comes from the hydrolysis of triglycerides in the presence of a base such as NaOH.

Fats (solid triglycerides) and an oil (a liquid triglyceride).

Fats are solid triglycerides. **Oils** are liquid triglycerides.

The difference between fats and oils is a very practical one that can be expressed in simple terms. At ordinary temperatures **fats** are solids and **oils** are liquids. Since all triglycerides (both fats and oils) are triesters of glycerol, the only differences among their molecules—differences that determine whether their melting points are above or below room temperature (Sec. 12.2)—must lie in the structures of their fatty acid side chains. Thus the only differences between fats and oils, and among various kinds of fats and oils, lie in

- the number of carbons in their side chains, which determines the length of their side chains, and
- the number of carbon–carbon double bonds in their side chains, which affects their *degree of unsaturation* (Sec. 6.10).

Table 15.1 shows some typical fatty acids that form the triglycerides of common fats and oils. (The notations C-14, C-16, and C-18 refer to the number of carbon atoms in the chain of each of the acids. They don't indicate isotopes of carbon as in Chapters 4 and 5.) Notice that, like the melting points of triglyceride esters themselves, those of the fatty acids also depend on their degree of unsaturation. The greater the number of double bonds in the chain, the lower the melting point and the more likely that at room temperature the acid is a liquid rather than a solid.

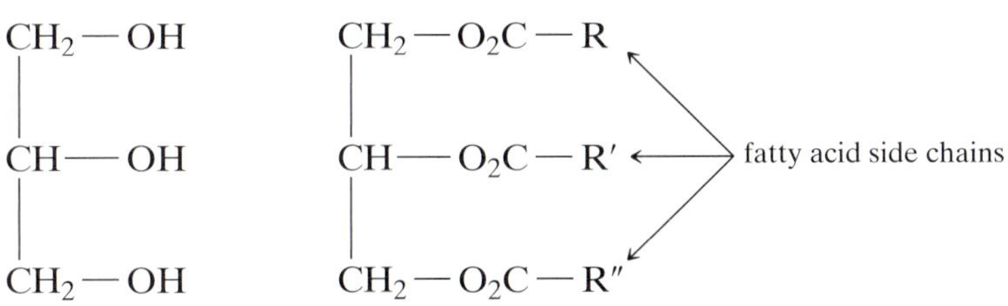

glycerol

Figure 15.2 The general structure of a triglyceride. R, R′, and R″ represent the long, fatty acid side chains.

Table 15.1 Representative Fatty Acids of Dietary Fats and Oils

Name	Melting Point (°C)	Class	Structure
Myristic acid	58	Saturated C-14	$CH_3—(CH_2)_{12}—CO_2H$
Palmitic acid	63	Saturated C-16	$CH_3—(CH_2)_{14}—CO_2H$
Stearic acid	71	Saturated C-18	$CH_3—(CH_2)_{16}—CO_2H$
Oleic acid	16	Monounsaturated C-18	$CH_3—(CH_2)_7—CH{=}CH—(CH_2)_7—CO_2H$
Linoleic acid	−5	Polyunsaturated C-18	$CH_3—(CH_2)_4—CH{=}CH—CH_2—CH{=}CH—(CH_2)_7—CO_2H$
Linolenic acid	−11	Polyunsaturated C-18	$CH_3—CH_2—CH{=}CH—CH_2—CH{=}CH—CH_2—CH{=}CH—(CH_2)_7—CO_2H$

Generally, a low melting point (and a tendency to exist as an oil) is favored by a short side chain and plenty of carbon–carbon double bonds; a high melting point is favored by a long side chain and little unsaturation. In terms of human nutrition, the presence of unsaturation in the side chains usually weighs more heavily with us than do the chain lengths.

In a saturated organic compound, each carbon is connected to its neighbors by single bonds; unsaturated compounds contain carbon–carbon double and/or triple bonds. (Triple bonds aren't important to the chemistry of fats and oils; we won't consider them further.) As we saw in Section 6.10, it's possible to saturate a double bond by adding a chemical reagent to it. In the presence of a catalyst, for example, hydrogen adds readily to the double bond of 2–butene to form butane. Appropriately enough, the reaction is known as **catalytic hydrogenation**:

Catalytic hydrogenation is the addition of hydrogen atoms to the double bonds of a molecule through the use of a catalyst. Hydrogen also adds to triple bonds if they are present.

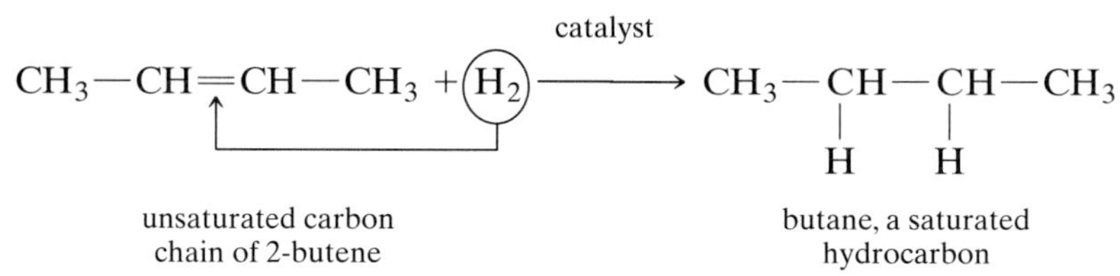

addition of a molecule of hydrogen to an unsaturated carbon chain through catalytic hydrogenation

The same kind of reaction leads to the addition of hydrogen to carbon–carbon double bonds of the side chain of a triglyceride, converting it to a saturated side chain and raising the melting point of the triglyceride. The notation $—(CH_2)_n—$ in the following structures represents a string of n $—CH_2—$ groups. We'll have more to say later about the commercial and nutritional importance of catalytic hydrogenation.

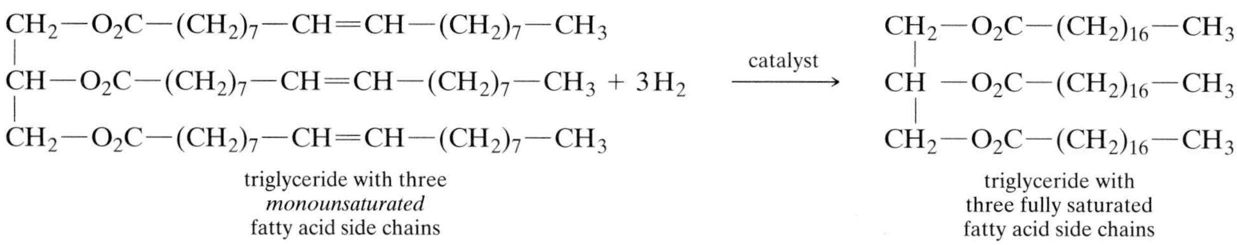

catalytic hydrogenation of an unsaturated triglyceride

In the unsaturated triglyceride shown above, each fatty acid side chain has a single carbon–carbon double bond located within its string of carbons. The side chain is therefore **monounsaturated**. (The prefix *mono-* comes from the Greek word *monos*, meaning alone or single.) Chains containing several C=C double bonds are more highly unsaturated than those with only one; carbon chains with two or more C=C double bonds are **polyunsaturated**. (*Poly-* is from the Greek *polus*, indicating many.) The polyunsaturated oils of foods such as margarines that claim to be "high in polyunsaturates" are simply triglycerides with two or more carbon–carbon double bonds in their fatty acid side chains. *Why* the manufacturers of margarines and other foods want us to know that they are "high in polyunsaturates" is a topic we'll take up after we understand a bit of the chemistry behind the claims.

A **monounsaturated** fatty acid is one with a single carbon–carbon double bond in its carbon chain. A **polyunsaturated** fatty acid contains two or more carbon–carbon double bonds.

■ **QUESTION** Does catalytic hydrogenation tend to convert fats to oils or oils to fats?

15.2 Additions to Carbon–Carbon Double Bonds: Iodine Numbers

Because of the importance of unsaturation in the chemistry of triglycerides, it's important to have a simple, convenient measure of the extent of unsaturation. The degree of unsaturation in a triglyceride is reflected nicely by a value known as the **iodine number**, which represents the number of grams of iodine that add to 100 g of the triglyceride. The addition of molecular iodine, I_2, which can be monitored closely and measured accurately in the laboratory through changes in the color of iodine solutions, allows us to compare the relative degrees of unsaturation in various fats and oils. The presence of molecular I_2 itself is responsible for the red–violet color of a tincture of iodine solution (the kind used as an antiseptic). Both molecular iodine and its color disappear as the iodine adds to a carbon–carbon double bond, producing a colorless product. For example 2, 3–diiodobutane, the product of the addition of iodine to 2-butene, is colorless. As I_2 adds to 2-butene, the reaction mixture loses its red-violet color:

A triglyceride's **iodine number** represents the number of grams of iodine that add to 100 g of the triglyceride.

$$CH_3{-}CH{=}CH{-}CH_3 + I_2 \longrightarrow CH_3{-}\underset{\displaystyle I}{\underset{|}{CH}}{-}\underset{\displaystyle I}{\underset{|}{CH}}{-}CH_3$$

2-butene	iodine	2, 3-diiodobutane
colorless	red-violet	colorless

The opening demonstration put this sort of chemistry into action as the I_2 added to the C=C double bonds in the fatty acid side chains of the triglycerides. With the transformation of the iodine atoms from the red-violet I—I of the tincture into the colorless iodine substituents on the chains, the color of the iodine–oil mixture lightened or vanished. The more pronounced loss of color of the sunflower oil indicates that the triglycerides of sunflower oil are more highly unsaturated—their carbon chains contain more double bonds—than those of peanut oil. For this reason the sunflower oil reacts with more iodine molecules than does an equal amount of peanut oil. With the right proportions of everything and the right temperature, the color persists in the peanut oil simply because of the residue of molecular iodine that remains in the peanut oil.

Since the size of the iodine number is a direct reflection of the number of double bonds in the side chains of the triglyceride, and since the melting point of the triglyceride drops with an increasing number of these double bonds, vegetable oils generally show larger iodine numbers than animal fats. As we might also expect, sunflower oil has a higher iodine number than peanut oil. Table 15.2 and Figure 15.3 present typical values and ranges for the iodine numbers and fatty acid contents of some of the fats and oils of our foods. (These iodine numbers, incidentally, are important only as indicators of the amount of unsaturation in the side chains. They have nothing whatever to do with the actual presence of iodine in our diets.)

As the data of Table 15.2 and Figure 15.3 indicate, nature isn't so neat and simple as to form animal fats exclusively from highly saturated fatty acids and to generate plant oils exclusively from unsaturated acids with plenty of double bonds. Instead, the fats and oils of our diets are complex mixtures of triglycerides containing a considerable variety of side chains: some long, some short, some fully saturated, some monounsaturated, some polyunsaturated. While fats are, as we've noted, generally richer in the saturated side chains, no fat consists entirely of fully saturated side chains and no oil is free of them. (The various kinds of side chains contained in our dietary fats and oils are generally scattered throughout their triglycerides; we rarely find any single triglyceride molecule containing three identical side chains.)

Table 15.2 Fatty Acid Content of Fats and Oils, by Percentage[a]

Fat or Oil	Iodine Number	Myristic Acid (Saturated) C-14	Palmitic Acid (Saturated) C-16	Stearic Acid (Saturated) C-18	Palmitoleic Acid (Monounsaturated) C-16	Oleic Acid (Monounsaturated) C-18	Polyunsaturated Acids C-18	Polyunsaturated Acids Larger than C-18
Animal Fats								
Beef fat	35–42	6	27	14	—	50	3	—
Butterfat[b]	26–48	11	29	9	5	27	4	1
Human fat	68	3	24	8	5	47	10	3
Lard	47–67	1	28	12	3	48	6	2
Vegetable Fat								
Cocoa butter	33–42	—	24	35	—	38	2	—
Animal (Fish) Oils								
Cod liver oil	135–165	6	8	1	20	(combined, 29)		35
Herring oil	140	7	13	—	5	—	21	53
Menhaden oil	170	6	16	1	16	—	30	31
Sardine oil[c]	185	5	15	3	12	(combined, 18)		32
Vegetable Oils								
Coconut oil[d]	6–10	18	11	2	—	8	—	—
Corn oil	110–130	1	10	3	2	50	34	—
Cottonseed oil	103–111	1	23	1	2	23	48	1
Linseed oil	180–195	—	6	3	—	19	72	—
Olive oil	80–88	—	7	2	—	85	5	—
Palm oil	50–60	1	40	6	—	43	10	—
Palm kernel oil[e]	37	14	9	1	—	18	1	—
Peanut oil	90–100	—	8	3	—	56	26	7
Safflower oil	145	—	4	3	—	17	76	—
Sesame oil	103–117	—	9	4	—	45	40	1
Soybean oil	120–135	—	10	2	—	29	57	1
Sunflower seed oil	125–135	—	6	2	—	25	66	1
Tung oil	168	(all saturated combined, 5)			—	4	91	—
Wheat germ oil	125	—	13	4	—	19	62	1

[a] Some entries total 99% because of the effects of rounding.

[b] Butterfat also contains about 11% saturated acids smaller than C-14 and 2.5% other saturated acids not shown on this table.

[c] Sardine oil also contains about 15% unsaturated C-14 acids.

[d] Coconut oil also contains about 60% saturated acids smaller than C-14.

[e] Palm kernel oil also contains about 57% saturated acids smaller than C-14.

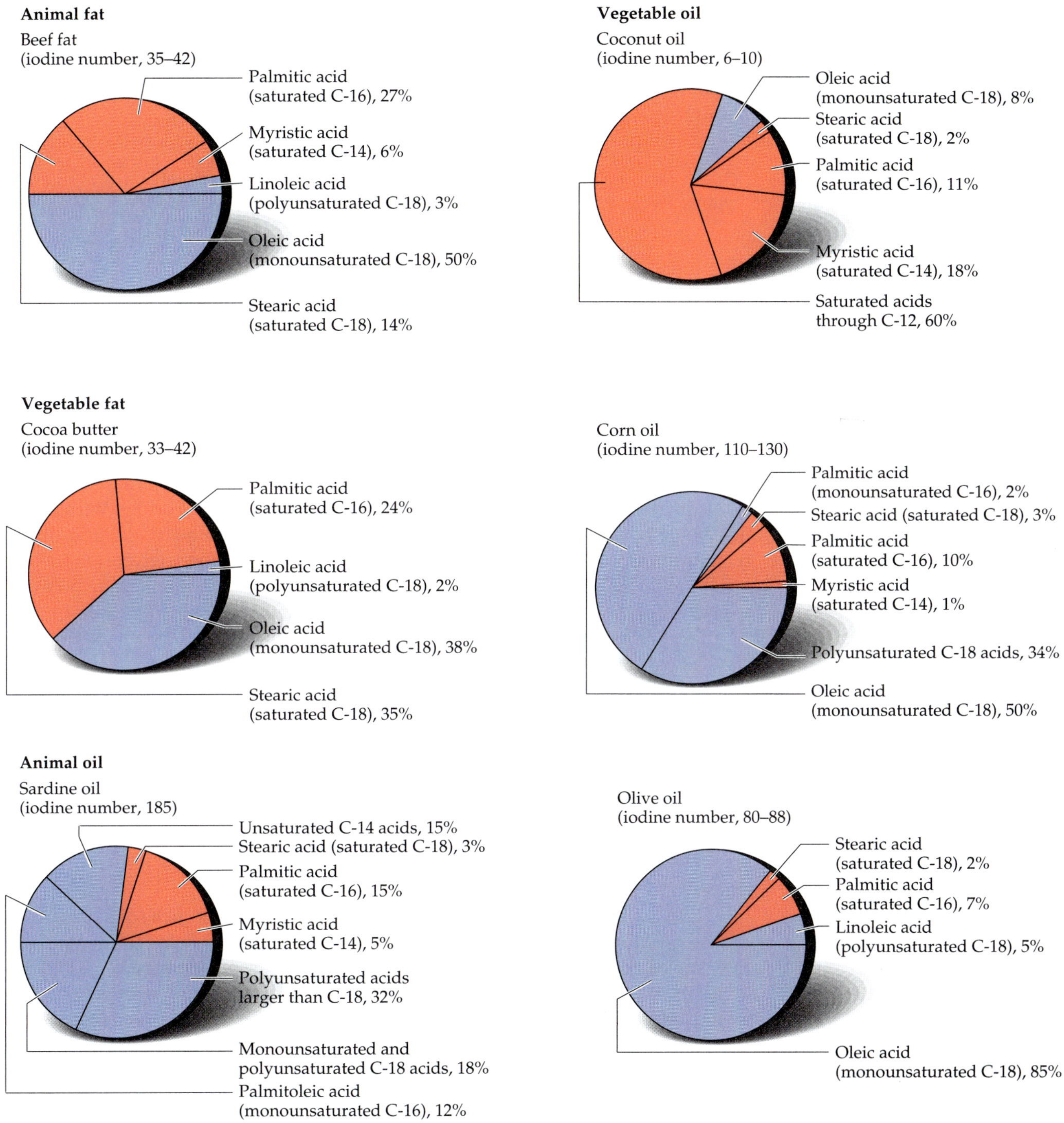

Figure 15.3 Fatty acid content of fats and oils (typical percentages).

In all these data we can see clear illustrations of several general trends. As a whole, vegetable oils and fish oils furnish a higher level of unsaturated side chains than do the fats. But there are exceptions. Lard, for example, carries a higher proportion of unsaturated side chains than either palm kernel oil or coconut oil. The iodine numbers associated with these three substances reflect the relative extent of unsaturation in each. (Both palm kernel oil and coconut oil are oils because their side chains are relatively short rather than because of any extensive unsaturation. Coconut oil is a major ingredient of many nondairy creamers and other prepared foods.) We'll examine some of the dietary implications of saturated and unsaturated triglycerides in Section 15.4 and in the Perspective.

■ QUESTION

Of the fats and oils of Table 15.2, which contains the most highly saturated side chains? Which contains the most highly *un*saturated side chains?

15.3 Additions to Carbon–Carbon Double Bonds: Catalytic Hydrogenation

While the addition of iodine to triglycerides is used only as an analytical tool for determining the degree of unsaturation in their side chains, the catalytic addition of hydrogen (Sec. 15.1) serves as an important chemical reaction in the manufacture of food products. Catalytic hydrogenation leads to a reduction in the number of double bonds, a decrease in the degree of unsaturation (an increase in saturation), and therefore an increase in the melting points of triglycerides. With few or no double bonds in its side chains, a triglyceride is likely to be a solid at ordinary temperatures. Catalytic hydrogenation thus can turn a liquid triglyceride (an oil) into a solid triglyceride (a fat). Because of its ability to raise the melting points of triglycerides, catalytic hydrogenation is also known as *hardening* of vegetable oils. Through this hardening process, for example, the liquid triglyceride *glyceryl trioleate* can be converted to the brittle solid *glyceryl tristearate*:

$$
\begin{array}{l}
CH_2{-}O_2C{-}(CH_2)_7{-}CH{=}CH{-}(CH_2)_7{-}CH_3 \\
\vert \\
CH{-}O_2C{-}(CH_2)_7{-}CH{-}CH{-}(CH_2)_7{-}CH_3 + 3H_2 \\
\vert \\
CH_2{-}O_2C{-}(CH_2)_7{-}CH{=}CH{-}(CH_2)_7{-}CH_3
\end{array}
\xrightarrow{\text{catalyst}}
\begin{array}{l}
CH_2{-}O_2C{-}(CH_2)_{16}{-}CH_3 \\
\vert \\
CH{-}O_2C{-}(CH_2)_{16}{-}CH_3 \\
\vert \\
CH_2{-}O_2C{-}(CH_2)_{16}{-}CH_3
\end{array}
$$

glyceryl trioleate, a liquid

glyceryl tristearate, a brittle solid not useful in processed foods

Catalytic hydrogenation is important to the production of such solid or semisolid processed foods as margarines, chocolate bars, and vegetable-based shortenings. The fully saturated side chains that result from complete hydrogenation of all the carbon–carbon double bonds of a typical vegetable oil produce a hard, brittle triglyceride with a consistency much like that of candle wax. Because of this harsh physical characteristic, fully saturated triglycerides have little or no value in the manufacture of processed foods. More useful are the partially hydrogenated vegetable oils that retain just enough unsaturation to produce our soft or semisolid food products.

The label on a can of a typical solid shortening used for frying foods states that it is "made from the finest vegetable oils (soybean, palm, and sunflower) which are partially hydrogenated for freshness and consistency." (We'll examine factors affecting the freshness of oils in Sec. 15.5.) On the carton of a popular margarine we read that it is "made from 100% pure corn oil" and that it contains, among other ingredients, "liquid corn oil and partially hydrogenated corn oil." It's the partial hydrogenation of the side chains that gives these products their desirable, semisolid consistency.

Partial hydrogenation of vegetable oils produces a mixture of triglycerides that soften and melt near body temperature.

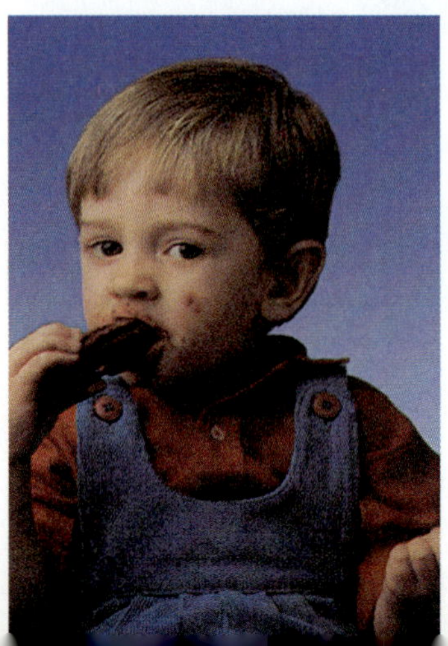

While it is true that these and similar products are, indeed, *made from* highly unsaturated vegetable oils, some of them no longer actually *contain* these same vegetable oils, at least not in significant amounts. The partial hydrogenation of the "100% pure corn oil" used in manufacturing the margarine converts the triglycerides of the liquid oil into a mixture of more highly saturated triglycerides with properties more like fats and, as a result, more suitable for the manufacture of the solid margarine. Margarine and similar solid and semisolid foodstuffs are hardly identical to the raw materials from which they are produced.

■ QUESTION

Which of the following triglycerides would you expect to be least useful in the manufacture of a chocolate candy bar: (a) glyceryl trilinoleate; (b) glyceryl tripalmitate; (c) glyceryl trioleate? Explain. (You may find it useful to refer to Table 15.1.)

15.4 Cholesterol

The use of these partially hydrogenated oils in foods brings with it a mixed story that centers on **cholesterol** (Fig. 15.4), a compound present in virtually all animal cells. Cholesterol falls into the class of *steroidal alcohols*, or *sterols* (a contraction of *ster*oidal alcoh*ols*). It's classified as a steroid because of its core molecular structure of three rings of six carbons each and a fourth ring of five carbons, connected as shown in Figure 15.5. All steroids share this same four-ring skeleton.

Cholesterol is a steroidal alcohol that contributes to the development of atherosclerosis (hardening of the arteries).

High levels of *serum cholesterol*, which is cholesterol present in the blood, are closely associated with atherosclerosis, a stiffening and thickening of the walls of the arteries sometimes known simply as hardening of the arteries. It's a condition that can lead to high blood pressure and heart disease. High levels of cholesterol in the blood are implicated as one cause of cardiovascular disease. Serum cholesterol appears to have two origins: cholesterol present in our diet and cholesterol manufactured by our bodies from other substances.

Cholesterol: a sterol that contributes to cardiovascular disease.

Controlling our intake of dietary cholesterol depends largely on avoiding foods rich in this sterol. Highest on the list are egg yolks. Red meat, animal organs, cream, butter, and many cheeses also contain high levels. On the other hand, egg whites, yogurt, and fat-free milk are among the animal products lowest in cholesterol. Since cholesterol occurs only in animal tissue, we can be certain that all fruits, vegetables, and vegetable oils are cholesterol-free. Table 15.3 lists the fat and cholesterol content of typical foods.

A set of complex, interacting factors appears to govern the production of cholesterol within our own bodies. These include genetics (over which we have no

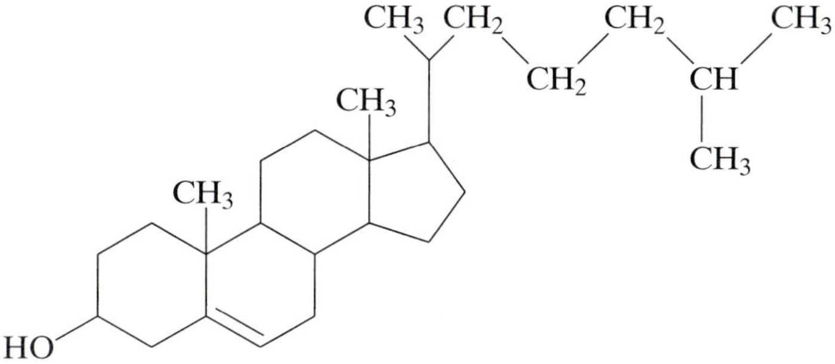

Figure 15.4 Cholesterol

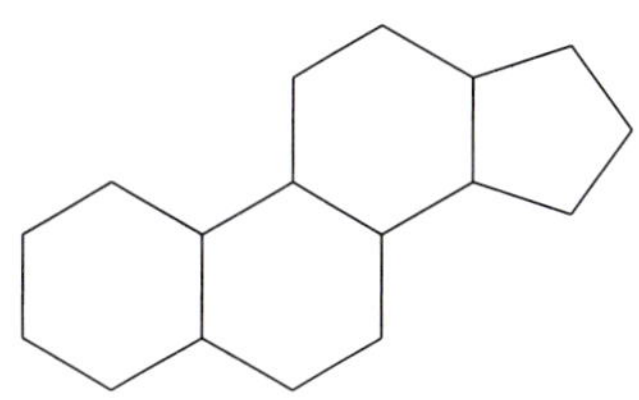

Figure 15.5 The core molecular structure of a steroid.

Table 15.3 Cholesterol and Triglycerides in Typical Foods (100-g servings, values are approximate)

		Fat and Oil Content (g)		
Food	Cholesterol	Saturated	Monounsaturated	Polyunsaturated
Beef	90	3	3	0.5
Butter	220	50	23	3
Cheese				
cheddar	105	21	9	0.9
cottage, dry	7	0.3	0.1	0.02
Swiss	90	18	7	1
Chicken, light meat, no skin	85	1	2	1
Corn oil	0	13	24	59
Eggs				
whole	430	3	4	1
yolk	1290	10	13	4
Frankfurter, all beef	50	13	15	1
Margarine, from corn oil	0	13	46	18
Milk				
skim	2	0.1	0.05	0.007
whole	15	2	1	0.1
Olive oil	0	14	74	8
Peanut butter	0	10	23	15
Peanut oil	0	17	46	32
Safflower oil	0	9	12	74
Salmon, pink, canned	35	1	2	3
Tuna, canned, in water	65	0.2	0.1	0.2
Turkey, light meat, no skin	70	1	1	1
Yogurt, plain, low fat	5	1	0.4	0.04

control), exercise, emotional stress, and still other influences, possibly including the amount and kinds of fiber in our diets. At any rate, the cholesterol generated within our bodies comes from the liver, which manufactures cholesterol from saturated fats. Reducing the amount of animal fats in our diets, especially those from red meats, is one recommended way of controlling serum cholesterol. Actually, animal fats pose a double threat: Not only do they provide the liver with the raw materials for its own production of cholesterol, but animal fats carry high levels of dietary cholesterol with them as they enter our digestive systems. Cholesterol, which has many of the properties of alkanes, doesn't dissolve in water to any appreciable extent. Like the hydrophobic tails of soap molecules (Sec. 13.3), cholesterol seeks out and dissolves in fatty, greasy substances, including animal fat. Thus animal fat serves to concentrate cholesterol within itself.

Diets rich in animal fats, then, are believed to promote high serum cholesterol levels both by carrying their own cholesterol with them and by providing our liver with saturated fats, which the liver uses to produce additional quantities of the sterol. It makes sense, then, to welcome manufacturers' boldly written claims that their products are CHOLESTEROL-FREE—until we read the small print on the labels that spell out, more quietly, the content of hydrogenated vegetable oils. The more highly hydrogenated these vegetable oils are, the more they become like animal fats, the liver's raw materials for putting its own cholesterol into our bloodstream.

Furthermore, while a dietary switch from animal fats to vegetable oils will certainly guarantee a reduction of our cholesterol intake, it will not necessarily ensure a nutritional turn to unsaturated side chains. If we want to replace the saturated triglycerides of our diets with their polyunsaturated counterparts, it's not enough simply to substitute vegetable oils for our dietary fats. We've got to be sure that the oils we choose are, indeed, polyunsaturated. Otherwise, as we see in the data of Table 15.3, we could end up with a diet of oils even richer in saturated triglycerides than the animal fats we abandon.

Plant products, including vegetables and vegetable oils, are all free of cholesterol.

■ QUESTION

Give two reasons why switching from a diet rich in red meat and animal products to one with plenty of fruits, vegetables, and other plant products might help lower serum cholesterol.

15.5 Oils That Spoil, Oils That Dry

Other chemical species, in addition to hydrogen and iodine, can also react with the double bonds of carbon chains. Atmospheric oxygen, for example, reacts with unsaturated chains of the triglycerides, fracturing them and converting them into a variety of smaller, foul-smelling molecules. Reactions of this sort cause cooking oils to turn rancid after long exposure to the air. Chemicals that react preferentially with oxygen, thereby sparing the polyunsaturated oils from oxidation, are sometimes added to cooking oils and other foods to increase their shelf lives. Among these *antioxidant* food additives are the synthetic compounds BHA and BHT and the natural antioxidant vitamin E. We'll examine these and other food additives more extensively in Chapter 19.

In addition to raising the melting and softening points of triglycerides, the partial hydrogenation used in the manufacture of shortening, margarines, and similar products also increases the shelf lives of the resulting triglycerides by decreasing the number of double bonds in their side chains and lowering their tendency toward oxidation. In this way the partial hydrogenation of triglycerides helps keep foods fresh.

The ability of polyunsaturated side chains to react with oxygen is put to good use in *drying oils*. These highly unsaturated vegetable oils are added to lacquers, varnishes, oil-based paints and some latex paints to produce tough, protective films as they dry. The drying of an oil-based paint isn't simply a matter of the evaporation of its solvent, usually mineral spirits or turpentine. In addition, the drying oil that remains on the painted surface reacts with atmospheric oxygen to form a tough web of triglyceride molecules linked to each other through oxygen atoms. Linseed oil and tung oil are especially useful as drying oils.

■ QUESTION

You are a manufacturer of varnishes that have drying oils as major ingredients. One of your laboratory chemists suggests that you add an antioxidant to the varnishes to improve their shelf life. How would you respond to this suggestion? Explain.

15.6 Double-Bond Geometry

We've seen repeatedly in this chapter the close connection between a triglyceride's physical state—liquid or solid—and the degree of unsaturation in its side chains. With a few easily explained exceptions, such as coconut oil and palm kernel oil (Sec. 15.2), carbon–carbon double bonds occur more often in the molecular structure of a liquid oil than in a solid fat. The connection between carbon–carbon unsaturation and melting point or softening temperature rests on a simple geometric requirement of the atoms associated with the double bond: the two carbons of the double bond and the four atoms attached to them (two to each carbon) must all lie in the same plane (Fig. 15.6). This particular bit of chemical geometry is a necessary result of the way the carbon–carbon double bond forms. We'll soon see how this geometry affects both the shape of the molecule and the melting point of the triglyceride.

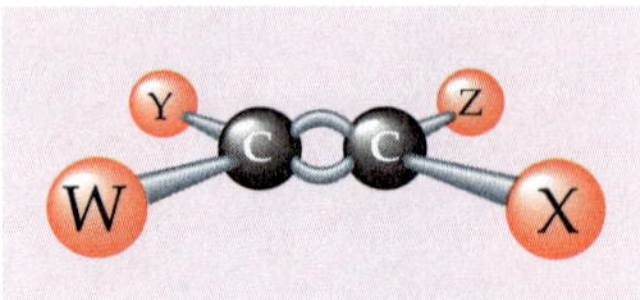

Figure 15.6 Planarity and the carbon–carbon double bond. The two carbons and atoms W, X, Y, and Z all lie in the same plane.

The required planarity of the set of six atoms—the two carbons of the double bond and the four atoms bonded to them—produces a new kind of isomerism. As we saw in Section 6.6, isomerism occurs when two different compounds share the same molecular formula. Here we encounter a form of isomerism, *geometric isomerism*, that results from the planar geometry of the C=C double bond. **Geometric isomers** are different compounds in which the same four groups are bonded to their C=C carbons, but in different geometric arrangements.

Geometric isomers are different compounds that have the same four groups bonded to the carbons of their double bonds, but with different geometries.

The alkene 2-butene (Sec. 6.10) illustrates geometric isomerism nicely. All four carbons of this compound, as well as the two hydrogens of the carbon–carbon double bond, lie in the same plane. Moreover, each carbon of the double bond carries two different substituents: a hydrogen and a methyl group. In one geometric isomer of 2-butene, *cis*-2-butene, both methyl groups lie on the same side of the double bond. (Both hydrogens also lie on the same side of the molecule—the side opposite the two methyl groups.) In the other geometric isomer, *trans*-2-butene, the two methyl groups lie on opposite sides of the double bond, as do the two hydrogens:

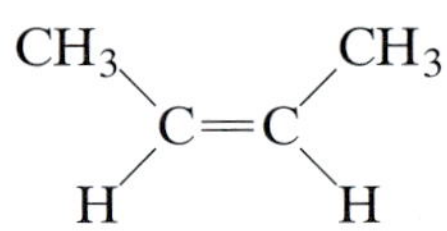

cis-2-butene

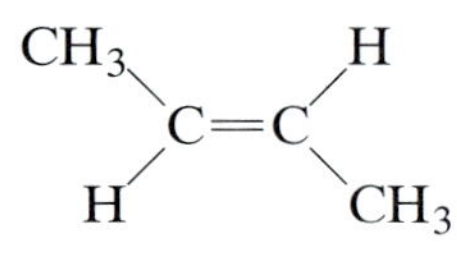

trans-2-butene

geometric isomers of 2-butene

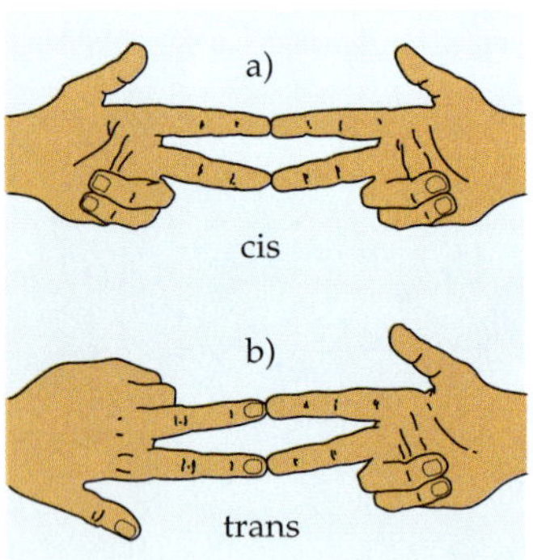

Figure 15.7 Geometric isomers.

As isomers, *cis*-2-butene and *trans*-2-butene are two different covalent compounds, with different chemical and physical properties: *cis*-2-butene melts at −138.9°C, *trans*-2-butene at −105.5°C; the *cis* geometric isomer boils at 3.7°C, the *trans* at 0.9°C. *cis*-2-Butene and *trans*-2-butene are geometric isomers. In *cis* isomers, similar groups (here the two hydrogens and the two methyl groups) lie on the same side of the C=C double bond; in *trans* isomers the similar groups lie on opposite sides. You can visualize this in two different ways. Figure 15.7 shows how you can put your hands together with two fingers representing the two covalent bonds of a double bond. At the top, part (a) of Figure 15.7, your extended thumbs are on the same side of the double bond and represent *cis* substituents; at the bottom, part (b), they're *trans*. Figure 15.8 represents the two isomers of a molecule of 2-butene itself.

Notice a peculiarity here. Geometric isomerism can occur *only* when each carbon of the double bond bears two different groups. Geometric isomerism is impossible if either one of the C=C carbons holds two identical groups. In 1-butene, for example, the end carbon bears two (identical) hydrogens. Switching the geom-

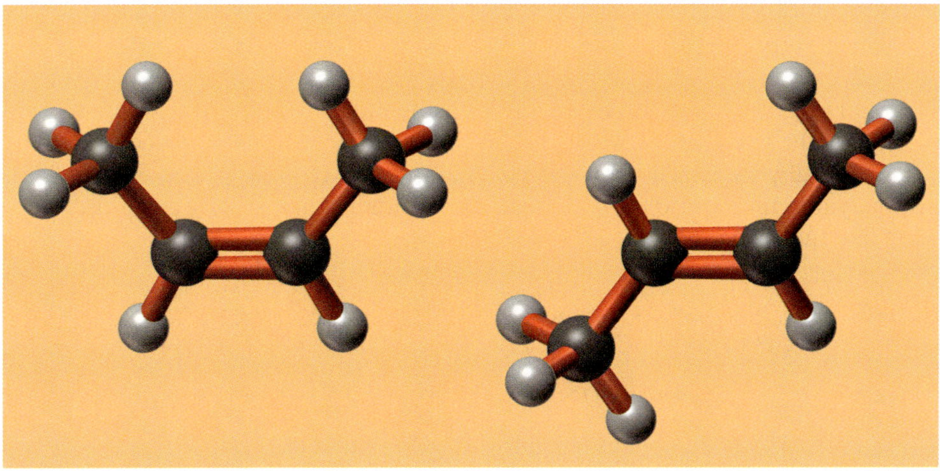

Figure 15.8 *cis*-2-Butene (left) and *trans*-2-butene (right).

etry of the double bond—transposing the positions of the two hydrogens—doesn't change a thing. The two hydrogens of the end carbon are identical, so there are no geometric isomers of 1-butene.

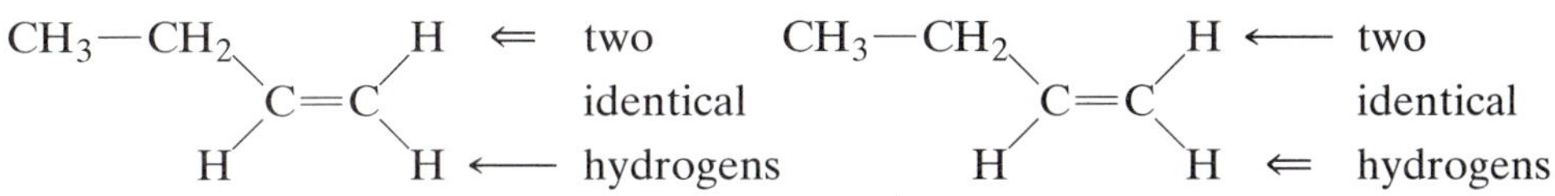

This molecule of 1-butene is identical to *this* molecule of 1-butene, even though the two hydrogens have been switched.

■ QUESTION

Label each of the following as *cis* or *trans* or incapable of geometric isomerism:

a. 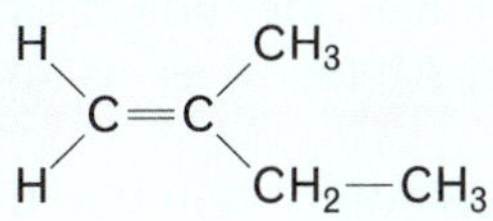

b. 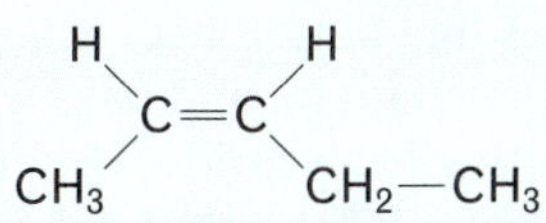

c. 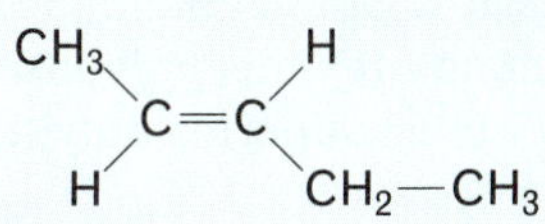

d. 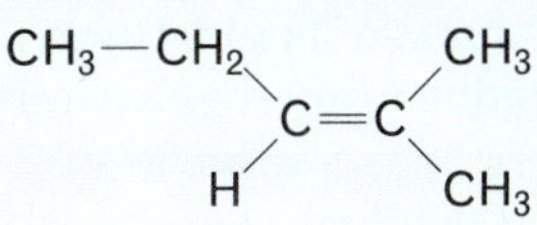

15.7 Side Chains and Human Nutrition

Early in Section 15.6 we noted the association between the melting or softening point of a triglyceride and the geometry of its double bonds. Here's how they're connected.

As we saw in Section 12.2, crystalline solids melt as their chemical particles leave the orderly arrangement of a crystal lattice and begin to move about more freely in the liquid phase. In contrast, the molecules of our dietary fats don't form the symmetrical, highly organized lattices that we find in the nicely defined crystals of, say, sodium chloride and sucrose. Nevertheless, short-range intermolecular attractive forces within the solid and semisolid triglycerides do hold their molecules together tightly, in a compact, somewhat orderly arrangement. These are the same sort of forces that bind together, for example, the large alkane molecules of candle wax, and they depend on the ability of the hydrocarbon chains to pack snugly against each other.

Long, fully saturated hydrocarbon chains, those of alkanes and triglycerides, normally fold nicely into fairly regular, three-dimensional zig-zag patterns, somewhat like the following:

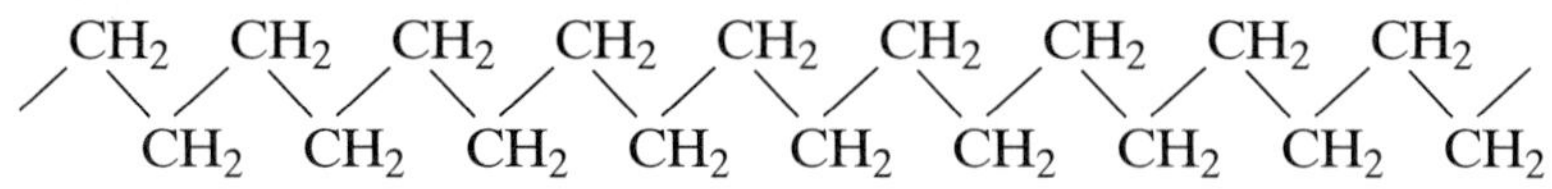

a portion of the regular pattern of fat's side chain

These symmetrical arrangements allow the saturated side chains, as well as the triglyceride molecules themselves, to fit compactly next to each other. Since they lie so close to each other, the short-range attractive forces effectively keep them packed together and keep them from melting—keep them in the solid phase—even at temperatures as high (relatively) as normal room temperature.

Things are different, though, in the polyunsaturated oils. All the double bonds of our *unprocessed* dietary triglycerides exist in the *cis* geometry. (We'll have more to say about *processed* foods in a moment.) The plentiful *cis* units of unsaturation in a polyunsaturated vegetable triglyceride introduce repeated kinks into the zig-zag arrangement of the carbon chains that, in effect, increase the bulk of the chains and prevent the close, orderly association of molecules that we find in the more highly saturated solid and semisolid animal fats. Polyunsaturation, as it occurs in most plant triglycerides, tends to increase the effective size of the triglyceride molecules, moving them farther apart, lowering the strength of their intermolecular attractive forces, and thereby lowering their melting points. All this gives us liquid triglycerides, the oils (Fig. 15.9).

That's why vegetable oils, most of which have many *cis* double bonds in their side chains, are liquids at room temperature. Substances like butter, beef fat, and lard, with fewer *cis* double bonds, are solids. Except for coconut oil and palm kernel oil (with their relatively short side chains), low-melting triglycerides generally consist of molecules that don't pack together well because of the abundance of the *cis* double bonds that produce kinks in their side chains. It's this polyunsaturation, resulting in their existence as liquids, that makes them superior nutritionally to the higher-melting, better organized fats (Sec. 15.4).

The story is a little different for the double bonds that remain standing even after partial hydrogenation of vegetable oils. Partial hydrogenation of these oils saturates many, but not all, of their carbon–carbon double bonds. But it does more as well. Affected by the reaction conditions, including the presence of the catalyst itself, many of the C=C double bonds that remain untouched by the hydrogen molecules isomerize from their natural *cis* geometry into a newly acquired *trans* geometry. Along with an increase in saturation, the geometry of the remaining double bonds, now *trans*, also allows neighboring chains to pack more tightly. As a result, the transformation from *cis* to *trans* also contributes to an increase in the triglyceride's melting point.

Partial hydrogenation, then, produces not only a decrease in unsaturation—an increase in saturation—but also an introduction of *trans* double bonds into the partially hydrogenated product. Some studies indicate that these *trans* double bonds add to the health-related threat posed by saturation itself. Unlike their *cis* isomers, these *trans* isomers appear to increase levels of a particularly dangerous

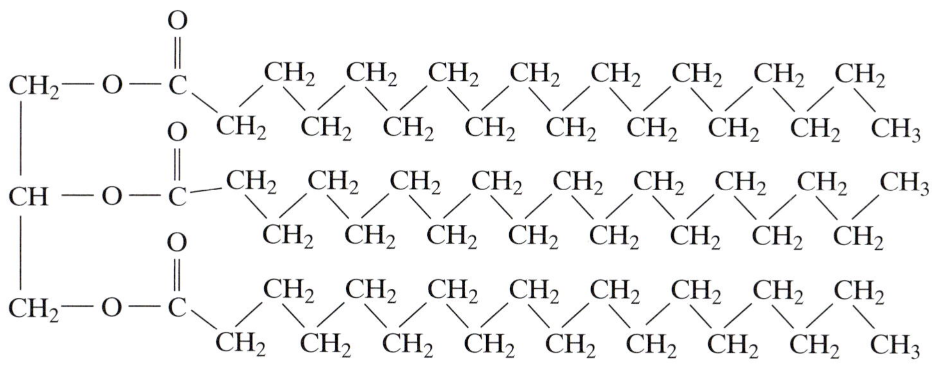

Glyceryl tristearate; chains are packed closely

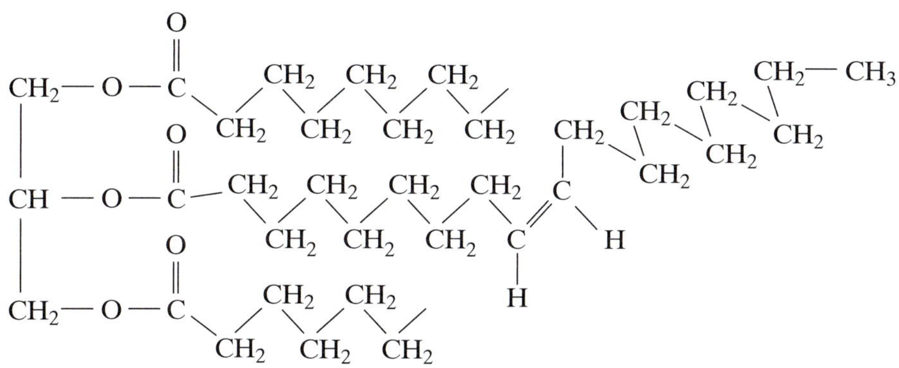

Glyceryl trioleate; double bonds prevent close packing
(For clarity, only the center chain is shown fully)

Figure 15.9 Molecular packing in glyceryl tristearate and glyceryl trioleate. Polyunsaturation increases the space occupied by fatty acid side chains, causing polyunsaturated triglyceride molecules to lie farther from each other than molecules with more fully saturated side chains. This increased distance of separation decreases the attractive forces the polyunsaturated molecules exert on each other and lowers the melting points of polyunsaturated triglycerides.

form of serum cholesterol, the LDL, which we'll examine in Section 17.10. In the next section we'll find that, in addition to its geometry, the location of the double bond within the chain also has implications for human health.

■ QUESTION

Why does the hydrogenation of a polyunsaturated triglyceride raise its melting point?

15.8 The Omega Acids

We use the fatty acids of triglycerides not only as a source of biochemical energy, but also as raw material for many of our bodies' physical structures and biochemical catalysts, including cellular membranes and the hormones that regulate a host of bodily functions. In generating these substances from our food chemicals, we can produce almost all of the necessary modifications ourselves, converting one molecular structure into another or joining several into larger, more complex forms.

We're often flexible in our approach. If we find a needed molecular structure among our food's nutrients, we use it. If not, we generate it from other components of our foods. But there are some chemicals our bodies simply don't have the ability to create, including the molecular structures of linoleic acid and linolenic acid (Table 15.1), the *essential fatty acids* of human nutrition. We can't generate either linoleic acid or linolenic acid by modifying the molecular structures of other fatty

The **essential fatty acids** are the fatty acids our bodies cannot synthesize from other chemicals and that we must obtain from our foods.

acids. Both linoleic and linolenic acid must be present in the foods we eat. Because we can't produce either of them by our own bodily processes, not even from closely related acids, we label these two critical acids as **essential fatty acids**—*essential* because they must be part of our diets.

Linoleic and linolenic acids are members of two more general categories of fatty acids critical to human well-being, the *omega-3* and the *omega-6* fatty acids. The defining characteristic of these categories is the location of the C=C double bond most remote from the carbonyl carbon (Sec. 10.8).

In dealing with these groups of acids, it's particularly convenient to label the carbons of the fatty acid chain with Greek letters. We identify the first carbon beyond the carbonyl group with the first letter of the Greek alphabet. It becomes the α (*alpha*) carbon. The second is the β (*beta*) carbon, the third is the γ (*gamma*), etc. If we're considering a set of acids with long chains of various lengths and we're interested in a structural feature near the most remote carbon—the one farthest from the carbonyl group—we can label that most remote carbon with the letter that's in a parallel location in the Greek alphabet, at the end: ω. That methyl group at the end of the chain becomes the *omega* carbon, regardless of the length of the chain:

$$\underset{\omega}{CH_3}-CH_2-CH_2-CH_2- \quad \cdots \quad -\underset{\gamma}{CH_2}-\underset{\beta}{CH_2}-\underset{\alpha}{CH_2}-\overset{\overset{\large O}{\|}}{C}-OH$$

Working backward from the methyl group at the end of the chain, with its omega carbon, the C=C double bond of linolenic acid begins at the third carbon from the end. Reflecting this, linolenic acid is a member of the *omega-3* group of fatty acids. Similarly linoleic acid is an *omega-6* fatty acid:

$$CH_3-CH_2-CH{=}CH-\cdots-CH_2-CO_2H$$

the two ends of linolenic acid, an omega-3 fatty acid

$$CH_3-CH_2-CH_2-CH_2-CH_2-CH{=}CH-\cdots-CH_2-CO_2H$$

the two ends of linoleic acid, an omega-6 fatty acid

In the same sense that the Perspective of each chapter is the terminal, or omega, section, regardless of the chapter's length, the methyl group of each fatty acid contains its omega carbon, regardless of the chain's length.

■ QUESTION

Considering the Perspective as the terminal section of this chapter, what omega designation would you give to this current section?

15.9 Storing Energy: Adipose Tissue

In Chapter 8 we saw that the sun serves as the ultimate source of our bodily energy. Photosynthetic reactions convert the sun's radiant energy into the chemical energy of plant nutrients. When we eat these plants, or products of the animals that feed on them, we convert this nutrient, chemical energy into our own biological energy. Yet although we expend biological energy constantly, we don't replenish it constantly.

In the Energy Equation of Section 8.8,

$$\text{Energy In} = \text{Energy Out} + \text{Energy Stored}$$

The body stores energy as the triglycerides of fat or adipose tissue.

our Energy Out is continuous—through basal metabolism (even while we're sleeping) and virtuously continuous exercise of various intensities throughout our waking hours—but Energy In is far from continuous. Several hours normally elapse between meals, sometimes with vigorous exercise. In fact, it's possible to fast for many days yet still have the energy to do most of the things we normally do. Very little of the energy we expend as exercise comes immediately from the food we eat, and what does often goes largely toward specific dynamic action and a portion of our basal metabolism.

If we didn't store the unused chemical energy of our food—if the only metabolic energy available to us and to our ancestors were the energy immediately available from the macronutrients of a recent meal—none of us would be alive today. The human race would have vanished eons ago as a result of mass starvation. Without some means for storing energy in the body, a temporary and minor shortage of the animals hunted by our cave-dwelling ancestors or a brief scarcity of the grains, fruits, or vegetables they ate would have been catastrophic; it would have wiped out the human race.

Even now, with food in plentiful supply and continuously available in many regions of the world, energy storage within our bodies is literally a matter of life and death. Our hearts, lungs, and other vital organs would run out of energy between meals, even as we sleep, if it were not for our body's ability to store the excess fuel of each meal and then use that stored fuel to provide a continuous supply of energy between meals. The principal mode of long-term energy storage available to our ancestors and to us operates through the formation, storage, and metabolism of body fat. (Short-term energy storage, usually from one meal to another, occurs through a starchlike substance called *glycogen*, which we'll examine in Chapter 16.)

We can think of body fat as stored energy. The body converts much of the unused carbohydrates, proteins, and triglycerides that make up our macronutrients into small globules of fat that end up in the specialized cells of **adipose tissue**, the fatty tissue of the body. One pound of adipose tissue stores (and provides when needed) roughly 3500 Cal of energy.

Adipose tissue is the fatty tissue of the body. It stores chemical energy at about 3500 Cal/pound.

Converting an excess of a macronutrient into body fat and then converting this fat into energy later, when we need it, is a particularly effective means of long-term energy storage. With a gram of carbohydrate or protein providing only about 44% (four-ninths; Sec. 8.10) of the amount of energy of a gram of fat, storing excess energy in the form of either carbohydrates or proteins would raise our bulk and weight considerably and make us far less mobile. The high *energy density* of fat—its ability to store energy compactly in relatively little space and with relatively little weight compared with carbohydrates and proteins—allows us to carry our stores of energy with us. It gives us and other animals the mobility and freedom necessary for survival in a world that was, and can still be, unforgiving to those poorly equipped for survival.

■ QUESTION

Suppose you're 15 pounds overweight. That is, you are carrying around, at all times, an excess of 15 pounds of adipose tissue. Translate this into the equivalent number of excess stored Calories you are carrying with you at all times.

PERSPECTIVE

Fats and Oils: The Perfect Examples

In this chapter we have examined the triglycerides of our food as chemicals that provide us with both the substance of many of the hormones, tissues, and other physical structures of our bodies and the energy to use them. They are esters that follow all the laws of chemistry and contain, scattered within their molecules, the same carbon–carbon double bonds that occur in the unsaturated hydrocarbons we know as alkenes. Hydrogen and iodine add to their carbon–carbon double bonds. They are the raw materials of the soap industry. They are chemicals we eat, chemicals we manufacture within our own bodies, and chemicals that provide the raw materials for major industries.

Fats and oils are also substances that work both for and against life and good health. We've already seen, for example, that body fat stores energy for us so that if we fast, miss a meal, or even sleep through a particularly long night, we still have the energy to breathe and to keep our hearts beating. Our adipose tissue also forms cushioning shields around our major organs, protecting them against damage from physical shock when we miss a curb or don't see the last step of a staircase. This tissue also provides thermal insulation to our bodies, guarding against a rapid loss of body heat to the external environment.

Fats also carry the flavors and vitamins of many of our foods. Although fats have no flavors of their own, many of the substances that do add flavor and enjoyment to eating are far more soluble in fats than in the more watery substances of food. The fat in meat, for example, carries the flavor we associate with the meat. Without fat meats would be tasteless. More importantly, we'll see in Chapter 19 that many of our vitamins are much more soluble in fats than in water. It's fat that carries vitamins A, D, E, and K from our foods to our tissues. And as we've already seen, our bodies use fatty acids in forming not only physical structures such as cell walls but many of our hormones and other regulatory substances as well.

Fats and oils clearly contribute to our life, health, and well-being. On the other hand, in excess they can be dangerous. Obesity is linked to both high blood pressure and diabetes. Dietary fat is implicated in heart disease and some forms of cancer. The connection among dietary fat, cholesterol, and atherosclerosis is well established (Sec. 15.4). Although several health and scientific organizations recommend that we obtain a maximum of 30% of our dietary calories from fats and oils, an average of about 37% of our calories now come from these triglycerides. While statistical correlations showing a connection between dietary fat and both heart disease and cancer don't provide proof of a cause-and-effect relationship, cutting fatty calories to well below 30% of our total caloric intake seems desirable.

Fats and oils provide us with almost perfect examples of the chemical basis of our everyday world; of the risks, hazards, and benefits in chemicals; of the contrast between the benefits that can be derived from moderate amounts of a particular chemical and the damage that can be done by an excess; and of our need for a knowledge and understanding of chemistry so that we can make intelligent and productive choices from among the many options we face each day. In many ways fats and oils represent virtually the ideal chemicals for an examination of the extraordinary chemistry of ordinary things. ■

EXERCISES

For Review

1. Following are a statement containing a number of blanks and a list of words and phrases. The number of words equals the number of blanks within the statement, and all but two of the words fit correctly into these blanks. Fill in the blanks of the statement with those words that do fit, then complete the statement by filling in the two remaining blanks with correct words (not in the list) in place of the two words that don't fit.

The fats and oils of the human diet are __________, or triesters of __________ and various __________. Those that are solids under ordinary conditions are called __________, while those that are liquids are the __________ of our diets. At the molecular level, the melting point of a triglyceride is determined by two factors: the __________ of the fatty acids' __________ and the degree of __________. Triglycerides undergo several important chemical reactions, including __________, which leads to soaps when it's carried out in the presence of a base, and the addition of __________ and of the halogens. __________ leads to an increase in the level of saturation of the side chains and an increase in the melting point of the triglyceride. The quantitative addition of iodine yields the __________ of a triglyceride, which is a measure of the extent of unsaturation in its molecules. It's undesirable to include large amounts of animal fats in a diet because these fats are rich in __________ and in the sterol __________, which can produce atherosclerosis, or hardening of the arteries.

carbon chains	hydrolysis
catalytic hydrogenation	iodine number
cholesterol	oils
cis double bonds	saturated side chains
fats	triglycerides
fatty acids	unsaturation
hydrogen	water

2. Identify or define each of the following:

adipose tissue	geometric isomerism
antioxidant	hydrogenation
drying oil	polyunsaturation

3. How does an increase in the number of $-CH_2-$ groups in a saturated fatty acid affect its melting point?

4. How does an increase in the number of carbon–carbon double bonds in a fatty acid, with no change in its carbon content, affect its melting point?

5. Given a sample of a triglyceride, how would you determine whether it is a fat or an oil?

6. Name the major saturated fatty acid obtained by the complete hydrogenation and then the hydrolysis of each of the following oils: (a) corn, (b) cottonseed, (c) linseed, (d) olive, (e) peanut, (f) safflower, (g) sesame, (h) soybean, (i) sunflower seed, (j) tung, (k) wheat germ.

7. Of beef fat, cocoa butter, and coconut oil, which has the highest total percentage of fully saturated fatty acids? Which has the lowest?

8. What nutritional hazard to human health is present in beef but not in cocoa butter or coconut oil?

9. While vegetable oils provide the same number of calories as animal fats, replacing animal fats with vegetable oils usually decreases the dietary levels of two undesirable substances. What are they?

10. Food labels occasionally claim that the product is made with partially hydrogenated vegetable oils "to preserve freshness." How does partial hydrogenation of an oil help preserve freshness?

11. Describe two undesirable dietary consequences of the partial hydrogenation of vegetable oils.

12. Why would a food product made of fully saturated triglycerides have little commercial value?

13. Name two fatty acids that the body uses but cannot synthesize from other substances.

14. An important group of fatty acids—a group that our bodies *can* synthesize—have chains that end in the structure

$$CH_3-CH_2-CH_2-CH_2-CH_2-CH_2-CH_2-CH_2-CH=CH-$$

Using the *omega* system of naming fatty acids, what designation would you give to this group of acids?

15. Draw the four-ring skeleton of a steroid.

16. What major class or category of foods contains no cholesterol whatever?

17. In what kinds of consumer products are drying oils used? Why?

18. What condition must exist for geometric isomerism to occur as a result of the presence of a carbon–carbon double bond?

19. What *two* changes in the molecular structure of a triglyceride occur as a result of partial hydrogenation?

20. What chemical reaction results in the *hardening* of a triglyceride?

21. Where do we find the *omega* carbon of a long-chain carboxylic acid?

22. Explain the difference between *fat* and *adipose tissue*.

A Little Arithmetic and Other Quantitative Puzzles

23. Which food in each of the following sets contains the highest percentage of saturated fats? (a) butter, Swiss cheese, whole milk; (b) olive oil, peanut oil, peanut butter; (c)beef frankfurter, white meat of chicken, white meat of turkey; (d) corn oil, egg yolk, whole milk.

24. Which food in each of the following sets contains the highest percentage of cholesterol? (a) butter, Swiss cheese, whole milk; (b) skim milk, whole milk, yogurt; (c) beef frankfurter, white meat of chicken, white meat of turkey; (d) beef, cheddar cheese, margarine?

25. Corn oil contains about 58.7% polyunsaturated triglycerides, but margarine, which is made from corn oil, contains only about 18% polyunsaturated triglycerides. Explain the difference.

26. Based on the degree of unsaturation in their side chains (Table 15.1), which one of the following would you expect to have the *lowest* melting point? The *highest* melting point? (a) glyceryl trioleate (the triester of glycerol and oleic acid); (b) glyceryl trilinolenate; (c) glyceryl tristearate; (d) glyceryl trilinoleate.

27. What effect, if any, does catalytic hydrogenation have on iodine numbers? Explain.

28. Neglecting any losses due to exercise, specific dynamic action, or basal metabolism, how many grams of each of the following would you have to eat to gain 1 pound of adipose tissue? (a) protein, (b) carbohydrate, (c) fat or oil.

29. Many food labels now proclaim that they are fat-free to a specified percentage. If a label claims that a processed lunch meat is "80% fat-free," what percentage of the calories of that lunch meat comes from fat? (Assume that if the meat is 80% fat-free, it contains 20% fat and the rest of it is a combination of carbohydrate and protein.)

30. Adipose tissue consists of fat globules and cellular protoplasm, which is mostly water. Virtually all of the energy stored in adipose tissue, about 3500 Cal/pound, comes from the chemical energy of its fat globules. Given the relationships

$$1 \text{ pound} = 454 \text{ g}$$
$$1 \text{ g of fat} = 9 \text{ Cal}$$

you can calculate the percentage of fat in adipose tissue. For this calculation, recognize that all of the energy stored in adipose tissue is stored in the triglycerides of its fat globules, determine the energy content per gram of adipose tissue, and compare this value with the energy content of fat.

a. How many Calories are stored in 1 g of adipose tissue?
b. How many Calories are stored in 1 g of fat triglycerides?
c. What is the percentage of fat in adipose tissue?

Think, Speculate, Reflect, and Ponder

31. Table 15.3 shows that margarine, which is made from corn oil, contains a higher percentage of saturated and monounsaturated fats and oils than corn oil, but a lower percentage of polyunsaturated fats and oils. Explain why.

32. Medical studies of connections between diets and diseases of the heart and circulatory system (cardiovascular disease) indicate that diets containing a large percentage of fish oils seem to protect against cardiovascular disease. This apparent protective effect of fish oils may come from some special characteristic of the fatty acid content of their triglycerides. Using the data of Table 15.2, identify what is unique about the side chains of the triglycerides of fish oils.

33. Why is iodine particularly useful for establishing the degree of unsaturation in a triglyceride?

34. The oils obtained from fish found in very cold waters are generally more highly unsaturated than the oils of fish whose natural habitats are in warmer waters. In what way would an increase in unsaturation benefit fish living in very cold waters?

35. Compare the health aspects of using butter, margarine, and vegetable cooking oils for frying an egg. (a) What health advantage does margarine have over butter? (b) What health advantage does cooking oil have over margarine?

36. Even though cocoa butter comes from a plant, it is a fat. After comparing cocoa butter's content of the various fatty acids with those of other vegetable triglycerides (Table 15.2), suggest a reason for its unusually high melting point.

37. Would you expect an omega-1 fatty acid to show geometric isomerism? Explain.

38. Which would you expect to be the better drying oil, an oil with a very high iodine number or one with a very low iodine number? Why?

Web Questions

Use the Web to answer or accomplish each of the following. Report your Web source for each response.

1. Give the name, structure, and biological importance of a steroid other than cholesterol.

2. What are some good dietary sources of omega-3 fatty acids? Of omega-6 fatty acids?

3. What are *prostaglandins*? What are some of their functions in the human body? How are they related to fatty acids?

4. Who was Michel Eugene Chevreul? What is his connection to the topics of this chapter?

5. What are *lipids*? Why are they important to human health? Which of the compounds discussed in this chapter can be classified as lipids?

6. Find a fatty acid that is important to human health, other than those listed in Tables 15.1 and 15.2. Give its name, molecular structure, and dietary sources, and describe its importance to our health.

7. What is meant by *obesity*? What is the *Body Mass Index (BMI)*? How is the BMI calculated and how is it related to obesity?

Carbohydrates

FOOD FOR THOUGHT

Sugar, our sweet carbohydrate.

How to Tell a Potato from an Apple, the Hard Way

Figure 16.1 The color produced by an iodine solution distinguishes the starch of a potato from the cellulose of an apple. The iodine turns blue–black with starch but remains reddish brown with cellulose.

You can tell a potato from an apple by smelling or tasting them. Of course, you could just look at them. But there's a harder way to do it. It's more fun, too.

The interior pulp of an apple looks very much like the interior pulp of a potato. They're both pale and appear to have the same consistency. If you slice out a piece of the interior of a raw potato and an apple it's hard to see any difference between them. But announce that you can tell which is the section of an apple and which is the section of a potato without smelling, tasting, or touching either one. Leave the room while a friend cuts out a piece of a raw potato and a similar piece of an apple (a ripe Delicious apple works well) and places them on a small dish, side by side, without any skin or other identifying characteristics on either one.

Now return to the room with a small bottle of tincture of iodine, the kind used for demonstrations in earlier chapters. Put a few drops of iodine on each. When the iodine contacts the potato it turns a very dark purple, nearly black, as you can see in Figure 16.1. While the apple might show a few small, dark spots too, for the most part the iodine on the apple retains the same reddish-brown color it produces on skin when it's used on a cut. The piece that turns almost black is the potato; the one that's mostly reddish brown is the apple. Why? We'll soon learn the secret.

In this chapter we'll examine carbohydrates, the second of the three classes of macronutrients. We'll find that carbohydrates form the digestible sugars and starches of our food and an indigestible component of our diet we call *fiber*. We'll also learn of two new classes of organic compounds—aldehydes and ketones—and we'll find a new form of isomerism that produces the same kind of differences in the shapes of molecules that we see in our right and left hands.

16.1 The Brain's Own Fuel

While you were reading the details of the opening demonstration and the brief summary of what follows, you were burning up energy. You still are, even as you read these words. Aside from the fundamental workings of basal metabolism and the possibility of specific dynamic action, your brain and its associated network of nerves are hard at work interpreting these short, straight, and rounded lines you see before you, converting them to recognizable letters, words, and thoughts, and absorbing and digesting them. All this nibbles away at your stores of energy.

You do very little actual work in this sort of mental activity, but you do consume a small, measurable amount of energy. The fuel that supplies this energy to your brain is the carbohydrate *glucose*, also known to us as *dextrose* or *blood sugar*. Glucose is the specific and exclusive fuel of the brain and the nervous system. It is quite literally our food for thought. Glucose also supplies the energy that keeps

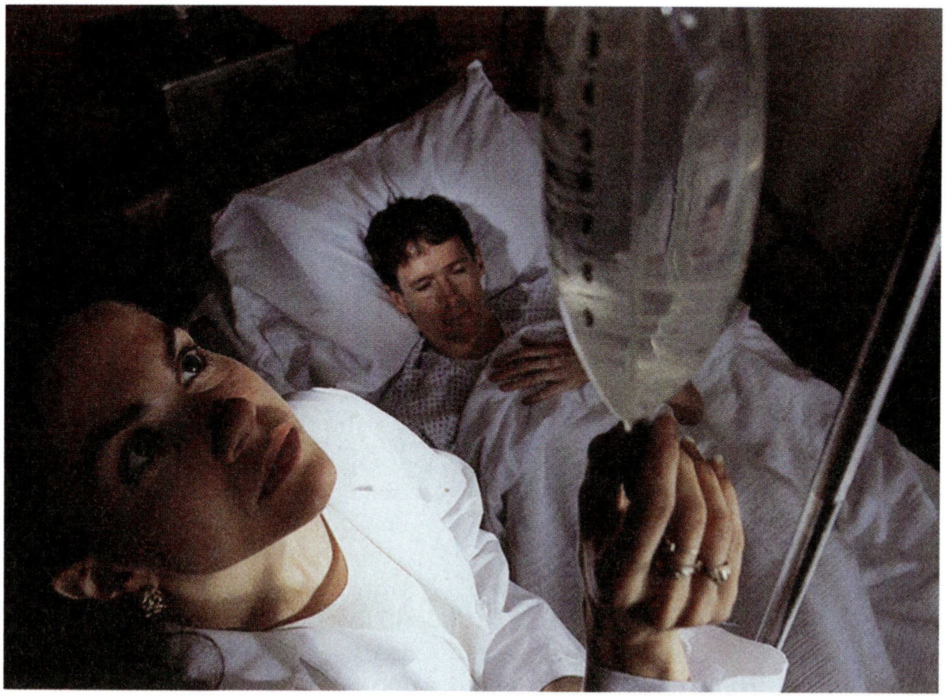

A 5% solution of glucose, a carbohydrate also known as dextrose and blood sugar, supplies the body with energy as it flows intravenously into the bloodstream.

our bodies at a constant temperature, moves our muscles, and keeps our digestive and respiratory systems running.

To provide a continuous supply of this fuel to the brain, nerves, muscles, and other bodily systems, the body maintains a concentration of about 0.06 to 0.11 weight-percent glucose in the blood. Somewhere below this range the brain begins to lose its ability to function effectively. Above concentrations of about 0.16%, glucose begins to seep through the kidneys into the urine, producing one of the symptoms of diabetes. We'll begin our examination of carbohydrates by focusing on glucose, which is one of the simplest of the common carbohydrates and one of the most important.

■ QUESTION

Blood makes up about 8% of the body weight of a healthy adult. How many grams of glucose circulate through the body of a typical 110-pound person?

16.2 A Case of Mistaken Identity

Unlike our nutritional fats and oils, all of which are esters of glycerol and which differ from one another entirely through the lengths and the degrees of unsaturation of their side chains, the carbohydrates come in a great variety of shapes and sizes. All carbohydrates, though, share certain key molecular characteristics:

- Carbohydrates are composed exclusively of three elements: carbon, hydrogen, and oxygen.
- Although carbohydrate molecules themselves occur in various shapes and sizes, the carbon atoms that form them are almost always strung along next to one another in unbranched chains.
- The ratio of hydrogen atoms to oxygen atoms in all carbohydrate molecules is almost always exactly 2 : 1, as it is in water.

Adding a little concentrated sulfuric acid to powdered sucrose produces heat, steam, and a black, brittle solid that resembles badly charred wood.

This virtually unvarying ratio generates molecular formulas for carbohydrates that suggested to the earliest workers in this field that these molecules are merely hydrates of carbon, or combinations of carbon and water. (The word *hydrate* and the latter portion of *carbohydrate* come from a Greek word for water.) The molecular formula of glucose, for example, $C_6H_{12}O_6$, can be written as though the molecule were composed of six carbon atoms bonded to each other and also to six water molecules, as in $C_6(H_2O)_6$. Although this interpretation of the molecular formulas proved to be completely incorrect, the name *carbohydrate*, meaning *hydrate of carbon*, has stuck. Now it applies not only to the carbohydrates themselves but also to other, closely related compounds with hydrogen-to-oxygen ratios varying just a bit from 2:1.

■ QUESTION

The carbohydrate sucrose (table sugar) has the molecular formula $C_{12}H_{22}O_{11}$. Rewrite this molecular formula as it would appear if sucrose were a true hydrate of carbon.

16.3 From Monosaccharides to Polysaccharides

Carbohydrates are easily sorted into several categories according to their size and the functional groups they carry. (As we saw in Section 6.12, a *functional group* is a portion of the molecular structure that confers a characteristic chemical reactivity to a molecule.) The smallest, simplest carbohydrates, those that form the foundation of carbohydrate chemistry, are the *monosaccharides*. The root *-sacchar-* comes from the Latin *saccharum*, for sugar. We see it also in *saccharin*, the first commercially successful, synthetic sugar substitute. *Monosaccharide*, then, literally means *one sugar*. Chemically, the **monosaccharides** are the smallest units of the carbohydrates. We can split any carbohydrate molecule into smaller and smaller units until we come to the monosaccharides that compose it. When we split apart a monosaccharide molecule, the products are almost never carbohydrates themselves. Glucose is a monosaccharide; it's a carbohydrate that can't easily be converted into a smaller carbohydrate.

A **monosaccharide** is the smallest molecular unit of a carbohydrate. Glucose is a monosaccharide.

Strings of monosaccharides, joined by covalent bonds, form larger and more complex carbohydrates, ranging in size from a combination of two monosaccharides to the enormously long molecules of the edible starches of plants and the indigestible cellulose of dietary fiber. Join two monosaccharide molecules together chemically and you get a **disaccharide**. Sucrose, known more commonly as table sugar, is a disaccharide formed by the combination of glucose with *fructose* (another monosaccharide). Lactose, also known as milk sugar, is a disaccharide formed by glucose and the monosaccharide *galactose*. Three monosaccharides joined by covalent bonds produce a trisaccharide; four give a tetrasaccharide; and so on. Linking very large numbers of individual monosaccharides into very long chains gives **polysaccharides**, such as starch and cellulose. Molecules of these complex carbohydrates consist of chains containing 300 to often more than 1000 glucose molecules joined, one after another, in long molecular strings.

A **disaccharide** is a molecule formed from a combination of two monosaccharides. A **polysaccharide** is a molecule that exists as a chain of hundreds of monosaccharides.

In all these cases, the carbon skeletons of the individual monosaccharides are connected to each other through oxygen atoms, via carbon–oxygen–carbon bonds. While degrading *monosaccharides* into smaller molecules is a tricky and complex job that requires breaking the carbon–carbon bonds of their individual molecular

Table 16.1 Common Carbohydrates

Carbohydrate	Molecular Formula	Source or Origin	Monosaccharide(s) Produced on Hydrolysis
Monosaccharide			
Glucose (also known as blood sugar, grape sugar, and dextrose)	$C_6H_{12}O_6$	Blood, plant sap, fruit, honey	
Fructose (also known as levulose)	$C_6H_{12}O_6$	Plants, fruit, honey	
Galactose	$C_6H_{12}O_6$	From the hydrolysis of lactose	
Disaccharide			
Sucrose (also known as table sugar, beet sugar, and cane sugar)	$C_{12}H_{22}O_{11}$	Sugar cane, sugar beets, maple syrup, and various fruits and vegetables	Glucose and fructose
Maltose (also known as malt sugar)	$C_{12}H_{22}O_{11}$	Partial hydrolysis of starch	Glucose
Cellobiose	$C_{12}H_{22}O_{11}$	Partial hydrolysis of cellulose	Glucose
Lactose (also known as milk sugar)	$C_{12}H_{22}O_{11}$	Makes up about 5% of milk	Glucose and galactose
Polysaccharide			
Starch		Potatoes, corn, various grains	Glucose
Cellulose		Cell walls of plants	Glucose

skeletons, the more complex carbohydrate chains of *polysaccharides* can be clipped easily by breaking the carbon–oxygen bonds that hold them together. Water does the job nicely through hydrolysis, especially if there's a little acid and perhaps a specific enzyme present. We'll have more to say about the catalytic action of enzymes in Section 16.12.

Table 16.1 presents some of the more common carbohydrates, their common names, and some of their chemical characteristics.

■ QUESTION

Give the chemical names of two: (a) monosaccharides; (b) disaccharides; (c) polysaccharides.

16.4 Glucose: An Aldohexose

It's useful to classify the monosaccharides themselves according to

- the lengths of their carbon chains, and
- which one of two important functional groups they contain.

We'll take the matter of chain lengths first. A monosaccharide with a molecular skeleton three carbons long is a *triose*. The suffix *-ose* tells us that we are dealing with a carbohydrate, while the prefix *tri-* indicates the number of carbons, three in this case. A *tetrose* is a four-carbon monosaccharide, a *pentose* contains five carbons, a *hexose* has six. The *-ose* ending of these names comes from the word *glucose*, which was itself adopted as a chemical term in 1838, by a committee of French scientists, from a Greek word for fermenting fruit juice or sweet wine, or simply for the quality of sweetness.

As for their functional groups, all common monosaccharides consist of a chain of carbons in which all but one are bonded to a hydroxyl group (C—OH). The single exception is a carbon doubly bonded to an oxygen, thereby forming a carbonyl group (Sec. 10.8).

carbonyl group

An **aldehyde** contains a carbonyl group bonded to at least one hydrogen.

The two atoms (other than the oxygen) that are bonded directly to the *carbonyl carbon* determine whether it is part of an *aldehyde* or a *ketone*. In **aldehydes**, at least one of the substituents on the carbonyl carbon is a hydrogen. Glucose contains just such a functional group. Among the simpler aldehydes are *formaldehyde* (in which *both* substituents are hydrogens), *acetaldehyde*, and *benzaldehyde*.

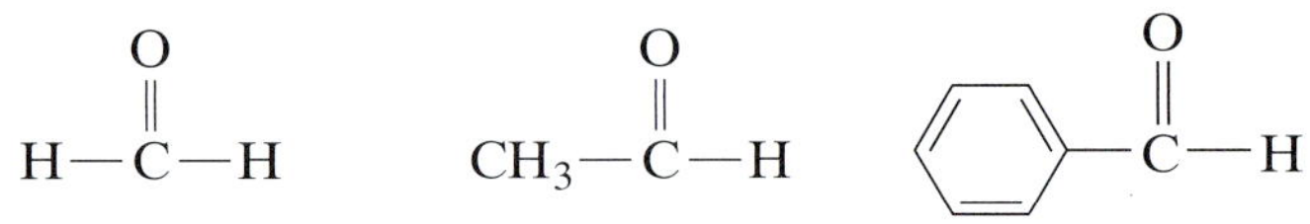

formaldehyde acetaldehyde benzaldehyde

Formaldehyde is a water-soluble gas. A 37% solution in water, known as *formalin*, is an effective biological preservative and is used in preparing embalming fluids. Benzaldehyde, also known as oil of bitter almond, is useful as a food flavoring and is an ingredient of many perfumes.

A **ketone** contains a carbonyl group bonded to two carbon atoms.

In a **ketone**, both of the atoms bonded to the carbonyl carbon are carbons. Typical simple ketones include the solvents *acetone* and *methyl ethyl ketone*. (Methyl ethyl ketone, which is known commercially as *MEK*, also has the more formal, chemical name, *2-butanone*.) Both acetone and methyl ethyl ketone are liquids at room temperature. Acetone serves as the major solvent in some nail polish removers. (You can usually find it listed among the ingredients on the label.) Methyl ethyl ketone is a common component of nail polish, nail polish removers, and commercial paint thinners and removers.

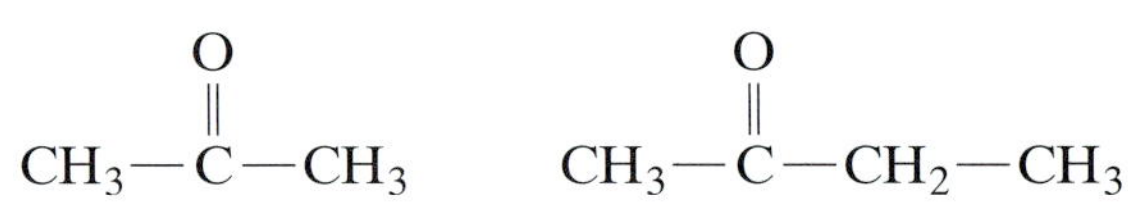

acetone methyl ethyl ketone or 2-butanone

The solvent acetone is a major component of nail polish remover.

All common monosaccharides contain one or the other of these two carbonyl groups. They are either polyhydroxy aldehydes or polyhydroxy ketones, or compounds very closely resembling polyhydroxy aldehydes and ketones. (The *poly-* of *polyhydroxy* reflects the presence of many hydroxyl groups on the molecular structure.) Monosaccharides with an aldehydic carbonyl group are the *aldoses*; those with a ketonic group are the *ketoses*. The classification of monosaccharides can indicate, simultaneously, both their carbon chain lengths and the kinds of substituents on their carbonyl groups. Glucose, for example, with a six-carbon skeleton topped off by an aldehydic group, is an *aldohexose*. Here the prefix *aldo-* tells us that glucose contains the aldehyde carbonyl group (*aldehyde*, because there's a hydrogen bonded to the carbonyl group's carbon), the stem *-hex-* refers to the six carbons present, and the suffix *-ose* places the compound in the carbohydrate family.

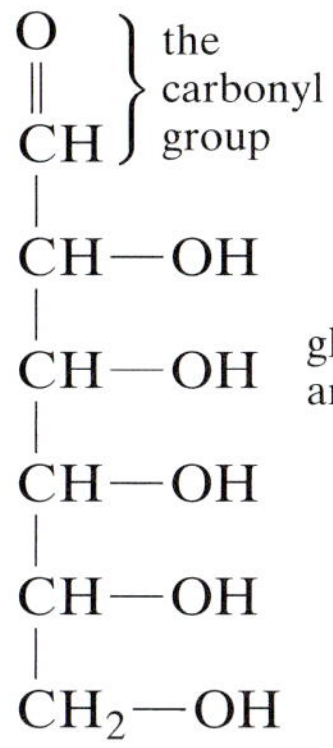

glucose, an aldohexose

E X A M P L E

Sweet Heredity

In Chapter 18 we'll learn about *ribonucleic acid (RNA)*, a molecule that uses the genetic information carried in all our cells to guide protein formation within cells. An important part of the molecular structure of RNA is the monosaccharide *ribose*, an aldopentose of molecular formula $C_5H_{10}O_5$. Draw a molecular structure for ribose.

■ The prefix *aldo-* tells us that ribose contains an aldehyde functional group, so we can begin by writing the structure:

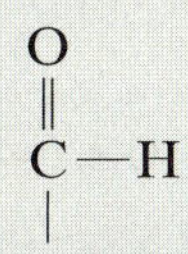

The next portion of the name, *-pent-*, indicates the presence of five carbons in the molecule. We'll add four more carbons to produce a total of five:

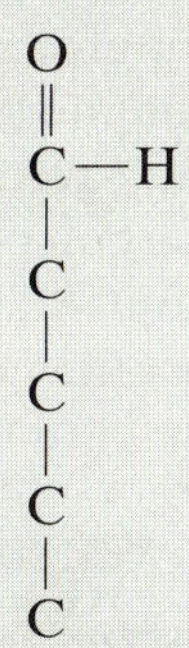

The suffix *-ose* emphasizes that ribose is a carbohydrate, so we'll add hydrogens and hydroxyl groups to produce a structure in which each carbon (except the carbonyl carbon) is bonded to one —OH group:

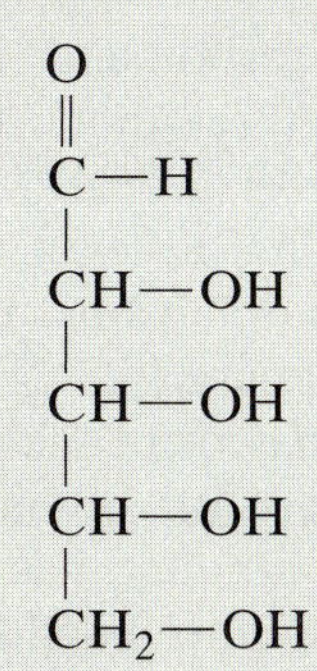

This structure now represents a molecule of ribose.

■ **QUESTION**

As we've just seen, glucose is an example of an *aldohexose* and ribose is an *aldopentose*. What would you call the category of monosaccharides that includes *fructose*?

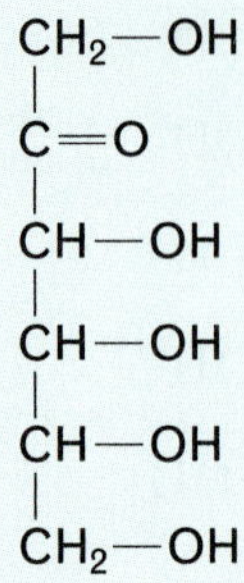

fructose

16.5 Chirality: Your Right Hand, Your Left Hand, and a Mirror

Two structural characteristics dominate the properties and behavior of monosaccharides that are important to human health and nutrition. All of them

- contain at least one carbon that's bonded to four different groups and
- form five- and six-membered rings with extraordinary ease.

After we examine each of these two characteristics individually, we'll see how they interact to produce remarkable differences in the properties of, for example, the starch of a potato and the cellulose of an apple.

Carbons bonded to four different groups are capable of a new kind of isomerism (Sec. 6.6), one that depends on the particular arrangement of these four groups around the carbon atom. As we saw much earlier, in Section 6.2, a carbon of an alkane or, more generally, any carbon bonded to four groups lies at the center of a *tetrahedron*, a symmetrical, four-sided structure. With the carbon at its center, each of the four substituents lies at one of the four corners or apexes of the tetrahedron. If we have before us a carbon bonded to four different groups, such as the generalized molecule CWXYZ,

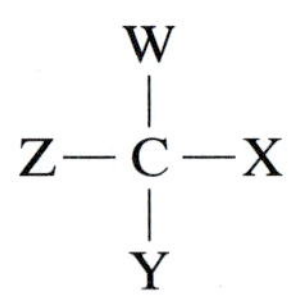

we can represent the shape of the carbon and its four bonded groups as a tetrahedron, with C at the center of the three-dimensional figure and the substituents W, X, Y, and Z occupying the apexes (Fig. 16.2). (This isn't to suggest that a carbon atom actually exists as a tetrahedron. Rather, if we imagine the carbon nucleus at the center of a tetrahedron, the four substituents would lie at the apexes of the imaginary, symmetrical geometric figure.)

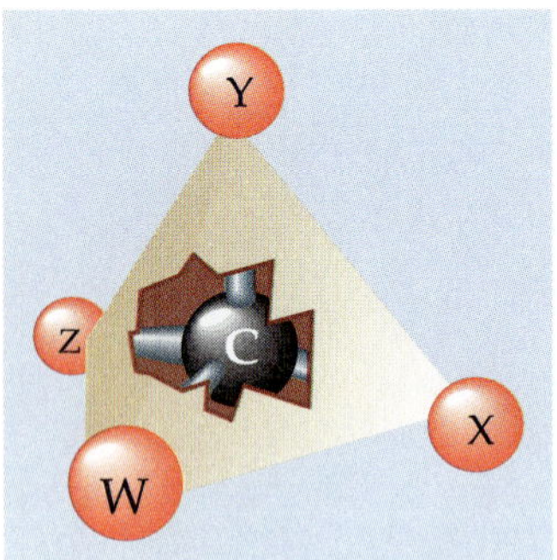

Figure 16.2 A tetrahedral carbon bonded to four different groups: W, X, Y, and Z.

We'll examine the unusual behavior of the CWXYZ carbon through the aldotriose *glyceraldehyde*, an aldehyde that contains three carbons and is a member of the carbohydrate family. This compound has no direct consumer applications whatever; its significance to us lies in its small size and its simplicity. It's a very good place to start:

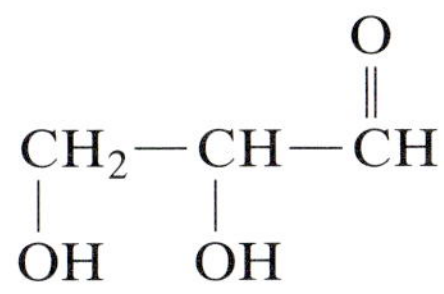

glyceraldehyde

With its central, tetrahedral carbon bonded to four different groups,

glyceraldehyde clearly represents a CWXYZ molecule. The specific orientation in space of the four groups surrounding the central carbon represents the carbon's *stereochemistry* (Sec. 6.2; Fig. 16.3). This stereochemistry represents an important characteristic of a CWXYZ molecule since *any tetrahedral carbon bonded to four different groups* takes on a peculiar property also found in your right and left hands: *it is not superposable on its mirror image*. We consider two structures to be superposable if we can merge them in space so that each and every point on one of the structures coincides exactly with its equivalent point on the other structure. If such point-for-point merging in space is not possible, the two are *not* superposable. Thus the glyceraldehyde molecule of Figure 16.3 is not superposable on its mirror image.

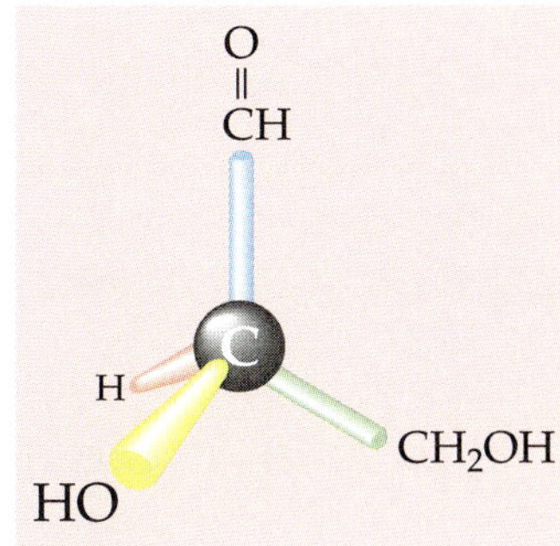

Figure 16.3 Glyceraldehyde stereochemistry.

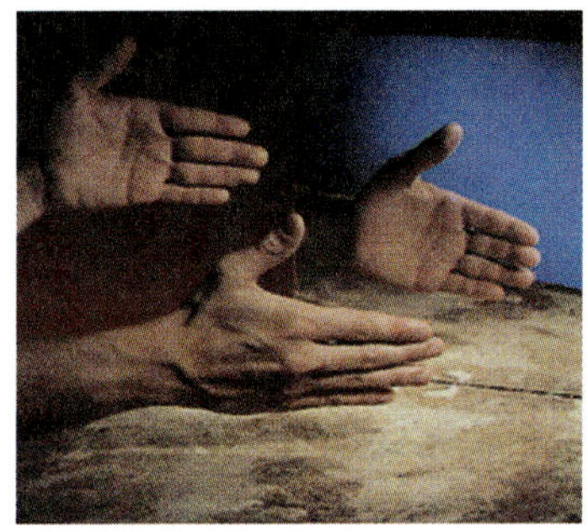

A left hand is a mirror image of a right hand.

Your right and left hands are not superposable. When you place your *right* hand before a mirror, the mirror produces an image that is essentially identical with your *left* hand. Just as your right hand cannot be superposed, point for point, on your left hand, your right hand cannot be superposed on its own mirror image. Nor can you fit your right hand into a left-hand glove.

The three-dimensional glyceraldehyde molecule of Figure 16.3 can no more be superposed on its mirror image than your right hand can be superposed on its own mirror image, which corresponds to your left hand. Molecules with this property of *handedness* are called **chiral molecules** (pronounced KI-rul), from a Greek word for a hand. Chiral molecules exist as two different, isomeric structures that are alike in every way except for their stereochemistry and its consequences. Each of these two isomers, the "right-handed" and the "left-handed" molecule, is an **enantiomer** of the other. Together they form an *enantiomeric* pair. The two enantiomers of glyceraldehyde appear in Figure 16.4. Notice that they are mirror images of each other, like your right and left hands. Notice also that they are not superposable on each other, again like your right and left hands. It's important to recognize that the two different molecules of Figure 16.4 represent different compounds, with different properties and different names, just as the two geometric isomers of 2-butene represent two different compounds (the *cis-2-butene* and *trans-2-butene* of Sec. 15.6). The most striking

Chiral molecules have the characteristic "handedness" of right and left hands. **Enantiomers** are nonsuperposable mirror images.

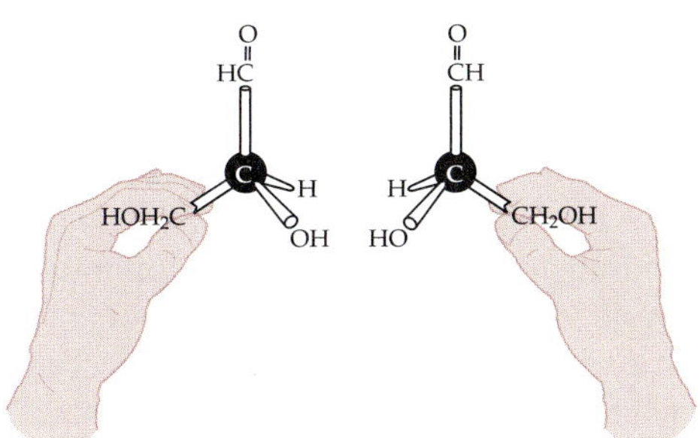

Figure 16.4 The two enantiomers of glyceraldehyde. The two molecules are nonsuperposable mirror images of each other.

difference in the properties of two enantiomers lies in their effect on light, as we'll see in Section 16.8. This effect allows us to give each a unique name.

We'll close here with a comment on an important structural requirement for chirality. We've seen that every carbon bonded to four different groups, as in CWXYZ, is chiral. Notice, though, that if two or more of the four substituents are identical, as in CWX*YY*, then the molecule *is* superposable on its mirror image and *cannot* be chiral. Remember that when we look at the substituents to determine whether any two are identical, we have to look at each substituent group as a whole, not simply at the four individual atoms bonded directly to the carbon.

■ QUESTION

Suppose that the carbon of CWXYZ were *square planar* rather than tetrahedral. That is, suppose all five of the atoms lay in the same plane, with the carbon at the center and each of the substituent atoms at one of the corners of the square. Would a molecule with this stereochemistry be chiral? Explain.

16.6 The Fischer Projection

A **Fischer projection** is used to show the stereochemistry of a chiral carbon.

Drawing the structure of a chiral molecule in as much detail as in Figure 16.3 is time consuming and requires a bit of artistic skill. A simpler and much more convenient approach to showing stereochemistry, the **Fischer projection**, appears in Figure 16.5. Emil Fischer, a German chemist, received the Nobel Prize in 1902 partly for his studies of the chemistry of sugars. To represent the *three-dimensional* stereochemistry of a chiral carbon on a *two-dimensional* surface, such as the pages of a book or a periodical, in as simple a way as possible, Fischer proposed that each chiral carbon be represented by a cross, following a few simple rules.

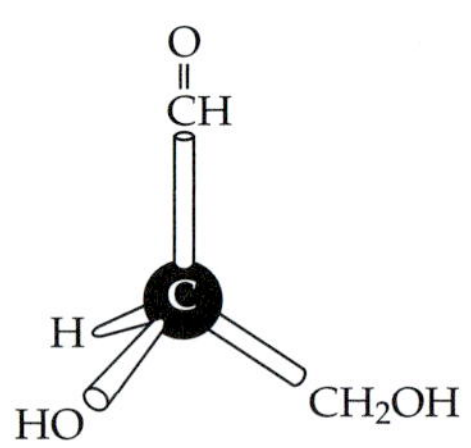

Figure 16.5a This is enantiomer held in the right hand of Figure 16.4.

To illustrate Fischer's procedure we'll use the enantiomer of glyceraldehyde that's held in the right hand of Figure 16.4. In Figure 16.5*a* you see this same enantiomer, but with the hand removed for clarity. Next, in Figure 16.5*b*, you see this enantiomer rotated slightly so that the top and bottom carbons appear to lie behind (or below) the paper and the H— and HO— substituents on the chiral carbon appear to lie in front of (or above) the paper. *With the vertical substituents behind (or below) the plane of the paper and with the horizontal substituents in front of (or above) the plane*, draw lines connecting the two vertical and the two horizontal substituents (Fig. 16.5*c*). The result is the Fischer projection of this particular enantiomer of glyceraldehyde, with the chiral carbon at the intersection of the two lines you've just drawn.

You now have the Fischer projection of one of the two enantiomers of glyceraldehyde. To generate the Fischer projection of the other enantiomer you can apply the same procedure to the model held in the left hand of Figure 16.4, or you can simply draw the mirror image of the Fischer projection of the right-hand model (Fig. 16.5*d*).

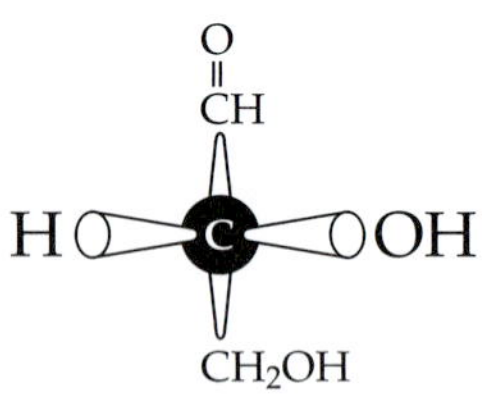

Figure 16.5b Here the molecule is rotated so that the top and bottom carbons lie behind or below the central, chiral carbon and the H— and —OH substituents lie in front of or above the chiral carbon.

Because of the way the Fischer projections are generated, we are not permitted to (mentally) remove them from the paper on which they're printed, nor are we permitted to rotate them except for a rotation of 180°, wholly within the plane of the paper. But we can slide them about on the paper to determine whether they represent structures that can or cannot be superposed. If the Fischer projections can be superposed, within the restrictions of movement we've just described, then the three-dimensional molecules they represent can also be superposed. The converse is also true: If the Fischer projections can't be superposed, neither can the three-dimensional molecules. Figure 16.6 shows the Fischer projections of the two enantiomers of 2-chlorobutane, which cannot be superposed on each other, and two Fischer projections of the same enantiomer of 2-bromobutane, which can be superposed on each other.

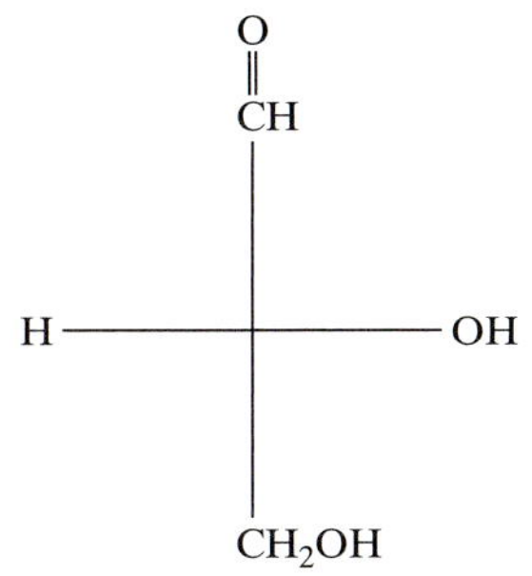

Figure 16.5c The Fischer projection of the enantiomer of glyceraldehyde held in the right hand of Figure 16.4.

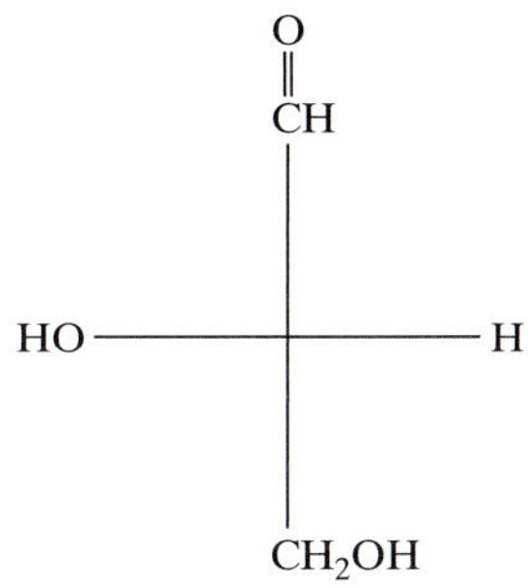

Figure 16.5d The Fischer projection of the enantiomer of glyceraldehyde held in the left hand of Figure 16.4.

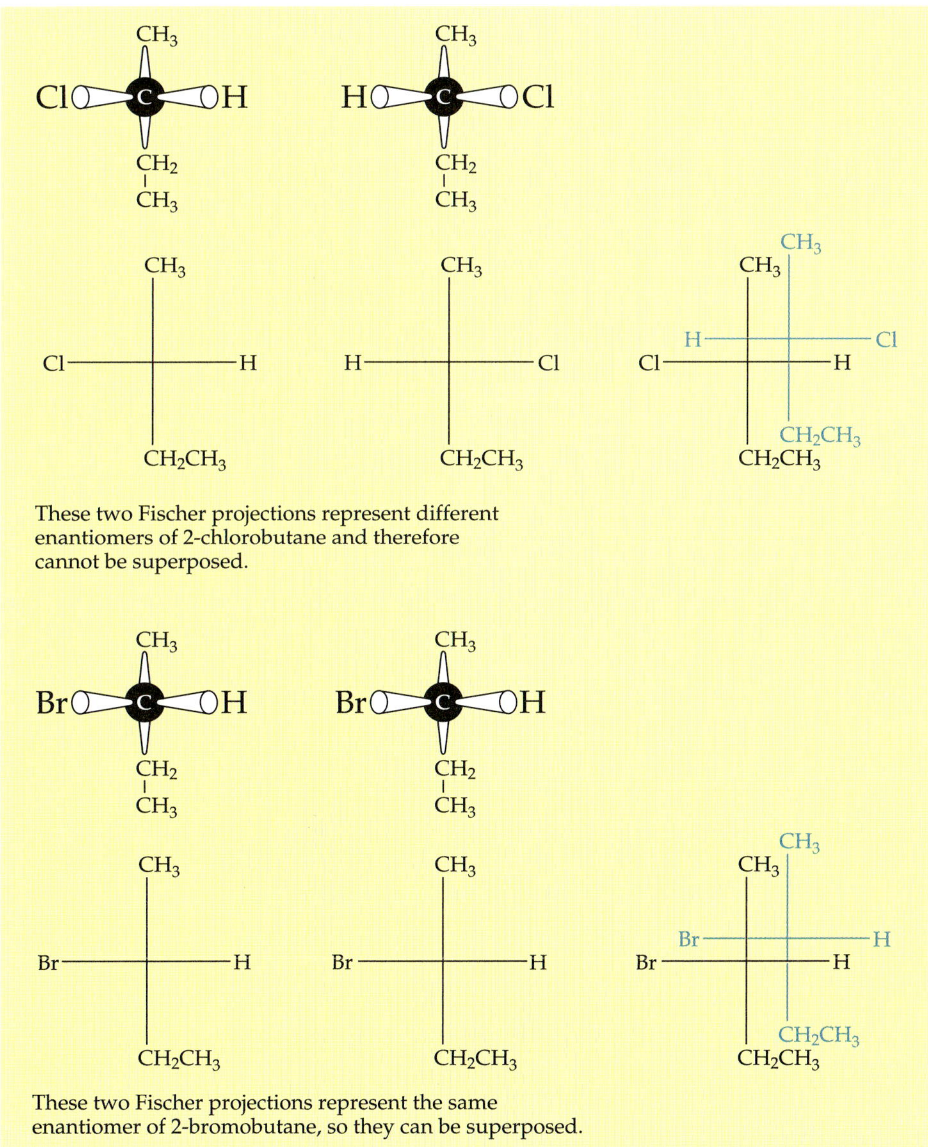

Figure 16.6 Superposition and nonsuperposition of Fischer projections.

EXAMPLE

Fischer Projection

Do the following two Fischer projections of 1,2-dibromopropane represent enantiomers?

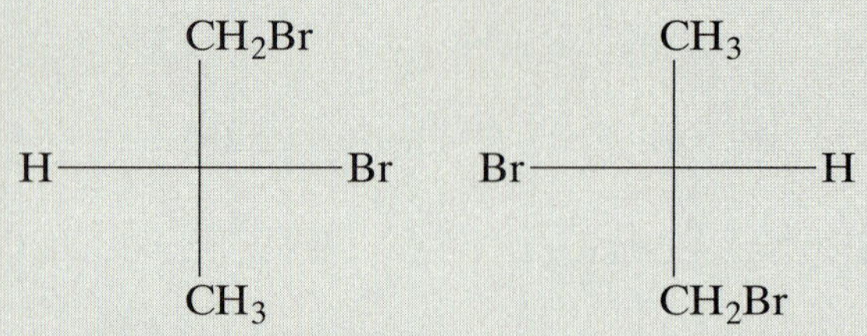

■ If we keep the first of these two Fischer projections fixed as shown above but rotate the second by 180°, as we are permitted by the rules described above, we get

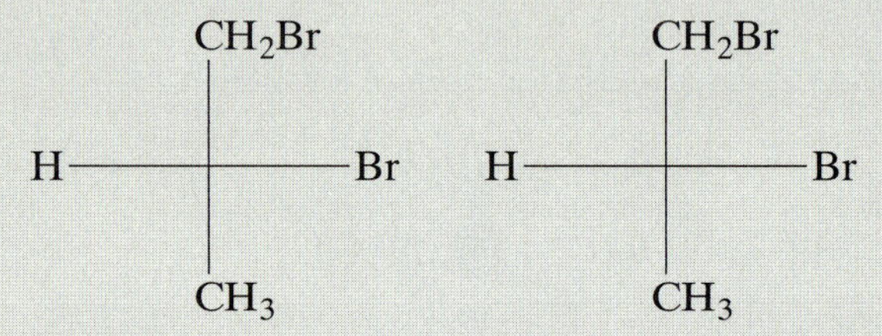

Clearly, if we slide one structure onto the other, all the groups will correspond. These are superposable and are therefore identical structures rather than enantiomers.

■ **QUESTION**

Draw the Fischer projections of both enantiomers of

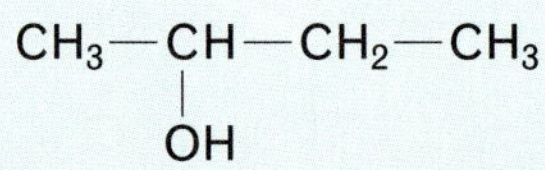

and show that they cannot be superposed on each other.

16.7 Polarized Light, Glare, and Sunglasses

The most remarkable property that chirality confers on a molecule lies in its effect on *polarized light*. To understand what polarized light is and how chiral molecules affect it, we'll start by examining the phenomenon of light itself. We'll treat radiant light as if it consisted of waves with crests and troughs, much like those we see on large bodies of water (Fig. 16.7).

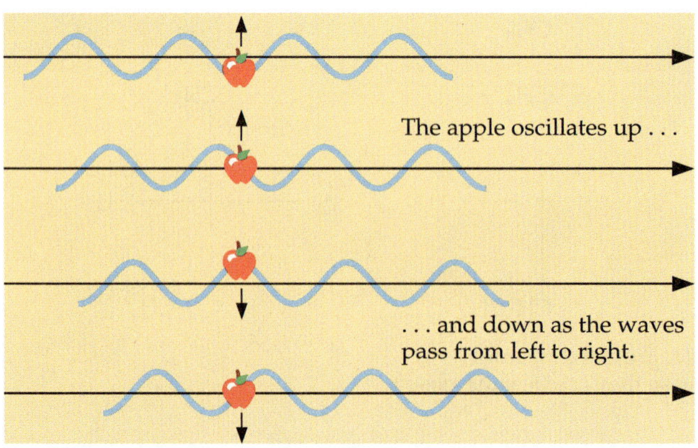

Figure 16.7 An apple bobbing in water.

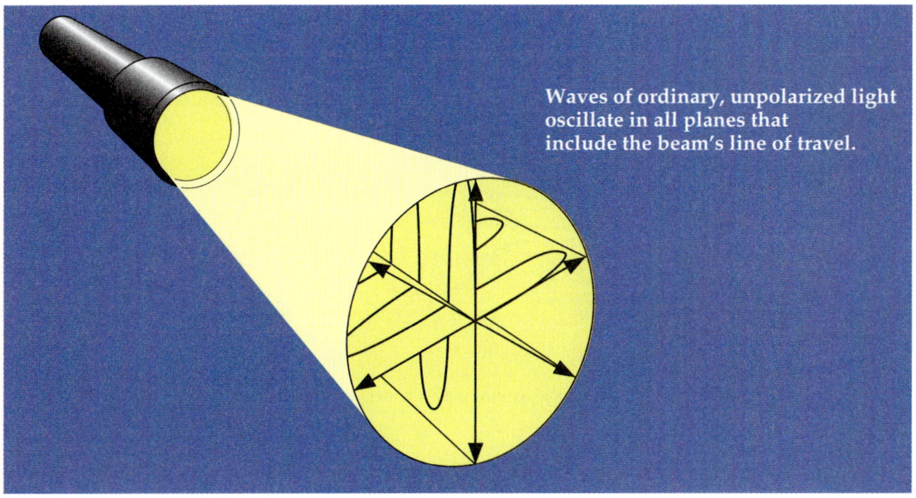

Figure 16.8 Unpolarized light.

In water waves any molecule of water, or any object floating on the water, moves largely up and down—from trough to crest and back down again—in a direction perpendicular to the direction in which the wave is traveling. While the wave itself appears to travel horizontally, the water itself oscillates vertically.

We can think of light in much the same way. According to one useful physical model, a light beam moves along with its waves undulating, like water's waves, in a direction perpendicular to the direction of the light's travel. Unlike water waves, though, the undulations of light oscillate in *all* directions perpendicular to the direction of the beam. If we could place a sheet of some special material into a light beam's path so that we could watch these vibrations, we'd see something that looks like Figure 16.8. This represents ordinary or *unpolarized* light.

Polarized light differs from unpolarized light in one critical respect. While unpolarized light is made up of oscillations in all planes that include the beam's line of travel, the oscillations of polarized light are limited to only one, single plane, the **plane of polarization**. For emphasis, this form of light is sometimes called *plane-polarized* light (Fig. 16.9).

Polarized light is light consisting of oscillations in only one plane, which is known as the **plane of polarization**.

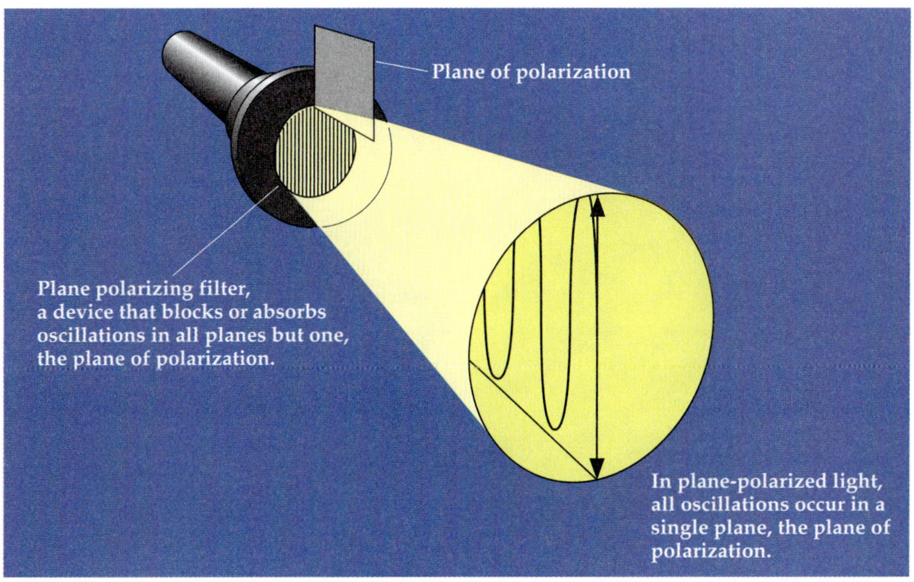

Figure 16.9 Plane-polarized light.

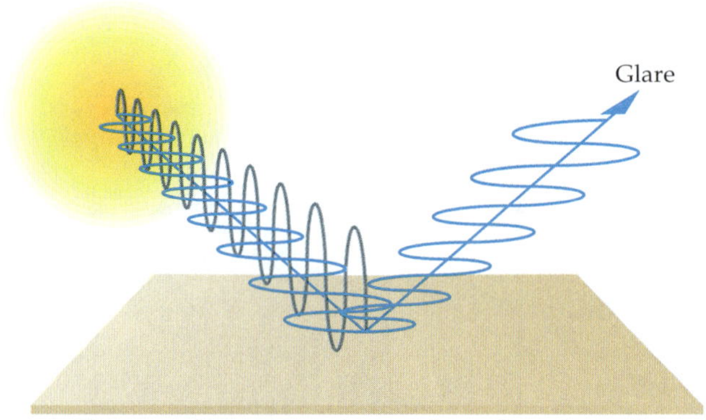

Figure 16.10 Reflected glare is plane-polarized light.

Several devices and some physical phenomena and processes can convert unpolarized light into plane-polarized light by blocking or removing all of the vibrations except those occurring in what becomes the plane of polarization. Light reflected as glare from water, sand, snow, ice, roads, or other surfaces becomes polarized, with the plane of polarization parallel to the reflecting surface. Glare, then, is an example of plane-polarized light (Fig. 16.10).

Polarizing sunglasses reduce glare because their lenses are made of polarizing filters that block out all planes of vibration except one. Setting the lenses into the frames so that their polarizing planes are vertical orients them perpendicular to the horizontal surfaces that cause glare. The lenses now block out all the horizontally polarized light that forms the glare. Hold a pair of these sunglasses in front of you, look through the lenses at surface glare, and rotate the glasses. You'll see the glare alternately brighten and grow dim as the lenses rotate (Fig. 16.11). That's because the lenses of the sunglasses let all the polarized light of the glare through only when the plane of polarization of the lenses matches the direction of the light's oscillations. When the two planes are crossed, none of the glare gets through. As you rotate the lenses their plane of polarization also rotates, alternately perpendicular to the plane of the glare and parallel to it.

In fact, you can block out virtually *all* light from any source of relatively low intensity by crossing the lenses of two pairs of polarizing sunglasses as shown in Figure 16.12. The first lens effectively converts ordinary, unpolarized light into

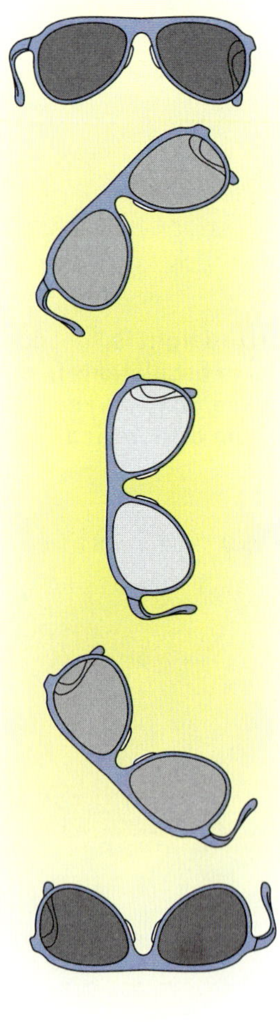

Figure 16.11 Polarizing sunglasses versus glare.

Figure 16.12 The effect of polarizing lenses on unpolarized light.

plane-polarized light; the second lens, with its plane of polarization perpendicular to that of the first lens, blocks out almost all the plane-polarized light coming through the first lens. The result is a dark spot.

■ QUESTION

Suppose you had a pair of polarizing sunglasses with perfectly circular lenses. One of the circular lenses suddenly popped out and rolled a bit on the ground. You want to repair the glasses by snapping the lens back into the frame but you're not sure what orientation to use. How would you go about determining the proper orientation of the lens before reinserting it into the frame?

16.8 Optical Activity

We can use polarized light and an instrument that operates much like the two polarizing lenses of Section 16.7 to distinguish between enantiomers and to examine an important property of carbohydrates. The instrument, a *polarimeter*, consists of a hollow tube placed between two sheets of polarizing material called *polarizing filters*. The filter you look through, the nearer filter, has a pointer attached to it and can be rotated clockwise and counterclockwise. An angular scale allows you to measure the degree of rotation. The filter at the far end of the tube is fixed in position; it can't be rotated. Behind the fixed filter is a light source (Fig. 16.13).

If you set the nearby, movable analyzing filter so that its plane of polarization is exactly perpendicular to that of the fixed, polarizing filter at the other end, you see only a dark or very dim field as you look through the empty polarimeter tube. But fill the polarimeter tube with any substance made up of chiral molecules of a single stereochemistry, either of the pure enantiomers of glyceraldehyde for example, and the image brightens. *Chiral molecules have the remarkable property of rotating the plane of polarized light.* To return to the darkest possible image you have to readjust the analyzing filter by rotating it through a measurable angle.

The actual, quantitative angle through which you have to rotate the filter to return to the darkest setting depends on several factors, including the molecular structure of the chiral compound, its concentration (if you were using a solution), the temperature, and the length of the polarimeter tube. But under any specific set of conditions, the *direction* of rotation depends only on which one of the two enantiomers you were examining. Under identical conditions, the two enantiomers of a chiral compound rotate the plane of polarized light to exactly the same degree, but in opposite directions. We call the enantiomer that requires the analyzing filter to be rotated clockwise *dextrorotatory*, from the Latin *dextra* for "to the right" (or, in a related sense, clockwise).

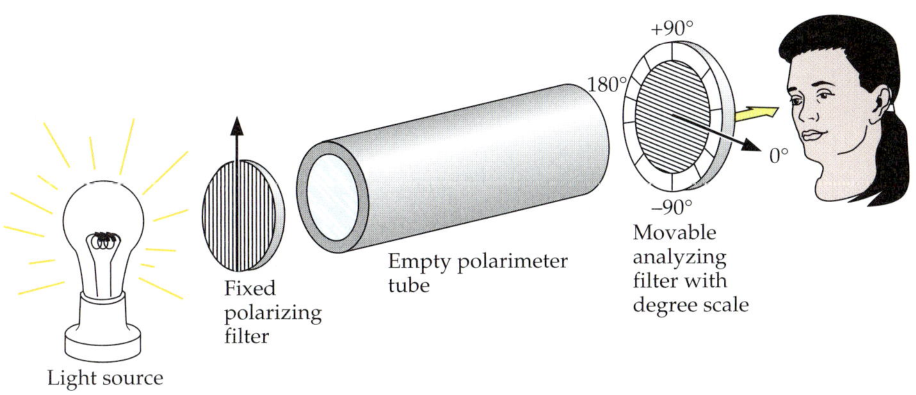

Figure 16.13 The essentials of a polarimeter.

The other enantiomer, the one that generates a counterclockwise rotation, is levorotatory, from the Latin *laevus*, for "to the left" (counterclockwise). Anything capable of rotating the plane of polarized light (in either direction) is **optically active**; substances that don't or can't produce such rotation are *optically inactive*.

An **optically active** substance is capable of rotating the plane of polarized light.

On examining each of the two enantiomers of glyceraldehyde we'd find that one of them is dextrorotatory and the other is levorotatory. While we could use these two terms to distinguish the two enantiomeric glyceraldehydes from each other, it's much simpler to use an algebraic "+" to indicate dextrorotation and a "−" to indicate levorotation. Encasing the sign in parentheses, we write the name of the dextrorotatory glyceraldehyde as (+)-*glyceraldehyde*; the levorotatory enantiomer is (−)-*glyceraldehyde*.

■ QUESTION

Knowing that the two enantiomers of a chiral compound rotate polarized light to exactly the same extent, but in opposite directions, what effect do you think *a mixture* of equal amounts of the two enantiomers of a chiral compound would have on plane-polarized light? Mixtures of this kind are called *racemates*. Is a racemate optically active? What term used in Section 16.8 can be applied to a racemate to describe its effect on polarized light?

16.9 Dextrose, Levulose, Honey, and Invert Sugar

We'll move now from the relatively simple glyceraldehyde, with its single chiral carbon, to the monosaccharides glucose and fructose, which contain several such carbons. The Fischer projections of Figures 16.14 and 16.15 reveal that glucose contains four chiral carbons and fructose three. As we might expect, each of these monosaccharides exists as enantiomers that are optically active. Virtually all naturally occurring glucose exists as the dextrorotatory enantiomer, (+)-*glucose*, shown in Figure 16.14. Another frequently used name for this monosaccharide, *dextrose*, emphasizes its dextrorotatory character. In contrast to glucose, naturally occurring fructose is levorotatory. It's often referred to as (−)-*fructose* or, emphasizing its levorotation, *levulose*.

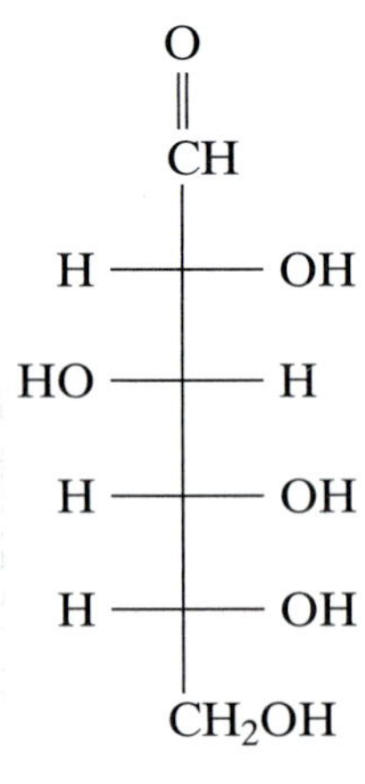

Figure 16.14 The Fischer projection of the naturally occurring enantiomer of glucose.

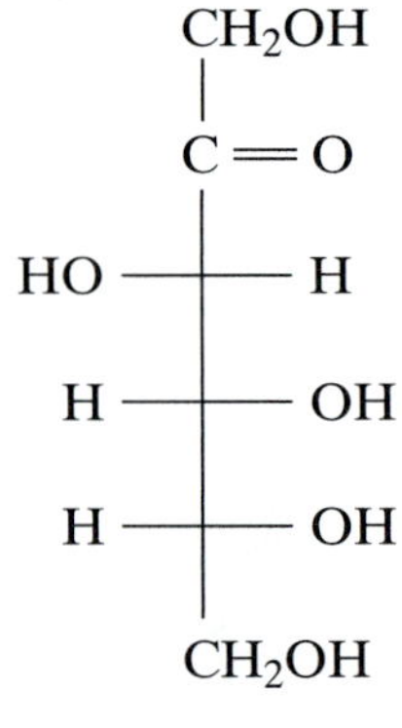

Figure 16.15 The Fischer projection of the naturally occurring enantiomer of fructose.

The difference in the direction of rotation caused by glucose and fructose produces a curious result in the hydrolysis of sucrose, a disaccharide. All naturally occurring sucrose is dextrorotatory and is composed of a chemically bonded combination of one unit of glucose and one unit of fructose. Hydrolysis of (+)-sucrose thus produces equal amounts of (+)-glucose and (−)-fructose. Since (−)-fructose rotates plane-polarized light through a larger angle, counterclockwise, than (+)-glucose does in a clockwise direction, a mixture of equal amounts of the two monosaccharides is levorotatory. As a result, the hydrolysis of sucrose proceeds with an inversion of the direction of rotation, from clockwise [because initially only (+)-sucrose is present] to counterclockwise [because at the end of the hydrolysis only a levorotatory mixture of equal amounts of (+)-glucose and (−)-fructose is present].

Honey is largely invert sugar, a mixture of glucose and fructose.

Because of this inversion in the direction of rotation, a mixture of equal amounts of glucose and fructose is known as **invert sugar**. Invert sugar is the principal component of honey. It's formed as bees, using an enzyme generated in their bodies, hydrolyze the sucrose of the nectar they gather from flowers and convert it into its two component monosaccharides. Because of the ability of invert sugar to decrease the volatility of water and slow its rate of evaporation (Sec. 7.1), this mixture of glucose and fructose is sometimes added to foods and candies to help keep them moist.

Invert sugar is a mixture of equal quantities of glucose and fructose.

■ QUESTION

You want to convert a portion of invert sugar into an optically inactive substance by increasing the amount of one of its monosaccharide components. Would you add naturally occurring glucose or fructose to the invert sugar? If you added just the right amount of the selected monosaccharide to achieve optical inactivity, would the resulting mixture be a racemate? Explain. (See the Question following Sec. 16.8 for additional information.)

16.10 Cyclic Monosaccharides

Now, with an understanding of the consequences of chirality, we'll turn to the cyclic structures of our most important monosaccharides, particularly the pentoses (five-carbon monosaccharides) and the hexoses (six-carbon monosaccharides). These monosaccharides form rings through the addition of an —OH group of one of their chiral carbons to the carbonyl group of the aldose or ketose. The reaction is much like the addition of water to the double bond of an alkene, which generates an alcohol (Sec. 6.10, Fig. 6.16). Ring formation in ketoses occurs in a manner similar to that shown for the aldose of Figure 16.16.

Ring formation of this kind occurs simply because it transforms the molecule into a more stable form. Chemically, these six-membered rings (and, in some cases, five-membered rings) are more stable than the straight chains of the monosaccharides.

It's customary to draw the cyclic monosaccharides as though we were viewing them almost edge-on, much as you see in Figure 16.16. With the rings drawn like this, the —H and —OH substituents that extended to the right and to the left in the Fischer projections now point upward and downward. The direction from which the —OH adds to the carbonyl group determines which one of two possible cyclic structures forms (Fig. 16.17).

[The cyclic representations of monosaccharide rings shown in Figure 16.17 are called *Haworth structures*, in honor of Walter Norman Haworth, a British chemist who shared the Nobel Prize in 1937 for his research on carbohydrates and vitamin C. Haworth devised methods for determining the size and stereochemistry of carbohydrate rings and played a major role in determining the chemical structure of vitamin C. Much of the chemistry described in this chapter was discovered by either Haworth or Emil Fischer (Sec. 16.6).]

Figure 16.16 Formation of the cyclist structure of glucose.

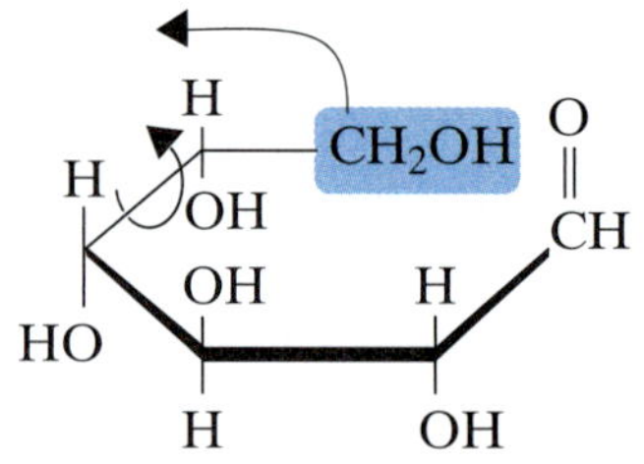

A side view of a glucose molecule, showing the arc of the carbon skeleton. Notice that the –OH groups alternate above and below the ring in the same order as in the Fischer projection. Rotation around the bond, as shown, produces the structure below.

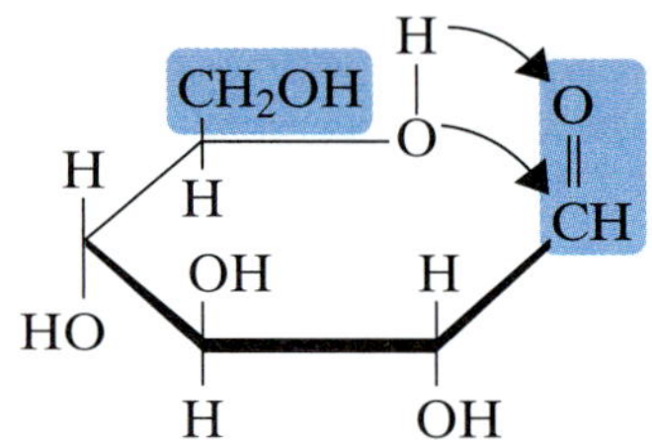

Addition of the O and the H to the carbonyl group produces the cyclic structure below.

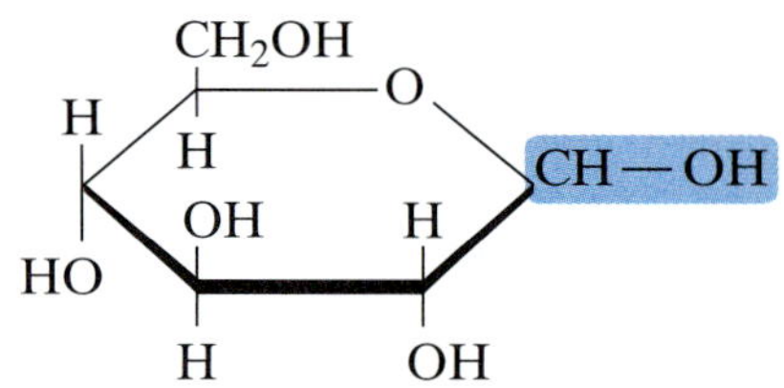

The cyclic structure of glucose.

Planar, and with only three groups bonded to its carbon, the carbonyl group of glucose can't be chiral as long as the molecule remains in the form of an aldehyde. (Anything that's two-dimensional or planar is always superposable on its own mirror image.) But with the addition of the —OH group in ring formation, the carbonyl carbon not only becomes tetrahedral, but it now bears four different substituents:

1. an —OH group,
2. —H,
3. the oxygen of the ring, and
4. a carbon of the ring.

Figure 16.17 The two glucose rings.

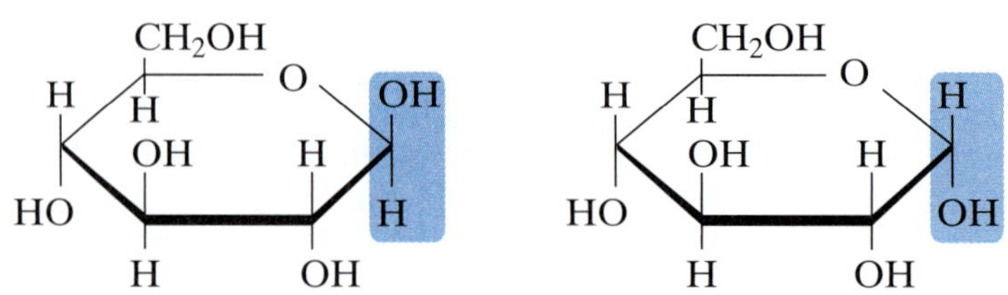

β-glucose   *α*-glucose

This carbon is now chiral and can hold its new —OH group pointing either up or down in the Haworth projection. If the new —OH group points down, we have an *alpha* (α) ring; if it's up, the ring is *beta* (β). There are, then, *two* cyclic forms of glucose: α-glucose and β-glucose (Fig. 16.17). Although the two isomers rotate the plane of polarized light to different extents, each one is dextrorotatory. They are, therefore, (+)-α-glucose and (+)-β-glucose.

■ QUESTION

How many chiral carbons are there in a molecule of the noncyclic form of glucose? How many are in a molecule of the cyclic form? What's the reason for this difference?

16.11 Starch and Cellulose: You Eat Both, You Digest One

The existence of these two forms of glucose, α and β, explains nicely the difference between the two polysaccharides that make up an important part of our diet:

- the *starch* of the seeds and roots of plants (such as wheat, corn, and potatoes), and
- the *cellulose* of their more fibrous structures (such as bran or seed husks of oats, rye, and wheat; celery stalks; lettuce leaves).

In their molecular structures, starch and cellulose resemble each other closely. Each consists of very long polysaccharide molecules that give only glucose on complete hydrolysis. Each is made up exclusively of glucose rings joined to each other to form long molecular strings. Yet while we digest and absorb the starch of our foods easily, their cellulose passes through our bodies largely unchanged and unabsorbed. Starch serves as a macronutrient; cellulose provides us with dietary fiber, sometimes called roughage or indigestible carbohydrate (Sec. 16.12). This difference between the two polysaccharides originates in their molecular structures: Starch is made up exclusively of rings of α-glucose; cellulose is composed exclusively of rings of β-glucose.

In forming any of the larger carbohydrates, ranging from the simple disaccharides of sucrose and lactose to the polysaccharides of starch and cellulose, individual cyclic molecules of monosaccharides combine and join together through the loss of water molecules. We'll now see how this combination occurs and how it produces the important difference between starch and cellulose. We'll look first at cellulose since it provides us with the simpler example.

There's probably more cellulose on this planet than any other single organic compound. By one estimate, growing plants produce something like 100 billion tons of cellulose worldwide each year. This polysaccharide, which doesn't occur at all in animals, is the principal component of plant cells. Cellulose gives plants their structural strength. The molecular chain of cellulose consists of molecules of β-glucose, joined together through oxygens that link carbon 1 of one of the rings with carbon 4 of another. As the rings join they lose a molecule of water.

Figure 16.18 illustrates a combination of this sort as two glucose rings join to form the disaccharide *cellobiose*. The ring carbons are numbered in the figure for easy reference. Notice that the ring on the left is β-glucose since the oxygen on carbon 1 lies above the ring. As a result the two glucose rings of cellobiose are joined

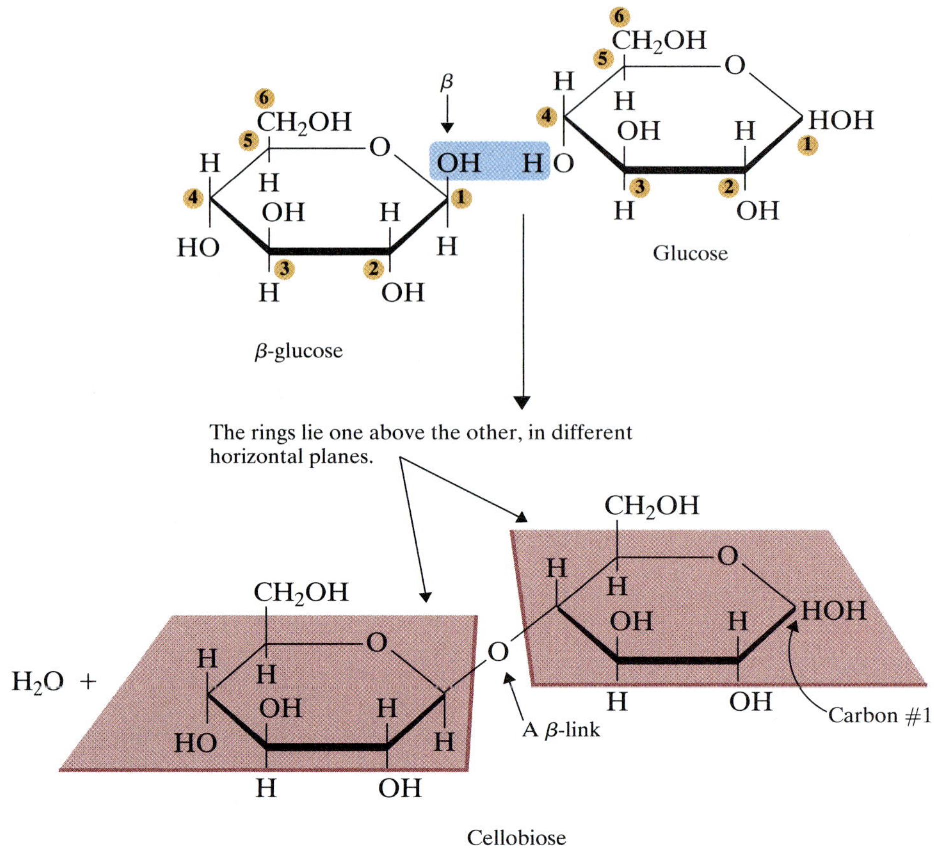

Figure 16.18 Two glucose molecules (one of them β-glucose) combine to form cellobiose and water.

through a β link. As they are shown in the figure, one of the rings lies slightly above the other, in a different horizontal plane. In the formation of the disaccharide cellobiose, it doesn't matter whether the glucose ring on the right is α or β. Cellobiose forms in either case. What *does* matter is that the oxygen joining the two rings lies above the left glucose ring, thereby forming a β link.

Except that the cellobiose molecule consists of two glucose rings joined through a β link, this disaccharide is of little interest to us. What's important is that as more and more glucose rings add to the chain in this way, *but always with the oxygen of the No. 1 carbon lying above its ring*, they eventually form a chain long enough to be called cellulose (Fig. 16.19). Cellulose molecules vary in length from several hundred to several thousand consecutive glucose rings, depending on their source. In cellulose all of the oxygens that link one ring to the next are in the β orientation.

If the oxygen joining the two glucose rings is in the α orientation, the resulting link is α and the disaccharide that forms is *maltose* (Fig. 16.20). Note that because of this α orientation the glucose rings of this figure all lie in the same plane. As with cellobiose, the maltose that's formed here is of interest to us only because of its relationship to a larger polysaccharide. The addition of more α-glucose rings in this same way eventually produces starch (Fig. 16.21). In starch, all of the linking oxygens are α.

Unlike cellulose, starch consists of two different kinds of polysaccharide molecules. A long, unbranched, linear chain with links exclusively between the No. 1 carbon of one ring and carbon 4 of the next, as shown in Figure 16.21, is

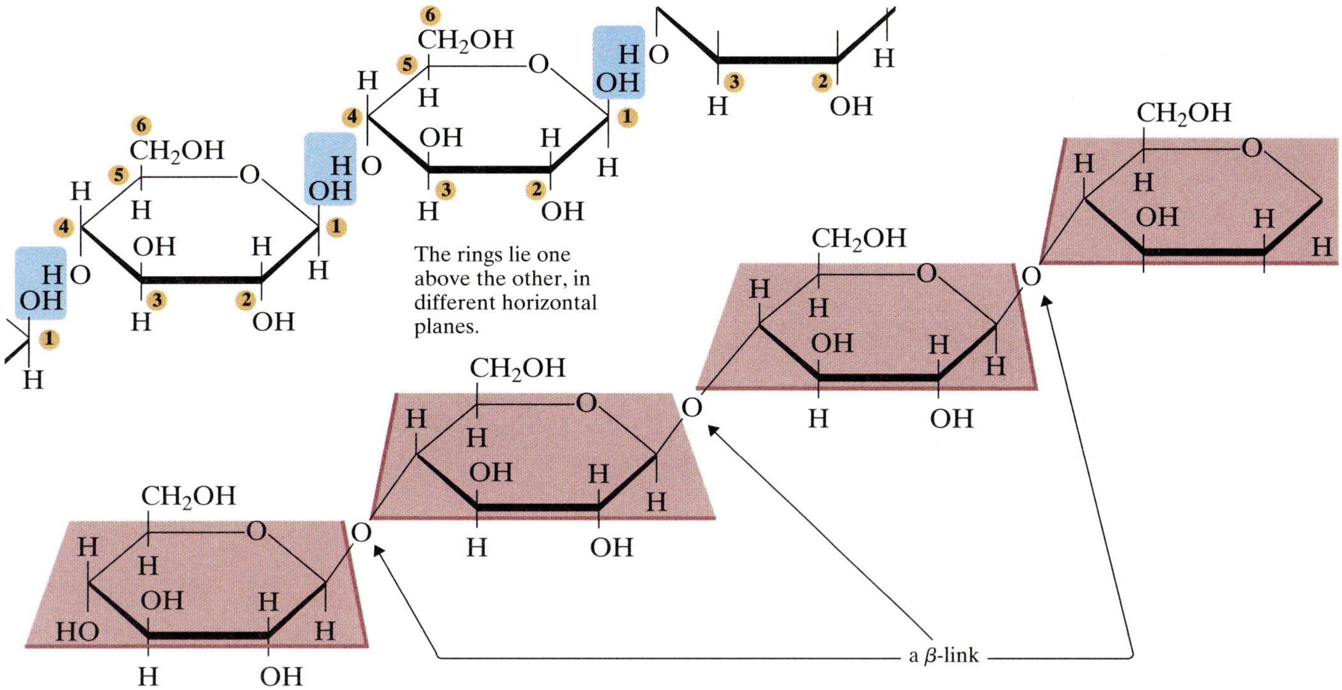

Figure 16.19 Beta-glucose and the chain of the cellulose molecule.

called *amylose*. These amylose molecules are strings of α-glucose rings averaging several hundred rings each. The chains of another type of starch, *amylopectin*, run several hundred thousand glucose rings long, with branches occurring from carbon 6 at about every 25th glucose ring. These branches lead to still other α-glucose chains of varying lengths. The amylose molecule looks like a long string with plenty of twists and turns; the amylopectin resembles the branches of a tree. The relative proportions of amylose and amylopectin in

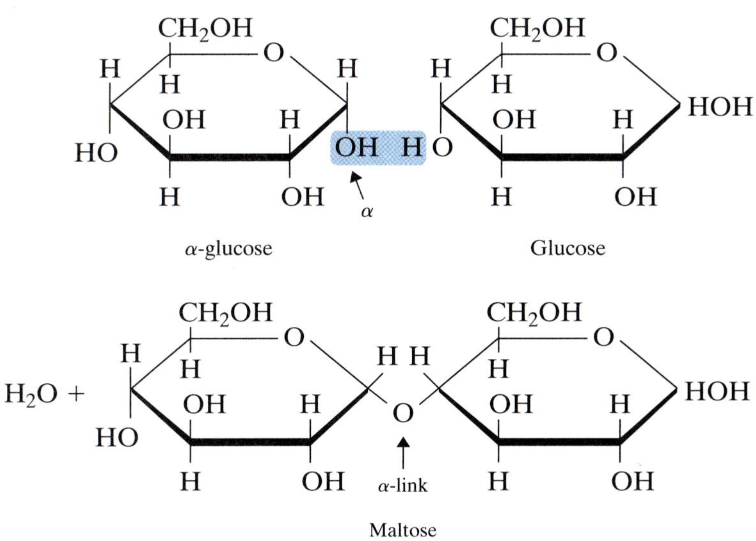

Figure 16.20 Two glucose molecules (one of them α-glucose) combine to form maltose and water.

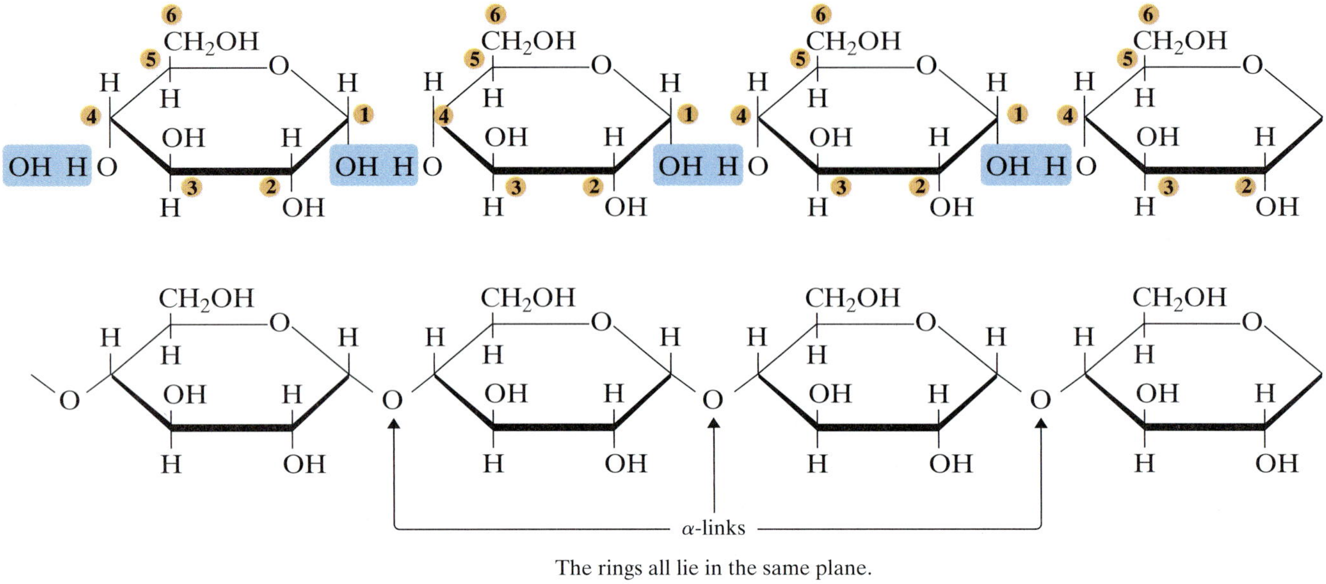

Figure 16.21 Alpha-glucose and the chain of the starch molecule.

Two polysaccharides of food: the starch of potatoes and the cellulose of cabbage.

food starches vary; the starches of corn and cereal grains run about three parts amylopectin to one part amylose.

One more polysaccharide is of interest to us here: *glycogen*. This carbohydrate resembles amylopectin in molecular structure except that its molecules are more highly branched and smaller, with fewer glucose rings in each. Animals manufacture and store glycogen to serve as a reservoir of readily available glucose. Deposited largely in the liver and muscles, the glycogen molecule can be cleaved quickly to individual glucose molecules, thus providing the body with a ready supply of energy. Because of its storage in the animal body as a reserve of glucose, glycogen is sometimes called *animal starch*. We don't store much glycogen, though. We use up the liver's supply each night as we sleep and must replenish it the next day. A longer fast causes us to eat into our stores of fat.

Neither starch nor cellulose has any significant taste, although many naturally occurring mono- and disaccharides are distinctly sweet. Table 16.2 lists the relative sweetness of several mono- and disaccharides, with table sugar (sucrose) assigned a rating of 100 for comparison.

Table 16.2 Relative Sweetness of Sugars

Sugar	Sweetness (Relative to Sucrose)
Fructose	173
Invert sugar (mixture of equal parts glucose and fructose)	130
Sucrose	100
Glucose	74
Maltose	32
Lactose	16

In what single way does the molecular structure of maltose differ from the molecular structure of cellobiose?

■ QUESTION

16.12 How We Digest Carbohydrates

While any of the monosaccharides—glucose or fructose for example—easily penetrates the intestinal wall to enter the bloodstream, neither disaccharides nor the larger carbohydrates normally get through the intestinal barrier. They are simply too large. To assimilate these larger carbohydrates, ranging from maltose and sucrose to starch, we must first clip them down to their component monosaccharides, which are able to pass through the intestinal wall and into the bloodstream. As we've seen, in forming the larger carbohydrate molecules the individual monosaccharide rings connect to each other with loss of a molecule of water. To reverse this process, to cleave the disaccharides and the polysaccharides to their component monosaccharides, we must return the water to the molecules through hydrolysis.

Our bodies carry out this hydrolysis through the catalytic action of our digestive *enzymes*. Enzymes, you'll recall, act as biological catalysts that allow chemical reactions to take place more rapidly or under milder conditions than they might otherwise, but without being consumed themselves. In cleaving the nutrient polysaccharides quickly and efficiently in the relatively mild chemical environment of our digestive systems, these molecular catalysts act very much like the platinum and palladium catalysts that help remove unburned hydrocarbons from automobile exhausts (Sec. 7.7).

The **fiber**, **bulk**, or **roughage** of our foods is largely cellulose, a carbohydrate we find indigestible because of our lack of the enzyme **cellobiase**.

An enzyme that enables us to digest both maltose and starch is *maltase*, which our bodies produce in sufficient quantities to allow us to digest the starch we eat. As you might infer from this example, an enzyme's name usually resembles the name of the substance it acts on. For the simple disaccharides, just replace the *-ose* ending of the sugar with *-ase* to get the name of the enzyme that hydrolyzes the disaccharide. Maltase helps us hydrolyze the links of maltose and starch.

Fruits and vegetables provide us with plenty of fiber.

We've seen that the combination of two glucose units through a β link produces *cellobiose*, which resembles two links of the cellulose chain. Since we don't produce **cellobiase**, the digestive enzyme that hydrolyzes the link, we can't digest either cellobiose or cellulose.

For humans, starch constitutes a *digestible carbohydrate*, while cellulose is one of the *indigestible carbohydrates* that form a large part of the **fiber**, **bulk**, or **roughage** of our diets. Grass, leaves, and other plant material, all of which are indigestible in our own intestines, provide metabolic energy to cows, goats, sheep, and other ruminants, and to termites and similar insects, simply because the digestive systems of these animals harbor microorganisms that produce the needed cellobiase. We don't have the required enzyme, so we can't live on grass and wood. For humans, fiber-rich foods include fruits, vegetables, bran, and nuts.

What monosaccharide is produced by the action of both cellobiase and maltase on polysaccharides?

■ QUESTION

16.13 Lactose Intolerance: Why Large Quantities of Dairy Products Can Cause Digestive Problems in Two Adults Out of Three

Lactose (from the Latin *lac*, for milk), the principal carbohydrate of milk (Table 16.1), offers a good example of the problems we can encounter through defects in our enzyme-catalyzed digestive processes. Lactose is a disaccharide that consists of a galactose ring joined to a glucose ring through a β link. The monosaccharide galactose differs only subtly from glucose, through the stereochemistry of carbon 4. If the —OH on carbon 4 of the monosaccharide ring points down, we have glucose; if it points up, we have galactose (Fig. 16.22). Since virtually no disaccharide of any kind gets through the intestinal wall and into the bloodstream, we need the enzyme *lactase* to hydrolyze lactose into its components, galactose and glucose. The more milk and milk products we consume, the more lactase we need.

Normally, there's plenty of lactase in the digestive systems of infants and children. That's fortunate, since nursing babies get about 40% of their calories from the lactose of their mother's milk. By and large, older children have more than enough lactase to digest all the lactose of the milk, cheese, ice cream, and other dairy products they eat. But gradually, as we grow from infancy to adolescence, most of us lose the ability to produce lactase in large quantities, so we're largely *lactase-deficient* as adults. In this condition we produce very small amounts of lactase, too little to handle more than a glass or two of milk at a time. It's been estimated that about 70% of the world's adult population, or about two out of three adults throughout the world, have very low levels of lactase in their digestive systems. More specifically, estimates for the percentage of lactase-deficient adults range from about 3% for Scandinavians and northern Europeans to 80 to 100% for some Asians. In North America the estimates for adults run from 5 to 20% for the white population to 70% for the nonwhite population.

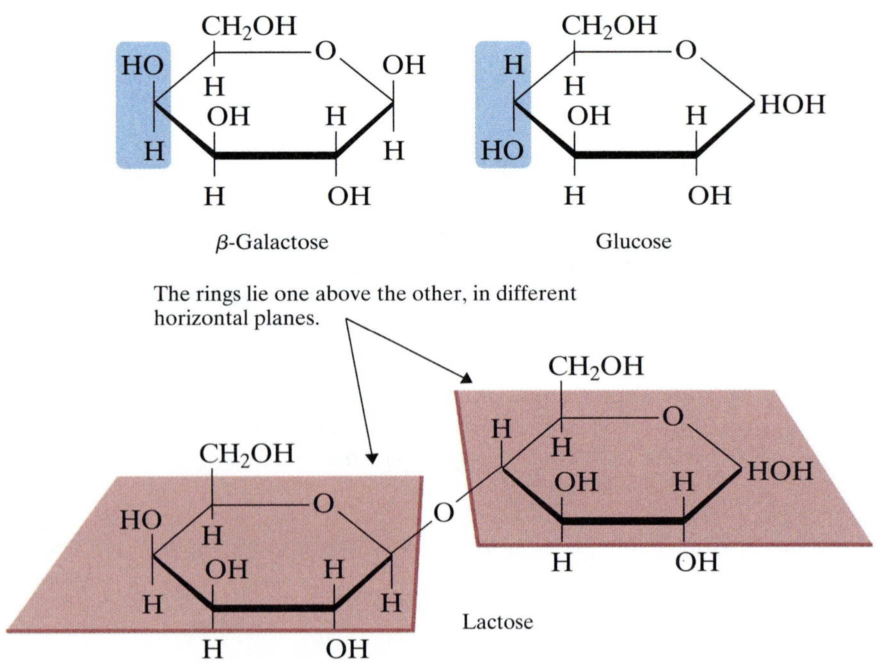

Figure 16.22 Galatose, glucose, and lactose.

Without enough lactase in the digestive fluids, the lactose of milk and milk products isn't hydrolyzed effectively. Whenever large quantities of this disaccharide are consumed, much of it, unhydrolyzed and unabsorbed, passes along the intestinal path to a region of the lower intestine where it undergoes fermentation to gases such as carbon dioxide and hydrogen, and to lactic acid, a bowel irritant. The combination easily produces gastric distress and diarrhea.

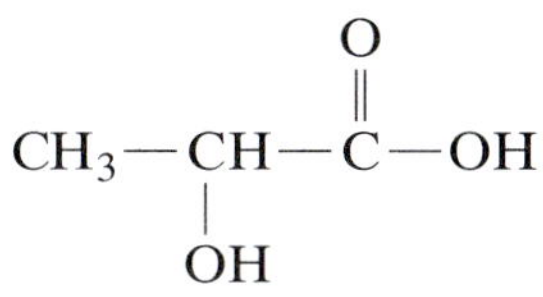

lactic acid

Milk and milk products, such as cheese and ice cream, are rich sources of lactose.

Not everyone is lactase-deficient and, for those who are, the amount of lactose needed to cause distress varies considerably. Nonetheless, gastric problems produced by drinking large amounts of milk or eating large quantities of cheese and ice cream could result from the simple inability to hydrolyze a β linkage connecting a galactose ring to a glucose ring in the disaccharide lactose.

■ QUESTION

Why might a baby who is born lactase-deficient not be able to obtain sufficient nourishment through nursing alone, without supplemental nutrition?

16.14 Enzymatic Locks and Molecular Keys

It's clear from what we've seen in this chapter that our digestion of carbohydrates requires the presence of specific enzymes. (The digestion of triglycerides and proteins, our other macronutrients, also requires the action of specific enzymes.) We've seen that biological enzymes allow certain chemical reactions, including those of our digestive processes, to occur under the relatively mild conditions present within living systems. We've also seen that enzymes can be highly specific in their actions. Maltase, for example, the enzyme that allows the easy hydrolysis of the α linkage of starch, has no effect on the hydrolysis of the β linkage of cellobiose or cellulose.

A similar sort of situation occurs with lactase, the enzyme that aids in the digestion of lactose. As we see in Figure 16.22, the link joining the two monosaccharide rings of lactose has the β geometry, as it does in the cellobiose molecule in Figure 16.18. Yet lactase, like maltase, has no effect on the hydrolysis of cellobiose or cellulose. Nor does cellobiase have any effect on the hydrolysis of lactose. As we can see in a comparison of Figures 16.18 and 16.22, the difference in molecular structure that distinguishes cellobiose from lactose lies in the stereochemistry of carbon 4 of the ring shown to the lower left in the two disaccharides. The —OH group lies above the ring in lactose, below in cellobiose.

To account for major differences in enzyme activity caused by small differences in molecular structure, a "lock-and-key" analogy has evolved. Enzymes are huge protein molecules (Chapter 17) with intricate and well-defined shapes. For effective catalytic activity an enzyme and the molecule on which it works must fit each other much as a key must fit into a lock. When the key's shape exactly matches the tumblers of the lock, the lock opens. If there's any change in the key's shape or in the tumblers, the combination fails to work. In this sense there must be a perfect match between the shapes of carbohydrates such as cellobiose, lactose, and maltose and the enzymes that work on them (Fig. 16.23).

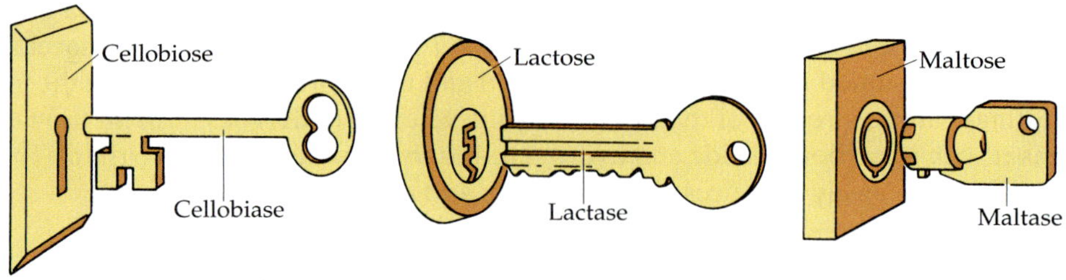

Figure 16.23 The lock-and-key analogy of enzyme action. Each key fits only its own lock.

16.15 Potatoes, Apples, and Iodine

We can now return to the opening demonstration to learn how iodine lets us distinguish easily between a slice of apple and a slice of potato. The test hinges on the difference in shapes of the carbohydrates of starches (in the potato) and cellulose (in the apple). The β links in the cellulose chain cause the long polysaccharide molecules of cellulose to stretch out in extended strings. Bundles of these strings form the tough structural material of plants. There's very little interaction between I_2 molecules and these molecular strings of cellulose. When we placed the iodine solution on the surface of the apple, we saw the color of the iodine solution simply lighten a bit as it spread out over the surface.

Starch produced a different result. The geometry of the α links bonding the individual glucose molecules together in starch causes the long amylopectin molecules to bend into a helix, with the three-dimensional shape of a spring or a corkscrew. The empty space inside the helix is just the right size to accommodate diatomic iodine molecules, which move into the empty central core of the amylopectin helix. The interaction of the interior I_2 molecules and the surrounding amylopectin helix produces the very dark color.

This deep color is useful for detecting the presence of small quantities of oxidizing agents, at least those capable of oxidizing iodide anions to molecular iodine. In a chemical procedure known as the *starch-iodide* test, a few drops of a solution under examination are added to a colorless solution of starch (or amylose) and potassium iodide. The appearance of a blue color indicates the presence of an oxidizing agent—at least one capable of oxidizing I^- to I_2—in the solution.

■ QUESTION

Suppose that you added a few drops of a solution of an oxidizing agent, one that gives a positive test with a starch-iodide solution, to a solution of potassium iodide and a soluble form of cellulose. Based on the results of the opening demonstration, what do you think you would see?

PERSPECTIVE

Can This Textbook Help End World Hunger?

Our examination of carbohydrates and of the enzymes that allow us to digest them raises an intriguing question: Would it be possible to eliminate human starvation by introducing cellobiase into our diets in a convenient, economical, and safe way? If humans could cleave the β links of the long cellulose chains and convert the cellulose into readily absorbable glucose, we might then be able to survive on grass, leaves, the bark of trees, and even paper. With a good source of cellobiase at hand, you might have a new option for a textbook at the end of a course: You might be able to put it between two pieces of bread, along with some lettuce and tomato, and enjoy it as a nutritious lunch.

The answer to our question is: "It depends." If starvation could be ended merely by adding energy (calories) to our daily diets—energy that provided at least the minimal calories needed for basal metabolism—then the answer could be "yes." But healthful and sufficient diets require more than calories alone. As we'll see in Chapter 17, they require, among other things, proteins (more specifically, proteins containing certain *essential amino acids*) to help form the physical structures of our bodies and our enzyme systems. Healthful diets also require certain minerals and vitamins to keep the chemical reactions of life running smoothly. (We'll examine these in more detail in Chapter 19.) What's more, for good health our diets even require the very indigestible dietary fiber that's made up of the β links we *can't* turn into energy. If we digested the β-linked polysaccharides as well as those that are α linked, roughage and fiber would disappear from our diets and our intestines would suffer the consequences.

Diets that provide sufficient energy do, indeed, protect against starvation. Yet the case against foods that provide *only energy*, to the exclusion of other nutrients, has been made so well that we now recognize many foods that are rich in calories but not in vitamins and minerals as "junk foods," filled with "empty calories." We'll have more to say about this in Chapter 19. ■

EXERCISES

For Review

1. Following are a statement containing a number of blanks and a list of words and phrases. The number of words equals the number of blanks within the statement, and all but two of the words fit correctly into these blanks. Fill in the blanks of the statement with those words that do fit, then complete the statement by filling in the two remaining blanks with correct words (not in the list) in place of the two words that don't fit.

> __________ are compounds of carbon, hydrogen, and oxygen, usually in a ratio of two hydrogens to one oxygen. Among the simplest, those called __________, are glucose (an __________) and __________ (a __________). Other common names for glucose include "blood sugar," so named because of its presence in the blood, and __________, which has its origin in the direction that glucose rotates __________. Common __________ include both maltose and cellobiose, in which two __________ rings combine with each other, and __________, a disaccharide that's known as "milk sugar" and that consists of a __________ ring joined to a glucose ring. Our most common __________ are __________, an important foodstuff that consists of long chains of glucose rings joined to each other through __________, and __________, which also consists of long chains of glucose rings. These are joined to each other through __________. We can digest starch because our bodies produce the enzyme __________, but fiber constitutes an indigestible carbohydrate in our diets since we lack __________. Those of us who suffer from a deficiency of __________ find that milk and milk products produce gastrointestinal distress.

α links	ketohexose
β links	lactase
aldohexose	lactose
carbohydrates	levulose
cellobiase	maltase
cellulose	monosaccharides
disaccharides	plane-polarized light
fructose	polysaccharides
glucose	starch
invert sugar	

2. Identify or define each of the following: (a) an aldehyde, (b) a chiral molecule, (c) dextrose, (d) invert sugar, (e) optical activity, (f) sucrose.

3. Give an alternative name for each of the following: (a) blood sugar, (b) grape sugar, (c) milk sugar, (d) cane sugar, (e) the principal carbohydrates of honey.

4. Give the name of a chemical compound that consists of a carbonyl group bonded to: (a) two methyl groups, (b) one methyl group and one hydrogen, (c) two hydrogens, (d) one methyl group and one ethyl group, (e) one phenyl group and one hydrogen.

5. Name (a) two different aldohexoses, (b) a ketohexose, (c) an aldotriose, (d) an aldopentose.

6. Classify each of the following as a monosaccharide, a disaccharide, or a polysaccharide: (a) amylose, (b) glucose, (c) lactose, (d) sucrose, (e) cellulose, (f) galactose, (g) cellobiose, (h) fructose.

7. What monosaccharide(s) do you get from the hydrolysis of: (a) table sugar, (b) lactose, (c) cellulose, (d) potato starch?

8. What is the linguistic origin of each of the following words, or of the italicized portion: (a) glucose, (b) mono*saccharide*, (c) *dextro*rotatory, (d) chiral, (e) *lac*tose, (f) carbohydrate?

9. Suppose you have some pure cellulose and some pure starch in separate containers, but you don't know which is which. What simple test could you perform to decide which is the starch and which is the cellulose?

10. Of what value is glycogen to us?

11. Name two foods that contain large amounts of indigestible carbohydrate.

12. Cows, sheep, and goats can digest grass and the leaves and stems of plants. Why can't we?

13. What is the single variation in molecular structure that differentiates cellobiose (Fig. 16.18) from lactose (Fig. 16.22)?

14. In what way does a cellulose molecule resemble an amylose molecule? In what way do they differ? In what way does an amylose molecule resemble an amylopectin molecule? In what way do they differ?

15. Given a molecular structure, how can you identify any chiral carbons that may be present?

16. Name an aldehyde that is not a carbohydrate but that is commercially useful. Describe its use.

17. Name a ketone that is not a carbohydrate but that is commercially useful. Describe its use.

A Little Arithmetic and Other Quantitative Puzzles

18. Each chiral carbon of a molecule can exist in a right-handed form and in a left-handed form. How many different isomers, differing only in the stereochemistry of their chiral carbons, can there be of a compound containing: (a) one chiral carbon? (b) two chiral carbons? (c) three chiral carbons?

19. Develop a formula that will give you the number of different isomers that can exist for a compound that contains n chiral carbons.

20. How many chiral carbons are there in the *enantiomer* of (+)-glucose? Explain.

21. How many of glucose's six carbons are chiral? How many of fructose's six carbons are chiral?

22. Draw the structure of a compound that can be classified as a ketotriose. How many chiral carbons does it have?

Think, Speculate, Reflect, and Ponder

23. Adding concentrated sulfuric acid, a very strong acid and a very powerful *dehydrating* agent, to powdered table sugar effectively and dramatically dehydrates the sucrose, forming water and a black, brittle substance resembling badly charred wood. The reaction is vigorous and evolves considerable heat—enough heat, in fact, to convert the newly removed water into steam. What is the brittle, black solid formed from the sucrose by this dehydration?

24. Which of the following are chiral? (a) a glove, (b) a shoe, (c) a tennis ball, (d) a molecule of methane, (e) the molecule

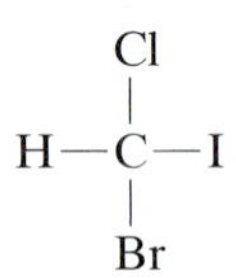

(f) CH_2BrCl.

25. Which of the following alcohols, if any, are chiral? (a) methyl alcohol, (b) ethyl alcohol, (c) propyl alcohol, (d) isopropyl alcohol?

26. What would you get if you rotated the Fischer projection of glyceraldehyde shown in Figure 16.5d by 90°? Explain. (Remember that all the rules governing the orientation of substituents in Fischer projections must be maintained even after the rotation.)

27. Are geometric isomers also enantiomers? Explain with an example.

28. Is (+)-fructose an enantiomer of (−)-fructose? Explain.

29. Is (+)-glucose an enantiomer of (−)-fructose? Explain.

30. Are α-glucose and β-glucose enantiomers of each other? Explain.

31. Consider the oxygen of the carbonyl group of the noncyclic form of glucose. During ring formation, does this carbonyl oxygen become the oxygen that's part of the new, six-membered ring, or does it become the oxygen of the newly formed —OH group?

32. Describe how cellulose and starch differ in: (a) their molecular structures, (b) the chemical behaviors, (c) the way our bodies use them.

33. Is the following statement true or false? "All of the carbohydrates of our diets provide us with 4 Cal per gram." Explain.

34. Why do glucose molecules exist primarily as rings rather than as straight chains?

35. What is the name of the enzyme our bodies use to hydrolyze table sugar?

36. A pair of clear, colorless, untinted polarizing sunglasses would reduce the intensity of most light very little, but they would still reduce glare sharply. Why?

37. Suppose you have a pair of heavily tinted sunglasses, but you don't know whether they contain polarizing lenses. What simple experiment could you perform to find out?

38. Suppose you have symptoms that you think may be caused by your sudden development of a lactase deficiency. You make an appointment with a physician to have the symptoms diagnosed, but meanwhile you make a small change in eating habits so that you can gain some evidence to confirm or deny your suspicions. What do you do to obtain this evidence, and how do you interpret it?

39. What is the major biological function of the glucose that circulates in your blood? Suppose you learn that your body will soon stop generating the enzymes needed to hydrolyze disaccharides, such as sucrose and the higher carbohydrates as well, and that medical assistance will be unavailable for several days. What would you do to ensure a sufficient concentration of glucose in your body?

40. How would your diet change if your body started generating cellobiase but stopped generating maltase and any other enzyme that could catalyze the hydrolysis of α-linked polysaccharides? (a) What foods, rich in carbohydrates, would you eat for energy? (b) What foods would you eat for roughage?

Web Questions

Use the Web to answer or accomplish each of the following. Report your Web source for each response.

1. Describe the symptoms, causes, and treatments for diabetes. Distinguish between Type 1 and Type 2 diabetes.

2. Write a brief description of *raffinose*. Discuss how it is related to the carbohydrates discussed in this chapter, how it differs from them, where it is found, and why it is important.

3. What are *diastereomers* and what is their relationship to enantiomers? What are *epimers*? Find two compounds discussed in this chapter that are epimers of each other.

4. Find two naturally occurring chiral compounds, other than carbohydrates, whose enantiomers have different biological properties or effects.

5. Write a brief biography of Emil Fischer describing his life, his education, some of his major contributions to science, and the work for which he received the Nobel prize.

6. Ingredient labels of processed foods sometimes list "high fructose corn syrup" as an additive used for sweetening. Discuss the use of this additive as a sweetener. What are its component carbohydrates? What special advantages does it supply as a sweetener?

7. The Example, "Sweet Heredity," of Section 16.4 describes the ribose of the body's RNA. (The *R* of RNA represents ribose.) What does the *D* of DNA, the carrier of the body's genetic information, represent? Describe how this structure, represented by *D*, is related to ribose. How are the two structures similar? How do they differ?

CHAPTER 17

Proteins

FIRST AMONG EQUALS

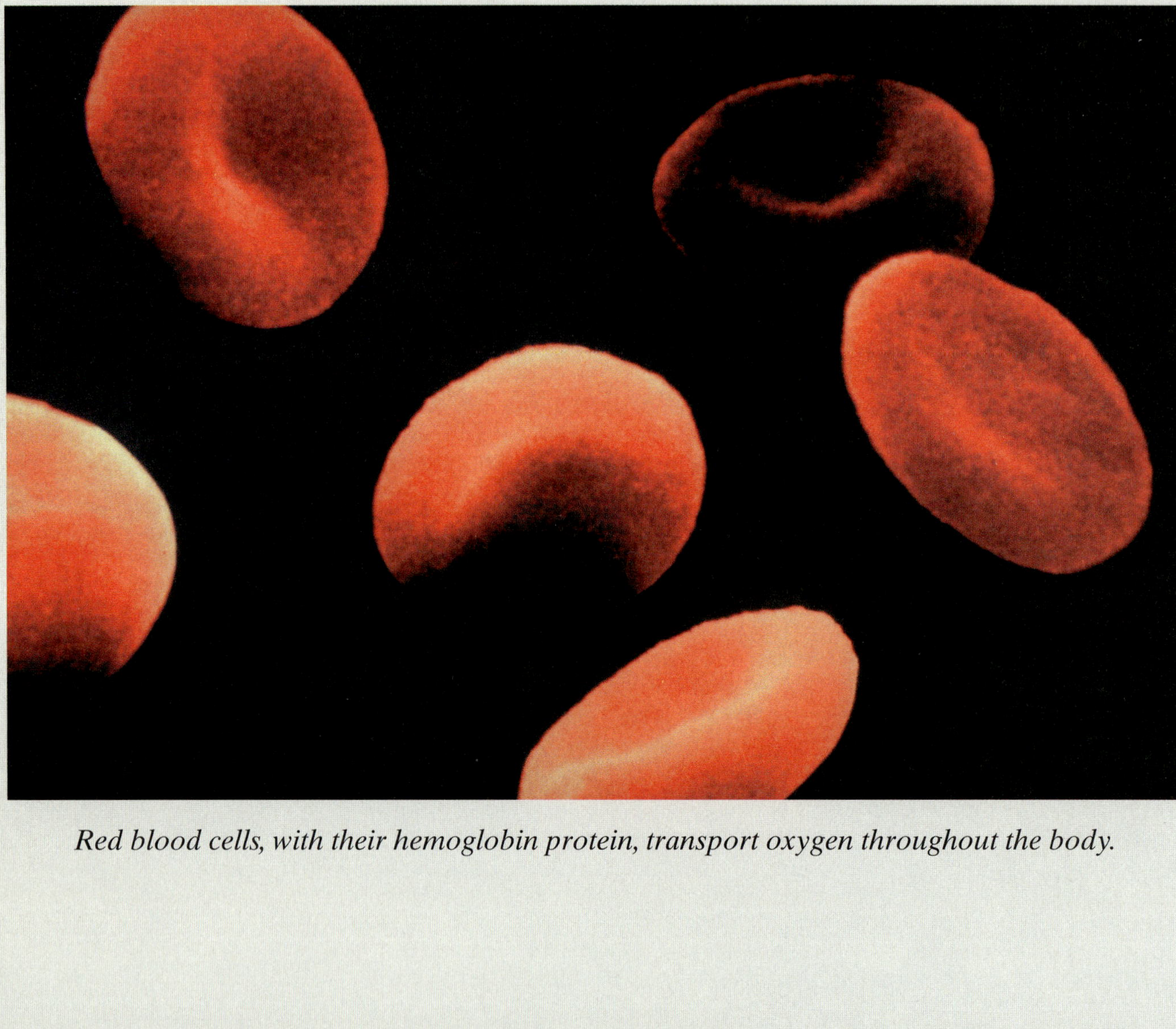

Red blood cells, with their hemoglobin protein, transport oxygen throughout the body.

How to Turn an Egg White White

Heating denatures the protein of egg white.

As we're all aware, it's possible to turn an egg white white by heating it. Actually, the clear, gelatinous part of a raw egg, the part we call the "white" (as distinct from the yolk), isn't white at all. It's colorless. But you can make it turn white by heating it or by doing a few other simple things we'll describe here.

An egg white is a combination of about 90% water and 10% proteins. Within the raw egg, the proteins of this mixture are distributed in small globules throughout the water. As they occur naturally, both in the egg white and more generally throughout the tissues and fluids of plants and animals, proteins are in their native state. That is, their molecules take the particular shapes and forms suited to each protein's function within the organism. In the egg white each long protein molecule folds and wraps around itself to form a small sphere. It's the dispersal of groups of these spheres through the water of the white that produces its characteristic transparency and gelatinous texture.

But when you heat an egg white, the protein molecules begin unfolding and assuming other shapes in a process called *denaturation.* (We'll have more to say about denaturation later. For now, we'll recognize it simply as a change in the shape of the polymeric string of a native protein molecule.) As denaturation progresses, the proteins of the egg white begin assembling into combinations with each other. The clear, transparent "white" actually turns white and becomes opaque and firm. The longer you cook an egg, the greater the change in the proteins. The white becomes more opaque and more firm as it progresses from soft boiled to poached to fried and as its native proteins become increasingly denatured.

You can denature some of the white's proteins even without heat. Break open a raw egg and separate the white from the yolk. (We'll use only the white here.) Put the white into a small glass and sprinkle a little table salt on it. Within a few minutes you'll see the egg white turn opaque and coagulate around the salt grains. The salt is denaturing the protein. You can do the same thing with some baking soda, which is the common name of sodium bicarbonate, $NaHCO_3$. It's a basic substance and it raises the pH of the protein just enough to denature it.

An even more dramatic effect involves use of the scientific method. Divide an egg white between two different glasses. Put about a tablespoon of rubbing alcohol into one of the glasses and swirl it slightly. The egg soon begins to turn white and coagulate, as if it were being heated. You'll recall from Section 9.7 that rubbing alcohol is a solution of 70% isopropyl alcohol and 30% water. Is it the isopropyl alcohol or the water that's denaturing the egg protein?

We can answer this question with the portion of the white in the other glass. Since we can't easily obtain pure isopropyl alcohol to determine its effect on the protein, we're left with an examination of water's effect. Add about a teaspoon of water, roughly the amount of water that's in a tablespoon of rubbing alcohol, to the white in the other glass. This may produce a little coagulation, but not nearly on the scale produced by the rubbing alcohol. We conclude that it was the isopropyl alcohol, not the water, that denatured the protein of the egg white (Fig. 17.1).

Figure 17.1 Adding rubbing alcohol, a mixture of isopropyl alcohol and water, to egg white denatures the protein and causes the egg white to curdle.

In this chapter we'll examine proteins in detail. We'll learn what they are made of, how they behave, what foods provide them, and why they are important to us. In Chapter 18 we'll examine the role proteins play in the chemical transfer of genetic information from one generation to the next.

17.1 Proteins: More Than Macronutrients

Proteins form the substance of our skin, nails and hair.

Proteins themselves form the third of the three classes of macronutrients. Like the polysaccharides starch and cellulose, proteins exist as long, threadlike molecules formed by linked sequences of smaller units. But while each unit of starch and cellulose is a cyclic glucose molecule, protein molecules consist of long sequences of nitrogen-containing units known as *amino acids*, which we'll examine in the next section.

The word *protein* comes to us from a 19th-century Dutch chemist and physician, Gerardus Mulder. In 1838 Mulder wrote of discovering a substance he believed to be the chemical essence of all life and that (he thought) permeates all parts of every living thing. To reflect what he imagined to be this new substance's fundamental importance, he chose its name from the Greek word *proteios*, meaning primary or first. Although proteins do, indeed, occur in every living cell, Mulder's enthusiasm and the limits of chemical science in the early 19th century overstated the case for proteins as the fundamental chemical substances of all life. Yet proteins are, indeed, vital to our lives and furnish us with virtually every atom of nitrogen that's found in our tissues.

While our brain, nerves, and muscles depend on the carbohydrate glucose for energy, and although we might be as immobile as plants if it weren't for the compact storehouse of energy that fat provides, it's protein that gives the very shape to our bodies. Proteins form our hair and our nails. Along with water, they are the principal substances of our muscles, organs, blood, and skin. Proteins form the collagen of the connective tissue that holds our bones together in a cohesive skeleton and wraps our bodies in flesh. They are the molecules of our enzymes, those catalysts so effective in promoting the chemical reactions that digest our food, provide us with energy, and manufacture the tissue of our bodies. Even the secret of life itself seems to be sealed in protein molecules.

In human nutrition, proteins stand along with carbohydrates and fats and oils as one of our three macronutrients. Yet in a broader sense, as we consider their importance in the physical structures of our bodies, in the formation and operation of our enzymes, and in the transmission of genetic information, proteins surely stand as the first among equals.

■ QUESTION

Fats and oils are chemical combinations of long-chain fatty acids and glycerol. Polysaccharides consist of chemically linked chains of monosaccharides. What chemical units form the chains of proteins?

17.2 Amines and Amino Acids

The long chains of protein molecules range in molecular weight from about 10,000 to well over 1,000,000 amu. For comparison, the molecular weight of a typical fat molecule runs between 500 and 1000 amu. Unlike the more common carbohydrates, though, whose individual links consist of one or another of only three different monosaccharides (glucose, fructose, and galactose), the protein molecules—at least those important to human life—are formed from an assortment of some 20 different amino acids. (Some authorities recognize as many as 23 amino acids in human proteins; others hold that the additional 3 are simply variants of the central set of 20. We'll stay with the fundamental 20 in this discussion.) By whatever criteria we may use, though, the simplest of all the amino acids is *glycine*:

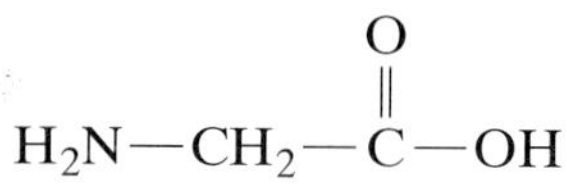

glycine

The amino acids themselves, these individual links of the protein chain, are unusual molecules that contain within their own structures two functional groups of contradictory properties. We see in glycine a methylene group, $-CH_2-$, bonded to both an acidic *carboxyl* group ($-COOH$, which can also be written as $-CO_2H$) and a basic *amino* group ($-NH_2$). The glycine molecule could be constructed by replacing one of the hydrogens of a basic ammonia molecule with an acidic $-CH_2-COOH$ unit,

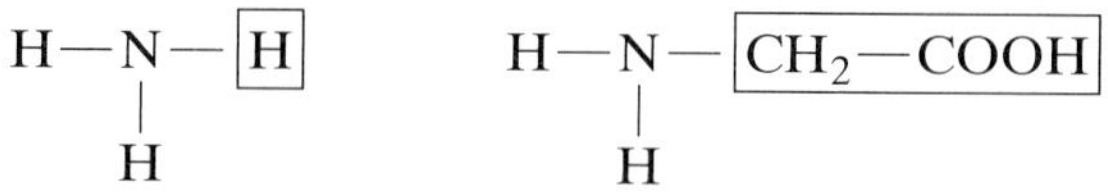

ammonia glycine

or by replacing one of the methyl hydrogens of acetic acid with a basic $-NH_2$, group,

acetic acid glycine

We've already seen that molecules bearing a carboxyl group as their only functional group are acids (Sec. 10.8). Molecules bearing only the amino function are **amines** and are bases, as defined by Brønsted and Lowry (Sec. 10.3). Because of the importance of the $-NH_2$ group to the structure and reactions of amino acids, we'll look briefly at the chemistry of amines.

An **amine** is an organic base whose molecules bear the *amino* group $-NH_2$.

We can think of amines as molecules in which one or more of the hydrogens of an ammonia molecule, NH_3, have been replaced with an alkyl group (Sec. 13.11). Replacing one of ammonia's hydrogens with a methyl group, for example, gives us methylamine. Like all bases, including sodium hydroxide and ammonia, amines can combine with an acid to form a salt:

$$\underset{\text{sodium hydroxide}}{NaOH + HCl} \longrightarrow \underset{\text{sodium chloride}}{Na^+Cl^- + H_2O}$$

$$\underset{\text{ammonia}}{NH_3 + HCl} \longrightarrow \underset{\text{ammonium chloride}}{NH_4^+Cl^-}$$

$$\underset{\text{methylamine}}{CH_3-NH_2 + HCl} \longrightarrow \underset{\text{methylammonium chloride}}{CH_3-NH_3^+Cl^-}$$

Generally, amines are less basic than ammonia. They are not commonly found in consumer products. For many of us our principal contact with amines comes through the odor of the sea and of fresh seafood. Since amines are released by all decaying animal flesh, the continual cycle of birth and death of marine organisms provides a constant supply of amines to seawater and to the air above the oceans. These amines are partly responsible for the characteristic odor of the sea and its inhabitants. This macabre aspect of amine chemistry is reflected by the common names of two amines, *cadaverine* (from *cadaver*, for a dead body) and *putrescine* (from *putrefy*, meaning "to decay with an odor"). Both these amines occur in decaying animal tissues. Ironically, putrescine not only serves as a mark of decay, but is also indispensable to life. It is essential to the growth of cells and occurs in all thriving animal tissue. Typical amines include the following:

$CH_3{-}NH_2$ methylamine

$CH_3{-}CH_2{-}NH_2$ ethylamine

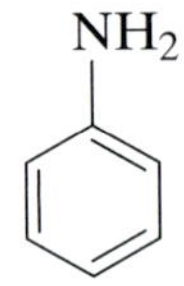

aniline

$H_2N{-}CH_2{-}CH_2{-}CH_2{-}CH_2{-}NH_2$

putrescine, also known as 1,4-diaminobutane

$H_2N{-}CH_2{-}CH_2{-}CH_2{-}CH_2{-}CH_2{-}NH_2$

cadaverine, also known as 1,5-diaminopentane

Typical Amines, Organic Bases

An **amino acid** is a compound whose molecules carry both a carboxyl group and an amino group. Amino acids have properties of both acids and bases.

Since amino acids bear both the carboxyl group of carboxylic acids (a class of acids) and the amino group of amines (a class of bases), the amino acids have properties of both amines and acids, and can therefore act as both acids and bases. That's why they are called **amino acids** and that's why they show some unusual chemical behavior. With both acidic and basic functional groups present in the same molecule, every amino acid possesses properties of both an acid and a base.

As in any other acid, the $-CO_2H$ group can release a proton to a base; as in any other base, the $-NH_2$ group can accept a proton. What's more, the overall proton transfer can occur internally, with the $-COOH$ losing its proton to the $-NH_2$ of the very same molecule. The product of this internal neutralization is an ion bearing both a positive (cationic) and a negative (anionic) charge. It's known as a *zwitterion* from a German word meaning hybrid, doubled, or twinned.

■ QUESTION

Write a balanced chemical equation for the reaction of glycine with (a) HCl and (b) NaOH. For simplicity, assume that the glycine molecule is un-ionized; that is, *not* in the form of its zwitterion.

17.3 The α-Amino Acids

The other amino acids, those that are larger and structurally more complex than glycine, carry longer carbon chains and, in many cases, additional functional groups. In *alanine,* for example, a methyl group replaces one of the hydrogens of the $-CH_2-$; in *serine* a $-CH_2{-}OH$ group replaces a hydrogen:

$$\boxed{H}{-}\underset{\displaystyle NH_2}{\underset{|}{CH}}{-}COOH \qquad \boxed{CH_3}{-}\underset{\displaystyle NH_2}{\underset{|}{CH}}{-}COOH \qquad \boxed{HO{-}CH_2}{-}\underset{\displaystyle NH_2}{\underset{|}{CH}}{-}COOH$$

glycine alanine serine

In every case, from glycine to the largest and most complex amino acid found in proteins, the nitrogen of the amino group bonds to the acid's *α carbon* (*alpha car-*

bon), the carbon that also bears the carboxyl group. Although we customarily number the carbons of a chain sequentially, as the IUPAC system instructs (Sec. 6.9), the Greek alphabet offers us a convenient alternative. In discussing the structure and chemistry of compounds bearing a carbonyl group,

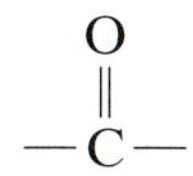

such as carboxylic acids, aldehydes, and ketones (Sec. 16.4), it's useful to refer to the carbon bonded directly to the carbonyl group as the α carbon. The next carbon becomes the β *carbon* (*beta carbon*); next to that is the γ *carbon* (*gamma carbon*), then the δ *carbon* (*delta carbon*), and so on, down the chain and through the Greek alphabet. Since we can view amino acids as carboxylic acids that contain an amino group on their carbon chains, we can use this same system in naming them.

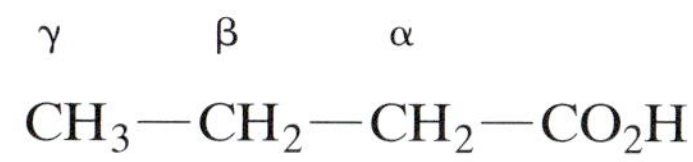

Greek Letter Designations of the Carbons of Butyric Acid

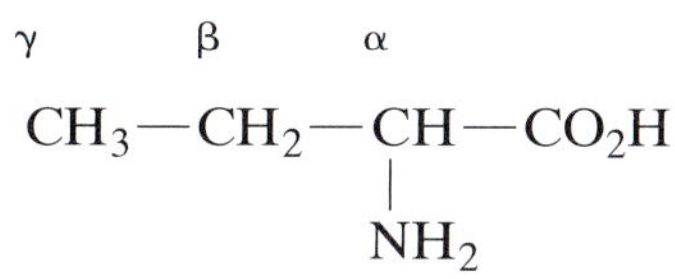

α-aminobutyric acid

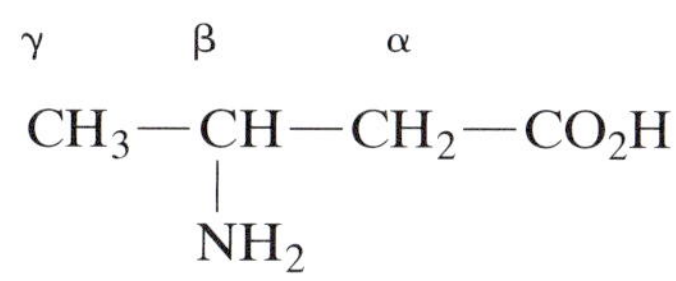

β-aminobutyric acid

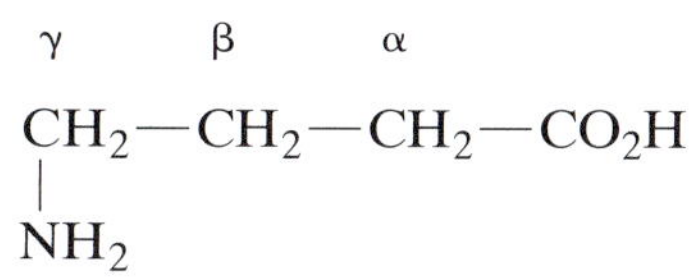

γ-aminobutyric acid

Greek Letter Names of the Aminobutyric Acids

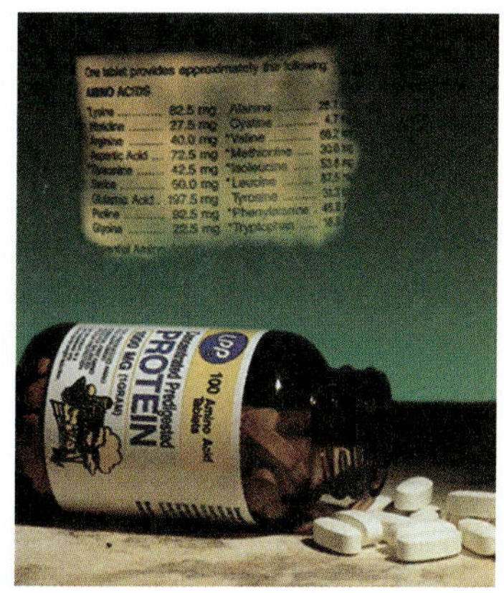

A food supplement containing many of amino acids of human protein.

All but one of the amino acids that occur in proteins contain an α carbon bearing four different groups, so all (except that one) are chiral (Sec. 16.5). Moreover, the chiralities of virtually all of these naturally occurring α-amino acids give their Fischer projections a common characteristic: With the carbonyl group at the top, the amino group lies to the left and the hydrogen lies to the right. This places them in a category of chiral compounds known as the *L-series.* Reverse the two—place the hydrogen to the left and the amino group to the right—and you have the *D-series.* At any rate, all the chiral amino acids that form protein molecules do belong to the L-series. (The designations D and L come from the two enantiomers of glyceraldehyde, the one that's dextrorotatory and the one that's levorotatory.)

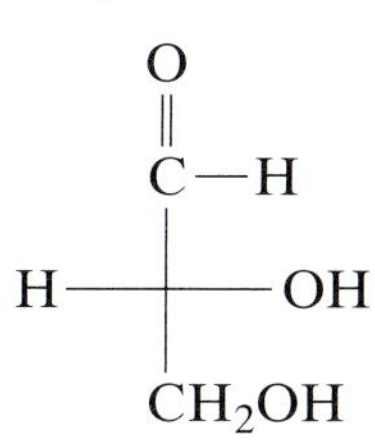

D-glyceraldehyde

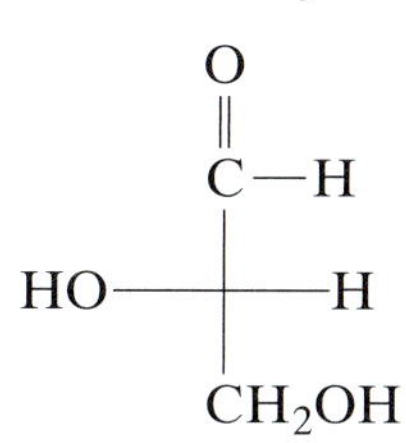

L-glyceraldehyde

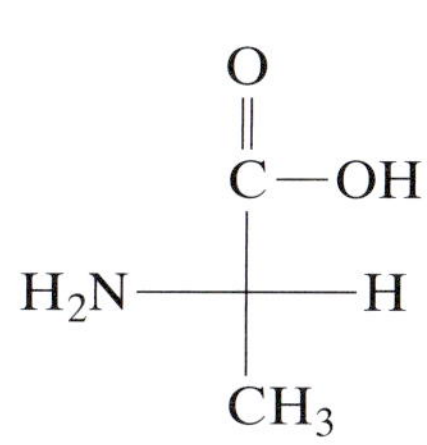

L-alanine
(The chiral amino acids
of protein
all have this same
kind of chirality)

■ **QUESTION**

Are all the aminobutyric acids—α-aminobutyric acid, β-aminobutyric acid, and γ-butyric acid—chiral? Explain.

17.4 Essential Amino Acids

Since all amino acids found in protein are α-amino acids, we can view each one of them as a variant of glycine in which an α-hydrogen is replaced by some other substituent. Table 17.1 presents the 20 amino acids of human protein, with emphasis on the structural group that differentiates each from the others. The table also gives the common abbreviations for the amino acids and specifies whether each is a necessary

Table 17.1 The 20 Amino Acids of Human Protein

Structure	Name	Abbreviation	Dietary Requirement[a]
$\boxed{H}-CH(NH_2)-CO_2H$	Glycine	Gly	Nonessential
$\boxed{CH_3}-CH(NH_2)-CO_2H$	Alanine	Ala	Nonessential
$\boxed{H_2N-C(=NH)-NH-CH_2-CH_2-CH_2}-CH(NH_2)-CO_2H$	Arginine	Arg	Essential for infants
$\boxed{H_2N-C(=O)-CH_2}-CH(NH_2)-CO_2H$	Asparagine	Asn	Nonessential
$\boxed{HO_2C-CH_2}-CH(NH_2)-CO_2H$	Aspartic acid	Asp	Nonessential
$\boxed{HS-CH_2}-CH(NH_2)-CO_2H$	Cysteine	Cys	Nonessential
$\boxed{HO_2C-CH_2-CH_2}-CH(NH_2)-CO_2H$	Glutamic acid	Glu	Nonessential
$\boxed{H_2N-C(=O)-CH_2-CH_2}-CH(NH_2)-CO_2H$	Glutamine	Gln	Nonessential
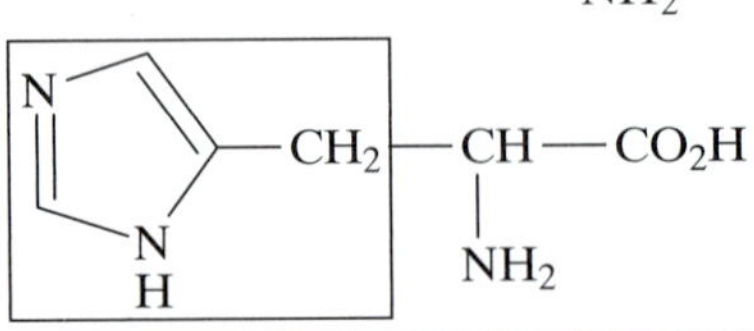	Histidine	His	Essential for infants

(continued)

Table 17.1 *(continued)*

Structure	Name	Abbreviation	Dietary Requirement[a]
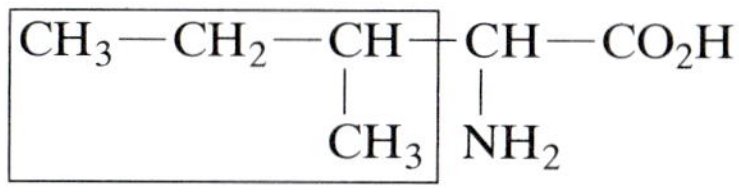	Isoleucine	Ile	Essential
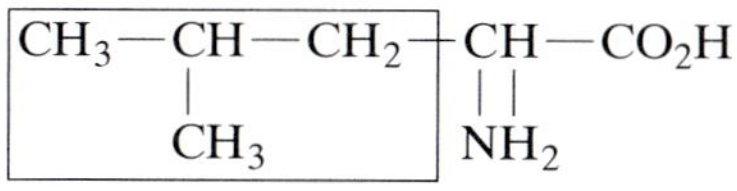	Leucine	Leu	Essential
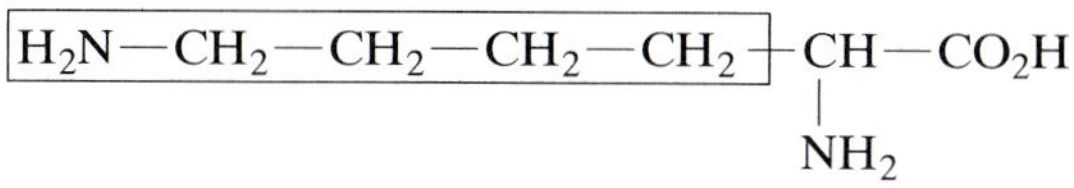	Lysine	Lys	Essential
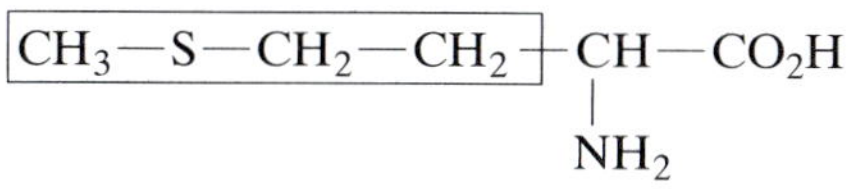	Methionine	Met	Essential
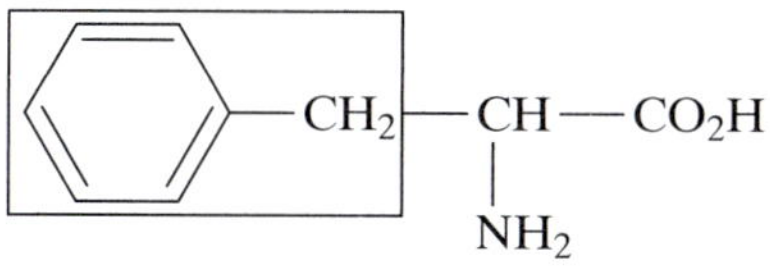	Phenylalanine	Phe	Essential
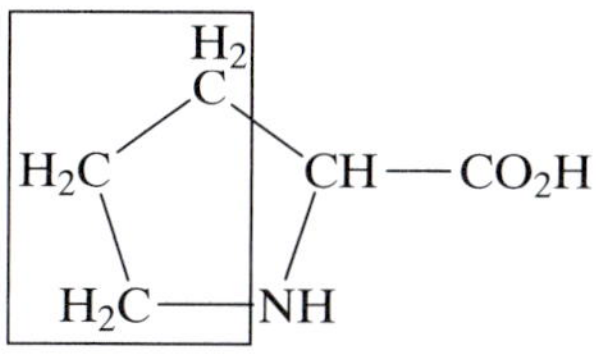	Proline	Pro	Nonessential
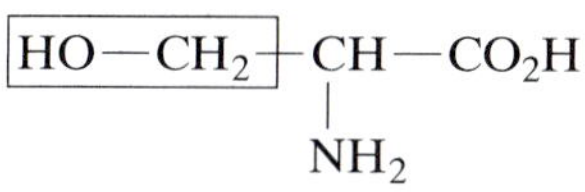	Serine	Ser	Nonessential
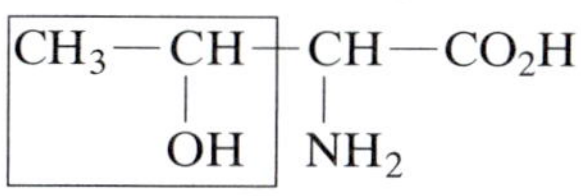	Threonine	Thr	Essential
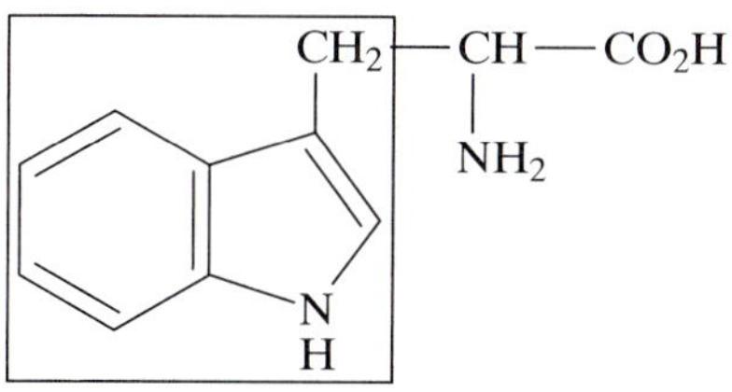	Tryptophan	Trp	Essential
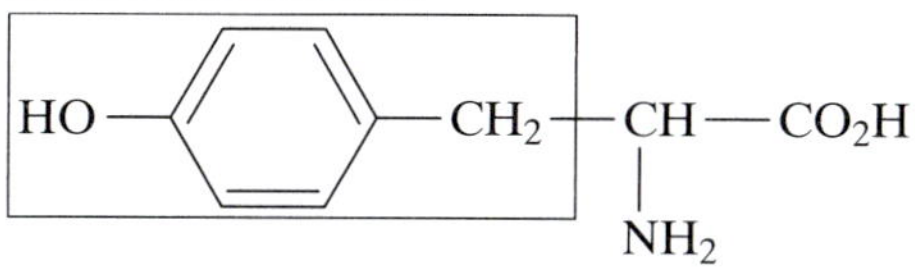	Tyrosine	Tyr	Nonessential
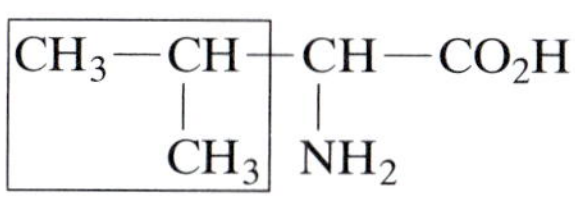	Valine	Val	Essential

[a] The essential amino acids are those that our bodies cannot synthesize from other chemicals and that we must obtain from our foods. The nonessential amino acids are those that our bodies can produce from other chemicals.

part of our diets. The question of whether any one of these amino acids *must* be part of the foods we eat, or whether we can form it within our own bodies (from other nutrients) as we need it, is important in assessing the quality of the proteins in our diet.

The **essential amino acids** are the amino acids that our bodies cannot synthesize from other chemicals and that we must obtain from our foods.

The **nonessential amino acids** are the amino acids our bodies can synthesize from other chemicals.

Of these 20 amino acids, our bodies lack the ability to synthesize about half, all of which are known as the **essential amino acids**. We must get these essential amino acids from the proteins of our foods so that we can use them in forming our own bodily protein. The remainder, those that our own bodies *can* produce as needed, are the **nonessential amino acids**.

The dividing line between the two classes isn't always completely clear. As shown in Table 17.1, eight of the amino acids appear to be essential for everyone; two more, arginine and histidine, probably can be generated in sufficient supply by an adult's body but not by the body of a young child. What's more, a couple of the nonessential amino acids seem to be able to substitute for at least a portion of two essential amino acids. Cysteine seems to be able to take up part of the slack left by a deficiency of dietary methionine (another sulfur-containing amino acid), while tyrosine can stand in, at least partially, for phenylalanine.

■ QUESTION

Is the following statement true or false? "A nonessential amino acid is one that is not essential for the formation of human protein." Explain.

17.5 The Quality of Our Protein

The quantities of total protein contained in various foods appear in Table 17.2. But total protein isn't enough. The essential amino acids must be there in the protein of the food we eat. They must *all* be there, simultaneously, in sufficient amounts and in the right ratios if our bodies are to use them in our own protein formation.

Table 17.2 Protein Content of Foods

Food	Percent Protein (average or range)
Cheese	Cream cheese 9 ↔ 36 Parmesan
Peanuts	27
Chicken	21
Fish	Flounder 15 ↔ 21 Sardines
Beef	Hamburger 16 ↔ 20 Round
Eggs	13
Wheat flour	13
White flour	11
Cornmeal	9
Lima beans	8
Milk	Whole 4 ↔ 8 Condensed
Rice	8
Peas	7
Corn	4
Dates	2
Potatoes	2
Bananas	1
Carrots	1
Orange juice	1

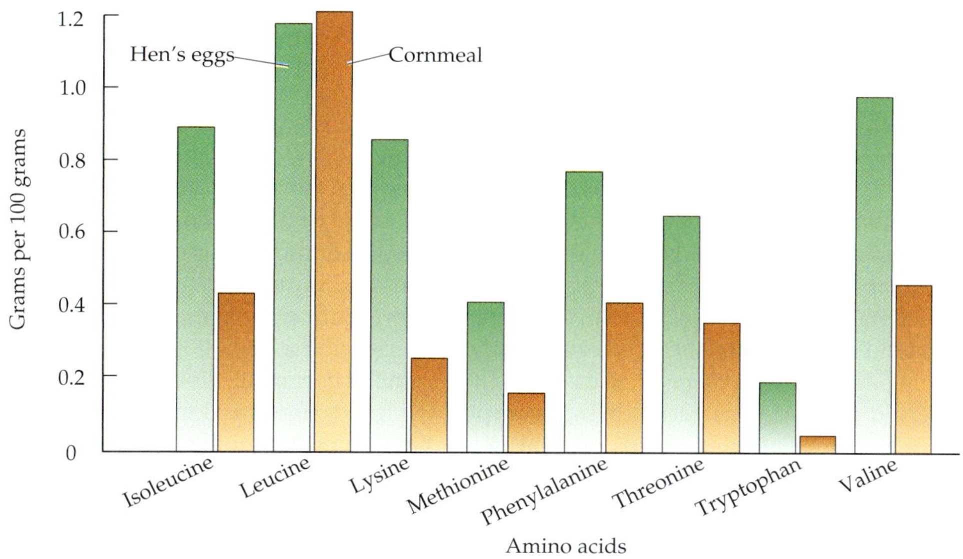

Figure 17.2 Grams of the essential amino acids per 100 g of hens' eggs and of cornmeal (approximate values).

Since we can't store significant amounts of any of the essential amino acids for future use in the same way we store fat, for example, all of them must be present as part of any particular meal. When our cells start to produce proteins, all the necessary components must be present in good supply, then and there, or the synthesis simply doesn't take place. Instead, the cells metabolize the unused acids that do happen to be present for the 4 Cal/g of energy that they yield.

To do us any good, then, the proteins of our foods must provide us with substantial amounts of all the essential amino acids, and in about the same ratio as they occur in our own proteins. The dietary proteins that provide us with something close to that ratio are the **complete** or **high-quality proteins**; those deficient in one or more of the nutrients are the **incomplete** or **low-quality proteins**.

Complete or **high-quality protein** contains all the essential amino acids in about the same ratio as they occur in our own proteins. **Incomplete** or **low-quality protein** is deficient in one or more of the essential amino acids.

Of all common foods, eggs furnish substantial amounts of the essential amino acids in a proportion closest to the human average. They provide us with the highest-quality dietary protein. This shouldn't surprise us since chickens are vertebrates, with proteins not far different from ours, and each egg must contain all of the amino acids that eventually become the proteins of a whole chick. In Figure 17.2 we see the amounts of each of the eight amino acids known to be essential for adults, as they occur in the protein of hens' eggs (and in the protein of cornmeal as well, for comparison in a few moments). All this, incidentally, isn't meant to suggest that eggs are the perfect food. Their very high cholesterol content (Sec. 15.4) limits their use by people who must avoid this substance.

The high-quality protein of eggs furnishes us with essential amino acids in nearly the same ratios found in human protein. The low-quality protein of cornmeal is deficient in several essential amino acids, especially lysine, methionine, and tryptophan.

To sum up, then, the ratio of the essential amino acids in any particular food is every bit as important as the actual quantities of each. Although the cornmeal of Figure 17.2, for example, contains almost as much total protein as eggs, and even a bit more leucine per 100 g, this grain serves as a comparatively poor source of dietary protein. Much of its leucine is lost through conversion to energy since cornmeal's deficiencies in the other essential amino acids don't allow very much protein formation. Most other grains and grain products also provide inferior dietary protein.

■ QUESTION

What dietary advantage do eggs have over cornmeal? What dietary advantage does cornmeal have over eggs?

17.6 The Vegetarian Diet

Dates, lima beans, cornmeal, and peanuts represent plant foods that are deficient in one or more essential amino acids and that furnish low-quality protein.

Finding the actual content of the essential amino acids in various foods and comparing their ratios with those found in eggs can be laborious and tricky. The approach used in Figure 17.3 makes the comparisons easier. Each bar in this figure reflects the amount of the amino acid (in a particular food protein) as a percentage of that amino acid's contribution to egg protein. A bar that reaches 100% represents an amino acid that's present in the same ratio to the others as it is in eggs. For any particular foodstuff that contains all the essential amino acids in exactly the same ratio as egg protein (or human protein) and is therefore of the same nutritional quality as egg protein, all the bars representing the amino acids would extend uniformly to the 100% mark. Bars that rise above the 100% mark represent amino acids present in excess. Like the leucine of cornmeal, they would be converted to energy instead of being incorporated into a protein. Bars falling below 100% represent amino acids present in smaller ratios than in egg protein. Since, as we saw in the previous section, all the necessary amino acids must be present for protein formation to occur, deficiencies in any of the essential amino acids in our diets limit the amount of protein our bodies can form. For each food protein, the essential amino acid that falls farthest below the mark serves to limit the amount of human protein that can be formed from the entire combination and is therefore known as the *limiting amino acid* of that particular food.

In Figure 17.3*a* we see clearly the deficiencies of lysine, methionine, and tryptophan in cornmeal. Because of these and similar deficiencies in other essential

Figure 17.3 Essential amino acids of the typical plant (*a*) and animal (*b*) proteins as compared with egg protein.

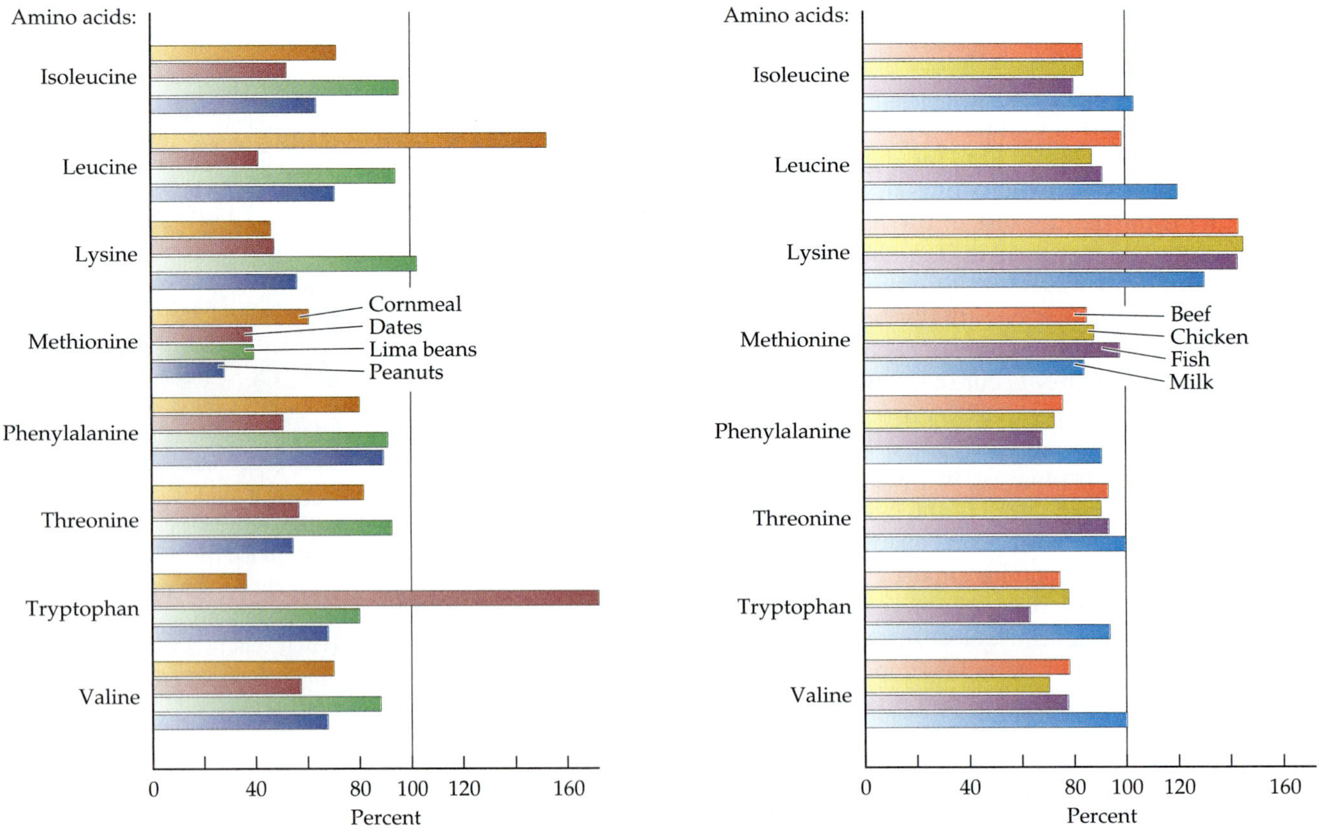

amino acids, the common food grains, fruits, and most vegetables and other plant products are poor sources of complete dietary protein. Of the common plant foods only peanuts provide substantial amounts of almost all the essential amino acids (with methionine the principal exception). It shouldn't surprise us that animal foods in general—beef, poultry, seafood, dairy products, and, of course, eggs—provide plenty of high-quality protein (Fig. 17.3*b*) while plants serve as poorer sources. The bodies of animals and the materials they produce to nourish their young are all chemically and biologically more nearly akin to our own physical bodies than are the products of the plant kingdom.

Animal products such as milk, fish, beef, and chicken typically provide more high-quality protein than do plant products.

While the high-quality proteins of meat, milk, and eggs provide us with our most efficient sources of the essential amino acids, they are by no means absolutely necessary to a healthful and perfectly satisfactory diet. After all, we're looking for only 8 to 10 essential amino acids, regardless of what sort of dietary proteins or combination of proteins furnish them. Consuming a combination of proteins from two or more different foods at the same meal can make up for a marked lack of protein quality in any one of the foods.

Different low-quality proteins that complement each other to mimic high-quality protein are known as **complementary proteins** and are found in many of the traditional dishes of various cultures. The breakfast of cereal and milk provides a good illustration. The plentiful lysine of the milk provides a nice complement to the lysine deficiencies of our cereals. Other sets of complementary proteins come in the combinations of peanut butter and bread, beans with rice or corn, and rice with peanuts or soybean curds. Generally, combinations of legumes (including peas and beans) with grains provide complementary proteins.

Complementary proteins are combinations of incomplete or low-quality proteins that, taken together, provide about tha same ratio of essential amino acids as do high-quality proteins.

Beef, veal, pork, and other meats are among our best sources of high-quality protein. Those of us who abstain from them (whether because of their potentially hazardous saturated fats and cholesterol or because of ethical, religious, or other dietary restrictions) can find good sources of protein in fish and poultry, as we see in Figure 17.3*b*. Eggs and dairy products provide rich protein for diets from which all forms of meat have been banished. Vegetarians who refuse to eat any animal products at all have a much harder time of it, but they can still obtain all the essential amino acids by careful dietary choices that include complementary plant proteins.

Failure to secure a good dietary supply of the essential amino acids, in the proper proportions, can produce devastating results, especially among the young. *Kwashiorkor* (kwash-e-OR-kor), a potentially fatal condition, develops from a protein deficiency due to the virtually complete absence of essential amino acids from the diet. The word *kwashiorkor* comes from the language of Ghana, a country on the west coast of Africa. There, nursing infants normally receive sufficient protein from their mother's milk. With the coming of a second child, though, the first is commonly weaned and placed on a diet consisting almost exclusively of readily available fresh fruits, potatoes, and other plants that may be rich in many nutrients but are poor in protein. This new diet contains only small quantities of protein, and low-quality protein at that. The first born soon suffers from a severe protein deficiency and develops a variety of symptoms, including a swollen abdomen, loss of color from the hair, patchy and cracked skin, loss of appetite, and general listlessness. Death often follows. The word *kwashiorkor* originated as the name of an evil spirit that, tradition held, afflicted the first child after the second was born. While kwashiorkor occurs principally in the tropics, its effects can be found worldwide.

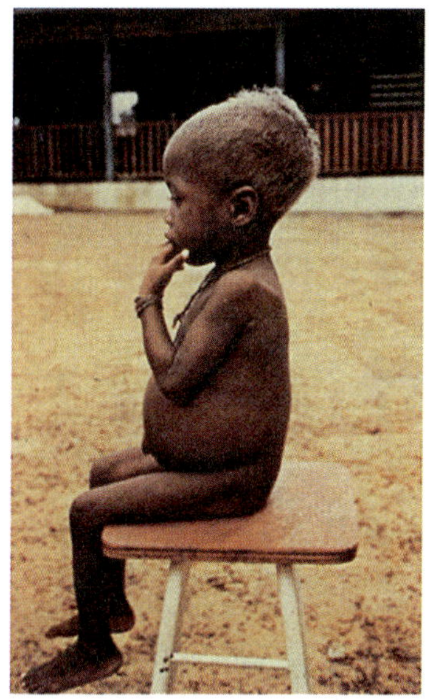

The loss of hair coloring and a swollen abdomen are typical symptoms of the protein deficiency known as *kwashiorkor*.

■ QUESTION

What is the limiting essential amino acid of both peanuts and lima beans?

17.7 Amides, Peptides, and Proteins

Up to now we've been considering what amino acids we need for protein formation and how we get them. We'll now look briefly at how we use these amino acids to construct the complex protein molecules that compose and regulate our bodies. In fashioning their various protein molecules, our bodies link the individual α-amino acids, both the essential and nonessential, into enormously long sequences, much as individual monosaccharides join with each other to form polysaccharides.

As a simple illustration of this combination, we can imagine two molecules of glycine joining together to form a new (and very short) molecule, *glycylglycine.* In this process, the —OH of the carboxyl group of one glycine molecule and one of the amino hydrogens of the other glycine molecule combine and leave as a water molecule (Fig. 17.4). The two glycines join through a covalent bond connecting the carbonyl carbon of one of them with the nitrogen of the other, thereby forming an **amide** functional group,

The **amide** functional group consists of a carbonyl group covalently bonded to a nitrogen.

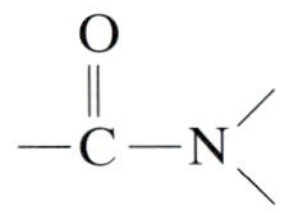

This amide group isn't limited to organic compounds related to amino acids. Simple examples include acetamide and benzamide. More important to us as consumers are the more complex amides such as *acetaminophen,* which is the *analgesic* or painkiller of several over-the-counter medications, including Datril and Tylenol. In acetaminophen, which is an amide despite the "amino" of acet*amino*phen, we see an example of an *N-substituted* amide, a compound in which one of the hydrogens of the amide nitrogen has been replaced by an organic group:

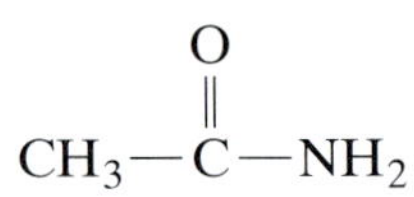

acetamide

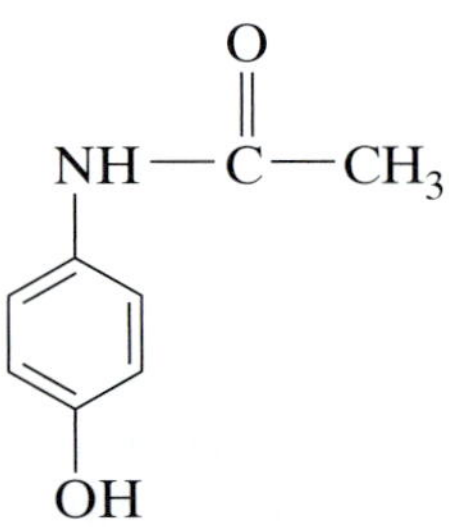

acetaminophen

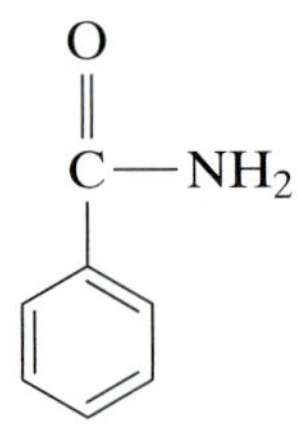

benzamide

A **peptide bond** or a **peptide link** is a carbon–nitrogen bond that links two amino acids. The resulting compound is a **dipeptide**, a **tripeptide**, a **tetrapeptide**, and so on, depending on the total number of amino acids jointed by peptide links. **Polypeptides** consist of chains of 10 or more amino acids.

Proteins are polypeptides with molecular weight greater than about 10,000 amu.

Although this amide link or functional group occurs in a variety of organic compounds, it's particularly important in connecting amino acids to each other. When it serves this function the carbon–nitrogen bond of the amide is called a **peptide bond** or a **peptide link**. The resulting compound, glycylglycine in the case we've just examined, is a *peptide*. A sequence of two amino acids joined through a peptide link forms a **dipeptide**. (Glycylglycine, then, is a dipeptide.) Three amino acids, joined by two peptide links, make up a **tripeptide**; four, joined by three peptide links, form a **tetrapeptide**; and so on. Chains of 10 or more amino acids are **polypeptides**. It's only when the molecular weight of a polypeptide exceeds about 10,000 amu that it becomes a **protein** (Fig. 17.5).

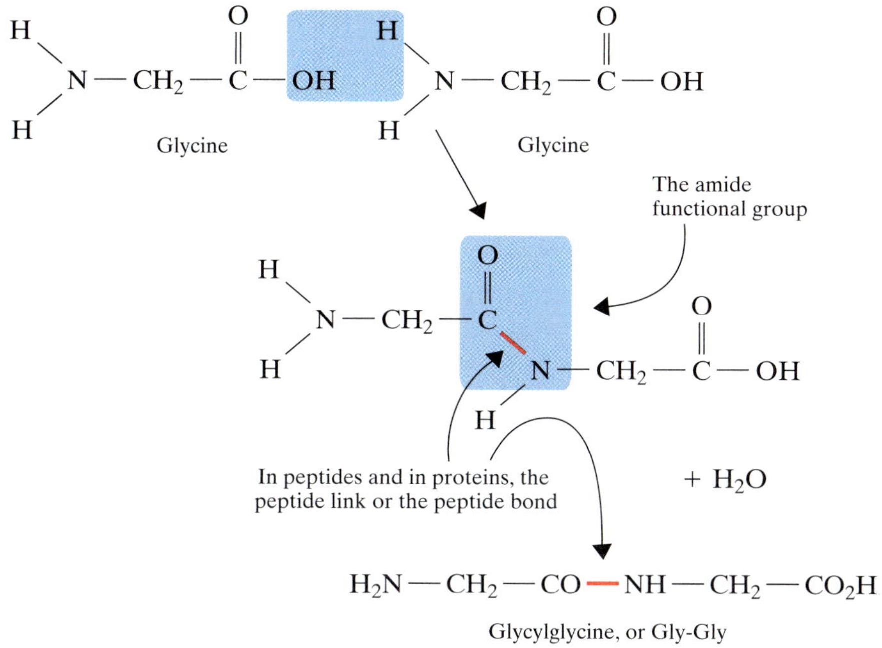

Figure 17.4 Peptide formation.

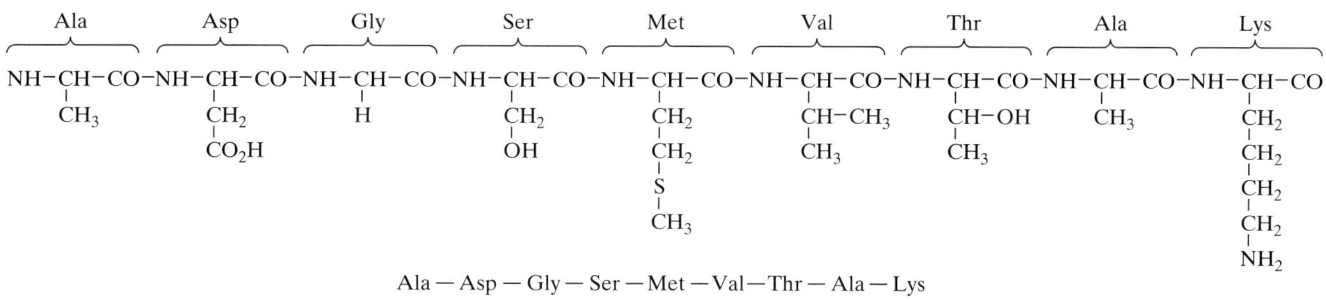

Figure 17.5 A typical segment of the chain of a polypeptide or a protein.

■ QUESTION

What name would you give to the following compound? (Refer to Table 17.1.)

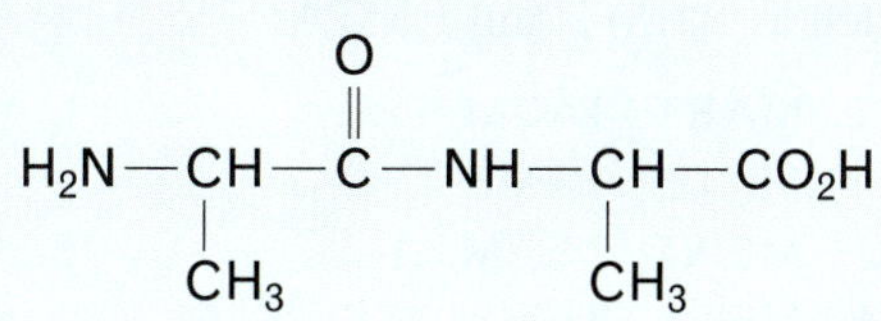

17.8 Primary Structures

To understand how the set of 20 amino acids of Table 17.1 can generate the enormous variety of peptides and proteins found in the human body, we'll begin by examining how many dipeptides we can construct from just two amino acids, alanine and glycine. We've seen two of the dipeptides already, glycylglycine in Section 17.7 and alanylalanine in the question at the end of that section. Two additional dipeptides are glycylalanine and alanylglycine or, using the abbreviations in Table 17.1, Gly–Ala and Ala–Gly:

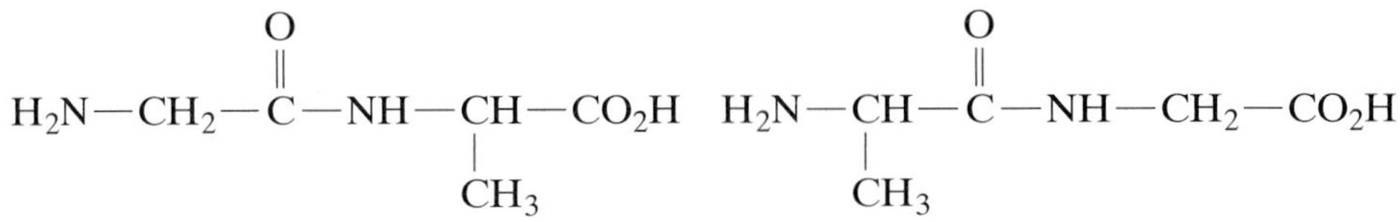

glycylalanine, Gly–Ala   alanylglycine, Ala–Gly

These are, indeed, two different compounds. Notice that the part of the structure bearing the carboxyl functional group ($—CO_2H$) comes from an alanine molecule in Gly–Ala but from a glycine molecule in Ala–Gly. (In writing the molecular structures of the peptides, it's customary to place the amino acid with the free amino group on the left and the one with the free carboxyl group on the right and to name the various amino acids from left to right.)

With just two amino acids, then, we can devise four different dipeptide chains: Ala–Ala, Ala–Gly, Gly–Gly, and Gly–Ala. As we increase the number of amino acids available for use and as we lengthen the chain we construct, the number of peptides that result grows almost beyond comprehension.

To grasp the enormous variety of polypeptide and protein structures that can exist, consider a simple analogy. We'll represent each amino acid by a letter of the alphabet and, in something resembling a game of anagrams, determine how many words can be formed by rearranging the sequence of letters. With *N*, for example, representing alanine, and *O* representing glycine (and limiting ourselves to using each letter once and only once in each word), we can form the two words NO (for alanylglycine) and ON (for glycylalanine):

NO   ON

In the next step, let the three letters A, E, and T represent three different amino acids. With these we can generate the four words ATE, EAT, TEA, and ETA (the seventh letter of the Greek alphabet), and two nonsense words, AET and TAE, for a total of six three-letter combinations:

AET	EAT	TAE
ATE	ETA	TEA

Adding an M to our supply gives us a total of 24 four-letter words, including such English words as MEAT, TEAM, MATE, and TAME and a long list of nonsense combinations, such as AEMT and EMTA:

AEMT	EAMT	MAET	TAEM
AETM	EATM	MATE	TAME
AMET	EMAT	MEAT	TEAM
AMTE	EMTA	META	TEMA
ATEM	ETAM	MTAE	TMAE
ATME	ETMA	MTEA	TMEA

The **primary structure** of the peptide is the sequence in which the amino acids are bonded to each other in the peptide chain.

The sequence of amino acids indicated by the letters of each of these words represents the **primary structure** of the peptide, which is simply *the sequence in which the amino acids are bonded to each other in the peptide chain.* By using each of four different amino acids once and only once in the chain, we can generate 24 different tetrapeptides, each with its own, unique primary structure. Of course, we're limiting ourselves here by using each amino acid just once in the sequence. We could form far more by omitting one or more of the amino acids and using any of the others more than once.

■ QUESTION

Our exercise with the letters A, E, and T demonstrates that three different amino acids can form six different tripeptides, with each amino acid represented once and only once. If we remove this restriction of once and only once—if we permit any of the three amino acids to be used more than once, and if we don't require that all three amino acids be represented in the tripeptide—what's the total number of tripeptides that can be formed from three different amino acids?

17.9 Oxytocin, Vasopressin, and Insulin

Now we can leap ahead to the entire assortment of 20 amino acids of human protein. If we were to assign a letter to each of these 20 amino acids and limit ourselves to constructing as many 20-letter words as possible, with each letter appearing once and only once in every sequence, much as we did in Section 17.8, we'd have more than 2,430,000,000,000,000,000 (or 2.43×10^{18}) different words, each representing a different protein.

Of course, this is only an analogy, and a very simple one at that. We've limited ourselves to 20 amino acids in the primary structure, with each acid used once and only once. Our bodies don't have any such limitations. Peptides and proteins of any length can be formed, and any of the 20 amino acids can be dropped from any particular primary structure, or repeated here and there anywhere within the sequence.

Yet each of the peptide chains that our bodies do, in fact, produce must be a rational, chemical "word" that performs a specific and important function in sustaining our lives. If the "word" is "misspelled," if just the right sequence of just the right amino acids isn't present in the polypeptide, the results can be tragic, as we'll see for *sickle cell anemia* in Section 17.10.

The nonapeptides *oxytocin* and *vasopressin* illustrate the importance of the primary structure of a peptide in determining its biological activity. Each is a hormone that regulates specific activities of the body. (The italicized amino acids in the following structures represent amino acids that differ in the two structures.)

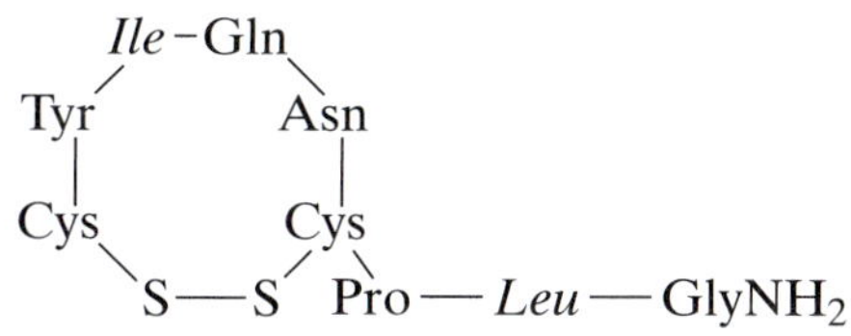

oxytocin

Oxytocin plays a vital role in the release of milk during lactation. Notice that the sulfur atoms of two cysteine units of the peptide are joined, forming a ring. Also note that an amide functional group of a glycine terminates the short chain:

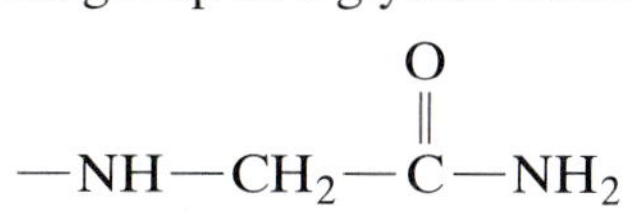

the $GlyNH_2$ that terminates the short chain of oxytocin

In vasopressin

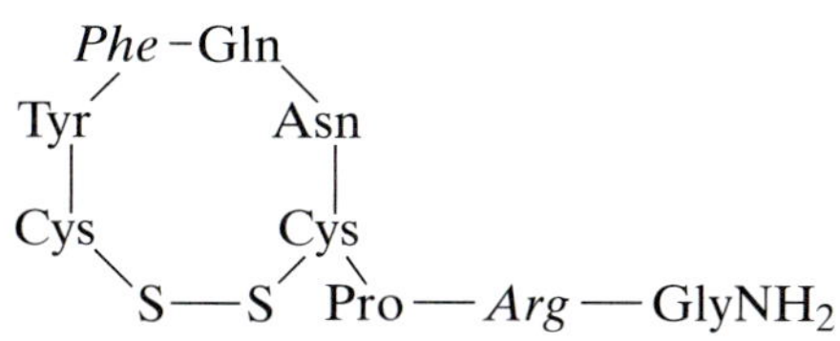

vasopressin

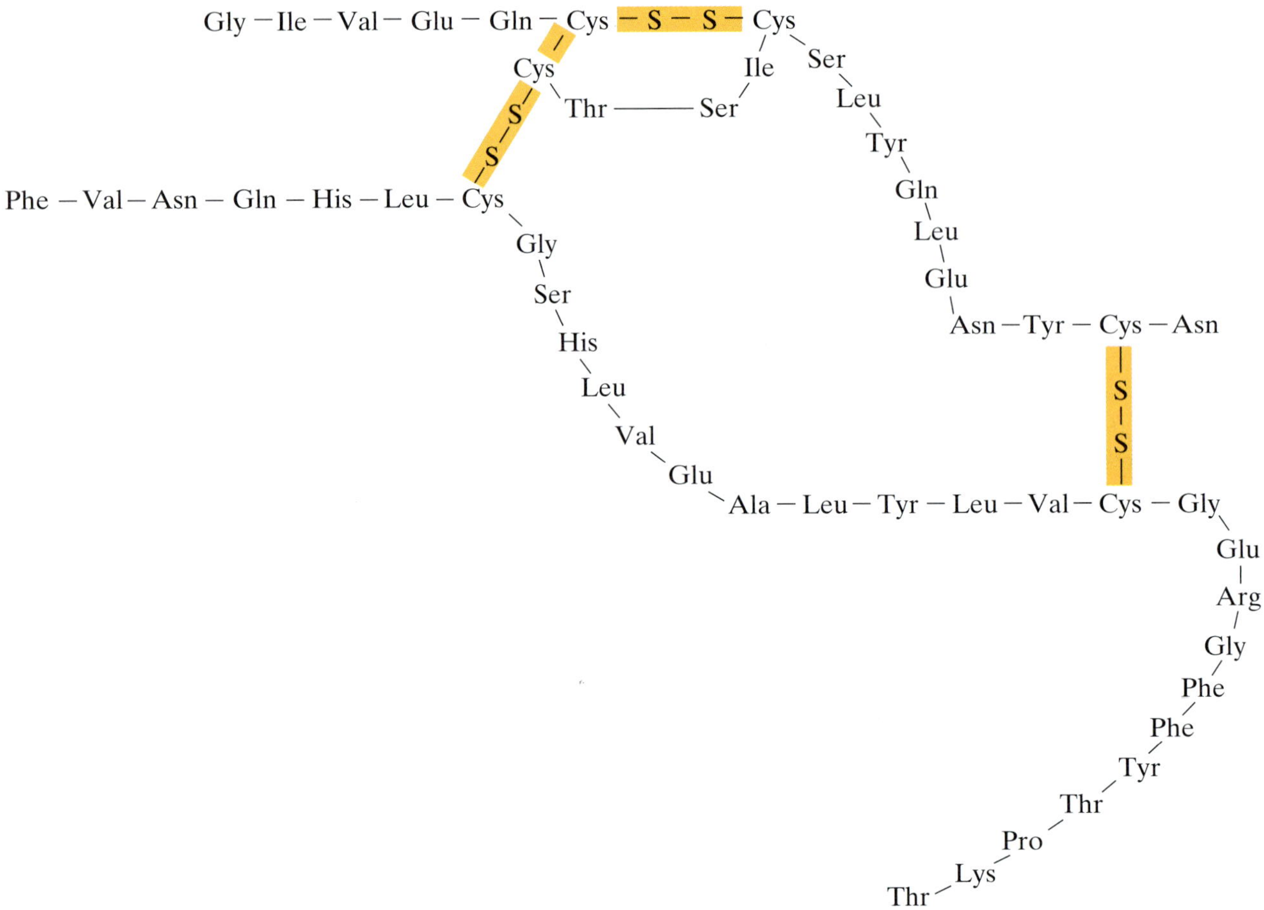

Figure 17.6 The primary structure of human insulin. Note the sulfur–sulfur bonds that connect pairs of cysteine groups.

phenylalanine and arginine, respectively, replace the isoleucine and leucine of oxytocin. As a result of this change in the primary structure of the peptide, vasopressin's principal function is to regulate the amount of water retained by the body. With this water-regulating action vasopressin also affects the blood pressure since excessive retention of water can lead to an increase in blood pressure.

Insulin is a much larger, more complex hormone, composed of 51 amino acids and important to the metabolism of glucose. Insufficient production of this polypeptide by the pancreas results in the condition of ineffective glucose metabolism known as *diabetes* or, more specifically, *diabetes mellitus* (Sec. 16.1). When the pancreas synthesizes insulin, it forms the molecule shown in Figure 17.6, with exactly that sequence of amino acids and no other.

■ **QUESTION**

What are the biological functions of oxytocin and vasopressin? In what ways do their molecules resemble each other? In what ways are their molecules different?

17.10 Higher Structures, More Complex Functions

In earlier examinations of triglycerides and carbohydrates we found that relatively small changes in molecular structure can produce major effects in properties and behavior. Converting carbon–carbon double bonds of triglyceride side chains from *cis* to *trans*, for example, can raise the melting point of a triglyceride, transforming an oil to a fat (Sec. 15.7). In carbohydrates, the geometric difference between an α link and a β link affects the three-dimensional shapes of polysaccharide chains, causing starch and cellulose to interact differently with molecular iodine. As we saw in Section 16.15, the long, α-linked carbohydrate chains of starch bend into helices that accommodate diatomic iodine molecules and that are thereby responsible for the dark blue color produced on starch by iodine.

In protein chemistry, these effects on molecular structure are magnified. The bending, twisting, and folding of the long molecular strands of primary protein structures result in the formation of increasingly complex shapes known as the higher structures of proteins, their *secondary*, *tertiary*, and *quaternary structures*. Among the forces that hold these higher structures in place are **hydrogen bonds**, weak bonds formed by the attraction of a nitrogen, oxygen, or fluorine atom for a nearby hydrogen that's bonded to a different nitrogen, oxygen, or fluorine (Fig. 17.7). Since protein molecules don't contain fluorine, only hydrogen bonds to nitrogen and oxygen are important in proteins.

A **hydrogen bond** is a weak bond formed by the attraction of a nitrogen, oxygen, or fluorine atom for a nearby hydrogen bonded to a different nitrogen, oxygen, or fluorine atom.

Hemoglobin, the complex protein of the red blood cells that transport oxygen from the lungs to tissues throughout the body, provides an illustration of the consequences of a change in even a single amino acid in a primary structure. Figure 17.8 presents schematic diagrams of a hemoglobin molecule itself: its primary structure (the amino acid chains); the helical secondary structure; the tertiary structure formed as the secondary structure bends in upon itself at several points; and hemoglobin's quaternary structure, which forms as a combination of two or more tertiary structures. The shapes of the secondary and higher structures are maintained by hydrogen bonds and sulfur–sulfur bonds formed by cysteine molecules, similar to those of oxytocin, vasopressin, and insulin (Sec. 17.9). All this refers to a normal hemoglobin molecule in a healthy human.

Hemoglobin takes a different shape in someone suffering from *sickle cell anemia,* a genetic disease that destroys the blood's capacity to carry oxygen. The disease occurs principally in people of West African ancestry. It gets its name from its distortion of the red blood cells from their normal disklike form into the shape of a sickle (Fig. 17.9).

Normally, the hemoglobin molecule consists of two pairs of polypeptide chains, with 141 amino acids units on each chain of one pair and 146 on each chain of the other pair, for a total of 574 individual amino acid units in all four chains. During hemoglobin's synthesis within the body, each amino acid is placed at a specific point within these chains, producing a primary structure that gives rise to normal secondary and higher structures, and normal red blood cells.

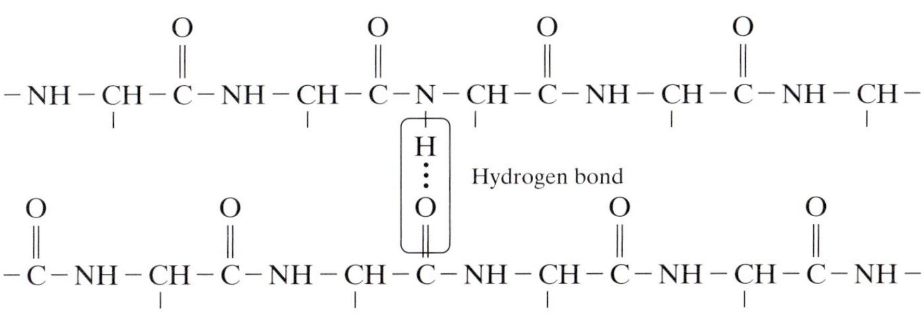

Figure 17.7 A hydrogen bond between two protein chains.

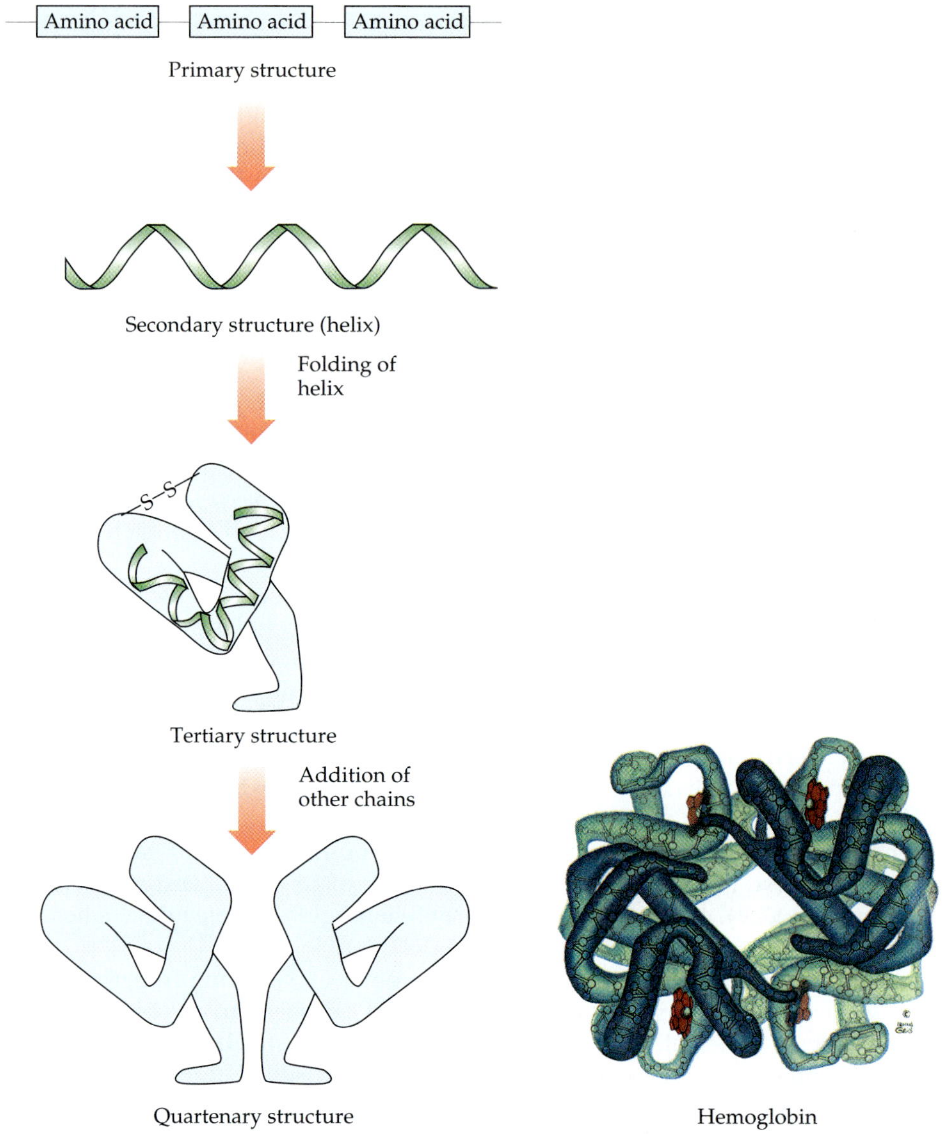

Small spheres and rectangles (in red) represent iron atoms held by small organic rings imbedded within the folds of hemoglobin.

Figure 17.8 The primary, secondary, tertiary and quaternary structures of the proteins of hemoglobin.

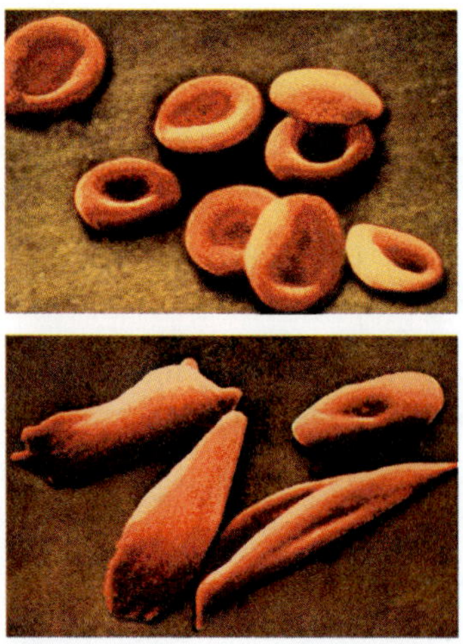

Figure 17.9 Normal red blood cells and sickle cells.

Sickle cell anemia results from a genetic error that places valine rather than the required glutamic acid at a precise point within one of the pairs of hemoglobin's polypeptide chains. As a result of this small change in primary structure, the hydrogen bonds that form the higher structures of hemoglobin produce an abnormal molecule. The hemoglobin protein becomes distorted, forcing the red blood cells into a characteristic sickle shape and impairing their ability to carry oxygen. Although often fatal, the disease can be treated with bone marrow transplants and with various forms of medication.

In addition to their roles as hormones and in giving shape and substance to various structures of the body, proteins perform other functions as well. They play a part, for example, in the control of serum cholesterol (Sec. 15.4). Cholesterol is a

	PROSTATE SPECIFIC AG	*(01)	0.96	MCG/L	0.50-4.00
*	GLUCOSE, FASTING		154.0	MG/DL	65.0- 115
*	PROTEIN BOUND GLUCOSE	*(02)	1.36	MG/GM	0.00-1.19
*	CHOLESTEROL		262.0	MG/DL	120- 240
*	HDL CHOLESTEROL		37.0	MG/DL	39.0-76.0
	CHOL./HDL RATIO	*(03)	7.10		
*	LDL CHOL.,CALCULATED		203.0	MG/DL	65.1- 171
	TRIGLYCERIDES		108.0	MG/DL	50.0- 200

Results of a blood analysis showing values for cholesterol, high-density lipoproteins and low-density lipoproteins, and ranges of desirable values.

hydrophobic molecule (Sec. 13.3) that has very little tendency to dissolve in water. To transport it through the blood, our bodies wrap packets of cholesterol in a sheath of proteins and varying amounts of triglycerides that present a more hydrophilic surface to the water of the blood. The resulting bundle, known as a **lipoprotein**, resembles the micelles of detergents.

A **lipoprotein** is a combination of a protein, cholesterol, and triglycerides, in varying proportions, that carries cholesterol through the blood.

The lipoproteins are classified by their density: the HDLs are high-density, the LDLs are low-density, and the VLDLs are very-low-density lipoproteins. The density of the lipoprotein appears to affect the fate of the cholesterol it carries. While only about a quarter of the cholesterol of the blood is carried in the form of HDLs, these high-density lipoproteins seem to have a protective effect on the body. They appear to remove cholesterol from deposits on the walls of arteries and transport it to the liver for disposal. LDLs and VLDLs, on the other hand, seem particularly effective in depositing cholesterol on arterial walls. There also seems to be a dietary connection here. Studies have connected large amounts of saturated fats in the diet with formation of the LDLs and VLDLs, while polyunsaturated fats seem to stimulate formation of the HDLs.

■ QUESTION

What is the chemical basis for sickle cell anemia?

17.11 Egg White Revisited

We've seen that the primary structure of a protein determines what protein it is and how it functions. We've also seen that hydrogen bonds can hold the primary structure in specific shapes of secondary, tertiary, and higher structures. The shape of a protein's higher structures determines how well the protein can do its job. Two types of proteins, *fibrous* and *globular*, have shapes that suit them particularly to several of the functions they perform.

Our hair and nails, and the claws of animals, for example, are tough and strong because of their fibrous proteins. Their strength comes from the organization of their protein molecules into parallel strands that entwine much like strands of fiber twisted into strong rope. The primary structures of fibrous proteins align and intertwine with each other, providing strength to the parts of the body they form.

Other peptide molecules remain largely aloof from their neighbors, twisting and coiling their own strands back into themselves, forming small spheres. These globular proteins take shapes similar to the micelles that soap and detergents form in water (Sec. 13.5). As a result they move around easily in the blood and other fluids, available to do their chemical work where they are needed. Enzymes are globular proteins, as are the proteins of egg white. The hydrogen bonds of egg white's proteins hold the long molecular strands in their globular form, giving the white its characteristic consistency and transparency.

Heating an egg white, as when we cook the egg, overcomes the hydrogen bonds that hold the protein in its globular form and allows the molecule to unravel, still keeping its original primary structure. As the primary structure unravels it comes in contact with other protein molecules, which have also lost their globular shapes. These denatured proteins now gather into shapes more like those of fibrous proteins and form the familiar tough white of the cooked egg. Adding salt or the isopropyl alcohol of rubbing alcohol to the protein, as we did in our opening demonstration, also disrupts the sequence of hydrogen bonds in the native protein and denatures it. Cooking an egg, then, is one way to denature the native proteins of an egg white. The heat breaks the hydrogen bonds holding the protein molecules in their native secondary and higher structures. The proteins then regroup in other ways that produce the texture and appearance of the cooked egg.

■ QUESTION

What is the advantage of a globular rather than a fibrous structure for the enzymes of the blood?

PERSPECTIVE

Protein and Health

Our first contact with proteins might lead us to believe—along with Gerardus Mulder (Sec. 17.1)—that they are, indeed, the foremost chemicals of life. Proteins build the tissues of our organs and our muscles. All our enzymes are proteins. We rely on proteins for

- digesting food;
- carrying oxygen to our tissues;
- growing hair and nails;
- healing cuts, wounds, and broken bones; and
- fighting off the invading microorganisms of disease.

"Surely," we might think, "the more protein we eat, the better our lives." Not quite. Remember that in many ways proteins are just another macronutrient, furnishing the same 4 Cal/g as carbohydrates. We don't use them as brain fuel, as we do glucose, nor store them as readily portable supplies of energy, as we do fats. While protein is being processed by our digestive systems, our bodies look at it as a collection of amino acids, to be used as needed.

If there's no demand for the special talents of the protein molecules we can make from amino acids—if we aren't ill, generating body tissue, recovering from an injury, running short of insulin or some other enzyme, or under some sort of physical stress—we use them for energy or convert them to body fat as we do our other macronutrients. We can't simply decide that this morning's fried eggs are going into muscle tissue, or antibodies, or hair. If there's no demand for muscle tissue, generated by exercise, or for antibodies, generated by an invasion of microorganisms, or for some other specialized molecules that only the amino acids can form, we either use them for energy or store them by converting them to fat.

How much protein do we need? As a general recommendation, only about 10 to 15% of our calories should come from protein. (Estimates for the remainder of our calories vary, but 20 to 25% has been recommended as a healthful maximum from fats and oils. For good health, most of our calories ought to come from the large carbohydrate molecules of starches.) For now we can say that, as a rough rule of thumb, most Americans consume at least twice as much protein as they need. Of course, the term *most* isn't the same as *all*. There are special demands and stresses brought on by illness, on the one hand, and pregnancy and lactation, on the other. Because of the need to repair and create tissue, these stresses require extra quantities of protein. Other groups who may not be getting enough protein include dieters with severely restricted food intake and those who simply can't obtain sufficient protein because of poverty or disability.

While both plants and animals can supply us with protein, each source provides its own advantages. Animal protein is high-quality protein, rich in the essential amino acids. Yet animal protein brings with it cholesterol and saturated fatty acids. Vegetable protein spares us these, but by and large it's poor in some vital minerals, notably calcium, iron, and zinc. On the other hand, plant protein is less expensive than animal protein and supplies us with fiber as a bonus. For most of us, fortunately, the answer is simple: A varied diet including our major sources of protein—meat, fish, poultry, vegetables, and legumes, along with milk, milk products, and some eggs—supplies us with adequate amounts of the nutrient.

In the chemical and biological mechanics of how we disassemble the molecules of our foods and recombine their components into the proteins and other substances that form and govern our human bodies lies one of the secrets of life itself. It's a secret that science is illuminating slowly, painstakingly, step by step. With each discovery we learn more clearly how living things form, how they grow and take their own shapes, and how they use their foods to build and sustain their bodies. Part of the solution to these mysteries lies in understanding how, through the operations of the DNA molecule, all living things pass their undiscovered mysteries along to generations yet unborn and unimagined. We'll examine this more closely in Chapter 18. ■

EXERCISES

For Review

1. Following are a statement containing a number of blanks and a list of words and phrases. The number of words equals the number of blanks within the statement, and all but two of the words fit correctly into these blanks. Fill in the blanks of the statement with those words that do fit, then complete the statement by filling in the two remaining blanks with correct words (not in the list) in place of the two wordss that don't fit.

> Dietary __________ provides us with the __________ we need to form our bodily enzymes and structures such as skin, hair, and nails. In generating our own protein, we use some 20 amino acids. Those we can form ourselves, through chemical reactions that take place in our own bodies, are called __________ amino acids. An example of this type of amino acid is __________. Those we cannot form but must obtain from our foods are the __________ amino acids. An example of this type of amino acid is __________. With the single exception of __________, all of the naturally occurring amino acids are __________ and are classified as members of the __________.
>
> The actual sequence of the amino acids along a protein or polypeptide chain represents the chain's __________. In this structure the individual amino acids are joined to each other through __________ links, which are incorporated within the __________ functional groups of simpler molecules. Two amino acids joined together in this way form a __________, three form a __________, and so on, up to chains of nine amino acids. Beyond nine and up to molecular weights of about 10,000, the chains are known as __________; beyond that size they are classified as proteins.

alanine	glycine
amide	leucine
amino acids	L-series
chiral	peptide
complementary protein	polypeptides
denatured	primary structure
dipeptide	protein
essential	

2. Define, identify, or explain each of the following:

α-amino acid	insulin
acetaminophen	kwashiorkor
amine	LDL
HDL	oxytocin
high-quality protein	zwitterion
hydrogen bond	

3. What happens when a protein is denatured?

4. Describe four ways to denature the protein of egg white.

5. What's the difference between a globular protein and a fibrous protein?

6. Name three foods that provide large amounts of high–quality protein.

7. Why do animal products generally provide better quality protein than plant products?

8. Assuming that the composition of egg protein accurately reflects our own, which essential amino acid is present in human protein in the smallest quantity?

9. Which, if any, of the following are essential amino acids for adults? (a) tryptophan, (b) valine, (c) isoleucine, (d) serine, (e) leucine.

10. What amino acids are essential for children but not for adults?

11. Why do vegetarians who avoid all animal products, including eggs, milk, and cheese, have to be particularly careful about their choice of the foods they eat?

12. Name each of the following α–amino acids:

a. $CH_3-CH(NH_2)-COOH$

b. $CH_2(NH_2)-COOH$

c. 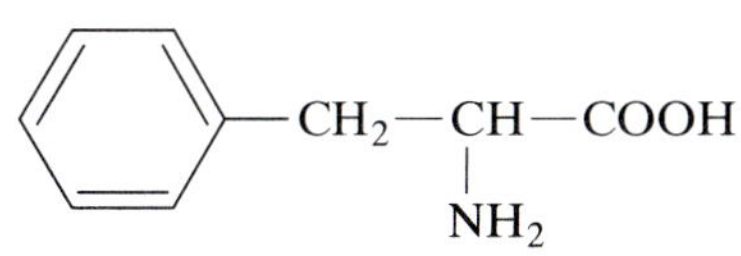

d. $HO-CH_2-CH(NH_2)-COOH$

e. $HOOC-CH_2-CH(NH_2)-COOH$

f. $HS-CH_2-CH(NH_2)-COOH$

13. Give the molecular structures of each of the following α-amino acids: (a) glutamic acid, (b) threonine, (c) arginine, (d) tyrosine.

14. Name the single, naturally occurring α-amino acid that is achiral (not chiral).

15. What's the difference between an α-amino acid of the D-series and one of the L-series?

16. In what way are cysteine and methionine different from the other amino acids of Table 17.1?

17. What is the difference between an essential amino acid and a nonessential amino acid? Name two in each category.

18. What differentiates a polypeptide from a protein?

19. Acetaminophen contains the amide functional group. Is acetaminophen a peptide? Explain.

20. Name (a) a genetic disease produced by an error in the primary structure of a protein, (b) a condition produced by a marked deficiency of essential amino acids in the diet, and (c) the structures that carry cholesterol through the blood.

A Little Arithmetic and Other Quantitative Puzzles

21. How many *different* amino acids make up the primary structure of the nonapeptides (a) oxytocin and (b) vasopressin?

22. With no restrictions on how many times you may use each, how many *tripeptides* can you generate from *two* different amino acids? Using the abbreviations for alanine and glycine, write the primary structures of all the tripeptides these two amino acids can form.

23. With no restrictions on how many times you may use each, how many *dipeptides* can you generate from *three* different amino acids? Using the abbreviations for lysine, methionine, and phenylalanine, write the primary structures of all the dipeptides these three amino acids could form.

24. Using the abbreviations of Table 17.1, name all the tripeptides that could possibly be formed from alanine, glycine, leucine, and any of their combinations. Each tripeptide may contain one, two, or all three of these amino acids. How many of these tripeptides are there?

Think, Speculate, Reflect, and Ponder

25. After a fried egg cools, why don't the fibrous proteins of the white return to their original clear, colorless, gelatinous form?

26. Write a chemical reaction in which alanine (a) acts as a base; (b) acts as an acid.

27. Most of the proteins that occur in blood serum (the fluid portion of the blood) are globular proteins. Why?

28. Explain the significance of the limiting amino acid in food protein.

29. What is the limiting amino acid in (a) cornmeal, (b) lima beans, (c) peanuts?

30. Of the following foods, which one is the best source of dietary lysine: bananas, carrots, corn, dates, fish, orange juice, potatoes?

31. Is each of the following statements true or false? Explain.

a. The combination of cereal and milk provides complementary protein.

b. The combination of eggs and cornmeal provides complementary protein.

32. Is each of the following statements true or false? Explain.

a. If, without changing any other aspect of your daily living, you eat more fatty foods, you will gain weight because the additional triglycerides of your diet will add adipose tissue to your body.

b. If, without changing any other aspect of your daily living, you eat more foods rich in high-quality protein, you will gain weight because the additional amino acids will add muscle tissue to your body.

33. Table 17.2 shows that peanuts contain about 27% protein. Yet Figure 17.3*a* shows that they contain very little of the essential amino acid methionine. Suppose someone suggested a diet in which peanuts provided for all of your protein needs. According to this diet, you would eat enough peanuts every day so that even with methionine as the limiting amino acid, you would get enough methionine to allow the syntheses of all the needed proteins in your body. How would you respond to this suggested diet? Describe your reasoning.

34. According to Table 17.2, peanuts provide about twice the amount of protein, gram for gram, as eggs. Yet (except for their cholesterol content) eggs are a better source of dietary protein than peanuts. Why?

35. Give the primary structure of the following peptide by using the abbreviations of the amino acids it contains:

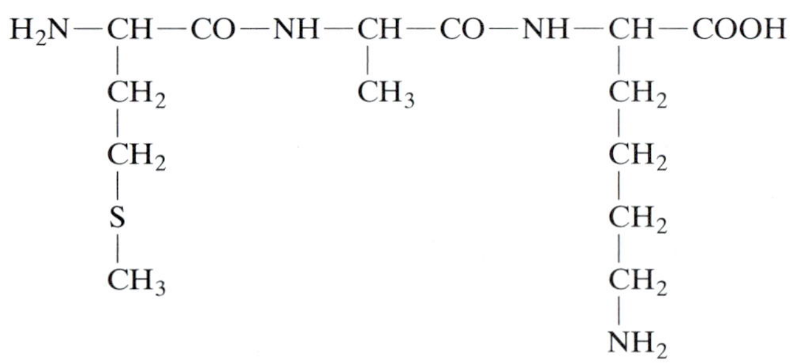

36. Is the compound in Exercise 35 a monopeptide, a dipeptide, a tripeptide, or a tetrapeptide?

37. How many peptide bonds does the compound in Exercise 35 contain?

38. Are all amides also peptides? Are all peptides also amides? Explain.

39. What two chemical elements occur in proteins but not in carbohydrates, fats, or oils?

40. Draw the molecular structure and give the full chemical name of the compound represented by the abbreviation Val–Pro–Gly.

41. Name an amino acid in Table 17.1 that has more than one chiral carbon.

42. Would you expect a zwitterion to be a good electrolyte? Explain.

Web Questions

Use the Web to answer or accomplish each of the following. Report your Web source for each response.

1. Summarize the life and work of Gerardus Mulder. Describe how he discovered the existence of proteins.

2. Many α-amino acids occur naturally, but only one β- amino acid is known to occur in nature. Identify this naturally occurring β-amino acid and describe its occurrence and importance.

3. What is *cystine*? How is cystine related to the amino acids of Table 17.1?

4. Identify *lypressin*. Using amino acid abbreviations, give its chemical structure. Describe its function as a consumer product?

5. What is *myoglobin*? Describe myoglobin's importance to human health and its relationship to hemoglobin.

6. Identify several amides, other than peptides, that are components of widely used consumer products. Describe their uses.

7. Describe several therapies currently being used or studied for the treatment of sickle cell anemia.

8. Summarize some foods and food combinations recommended for people who want to avoid animal products such as meat, milk, and eggs.

The Chemistry of Heredity

MOLECULAR GENETICS

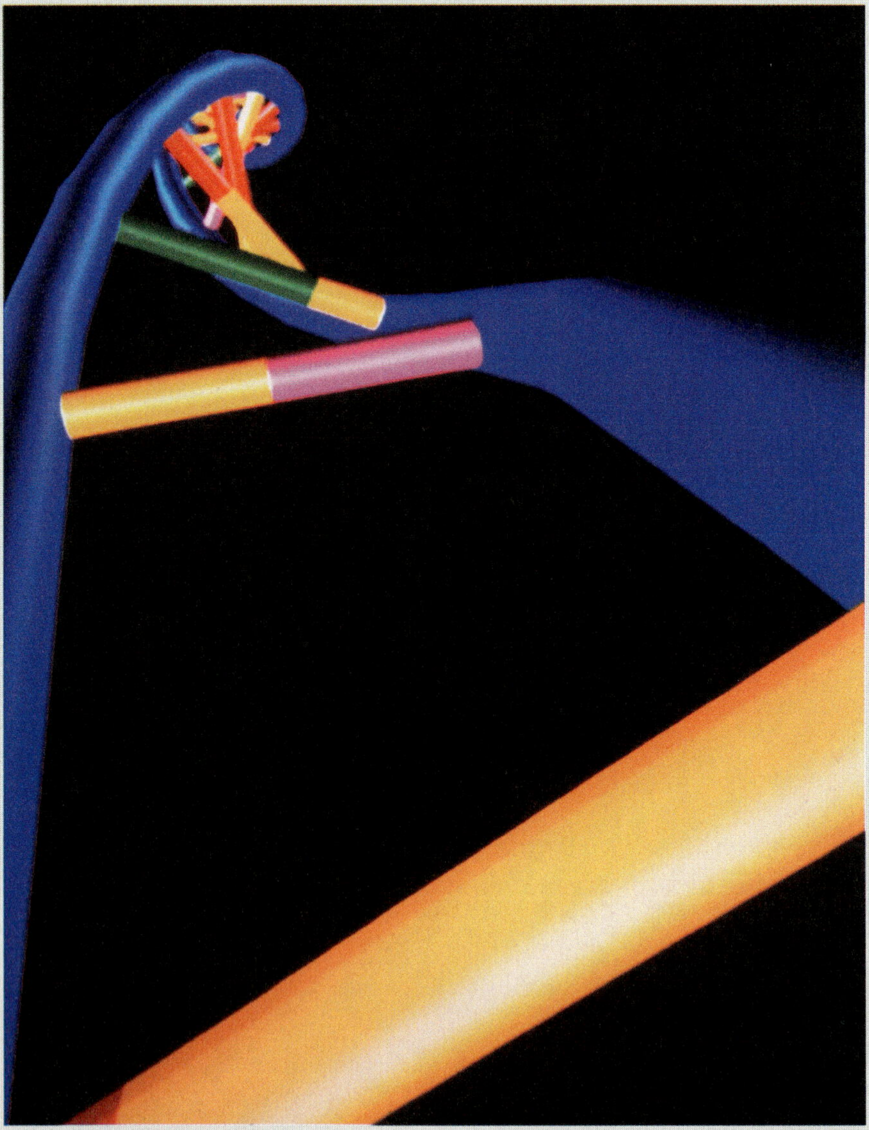

A view along the major axis of a schematic model of the DNA double helix.

Pineapple Gelatin

If you want to make some pineapple gelatin for dessert, not simply pineapple-flavored gelatin but gelatin holding suspended chunks of the fruit itself, you have to use canned pineapple. Fresh pineapple won't work. Here's proof.

Get a package of a simple gelatin dessert mix. Jell-O, Knox, or a similar powdered mix will do. It can be any flavor or no flavor at all, either with sugar or sugar-free. Also pick up a can of pineapple, crushed or in small chunks, and a piece of fresh pineapple as well. In addition you'll need five small bowls or containers, roughly half a cup or about 100 mL each.

Before beginning work on the gelatin, cut or crush both the canned pineapple chunks and the fresh pineapple into very small pieces. But keep the two samples separate from each other. Use different utensils for each set. Don't allow the two pineapple samples to contaminate each other in any way.

Now follow the directions on the gelatin package. Add the specified amount of boiling water, stir until the gelatin powder dissolves, add cold water (and perhaps ice cubes) as directed. Pour equal amounts of the liquid mixture into each of the five containers.

For the scientific analysis, stir a teaspoon of the canned pineapple into one bowl of liquid gelatin and a teaspoon of the fresh pineapple into another. In the same way add a teaspoon of table sugar to the third and a teaspoon of table salt to the fourth. Add nothing at all to the fifth. The four bowls that receive the pineapple, sugar, and salt should get nearly equal amounts of each. Label or mark the containers so you know what each contains. The bowls with the sugar, the salt, and with no additive at all represent applications of the scientific method of Chapter 1.

Examine the bowls after you have refrigerated them as directed or overnight. You're likely to find that all have gelled nicely except the one containing the fresh pineapple. Success with the bowl containing no additive tells us that an additive, in itself, isn't needed to produce a gel—sometimes the scientific method serves only to confirm our preconceptions, but even that can be valuable —while those containing added sugar and salt prove that neither electrolytes nor nonelectrolytes (Sec. 1.1) inhibit gel formation. The difference comes down to whether we used canned or fresh pineapple. The mixture in the bowl containing canned pineapple does form a gel, but we find only a cold, watery pineapple soup in the bowl with the fresh fruit (Fig. 18.1). The explanation for this difference involves both the chemistry of proteins and the biology of survival.

Figure 18.1 (1) Preparing the gelatin. (2) Adding canned pineapple to the liquid gelatin. (3) Adding fresh pineapple to the liquid gelatin. (4) The gelatin mixture does not solidify in the presence of fresh pineapple.

18.1 Pineapple's Enzymes vs. Gelatin's Proteins

Examining the effect of pineapple on the gelatin begins with the protein chemistry of Chapter 17. Gelatin is a protein derived from *collagen*, a component of connective tissue found in the bodies of animals. Meat processors remove collagen from the hooves, skins, bones, and other parts of slaughtered animals and convert the raw collagen to the purified form of collagen protein we know as gelatin. The powder we pour out of a box of commercial gelatin consists of a mixture of fibrous protein filaments associated with each other in the form of triple helices.

[A *helix* is a three-dimensional structure resembling the thread on a screw. A double helix forms when a pair of parallel strands assume the shape of a helix. The triple helix of gelatin proteins forms from a set of three parallel strands (Fig. 18.2). We'll examine the double helix of DNA in Sec. 18.10.]

These helical proteins separate from each other and become disorganized, individual protein strands when they dissolve in hot water and are denatured by the high temperature. Among other changes, they lose their helical structures. The hot mixture of gelatin and water is essentially a dispersion of individual protein molecules in water. As the water cools, the protein molecules reassociate with one another, trapping water molecules among themselves as they do. Unlike the *irreversible* denaturation of the native proteins of egg white as they are heated when we fry an egg, the denaturation of gelatin's proteins by hot water is *reversible*. Cooling restores their original characteristics.

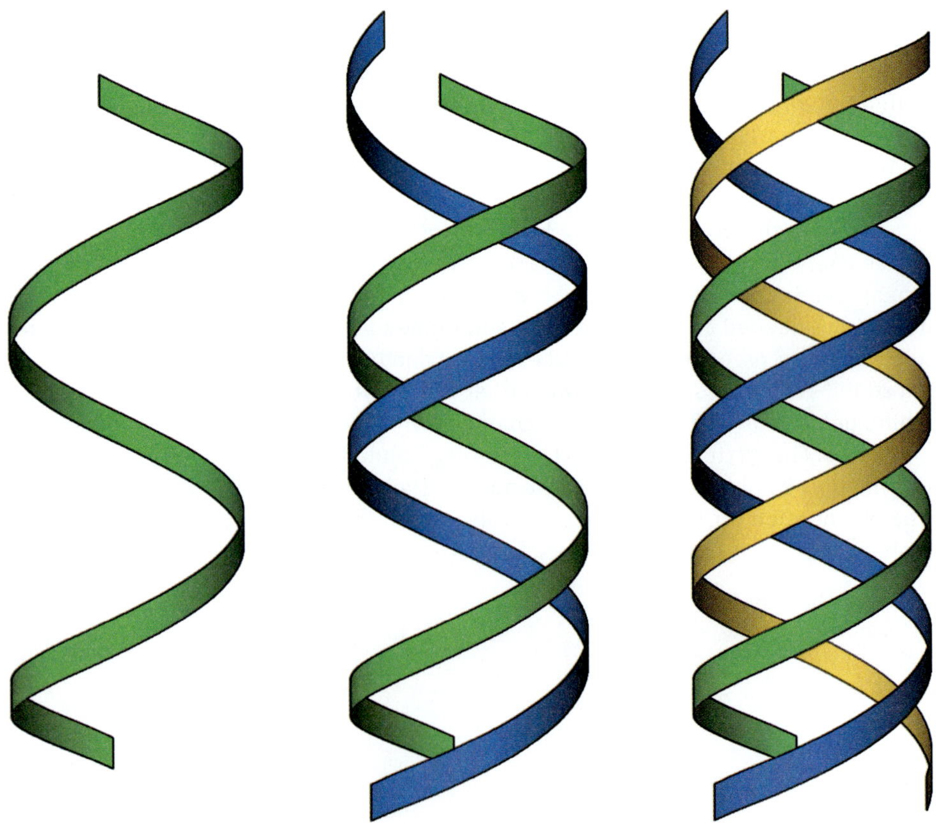

Figure 18.2 The geometry of a single, double (DNA), and triple (gelatin) helix.

The cold gel is essentially a network of water molecules trapped within a framework of associated, organized, helical protein molecules. This cold, semisolid gelatinous network holds fruit, nuts, and other small pieces of food nicely—including even the canned bits of pineapple—as the protein molecules and trapped water molecules wrap around the suspended food particles and suspend them in the semisolid protein–water network. This gives us our pineapple gelatin dessert when we use the canned fruit.

But as we've seen, something very different happens if we use fresh pineapple instead. Here the failure of gel formation reflects the ability of the fresh pineapple to destroy the protein network. The culprit here is the *bromelin* of the fresh pineapple.

Many plants, including pineapple, produce within their leaves, stems, fruit and other structures various substances that serve as poisons, toxins, irritants, or other kinds of chemical deterrents to attack by predators. Among these is the bromelin of the pineapple plant. Bromelin is a set of closely related *proteolytic enzymes*, biological catalysts that promote the degradation, cleavage, or destruction of proteins. Both the sap and the fruit of the pineapple plant contain these enzymes, which effectively inactivate or destroy many of the proteins they contact. They have been used for a variety of purposes, including tenderizing meat (by destroying the protein of the collagen and other connective tissue that strengthen the physical structure of the meat) and reducing some forms of pain and swelling (by removing or inhibiting the formation of proteins that produce these effects). They can also degrade structural and enzymatic proteins of insect predators, thus providing some protection against attack by insects. The ability to produce bromelin enzymes, then, can be viewed as a survival trait of the pineapple plant.

Extended exposure to the high temperatures of the canning process denatures and deactivates the bromelin enzymes of the canned pineapple. Under these conditions the bromelin loses its ability to decompose the helical proteins of the gelatin. The gelatin proteins, then, survive in the environment of the canned pineapple but are destroyed by the active enzymes of the fresh pineapple.

■ QUESTION

Classify as reversible or irreversible the effects of (a) hot water on the proteins of gelatin, (b) the high temperature canning process on the proteins of bromelin, and (c) the high temperatures of frying on the proteins of egg white.

18.2 Biochemistry

The proteins of gelatin and bromelin are chemicals produced by living organisms. Although—given enough time, chemicals, and other resources—we could replicate their molecular structures in the laboratory, their actual generation through the chemical processes of life is an illustration of **biochemistry**, the chemical reactions that take place within a living organism. As we saw in Section 6.1, scientists at one time believed that compounds generated by life processes could be produced only through the operation of a "vital force," a mysterious power that (it was thought) operates in living organisms but is unavailable to scientists working with the apparatus of a chemical laboratory. Friedrich Wöhler's chemical synthesis of urea helped dispel this idea.

Biochemistry is the branch of science devoted to the study of the chemical reactions taking place within a living organism.

Nonetheless, a complete understanding of how biochemical reactions produce the extraordinarily complex molecules of the organisms that generate them presents science with one of its greatest challenges. In this chapter we'll examine a closely related question: How do plants and animals transmit their own ability to create these chemicals, such as the bromelin of pineapples and the collagen of animals, from generation to generation? We'll examine this and related questions as we consider the biochemistry of heredity.

■ QUESTION

Identify a polypeptide produced by the human body that helps ensure the effective nourishment of newly born children. (See Sec. 17.9.)

18.3 Apples and Oranges; Pineapples and Potatoes; Cows and Humans

As the common saying points out, "You can't mix apples and oranges." They're clearly different fruits. But beyond their obvious differences—appearance, taste, odor, texture, and so on—lie more fundamental, chemical differences. Oranges, for example, contain far more vitamin C than apples; pineapples contain far more bromelin than either apples or oranges. These represent only two chemical illustrations of the often complex and extensive chemical differences that exist among all species of plants and animals.

We have already examined a few of these differences. In Chapter 16 we examined properties of apples and potatoes that stem from differences in the structures of their molecules. In the opening demonstration we saw that the color of iodine on potatoes is much deeper and darker than on apples, and we found in Section 16.15 that this difference in color originates in differences in the way apples and potatoes join the glucose units of their carbohydrate molecules. Apples connect theirs through β links, potatoes through α links. We saw that this fundamental chemical difference also explains why humans can use the starch of a potato as an energy-producing nutrient, while the cellulose of an apple serves as an indigestible carbohydrate.

These chemical differences and their consequences arise from the biochemical processes that connect glucose units one way in apples, another way in potatoes. We've seen this same kind of chemistry at work in animals as well. Through its biochemical operations, the human body generates the maltase needed for digestion of starch but not the cellobiase necessary for the digestion of cellulose. On the other hand, cellobiase-generating biochemical reactions do take place within the bodies of cows, sheep, goats, and similar animals. As a result, these animals feed successfully on grass and other cellulose-containing plants.

Oranges produce vitamin C, apples produce cellulose, potatoes produce starch, humans produce maltase, and cows produce cellobiase, all because of the genetic programming of their biochemical reactions. The chemical behavior of each living organism is a function of the organism itself and helps define the organism. And the origin of that chemical behavior lies in the progenitors of each plant and animal. It comes to them through the process of heredity.

■ QUESTION

Name another chemical, in addition to vitamin C, that is found in much higher concentrations in oranges (and lemons, as well as other citrus fruit) than in apples. (See Chapter 10.)

18.4 Gregor Mendel

Observations that offspring of plants and of animals resemble their parents go back to the beginnings of civilizations. They formed the basis for agricultural and farming methods that sought to reproduce the best characteristics of domesticated plants and animals by encouraging the reproduction of the most desirable specimens, and they supported royal lines of succession in which the descendants of great rulers were themselves believed destined to be great rulers as well.

The fact of inherited traits was well established by observation, but the mechanism of inheritance was unknown. Speculation gave rise to a variety of theories, each deficient and unproductive, until the time of Gregor Mendel (Fig. 18.3). Mendel (1822–1884), an Austrian monk, studied the effects of cross-breeding on the traits of garden peas and other plants. He carefully recorded and analyzed connections between the characteristics of parents and their offspring. Although he published his observations and conclusions in 1866, his writings remained dormant and unknown to the rest of the scientific world until they were discovered independently by three European botanists in 1900, 16 years after Mendel's death.

Figure 18.3 Gregor Mendel laid the foundation for our understanding of how individual characteristics are passed from parents to offspring.

Mendel's writings revealed an accurately and brilliantly developed mathematical model for the transmission of individual characteristics from an organism to its offspring. He concluded that an observable characteristic in offspring, a *phenotype*, must have its origin in some obscure genetic feature or structure carried by a parent, a *genotype*. Although Mendel had no idea what this genotype might be and could not explain how it might be transmitted to the offspring, his writings opened the path to our current understanding of the molecular basis of heredity.

Briefly, we now recognize that all organisms—both plants and animals—resemble their parents because each generation transfers to its offspring its own set of biochemical information. That is, all preceding generations transfer to succeeding generations the biochemical instructions needed for looking and acting—in the broad, biological sense—very much like themselves. We are humans because we received from our parents all the biological information necessary to make us humans. And the details are carried from one generation to the next by a molecular messenger, a courier of genetic information inscribed in molecular structure.

■ QUESTION

Is red hair classified as a phenotype or a genotype? What term do we give to the carrier of the genetic information that transmits the characteristic of red hair from parent to child?

18.5 Mitosis and Meiosis: Information Transfer Through Cell Division

The cell is the fundamental structural unit of all living things. With a few exceptions, such as red blood cells (Sec. 17.10) and single-celled bacteria, each cell consists of a well-defined nucleus embedded in cytoplasm, and a surrounding membrane (Fig. 18.4). The membrane encloses and protects the cellular material. Structures within the cytoplasm do most of the work of the cell, including photosynthesis (in plants) and the conversion of nutrient chemicals into biological energy (in animals). Our interest lies in the nucleus, which contains the organism's genetic information.

Within the nucleus are the **chromosomes**, long strands of material that (as we'll see) serve as the carriers of this information. The term *chromosome* reflects the ability of these bodies to accept biological dyes and show distinctive colors.

Chromosomes are structures within a cell's nucleus that carry the organism's genetic information.

Figure 18.4 A typical animal cell.

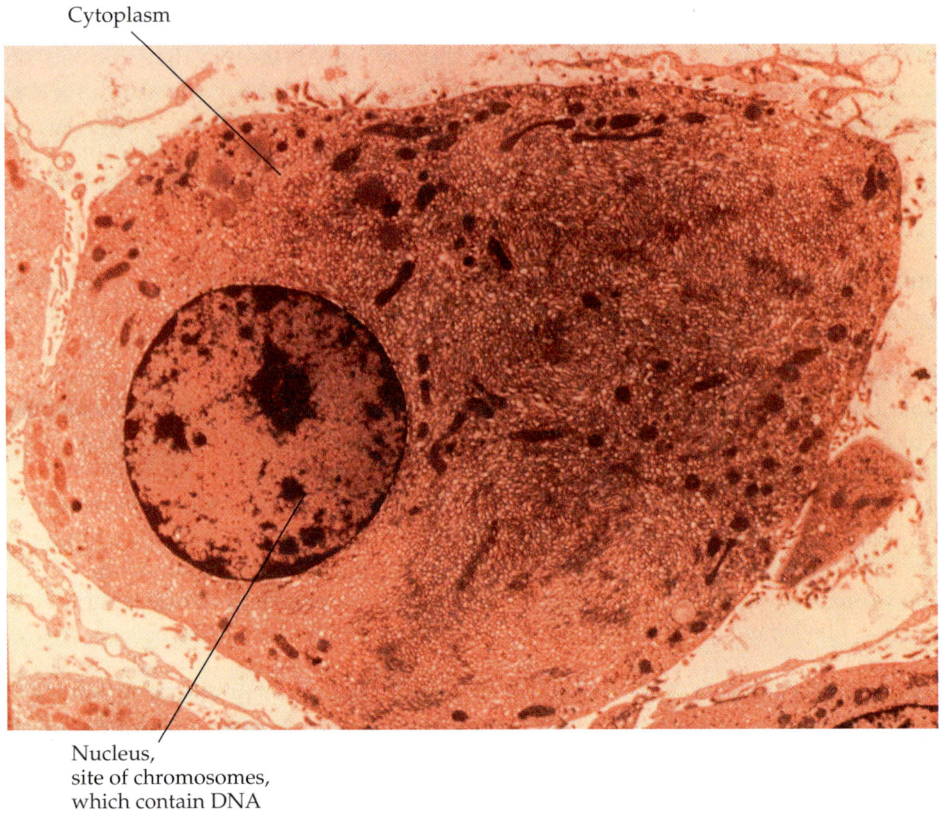

As long as the cell is not in the process of dividing, the material of the chromosomes lies diffused within the nucleus in threadlike filaments. But as the cell prepares to divide, these filaments coalesce into more highly structured coils that twist about themselves to produce the forms of Figure 18.5. All the cells of each organism of any given species hold the same number of chromosomes, but this number varies

Figure 18.5 Chromosomes and DNA (Sec. 18.6).

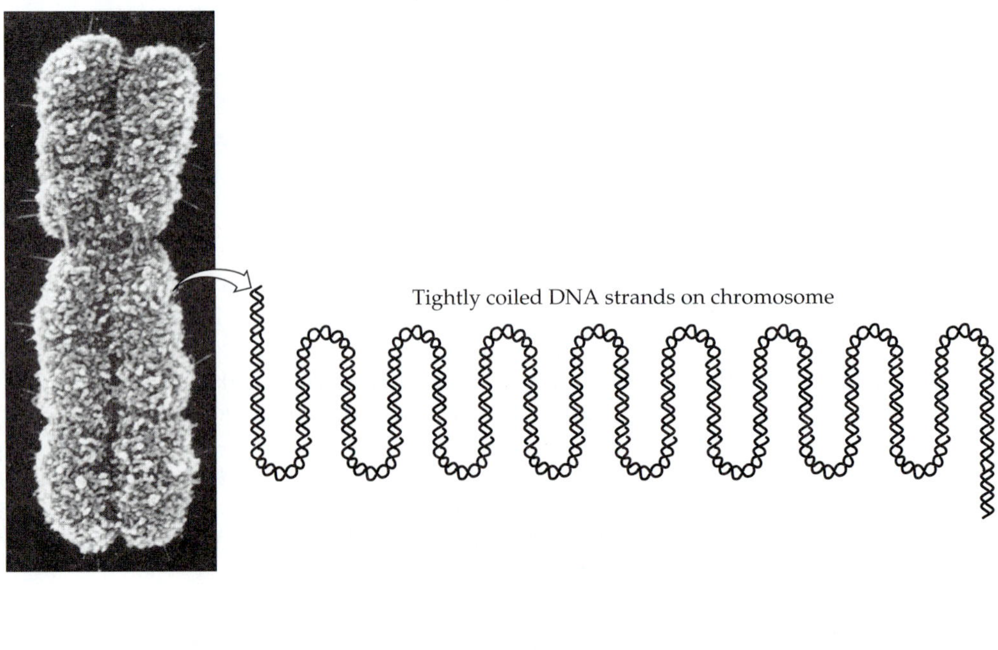

from one species to another. Each cell of a fruit fly holds 8; for horses, 64; for both potatoes and red ants, 48; dogs, 78; sugar cane, 80; carp, 104. Each human cell contains 46. In most species, including humans but excluding the cells involved in reproduction, the chromosomes form pairs. Human cells, then, except for the single-celled sperm and eggs, contain 23 pairs of chromosomes.

As the cells of the body divide, the chromosomes divide as well, with their genetic information transferred equally to the newly produced cells. In *mitosis*, a cell divides to produce two identical copies of itself. The chromosomes of the cell duplicate themselves, with one copy going into each of the new cells (Fig. 18.6). Where we previously

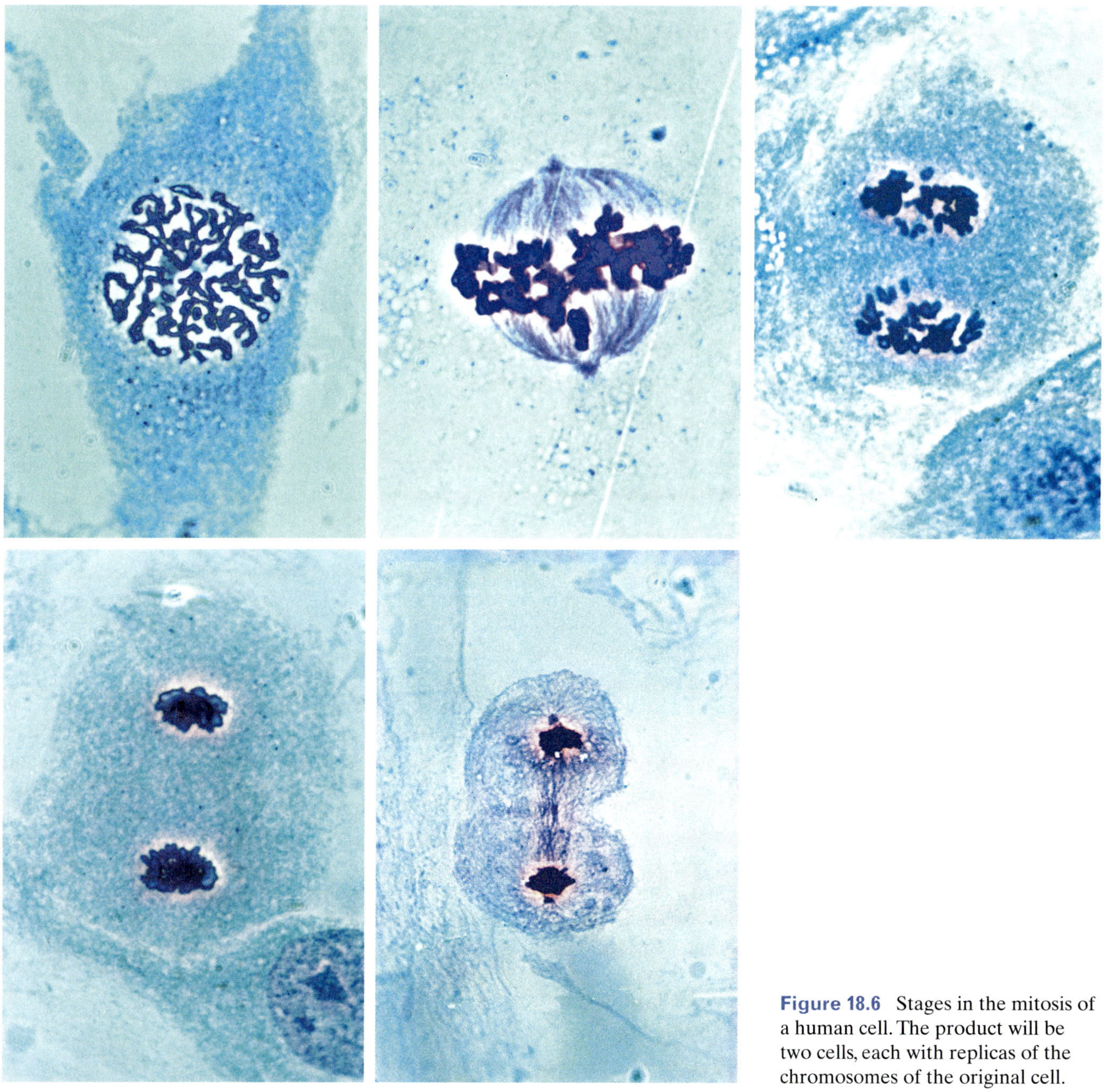

Figure 18.6 Stages in the mitosis of a human cell. The product will be two cells, each with replicas of the chromosomes of the original cell.

had a single cell, we now have two cells, identical to each other and to the original cell. Since the chromosomes of the original cell have simply replicated themselves in the formation of the new cells, each new cell holds the same genetic information as the original cell. This form of cellular reproduction is sometimes known as asexual reproduction. It's the mode by which most body cells, those of the skin, for example, divide. In effect, cellular division by mitosis simply leads to more of the same.

Meiosis, the kind of division important in sexual reproduction, represents a more complex process. Here half the chromosomes (one from each pair) move to one of the new cells, half (the other member of each pair) go to the other. This kind of division occurs in the formation of sperm and egg cells. To produce a cell containing the full complement of chromosomes, 46 in the case of humans, two of these cells—one sperm cell and one egg, each containing half the required number of chromosomes—must unite. In human reproduction this produces again the characteristic 23 pairs of chromosomes. Half the new set of 46 chromosomes (and half the genetic information) comes from one parent, half from the other. Unlike mitosis, meiosis, with its resultant mixture of the genetic properties of two parents, produces something new rather than more of the same. Meiosis produces a new combination of chromosomes and a new set of genetic information, with half of each coming from each parent.

■ QUESTION

Based on the limited data of this section, what correlation, if any, appears to exist between the number of chromosomes in the cells of mammals and nonmammals? Of plants and animals?

18.6 The Gene

Figure 18.7 An early (1953) model of a portion of a DNA molecule, designed and built by James Watson, left, and Francis Crick, right (Sec. 18.10).

Chemically, the structures characterized as chromosomes consist of long, tightly coiled molecules known as *DNA*, an abbreviation for *deoxyribonucleic acid*, an acid found almost exclusively in the nuclei of our cells (Fig. 18.7). This combination of biological location (the nucleus) and chemical characteristic (acid) accounts for the *-nucleic acid* portion of the term. We'll examine the significance of the first part of the name, *deoxyribo-*, in the next section. These DNA molecules carry all the genetic information of every living organism. Moreover, the DNA in virtually every cell of any individual organism is identical with the DNA in every other cell of that same organism. With very few exceptions, then, each cell of every organism carries within its DNA the complete genetic blueprint of the entire organism.

An organism's genetic information resides in the DNA molecule, within sequences of *genes* strung out along the length of the DNA. Each gene coincides with a specific, short section of DNA that (generally) determines the amino acid sequence of a specific polypeptide. Each gene represents a chemical blueprint for the formation of the globular and fibrous polypeptides and proteins that serve as enzymes, hormones, and structural and regulatory proteins that determine the form, appearance, and functions of an organism. They hold (and transmit from generation to generation) the directions for the formation of the bromelin, collagen, insulin, oxytocin, vasopressin, hemoglobin, and all other polypeptide structures that define the organism. In summary, **gene** is the term used to describe a segment of a chromosome that carries the information needed to create a specific protein. With this, genes hold the keys to the biochemistry of life.

A **gene** is a segment of a chromosome that carries information necessary for the creation of a specific protein or polypeptide.

■ QUESTION

What is the full chemical term represented by the letters of DNA?

18.7 DNA: The Storehouse of Genetic Information

The deoxyribonucleic acid molecule that coils into a chromosome consists of three chemical segments:

- a series of four cyclic amine bases (Sec. 17.2), *adenine*, *cytosine*, *guanine*, and *thymine*, that are bonded to a backbone consisting of
- *2-deoxyribose* and
- *phosphoric acid*, H_3PO_4.

In its chemical structure, 2-deoxyribose resembles the cyclic form of the aldopentose ribose (Exercise "Sweet Heredity," Sec. 16.4). The *2-deoxy-* portion indicates that it's identical with ribose except for the absence of an oxygen at the number 2 carbon. The structures of β-ribose, β-2-deoxyribose, and phosphoric acid appear in Figure 18.8; adenine, cytosine, guanine, and thymine are shown in Figure 18.9.

Combining 2-deoxyribose, phosphoric acid, and any one of the four organic bases of Figure 18.9 produces a *nucleotide* (Fig. 18.10), the repeating structural unit of DNA. A long sequence of one nucleotide unit after another, each bearing one of the bases of Figure 18.9, produces the primary structure of a DNA molecule. Figure 18.11 represents a typical segment of the primary structure of DNA. To sum up, the DNA chain is simply a set of alternating 2-deoxyribose and phosphate units, with four amines—adenine, cytosine, guanine, and thymine—strung out as substituents along the chain.

The *sequence* of these amine bases determines the sequence of the amino acids that make up the primary structures of the body's proteins. That is, the primary structure of each of the body's proteins is determined by the genetic information passed from generation to generation through the sequence of bases on the DNA molecule. A gene is simply a portion of the DNA chain with a large number of bases arranged in the right order to direct the formation of a specific polypeptide.

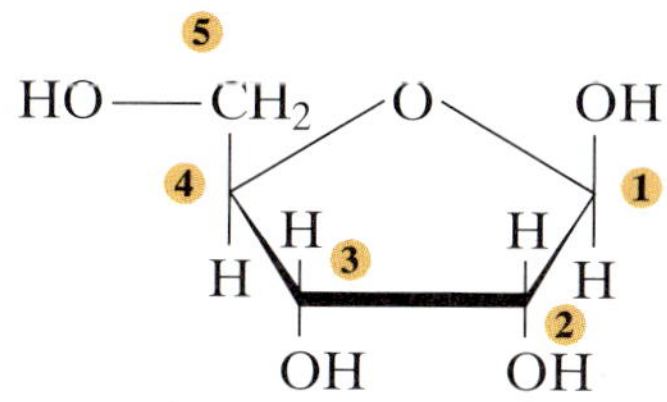

β-Ribose

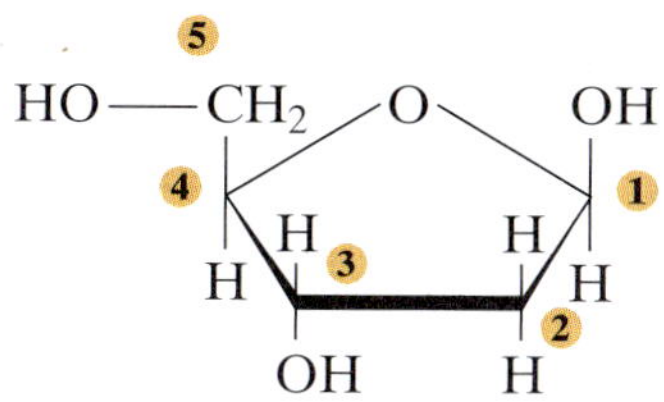

β-2-Deoxyribose

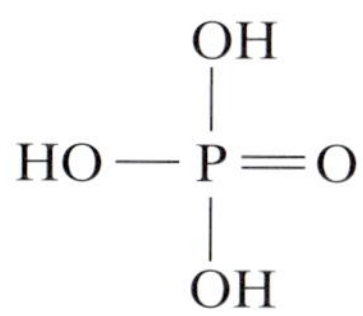

Phosphoric acid

Figure 18.8 Ribose, 2-deoxyribose, and phosphoric acid.

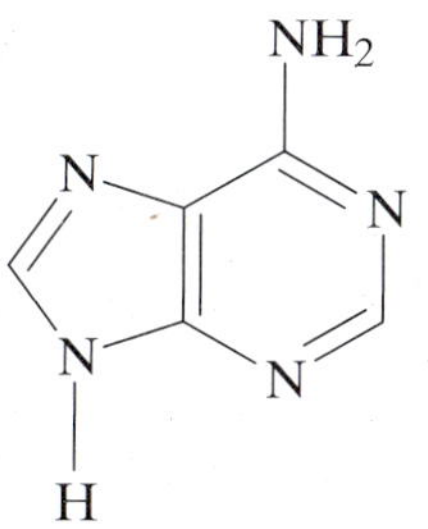

Adenine

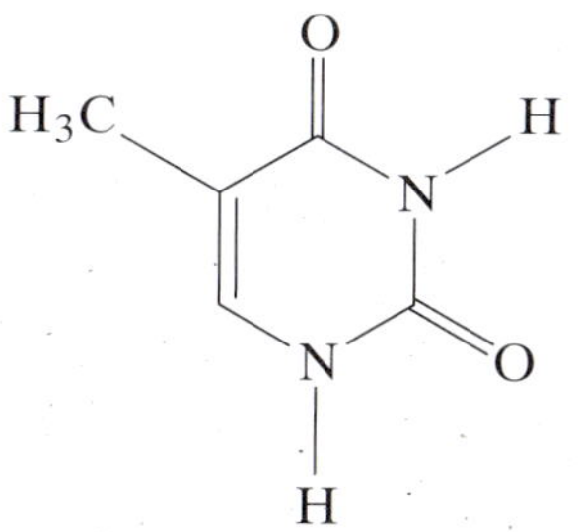

Thymine

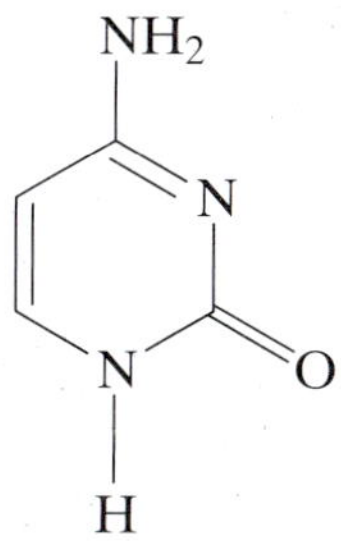

Cytosine

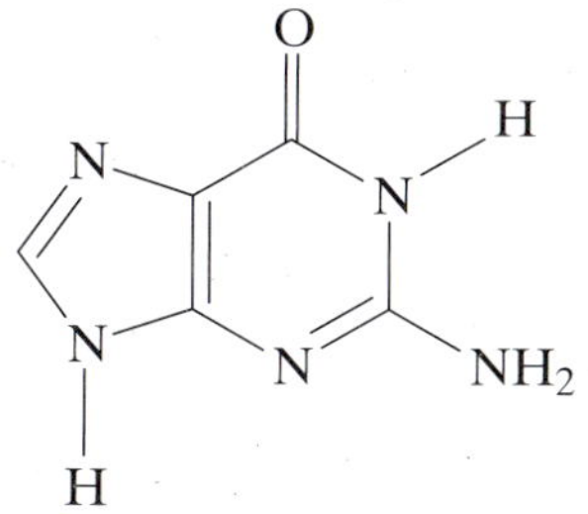

Guanine

Figure 18.9 The amine bases of DNA.

Figure 18.10 The combination of 2-deoxyribose, cytosine, and phosphoric acid produces a typical nucleotide.

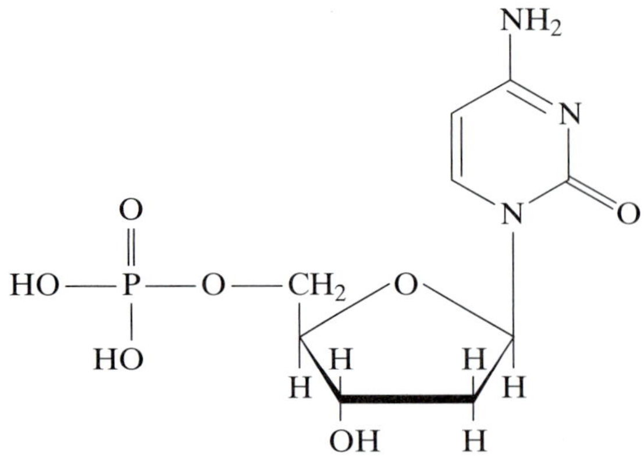

Figure 18.11 A typical segment of the DNA chain.

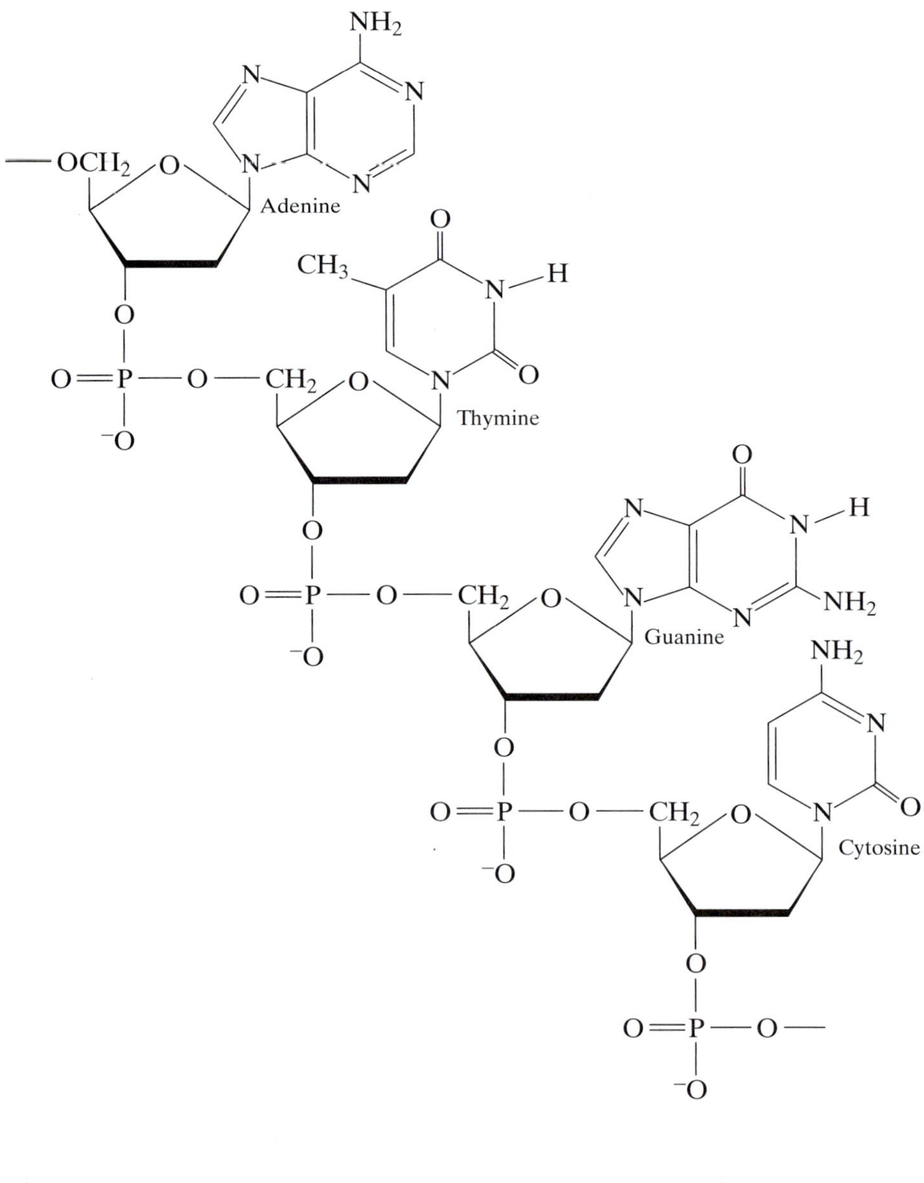

How would you define a gene in biological terms? In chemical terms?

■ QUESTION

18.8 RNA: Transforming Genetic Information into Biochemical Action

Protein synthesis itself takes place outside the nucleus, in the cytoplasm (Sec. 18.5). As a result, the information contained in the DNA molecule, which remains confined to the nucleus, must be transferred out of the nucleus into the cytoplasm.

The transfer begins within the nucleus as the information contained on the DNA molecule is transcribed, in portions, onto another kind of molecule, *RNA* (an abbreviation for *ribonucleic acid*). RNA, formed by biochemical reactions within the nucleus, differs from DNA in four important ways:

- RNA molecules can pass from the nucleus into the cytoplasm.
- *Ribose* (rather than 2-deoxyribose) forms the carbohydrate ring of the RNA chain.
- In RNA the amine *uracil* replaces the thymine of DNA (Fig. 18.12).
- RNA molecules are much shorter than DNA; each RNA carries only a small segment of the genetic information stored on the entire strand of DNA.

RNA occurs in several different forms, each with a specific function. One form, called *mRNA* (for *messenger RNA*), carries information stored in DNA (the message) to the cytoplasm, where the protein synthesis occurs. For the protein synthesis itself, still another kind of RNA, *tRNA* (*transfer RNA*), finds and transports (transfers) each required amino acid to the site where the peptide bonds are formed under the direction of the mRNA. This peptide synthesis takes place within the cytoplasm, in bodies known as *ribosomes*. With biochemical efficiency, one mRNA molecule can simultaneously guide polypeptide synthesis in several different, associated ribosomes.

To summarize the entire process, genetic information is stored as a sequence of four different amine bases in molecules of DNA, which lie within the nucleus of the cell. The information is transcribed, in portions, into a sequence of bases of newly formed mRNA molecules. Three of these bases on the mRNA are identical with those on DNA; one is different. This mRNA leaves the nucleus for the ribosomes of the cytoplasm and there serves as a guide for joining individual amino acids, brought to it by tRNA, into useful polypeptides.

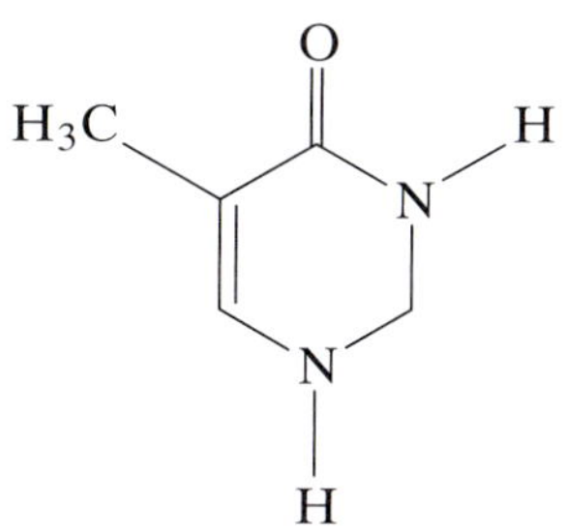

Thymine
occurs in DNA

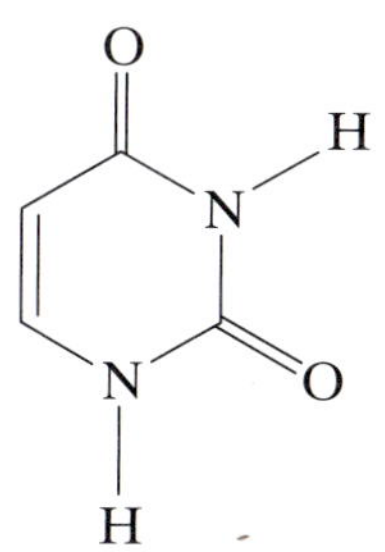

Uracil
occurs in RNA

Figure 18.12 Thymine and uracil.

In what way(s) are mRNA and tRNA similar? How do they differ?

■ QUESTION

18.9 The Genetic Code

The sequence of the amines strung out along the RNA molecule determines the primary structure of the polypeptide it forms. Each set of *three* sequential amines on RNA either identifies one particular amino acid that fits into the polypeptide chain or acts as signal for the chain-forming process itself, such as "start" or "stop". To reflect this coding function, sets of three successive amine bases are called *codons*. The correspondence between any particular sequence of three RNA bases and the specific amino acid or function it represents is known as the *genetic code*. Table 18.1 presents the correlation between RNA codons and amino acids or synthetic functions.

Table 18.1 Codons of mRNA

Amino Acid	Codon[a]	Amino Acid	Codon[a]
Ala	GCA GCC GCG GCU	Lys	AAA AAG
Arg	AGA AGG CGA CGC CGG CGU	Met	AUG
Asn	AAC AAU	Phe	UUU UUC
Asp	GAC GAU	Pro	CCA CCC CCG CCU
Cys	UGC UGU	Ser	AGC AGU UCA UCC UCG UCU
Gln	CAA CAG	Thr	ACA ACC ACG ACU
Glu	GAA GAG	Trp	UGG
Gly	GGA GGC GGG GGU	Tyr	UAC UAU
His	CAC CAU	Val	GUA GUG GUC GUU
Ile	AUA AUC AUU	Initiate peptide chain	AUG GUG
Leu	CUA CUC CUG CUU UUA UUG	Terminate peptide chain	UAA UAG UGA

[a] A = adenine; C = cytosine; G = guanine; U = uracil.

With *four* different bases arranged in codons of *three* bases each, a total of 64 different sequences (codons) is available. This allows for plenty of redundancy. For example, with adenine, cytosine, guanine, and uracil represented by A, C, G, and U, respectively, each of the following four codons represents the amino acid alanine: GCA, GCC GCG, and GCU. The signal to terminate the protein chain is given by each of the codons UAA, UAG, and UGA.

■ QUESTION

What amino acid does the codon GGC correspond to?

18.10 The Double Helix of Watson and Crick (and Wilkins and Franklin)

As we saw for proteins in Chapter 17, the primary structures of long molecular chains—the sequence of the individual units that form them—tell only part of the chemical story. And as with proteins, the shape taken by the primary structure of DNA influences its behavior and biochemical effects.

The discovery in 1953 that DNA exists as a *double helix* led to rapid advances in understanding the chemical basis of biological heredity. Working in Cambridge, England, the biologists James D. Watson (born 1928) and Francis Crick (born 1916), American and English respectively, elucidated the structure of DNA largely through the construction of models based on X-ray studies carried out in London by Rosalind E. Franklin, an English biophysicist (1920–1958). In 1962 Watson and Crick shared the Nobel Prize in medicine with Maurice H. E Wilkins (born 1916, New Zealand), who had been a coworker of Franklin. Because Rosalind Franklin died before the award of the Nobel Prize, her name does not appear on it. Nobel Prizes may not be awarded posthumously.

Watson and Crick found that the DNA molecule consists of two nucleotide strands held together in a double helix by hydrogen bonds (Sec. 17.10) formed between the amine bases of the adjacent strands. Although the hydrogen bond is a weak bond, the huge numbers of these bonds that can form between the two strands hold them together effectively in the double helix (Fig. 18.13).

The two deoxyribonucleic acid strands that form the double helix are not identical. They are *complementary* strands, which means that *every cytosine of one strand lies opposite a guanine of the other strand; every adenine of one strand lies opposite a thymine of the other strand.* The hydrogen bonds that hold the strands together form between these base pairs, which occupy the central portion or core of the helix, as shown in Figure 18.13. An estimated 6 billion of these base pairs exist in the DNA of each human cell. (In contrast, RNA consists of only a single strand of nucleotides, with genetic information copied selectively from only one of the two DNA strands.)

When a cell divides, replicas of its DNA form as the double helix uncoils and a complementary strand forms adjacent to each one of the original strands (Fig. 18.14). Each of the two new cells receives one of the newly formed double helixes. This accounts for the identity of the DNA in each cell of an organism.

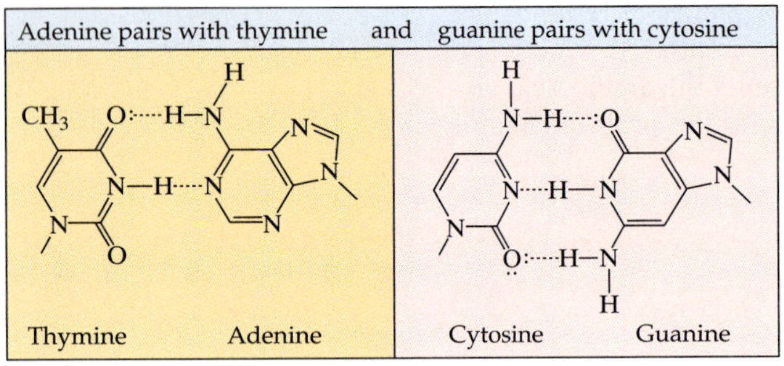

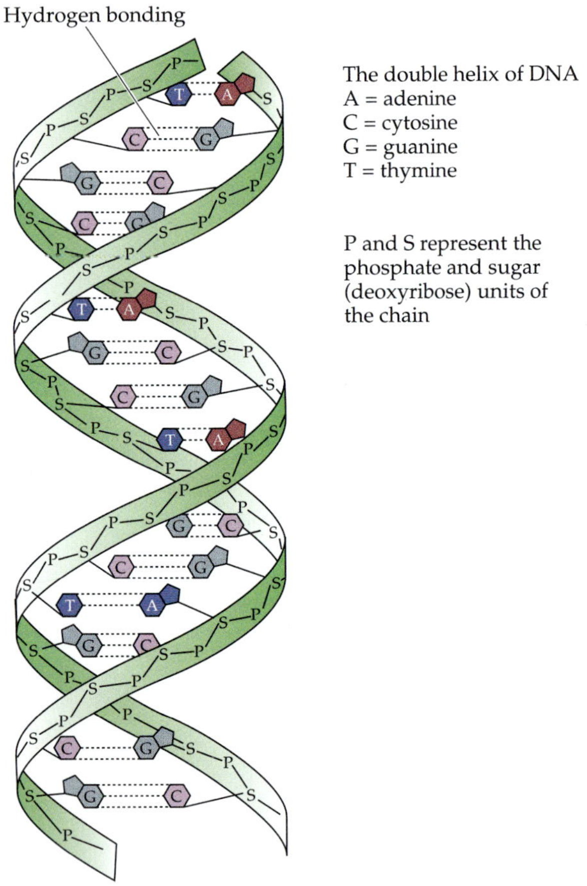

Figure 18.13 Hydrogen bonding and the double helix of DNA.

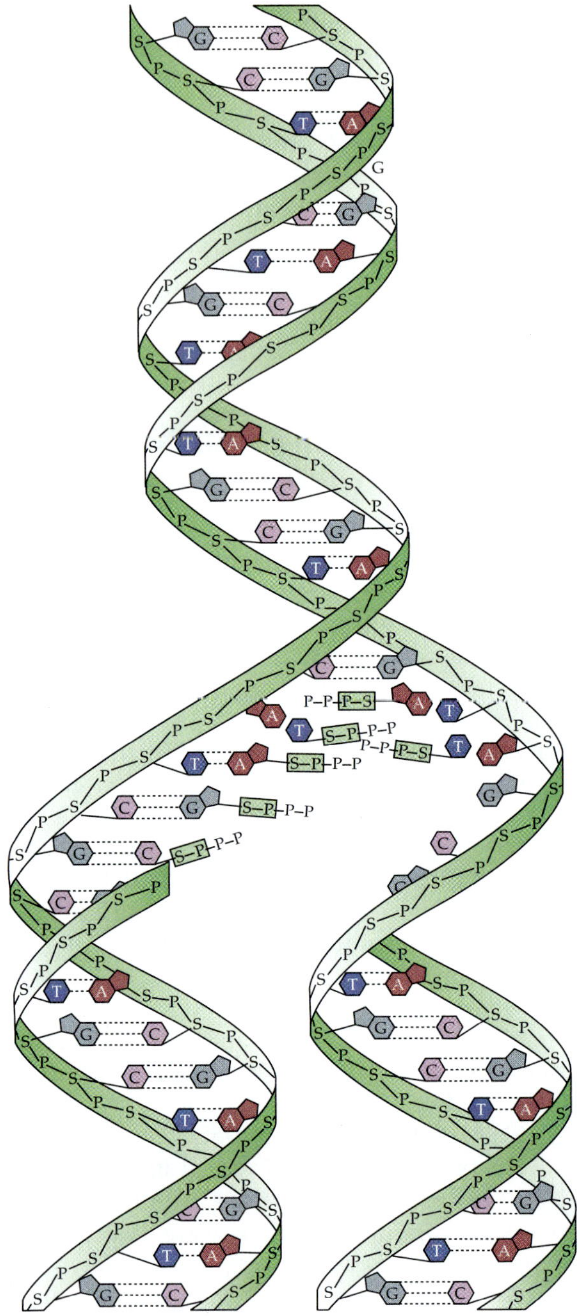

Figure 18.14 DNA replication.

■ **QUESTION**

Why are the two DNA strands of a double helix regarded as complementary rather than identical?

18.11 The Human Genome

Given the importance of our genetic information—it's all that distinguishes you from a yeast cell or an earthworm—an enormous amount of time and effort have gone into determining the precise sequence of genes and their bases in our DNA.

This sequence for any particular species or individual is a **genome**. The sequence responsible for our human characteristics represents the *human genome*. In February 2001, initial results of the first analysis of the human genome were made public, with additional detailed results expected to follow in a few years.

A **genome** is the complete set of genes responsible for the hereditary characteristics of an organism or a species.

Current research indicates that the human genome consists of somewhere between 32,000 and 70,000 genes. The smaller number is based on computer analyses of the data; the larger is suggested by laboratory work. The total number of genes in human chromosomes ranges from 231 in the smallest to 2968 in the largest. It's estimated that the entire human genome contains more than 31 million amine bases, with about 3000 in the average gene. The largest gene, the one responsible for the protein *dystrophin*, contains 2.4 million bases. (Dystrophin is required for healthy muscle tissue. Defects in dystrophin formation are related to the condition known as *muscular dystrophy*.)

Of all the genes that have been found strung out on the DNA molecules, the roles of more than half still aren't understood. In fact, large portions of the DNA molecules themselves, over 95%, have repeated sequences of bases that seem to serve no function at all. These apparently useless segments are known as *junk DNA*. Some of this junk DNA lies between genes, some takes the form of useless sequences within genes. The code-carrying segments of DNA are known as *exons*. These are the portions that are transcribed into the mRNA that guides protein synthesis in the cytoplasm. The exons contain the information that is expressed as proteins. Between the exons lie the *introns*, the intervening segments that hold no (known) code and that serve no (known) function in protein synthesis.

Although each of us is an individual, different in substantial ways from anyone else now alive or who has ever lived, we are all members of the human race. We all share the characteristics of all the biological organisms known as humans. We might wonder, then, how much of our DNA belongs to us alone, making us the individuals we are, and how much we share with all other members of our species. The answer is that of the sequence of bases in our DNA, about 99.9% is shared by all other humans. It takes only your own, unique one-tenth of one percent of your DNA to make you the individual you are.

■ QUESTION

What is the difference between a gene and a genome?

18.12 Cloning

Reproduction in humans and other higher animals begins when a sperm cell, with its set of the male organism's DNA (Sec. 18.5), enters an egg cell, with its set of the female's DNA, and the two sets of DNA combine to form the full complement of DNA that characterizes the offspring. As the fertilized egg begins dividing, each of the newly formed cells of the developing embryo carries this same complete set of DNA. As a result, virtually every cell of the new organism has this same complete set of DNA that was created in the newly fertilized egg and that characterizes the organism (Sec. 18.6).

It's possible to carry out this reproductive process with a significant twist. In *cloning*, the nucleus of an egg cell, with its half set of chromosomes, is removed and replaced by a nucleus extracted from a body cell of another adult animal, a nucleus with a full set of the adult organism's chromosomes. The egg now contains a nucleus with a full complement of chromosomes, all of which came from the adult, donor organism. Genetically, the egg is now the equivalent of a freshly fertilized

Dolly, the first successful clone of an adult mammal.

egg, but with all of its genetic information coming from a single existing organism rather than from two parents.

The egg is now induced to begin dividing. When it reaches an embryonic stage it is implanted into the reproductive system of a female, a surrogate mother, where it develops into a fetus and in time is born as a new infant, genetically identical to the adult whose DNA was inserted into the egg. The adult individual that served as the source of the DNA has been cloned.

Cloning is a tricky process, with much that can go wrong. It has a high failure rate. The first report of the cloning of an adult mammal came in February 1997, when Ian Wilmut of the Roslin Institute, Edinburgh, Scotland, announced that he and his colleague, Keith Campell, had successfully cloned a sheep, producing a lamb they named Dolly. The cloning had been accomplished the previous year, with Dolly born on July 5, 1996. The success came after 277 previous failures by the same team of scientists.

Following Dolly, cows, goats, mice, pigs, and sheep have also been cloned. In December 2001, a cloned cat, CC (for Carbon Copy), was born, the first success in a total of 87 failed attempts. Although CC is genetically identical to the adult DNA donor, a two-year-old female named Rainbow, the kitten's markings aren't the same as the donor's. The striking difference in fur patterns illustrates an important point about the connection between the molecular basis of heredity and the characteristics of each individual organism. Although CC and Rainbow are genetically identical, factors other than the codons on their (identical) DNA influence at least some aspects of their physical appearance. These include random and unpredictable events that may occur during the development of the fetus. Variations in blood flow to the fetus, for example, occurring at various stages in fetal development, could affect the patterns of pigmentation in the fur.

Cloning produces a new organism with the same genetic information as the cloned individual. As demonstrated by CC and Rainbow, cloning does not produce an exact replica of the cloned organism. Differences among all plants and animals extend beyond differences in the genetic information carried by their DNA. Factors ranging from the effects of haphazard events during fetal development to environmental influences extending beyond birth all place their imprint on each individual organism. The string of bases of a DNA molecule does not tell the entire story of any plant or animal.

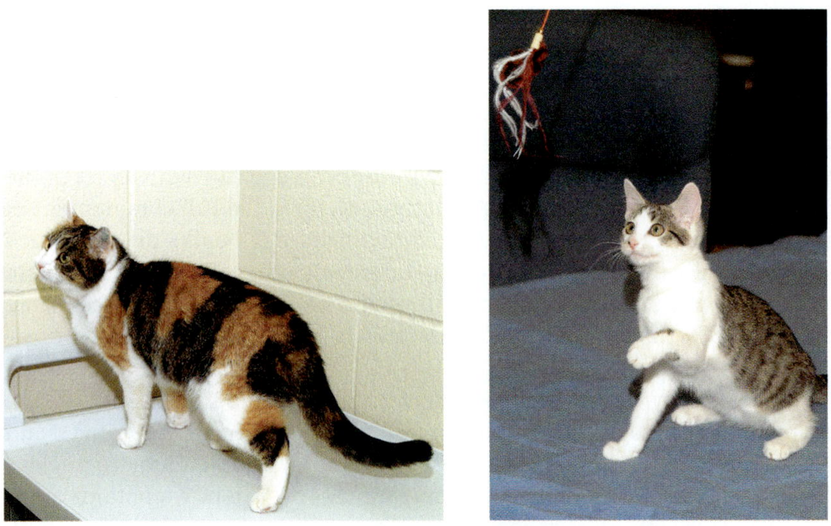

Rainbow (left), the cat, and CC (right), the cloned kitten. Although the two are genetically identical, factors beyond genetics influence fur patterns and colors.

■ QUESTION

What was the first adult mammal to be cloned successfully? When and where did this occur, and how many unsuccessful attempts preceded the first success?

PERSPECTIVE

Will There Ever Be Another You?

Cloning can originate from a variety of motivations. Initially, there is the scientific drive to demonstrate that our theoretical understanding of how cloning might be accomplished can be applied to a practical demonstration of the process. It's an illustration of the connection between the *science* involved in understanding how the universe operates and the *technology* that allows us to convert this understanding into practical applications in our society and in our daily lives (Perspective, Chapter 1).

In addition there is a motivation for material improvement in our own lives and in our society. As we saw in Section 18.4, an understanding of genetics evolved partly from the application of breeding methods to improve the quality of farm animals and agricultural crops, leading to gains for both the farmer and society. Cloning carries this pattern from traditional forms of sexual breeding of farm animals to the more direct cloning of the most desirable individual animals—cows, for example, that produce milk effectively.

The cloning of a cat opens still another possibility, that beloved pets might be replicated—genetically, at least—to provide additional years of comfort and satisfaction to their owners.

Lurking beyond all this, beyond the cloning of animals for scientific achievement, for material gain to individuals and to society, and for continuing companionship, is the possibility of cloning humans themselves. Pairs of humans with mutually identical DNA—identical twins—are well known. Yet despite having developed under nearly identical fetal conditions and having been raised in nearly identical environments, each twin still shows distinguishing physical and other characteristics. Here again, our own genes do not define us in total. And as with the cloning of the two-year-old Rainbow to produce the kitten, CC (Sec. 18.12), the cloning of an adult human might not produce an exact replica. Differences in fetal conditions and in the external environment may be expected to produce human distinctions at least comparable to the difference in cat-fur patterns.

Motivations for the cloning of humans are as complex as the process itself. And so are the ethical, social, religious, medical, and other issues involved. But even though we are, in a sense, defined by our genes, the expression of that definition depends on a multitude of complex, variable, unpredictable, and, to an extent, unknown factors. Although science might someday be able to clone any one of us just as it has already cloned several other animals, it's unlikely that there will ever be anyone else quite like any one of us.

Once again in the issue of cloning, we find that whatever choices we may make, as individuals and as a society, a knowledge of the science behind the issue, and of the possible and probable outcomes, adds reason, value, and confidence to our judgments. ■

EXERCISES

For Review

1. Following are a statement containing a number of blanks and a list of words and phrases. The number of words equals the number of blanks within the statement, and all but two of the words fit correctly into these blanks. Fill in the blanks of the statement with those words that do fit, then complete the statement by filling in the two remaining blanks with correct words (not in the list) in place of the two words that don't fit.

Nuclei within cells of most organisms contain long strands of deoxyribonucleic acid, or __________, that normally exist as threadlike filaments. This deoxyribonucleic acid exists as a __________, with the two parallel, __________ strands of the acid held together through __________.

During cell division these threads form highly structured coils known as __________. These molecular filaments consist chemically of a backbone formed by alternating __________ units and rings of the carbohydrate __________. Bonded to these carbohydrate rings are four amine bases: adenine, __________, guanine, and thymine. The sequence of these bases varies from species to species and organism to organism and carries __________ information from one generation to the next.

A sequence of three bases constitutes a __________, which codes for a specific __________ that will be placed in a __________ formed in the cellular __________. A __________ is a string of bases that carries enough information to control the biochemical synthesis of a specific polypeptide. The genetic information is transferred from the nucleus to cellular structures known as _________, which lie outside the nucleus, by molecules of _________, or more specifically, _________. In these molecules the amine base _________ replaces the _________ of DNA. The correspondence between a sequence of three bases on these molecules and the amino acid that the three-base sequence codes for is known as the _________.

2-deoxyribose	gene
amino acid	genetic code
bromelin	genetic
chromosomes	hydrogen bonds
codon	meiosis
complementary	mRNA
cytoplasm	polypeptide
cytosine	ribonucleic acid
DNA	ribosomes
double helix	thymine

2. Define, identify, or explain each of the following:

clone	mitosis
cytoplasm	nucleotide
Dolly	ribose
dystrophin	triple helix
GUC	tRNA
junk DNA	

3. What proteolytic enzyme is present in fresh pineapple but not in canned pineapple? Why isn't this enzyme present in canned pineapple?

4. What is a commercial or industrial source of collagen? What common product is derived or produced from collagen?

5. Describe or identify the kinds of cells produced by: (a) mitosis; (b) meiosis.

6. In what biological structures are chromosomes located?

7. Excluding cells that participate in sexual reproduction, identify a plant or an animal whose cells contain fewer chromosomes than human cells. Identify one whose cells contain more chromosomes.

8. What four amine bases occur in DNA? In RNA?

9. What holds the strands of the double helix next to each other?

10. What is the difference in structure and in function between DNA and RNA?

11. What six sequences of amine bases on RNA form the codons for serine?

12. What is the genetic code?

13. In addition to identifying the specific amino acids that form the primary structure of polypeptides, what other functions do codons serve?

14. Explain or describe the difference between:

a. a genotype and a phenotype
b. a gene and a chromosome
c. a double helix and a triple helix
d. DNA and RNA
e. mRNA and tRNA
f. meiosis and mitosis
g. nucleic acid and a nucleotide
h. exons and introns

15. Describe the contributions to the topics of this chapter made by:

a. Gregor Mendel
b. James Watson and Francis Crick
c. Rosalind Franklin
d. Maurice Wilkins
e. Ian Wilmut and Keith Campbell

16. Maurice Wilkins shared the Nobel Prize in medicine with James Watson and Francis Crick, but Rosalind Franklin did not. Why didn't she share the Nobel Prize?

A Little Arithmetic and Other Quantitative Puzzles

17. Show that the four amines of RNA, taken three at a time, give 64 different sequences (codons).

18. How many *different* nucleotides can a strand of DNA contain? Explain.

19. Section 18.11 tells us that the largest gene, the one that's responsible for the formation of the protein dystrophin, contains 2.4 million bases. Of these, 11,031 are located in exons. How many codons does this represent?

20. In Section 18.10 we found that an estimated 6 billion base pairs exist in the DNA of each human cell. Figure 18.13 shows that the thymine–adenine pair is connected by two hydrogen bonds, and the cytosine–guanine pair by three. Assuming that these two base pairs are present in equal numbers in the DNA of each human cell, calculate the total number of hydrogen bonds in the DNA of each human cell.

Think, Speculate, Reflect, and Ponder

21. Is DNA a protein? Explain.

22. Explain why virtually every cell of an organism contains the same DNA as every other cell.

23. In forming the double helix of DNA, thymine pairs with adenine rather than with guanine and cytosine pairs with guanine rather than with adenine. Suggest a reason why this is so.

24. Where else, in addition to the thread of a screw, might we find a helix in our everyday world?

25. In one of their early models of the double helix, Watson and Crick assumed that the amine substituents on each DNA molecule pointed outward, away from the central axis. They later discarded this model in favor of the one in which the amine substituents point inward, toward the central axis. Why do inward-pointing amine bases provide a more stable double helix?

26. A rabbit can run from a hungry tiger, but a plant can't run from a hungry locust. Describe what defense(s) a plant might have against a predatory insect.

27. The bromelin of pineapple is a set of proteolytic enzymes capable of inactivating and/or destroying the proteins of a variety of organisms. Yet although the bromelin enzymes are themselves proteins, they do not inactivate or destroy themselves. What conclusion(s) do you draw from this observation?

28. What function does cell division by meiosis serve in the reproduction of the organisms of a species? In this form of reproduction, why can meiosis be considered superior to mitosis?

29. What are some possible reasons that might have led the scientists who first cloned a sheep to wait approximately half a year before announcing their success?

30. Human cloning is a controversial topic. Suggest some ethical considerations in favor of human cloning. Suggest some in opposition to human cloning.

31. Give your own answers to the following questions and explain your reason for each.

a. Would you wish to have a pet of yours cloned?

b. Would you wish to have yourself cloned?

c. Are there circumstances under which a human might be cloned?

d. Should governments control animal and/or human cloning or should cloning be left to the judgments of individuals, physicians, and scientists.

Web Questions

Use the Web to answer or accomplish each of the following. Report your Web source for each response.

1. Find two other edible plants or plant products, in addition to pineapples, that produce proteolytic enzymes. Identify the plants and the enzymes.

2. Write a brief description of the life and work of Gregor Mendel.

3. Describe the contributions to our understanding of genetics and cellular division made by Walther Flemming and by Eduard Strasburger.

4. What are *purines* and *pyrimidines*? Which of the compounds discussed in this chapter are purines? Which are pyrimidines?

5. How many genes are there in the DNA of a fruit fly? How does this compare with the number of genes in human DNA?

6. Explain why the cloning of cats is easier to accomplish than the cloning of dogs.

7. Describe in detail the difference between the reproductive processes of cloning and *twinning*.

8. Identify Francis Collins and Craig Venter and describe their connection to the topics of this chapter.

Vitamins, Minerals, and Additives

MICRONUTRIENTS AND OTHER FOOD CHEMICALS

A shopper compares nutrition and ingredients panels of two soup cans.

The Power of Vitamin C

When we think of vitamin C we think of oranges, and with good reason. Oranges are rich in vitamin C. A half cup of orange juice, 4 oz, contains about 62 mg of the vitamin. That's enough to carry most of us through the day. Citrus fruit generally is rich in this vitamin. Other foods containing plenty of vitamin C include broccoli, Brussels sprouts, cantaloupe, cauliflower, collard greens, green beans, and potatoes.

Many of us regard vitamin C as a natural component of healthful foods that protects us against illness. That's true, but there's another way of looking at it that's equally valid. Vitamin C is also a chemical, consisting of a highly organized arrangement of protons, neutrons, and electrons, capable of reacting in well-defined ways with other chemicals. It's formed by a combination of carbon, hydrogen, and oxygen atoms held together by covalent bonds, with the molecular formula $C_6H_8O_6$ and the chemical name *ascorbic acid*.

We can see vitamin C in action with a little water, tincture of iodine, household bleach, and two vitamin C tablets. As we saw in the table of standard reduction potentials of Chapter 11, iodine is a good oxidizing agent, capable of being reduced to iodide ions as it oxidizes other chemicals (Table 11.1). Iodine is especially useful here because we can see clearly the loss of color as the red-violet molecular I_2 is reduced to colorless I^- anions by the vitamin.

You can see the vitamin's chemical prowess in either of two ways. In one approach, simply put a drop of the iodine solution on the surface of a vitamin C tablet. (A 250-mg tablet of the pure vitamin works nicely.) Watch closely, because the iodine quickly loses its color. The molecular iodine, I_2, is rapidly reduced to colorless iodide ions, I^-, by the powerful reducing activity of the vitamin. As part of this redox reaction (Chapter 11), the iodine oxidizes the vitamin C to a different compound.

After the color of the iodine has disappeared, add a drop or two of liquid bleach to the tablet. You'll see its surface darken immediately as the chlorine of the bleach reoxidizes the iodide ions back to molecular iodine. We saw something like this happen in the opening demonstration of Chapter 11, when we used galvanized tacks to decolorize iodine. In this current demonstration, the vitamin C is behaving chemically much like the zinc tacks of Chapter 11.

You can also carry out this demonstration by dropping a vitamin C tablet into about a quarter cup of water. You don't have to wait until the tablet dissolves. Swirl the tablet in the water for a few moments and then add a few drops of iodine. After the color disappears, add a few drops of the bleach. You'll see the color reappear, as it did with the dry tablet (Fig. 19.1). The speed of the appearance of the color depends on the temperature of the water, the amounts of iodine and bleach you add, and other factors. You may want to try this a few times to see the effects of different water temperatures and other conditions.

(After the red-violet color of the iodine reappears, both with the dry tablet and with the solution of vitamin C in water, it fades away again after a short time. This results

Figure 19.1 Adding a drop of iodine solution to a vitamin C tablet.

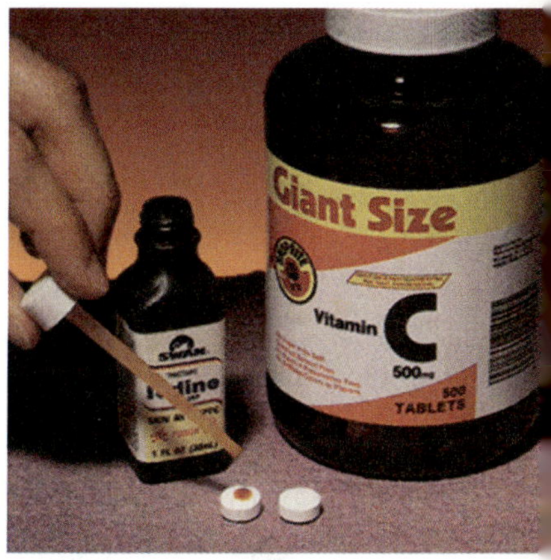

The deep color of a drop of iodine solution on a vitamin C tablet. . .

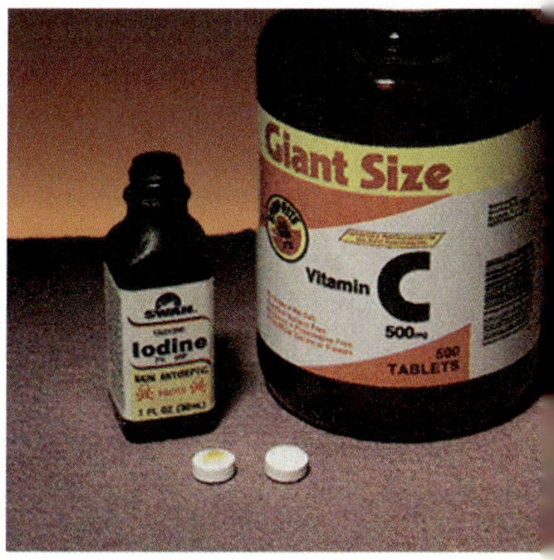

. . . quickly fades and soon vanishes as vitamin C reduces the deeply colored I_2 to colorless I^-.

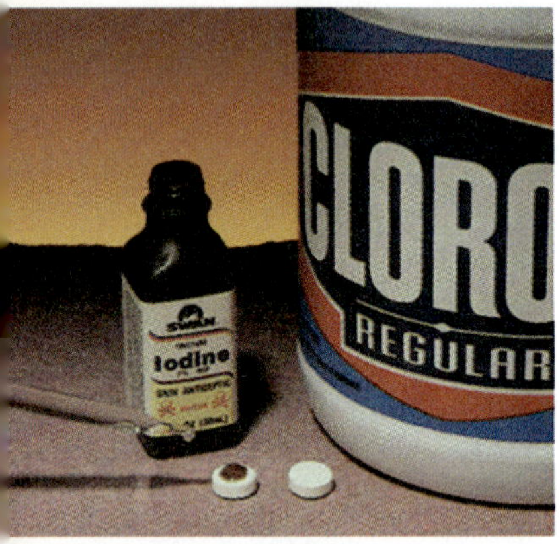

The color reappears as a drop of liquid bleach oxidizes the colorless I^- back to I_2.

Adding a drop of bleach to a fresh vitamin C tablet produces no color, thereby demonstrating that the reappearance of color is due to a reaction of the bleach and the I^-.

from continued oxidation of the I_2 to still other, colorless substances by the household bleach. You can demonstrate that this is happening by adding a drop or two of the tincture of the iodine solution to a little liquid bleach. The color of the iodine rapidly fades in the bleach because of this continued oxidation. The nature of this oxidation of I_2 by the bleach does not affect the results we obtain from the vitamin C or our discussion of them.)

You can use the scientific method in this demonstration to prove an important point. When you add the bleach to the vitamin C, after the original color has disappeared, you see a return of color. How do we know this comes from a reaction of the bleach with the iodide ions? Could the color now result from a reaction of the bleach with the vitamin C itself? We can answer this question by using a *control*, an experiment identical to the first in every way but a critical one. For the control experiment add a bit of liquid bleach to a fresh vitamin C tablet untouched by iodine. There's no color this time. We interpret this observation as proving that the color you saw on adding the bleach to the first tablet must have come from a reaction of the bleach with iodide ions that were present, not from a reaction with the vitamin. You can perform this same test with a tablet in a little water. We'll have more to say about the ease of oxidation of vitamin C later in this chapter.

This demonstration illustrates an important point: Aside from its presence in healthful foods and its value in protecting us against some forms of illness, vitamin C is a chemical. Like the other chemicals we've examined, vitamin C has a unique chemical name (ascorbic acid), a specific set of chemical properties, and a well-defined chemical structure.

In preceding chapters we examined our macronutrients: carbohydrates, proteins, and fats and oils. We saw that the macronutrients are also chemicals, that they take part in chemical reactions, producing specific, observable results. We saw that the macronutrients make up the major portion of the food we eat and provide us with energy and the materials of our physical bodies. In this chapter we'll examine the minerals and vitamins—both of which compose a far smaller portion of our foods than the macronutrients—and some of the chemicals (the *food additives*) we use to help preserve foods and make them more nutritious and more appealing.

Micronutrients are the components of food that we need in very small amounts for life and good health.

It's customary to combine minerals and vitamins into the category of **micronutrients**. These substances aren't converted into energy in any significant amounts. Nor, except for calcium and phosphorus, which make up almost the entire mass of our bones and teeth, do they provide the raw material for tissue formation. We do need the micronutrients, though, usually in minuscule quantities, to convert the macronutrients into energy and into the materials of the body. In this chapter we'll examine both the micronutrients and food additives.

19.1 The Major Minerals: From Calcium to Magnesium

Minerals are the chemical elements we need for life and good health, beyond the carbon, hydrogen, nitrogen, oxygen, and sulfur that are brought to us by our macronutrients.

We might think of *minerals* as hard, sometimes metallic or crystalline substances we obtain from the ground, ranging in beauty and value from lead to marble or to gems such as opals or rubies. Nutritionally, though, the term "minerals" has a different meaning. The classification **minerals** includes all the nutritionally important chemical elements of our foods, except for the carbon, hydrogen, nitrogen, oxygen, and (by some definitions) sulfur that come to us with the macronutrients. Of all the minerals, calcium is the most abundant in our bodies, ranking fifth among all of our atoms, just behind hydrogen, oxygen, carbon, and nitrogen (see Fig. 2.5). Between 1.5 and 2.0% of your body weight comes from calcium, and more than 99% of all this calcium lies in your bones and teeth. To translate this into more concrete terms, if you weigh 70 kg (154 pounds), you're carrying around some 1.0 to 1.5 kg (roughly 2 or 3 pounds) of calcium in your skeleton and in your mouth.

Next in rank comes the element phosphorus. Although calcium is the major mineral of our bones and teeth, it takes about half a gram of phosphorus to pack each gram of calcium firmly into the bony lattices and hold it there. Add to this the smaller amounts of phosphorus in our soft tissues, and a 70-kg person owns a total of almost a kilogram (about 2 pounds) of the element.

We have smaller amounts of potassium, chlorine, sodium, and magnesium inside us, in that order. These elements form the major ions of the fluids in and around our cells. (The fluids within the cells are *intracellular*; those outside are *extracellular*.) Potassium, the principal cation inside the cells, governs the activities of the cellular enzymes. Sodium, which is the dominant cation outside the cells, keeps the water content of the intracellular and extracellular fluids in a healthy balance. Both sodium and potassium cations regulate the distribution of hydrogen ions throughout our bodies and thereby keep the acidities of our various fluids within their normal ranges.

Along with these alkali metal cations, chloride anions also help regulate fluid balances and, in combination with protons, provide the hydrochloric acid of our gastric juices. They also serve to balance the electrical charges of the various cations.

Magnesium plays several secondary roles. It's second to calcium as the hard mineral of bones and teeth and second to potassium in regulating chemical activities within the cells. Magnesium helps control the formation of proteins inside the cells and the transmission of electrical signals from cell to cell.

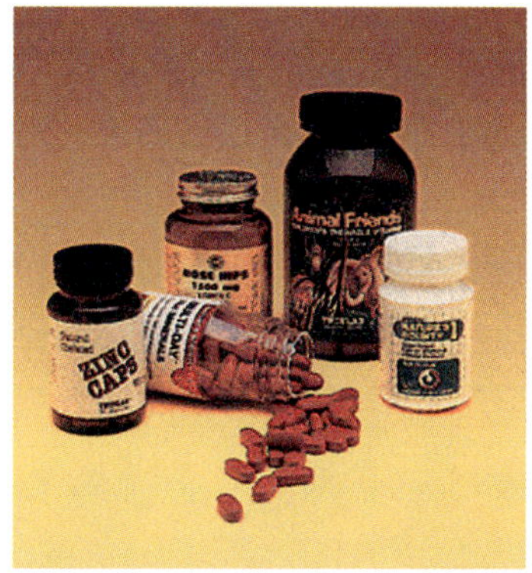

Commercially available mineral and vitamin supplements.

Sodium and potassium play major roles in regulating the water balance of the body, including the formation and secretion of sweat.

Calcium, phosphorus and magnesium are the major minerals of teeth and bones.

■ QUESTION

What structures hold most of the phosphorus of our bodies?

19.2 Trace Elements: From Iron on Down

The **trace elements** are the dietary elements we need at levels of less than 100 mg/day.

Together, just 11 elements—the carbon, hydrogen, nitrogen, oxygen, and sulfur of the macronutrients and of water, and the calcium, phosphorus, potassium, chlorine, sodium, and magnesium of the micronutrients—compose well over 99% of the mass of all living matter. All the remaining elements necessary for good health make up the category of **trace elements**. Generally, we consider as trace elements those needed in our diets at levels of less than 100 mg (0.100 g) per day. We'll examine the quantitative daily requirements more closely in the next section.

Among the more important of the trace elements are iron, fluorine, zinc, copper, selenium, manganese, iodine, molybdenum, chromium, and cobalt, arranged here in decreasing order of the amount of each in an adult body. They function in various ways, with most of them incorporated into the structures of enzymes, hormones, and related molecules.

Iron forms a critical part of the hemoglobin molecule of red blood cells. The iron of hemoglobin bonds to oxygen acquired as the blood passes through the lungs and carries the oxygen throughout the body. *Anemia*, a condition characterized by fatigue, results from a deficiency of iron and the resulting inability of the blood to carry sufficient oxygen to the cells. Fluorine, as the fluoride anion, helps harden the enamel of teeth, making them resistant to decay. Many communities add fluoride salts to drinking water to protect the teeth of children (Perspective, Chapter 14), and a compound of fluorine, often sodium fluoride, is an ingredient of most anticavity toothpastes. Zinc promotes growth, the healing of wounds, and the development of male sex glands. A deficiency of iodine leads to enlargement of the thyroid gland of the neck, a condition known as a *goiter*, and has been implicated in learning deficiencies. A single cobalt atom is incorporated into every molecule of vitamin B_{12}.

Other trace elements play various roles in the body. Deficiencies of copper, for example, can produce symptoms ranging from changes in hair color to anemia and bone disease and may impair the immune system. Manganese is essential for healthy bones, a well-functioning nervous system, and reproduction. Chromium plays an important part in the metabolism of glucose. Because of the abundance of some of the trace elements in our diets, our knowledge of their actions sometimes comes from studies of deficiencies artificially produced in the diets of animals rather than directly from human studies.

Some of the more common dietary sources of minerals appear in Table 19.1.

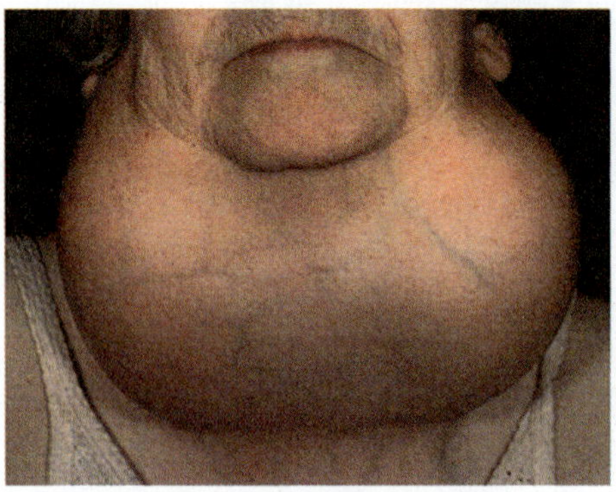

A deficiency of iodine can lead to an enlargement of the thyroid gland, a condition known as *goiter*.

Table salt is available with added iodide ion, usually as potassium iodide, KI, and without this micronutrient.

Table 19.1 Dietary Sources of Minerals

Mineral	Source
Calcium, Ca[a]	Milk and dairy products such as cheese and ice cream; fish, such as sardines, eaten with their bones; broccoli and other dark green vegetables; legumes
Chlorine, Cl[a]	Table salt
Chromium, Cr[b]	Brewers' yeast; meat (except fish); whole grains
Cobalt, Co[b]	Most animal products, including meat, milk, and eggs
Copper, Cu[b]	Liver and kidneys; shellfish; nuts; raisins; dried legumes; drinking water in some areas
Fluorine, F[b]	Drinking water (in some regions); tea; fish, eaten with their bones
Iodine, I[b]	Iodized salt; seafood; bread
Iron, Fe[b]	Liver and other red meats; raisins; dried apricots; whole-grain cereals; legumes; oysters
Magnesium, Mg[a]	Whole-grain cereals; nuts; green vegetables; seafood
Manganese, Mn[b]	Nuts; whole grains and cereals; leafy green vegetables; dried fruits; roots and stalks of vegetables
Molybdenum, Mo[b]	Animal organs; cereals; legumes
Phosphorus, P[a]	Nearly all foods
Potassium, K[a]	Nearly all foods, especially meat, dairy products, and fruit
Selenium, Se[b]	Various foods, including grains, meat, and seafood
Sodium, Na[a]	Table salt
Sulfur, S[a]	Dietary proteins of meat, eggs, dairy products, grains, legumes
Zinc, Zn[b]	Meat; eggs; seafood (particularly oysters); dairy products; whole grains

[a] A major element of our micronutrients.

[b] A trace element of our micronutrients.

■ QUESTION

Of the elements classified as *trace elements*, which one is most abundant in our bodies? What condition results from a severe deficiency of this element?

19.3 Vitamins

We also need small amounts of vitamins in our diets, in addition to minerals. But despite the nutritional similarities, there are important chemical differences between the two. First, the dietary minerals are simply the less abundant chemical elements of our foods. **Vitamins**, on the other hand, are organic compounds, with well-defined molecular structures. They are the organic compounds we need in small quantities in our foods for life and good health. We don't use them directly for energy or for the materials of our bodies.

Vitamins are organic compounds that are essential in very small amounts for life and good health.

The word *vitamin* came into our language in 1912, devised by the Polish biochemist (later, American citizen) Casimir Funk. He discovered these organic substances and believed that they were all amines (Sec. 17.2) that are vital to our health, or "vital amines." Hence he gave them the name *vitamines*, later shortened to our current *vitamins*. While we now know that vitamins are, indeed, vital to our health, we also know that some are amines and others are not.

The difference between minerals (nutritionally important elements) and vitamins (nutritionally important organic compounds) is reflected in their source or origin. Plants pick up minerals directly from the soil, and animals get them from their food. If any particular mineral doesn't occur in the soil in which a plant grows or in the food of an animal, there's no way in the world that it can be part of that plant's or animal's makeup. Living beings simply can't generate minerals in their own bodies.

All of our minerals became part of the earth at its birth and enter our bodies only from the earth, through the plants we eat or the animals that feed on the plants.

Vitamins, on the other hand, form biochemically, through the life processes of the plants and animals we eat. Plants and animals synthesize vitamins within their own bodies just as we synthesize protein and fat molecules within ours. Thus, the vitamins of plants come from the plants themselves, not from the soil or the air or any fertilizers. They're as much a part of the plant as a nose is part of your face. An orange contains vitamin C for the same reason it has a tough, orange-colored skin. Both the orange's vitamin C and its orange skin develop as the fruit grows and matures. Otherwise, it wouldn't be an orange. Table 19.2 presents the best known vitamins and brief dietary descriptions.

Table 19.2 The Vitamins

Chemical Name (and Letter Designation)	Dietary Source	Deficiency Symptoms or Disease
Water-Soluble Vitamins		
Ascorbic acid (vitamin C)	Fruits, especially citrus; many vegetables	Scurvy; degeneration of tissues
Biotin	Various foods	Rare; nausea, loss of appetite
Cobalamin (vitamin B_{12})	All foods of animal origin	Pernicious (fatal) anemia
Folic acid Folacin	Meat; green, leafy vegetables	Anemia
Niacin, or its equivalent in tryptofan (vitamin B_4)	Meat; legumes; grains	Pellagra; skin, digestive, and nervous system disorders; depression
Pantothenic acid (vitamin B_3)	Widely distributed; occurs in virtually all foods	Defects in metabolism
Pyridoxine Pyridoxal Pyridoxamine (vitamin B_6)	Various foods; widely distributed	Deficiency is rare; results in defects in amino acid metabolism with various symptoms
Riboflavin (vitamin B_2)	Meat, especially animal organs; milk and dairy products; green vegetables	Skin disorders
Thiamine (vitamin B_1)	Pork; animal organs; whole grains; nuts; legumes	Beriberi; muscular weakness; paralysis
Fat-Soluble Vitamins		
Cholecalciferol (vitamin D_3)	Liver and liver oils; fortified milk	Rickets; malformation of the bones; osteomalacia (adult counterpart of rickets)
Retinol (vitamin A)	Liver and liver oils; carrots and other deeply colored vegetables	Night blindness; degenerative diseases of the eyes leading to total blindness
α-Tocopherol (vitamin E)	Various foods, especially grain oils	Deficiency disease unknown in humans
Vitamin K	Plants and vegetables; produced by intestinal bacteria and absorbed through the intestinal wall	Deficiency disease unknown in adults; needed for blood clotting

Notice the two categories of data in Table 19.2. The **fat-soluble vitamins** are much more soluble in fats, hydrocarbons, and similar solvents than they are in water; the **water-soluble vitamins** exhibit the opposite property. Vitamins A, D, E, and K constitute the fat-soluble class; the B complex of vitamins and C are water-soluble. The B complex actually consists of several vitamins, including

- thiamine (B_1),
- riboflavin (B_2),
- pantothenic acid (B_3),
- niacin (B_4),
- pyridoxine (B_6), and
- cobalamin (B_{12}).

The **fat-soluble vitamins** are much more soluble in fats, hydrocarbons, and similar solvents than in water. The **water-soluble vitamins** are much more soluble in water than in these other solvents.

All of these occur in minute quantities in a great variety of foods and are often considered as a group. There is no single "vitamin B."

This classification (water-soluble vs. fat-soluble) doesn't tell us much about what kinds of foods contain any particular vitamin. Some fatty foods can be good sources of the water-soluble vitamins, and the moist, green, leafy vegetables often provide ample supplies of fat-soluble vitamins. But it does explain some of the properties of the vitamins and their effects on our health. A full description of the chemistry and the nutritional importance of all the vitamins is beyond the scope of this discussion. Instead, we'll emphasize three of the better known vitamins: the fat-soluble A and D and the water-soluble C.

■ QUESTION

What do vitamins A and D have in common? What do the B complex of vitamins have in common with vitamin C?

19.4 Vitamin A: Polar Bear Livers and Orange-Tinted Skin

Vitamin A maintains the health of the eyes, skin, and mucous membranes and is particularly important for good vision in dim light. A deficiency can produce "night blindness," an inability to see in the dark. More severe deficiencies can cause a failure of the tear ducts and eventual blindness.

As one of the fat-soluble vitamins, vitamin A accumulates in the body's fat cells, particularly in the liver. A moderate excess of vitamin A in our bodies normally does no harm. We simply store it, largely in our livers, for use later, when it's in short supply. Large excesses, though, can overwhelm the body's storage capacity and can generate toxic symptoms, including headaches, dizziness, nausea, loss of appetite, sore muscles, blurred vision, loss of hair, and worse, including death. Moreover, even though vitamin A is required for normal development of the embryo, excesses taken in early stages of pregnancy—and even shortly before conception, since the vitamin can be stored—can produce serious birth defects.

Toxic doses of the vitamin can result from overzealous use of vitamin supplements and from eating certain rare and bizarre foods that are particularly rich in vitamin A. Polar bear livers, for example, contain unusually large quantities of the substance; a 3-oz portion provides about 200 times the currently recommended daily intake for an adult. There have been reports of extreme toxicity and death among early Arctic explorers who ate large quantities of polar bear livers.

While animal livers can store very large concentrations of this vitamin, plants don't contain any. Carrots, which are supposed to be a very good source of the nutrient (and indeed *are*), have none whatever. This strange-sounding contradiction makes more sense when we examine the chemistry of the vitamin.

Our bodies can use vitamin A (known chemically as *retinol*) itself, or we can easily convert several chemical structures closely related to retinol into the vitamin, or we can even use these closely related compounds directly as retinol's equivalents. Several of retinol's esters (Sec. 13.7), for example, are every bit as effective as retinol itself and are *physiological equivalents* of the vitamin. For this reason vitamin levels in foods or supplements are sometimes stated in *Retinol Equivalents* of the vitamin, *REs*, or in *International Units, IUs*. REs and IUs represent not only the quantity of the vitamin itself but also the total amount of the vitamin and all its physiological equivalents that can be converted into or used as the vitamin.

(One RE of vitamin A is equivalent to 1 μg—one *microgram*, or 10^{-6} g—of retinol. One IU of vitamin A represents 0.3 μg of retinol. A little arithmetic shows that 3.33 IUs of retinol constitute one RE.)

Although carrots contain no retinol, they do provide large amounts of β-carotene, a deeply colored material that we easily convert to the vitamin itself (Fig. 19.2). With enough of the vitamin stored away in the liver, our bodies stop converting the β-carotene into retinol. As a result we can't very well poison ourselves by overeating carrots and other deeply colored red, orange, and yellow fruits and vegetables, all of which provide plenty of β-carotene in one form or another.

When we consume more β-carotene than we need, some of it reaches the surface tissues, where it's stored and begins imparting its own color to the skin and eyes. There are reports of people whose skin acquires a yellow-orange tint from consuming enormous quantities of carrots and tomato juice, which is also rich in carotenes, over a long period. It's a harmless and reversible condition that disappears when carrots and tomatoes are removed from the diet.

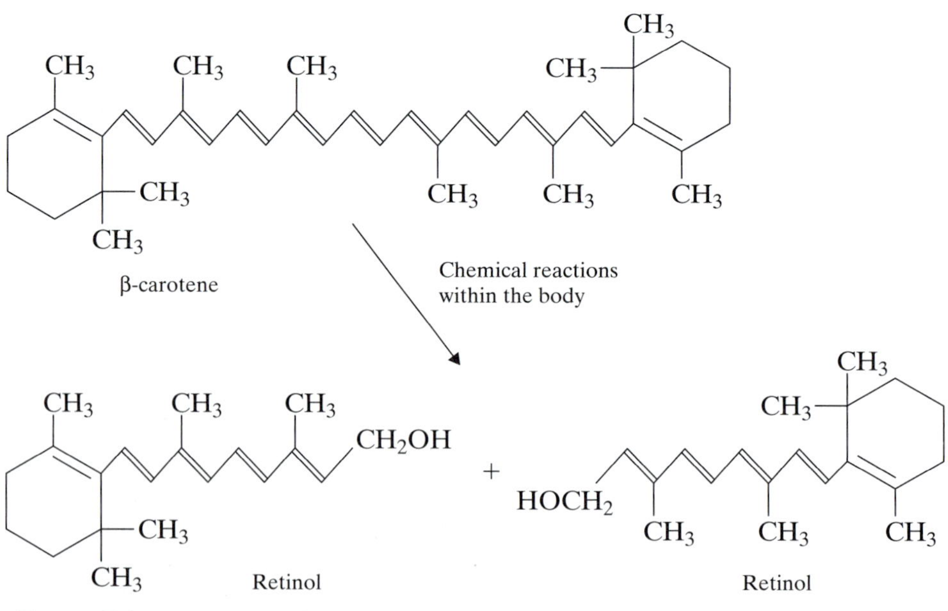

Figure 19.2 Human metabolism converts one molecule of β-carotene into two molecules of vitamin A.

■ QUESTION

Carrots are considered to be excellent sources of vitamin A, yet they contain no vitamin A. How do you explain this apparent contradiction?

19.5 Vitamin D ... Or Is It?

Vitamin D promotes the absorption of calcium and phosphorus from our foods through the intestinal wall and into the bloodstream, thereby providing the raw materials for forming and maintaining healthy bones. Without vitamin D, children's bones develop poorly, resulting in the severely bowed legs and other skeletal deformations that characterize the physical condition known as *rickets*. Fortifying milk with added vitamin D makes sense since milk is a major food of young children, it's rich in calcium, and the fat-soluble vitamin dissolves readily in the milk fats.

But vitamin D, like A, has its own story. Both rickets and the vitamin itself may be largely the products of the way civilization has developed in the northern and southern latitudes. For some of us the chemical may not be a vitamin after all. Part of the reason for questioning whether it's really a vitamin is that, unlike other vitamins, D forms in our own bodies. Under the right conditions we don't need it in our foods; instead, it forms in our skin, under the action of the sun's ultraviolet rays.

Actually, vitamin D isn't even a single organic compound. The term applies to a set of very closely related molecular structures, differentiated from each other by subscripts. There are vitamins D_1, D_2, D_3, and so on, all with the same physiological function. As ultraviolet radiation strikes our skin, it converts a bodily substance named *7-dehydrocholesterol* into the form of the vitamin known as vitamin D_3 or, by its chemical name, *cholecalciferol* (Fig. 19.3). With plenty of sunshine, lots of exposed skin, and long hours in the outdoors, cholecalciferol would be in plentiful supply in all our bodies, rickets would be virtually unknown, and vitamin D wouldn't be a vitamin at all. But with the limited sunshine, copious clothing, and indoor living and working conditions of the more extreme northern and southern latitudes, the chemical becomes scarce. This could be especially tragic for children, whose bones are still growing. Under these conditions rickets would be common if it weren't for the addition of vitamin D to children's diets, usually in the form of fortified dairy products or, in their absence, cod liver oil or vitamin supplements.

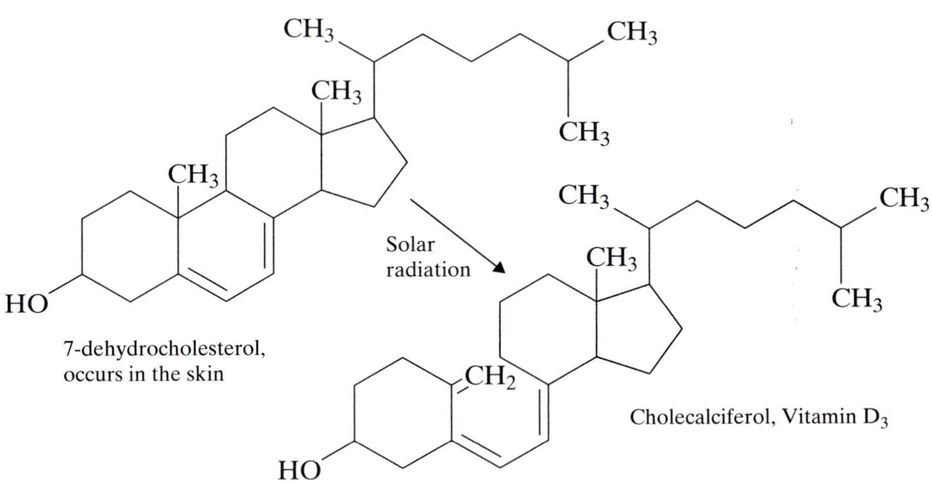

Figure 19.3 The formation of vitamin D.

It's probably impossible to produce dangerously high levels of vitamin D in the body through exposure to the sun, since the natural tanning of the skin blocks out the rays and stops the chemical conversion. Unfortunately the same can't be said about the excessive use of vitamin supplements. Overzealous use of vitamin D supplements has led to excessive absorption of calcium and phosphorus and the formation of calcium deposits in the soft body tissues, including the major organs such as the heart and the kidneys. Extreme cases have resulted in death.

■ QUESTION

What are the name and the symptoms of the physical condition brought about by a deficiency of vitamin D?

19.6 Vitamin C: Bleeding Gums and the Bulbul Bird

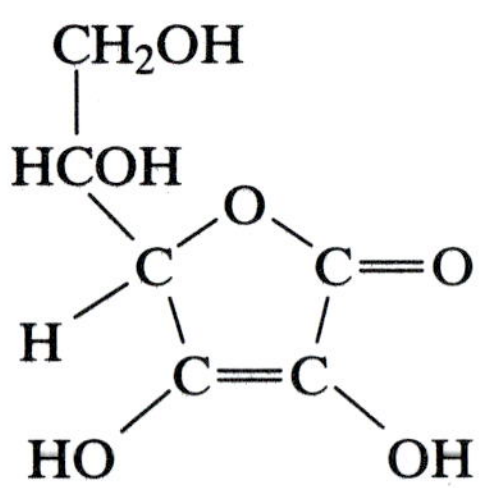

Figure 19.4 Vitamin C, ascorbic acid.

Water-soluble vitamin C, the best known of all the vitamins, is taken as a dietary supplement by more Americans than any other. As *ascorbic acid* (Fig. 19.4), its chemical name represents two properties of the substance, one chemical, the other biological. First, it's an acid, although it clearly doesn't belong to the class of carboxylic acids (Sec. 10.8). Moreover, the word *ascorbic* reflects its biological value in protecting against the disease *scurvy*. (The words *scurvy* and *ascorbic* come from the Latin word for the symptoms of the disease, *scorbutus*, described below.)

Humans require vitamin C for formation of the collagen of our connective tissues (Sec. 18.1). Tough fibers of this protein are important components of our skin, muscles, blood vessels, scar tissue, and similar bodily structures. Since the gums are rich in blood vessels and are subject to wear and abrasion through eating and by brushing the teeth, the bleeding of gum tissue weakened by deterioration of its collagen becomes the first visible symptom of scurvy. As the disease progresses, the gums weaken and decay, the teeth fall out, and the body bruises easily. Scurvy itself isn't fatal, but as it progresses it leads to a decline in the body's resistance to other, lethal diseases.

Until the last few hundred years, scurvy had been the scourge of sailors, explorers, and those on long military expeditions, people who had to survive for months or years on stored provisions. Although some, especially the sailors, were well stocked with meat preserved by salting and smoking, and with carbohydrate-rich grains, many of them lacked the fresh fruits and vegetables that could supply them with vitamin C. Descriptions of diseases affecting the Crusaders suggest that they suffered from scurvy. Vasco da Gama, the Portuguese navigator who, in 1497, was the first European to sail beyond the Cape of Good Hope at the southern tip of Africa, lost more than half his men to diseases probably brought on by the debilitation of scurvy.

Although some Dutch and English seamen of the 1500s knew of the value of fresh fruit and lemon juice in preserving sailors' health on long voyages, it wasn't until 1795 that the British navy officially made lemons part of the required stores of its fleet at sea. It took another 70 years for lemons to become an official part of the diet of the British merchant marine. Because lemons were known as "limes" in those days, British sailors, who were required to eat the "limes," became known as *limeys*.

Curiously, humans are one of the few species of animals for which ascorbic acid is actually a vitamin. We're among the few that can't produce the acid within our own bodies but have to obtain it from our foods. The story seems to be connected to the trade-offs of evolution. Neither the newest of the animal species, we

Like humans, guinea pigs need a dietary source of vitamin C.

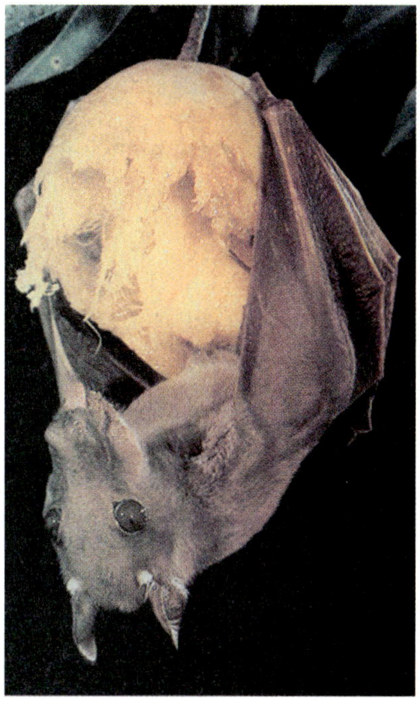

The mango is a source of vitamin C for the Indian fruit bat.

The bulbul bird also requires a dietary source of vitamin C.

humans, nor the oldest, including the invertebrates and fishes, can produce ascorbic acid. The intermediate orders in the chronology of evolution, including amphibians, reptiles, and the more ancient birds, developed the ability to form ascorbic acid themselves. Then, as the thread of evolution lengthened, still higher orders lost the ability, perhaps in making way for more advantageous biochemical processes. The other primates, along with humans, also need to obtain ascorbic acid from food, as do the Indian fruit bat, the guinea pig, and a favorite songbird of the Orient, the bulbul bird.

■ QUESTION

What is the first visible symptom of a severe deficiency of vitamin C? What is the origin of this symptom?

19.7 Vitamins: Myths and Realities

Although remarkable curative powers have been ascribed to various vitamins, few have actually been demonstrated with any kind of rigor. The major medical use of the vitamins lies largely in curing the deficiency diseases, those caused by their absences from the diet. Vitamin A certainly alleviates the night blindness caused by a deficiency of vitamin A; vitamin D unquestionably prevents rickets; vitamin C dramatically cures scurvy.

Orange juice is one dietary source of vitamin C.

But other medical claims that have been made for the vitamins are spurious or unproven. Allegations that vitamin C cures the common cold or is effective in treating cancer have never been demonstrated clearly and unambiguously. Fortunately for anyone who takes massive doses of ascorbic acid as a dietary supplement, the vitamin is relatively nontoxic and is highly soluble in water, with 1 g dissolving in just 3 mL of water. Moreover, ascorbic acid isn't very soluble in fat so it isn't stored in the body. When taken in amounts that far exceed the daily requirement, ascorbic acid leaves through the urine almost as soon as it enters through the mouth. The consequences of taking large supplements of the vitamin over a long period aren't known with any certainty.

On the other hand, the common belief that cooking foods for a long time destroys vitamin C is certainly true. In the opening demonstration you saw the ease with which vitamin C reduces iodine. By looking at the same reaction from another chemical perspective, we can say that iodine easily oxidizes the vitamin. So does the oxygen of air. As you can demonstrate by using water of different temperatures, the oxidation of vitamin C takes place more rapidly at high temperatures than at low temperatures. Cooking food for extended periods can lower its content of the vitamin through air-oxidation. Simply dissolving a powder gelatin dessert mix in boiling water before chilling it, for example, results in oxidation of a quarter to a third of its vitamin C. Even refrigerating it for long periods leads to a decrease through oxidation. As another illustration of the fragility of vitamin C, orange juice refrigerated in paper cartons reportedly loses half its vitamin C in just three weeks.

An **antioxidant** is a chemical so easily oxidized itself that it protects others from oxidation.

The exceptional ease with which vitamin C is oxidized makes it a fine **antioxidant**: a compound that can protect other compounds from oxidation by sacrificing itself instead. We saw something of this sort in Section 10.16, with the use of galvanizing to protect iron.

■ QUESTION

In its chemical reactions, is ascorbic acid likely to be a good source of electrons to be transferred to other reagents, or a good acceptor of electrons transferred from other reagents? Explain.

19.8 Natural versus Synthetic

All pure ascorbic acid is identical, regardless of its source. The vitamin C of food is indistinguishable from vitamin C prepared in the laboratory.

Another set of myths and realities focuses on the distinction between "natural" and "synthetic" vitamins. We'll examine this topic through a set of questions: Does it matter whether we get our micronutrients principally from our foods or largely from a chemical processing plant? Is the calcium in a vitamin and mineral supplement any different from the calcium that's in milk? Is the vitamin C that comes out of a chemist's laboratory as good as the vitamin C in a real orange? The answers to these questions depend on a bit of chemistry.

The answer to the question "Are there different kinds of calcium?" is *no* (except for insignificant differences among its isotopes). As we saw in Chapter 2, all atoms that contain exactly 20 protons in their nuclei are atoms of calcium. That holds true no matter what their source. A calcium atom, in short, is a calcium atom no matter where it comes from.

But what comes along with it *could* matter. Recall that we need vitamin D to absorb calcium into our bloodstream and phosphorus to add the calcium to our bones and teeth. So while all calcium atoms are identical in their chemical behavior, the calcium of food is much more likely to carry along with it any vitamin D and phosphorus that we might need than is the calcium of, say, an antacid tablet, or a piece of chalk, or a poorly formulated supplement.

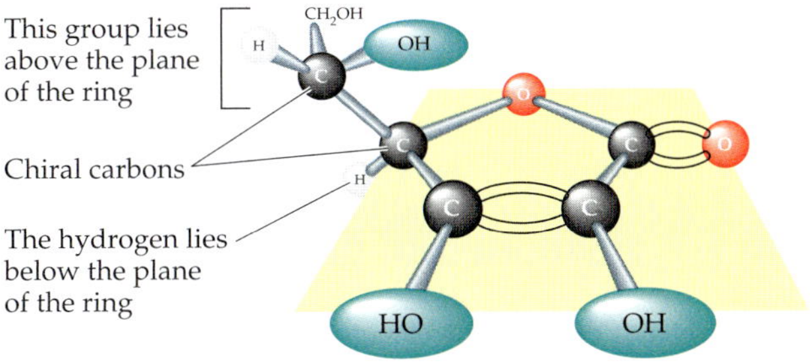

Figure 19.5 The stereochemistry of vitamin C. All molecules of vitamin C have this structure and stereochemistry, regardless of their source.

Like all atoms of one isotope of any particular element, all molecules of any particular compound are mutually identical. For example, it takes exactly 20 protons, 20 electrons, and 20 neutrons, arranged so that all the protons and neutrons form the nucleus and all the electrons are organized into shells surrounding the nucleus, to make up the most common isotope of a calcium atom. Similarly, it takes exactly 6 carbon atoms, 6 oxygen atoms, and 8 hydrogen atoms to make up a molecule of vitamin C. Moreover, the atoms must all be organized into a precise arrangement, even to the exact arrangement of the atoms bonded to the chiral carbons (Sec. 16.5; Fig. 19.5).

With this specific molecular structure, including the necessary chirality, we have vitamin C; without it we don't. Any molecule that *does* have the structure and chirality shown in Figure 19.5 *is* vitamin C; any other molecule *isn't* (Fig. 19.6). It doesn't matter whether the molecule is formed inside an orange as it grows and ripens or whether it's formed in the apparatus of a chemical laboratory. It all depends on the molecular structure and *only* on the molecular structure. Any molecule that has this particular arrangement of atoms and chirality is indistinguishable from any other of exactly the same structure and chirality, regardless of its origin, and has all the chemical, physical, and biological properties of vitamin C. *Pure* ascorbic acid obtained from living things is absolutely identical in every way, but one, with *pure* ascorbic acid (of the same chirality) synthesized in the laboratory. The *single* difference between the two is the (usually) higher cost of the "natural" vitamin. The same identity of "natural" and "synthetic" holds true for all other vitamins as well.

Figure 19.6 Vitamin C and molecular structure.

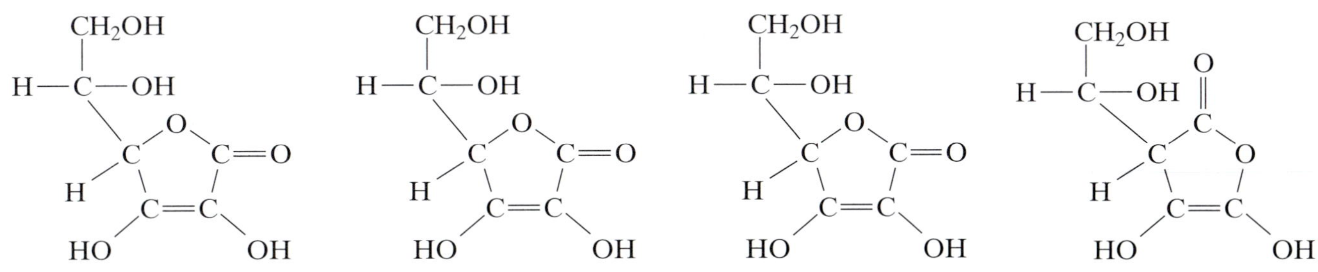

A molecule of vitamin C isolated from an orange, lemon, or lime

A molecule of vitamin C from a kilogram batch synthesized in large vats by a major pharmaceutical firm

A molecule of vitamin C of unknown origin, found in an ancient bottle, in an old attic

A molecule of a substance that may or may not have useful properties, but that is definitely *not* vitamin C

■ **QUESTION**

What is the single factor that determines whether or not a given molecule is vitamin C?

19.9 Foods versus Pills

The discussion of "natural" versus "synthetic" vitamins brings us to a related question: If all molecules of any particular vitamin are mutually identical, regardless of their source, does it matter whether we get our vitamins from a well-balanced diet, or can we eat what we please and rely on vitamin and mineral supplements for our micronutrients? The answer seems to be that it matters very much whether we get our vitamins from food or pills, that a well-balanced diet is far superior to reliance on vitamin and mineral supplements. The following explains why.

While there's no solid evidence that vitamins can *cure* any disease (other than deficiency diseases), there is circumstantial evidence that some vitamins, particularly C and E, and the β-carotene that forms vitamin A, might *prevent* certain diseases, especially heart disease and some forms of cancer. The problem with this kind of evidence is that it comes largely from two different kinds of studies, involving people

- whose diets are rich in fresh fruits and vegetables, and
- others who report taking high doses of vitamin supplements.

The correlation between diets containing plentiful amounts of vitamin-containing foods—especially fresh fruits and vegetables—and a low incidence of cancer and other diseases is well established. Based on evidence of this sort, government agencies and health organizations have been urging us for many years to increase the proportions of fresh fruits and vegetables in our diets as protection against heart disease, cancer, and perhaps other illnesses.

Yet even an established connection between consumption of foods that contain plenty of vitamins and low cancer rates doesn't prove that the vitamins are the active agents in disease prevention. There are many uncertainties. Perhaps specific, as-yet-unidentified components of these foods, other than vitamins, guard against the disease. Or maybe there are other, as-yet-unrecognized protective characteristics common to people who eat lots of fresh fruits and vegetables. The same sort of reasoning also clouds any conclusions about those who take vitamin supplements.

Several *double-blind* studies (Perspective, Chapter 23) have been carried out to clarify these ambiguities and to pinpoint the effects of the vitamins themselves. One, which ended in 1996, examined a group of about 18,000 men and women who were considered to have a high risk of lung cancer because of their history of smoking or exposure to asbestos. Those who took large doses of β-carotene and of vitamin A daily for at least four years showed a 28% increase in incidence of lung cancer and a 17% increase in overall death rate, compared to those in the group who did not take these large doses of the supplements. These results echo the conclusions of a study of 30,000 male smokers in Finland that ended in 1994. Smokers in that study who took supplements of vitamin E over periods of five to eight years were *more* likely to die of strokes than those who did not take the supplements, and those taking β-carotene were more likely to die of lung cancer or heart disease. (In still another recent study of 22,000 male physicians, taking β-carotene supplements every other day for 12 years produced no benefits and did no harm.)

One possible interpretation of the results of these studies is that plant chemicals (*phytochemicals*, from the Greek word *phuton*, for plants) other than vitamins might produce the protective effects noted in previous studies. We'll have more to say about this in the Perspective of this chapter.

■ QUESTION

What evidence suggests that plant chemicals other than vitamins may offer protection against cancer, heart disease, and strokes?

19.10 Health Foods versus Junk Foods

To this point in our examination of the chemicals of our foods, we've considered the chemicals that appear in foods without any human intervention: the macronutrients and vitamins that plants and animals form through their own biochemical processes and the minerals that plants receive directly from the soil and animals get from the foods they eat. We'll soon look at some of the chemicals we ourselves add to our foods as we prepare, cook, or process them for our tables and that food manufacturers and processors add to foods they package.

Before moving to these food additives we'll examine one final pair of categories: *health foods* and *junk foods*. We sometimes speak of highly processed foods, sweet foods, snacks, and/or foods served by fast-food restaurants as junk foods. On the other hand, we sometimes think of unprocessed or uncooked fresh fruits and vegetables or foods with little or no sugar, fat, or cholesterol as health foods. Descriptions like these can be a bit vague and hard to apply in specific instances.

It's clear that one of the difficulties in discussing health foods and junk foods stems from the lack of a general agreement on what these terms mean. One way of sorting one from the other is by balancing the number of calories in any particular food against its content of minerals and vitamins. Since we need energy, in moderate amounts, to maintain life and health and since *all* foods provide energy, we can differentiate among our foods by focusing on the relative proportions of their calories and their micronutrients. (Since the micronutrients don't supply us directly with energy, we are dealing with two distinct categories here, with no overlap.) Using this approach and applying what we have learned about the chemistry of food, we can describe at least one of the two classes of foods in chemical terms. A **junk food**, by one popular definition that also makes good chemical sense, is a food that supplies a large number of calories but few micronutrients.

By one definition, a **junk food** is a food that supplies a large number of calories but few micronutrients.

A "health food" store, where we can find "organic foods," "natural foods," and foods "without chemicals."

Sometimes the term *empty calories* enters into these discussions, often with reference to a food, such as highly refined sugar, that provides calories but few or no micronutrients to accompany them. Highly refined sugar is, as we know, purified sucrose, a carbohydrate. As such it's one of the three classes of macronutrients and furnishes us with energy at the rate of 4.0 Cal/g. We saw in Chapters 8 and 16 that sucrose and other carbohydrates provide the energy we need for our basal metabolism, physical activities, and specific dynamic action. Consumed in excessive amounts, sugar, or any of the other macronutrients for that matter, also provides the energy we store as fat.

Turning now to health foods, we find the label used so loosely and with so little general agreement that it's nearly impossible to define. Nonetheless, using the idea that the relative proportions of calories and micronutrients are important to good health, we can conclude that, in general, the larger the variety and amount of micronutrients that a food provides and the fewer its calories, the more likely we are to consider it a **health food**. While this doesn't provide a precise definition or description of a "health food" and doesn't take into consideration cholesterol, saturated fats, fiber, and other factors that affect our state of health, it does give us a basis for examining various foods and assessing their nutritional value. In the next section we'll look at some quantitative ways to estimate the levels of these components.

By one definition a **health food** is a food that supplies a large number of micronutrients compared with its calories. This definition does not take into consideration the food's content of saturated fats, cholesterol, or other factors that can affect our health.

■ QUESTION

Is there any processed, packaged food sold in grocery stores or supermarkets that can be considered a health food? If your answer is *yes*, identify one. If it's *no*, explain why not.

19.11 Nutrition Facts, Daily Values, and Ingredients Labels

Tables 19.1 and 19.2 show us what kinds of foods provide the minerals and vitamins we need. But this leaves us with questions of how much we need of each and what quantities our foods provide.

To help answer the question of how much we need, since 1943 the National Academy of Sciences and the National Research Council, organizations of distinguished U.S. scientists and engineers, have published periodically the recommendations of their Food and Nutrition Board. These have been summarized as the Recommended Dietary Allowances, or RDAs, and, as stated in the publication, are "the levels of intake of essential nutrients that, on the basis of scientific knowledge, are judged by the Food and Nutrition Board to be adequate to meet the known nutrient needs of practically all healthy persons." The RDAs are, in brief, the average daily amounts of various nutrients that, in the opinion of a group of well-informed scientists, most of us should consume to maintain good health. They're considered to be at least adequate for nearly everyone and certainly more than enough for most of us. Table 19.3 presents a representative set of RDAs for minerals and vitamins selected from the 1989 publication.

The RDAs serve a valuable role as quantitative standards for recommending daily dietary intakes of our micronutrients, yet they suffer from a few weaknesses. Although establishing RDAs for various categories of gender, age, weight, height, and conditions such as pregnancy and lactation increases the statistical and

Table 19.3 Selected Recommended Dietary Allowance (RDAs) for Minerals and Vitamins[a]

		Weight		Height		Minerals				Vitamins		
Gender	Age	kg	lb	cm	ft′in.″	Ca (mg)	P (mg)	Fe (mg)	Mg (mg)	A (RE[b])	C (mg)	D (μg[c])
Both	0–0.5	6	13	60	2′	400	300	6	40	375	30	7.5
	0.5–1	9	20	71	2′4″	600	500	10	60	375	35	10
	1–3	13	29	90	2′11″	800	800	10	80	400	40	10
Male	15–18	66	145	176	5′9″	1200	1200	12	400	1000	60	10
	19–24	72	160	177	5′10″	1200	1200	10	350	1000	60	10
	25–50	79	174	176	5′10″	800	800	10	350	1000	60	5
	51+	77	170	173	5′8″	800	800	10	350	1000	60	5
Female	15–18	55	120	163	5′4″	1200	1200	15	300	800	60	10
	19–24	58	128	164	5′5″	1200	1200	15	280	800	60	7.5
	25–50	63	138	163	5′4″	800	800	15	280	800	60	5
	51+	65	143	160	5′3″	800	800	10	280	800	60	5
—Pregnant						1200	1200	30	320	800	70	10
—Lactating												
(First 6 months)						1200	1200	15	355	1300	95	10
(Second 6 months)						1200	1200	15	340	1200	90	10

[a]Condensed version of Recommendations by the Food and Nutrition Board of the National Academy of Science, National Research Council. Published in 1989.

[b]RE represents retinol equivalents.

[c]1 μg (one microgram) = 10^{-6} g = 10^{-3} mg.

scientific value of the data, a direct application to the population at large, through printed nutrition labels of processed foods for example, becomes confusing and impractical. Furthermore, nutritional work has extended our knowledge of dietary requirements (and benefits of related recommendations) beyond the micronutrients to categories including macronutrients and fiber.

To accommodate these and related concerns, two new categories have been established. The *Reference Daily Intake* (*RDI*) represents the highest RDA for each micronutrient (excluding pregnancy and lactation). This allows easy reference to both genders and all age groups. In addition, the *Daily Reference Value* (*DRV*) covers a related set of data for the calorie-providing macronutrients.

At the expense of adding still another set of values, both the RDI and the DRV have been consolidated into a single value, more convenient to use, to print on packaging labels, and to understand: the *Daily Value* (*DV*). The DV represents the daily recommended level of a micronutrient or a macronutrient, based on a diet of a specific number of calories. On nutrition labels it's generally represented as a "% Daily Value" for a diet of 2,000 Calories per day (Fig. 19.7). (As we might expect, the DV for fat and other calorie-providing macronutrients depends on the number of Calories in your individual diet; the DV for the micronutrients doesn't.)

This "% Daily Value" appears in a section titled "Nutrition Facts" on packages of processed foods. Below this you can usually find a list labeled "Ingredients" that presents the ingredients of the packaged food, in order of decreasing weight. Except for the plant or animal materials that may form the bulk of the food, these ingredients represent the chemicals we think of as food additives. We'll examine these in the remainder of this chapter.

Figure 19.7 Nutrition labels showing % Daily Values.

■ **QUESTION** What's the value of the RDI for vitamin D?

19.12 How the Search for Food Additives Led to Marco Polo's Adventures in the Orient, Columbus's Voyage to America, and Dr. Wiley's Poison Squad

Food additives of one sort or another have been with us for a long time. They've been in use since prehistoric cooks began salting and smoking meat to preserve it against decay. Food dyes were used to improve the appearance of food by the Egyptians as early as 3500 years ago. Adding herbs, spices, and sweeteners is also an ancient practice.

Another ancient practice, especially among the manufacturers and vendors of food, was the use of additives to make bad food more palatable. Covering the foul tastes and odors of spoiled food was the principal function of spices in the Middle Ages. Indirectly, the lack of good methods for preserving food in those times led to the opening of trade routes to the Orient and to the European discovery of the Western Hemisphere. It was at least partly the search for foreign spices that led Marco Polo eastward to the Orient and Columbus westward to the New World.

As the centuries passed and people throughout the world moved from the countryside to the cities in more modern times, the distance from the farm and the dairy herd to the dinner table lengthened and so did the time it took to transport food from where it was grown to where it was to be eaten. Because of a lack of effective methods for preserving foods—efficient and economical mechanical refrigeration is a product of the 20th century—food spoilage became widespread and so did the use of chemicals to preserve food and to mask the taste and odors of decay, just as in the Middle Ages.

With the development of chemical technology and food processing late in the 19th century, the use of chemical additives grew enormously. By 1886 the U.S. Patent Office had issued its first patent for a food additive, a combination of sodium chloride and calcium phosphate designed for use as a food seasoning. As the use of additives grew, the variety of the chemicals used as additives also increased, and so did their hazards.

In 1902, Dr. Harvey W. Wiley of the United States Department of Agriculture organized a Poison Squad of volunteers to begin the first tests of the effects of food additives on humans.

We now have effective governmental regulations to ensure the safety of the chemicals added to our foods. That wasn't always the case. In the United States of the late 1800s, the uninformed and indiscriminate use of chemicals as food additives led to frequent and sometimes fatal outbreaks of illness and, eventually, to detailed regulation by the federal government. The first step, taken in 1902, was the formation of a group within the U.S. Department of Agriculture to examine the usefulness and the dangers of food additives. Headed by Dr. Harvey W. Wiley, an American chemist who was at the time the chief of the Department's Bureau of Chemistry, the group became known as Dr. Wiley's Poison Squad. His group of a dozen healthy young volunteers taken from the Department actually ate the additives under investigation, along with their regular, carefully supervised meals, while Wiley watched for signs of illness.

Wiley's investigations and his vigorous public support of effective control of food additives, coupled with public indignation at the filth of slaughterhouses and meat products depicted in Upton Sinclair's novel *The Jungle*, stimulated Congress to pass both the Meat Inspection Act of 1906 and the Pure Food and Drug Act of 1906. Since 1906, the enforcement of the Pure Food and Drug Act and its subsequent revisions has been the responsibility of organizations bearing a variety of names and titles. In 1931 Congress created the Food and Drug Administration (FDA) to ensure the safety of the chemicals added to the food and drugs sold to the public. The FDA is now part of the U.S. Department of Health and Human Services.

■ QUESTION

What chemicals were used to formulate the first patented food additive? What was the function of the additive?

19.13 The Law of Additives

Several amendments and other changes since 1906 have extended and strengthened the Pure Food and Drug Act. Currently, the law also covers cosmetics and is known as the Federal Food, Drug, and Cosmetic Act. In effect, this law protects the public from unsafe food additives by declaring that *all* substances that might be added to food, in any way, are legally unsafe unless they are specifically exempted from the Act or unless they are used in ways specifically permitted by the Act. As we might expect, the effective operation of the law requires a definition of a food additive (and of the various exemptions) more attuned to the ears of lawyers, administrative officials, and judges than to the ears of the general public or scientists. The Act's definition of a food additive has thus been embellished into the legal jargon of the *single sentence* of Figure 19.8.

To paraphrase this elaborate, legal definition, a food additive is any substance added to food, directly or indirectly, except for

- those shown by scientific studies to be safe, at least under the conditions of their use in foods;
- those used as additives before January 1, 1958 (the date this part of the Act became effective), and shown to be safe either by scientific studies or by our common experience;
- pesticides;
- color additivcs;
- substances approved by earlier acts of Congress; and
- new animal drugs.

To work backward through this list of exemptions, the bottom four are defined and regulated either in other laws or in some other section of this Act,

Figure 19.8 The legal definition of a food additive as it appears in the current version of the federal Food, Drug, and Cosmetic Act.

The term "food additive" means any substance the intended use of which results or may reasonably be expected to result, directly or indirectly, in its becoming a component or otherwise affecting the characteristics of any food (including any substance intended for use in producing, manufacturing, packing, processing, preparing, treating, packaging, transporting, or holding food; and including any source of radiation intended for any such use), if such substance is not generally recognized, among experts qualified by scientific training and experience to evaluate its safety, as having been adequately shown through scientific procedures (or, in the case of a substance used in food prior to January 1, 1958, through either scientific procedures or experience based on common use in food) to be safe under the conditions of its intended use; except that such term does not include—

1. a pesticide chemical in or on a raw agricultural commodity; or
2. a pesticide chemical to the extent that it is intended for use or is used in the production, storage, or transportation of any raw agricultural commodity; or
3. a color additive; or
4. any substance used in accordance with a sanction or approval granted prior to the enactment of this paragraph pursuant to this Act, the Poultry Products Inspection Act (21 U.S.C. 451 and the following) or the Meat Inspection Act of March 4, 1907 (34 Stat. 1260), as amended and extended (21 U.S.C. 71 and the following); or
5. any new animal drug.

so they aren't actually considered to be food additives by this particular section of the law.

The second category results from a combination of expediency and common sense. By January 1, 1958, the date this particular section of the Act went into effect, so many substances had been used as food components regularly and for so long that testing all of them would have been entirely impractical. Instead, panels of scientists were set up to evaluate the safety of these materials that had been in common and long-term use. Those that were generally recognized by the panelists to be safe, on the basis of common experience (but without laboratory testing), were compiled into a list of substances *G*enerally *R*ecognized *A*s *S*afe, the **GRAS list**. In effect, the GRAS list constitutes the second category of exemptions. (Legally, GRAS substances aren't food additives either.)

The **GRAS list** is a list of food additives that are generally recognized as safe by a panel of experts but that have not been subjected to laboratory testing.

This GRAS list is by no means fixed and unchangeable, nor are all the substances on the list necessarily free of hazard. They are simply generally recognized as safe, and they are reviewed periodically to determine their continued suitability for GRAS status. Table 19.4 contains examples of chemicals and other substances currently on the list, together with brief descriptions of their functions in food. We'll look at some of their uses in greater detail in the sections that follow.

The remaining exemption, the first one of the series, requires that newly developed chemicals, which are by no means GRAS, be shown by scientific studies

Table 19.4 Examples of GRAS Substances

Substance	Structure or Chemical Formula	Use or Classification
Acetaldehyde	$CH_3-\overset{\overset{\displaystyle O}{\|}}{C}H$	Spices, seasonings, and flavorings
Anise		
Cinnamon		
Ethyl acetate	$CH_3-\overset{\overset{\displaystyle O}{\|}}{C}-O-CH_2-CH_3$	
Aluminum calcium silicate	$CaAl_2(SiO_4)_2$ and $Ca_2Al_2SiO_7$	Anticaking agents
Sodium metabisulfite	$Na_2S_2O_5$	
Sorbic acid	$CH_3-CH=CH-CH=CH-CO_2H$	
Ascorbic acid	$C_6H_8O_6$	Dietary supplements and nutrients
Calcium phosphate	$Ca_3(PO_4)_2$	
Ferrous sulfate	$FeSO_4$	
Linoleic acid	$CH_3-(CH_2)_4-CH=CH-CH_2-CH=CH-(CH_2)_7-CO_2H$	
Zinc oxide	ZnO	
Dipotassium hydrogen phosphate	K_2HPO_4	Sequestrants (Sec. 19.18)
Trisodium citrate	$\begin{matrix} & CH_2-CO_2Na \\ & \| \\ HO- & C-CO_2Na \\ & \| \\ & CH_2-CO_2Na \end{matrix}$	

to be safe before they can be used in foods. Of course once they are demonstrated to be safe and may legally be added to food, they're no longer legally defined as food additives. What we mean by "safe" and how chemicals are demonstrated to be safe for addition to our foods are topics we'll examine in Chapter 20, after we take a brief look at the more practical side of additives.

There are other good and workable (and certainly simpler) definitions of food additives, aside from the legal one of the Federal Food, Drug, and Cosmetic Act. A more practical definition holds that a **food additive** is simply *anything intentionally added to a food to produce a specific, beneficial result*, regardless of its legal status. Although the extended, legal definition of the Act has the force of U.S. federal law behind it and has the close attention of anyone who processes food for sale to the public, we'll use this simpler definition as we examine what chemical additives are doing in our food and why we use them at all.

By a practical definition, a **food additive** is anything intentionally added to a food to produce a specific, beneficial result.

■ QUESTION

Under what circumstances is a GRAS substance considered to be a food additive? Under what circumstances is it not considered to be a food additive?

19.14 Additives at Work

We use additives for a remarkably simple reason: to improve or maintain the quality of our foods. For many of us, adding some combination of salt, sweeteners such as sucrose or corn syrup, and seasonings like mustard, pepper, and other condiments improves the taste of an otherwise bland dish. Food colorings, another class of additives, enhance the appearance of our foods by converting a pallid frosting, for example, into the colorful decorations of a birthday cake.

Other additives function in less obvious ways. Adding vitamin D to milk doesn't add any appeal to the milk through taste, odor, or color, but it does improve the absorption of calcium into the body, so it helps protect against rickets, as we saw in Section 19.5. Similarly, the potassium iodide added to iodized table salt acts to prevent the formation of goiter (Sec. 19.2). Ascorbic acid (vitamin C), on the other hand, not only improves the nutritional qualities of food but also protects other food components from oxidation by contact with air. The ascorbic acid is itself so easily oxidized that it reacts preferentially with atmospheric oxygen. As a food additive, then, ascorbic acid acts both as a nutrient and as a preservative (specifically, an *antioxidant;* Sec. 19.7).

Although additives serve a large variety of purposes in foods, we can place them into four major groups, according to their function. Generally, chemicals are introduced into foods in order to

- make them more appealing,
- make them more nutritious,
- preserve their freshness and keep them unspoiled, and
- make them easier to process and keep them stable during storage.

These categories aren't exclusive, and they even overlap a bit. For example, we've just seen that through its own preferential oxidation ascorbic acid can act as a preservative as well as a nutrient. In this same sense, any one chemical additive can function in two or more of these categories. Nonetheless, we'll examine the categories individually in the following sections.

■ QUESTION

What food additive acts both to make foods more nutritious and to preserve their freshness and keep them unspoiled?

19.15 Making Food More Appealing ... with Nail Polish Remover!

Among the additives that make our foods more appealing to our sight, smell, and taste are the natural and synthetic sweeteners, colorings, flavor extracts, and flavor enhancers (Table 19.5). Naturally occurring flavorings and fragrances include the essential oils and extracts of plants, such as oil of orange and vanilla extract, all of which contribute a bit of their own tastes and odors to foods. Some of the synthetic organic compounds used for these purposes also occur naturally in the plant essences. A particularly simple example is *ethyl acetate*, a major commercial solvent and a naturally occurring ester (Sec. 13.7) widely distributed in the plant kingdom. Its fruity odor and pleasant taste (when it's considerably diluted) make ethyl acetate valuable in the formulation of synthetic fruit flavorings.

As with so many other chemicals, ethyl acetate has properties that serve very well in a variety of functions. In addition to its pleasant taste and odor, its fine characteristics as a solvent and very low hazard to humans make it an excellent solvent for such consumer products as nail polishes and nail polish removers. Its ability to do either one of these jobs—to add its flavor to foods or to remove nail polish—doesn't detract in the least from its value in the other.

While these natural and synthetic flavorings and fragrances add their own flavors and aromas to our foods, the *flavor enhancers*, such as MSG (the *monosodium glutamate* of Table 19.5), sharpen some of the weaker flavors already present in the food without adding any significant taste of their own. MSG is used in preparing some processed canned and frozen foods, and it's often used in large amounts in Chinese foods. Some people develop symptoms including lightheadedness, headaches, an uncomfortable sense of warmth, and difficulty in breathing after eating a meal containing MSG. Because of the considerable quantities of MSG often used in meals served at Chinese restaurants, the condition has been named the *Chinese restaurant syndrome*. Many Chinese restaurants will omit MSG from a meal on request.

A menu from a Chinese restaurant advising that it does not use the flavor enhancer MSG.

Table 19.5 Representative Substances Used to Increase the Appeal of Foods

Substance	Structure or Chemical Formula	Function
β-Carotene	$C_{40}H_{56}$	Colorant
Ethyl acetate	$CH_3-C(=O)-O-CH_2-CH_3$	Flavoring and fragrance
Ferric oxide	Fe_2O_3	Colorant
Glucose	$C_6H_{12}O_6$	Sweetener
Monosodium glutamate, MSG	$HO_2C-CH_2-CH_2-CH(NH_2)-CO_2Na$	Flavor enhancer
Paprika		Flavoring and colorant
Sucrose	$C_{12}H_{22}O_{11}$	Sweetener
Titanium dioxide	TiO_2	Colorant

■ QUESTION

Monosodium glutamate is a salt of *glutamic acid*. To what class of compounds, described in Chapter 17, does glutamic acid belong?

19.16 Preserving and Protecting Food against Spoilage

The added minerals and vitamins that make our foods more nutritious constitute the next category of additives. Adding vitamin D to milk, for example, protects children against rickets. Canned fruit drinks often contain added vitamin C for nutritional quality. Some of the chemicals we add to food to increase its nutritional value are illustrated in Table 19.6.

Since we've already described the micronutrients in detail earlier in this chapter, we'll move along to the next of the categories of Section 19.14, the preservatives. These keep foods fresh and slow down the process of spoilage and deterioration. Without these preservatives many of our foods would become unpalatable, completely inedible, or even hazardous to our health as they travel from the farm, the dairy, or the processor to our tables. As we saw earlier, chemicals were first added to food as preservatives with the smoking and salting of meat in prehistoric times. Even in those times a rudimentary form of food processing was needed to maintain the stability of the food supply.

Today mechanical refrigeration represents our primary nonchemical mode of preservation. To appreciate the important role of food preservation of every sort to our current way of living, imagine the chaos of a modern world without the freezers and refrigerators of our stores and homes. Even though chemical preservatives are far less visible to us than these freezers and refrigerators, the abandonment of chemical preservatives could produce as great a catastrophe as the disappearance of all forms of refrigeration.

The preservatives we rely on so heavily make up the group of compounds that protect against

- oxidation by air and
- spoilage by bacteria and other microorganisms.

Table 19.7 shows some of these chemical preservatives.

We've already seen that ascorbic acid, by oxidizing preferentially, can slow the reaction between food chemicals and atmospheric oxygen (Sec. 19.14). A pair of widely used antioxidants on the GRAS list are *BHA*, which is chemically a mixture of the two isomers (Sec. 6.6) of *b*utylated *h*ydroxy*a*nisole, known as *2-tert-butyl-4-methoxyphenol* and *3-tert-butyl-4-methoxyphenol*, and *BHT*, also known as *b*utylated *h*ydroxy*t*oluene or *2,6-di-tert-butyl-4-methylphenol*.

The additives that guard food against the growth of bacteria and other microorganisms not only preserve the taste and appearance of food but also

Table 19.6 Representative Substances Used to Improve Nutrition

Substance	Chemical Formula	Function or Classification
Ascorbic acid	$C_6H_8O_6$	Vitamin C
β-Carotene	$C_{40}H_{56}$	Provitamin A[a]
Ferrous sulfate	$FeSO_4$	Mineral
Potassium iodide	KI	Prevents goiter
Riboflavin	$C_{17}H_{20}N_4O_6$	Vitamin B_2
Zinc sulfate	$ZnSO_4$	Mineral

[a] A provitamin is a substance that can be converted easily into a vitamin.

Table 19.7 Representative Substances Used as Preservatives

Substance	Structure or Chemical Formula	Function
Ascorbic acid	$C_6H_8O_6$	Antioxidant and antimicrobial
Butylated hydroxyanisole, BHA	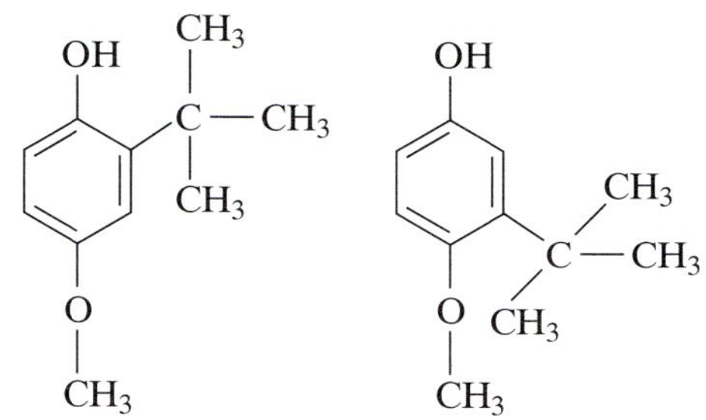	Antioxidant
Butylated hydroxytoluene, BHT	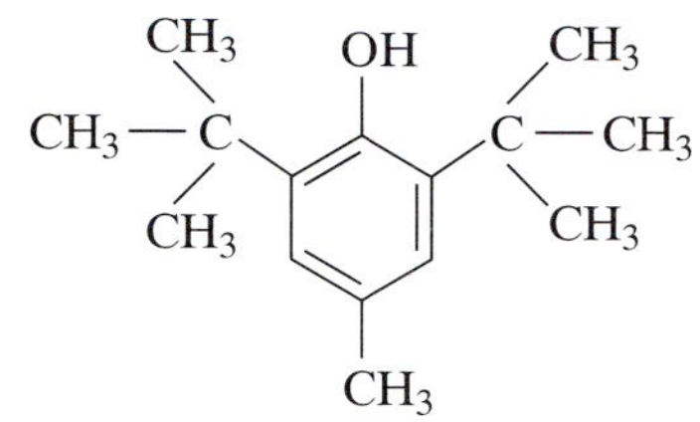	Antioxidant
Calcium propionate	$(CH_3{-}CH_2{-}CO_2)^- \; Ca^{2+} \; ^-(O_2C{-}CH_2{-}CH_3)$	Inhibits growth of molds and other microorganisms
EDTA (ethylenediaminetetraacetic acid)	(see Fig. 19.8) $C_{10}H_{16}N_2O_8$	Antioxidant
Sodium benzoate	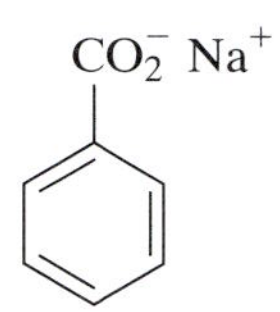	Inhibits growth of microorganisms in acidic foods
Sodium nitrite	$NaNO_2$	Inhibits growth of microorganisms in meat
Sorbic acid and its salts	$CH_3{-}CH{=}CH{-}CH{=}CH{-}CO_2H$	Inhibits growth of molds and yeast, especially in cheese

protect us against microbiologically generated toxins. Labels on breads and other baked goods usually list *calcium propionate* (or, less often, *sodium propionate* or *propionic acid* itself) among their ingredients. Propionic acid occurs naturally in cheeses and other dairy products, and both the acid and its salts effectively inhibit the growth of mold, a major cause of the spoilage of both dairy products and baked goods. Packaged acidic fruit drinks, especially those that don't have to be refrigerated before being opened, such as canned orange juice and grapefruit juice, usually contain added *sodium benzoate*, sometimes shown on the ingredients label as *benzoate of soda*. When present in amounts a little less than 0.1%, the sodium benzoate serves as an effective antimicrobial ingredient.

■ QUESTION

What naturally occurring chemical is used as an additive both to improve nutrition and to add color?

19.17 Chemistry in a Crab's Claw

Some additives protect food against spoilage through complex and indirect but nonetheless intriguing chemical actions. One of these, EDTA or *e*thylene*d*iamine*t*etraacetic *a*cid (sometimes called *edetic acid*), owes its prowess as an antioxidant to a peculiarity of its molecular structure and to a subtle aspect of oxidation chemistry. As shown in Figure 19.9, the intricacy of EDTA's molecular structure rivals the complexity of its name.

Behind EDTA's ability to protect foods against oxidation lies a bit of chemistry involving the catalytic effects of metal ions. Trace quantities of the ions of various metals—aluminum, iron, and zinc, for example—easily enter our processed foods as they go through their many stages of preparation in an assortment of ovens, vats, kettles, and other metal equipment. While these exceedingly small quantities of metal ions don't affect us or our foods directly, they can very effectively catalyze the air oxidation of many of the compounds that are present in food and thereby lead to spoilage. It's important to remove even the smallest traces of metal ions from some foods to protect against this catalytic oxidation. That's where EDTA comes in.

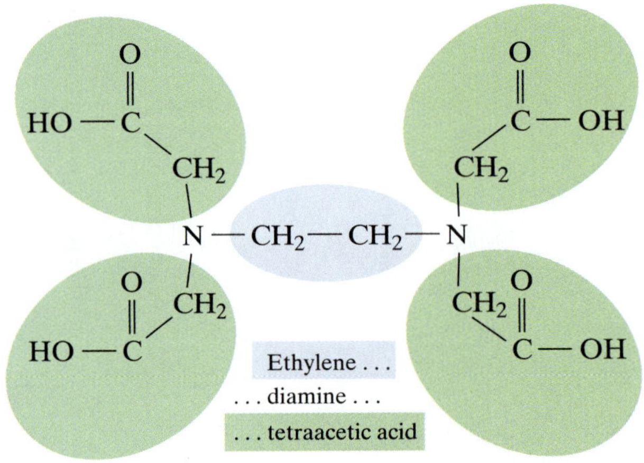

Figure 19.9 Ethylenediaminetetraacetic acid, or EDTA.

A close look at Figure 19.9 shows that the molecule consists of an *ethylene* core (the $—CH_2—CH_2—$ structural unit is the *ethylene* group, just as $—CH_2—$ is the *methylene* group; Sec. 6.5) to which two *amine* nitrogens are attached. Each of these amine nitrogens is bonded to two groups derived from *acetic acid*, for a total of four acidic structural units. Putting all these groups together—one ethylene, two amine, and four acetic acid units—gives ethylenediaminetetraacetic acid, a long name but one that reflects very nicely the architecture of the molecule.

Note that six of the atoms of the EDTA molecule—the two amine nitrogens and four of the acetic acid oxygens—are relatively basic. The amine nitrogens have free, nonbonding electrons that they can use to form covalent bonds to other atoms (Sec. 17.2), and although the four basic oxygens of the acetic acid groups are already bonded to hydrogens, those four oxygens can also form similar bonds as each of their hydrogens is displaced by some other chemical species.

When EDTA or, more commonly, one of its sodium salts is added to a food, the molecule surrounds extraneous metal ions that may be present and bonds tightly to the metal through covalent bonds formed by the free electrons of its nitrogen and oxygen atoms. With this bond formation the EDTA wraps itself around the metal and holds it in a molecular grasp, much as a crab or a lobster holds a pebble in its claw (Fig. 19.10). This analogy gives these combinations of organic molecule and metal ion the name *chelates*, taken from the Greek *khēlē* for a crab's claw. What's more, with their firm hold on the metals, these chelates, in effect, remove them chemically from contact with the foods. As a result they prevent any traces of the metallic impurities from catalyzing a reaction of the food with air, as the metals might in the absence of the additive. As the chelating agents incorporate the metals into their own structures and remove them from chemical contact with the food, the chelating agents become **sequestrants**, from the English word *sequester*, meaning to remove or to set apart.

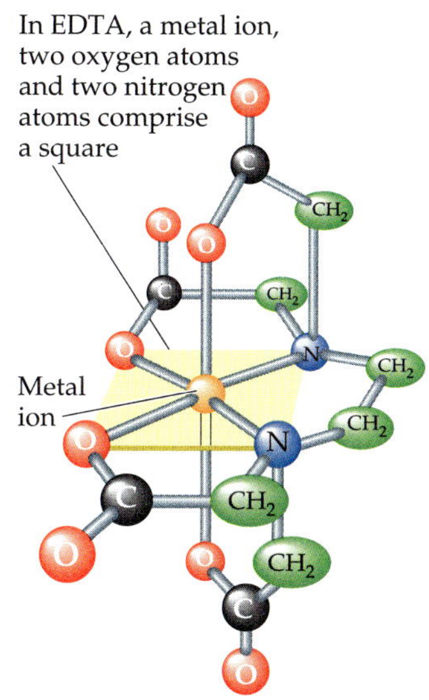

Figure 19.10 The chemical claws of EDTA. The positions of the nitrogens and the oxygens within the molecule allow it to bond tightly to a metal through covalent bonds, thereby preventing the metal from catalyzing the oxidation of components of the food.

A **sequestrant** is an agent that is able to bond with a metal ion so firmly that it removes the metal from chemical contact with other substances.

To summarize, the chemical significance of this sequestering action lies in the ability of even trace amounts of free metal ions to catalyze the air oxidation of many of the compounds of foods. Isolating the ions into the chelate molecule effectively eliminates their catalytic activity and preserves food quality.

EDTA and other sequestrants, including salts of citric acid and phosphoric acid, are used as additives in fats and oils to protect the unsaturated side chains of triglycerides against the oxidation reactions that turn the fats and oils rancid. They're also added to foods to protect various vitamins, including the easily oxidized vitamin C, and to guard against air-induced discoloration of canned and processed corn, potatoes, and other vegetables, as well as fish, shellfish, dairy products, vinegar, meats, and many other kinds of foods.

■ QUESTION

What characteristic of the nitrogen and oxygen atoms of EDTA allows the molecule to act as a chelating agent?

19.18 Stabilizers: Moist Coconut, Soft Marshmallows, and Creamy Peanut Butter

Finally, there's the class of processing aids and stabilizers, which make it easier to prepare and use processed foods and also help prevent undesirable changes in the appearance or physical characteristics of foods while they're being stored. Among these are *humectants*, such as the *glycerine* that keeps shredded coconut moist and the *glyceryl monostearate* that softens marshmallows. *Anticaking agents*, such as

Table 19.8 Representative Substances Used in Processing and to Maintain Stability

Substance	Structure or Chemical Formula	Function
Acetic acid	$CH_3—CO_2H$	Control of pH
Calcium silicates	$CaSiO_3, Ca_2SiO_4, Ca_3SiO_5$	Anticaking agents
Glycerine	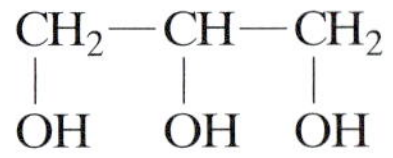	Humectant
Glyceryl monostearate	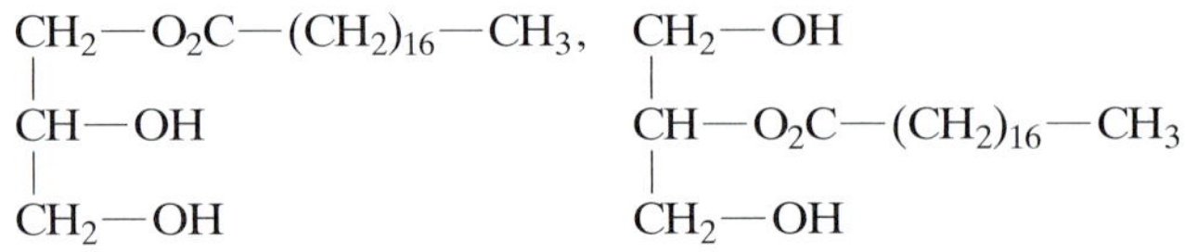	Humectant
Gum arabic	(Gummy plant fluid)	Thickener, texturizer
Mono- and diglycerides	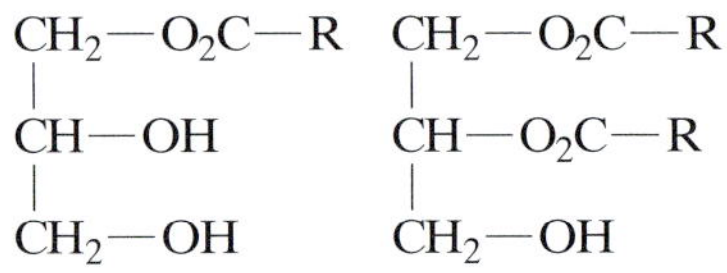	Emulsifiers
Phosphoric acid and its salts	$H_3PO_4, NaH_2PO_4, Na_2HPO_4, Na_3PO_4$	Control of pH
Silicon dioxide	SiO_2	Anticaking agent
Xanthan gum	Complex polysaccharide from corn fermentation	Emulsifier, thickener

silicon dioxide and the *calcium silicates*, keep table salt, baking powder, and other finely powdered food substances dry and free-flowing while they're being processed and as they stand on the store shelf or in the kitchen cabinet.

Glyceryl monostearate and other *mono-* and *diglycerides* also serve as *emulsifiers* that help whip peanut butter into a creamy smoothness and keep it that way until the jar is empty. *Xanthan gum* helps blend the oils and water that make up salad dressings and stabilizes the product as a smooth, homogeneous mixture. Chemically, emulsifiers are surfactants (Sec. 13.4) that make it possible to mix intimately two phases, oil and water for example, that don't dissolve in each other. Acting very much like the laundry detergents of Chapter 13, the emulsifiers convert one of the phases into micelles, disperse them in the other phase, and stabilize the mixture by keeping the micelles from coming together. If it weren't for its emulsifiers, peanut butter would soon separate into a thick mass of peanut solids topped off by an unappetizing peanut oil. Table 19.8 presents some typical processing aids and stabilizers.

■ **QUESTION**

Silicon dioxide is often listed as an additive on boxes of table salt. What function does it serve?

19.19 A Return to Vitamin C and Iodine

Vitamin C serves as an appropriate focus for this chapter. On the one hand, it's a vitamin that occurs naturally as a micronutrient in many foods and works to preserve good health. Because of its important role in protein synthesis, for example, it prevents scurvy. On the other hand, it's a food additive: a chemical deliberately added to processed foods to produce a desirable effect.

In the opening demonstration we witnessed the ease with which iodine oxidizes vitamin C and we saw evidence of this vitamin's ability to act as an antioxidant. Because vitamin C reacts so readily with a variety of oxidizing agents, including the oxygen of air, it's added to foods not only to increase their nutritional value but also as a preservative chemical, an antioxidant. This dual role illustrates once again the need to view chemicals as agents that serve us well or badly, not because of their intrinsic properties, but through the ways we choose to use them.

PERSPECTIVE

Sodium Nitrite: Balancing Risks and Benefits

Another widely used additive, *sodium nitrite*, $NaNO_2$, presents us with a fine illustration of the need to balance risks against benefits in the use of chemicals as food additives. To sum it up, the risk in sodium nitrite is that it may cause cancer; the benefit is that it protects processed meats effectively against what may be the most deadly of all poisons, the microbial toxin *botulism*, which is responsible for a particularly lethal form of food poisoning. There's also an ironic twist to this story that complicates the issues involved in making any judgments about the value of sodium nitrite. We'll examine that twist after we've considered the risks and the benefits.

First, the risk. There's evidence that, under the chemical conditions that occur in the human stomach during the digestion of food, sodium nitrite can transform some of the digestive and metabolic products of proteins into a class of compounds known as *nitrosamines*. As a class of compounds nitrosamines are among the most powerful **carcinogens** (cancer-causing agents) known. They've been found to produce cancer in every species of laboratory animal tested.

The conversion of sodium nitrite into a nitrosamine occurs as a part of a three-step process. It's known that when sodium nitrite and compounds like it dissolve in water they ionize to form a *nitrite anion*, NO_2^-:

$$\underset{\text{sodium nitrite}}{NaNO_2} \underset{\text{ionizes to}}{\longrightarrow} \underset{\text{sodium cation}}{Na^+} \underset{\text{and}}{+} \underset{\text{nitrite anion}}{NO_2^-}$$

It's also known that the nitrite anion can react with hydrochloric acid (which occurs in our gastric juices) to form *nitrous acid*:

$$\underset{\text{nitrite anion}}{NO_2^-} + \underset{\text{hydrochloric acid}}{HCl} \longrightarrow \underset{\text{nitrous acid}}{HNO_2} + \underset{\text{chloride anion}}{Cl^-}$$

Finally, it's known that nitrous acid, the product of this reaction, can react with certain types of amines (similar to those formed during the digestion of proteins) to form nitrosamines. In reacting with *dimethylamine* in the laboratory, for example, nitrous acid generates *N-nitrosodimethylamine*. (The *N-* in the name of the compound tells us that the *nitroso* group, —NO, is bonded directly to the nitrogen and not to a carbon of the methyl groups):

$$\underset{\text{nitrous acid}}{HNO_2} + \underset{\text{dimethylamine}}{CH_3{-}NH{-}CH_3} \longrightarrow \underset{N\text{-nitrosodimethylamine}}{CH_3{-}\overset{\overset{\large NO}{|}}{N}{-}CH_3}$$

Whether sodium nitrite, at the additive levels currently used, does actually generate nitrosamines in the human stomach and whether these nitrosamines (if they are, indeed, formed in our digestive system) do actually increase the risk of cancer in humans are uncertain at present.

We might think that any doubt and uncertainty about the connection between sodium nitrite and cancer is irrelevant. We might think that even the suspicion that sodium nitrite could generate a carcinogen would be enough to banish it as a food additive. In fact, though, the use of sodium nitrite as a food additive brings with it a very real and very important benefit: Sodium nitrite is very effective at preventing the growth of a microorganism known as *Clostridium botulinum*, which produces botulinum toxin and is responsible for botulism food poisoning. (We'll have more to say about botulism poisoning in Chapter 20.) While removing sodium nitrite from meats might reduce the risk of cancer, removing this preservative would also certainly increase the risk of a particularly lethal form of food spoilage.

As for the ironic twist we mentioned earlier, we'd find that even a total ban on the use of sodium nitrite and other nitrite salts as food additives wouldn't remove the potential hazard. Only about a third of the nitrite salts in our bodies comes from the use of these additives. The remaining two-thirds results from the action of bacteria that live quite normally in our mouths and digestive systems. These bacteria convert sodium *nitrate* ($NaNO_3$) and other nitrate salts that occur naturally in

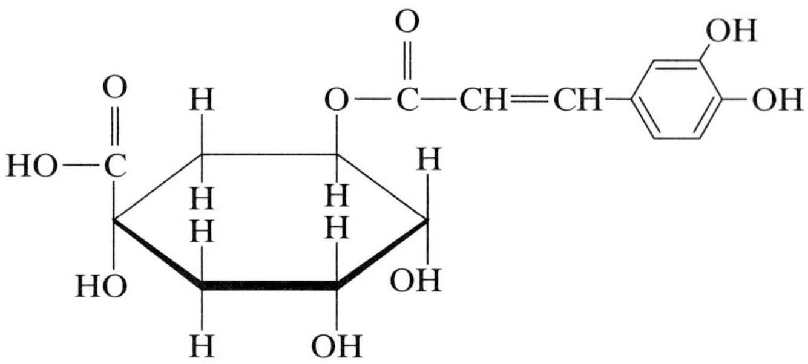

Chlorogenic acid

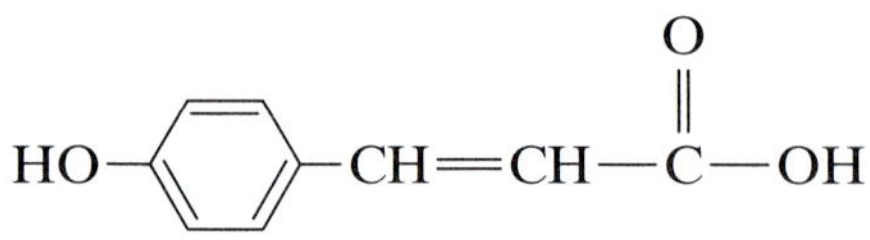

p-Coumaric acid

Figure 19.11 Phytochemicals that protect against nitrosamine formation.

the fresh fruits and vegetables of our diets into *nitrites*, as indicated by the following unbalanced equation:

$$\underset{\text{sodium nitrate}}{NaNO_3} \xrightarrow[\text{digestive system}]{\text{bacteria in the human}} \underset{\text{sodium nitrite}}{NaNO_2}$$

Thus nitrites might still be with us through the action of very natural processes even if they were never again used as food additives. Yet, in another twist, some fresh fruits and vegetables may provide their own protection against the hazards of the nitrites they bring with them. *Chlorogenic acid* and *p-coumaric acid* (Fig. 19.11), for example, two phytochemicals (Sec. 19.9) that occur in various fruits and vegetables—including carrots, green peppers, pineapples, strawberries, and tomatoes—can react with nitrous acid and thus inhibit its reaction with amines.

Instead of trying to assess the value of food additives by balancing the risks of their use against the benefits they provide, perhaps we can gain a clearer picture by balancing risks of one activity against risks of another. In the case of nitrites, we would balance the risk of generating a carcinogen within our own digestive systems (by using nitrites as additives) against a risk of the botulinum organism and its toxin contaminating our processed meats (by not using nitrites). Each is a risk of uncertain magnitude.

Here, as in so many other cases in which we must perform a balancing act of this kind, chemistry can provide us with an understanding of our options and their potential consequences. In the matter of food additives, as we have already seen for the use of nuclear energy and petroleum products and gasoline, and as we will see again in later chapters, we must ultimately apply our own judgment and sense of values to determine where the best balance lies. ■

A **carcinogen** is a cancer-causing agent.

EXERCISES

For Review

1. Following are a statement containing a number of blanks and a list of words and phrases. The number of words equals the number of blanks within the statement, and all but two of the words fit correctly into these blanks. Fill in the blanks of the statement with those words that do fit, then complete the statement by filling in the two remaining blanks with correct words (not in the list) in place of the two words that don't fit.

The __________ and minerals of our diet constitute our __________. Except for carbon, hydrogen, nitrogen, and oxygen (and, by some definitions, sulfur), the elements of our foodstuffs make up the class of __________. Of these, __________ is the major mineral of our body, located largely in our __________. Sodium and __________ are the major cations of our bodily fluids, with __________ ions as the major anion. Among our vitamins (conveniently divided into the two categories of the __________ A, D, E, and K and the __________ B and C) are vitamin C, which is known chemically as __________ and which protects against __________; vitamin D, which protects against __________; and vitamin A, which protects against a vision defect known as __________. Much of our vitamin A comes from the __________ of our yellow, orange, and green vegetables.

ascorbic acid	minerals
bones and teeth	micronutrients
calcium	night blindness
β-carotene	rickets

chloride	sodium
fat-soluble	vitamins
manganese	water-soluble

2. Following is a statement containing a number of blanks and a list of words and phrases. The number of words equals the number of blanks within the statement, and all but two of the words fit correctly into these blanks. Fill in the blanks of the statement with those words that do fit, then complete the statement by filling in the two remaining blanks with correct words (not in the list) in place of the two words that don't fit.

In its most general sense, a __________ is any substance deliberately added to a food to produce a __________. Among the additives that make our foods __________ are __________, which adds its flavor to foods; glucose and sucrose, which serve as __________; and ferric oxide and __________, which add __________. Increasing the __________ are additives such as __________, which helps prevent goiter formation, and __________. Several additives, including ascorbic acid and __________, help prevent the air oxidation of foods. __________ protects against oxidation by combining with traces of __________ and reducing their ability to __________ the process. __________ are surfactants that help __________ foods and keep them from separating during storage.

BHA and BHT	legal definition
catalyze	metal ions
color	mono- and diglycerides
EDTA	monosodium glutamate
emulsify	more appealing to the senses
ethyl acetate	nutritional value
ferrous sulfate	potassium iodide
food additive	titanium dioxide

3. Explain, define, or describe the significance of each of the following:

cholecalciferol	GRAS
collagen	retinol
DV	retinol equivalent
FDA	sodium benzoate
goiter	trace elements

4. What is the difference between a macronutrient and a micronutrient? Name two of each.

5. Describe the major function of (a) phosphorus in the bones, (b) sodium in the extracellular fluid, (c) fluoride in the teeth, (d) chloride in the gastric juices, and (e) cobalt in the body.

6. Name a vitamin that plays an important role in (a) protein synthesis, (b) the absorption of calcium.

7. Name two good sources of (a) vitamin A, (b) vitamin C, (c) vitamin D.

8. What is meant by the IU and the RE of vitamin A and what is the advantage of describing the RDAs of vitamin A in terms of its IUs and REs?

9. Why are bleeding gums one of the first symptoms of scurvy?

10. What nutritional characteristic do humans, guinea pigs, Indian fruit bats, and bulbul birds have in common?

11. What are four major benefits a chemical can provide when it is added to a food?

12. What does the term empty calories imply and how can it be used to distinguish between a "junk food" and one that provides substantial nourishment?

13. Describe at least one function each of the following chemicals provides when it is used as a food additive: (a) potassium iodide, (b) iron salts, (c) calcium propionate, (d) calcium carbonate, (e) mono- and diglycerides, (f) acetic acid.

14. Into which of the four groups of Section 19.14 does each of the following additives fall? (a) vitamin D, (b) potassium iodide, (c) pepper, (d) sugar, (e) food coloring?

15. What is the function of a humectant? Name a chemical used as a humectant and name a food to which it is added.

16. The glycerine and the mono- and diglycerides of Table 19.8 enter our bodies as food additives and as the result of the metabolism of one of our macronutrients. Which macronutrient produces these chemicals as it is metabolized?

17. What two functions does ascorbic acid serve when it is used as a food additive?

18. What connection existed between the lack of effective methods of food preservation and the travels of Marco Polo and Columbus?

19. Legally, food colorants are not food additives. Explain why not.

20. What function does sodium nitrite ($NaNO_2$) serve when it is used as an additive in processed meats?

21. What is the hazard of using sodium nitrite as a food additive?

A Little Arithmetic and Other Quantitative Puzzles

22. Using the data of Section 19.1, calculate the ratio of calcium atoms to phosphorous atoms in bones and teeth.

23. One average raw carrot provides us with 8000 IUs of vitamin A. Why is the amount of vitamin A given in terms of IUs rather than in a weight of retinol?

24. Using the information provided in Exercise 23 and in Section 19.4 and assuming that an average carrot weighs 3 oz, determine the weight-percent of retinol equivalent in an average carrot.

25. How many REs of retinol do 5 IUs of retinol represent? How many IUs of retinol does 1 μg of retinol represent?

26. Suppose that one cup of a certain cereal (30 g) provided 6% of the Daily Value of fat for a 2000-Calorie diet. If you wanted to maintain a daily diet of 3000 Calories, what Daily Value of fat would one serving of this cereal represent?

Think, Speculate, Reflect, and Ponder

27. Which of our macronutrients would you expect to be a major source of dietary sulfur?

28. What advantage is there in taking β-carotene rather than retinol as a food supplement?

29. Why is the vitamin A content of food better described in terms of International Units and Retinol Equivalents than in terms of actual weight?

30. Would a chemical analysis of the soil of a vegetable garden reveal (a) what minerals we might find in the vegetables that come from the garden? (b) What vitamins we might find in the vegetables that come from the garden? Explain.

31. Would you expect the vitamin C contained in an orange grown in soil of poor quality to be inferior to the vitamin C contained in an orange grown in high-quality soil? Explain.

32. In what chemical or physical way does the vitamin C produced by chemical reactions carried out in laboratory equipment differ from the vitamin C obtained from fruits and vegetables?

33. Vitamin B_3 is also known as pantothenic acid, which comes from the Greek word *pantothen*, meaning "from all sides" or "from every side." Suggest a reason why *pantothenic* is an appropriate name for this acid.

34. To provide fresh meat on long voyages, sailors of past centuries took with them live cattle, which they slaughtered as they needed meat. With no supplies of fresh fruit and vegetables, the sailors often succumbed to scurvy. Why didn't the live cattle also develop scurvy?

35. People living in the far northern latitudes receive little or no sunshine during the winter, wear a good deal of clothing for protection against the cold, and lack readily available vitamin supplements and fortified milk. Where does their vitamin D come from?

36. Vitamin D protects against rickets (Sec. 19.5). Suppose a child receives the vitamin D only through vitamin supplements. Under what conditions might this child develop rickets?

37. Is the decolorization of tincture of iodine a reliable test for the presence of vitamin C in a food? Explain your reasoning.

38. In cooking fresh vegetables in water, it's advisable to cook them for a short period in the minimum amount of water required. Why does this make good nutritional and good chemical sense?

39. Name a food that you would put into the category of (a) junk food, (b) health food.

40. Why were spices one of the most widely used food additives of the Middle Ages?

41. One definition regards a food additive as "anything intentionally added to a food in order to produce a specific, beneficial result," regardless of its legal status. In what way is this definition (a) more useful than the legal definition? (b) less useful than the legal definition?

42. Another possible definition of a "junk food," in addition to the one proposed in this chapter, is: "A junk food is any food that is so enjoyable that people tend to eat too much of it." Do you think this is an acceptable definition? Explain.

43. BHA and BHT are colorless. What do you think might happen if a drop of iodine solution were placed on some solid BHA or BHT? Explain.

44. Suppose someone began marketing packages of refined sugar that's enriched with vitamins A, the B complex, C, and D and that had added iron, phosphorus, and zinc. Would you consider this a health food or a junk food? Explain.

45. Why might EDTA be useful as an additive in a box of laundry detergent? (Refer to Sec. 13.12.)

46. Many chemicals used as food additives have never been subjected to scientific laboratory tests for safety, and yet their use is perfectly legal. What do all these legal, yet untested, additives have in common?

47. Assuming that it's desirable to test chemicals that are added to our foods to determine their effects on us, do you believe it is better to test them on laboratory animals or on human volunteers, such as those of Dr. Wiley's Poison Squad (Sec. 19.12)? Under what conditions, if any, would you volunteer to participate in such a test?

Web Questions

Use the Web to answer or accomplish each of the following. Report your Web source for each response.

1. Write a brief summary of the life of Casimir Funk. Include a description of his discovery of vitamins.

2. Summarize current knowledge about the effects of very large doses, or overdoses, of vitamins A, the B complex, C, and D. Which can produce damage to bodily organs? What are the symptoms of overdoses?

3. What health claims have been made for vitamins A, the B complex, C, and D, other than their ability to cure or prevent deficiency diseases such as night blindness, scurvy, and rickets? What justification has been made for these claims? Have any of them been verified by scientific studies?

4. Describe the operation of the thyroid gland. What functions does it perform? Why is iodine necessary to its operation? What are the consequences and symptoms of a diseased or defective thyroid? How is iodine used to treat thyroid disease?

5. What are the currently accepted values of Dietary Reference Intakes (DRIs) for calcium, chromium, copper, iodine, iron, magnesium, manganese, selenium and zinc? How are these values determined?

6. Summarize the life and contributions of Dr. Harvey W. Wiley. Describe the formation and activities of his Poison Squad.

7. Give the date of the latest amendment to the Federal Food, Drug, and Cosmetic Act and summarize its effects on the use of food additives.

8. Identify five chemicals currently on the GRAS list, other than those shown in the tables of this chapter, and describe their uses as food additives.

9. Find the most recent data on the amount of money Americans spend each year on vitamin and mineral supplements. Report this as a total and on a per-person basis.

10. Summarize current information about the sources and functions of vitamins E and K. Where are they found? What are the symptoms or effects of deficiencies? Are dietary deficiencies common or widespread?

Poisons, Toxins, Hazards, and Risks

WHAT'S SAFE AND WHAT ISN'T

Is rock climbing safe?

Spreading Poison All Around

Figure 20.1 When we spray an insecticide into the air, we release a toxin into our environment in order to eliminate insects that may carry diseases. As we use an insecticide, we balance the risk of one hazard against the risk of another.

For the opening demonstration of this chapter, you deliberately release a toxic chemical into your own environment, subjecting yourself, your friends, and others to its effects. Whether you actually follow the directions or pass this one by is, of course, up to you. But it's something many of us have already done many times. The toxic chemical is the active ingredient of a commercial insecticide; the directions are whatever instructions the manufacturer has printed on the label. Whenever you spray an insecticide in a room or outdoors you are, indeed, releasing a potent poison into your environment (Fig. 20.1).

To put this into a bit of perspective, we can use the same sort of risk–benefit analysis we discussed in the concluding section of Chapter 19. There we considered both the risks and the benefits of using nitrites to preserve processed meats. Applying the same sort of reasoning here, we can see that using an insecticide in our homes and gardens provides a clear benefit: It rids us of annoying, possibly destructive, disease-bearing insects and similar pests. There's also a risk involved: the possible hazard that the active ingredient in the insecticide—the chemical that does the killing—may put humans at risk of injury, poisoning, or illness.

In this chapter we'll examine some of the factors affecting the hazards that poisons of various kinds present to us. Before we begin, though, let's take a closer look at the risk of poisoning presented by the insecticide. We'll find, as we progress through this chapter, that the toxic risk depends on two factors:

- the intrinsic toxicity of the insecticide and
- the concentration of the insecticide in an animal's body.

As for the first factor, the more potent any particular insecticide is toward insects, the less we need to use and the less of it we, ourselves, are exposed to. For the second, we can compare the concentration of the poison in the insect's body with that in a human body. To simplify the comparison, we will assume we're dealing with a flying insect weighing about 0.7 g and a person who weighs about 70 kg. With this ratio of weights, our typical person weighs about 10^5(100,000) times as much as the insect. If the person doing the spraying or someone nearby absorbs the same weight of the active ingredient as the insect does, the weight percent of the poison in the human's body (Chapter 9) would be about 10^{-5}(1/100,000) as great as it is in the insect's body. Even if our assumptions are a little off, it's not hard to see that the insecticide can be lethal at its very high concentration in the insect's body yet produce no noticeable effect at its very low concentration in a human.

[Naturally, some of the poisons used in commercial insecticides are particularly hazardous and can harm humans and pets if enough of the poison is absorbed. This is particularly true of some chemicals used in lawn care (Sec. 14.14). Some of these can pass through the skin on contact and enter the bloodstream directly. Anything designed to kill insects or other pests has to be considered as a poison and must be used with caution. Always follow the manufacturer's directions.]

In this chapter we'll examine poisons and toxins of various kinds. We'll learn how some of them work and how we can assign quantitative values to their lethal effects. We'll explore some of the risks and hazards they present to us. One of our most important topics will be safety itself. We'll examine what we mean when we say something is "safe" or "unsafe," and we'll look at a definition of safety that's especially useful in dealing with the ordinary things of our everyday lives. Of particular interest will be the chemicals of our foods. ■

20.1 Chemical Poisons, Biological Toxins

Considering the number of hazards already present in our foods—the sugars that can decay our teeth, the saturated fats and cholesterol that can harden our arteries, the sodium that can aggravate high blood pressure, the minerals and the fat-soluble vitamins that can make us ill if consumed in excess—it's surely worth wondering about the value of all the chemical additives introduced into our foods. Are they hazardous? Do their benefits outweigh the increased risks they bring along with them? We'll focus our concerns about chemicals in our foods on two crucial questions:

- Are there actually poisons in our foods?
- Is our food supply safe?

Answering these questions requires, first, good definitions of the key words *poison* and *safe*. We begin with "poison" and its close relative, "toxin."

Generally, when we speak of a **poison** we're referring to a chemical substance that can cause illness or death when it enters our bodies in one way or another, usually through eating, drinking, or inhaling. The word **toxin** usually refers to a poison of biological origin, specifically a protein molecule produced by a plant or animal, with origins ranging from the secretions of microbes to the venoms of poisonous snakes. Toxins can reach us by way of spoiled foods, infectious diseases, or, more violently, through bee stings, snake bites, or other wounds.

A **poison** is a substance that can cause illness or death when it enters our bodies, usually as a component of our food or drink or of the air we breathe. A **toxin** is a harmful substance that has a biological origin.

The ways we use the terms reflect something of their origins. *Poison* comes to us from a Middle English word that was spelled either *poysoun* or *puison* and that referred to a potion or a poisonous drink. The word originated in the Latin *potare*, to drink, and still carries the implication of something hazardous entering our bodies through food or drink. Toxin was carried into English via the Latin *toxicum,* a poison, especially a poison smeared on arrowheads used in warfare. The ancient Romans got their *toxicum* from words for arrow poisons that trace back to *toxon,* an ancient Greek term for bows and arrows, poisoned or not. Our own sense of the word usually emphasizes the biological origin of the poison, especially when we speak of such substances as botulism, diphtheria, and tetanus toxins or the toxins of poisonous plants and animals.

Even with this distinction between a poison and a toxin, these words and others closely related to them are often used interchangeably. Despite the industrial origins of chemicals that pollute our environment, for example, we often speak of them as toxic wastes. And despite the biological origin of snake toxins, we call the snakes that produce them "poisonous." With this in mind, we'll consider the words *poison* and *toxin* virtually interchangeable in this discussion.

■ QUESTION

Using the definitions in this section, name a substance you would consider to be (a) a poison; (b) a toxin.

20.2 Degrees of Danger and Powers of Poisons

Sodium cyanide, aspirin (acetylsalicylic acid), table salt (sodium chloride), and water. In large enough quantities, each can be lethal.

The world, as we're well aware, is filled with dangers. But we're also aware that the dangers are unequal. Some hazards are more severe than others; dangers appear in various degrees, and poisons come in different shadings of potency. Hydrogen cyanide and sodium cyanide, for example, are both powerfully lethal. It takes only a very little of either to kill. Ethyl alcohol is also a poison, but it's much less potent than the cyanides. Aspirin, the most widely used nonprescription drug in the world, can also kill. Although the drug seems harmless enough in ordinary use, years ago it led all other chemicals in the accidental poisoning of young children in the United States. Its victims mistook aspirin for candy and swallowed the tablets, often in lethal doses. In 1970, to protect children from accidental poisoning of this sort—through misuse of aspirin and other ordinary household products—Congress enacted the Poison Prevention Packaging Act. Among other things, this act requires that hazardous household substances, including aspirin, be packaged in containers equipped with safety caps that are especially difficult for children to open. Before August 1972, the date this Act went into effect for aspirin, children under five years of age were dying at the average rate of about 46 per year from accidental aspirin poisoning. After 1972, with increased protection, the rate dropped abruptly to about 25 per year.

Ordinary table salt, the sodium chloride of chemists, is another common but very weak poison. Eating enormous amounts, far more than the average person would consider palatable, upsets the balance of the ions of the body and can cause illness and death. Several years ago, for example, a six-year-old boy with a history of insatiable desire for table salt reportedly died from what was called the "grossly excessive" amount of salt he had added to his meal.

Even water, the compound that makes up more than half of our weight and without which there can be no life at all, is harmful and even deadly in excess. Very much like eating huge quantities of sodium chloride, drinking enormous volumes of water can lead to intolerable and potentially fatal imbalances in the composition of the body's electrolytes and fluids. The result can be mental confusion, lethargy, stupor, coma, and ultimately death. Medical reports tell of endurance runners who, after drinking huge quantities of water during races, exhibit a set of symptoms called *water intoxication* and then lose consciousness. These runners were hospitalized with seizures resembling those of epilepsy. In another case, a Florida woman died tragically in 1977 from drinking as much as 4 gallons of water a day in an effort to rid herself of poisons she believed were accumulating in her body. It was the excessive water rather than her imaginary poisons that killed her.

The key word in all this is *excessive.* Anything consumed in large enough quantities, including household staples such as aspirin, table salt, and even water, can act as a poison. ("The poison," it's been said, "is in the dosage.") Since *anything* can be harmful to us if it's ingested in large enough quantities, we're forced to regard literally *everything* as poisonous. The question, then, isn't whether any particular substance can be lethal. The question is, how much does it take? How do we determine the lethal quantity of a chemical?

■ QUESTION

What effect did the introduction of safety caps on aspirin bottles have on the accidental aspirin poisoning of children? What were the annual death rates before and after the introduction of the caps?

20.3 LD_{50}s

The most sensible way to measure the virulence of a poison or a toxin is to go directly to the heart of the matter and determine the amount needed actually to kill a living thing: the **lethal dose**. There are some problems with this approach, though, even aside from any ethical issues involved in deliberately killing an animal to measure a chemical's lethal strength. As a practical matter, simply feeding the substance in question to a laboratory animal and determining the amount needed to kill the creature doesn't give a satisfactory measure of the lethal dose.

The **lethal dose** of a substance is the quantity that causes death.

Individual animals of the same species, even of the same litter, can show different responses to identical stresses and to identical poisons. Some unusually sturdy individuals, for example, can withstand relatively large amounts of any particular toxin while others, more susceptible, succumb to traces. To nullify the effects of these relatively rare individuals, it's customary to apply the test substance to large groups of animals, with each group containing a large, statistically realistic spectrum of individual susceptibilities. One useful measure that comes from studies of this sort is the amount of a chemical that kills exactly half of a large population of animals, usually within a week. It's known as the *L*ethal *D*ose for *50%* of the group, or the $\mathbf{LD_{50}}$.

The $\mathbf{LD_{50}}$ of a chemical is the amount that kills exactly half of a large population of animals.

Since a chemical's capacity to cause harm depends partly on its concentration in the animal's body, it's common practice to report the LD_{50} of a substance in terms of the weight of the poison per unit weight of the test animal. Moreover, because the method used for administering the substance often affects the results, that too is usually included. For example, the LD_{50} for aspirin fed orally to mice and rats is 1.5 g/kg. This means that feeding aspirin orally to a large group of mice or rats, at the level of 1.5 g of aspirin per kilogram of animal, kills half of the population (Fig. 20.2). From still another viewpoint, we could say there's a 50% chance that 1.5 g of aspirin per kilogram of body weight will kill any particular mouse or rat that eats it.

In contrast, the LD_{50} of sodium cyanide is only 15 mg/kg, fed orally to rats. Since LD_{50}s serve as a good guide to the potency of poisons, we can conclude that sodium cyanide is about 100 times as lethal as aspirin, at least when taken orally by rats. Notice that the smaller the value of the LD_{50}, the less it takes to kill and the more toxic the substance. Table 20.1 lists the LD_{50}s of some more familiar chemicals.

Figure 20.2 The LD_{50} for aspirin.

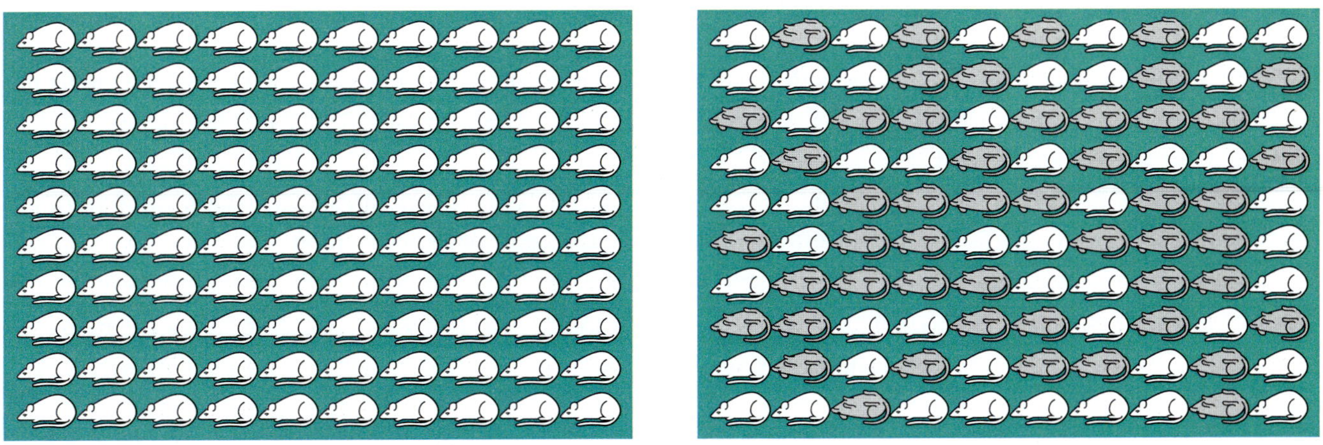

Feeding aspirin to a large group of mice at the level of 1.5 g of aspirin per kg of mouse . . .

. . . kills half of them.

Table 20.1 **LD_{50}s of Some Familiar Substances**

Substance	Animal[a]	LD_{50}
Acetaminophen (an analgesic in medications such as Excedrin and Tylenol)	Mice	0.34 g/kg
Acetic acid (major organic component of vinegar)	Rats	3.53 g/kg
Arsenic trioxide (poison of mystery stories)	Rats	0.015 g/kg
Aspirin	Mice, rats	1.5 g/kg
BHA (antioxidant food additive)	Mice	2 g/kg
BHT (antioxidant food additive)	Mice	1 g/kg
Caffeine	Mice	0.13 g/kg
Citric acid	Rats (abdominal injection)	0.98 g/kg
Ethyl alcohol	Rats	13 mL/kg
Glucose	Rabbits (intravenous)	35 g/kg
Niacin (vitamin B_4)	Rats (injection under skin)	5 g/kg
Nicotine	Mice	0.23 g/kg
Sodium chloride	Rats	3.75 g/kg
Thiamine hydrochloride (vitamin B_1)	Mice	8.2 g/kg
Trisodium phosphate (pH-adjusting food additive)	Rats	7.4 g/kg

[a]The substance is administered orally unless noted otherwise.

While these LD_{50}s serve as a useful guide to a poison's lethal power, they have to be interpreted with a bit of caution. Different animal species often respond differently to a particular chemical and, as we noted earlier, the method of introducing the substance into the animal influences the results. Even for the same species different methods of administration—orally, by injection under the skin (*subcutaneously*), directly into the abdominal cavity (*intraperitoneally*), or directly into the bloodstream (*intravenously*)—can produce different values for the LD_{50}. For example, the LD_{50} for nicotine administered to mice is 230 mg/kg orally, 9.5 mg/kg intraperitoneally, and only 0.3 mg/kg intravenously.

Considering that *anything* can cause harm if it's used in excess (a word whose meaning varies with the material we're talking about), it's clear that to be realistic we've got to put *everything* around us into the same category. This includes chemicals ranging from cyanide and the most lethal biological toxins to caffeine, nicotine, glucose, and sodium chloride. It even includes water and our food's macronutrients, micronutrients, and chemical additives. Each one of these is a potential poison.

Understanding this, we're now in a position to answer one of the questions that began this chapter: "Are there actual poisons in our foods?" The answer must be, "Yes, there are." But since everything presents a potential hazard, both the question and its answer have lost most of their meaning.

The significant question now is whether our foods, with all their chemicals, are actually *safe*. The answer to this one isn't quite as clear-cut. Before taking up the matter of safety we'll examine, in the next few sections, some of the most powerful chemical poisons known.

■ **QUESTION**

What are two experimental conditions of administration or measurement that must be defined along with an LD_{50} in order to give precise meaning to the value?

20.4 The Most Deadly Molecules

It's sometimes useful to discuss the hazards of chemicals in terms of the categories of their use. The chemicals that affect our everyday world can be described, for example, in terms of pesticides, food additives, industrial toxins, environmental pollutants, household hazards, and so on. While useful, categories such as these can also be misleading. Take ethyl acetate, for example. Consumers encounter this ester as a solvent in both nail polish removers and paint strippers (Sec. 13.7) and as a food flavoring (Sec. 19.15). It's also used as an industrial solvent in the manufacture of various consumer products. With all these functions, it's not particularly useful to categorize ethyl acetate as a cosmetic ingredient, a food additive, a household chemical, or an industrial solvent. It belongs in each of these categories.

Chlorine furnishes another illustration. This element is a disinfectant for swimming pools and drinking water, a household bleach, and an industrial chemical in the manufacture of products ranging from plastics to pharmaceuticals. Chlorine was also used as a war gas in World War I (Sec. 3.2). Nicotine, another example, is the major toxin of tobacco and a useful agricultural insecticide. Instead of looking at individual chemicals in terms of their use, we'll look at them in terms of their hazard, regardless of use.

We might expect that with all the ingenuity at a chemist's disposal and with all the power of molecular design, the most lethal of all the poisons around us would be those synthesized by chemists in a chemical laboratory. That simply isn't so. By far, the most powerful poisons and toxins are created by nature, as shown clearly in Table 20.2, which presents the LD_{50}s of the most deadly of all chemicals. (The

Table 20.2 Approximate LD_{50}s of the Most Lethal Poisons

Substance	LD_{50} (mg/kg)
Botulinum toxin A	3×10^{-8}
Tetanus toxin A	5×10^{-6}
Diphtheria toxin	3×10^{-4}
TCDD[a]	3×10^{-2}
Muscarine	2×10^{-1}
Bufotoxin[b]	4×10^{-1}
Sarin[a] (also known as isopropoxymethylphosphoryl fluoride)	4×10^{-1}
Strychnine	5×10^{-1}
Soman[a] (also known as pinacoloxymethylphosphoryl fluoride)	6×10^{-1}
Tabun[a] (also known as dimethylamidoethoxyphosphoryl cyanide)	6×10^{-1}
Tubocurarine chloride	7×10^{-1}
Rotenone	3
Isoflurophate[a] (also known as diisopropyl fluorophosphonate)	4
Parathion[a]	4 (female rats)
	13 (male rats)
Aflatoxin B_1	10
Sodium cyanide[a]	15
Solanine	42

[a]Manufactured; others occur in nature.

[b]Toxicity to cats; all others refer to mice or rats.

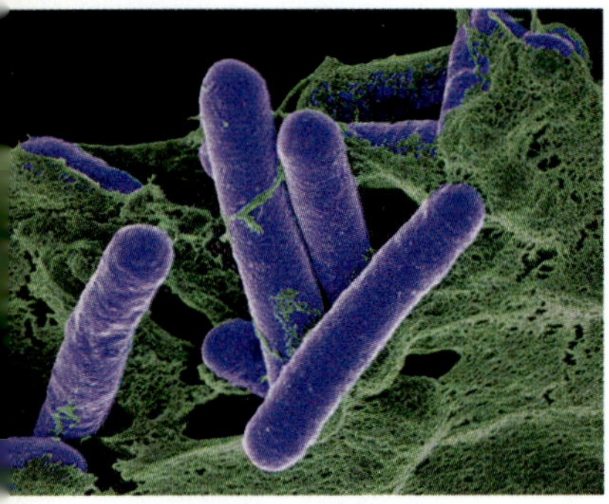

Microphotograph of *Clostridium botulinum*. These microorganism produce the deadly botulinum toxin, the most deadly poison known. It's the cause of the botulinum poisoning that results from eating spoiled food.

LD_{50}s in the table generally represent the most lethal method of administering the substance to mice or rats, whether orally or through injection.)

Leading the list as the deadliest poison of all is a work of nature known as *botulinum toxin,* especially a strain called *botulinum toxin A*. It's a polypeptide, much like those described in Chapter 17, produced by a common microorganism of the soil called *Clostridium botulinum.* This microbe thrives in many foods, especially improperly canned fruits and vegetables and poorly preserved meats. Fortunately, the microbe doesn't survive inside the human body, although the toxin it produces does. The microbe is easily killed by the high temperatures used in approved methods of canning. Moreover, formation of the toxin doesn't occur in acidic foods with a pH below 4.6. Whatever botulinum toxin may be present in contaminated foods is destroyed by cooking at high temperatures. Nonetheless, the toxin that the microorganism does produce is particularly lethal and dangerous.

Both *botulism* (the term for the illness) and *botulinum* (which applies to both the microorganism and the toxin it produces) owe their ultimate origin to the language of the ancient Romans. The first well-recorded outbreak of the disease occurred in 19th-century Germany with more than 200 cases of poisoning. The origin of the epidemic was finally traced to contaminated sausages, which eventually gave a variation of the Latin name for sausage, *botulus,* to the disease itself. Later, with the isolation and identification of the organism that produces the toxin, the sausage connection was transferred to it as well.

Since the beginning of the 20th century, the overwhelming majority of the cases of botulism poisoning in the United States (roughly 90%) have been caused by foods prepared and preserved in the home rather than by commercially canned foods. The chief offenders have been home-preserved foods that aren't heated to temperatures high enough or for periods long enough to kill the microorganism.

Using an LD_{50} of 3×10^{-8} mg/kg for botulinum toxin A (Table 20.2) and a body weight of 70 kg for the average adult (and assuming that humans respond to the toxin exactly as mice do), we can calculate that 5 g of the toxin, about a teaspoonful, would be enough to kill half of a group of 2,400,000,000 adults. Looking at the same numbers from a different angle, one teaspoon of botulinum toxin A could be enough to kill almost a quarter of the world's entire current population.

Another calculation shows that a teaspoon of tetanus toxin, the second deadliest toxin in the table and another one of nature's wonders, would be enough to kill half a population of 14,300,000 people, or about half the total population of Texas. Tetanus toxin is a secretion of the *Clostridium tetani* bacillus and enters the body through wounds, cuts and punctures.

The secretion of the microorganism that produces diphtheria comes in third in the roster, with 1/10,000 the lethal power of botulinum.

■ QUESTION

What is the major source of botulism poisoning in the United States? What is the most practical method for eliminating this source?

20.5 Some Lethal Products of the Laboratory ...

Only now, fourth on the list, comes a chemical produced by chemists in a chemical laboratory, rather than by one of nature's own creatures. With an LD_{50} of about 0.03 mg/kg, this synthetic chemical bearing the formal name *2,3,7,8-tetrachlorodibenzo-p-dioxin* (Fig. 20.3) is only about a hundredth as lethal as nature's

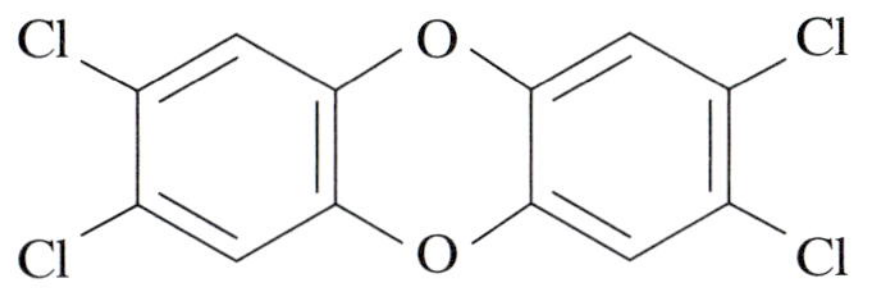

2, 3, 7, 8-tetrachlorodibenzo-*p*-dioxin
(TCDD)

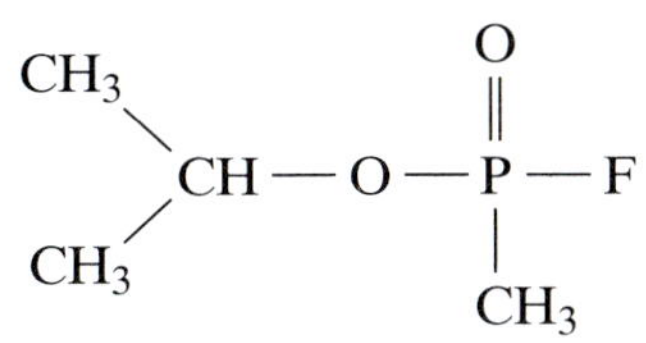

Sarin

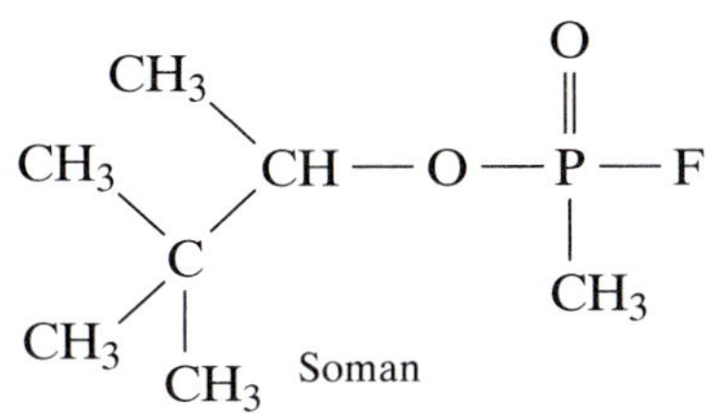

Soman

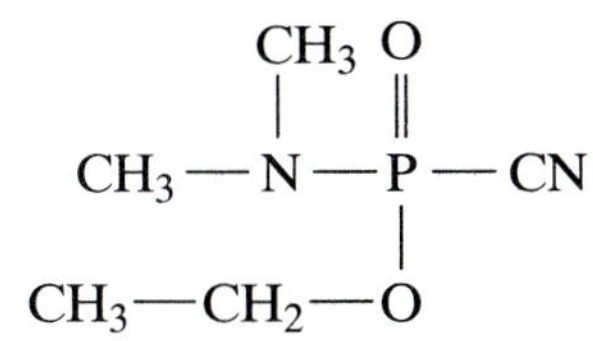

Tabun

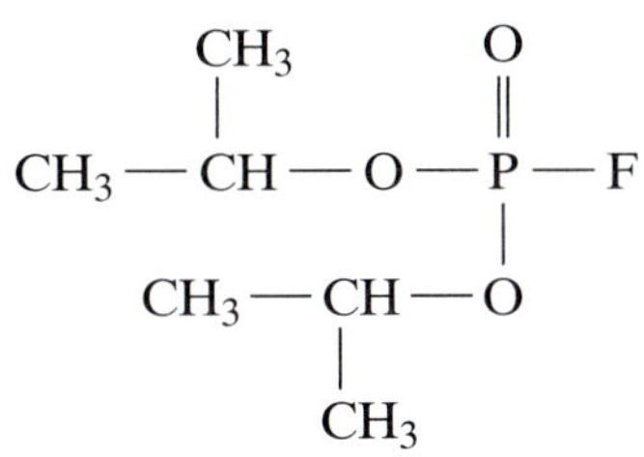

Isoflurophate

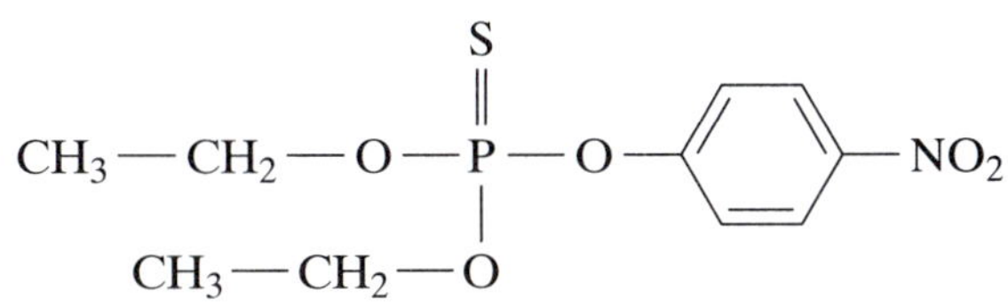

Parathion

NaCN

Sodium cyanide

Figure 20.3 Manufactured poisons.

own tetanus toxin, at least to a mouse. Often known simply as *TCDD*, this compound was formed as a by-product in a few (now discontinued) industrial processes and was implicated as an environmental contaminant in a herbicide used in Vietnam and in industrial disasters of the 1970s.

To describe the potency of TCDD in the same sort of terms we used for botulism and tetanus, terms based on studies with laboratory animals, a teaspoon of this substance should be lethal to about half of a population of 2500 people. TCDD represents the most lethal of the chemicals known to be produced by humans rather than nature.

Following TCDD we find a set of toxic chemicals ranging from *muscarine* (LD_{50}, 0.2 mg/kg) to *solanine* (LD_{50}, 42 mg/kg). Interspersed among these are some highly specialized, closely related, phosphorus-containing synthetic chemicals that are highly effective in keeping nerve cells from transmitting signals to each other. Among these are *sarin, soman,* and *tabun,* which arouse interest because of their potential as nerve gases in chemical warfare. With human LD_{50}s estimated to be as low as 0.01 mg/kg, these chemicals are particularly lethal.

Other compounds with molecular structures similar to the nerve gases have uses a bit closer to our everyday experiences. *Isoflurophate* is effective in contracting the pupil of the eye in veterinary medicine, and *parathion* serves as a particularly toxic agricultural insecticide. *Sodium cyanide*, the last of the synthetic chemicals on this list, has various important commercial and industrial applications, ranging from the extraction of gold from ores and the electroplating of metals to fumigation and pest control. The structures of these synthetic chemicals appear in Figure 20.3.

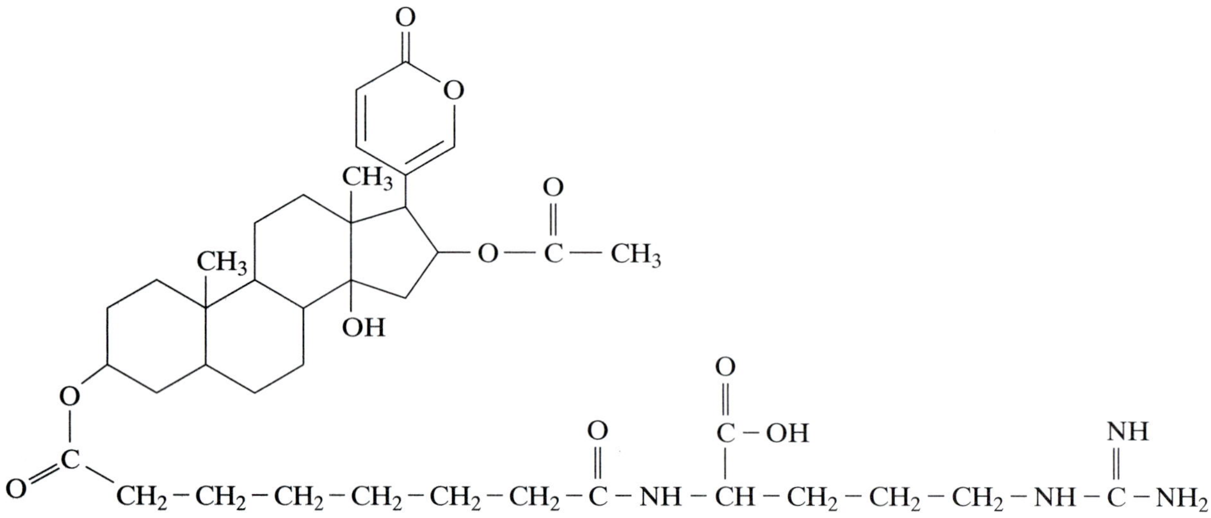

Bufotoxin

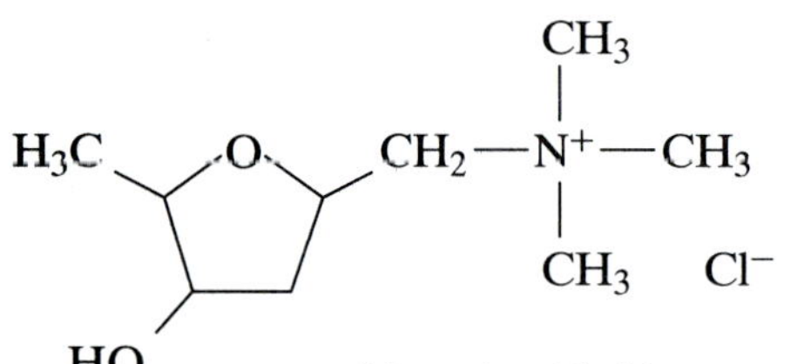

Muscarine chloride

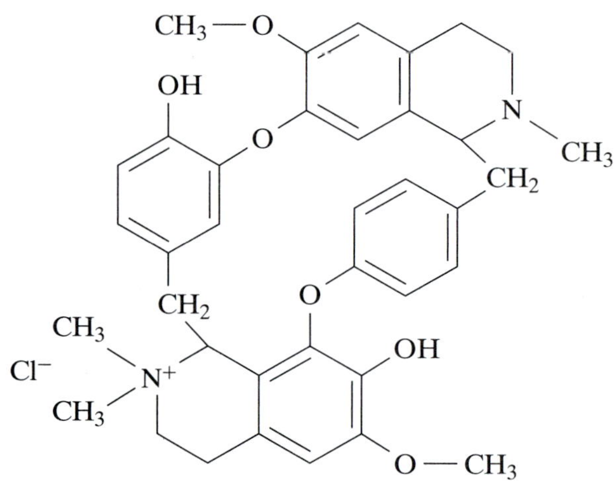

Tubocurarine chloride

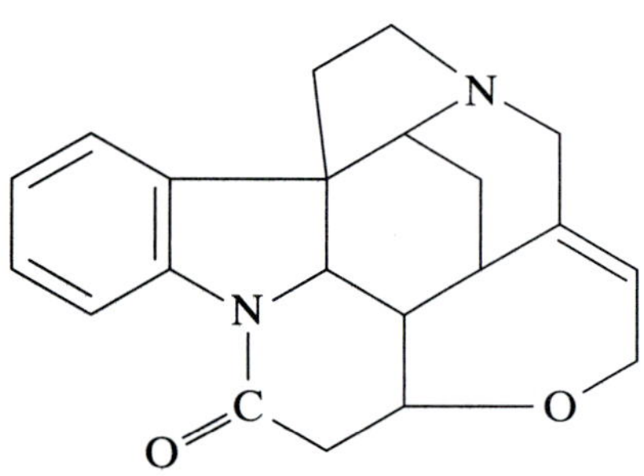

Strychnine

Figure 20.4 Nature's toxins.

The remaining natural toxins of Table 20.2 include, in order of decreasing toxicity, the following:

- *muscarine*, the lethal ingredient of deadly mushrooms such as the fly agaric or fly fungus;
- *bufotoxin*, the active component of the venom of the common toad, *Bufo vulgaris*;
- *strychnine*, which occurs in the seeds of trees native to Australia, India, and countries of the southern Orient and which has been used as a natural rat poison for centuries;
- *tubocurarine*, the toxin of curare, a plant extract used by South American Indians as an arrow poison;
- *rotenone*, a naturally occurring insecticide found in the roots of derris vines; and
- *aflatoxin* and *solanine*, both of which affect our foods, as we'll see in the next section.

Figure 20.4 presents the molecular structures of these natural toxins.

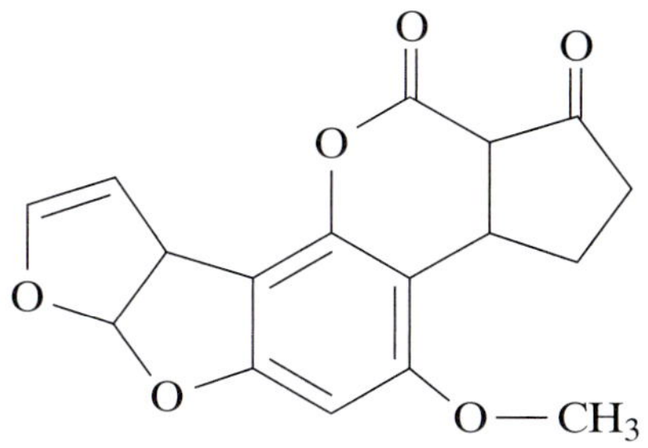

Aflatoxin B_1

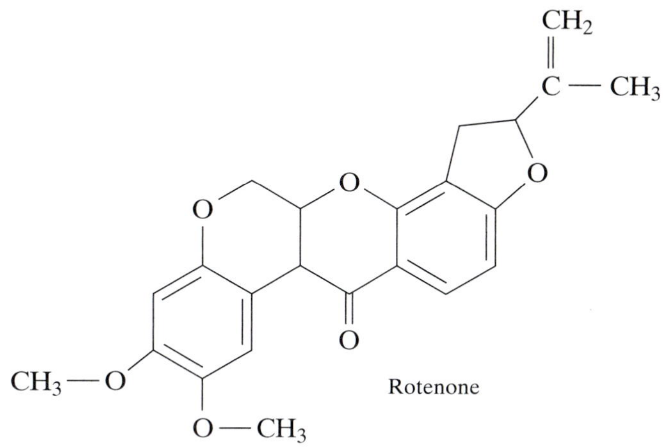

Rotenone

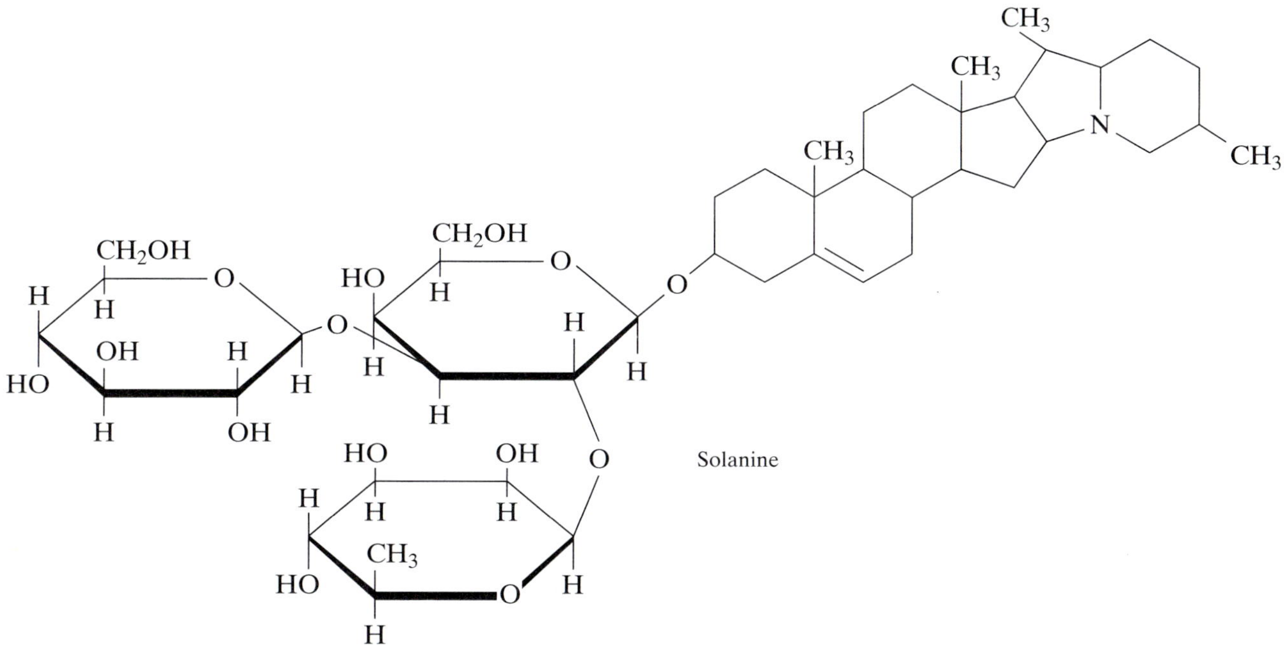

Solanine

Figure 20.4 *(cont.)*

Which of the compounds of Table 20.2 have useful applications unrelated to their toxicities? Explain.

■ QUESTION

20.6 ...And of Nature

The last two toxins on the list, nature's *aflatoxin* and *solanine*, bring us back to our questions about the safety of the food supply. Aflatoxins are a group of toxins produced by a mold that flourishes on various grains and legumes, including peanuts, wheat, and corn. Not only are the aflatoxins this mold releases highly toxic, they are among the most potent liver carcinogens known.

Differences in the effects of aflatoxin B_1 on newly born mice and on day-old ducklings illustrate very nicely the enormous variations in LD_{50}s that occur with

The toxin *solanine* accumulates in the green areas of a potato.

different species of animals and different methods of administration. For the mice, the LD_{50} is about 10 mg/kg for aflatoxin injected directly into their abdominal cavity; for the ducklings, oral feeding produces an LD_{50} of 18×10^{-3} mg/kg. The two values differ by more than a factor of 500.

Solanine, which occurs in potatoes, especially in or near green areas of the skin and near fresh sprouts projecting from the potato, inhibits the transmission of nerve impulses. In this respect it resembles the synthetic phosphorus-containing compounds described earlier. Solanine's potency, like that of the aflatoxins, depends on how it enters the body. Its LD_{50} by direct injection into mice is 42 mg/kg, while oral feeding of 1000 mg/kg produces no ill effects at all. (Direct ingestion can produce ill effects and death in many other species, though, including humans. It's best to cut green areas and fresh sprouts out of potatoes before using them.)

■ QUESTION

What hazard is associated with moldy peanuts? With potatoes containing many fresh sprouts?

20.7 Safety: Is It Freedom from Hazards or Simply a Matter of Good Judgment?

To sum up, we've seen that any chemical can be hazardous to our health and that lethal potencies can be measured as LD_{50}s. We've also seen that these LD_{50}s vary with the species of test animal used and with the method of administration. Nonetheless, it's still possible to estimate the relative toxicities of various chemicals by ranking them with respect to each other. Although both botulinum toxin and ordinary table salt can kill, for example, the protein that makes up the toxin is unquestionably more lethal than sodium chloride. Taken as a whole, nature's poisons outrank synthetic chemicals in both number and virulence.

Recognizing that there are, indeed, potentially harmful substances in our foods—poisons—we can now turn to the second question we asked at the beginning of this chapter: Are our foods safe? A reasonable answer requires a reasonable definition. Often, when we say that something is "safe" we imply that it is free of danger, hazard, or harm. But no substance is inherently and totally free of danger, hazard, or harm. The harm any substance can do to *us*, each one of us, depends on

- its chemical characteristics,
- how much of it we use,
- how we use it,
- how susceptible to its hazards we are as humans, and
- how susceptible we are as individuals.

By one useful definition, **safety** is the acceptability of risk

For a more realistic test of safety, we can ask whether, knowing the risks involved in using a chemical (or in anything else for that matter), we are willing to accept those risks. In this sense **safety**, as defined by William Lowrance, a contemporary American organic chemist, is the *acceptability of risk*. It's a realistic and practical definition of a subtle and sometimes ambiguous term, and it provides us with a convenient way to examine the safety of our food supply and of chemicals in general. With this definition in mind we can look to our own (informed) judgment in deciding matters of safety.

Marked crosswalks, traffic lights, and pedestrian signals help control the risks involved in crossing a busy street and make these risks acceptable.

In short, everything in life involves a risk of one sort or another. Those risks we find acceptable are the ones we, ourselves, define as safe. Those we find unacceptable we define as unsafe. All of us, collectively, make similar judgments about the risks we are willing to take as a society.

Viewing safety as a matter of acceptable risk shines a somewhat clearer light on any attempts to prove that something is indeed safe. It is simply impossible to prove that anything is safe in the sense that it presents no hazard whatever, to anyone, at any time, in any possible circumstances. We might think that if we tried to demonstrate in 10,000 different ways that a certain chemical, for example, is harmful to mice and failed each time, we might have shown it to be safe. Yet all this is negative evidence, which proves nothing at all. There's always the possibility that one ingeniously designed additional test, a 10,001st experiment, might show that the chemical does, indeed, do harm to mice.

A tragic illustration of this occurred in the late 1950s and early 1960s with the drug *thalidomide* (Fig. 20.5). The drug, which had been developed as a sleeping pill, passed several routine toxicity tests with no evidence that it might be harmful. With the belief that this negative evidence was a demonstration of safety, physicians in England and other countries began prescribing thalidomide as a sleeping pill for pregnant women. In Germany it was sold as an over-the-counter drug, with

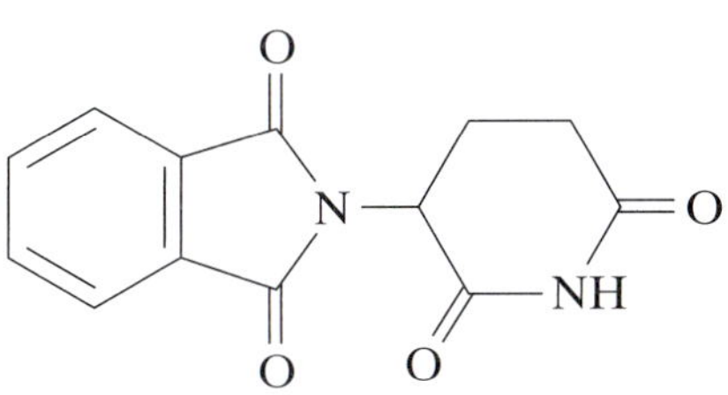

Figure 20.5 Thalidomide. Prescribed as a sleeping pill for pregnant women before its dangers were known, thalidomide acted as a teratogen in women who took it within their first 12 weeks of pregnancy.

no prescription required. (Doubts about the results of the animal testing kept it off the market in the United States.)

Within a few years, some 4000 babies in Germany, 1000 in Great Britain, and 20 in the United States (to women who had obtained the drug on trips to Europe) had been born with severe birth defects, including severely shortened or absent arms and legs. These defects were traced eventually to thalidomide that the women had taken within the first 12 weeks of pregnancy. Thalidomide is an example of a **teratogen**, a substance that produces severe birth defects. The word comes from the Greek *teras* for a monster or a marvel.

A **teratogen** is a substance that produces severe birth defects.

More extensive studies of thalidomide, spurred by the birth defects, showed that while it posed no apparent hazards to rats, dogs, cats, chickens, and other species, it did act as a teratogen in monkeys and rabbits, neither of which had been used in the original studies. It was all clearly a case of a 10,001st experiment that, tragically, had not been performed.

More recently thalidomide, which has a very low inherent toxicity and seems to be hazardous only to a developing fetus, has proven useful in treating severe inflammations and disorders of the immune system. Thalidomide or compounds closely related to it have shown promise in alleviating a variety of diseases and disorders, including mouth ulcers, rheumatoid arthritis, leprosy, AIDS, tuberculosis, and transplant rejection. Once again we see that chemicals, including even thalidomide, are neither good nor evil in themselves but appear to be so only in the ways we choose to use them (Sec. 1.2).

■ QUESTION

Under what conditions is thalidomide a hazardous substance? Under what conditions does thalidomide provide benefits that outweigh its hazards?

20.8 The Laws of Safety

With safety defined as the acceptability of risk, the question of which risks are acceptable and which are not falls on us as individuals and also as members of a larger society, one represented by our elected government and its laws. As individuals, for example, we're free to choose whether or not to use aspirin to relieve headaches and other minor pains, inflammations, and fevers. The choice is ours, individually, and the judgment of safety rests with each of us as individuals.

In 1970, though, with the passage of the Poison Prevention Packaging Act, Congress placed a larger, societal judgment on free access to aspirin. Aspirin, they said in effect, is safe enough to be handled freely by anyone who can open a simple but obstinate twist-and-snap cap, but it is unsafe for those very young children who can't. That judgment has become society's assessment of the safety of aspirin.

It also serves as a model for our decisions about the safety of chemicals in our foods. While each of us makes daily decisions about what to eat, our representatives at the centers of government decide what chemicals are safe enough to be used as additives in commercial foods and in what quantities these additives may be used. The executive branch of the government operates through the Food and Drug Administration (the FDA) and many other agencies to implement legislative judgments of the safety of various chemicals. In addition to the FDA, these other agencies include the

- Environmental Protection Agency (EPA), which regulates chemical pesticides, among other matters affecting the environment;

- Occupational Safety and Health Administration (OSHA), which is concerned with exposure to chemicals in the workplace;
- Bureau of Alcohol, Tobacco and Firearms (BATF), a division of the Department of the Treasury that has jurisdiction over the chemicals of beer, wine, liquor, and tobacco; and
- U.S. Public Health Service (USPHS), which, among its other activities, investigates outbreaks of illness due to food spoilage.

■ QUESTION

Which agency of the federal government has responsibility for: (a) chemicals used as food additives; (b) substances that may contaminate the environment; (c) chemicals that may prove hazardous in the workplace?

20.9 A Factor of 100

Through its rules and regulations, the FDA exercises a societal judgment over the safety of the chemicals that are added to our foods. In practice the agency must first determine that any proposed additive does, indeed, produce a desired, beneficial effect, which is one of the necessary characteristics of any food additive. The FDA then determines the acceptable level of an additive's use in foods through examination of

- the LD_{50} of the proposed additive in at least two (and often more) species of animals,
- the chemical's maximum "no-effect" level, and
- its long-term hazards, if there are any.

Values of LD_{50} represent the acute or immediate hazard in using a chemical. Naturally, a substance must produce a beneficial effect in foods at a level well below its acute toxicity if it's to be used as an additive. How much below, though, is a matter of judgment.

In making judgments of this kind the FDA requires that the substance be fed to at least two different species of animals for several months and that the animals be examined thoroughly throughout and at the end of this period for any changes the chemical may have produced. Feeding tests of this sort make it possible to determine the *maximum* daily amount of the chemical that produces *no observable effect* on the animals. This is the *no-effect level*, measured as milligrams of the additive per kilogram of the animal's body weight.

With a safety factor of 100, the FDA permits no more than 1/100, or 1%, of this no-effect level in commercially prepared foods, based on estimates of the average person's daily consumption. This represents the additive's **acceptable daily intake**, or **ADI**. Any evidence that humans are even more sensitive to the additive than the test animals are or that the additive can do its job at less than the 1% level can result in even lower ADIs. Then again, if there's evidence of harm detected through much longer feeding tests, sometimes lasting several years, the substance may be prohibited entirely.

The acceptable daily intake, or **ADI**, of a food additive is 1% of the maximum daily amount of additive that produces no observable effect on laboratory animals.

These long-term feeding tests reveal the chronic effects of proposed additives, including some of the more dreaded hazards we sometimes associate with synthetic chemicals. Long-term studies uncover the capacity of a chemical to affect an animal's ability to reproduce (fertility effects) and to affect offspring either through the development of severe birth defects (teratogenic effects) or through lesser genetic changes (mutagenic effects). Such studies also reveal a chemical's ability to produce cancer (to act as a carcinogen). It's this last hazard—the possibility of chemically induced cancer—that arouses some of our strongest feelings about the chemicals of our foods and our environment. We'll turn now to the question of carcinogens in our foods and to an examination of our search for safety in its strictest form: absolute safety.

■ QUESTION

What is an additive's ADI and how is it determined?

20.10 The Legacy of James J. Delaney

Through the original Food and Drug Act of 1906 and its many revisions, Congress has generally given federal agencies substantial leeway in setting the standards of safety for chemicals in our food, drugs, and cosmetics and in public areas such as the workplace and municipal drinking water. The matter of safety is, as we have seen, a matter of judgment; our lawmakers have been content, by and large, to delegate the finer, more technical matters of judgment to agency officials and their technical experts.

Yet with respect to one particularly fearsome hazard, the risk of cancer from food additives, federal law sets a *zero tolerance* level. That is, with the notable exception of saccharin, which we'll examine in detail in the next section, our society judges that there is *no* acceptable level of risk of cancer from a food additive. We'll now see what this zero tolerance implies, how it came about, and why an exception has been made for saccharin.

In 1958 Congressman James J. Delaney of New York introduced an amendment into a revision of the Food, Drug and Cosmetic Act which provided that

> no additive shall be deemed to be safe if it is found to induce cancer when ingested by man or animal, or if it is found, after tests which are appropriate for the evaluation of the safety of food additives, to induce cancer in man or animal.

In effect this portion of the Act, now known as the *Delaney Amendment* or the *Delaney Clause,* prohibits the use of any chemical as a food additive, at any level, if the chemical is found to produce cancer in any way, in any test, at any concentration, in any animal. Although the FDA has established acceptable daily intake levels for various chemical additives of known acute toxicity, including some of those listed in Table 20.1, the Delaney Amendment sets a zero tolerance for the deliberate addition to food of any chemical known to cause cancer in humans or laboratory animals.

While this legislative legacy has served as an effective bar against the addition of known carcinogens to our processed foods, it also brings into focus a larger need for balancing the potential hazards of any particular food additive against the very real and very desirable benefits that additive may provide. An illustration of this balance occurred a few years ago with the nonnutritive sweetener *saccharin.* (A nonnutritive sweetener is one that sweetens a food while introducing few or no calories per serving.)

■ QUESTION

Under what condition or circumstance does the Delaney Amendment take effect in regulating the use of a food additive?

20.11 The Story of Saccharin

In 1879 Ira Remsen, Professor of Chemistry at the newly founded Johns Hopkins University, and Constantine Fahlberg, a student of Remsen's, were studying some of the more esoteric aspects of the reactions of organic compounds. Remsen was later to become president of the University and one of the major influences on the teaching and practice of chemistry in the United States.

One evening at dinner, after having finished his laboratory work and cleaned up, Fahlberg noticed a peculiar sweetness in the bread he was eating. He traced the source of the exceptional taste to his hands and arms. Recognizing that it must have come from residues of chemicals he had been in contact with that day, he returned to his laboratory to find its origin. By using the sensitive but very risky technique of tasting remnants of his work here and there in the laboratory, he traced the unusual sweetness to his chemical apparatus and to the compounds on which he had been working. He and Remsen immediately published their discovery of the source of the sweetness, the newly prepared chemical *saccharin*. The name of the sweet compound comes from the Latin and Greek words for sugar itself.

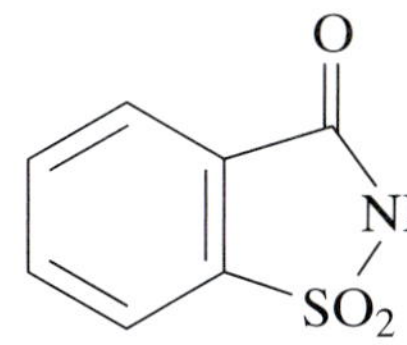

Figure 20.6 Saccharin. A nonnutritive sweetener about 500 times as sweet as sucrose, saccharin has been shown to be a weak carcinogen.

Saccharin (Fig. 20.6) provides neither energy nor the materials from which we produce our body tissues, so it has no food value. Yet its intense sweetness, about 500 times that of table sugar, lasts even to incredible dilutions. Dissolve one teaspoon of pure saccharin, about 5 g, in 145,000 liters of water—enough water to fill a cube just over 52.5 m on a side or enough to cover a football field about an inch and a quarter deep—and the sensation of sweetness is still present.

Saccharin has been used as a nonnutritive sweetener since around the beginning of the 20th century. Although it has been shown to cause cancer in mice, saccharin continues to be used as a food additive because of public demand. All products containing saccharin must carry a warning label.

Soon after Remsen and Fahlberg's discovery, saccharin was in use as a food preservative and as an antiseptic and then as a replacement for sugar. Around 1907 commercial food canners discovered its popularity among people with diabetes, and it has been with us ever since as a nonnutritive, noncaloric sweetener. With the passage of the 1958 Food, Drug and Cosmetic Act, saccharin, which had been in general use for about half a century, joined the GRAS list of approved food additives (Sec. 19.13).

Although sugar-free foods for sufferers of diabetes formed the original market for the sweetener, the demand for diet soft drinks that sprang up in the early 1960s brought saccharin into much wider use. By 1977 between 50 and 70 million Americans were regular users of saccharin, consuming about six million pounds each year, about three-quarters of it in soft drinks.

With this widespread public consumption, doubts arose about saccharin's safety. A few years earlier, in 1972, the FDA had removed it from the GRAS list and placed temporary restrictions on its further use pending the outcome of laboratory tests. Laboratory studies soon indicated that saccharin produces bladder cancer in mice. Even though the sweetener proved to be an extremely weak carcinogen, posing a trivial and perhaps undetectable threat of cancer to the general population (according to the experts who examined it), the Delaney Amendment came into operation through its zero-tolerance provision and forced the FDA, early in 1977, to announce its intention to ban saccharin as a food additive.

Before the FDA's order could take effect, a public outcry against the proposed prohibition overwhelmed Congress. Saccharin was, at the time, the only sweetener available to those who wanted sweetness without calories, taste without the accompanying prospect of weight gain. One physicist published scholarly calculations suggesting that for a person 10% overweight the health risk from the saccharin in one diet soft drink was less than the health risk from the calories of the sugar it replaced.

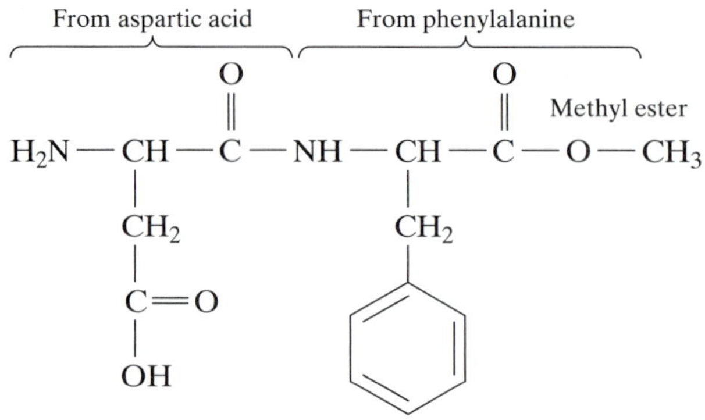

Figure 20.7 Aspartame, a dipeptide about 160 times as sweet as sucrose.

Aspartame is the principal ingredient of several non-nutritive sweetners, including NutraSweet.

Spurred by public pressure, Congress acted. Through the Saccharin Study and Labeling Act of 1977 it ordered additional studies of saccharin and temporarily prohibited the FDA from banning it as a food additive. Congress also required that a label appear on all commercial foods containing the additive, warning that it causes cancer in laboratory animals and that "use of this product may be hazardous to your health." The moratorium against FDA action has been extended every few years and continues even now.

The public had decided, through its elected representatives, that the benefits accompanying continued use of saccharin outweighed the risk of cancer. The risk was judged to be acceptable and, as a result, saccharin was to be considered safe.

Today saccharin stands in the shadow of still another synthetic sweetener, *aspartame*, a synthetic dipeptide (Sec. 17.7) of aspartic acid and the methyl ester of phenylalanine (Fig. 20.7). As a dipeptide, aspartame provides the same 4.0 cal/g as do proteins, yet its intense sweetness, about 160 times that of sucrose, allows it to provide the equivalent sweetness of table sugar with about 1/160 as many calories. Aspartame is the major ingredient of sweeteners such as NutraSweet and Equal.

■ **QUESTION**

Even though the application of the Delaney Amendment might have ended the use of saccharin as a food additive when the sweetener was shown to cause cancer in laboratory animals, its continued use was not prohibited. Why not?

20.12 An Ingredients Label for the Mango

We might speculate briefly on what our attitude would be toward the chemicals of our foods and toward the hazards they present if all foods, both processed and unprocessed, bore ingredients labels. Imagine what we might find if an orange growing on a tree or a potato freshly dug up from the ground carried a stamp showing all the chemicals it contained. Certainly one major difference between processed foods and those prepared by nature is that packages of processed foods must carry lists of their ingredients, while foods taken directly from the ground don't. To learn what's in unprocessed foods like oranges freshly off the trees or potatoes freshly out of the ground, or in apples, steaks, tomatoes, and the like, we're forced to turn to the chemists who analyze them and study their components.

Without exception, the lists of chemicals isolated from foods grown on the farm and in the orchard and sold with little or no processing far exceed those appearing on the ingredients panels of processed foods. The oil of an orange, for example, contains more than 40 different chemicals, grouped by organic chemists

Figure 20.8 A partial ingredients list for a typical mango. α-Terpinolene, ethyl butanoate, 3-carene, ethyl acetate, ethyl 2-butenoate, α-terpinene, α-thujene, dimethyl sulfide, limonene, β-phellandrene, myrcene, *p*-cymen-8-ol, β-caryophyllene, *cis*-3-hexene-1-ol, hexadecyl acetate, 5-butyldihydro-3H-2-furanone, *trans*-2-hexenal, ethyl tetradecanoate, α-humulene, sabinene, 2-carene, camphene, ethyl octanoate, 4-isopropenyl-1-methylbenzene, 1-hexanol, α-terpinene, hexanal, ethyl hexadecanoate, α-copaene, hexadecanal, ethanol, ethyl propionate, dihydro-5-hexyl-3H-2-furanone, carveol, geranial, ethyl decanoate, furfural, butyl acetate, methyl butanoate, dihydro-5-octyl-3H-2-furanone, *p*-cymene, octadecanal, 6-pentyltetrahydro-2H-2-pyranone, 2, 3-pentanedione, 1, 1-diethoxyethane, pentadecanal, butyl formate, 1-butanol, 5-methylfurfural, ethyl dodecanoate, 2-acetylfuran, 2-methyl-1-butanol, 4-methylacetophenone, acetaldehyde, cyclohexane.

into categories such as alcohols, aldehydes, esters, hydrocarbons, and ketones. A potato yields some 150 different compounds, each of which can be synthesized in a chemical laboratory or poured from a bottle on the chemist's shelf. Each is a chemical put there by nature, synthesized by the plant itself as it grows.

The mango, a particularly tasty tropical fruit, offers an excellent example of what an ingredients label on a piece of a natural, unprocessed fruit might look like. Mangoes are prized delicacies throughout much of the world, especially in India and the Far East. In terms of total tonnage consumed, the mango is the most popular of all fruits. Figure 20.8 presents a partial list of the chemicals responsible for the flavor of a typical mango. It represents only a small fragment of all the chemicals that make up the fruit. Within Figure 20.8 the chemicals appear in decreasing order by their relative abundance in the essential oil, just as they would on an ingredients label. Very few of them have been examined sufficiently for use as food additives.

Is the mango safe to eat? By any reasonable standard, and based on the masses of people throughout the world who eat mangoes daily without ill effects, the answer must be *yes*. (Of course, the answer could be *no* for those individuals who may be allergic to mangoes, for those with diabetes who may have to avoid the mango's sugars, and for others who may be sensitive to individual chemicals within the fruit.) Yet since its chemicals have not been tested for safety and have not been approved for use in foods, an identical mango that might somehow be manufactured in a food processing plant could not be sold legally as a food.

■ **QUESTION**

At least one of the major flavor ingredients of the mango (Fig. 20.8) has been approved for use in foods. Which one has this approval? (See Sec. 19.15 for help.)

20.13 The Questionable Joys of Natural Foods

The example of mangoes can be repeated with all of our other foods, many times over and sometimes with sinister implications. Many of the compounds that form in plants and that become part of our food supply are very effective insect poisons, rivaling commercial insecticides in toxicity (certainly to insects and probably

Examples of foods containing hazardous chemicals. Nutmeg contains myristicin, a chemical that can produce liver damage and hallucinations when taken in very large amounts. Large amounts of the glycyrrhizic acid of licorice can damage the cardiovascular system. The allyl isothiocyanate of horseradish, the symphytine of comfrey, and the estragole of basil can produce tumors.

to humans) and overwhelming them in number. For every *gram* of synthetic pesticide that we eat as an unwanted residue of a commercial, agricultural spray, an estimated *10 kg* of natural pesticides enter our bodies as natural poisons of natural foods. With some reflection, this shouldn't surprise us. With eons of evolution behind them, it's entirely reasonable that plants should develop powerful toxins as part of their chemical armory against insects and other predators. Unlike animals, plants can't run from their enemies; they can only poison them (Sec. 18.1).

The armaments of some plants include chemicals that are not only toxic in the more common sense of the word but are carcinogenic as well. *Safrole*, for example, makes up about 85% of oil of sassafras, which comes from the bark around the root of the sassafras tree and is also a minor component of cocoa, black pepper, and spices and herbs such as mace, nutmeg, and Japanese wild ginger. Safrole produces the taste of sassafras tea and was at one time used as the principal flavoring ingredient in the manufacture of root beer.

Both the oil of sassafras and safrole itself have been banned from use as food additives. Safrole produces liver cancer in mice and has been listed as a carcinogen by the Environmental Protection Agency. The FDA judges both safrole and the oil to be too dangerous for use in foods and considers any food containing either one of them to be adulterated.

Figure 20.9 shows safrole and several additional plant substances that produce tumors in mice:

- *allyl isothiocyanate*, a pungent, irritating oil that occurs in brown mustard as well as horseradish and garlic and is also known as mustard oil;
- *estragol*, a major component of oil of tarragon and occurring also in the herbs basil and fennel; and
- *symphytine*, a component of the comfrey plant and one of its natural defenses. (The comfrey plant is used for brewing herbal teas.)

The roster of toxic, carcinogenic, or otherwise hazardous chemicals in our natural food supply continues, seemingly without end (Table 20.3). Yet we do accept the hazards of our natural food supply, those that exist through the action of nature even in the absence of any processing or chemical additives. Despite the presence of an enormous variety of naturally occurring chemicals that may cause harm to us, we ordinarily consider the fruits, vegetables, grains, legumes, dairy products, eggs, meats, poultry, fish, and other foods that we eat in our daily meals to be safe, especially when we eat each in some measure of moderation.

Figure 20.9 Plant substances that produce tumors.

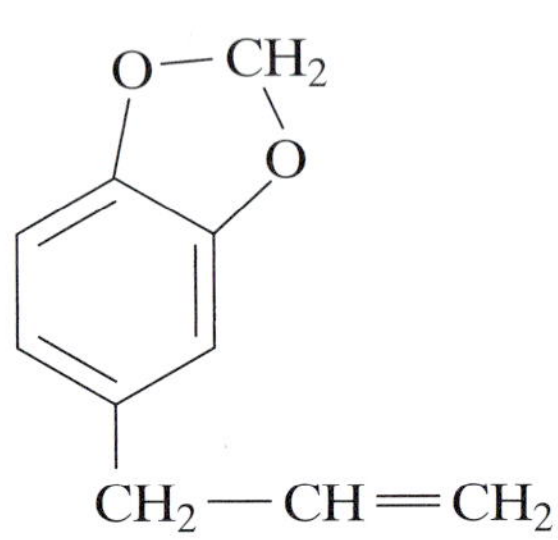

Safrole

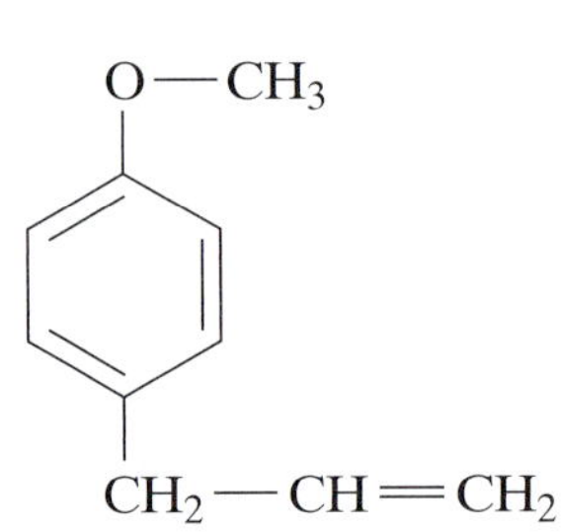

Estragole

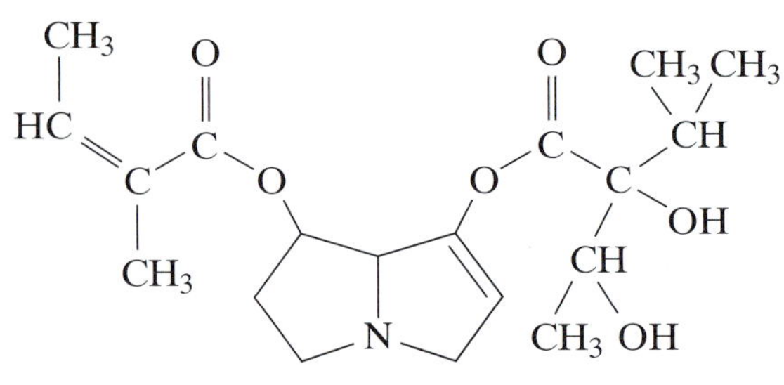

Symphytine

$CH_2{=}CH{-}CH_2{-}NCS$

Allyl isothiocyanate

Table 20.3 Toxic or Carcinogenic Chemicals Occuring Naturally in Foods or Produced through Cooking

Food Substance	Source	Potential Hazard
Allyl isothiocyanate	Brown mustard, horseradish, garlic	Tumors
Benzo(*a*)pyrene	Smoked and broiled meat	Gastrointestinal cancer
Cyanides	Oil of bitter almond cashew nuts, lima beans	General toxicity
Dimethylnitrosamine	Cooked bacon	Cancer
Estragole	Basil, fennel, oil of tarragon	Tumors
Glycyrrhizic acid	Licorice	Hypertension and cardiovascular damage
Hydrazines	Raw mushrooms	Cancer
Lactose	Milk	Gastrointestinal distress (Sec. 16.13)
Myristicin	Black pepper, carrots, celery, dill, mace, nutmeg, parsley	Hallucinations
Oxalic acid	Rhubarb, spinach	Kidney damage
Saxitoxin	Shellfish	Paralysis
Symphatine	Comfrey plant	Tumors
Tannic acid and related tannins	Black teas, coffee, cocoa	Cancer of the mouth and throat
Tetrodotoxin	Pufferfish	Paralysis

This word *moderation* holds the key to our own defenses against food toxins. For most of our ordinary foods, normal levels of consumption lie far below anything that might produce an acute illness. Although the *myristicin* of nutmeg (Table 20.3), eaten in very large quantities at one sitting, can produce hallucinations and liver damage, the amount of nutmeg ordinarily used in even the most heavily spiced meal runs about 1 or 2% of the quantity needed to produce anything more notable than flavorful food.

To suffer cardiovascular damage from the *glycyrrhizic acid* of licorice seems to require eating perhaps 100 g of the candy each day for several days. *Caffeine*, with its LD_{50} of about 130 mg/kg in mice, is the major physiologically active compound of coffee and a powerful stimulant of the central nervous system. At 100 to 150 mg of caffeine to a cup of coffee, it would take about 70 cups of coffee at one sitting to reach the mouse's LD_{50} in a 70-kg human. Surely other, urgent problems would arise long before the approach of a lethal dose. (Structures of caffeine, glycyrrhizic acid, and myristicin appear in Fig. 20.10.)

The body's most effective shield against molecular poisons lies in the operation of the liver. This organ can call up a great variety of metabolic reactions to change the chemical structures of poisons, usually rendering them harmless. As long as a healthy liver isn't overpowered by a massive dose of any one toxin, it can effectively protect the rest of the body. By using a variety of metabolic defenses, the liver far more easily disposes of, say, a tenth of a gram of each of 10 different toxins than the same total weight, one gram, of a single toxin. While small amounts of the 10 poisons might be metabolized effectively by 10 different enzymatic reactions, a large dose of a single poison could overwhelm the only biochemical path

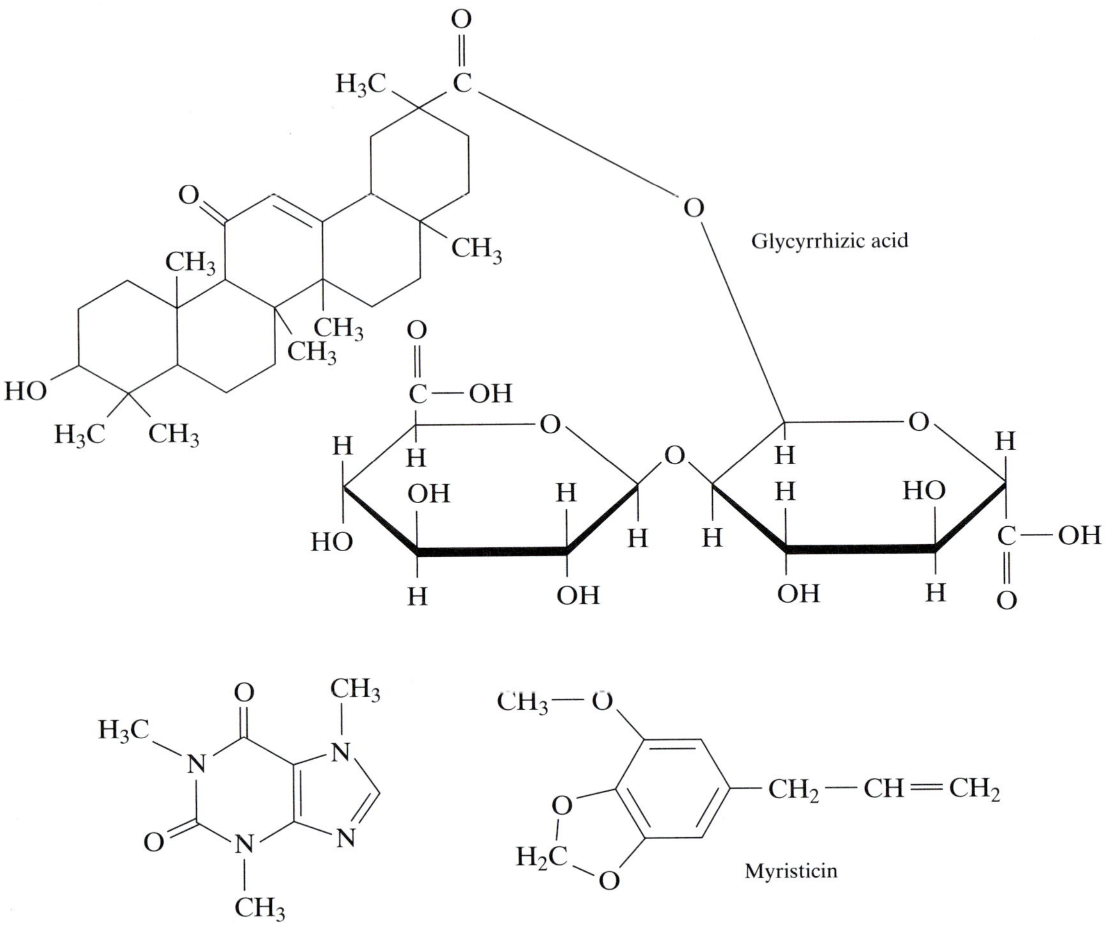

Figure 20.10 Caffeine, myristicin, and glycyrrhizic acid.

available for its removal. In this sense a varied diet, which implies eating foods containing a variety of natural toxins, each present in exceedingly small amounts, is far less hazardous than a diet that concentrates on large quantities of only one or two foods.

■ QUESTION

What hazardous chemical occurs in each of the following and what effects or bodily harm are associated with it? (a) nutmeg; (b) licorice; (c) rhubarb; (d) black tea; (e) horseradish.

20.14 Some Factors to Be Considered in Spreading Poisons All Around

In opening this chapter we demonstrated that an insecticide can be safe enough for use in the human environment even though it's lethal to insects. One factor we examined there is the quantitative difference between the concentration of the active ingredient in the human body and the insect body. Another important factor, which we only touched on in that discussion, involves differences in the effects of poisons on different species—insects and humans in the opening demonstration. As the chapter progressed we examined LD_{50}s, which allow us to quantify these differences in toxic effects, and we defined "safety" in terms of our own and society's judgments about potential hazards.

In an earlier discussion of Dr. Wiley's Poison Squad (Sec. 19.12) we saw an example of how poisons were tested near the beginning of the 20th century by feeding them to human volunteers. Ethical standards for testing chemical hazards have changed since then. Humans, even volunteers, no longer form Poison Squads, and the use of animals such as mice and rabbits is coming under critical examination. In the closing Perspective we'll see how technical advances in microbiology have provided a simple technique for testing the mutagenic effects of chemicals using nothing more advanced biologically than bacteria.

PERSPECTIVE

How Many Dead Mice Is a Bowl of Fresh Cereal Worth?

Today laboratory animals, usually mice, replace Dr. Wiley's human Poison Squad volunteers. Even now, though, both a heightened sense of ethics and the costs of large-scale animal testing spur the development of other, faster, less expensive, and more humane tests: laboratory studies of a type that would provide accurate information without the sacrifice of large numbers of laboratory animals.

New laboratory procedures such as the *Ames test,* developed by Bruce N. Ames of the University of California, Berkeley, may point the way to better, less expensive, and surer methods of toxicity testing. The Ames test operates on the assumptions that both cancer and mutations begin with genetic damage of some sort and that a chemical mutagen stands a very good chance of being a chemical carcinogen as well. The test uses a strain of bacteria known as *Salmonella typhimurium* that is modified biologically to make it very sensitive to genetic damage by chemical means and therefore sensitive to chemically induced mutation. What's more, the bacteria are altered further so that the amino acid *histidine,* which is normally a *nonessential* dietary amino acid for the bacteria, becomes an *essential* amino acid (Sec. 17.4). As a result of its newly acquired dependence on histidine as an essential amino acid, a colony of the biologically altered bacteria won't grow in the absence of this amino acid. But exposing it to a mutagen that switches it back to its more common form allows it to synthesize its own histidine and to flourish.

Bruce N. Ames, developer of the Ames test to determine whether chemicals produce mutations in certain bacteria. A positive Ames test suggests that the chemical may also cause cancer.

In practice the investigator places the altered bacteria in a medium that contains all the needed nutrients for growth, except histidine, as well as an extract taken from rat livers. Using the liver extract helps identify chemicals that may not be mutagens themselves but that are converted into other, mutagenic substances as the liver enzymes detoxify them. Adding the chemical under examination starts the test. Rapid growth of the bacterial colony, at a rate comparable to that of unaltered bacteria under the same conditions, indicates that the added chemical (or whatever it's being metabolized to) is a mutagen and causes the bacteria to switch back to the form for which histidine is a nonessential amino acid.

Because of a very good overlap between the list of chemicals known to produce mutations and those known to cause cancer—about 90% of the chemicals on each list appear on the other—the Ames test and similar techniques can serve as useful, rapid, and inexpensive screening tools for identifying chemical carcinogens. Yet laboratory animals, which stand far closer to humans in their genetic makeup than bacteria do, still offer the best and most reliable estimate of chemical dangers to humans. In this respect, questions of ethical and economical values in animal testing must be balanced against the reliability of the animal experiments in protecting humans against chemical harm.

In concluding, we return to the questions about safety that began this chapter. We've seen that all chemicals present hazards. We've seen that there are, indeed, poisons and carcinogens in our foods and that they are to be found in processed foods and in natural, unprocessed foods as well.

We've seen that various governmental agencies have the power to protect us from excessive, known chemical risks but that it is our own judgment of the acceptability of risks, both as individuals and as members of society, that ultimately determines the issue of safety.

We've seen, finally, that the idea of *absolute* safety is a phantasm. To the extent that we are well informed and that our judgments are sound, we can weigh the very real benefits that chemicals provide and balance them against the very real risks of their use. In this way, and only in this way, can we ensure that our world, while never free of hazards, is indeed safe. ■

EXERCISES

For Review

1. Following are a statement containing a number of blanks and a list of words and phrases. The number of words equals the number of blanks within the statement, and all but two of the words fit correctly into these blanks. Fill in the blanks of the statement with those words that do fit, then complete the statement by filling in the two remaining blanks with correct words (not in the list) in place of the two words that don't fit.

Since anything can be harmful if we consume it in excessive amounts or use it carelessly or improperly, everything we come in contact with presents some risk of __________. Because of this, we find it useful to define safety as the __________ of __________. One measure of lethal risk in the chemicals we consume is the __________, which is the weight of the substance (per unit of __________) that is __________ to __________ of a large population of laboratory animals. According to this measure, the deadliest chemical known is __________, produced by a common microorganism. Second and third on the list of deadly chemicals, respectively, are __________ and __________. The fourth most lethal chemical is one produced by laboratory reactions carried out by humans, __________, which is also known by the simpler term __________. Other risks, not immediately lethal, include the risk of severe birth defects, produced by __________; the lesser genetic changes induced by __________; and the risk of cancer, generated by __________. Among the carcinogens are natural products such as __________, which is a component of __________, and the synthetic sweetener __________. Both of these chemicals are banned from use as food additives by provisions of the __________, but the sweetener is in continued use because of the enormous public demand for it. The regulation of food additives is the responsibility of the __________, an agency of the federal government.

2, 3, 7, 8-tetrachlorodibenzo-*p*-dioxin	lethal
acceptability	mutagens
aspirin	oil of sassafras
body weight	risk
botulinum toxin A	saccharin
carcinogens	safrole
diphtheria toxin	sodium cyanide
half	TCDD
injury	teratogens
LD_{50}	tetanus toxin

2. Describe, define, or identify the following:

aflatoxin	OSHA
comfrey	parathion
isoflurophate	solanine

3. What contribution did each of the following make to the subject matter of this chapter: (a) William Lowrance, (b) James J. Delaney, (c) Bruce N. Ames, (d) Ira Remsen and Constantine Fahlberg, (e) Harvey W. Wiley?

4. Name a chemical affected by each of the following acts of Congress and describe how the law affects our use of the chemical: (a) the Poison Prevention Packaging Act of 1970, (b) the Saccharin Study and Labeling Act of 1977.

5. In what food does each of the following chemicals occur: (a) lactose, (b) glycyrrhizic acid, (c) myristicin, (d) oxalic acid, (e) benzo(*a*)pyrene, (f) tetrodotoxin?

6. What governmental agency is responsible for (a) investigating outbreaks of illness caused by food spoilage; (b) the chemicals of beer, wine, liquor, and tobacco; (c) chemical pesticides; (d) exposure to chemicals in the workplace; (e) chemicals used as food additives?

7. Give the titles of two acts passed by Congress that contain expressions of its judgment about the safety of chemicals used by consumers. (You may wish to review Sec. 19.13.)

8. Which of the natural toxins described in Section 20.5 is derived from a steroid? Which contains a glucose ring?

9. What was thalidomide used for before it was found to produce birth defects?

10. What is the difference between (a) a teratogen and a mutagen? (b) A mutagen and a carcinogen?

11. What is an additive's ADI and how is it determined?

12. On what factors does the harm that any particular substance can do to us depend?

13. What are natural sources of each of the following chemicals: (a) oxalic acid, (b) caffeine, (c) tetrodotoxin, (d) symphytine, (e) safrole?

14. What characteristic or property of a chemical does the Ames test reveal?

A Little Arithmetic and Other Quantitative Puzzles

15. Which is more toxic to rodents when administered orally: (a) arsenic trioxide or sodium chloride? (b) aspirin or trisodium phosphate? (c) caffeine or nicotine? (d) acetaminophen or BHA?

16. What three poisons of Table 20.1 are the most powerful when administered orally to mice or rats? What three are the least powerful when administered in the same way?

17. Commercial aspirin contains 325 mg of aspirin (acetylsalicylic acid) per tablet. Assuming that the LD_{50} for aspirin in mice and rats applies equally well to humans, how many aspirin tablets, taken all at once, would produce a 50% chance of a lethal dose of aspirin in a 70-kg person?

18. Canned fruit drinks often contain 0.1% of sodium benzoate as a preservative. Studies provide a value of 4 g/kg for the LD_{50} of sodium benzoate, orally in rats. Assuming that humans respond to this chemical as rats do, and assuming that the density of a commercially canned fruit drink is 1.0 g/mL, how many liters of canned fruit punch would a 70-kg person have to drink, all at the same time, to produce a 50% chance of a lethal dose of sodium benzoate?

19. Solanine has the lowest LD_{50} of the poisons listed in Table 20.2. What amount of solanine would a 70-kg person have to consume at one time to produce a 50% chance that it would be lethal?

20. Arrange the following in order of their ability to cause illness or death (from the most hazardous or lethal to the least): aspirin, sodium chloride, sodium cyanide, water.

21. Why is the quantity of a chemical that is lethal to 50% of a population a better measure of its hazard than the quantity that is lethal to 100% of a population?

22. Section 20.4 contains the statement that 5 g of botulinum toxin A would be enough to kill half of a group of 2,400,000,000 people. Using data presented in this section, show by a calculation that this population figure is correct.

23. What is the ADI of a chemical that produces no detectable effect on laboratory animals when they are fed it at a level of 1 mg/kg per day but produces bladder cancer at a level of 2 mg/kg per day? Explain.

24. Applying LD_{50}s obtained with mice and other laboratory animals directly and quantitatively to humans requires the questionable assumption that humans respond to poisons in exactly the same way as mice. For a more realistic illustration of the potency of botulinum toxin, calculate the total population of a large group of mice, half of which would be killed by a single teaspoon of botulinum toxin. Assume that the average mouse weighs 20 g.

25. Section 20.4 contains the statement that a teaspoon of tetanus toxin would be enough to kill half a population of 14,300,000 people. What assumptions are made in arriving at this figure? Using data of Section 20.4, demonstrate that this population figure is correct.

26. How would you go about determining a toxicity value we might term an LD_{75}? Would you expect an LD_{75} to be larger or smaller than the LD_{50} for the same substance?

27. As described in Section 20.9, a safety factor of 100 is applied to the results of feeding studies to determine the ADI for a newly proposed food additive. What is the numerical value of the safety factor that's applied if the proposed additive is discovered to be a carcinogen?

28. Assuming that an average of 60 million Americans were regular users of saccharin in 1977, what was their average daily consumption in grams?

Think, Speculate, Reflect, and Ponder

29. Give two definitions of safety described in this chapter. Is it possible to achieve absolute safety under either definition? If so, under which? Is it possible to prove that something is safe under either definition? If so, under which?

30. What do the molecular structures of the nerve gases sarin, soman, and taubin have in common with the molecular structure of the insecticide parathion?

31. With "safety" defined as the acceptability of risk, name three activities you would consider to be unsafe.

32. Give two different illustrations of how the hazard presented by a substance can depend on the species (and age) of the animal and the method by which it is introduced into the animal's body.

33. Under what circumstances would an ADI be lower than 1/100 of the no-effect level of a chemical?

34. Suppose that laboratory tests on a newly discovered chemical showed that it produced absolutely no effects on any animal tested, no matter how or at what level it was administered. Would you consider this new chemical to be "safe"? Explain.

35. A statement sometimes used about the hazards of medicines is: "The poison is in the dosage." Explain what this means.

36. Which of the following does our society (as represented by its laws) consider to be safe? (a) use of ethylenedi-

aminetetraacetic acid as a food additive (Sec. 19.17); (b) cigarette smoking by adults in the privacy of their own homes; (c) easy access to aspirin by children under age five (Sec. 20.2); (d) use of sodium nitrite as a food additive (Perspective, Chapter 19).

37. Describe your own thoughts about the safety of each of the following: (a) sodium chloride, (b) aspirin, (c) ethyl alcohol, (d) caffeine, (e) nicotine.

38. If we can say that anything is hazardous if it is used in excess, can we also say that everything is safe if used in very small amounts? Explain.

39. What conclusions, if any, could you draw from a positive Ames test (with bacterial growth) in an examination that did not contain added liver extract? What if this same examination produced a negative result?

Web Questions

Use the Web to answer or accomplish each of the following. Report your Web source for each response.

1. Find the LD_{50} for an example of each of the following, other than substances described in this chapter: (a) an insecticide or a pesticide, (b) a food additive, (c) a natural component of an unprocessed food, (d) a chemical element, (e) a vitamin.

2. Identify and describe the industrial disasters of the 1970s, mentioned in Section 20.5, that involved TCDD.

3. Summarize some current arguments for the modification, revision, or repeal of the Delaney Amendment.

4. What were the experimental results that led to the proposal to end the use of saccharin as a food additive? What is the current status of saccharin as a food additive?

5. Give the molecular formula and a brief description of the three most abundant flavor-producing ingredients of the mango, as shown in the caption of Figure 20.8.

6. Describe in detail an application of the Ames test. Identify five chemicals that have been subjected to this test and present the results obtained.

7. Both sodium cyclamate and calcium cyclamate, salts of cyclamic acid, have been used as artificial sweeteners. Give the molecular structure of cyclamic acid and describe the history and current status of its sodium and calcium salts as food additives.

Polymers and Plastics

THE PLASTIC AGE

Plastics, the material of our everyday world.

Plastics, Water, and Rubbing Alcohol

Figure 21.1 Separating plastics of different densities with a mixture of isopropyl alcohol (rubbing alcohol) and water. The more dense plastics sink; the less dense plastics float.

Plastics are remarkably useful materials, but one of their greatest advantages over other materials, their durability, is also one of their greatest handicaps. They don't disintegrate. Once we dump them into the environment, they just stay there. Some plastics seem to last forever. One way to keep discarded plastics from overwhelming us is to recycle them, to use them over and over again. But there are different kinds of plastics, and one type doesn't mix well with another. To reuse plastics we have to separate one kind from another so that we can combine all compatible plastics into one group and reprocess them as a whole. One way to sort them is by taking advantage of differences in the densities of different kinds of plastics (Sec. 13.1).

You can do this yourself by using some rubbing alcohol, water, and a few pieces of different kinds of plastics. For a good start, use the rather hard plastic of a milk bottle. (Be sure to use a plastic bottle, not a waxed paper carton.) Another good plastic to use here is a trash bag. Plastic milk bottles are made of a high-density polyethylene, while trash bags are made of a low-density polyethylene. Made by one process, the polyethylene forms as a hard, rigid, high-density material well suited to bottles and other containers. Made in a different way, the polyethylene forms as a soft, flexible, low-density plastic that gives trash bags their desirable characteristics.

A mixture of one part water to two parts rubbing alcohol, by volume, provides a good start for separating the two. The milk bottle plastic should sink in this mixture and the trash bag should float. If both sink, add a little water. If both float, add a bit of rubbing alcohol. Water is more dense than the isopropyl alcohol of the rubbing alcohol (Sec. 9.7). If both plastics sink, adding some water should force the less dense plastic to float. Adding the rubbing alcohol decreases the density of the liquid mixture, causing the more dense plastic to sink.

Either way, by adding water or rubbing alcohol to adjust the density of the liquid, you can separate plastics of different densities. Once you learn how with the polyethylenes, try separating pieces of plastic taken from other containers (Fig. 21.1). Techniques of this sort can help us keep used plastics out of our trash dumps and put them back into consumer products.

21.1 The Age of Plastics

If an era is known by the kinds of materials its people use to manage the world they live in, then the Stone Age, the Bronze Age, and the Iron Age have given way to our own Plastic Age. Plastics form much of our packing and wrapping materials, many of our bottles and containers, textiles, plumbing and building materials, furniture and flooring, paints, glues and adhesives, electrical insulation, automobile parts and bodies, television, stereo and computer cabinets, medical equipment, cellular telephones, compact disks, personal items, including pens, razors, toothbrushes, and hairsprays, and even the plastic bags we use to discard all our plastic trash. Except for our food, air, and water, almost every ordinary thing we come into contact with each day contains some sort of plastic somewhere in, on, or around it. So many of our throwaway goods are made of plastic that, despite its low density, the material makes up roughly 10% of the total weight of all solid municipal wastes, largely as plastic containers and packing materials. What's more, at about a quarter of the entire volume of our trash, plastic products form a highly visible part of what we discard. We'll have more to say about the problem of plastic wastes in the Perspective.

Plastics in everyday trash.

In 2000 the U.S. chemical industry produced over 37 million tons (roughly 34 billion kilograms) of unprocessed, raw plastics of all kinds. Table 21.1 describes some of the more important properties and uses of the most common commercial plastics, in declining order of the quantities manufactured. About half of all manufactured plastics goes into packaging and into building and construction materials; roughly 10% forms personal consumer products; the remainder is used in a variety of products, including furniture, parts for cars and other vehicles, and electrical and electronic equipment.

Table 21.1 Major Types of Commercial Plastics and Related Materials

Plastic	Characteristics	Typical Applications
Low-density polyethylene	Low melting; very flexible; soft; low density	Bags for trash and consumer products; squeeze bottles; food wrappers; coatings for electrical wires and cables
Poly(vinyl chloride), also known as PVC	Tough; resistant to oils	Garden hoses; inexpensive wallets, purses, keyholders; bottles for shampoos and foods; blister packs for various consumer products; plumbing, pipes, and other construction fixtures
High-density polyethylene	Higher melting, more rigid, stronger, and less flexible than low-density polyethylene	Sturdy bottles and jugs, especially for milk, water, liquid detergents, engine oil, antifreeze; shipping drums; gasoline tanks; half to two-thirds of all plastic bottles and jugs are made of this plastic
Polypropylene	Retains shape at temperatures well above room temperature	Automobile trim; battery cases; food bottles and caps; carpet filaments and backing; toys
Polystyrene	Light weight; can be converted to plastic foam	Insulation; packing materials, including "plastic peanuts"; clear drinking glasses; thermal cups for coffee, tea, and cold drinks; inexpensive tableware and furniture; appliance cabinets
Poly(ethylene terephthalate), also known as PET	Easily drawn into strong, thin filaments; forms an effective barrier to gases	Synthetic fabrics; food packages; backing for magnetic tapes; soft drink bottles
Phenol-formaldehyde resins	Strongly adhesive	Plywood; fiberboard; insulating materials
Nylon	Easily drawn into strong, thin filaments; resistant to wear	Synthetic fabrics; fishing lines; gears and other machine parts

Durable or fragile, rigid or flexible, sturdy or flimsy, dense or light, strong or weak, plastics provide us with inexpensive materials of virtually unlimited properties. With chemical ingenuity we can transform them into almost whatever shapes we wish with almost whatever properties we desire. And at their root, in the *polymeric* molecules that make up these extraordinary substances of our everyday world, lies one of the shining achievements of modern chemistry. We'll begin by examining the difference between the *plastics* of our world and the *polymers* that form them.

■ QUESTION

Plastic makes up about 10% of the mass of our discarded trash, but about 25% of its volume. What is the likely reason for the difference between these two figures?

21.2 Plastics and Polymers

Plastics, especially the plastics of our most common commercial products, are materials that we can shape into virtually any form we want. The word itself comes from the Greek *plastikos*, for a substance that's suitable for molding or shaping. We can form plastics into round, hard, resilient bowling balls, draw them out into the thin, flexible threads of synthetic fibers, mold them into intricately designed, durable machine parts, or flatten them into flimsy but tough sheets of clinging kitchen film. Today, the word **plastic** refers mostly to a property of a material: its ability to be shaped into the myriad forms of today's commercial and consumer products.

A **plastic** is a material capable of being shaped into virtually any form.

When we speak of a *polymer*, though, we return to the molecular level. All the plastics of our everyday lives, as well as all the proteins, starch and cellulose of our foods; the cotton, silk, and wool of our textiles; and even the DNA that carries the genetic code within the nucleus of the cell are formed of enormously large polymeric molecules. The combination of the Greek words *poly*, meaning many, and *meros*, parts, gives us the word for the molecules that compose these substances, *polymer*. A **polymer** is a molecule of very high molecular weight, formed through the repeated chemical linking of a great many smaller molecules.

A **polymer** is a molecule of very high molecular weight formed by the repeated chemical linking of a great many simpler, smaller molecules.

Monomers are the individual structures that are linked to each other to form a polymer.

As the word implies, polymers are extremely large molecules, sometimes called *macromolecules* to emphasize their very large size. The individual parts that combine to form them, their **monomers** (from the Greek *mono*, signifying one or a single unit), join to each other in enormously large numbers to produce polymers with molecular weights ranging from the tens of thousands to millions of atomic mass units.

In many polymers the monomers have joined to each other to form enormously long, linear molecular threads, very much like long chains we might find in a hardware store (with the individual links representing the monomers). In other polymers the chains may be branched to various degrees, or they may be interconnected at occasional junctions, or so frequently that they form a web or even a rigid, three-dimensional lattice (Fig. 21.2). In any event, a polymer is a substance composed of huge molecules, sometimes in the form of very long chains, sometimes as sheets, sometimes as intricate, three-dimensional lattices.

A plastic, on the other hand, is a material that can be molded readily into a variety of shapes. All of today's commercial plastics are polymers, even though some of our most important polymers—the starch and cellulose of our foods (Chapter 16), the proteins of our foods and bodies (Chapter 17), and the silicates of our earth (Sec. 21.15), for example—are not at all plastic.

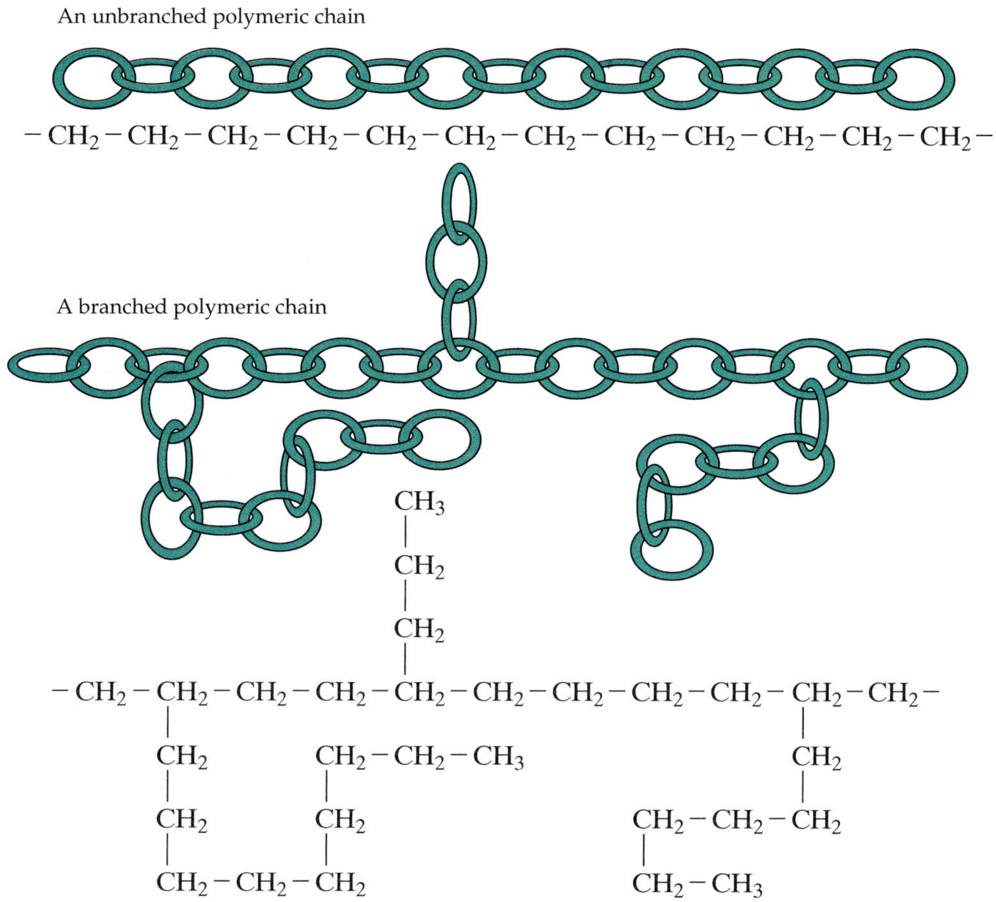

Figure 21.2 Chains and polymers.

We'll begin our review of plastics and polymers by examining their

- chemistry of formation,
- molecular structures, and
- physical properties.

Then we'll look briefly at a history of polymers and their discoverers and inventors.

■ QUESTION

What are three polymers found as major components of our foods?

21.3 Modes of Polymerization: Addition Polymers and Condensation Polymers

Polymerization is the process whereby individual monomers link together to form a polymer.

The actual linking of the monomers through covalent bonds occurs through **polymerization**, a chemical process easily divided into two broad categories: *addition polymerization* (which produces *addition polymers*) and *condensation polymerization* (which produces *condensation polymers*). Addition polymers, such as *polyethylene* (Sec. 21.13) and several of its close molecular relatives, dominate today's chemical economy. They account for more than half of all the plastics currently

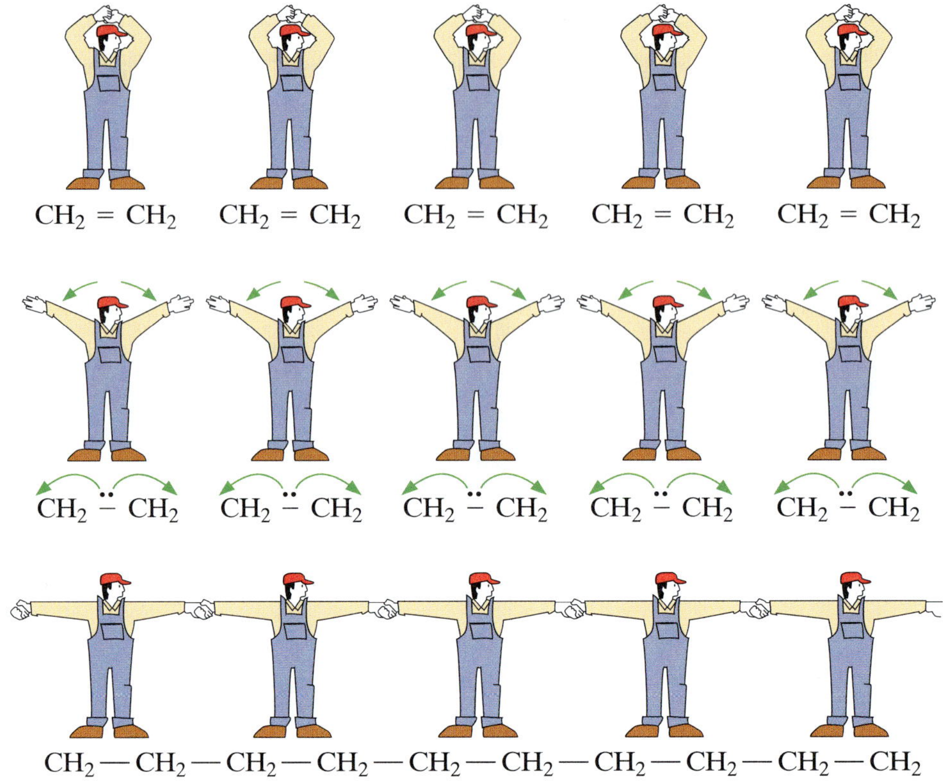

Figure 21.3 The addition polymerization of ethylene to polyethylene.

produced in the United States. Polyethylene and other addition polymers form as their individual, unconnected monomers join each other to form a polymeric chain in much the same way as people standing next to each other can form a human chain by holding the hands of those next to them.

Figure 21.3 shows a group of people lined up side by side, clasping their hands over their heads, and also a group of ethylene molecules arranged similarly, each with its two covalent carbon–carbon bonds. As we saw in Section 3.13, every covalent bond is actually a shared pair of electrons. In a rough analogy, each pair of clasped hands in the figure corresponds to a pair of bonding electrons. Just as each person in the row can unclasp his or her own hands and link up with those on either side to form a human chain, each ethylene molecule, under the right chemical conditions, can swing the electrons of one of its bonds toward each of its neighbors to form a molecular chain. The result is polyethylene, an addition polymer.

The analogy isn't perfect. For example, polymerizations aren't limited to the specific sort of electron movements shown in Figure 21.3. Under different chemical conditions, both electrons can move in the same direction, as in Figure 21.4.

Figure 21.4 Another route for the polymerization of ethylene.

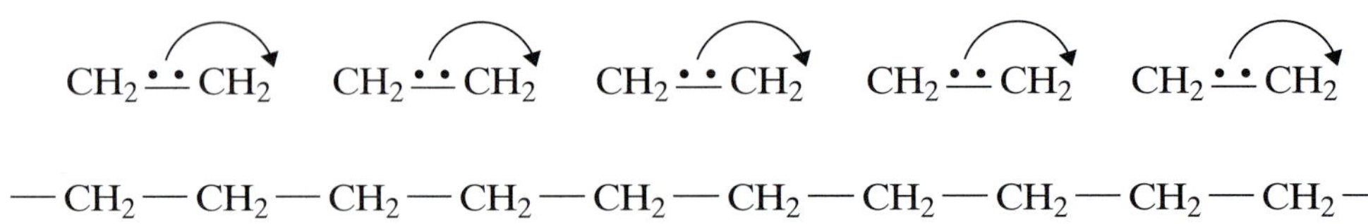

In either case, though, the process is still polymerization and the product is still polyethylene, an addition polymer.

In a *condensation reaction* (a general term for the kind of chemical reaction that takes place in condensation polymerization) two molecules combine with the formation and loss of another, smaller molecule, usually water or a simple alcohol or acid. [The general term "condensation reaction" probably originated as early chemists observed water or similar liquids forming droplets of condensate on the sides of flasks during this sort of reaction (Fig. 21.5).] Each of the condensing molecules contributes some portion of the smaller molecule being eliminated. In Figures 16.18 through 16.21 we saw how molecules of glucose can undergo simple condensation reactions to form cellobiose and maltose, and condensation polymerizations to form polymeric chains of cellulose and starch. In each case a molecule of water is lost in the formation of each bond linking two adjacent monomers. In Figure 17.4 we saw a corresponding condensation reaction in which two amino acids form a dipeptide with the loss of an H_2O molecule. The naturally occurring polysaccharides and proteins provide us with good examples of condensation polymers, even though they form through complex enzymatic reactions, far removed from the relatively straightforward industrial processes that produce our everyday polymers, such as *nylon* (Sec. 21.9) and *Dacron* (Sec. 21.10).

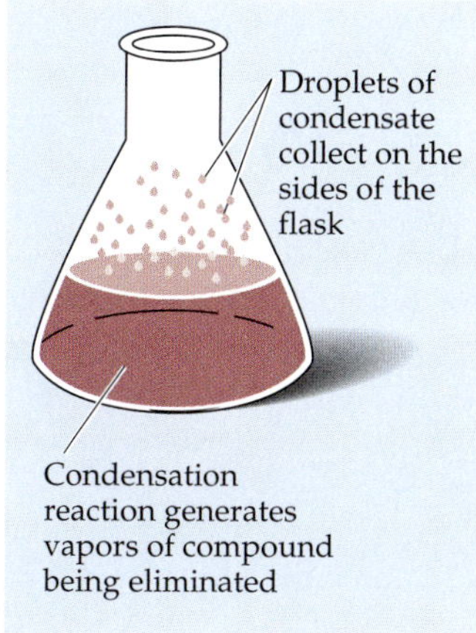

Figure 21.5 A condensation reaction.

■ **QUESTION**

Of the following, which are addition polymers and which are condensation polymers? (a) nylon, (b) starch, (c) protein, (d) cellulose, (e) polyethylene.

21.4 Molecular Structures: Homopolymers and Copolymers

Regardless how they may be formed, we can also divide polymers into two different categories to reflect the *structures* of their molecular chains:

- *homopolymers*—polymers consisting of chains in which each link (each monomer) is identical with every other link—and
- *copolymers*—polymers consisting of chains composed of two or more different kinds of links.

Since all links of homopolymers are mutually identical, we can represent the monomer of the chain as X and write the molecular chain of a homopolymer as

$$-X-X-X-X-X-X-X-X-X-X-X-X-X-X-X-X-X-X-X-$$

with the Xs running off both sides of the page and continuing to the full length of the chain. To simplify matters we can show the molecular structure just as well by writing *one* of these monomeric units in parentheses and adding a subscript n, which represents a very large but unspecified number of links. The short, horizontal lines running from X through each of the parentheses represent the covalent bonds linking X to its neighbors:

$$+\!\!(X)\!\!+_n$$

Using this approach, we can recognize the starch and cellulose of Chapter 16 as homopolymers, with glucose as their repeating monomeric unit. We can then draw

The cellulose of the leaves, stems, and trunks of plants is a homopolymer of glucose.

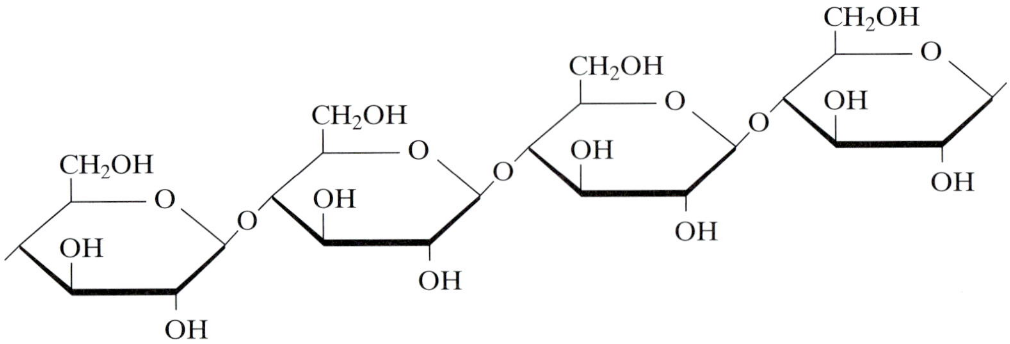

Figure 21.6 Cellulose as a homopolymer.

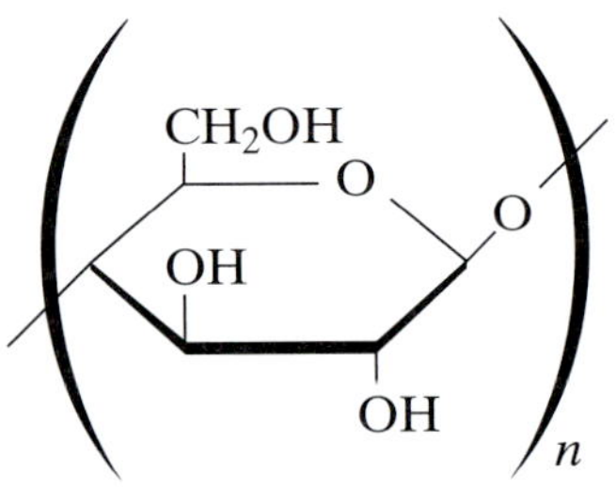

Figure 21.7 Cellulose.

their molecular structures as the full chain (Fig. 21.6) or simply as the repeating monomeric unit (Fig. 21.7).

Protein chains are also polymeric. But with an assortment of 20 different amino acids serving as their monomeric units and appearing in an elaborate sequence along the chain, proteins fall into the class of copolymers (Fig. 21.8). While it isn't possible to use a form of $-\!(\mathrm{X})\!-_n$, in writing the molecular structure of a typical protein, some other copolymers, especially those in which two different monomers alternate with each other throughout the chain, do lend themselves to this approach. For example,

$$-X-Y-X-Y-X-Y-X-Y-X-Y-X-Y-X-Y-X-Y-X-Y-X-Y-$$

can be written as $-\!(\mathrm{X}-\mathrm{Y})\!-_n$. In Section 21.9 we'll see how we can use this same kind of symbolism to write the structure of the copolymer nylon.

In summary, we'll note that any particular polymer or plastic can fall into several different categories. Both condensation and addition polymerization can produce homopolymers and copolymers. The terms *homopolymer* and *copolymer* refer to the molecular structures of the polymeric chains; *addition* and *condensation* refer to how the chains are formed.

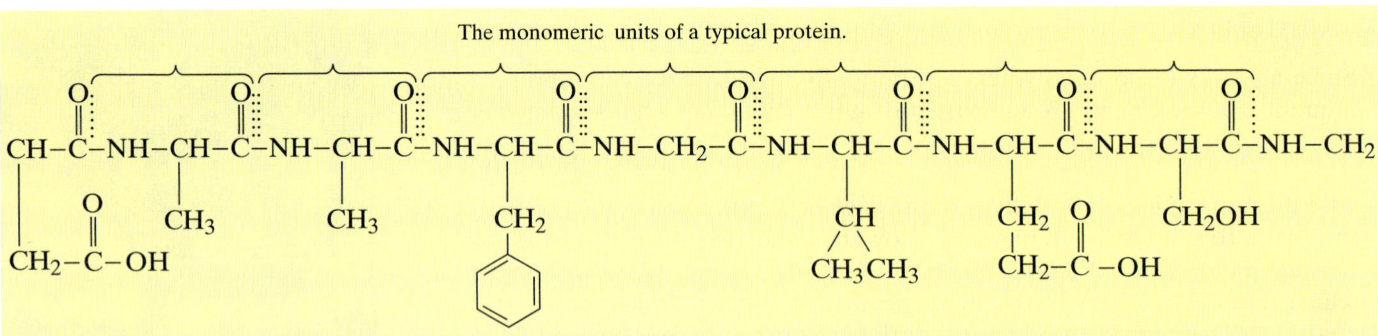

Figure 21.8 A typical protein as a copolymer.

■ **QUESTION** Of the following, which are homopolymers and which are copolymers? (a) starch, (b) $-\!(\mathrm{X}-\mathrm{Y})\!-_n$, (c) a typical protein, (d) cellulose, (e) $-\!(\mathrm{X})\!-_n$, (f) polyethylene.

21.5 The Properties of Polymers: Plastics That Act Like Fats and Plastics That Act Like Eggs; Polymers That Bounce and Polymers That Don't

We'll turn now to some of the properties of the plastics and other polymers we find in our everyday world. We'll examine what some of them do when we heat them and what others do when we toss them against a wall or drop them. If you've ever left something made of plastic in a hot location, perhaps next to a window of a car parked in the sun, or if you've ever watched an optician heat the plastic frames of eyeglasses before readjusting them, you know that some plastics soften as they are heated. On the other hand, if you've ever let the plastic handle of a frying pan stand over a hot element on a stove, you know that some plastics keep their shapes and remain rigid even though they get hot enough to decompose.

Plastics that soften when heated and become firm again when cooled, much like the fats of Chapter 15, are called **thermoplastics**. Another category of plastics, the **thermosets** or *thermosetting plastics*, are soft enough to be molded when they are first prepared, but on heating they firm up permanently. Reheating may cause them to decompose (if the temperature is high enough) but it certainly won't soften them again. Bakelite (Sec. 21.8) is a good example of a thermosetting plastic. Since Bakelite is a strong material and a poor conductor of heat and electricity, it's used in making handles for toasters and pots and pans, for molding common electrical goods such as wall outlets and adapters, and for such diverse items as buttons and billiard balls. Resins similar to Bakelite are used in fiberboard and plywood. While thermoplastics mimic fats in their response to heat, the thermosets are more like eggs. Heating produces irreversible changes in both thermosets and eggs.

Thermoplastics are plastics that soften when they are heated, then harden again as they cool.

Thermosets are plastics that hold their shape, even when they are heated.

Billiard balls made of a thermosetting polymer.

Polymeric molecules also make up another kind of material, one that returns to its original shape after it's stretched or squeezed. Materials of this sort bounce back when they're thrown against a wall or dropped on a floor. Christopher Columbus and his companions found the inhabitants of the newly discovered Western Hemisphere playing with a natural polymer that possesses this property. The

Recycling the thermoplastic polymers of soda bottles.

people of this hemisphere were fond of bouncing heavy balls made of a plant gum that Europeans eventually named *rubber* when, almost three centuries after Columbus's voyage, they found it was useful for rubbing out marks made by lead pencils. While rubber is plastic in the sense that we can mold, twist, stretch, and compress it into various shapes, rubber belongs to the class of polymers called **elastomers**, substances that stretch easily and return readily to their original shapes. It's the resilience of rubber that we value.

An **elastomer** is a substance that stretches easily and returns readily to its original shape.

Rubber itself is an addition polymer of a *diene* (a hydrocarbon molecule containing two double bonds) called *isoprene* or, applying the IUPAC rules of Sections 6.9 and 6.10, *2-methyl-1, 3-butadiene*. As shown below, we number the carbons of the unsaturated chain that contains the double bonds from 1 to 4, beginning at the end closer to the methyl substituent.

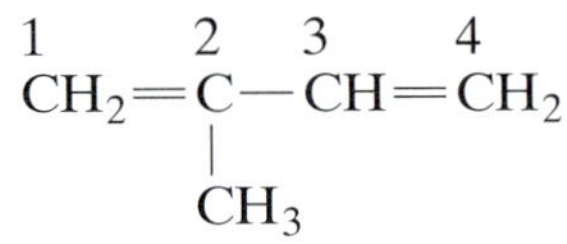

isoprene,
or
2-methyl-1,3-butadiene

Like the polymeric proteins and polysaccharides, the rubber polymer forms in nature through a complex set of reactions. Nevertheless, we can view it as *polyisoprene*, an addition polymer of isoprene, and we can imagine it forming with electron movements similar to those that convert ethylene into polyethylene (Figs. 21.3 and 21.4). Double bonds remain in the polymeric chain of rubber and, moreover, fix the chain into the *cis* configuration, one of two possible shapes available to it (Fig. 21.9).

As we saw in Section 15.6, a carbon–carbon double bond can secure an unsaturated molecule into a *cis* or a *trans* geometric configuration. The uniformly *cis* configuration of all the double bonds in rubber's polyisoprene chain keeps the

Figure 21.9 Rubber as *cis*-polyisoprene.

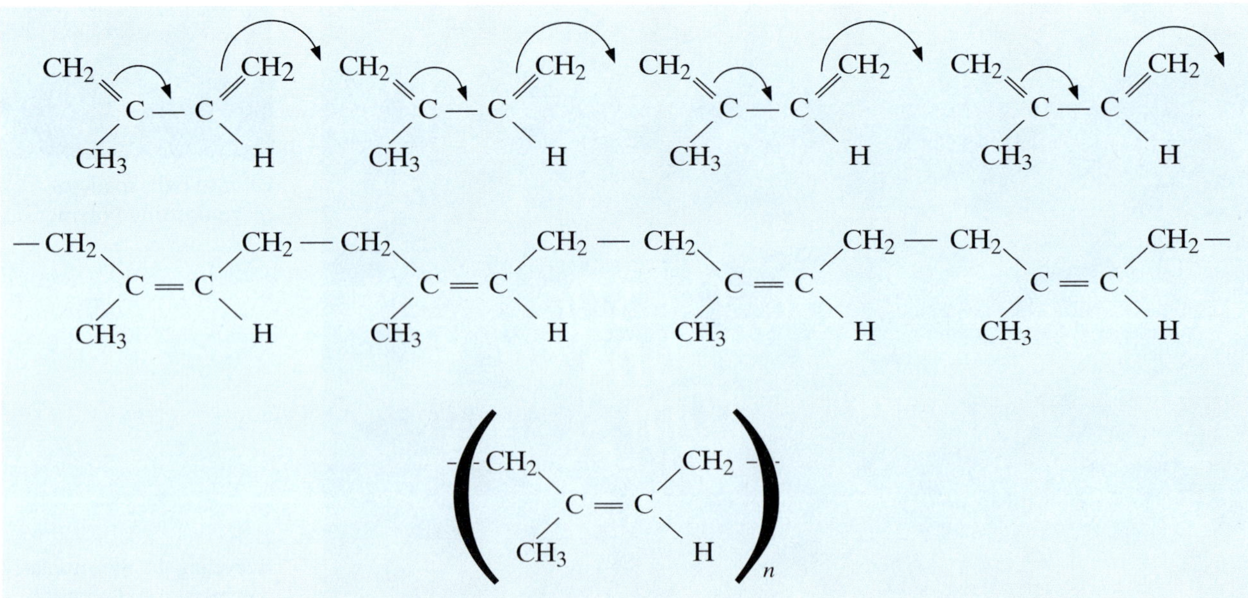

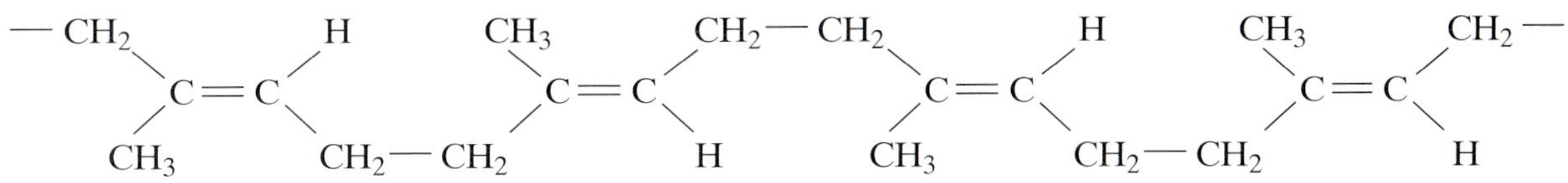

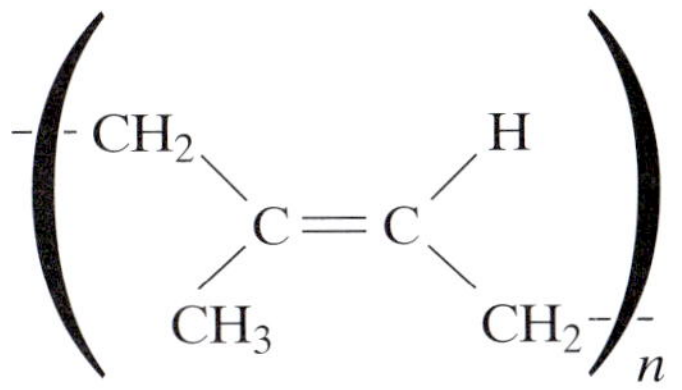

Figure 21.10 Gutta-percha as *trans*-polyisoprene.

carbons of the chain on one side of the average line of the double bonds and the methyl substituents on the other, as in Figure 21.9. In *gutta–percha*, another naturally occurring polymer of isoprene, all the double bonds are in the *trans* configuration (Fig. 21.10). With an all-*trans* molecular geometry, gutta–percha isn't nearly as elastic as rubber. It does find uses, though, as a covering for golf balls, in surgical equipment, and as an electrical insulator, especially for underwater cables.

Having completed a review of the structures of polymeric molecules, the ways they form from monomers, and some of their properties, we'll now examine some of their histories.

■ QUESTION

We've seen that polyisoprene can exist as geometric isomers. Can polyethylene also exist as geometric isomers? Explain.

21.6 A Brief History of Polymers and Plastics. Part I: The Roman God of Fire

In its elasticity, rubber illustrates the importance of molecular structure as a source of physical properties. Its elasticity originates in the way its molecules coil up, which allows them to stretch out when we pull on a piece of rubber and then to spring back when we release it. Heat a piece of pure, natural rubber, though, and you'll find that as it becomes warm it loses much of its resilience. Its usual bounce becomes sloppy and it turns sticky. That happens because at high temperatures the intertwined, threadlike polyisoprene molecules slide past each other a bit too readily when we stretch the rubber and they don't pull back to their original positions when we release it.

Charles Goodyear, born in 1800 in New Haven, Connecticut, solved the problem of sticky rubber partly by accident. Goodyear was an inventor and the son of an inventor, but he lacked the talents of a good businessman. He had already spent time in jail for his debts when he became obsessed with the idea of creating a rubber that retained its elasticity even when hot. He tried to perfect it for 10 years with little success until, one day in 1839, he accidentally dropped a mixture of crude rubber and sulfur onto a hot stove. When the charred mixture cooled a bit he found that it was nicely elastic, even though still warm. In 1839 neither Goodyear nor anyone else knew anything about the molecular structure of rubber. Goodyear knew only that heating rubber with sulfur worked; he had no idea why. Once again we see the importance of serendipity in science (Sec. 4.1).

Charles Goodyear discovers vulcanized rubber.

We know now that with heating, the sulfur and the polyisoprene molecules react to *cross-link* the polyisoprene molecules to one another. With this cross-linking the molecular structures became loosely bound to each other in a three-dimensional lattice. The sulfur links keep the long molecules from slipping past their neighbors at high temperatures and thereby keep the rubber resilient and prevent it from becoming sticky (Fig. 21.11).

Five years later Goodyear received a patent for the process, but financial success eluded him. Poor and debt ridden, he tried unsuccessfully to make his fortune by manufacturing rubber in both England and France. Close to his 55th birthday, he was thrown into a Paris jail for debt. He returned to the United States and died in New York on July 1, 1860, still poor and still in debt. Someone else named his process *vulcanization* in honor of the Roman god of fire, *Vulcan*. Today vulcanization forms the basis of a major industry. All natural rubber now used commer-

Figure 21.11 Vulcanized rubber.

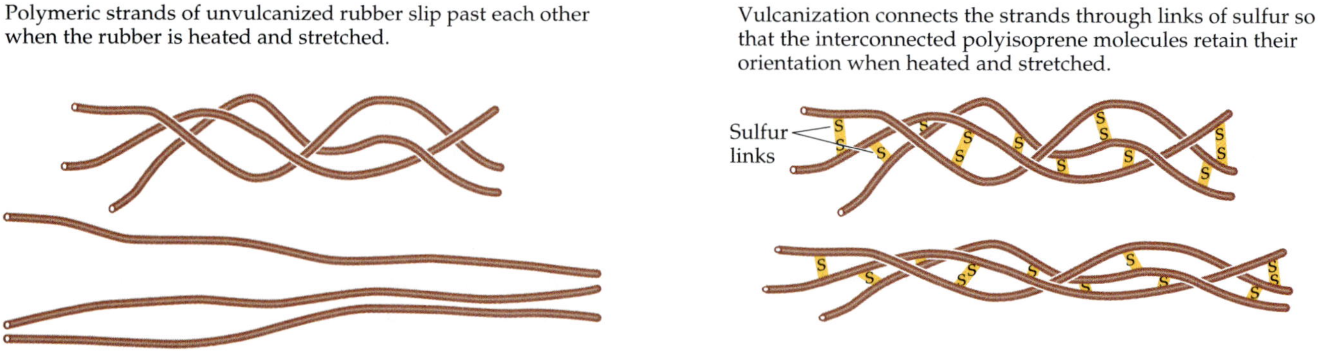

The elastomers *neoprene* and *styrene–butadiene rubber* make up the bulk of auto and truck tires.

cially is vulcanized, with most of it going into the production of auto tires. The name of Charles Goodyear is preserved in the Goodyear Tire and Rubber Company.

Several synthetic elastomers now supplement or replace rubber in various consumer applications, particularly in tire treads and engine hoses and belts. Today's leading synthetic elastomer, *styrene–butadiene rubber*, was developed by German chemists in the early 1930s and further refined by Americans during and after World War II. It results from the copolymerization of a mixture of 75% butadiene and 25% styrene (Fig. 21.12), and it serves as a good substitute for natural rubber in the manufacture of various consumer goods, especially the treads of tires.

Neoprene is a homopolymer produced by the polymerization of *chloroprene*, a monomer with a molecular structure resembling isoprene (Sec. 21.5), but with isoprene's methyl group replaced by chlorine (Fig. 21.13). Neoprene resembles vulcanized rubber except that this synthetic substitute is much more resistant to heat and to the action of hydrocarbon solvents like gasoline and automotive greases and oils. Its properties are especially valuable in belts and other parts used in and

Figure 21.12 A segment of the styrene–butadiene polymer.

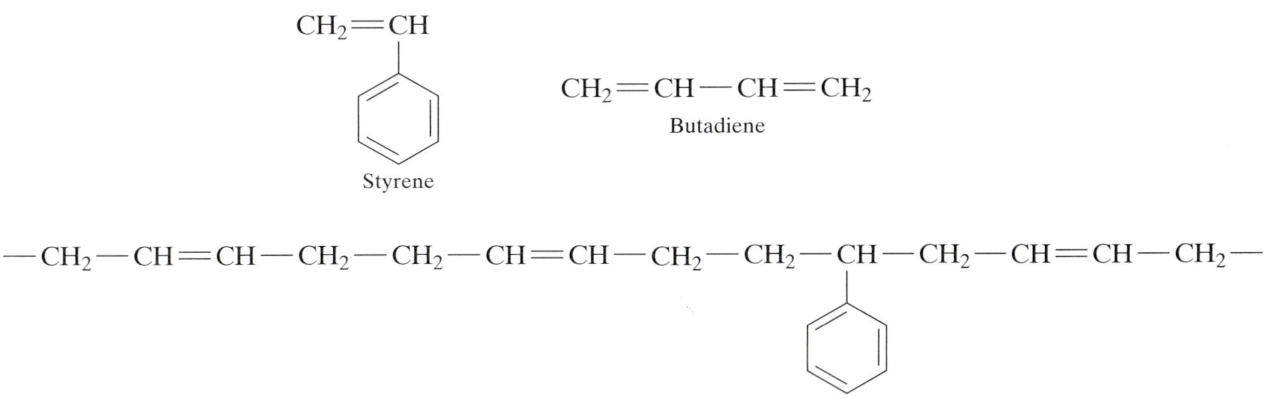

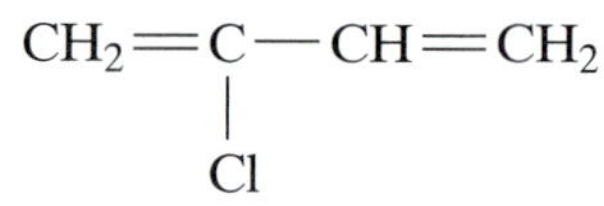

Chloroprene

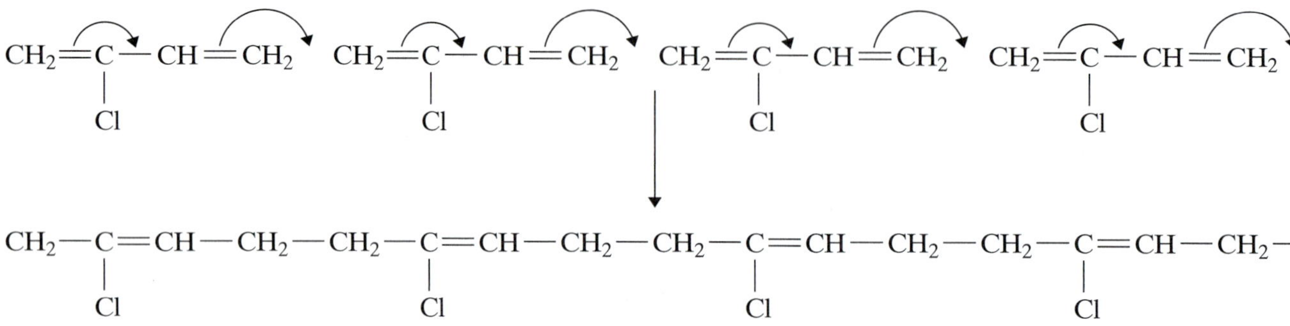

Figure 21.13 Neoprene, polychloroprene.

around auto engines, in the hoses of gasoline pumps, in gaskets and filling material used in the construction of highways and bridges, and in the rubber stoppers found in chemistry laboratories. Neoprene was developed for commercial production by Wallace Carothers, the inventor of nylon (Sec. 21.9).

■ **QUESTION**

What element, absent from natural rubber, is present in: (a) neoprene? (b) vulcanized rubber?

21.7 Part II: Save the Elephants!

Celluloid and other polymers that can be made to resemble ivory have helped reduce the demand for elephant tusks.

Charles Goodyear didn't invent a new polymeric molecule or a new elastomer when he spilled a mixture of rubber and sulfur onto a hot stove; he accidentally *modified* a natural, polymeric elastomer and came up with an enormous improvement in its properties. John Wesley Hyatt and his brother Isaiah, on the other hand, consciously and deliberately converted a modified natural polymer into the world's first new, commercially successful synthetic plastic. They produced a plastic that hadn't existed earlier.

By 1863, three years after Goodyear's death, the slaughter of the world's elephants for their tusks had become a serious matter, as it is even today. In that era the disappearance of elephants was threatening to disrupt the world's ivory supply. Ivory, a valuable luxury item of the 19th century used for jewelry, ornaments, piano keys, and various other items, was becoming scarce and very expensive. Perhaps to protect the elephants but certainly to ensure a source of raw materials, the firm of Phelan and Collander, a New England manufacturer of ivory billiard balls, offered \$10,000 to anyone who could devise a satisfactory substitute for the rapidly disappearing natural ivory. John Hyatt, a 26-year-old printer born in Starkey, New York, took up the challenge. He was helped by a startling discovery made years earlier by a chemist in Basel, Switzerland.

In 1846 a Swiss chemistry professor, Christian Schoenbein, had accidentally invented guncotton by spilling a mixture of nitric acid and sulfuric acid in the kitchen of his home and then wiping up the mess with his wife's cotton apron. He rinsed out the apron thoroughly with water and hung it up to dry near a hot stove. As it hung drying it disappeared in a sheet of flame.

The smokeless *guncotton* of Schoenbein's accident proved far superior to the very smoky *gunpowder* used at the time in warfare and it became a popular military item. More to the point, by inadvertently inducing a reaction between the mixture of nitric and sulfuric acids and the cellulose of the cotton apron, Schoenbein had successfully transformed the polymeric cellulose into *nitrocellulose*, a compound in which varying numbers of the hydroxyl groups ($-OH$) of the polymer are converted into nitrate groups ($-O-NO_2$). (The sulfuric acid serves to catalyze the reaction.)

John Hyatt won the $10,000 prize by combining *camphor*, a pungent substance obtained from the camphor tree, with a lightly nitrated form of Schoenbein's nitrocellulose. The mixture forms a thermoplastic so similar to ivory that it was known for some time as artificial ivory. We now call it *celluloid*. With the help of his brother Isaiah, Hyatt began manufacturing celluloid in 1870 and became more successful financially than Goodyear. His synthetic billiard balls proved a bit too brittle to be useful, but the plastic did make fine dental plates, photographic film, brush handles, detachable collars, ping pong balls, and a host of other small products. Its major defect in consumer products is its tendency to burst into flames. Movie film was once made of this highly flammable celluloid and often ignited from the heat of the projector.

A burning pong-pong ball made of celluloid illustrates the flammability of the plastic.

Although the Hyatts had produced celluloid, the world's first successful commercial plastic, they hadn't actually constructed a new polymeric chain. The plastic was, rather, a combination of camphor and chemically modified, naturally occurring cellulose. The credit for the world's first fully synthetic organic polymer goes to Leo Hendrik Baekeland.

■ QUESTION

What are the chemical names of the three reagents or starting materials needed for the production of guncotton?

21.8 Part III: The First Synthetic Polymer

Born in Belgium in 1863, the same year that Phelan and Collander offered its $10,000 reward for an ivory substitute, Leo H. Baekeland became an active, productive, and successful academic and industrial chemist. As a young man he emigrated to the United States and became a citizen; eventually he was elected president of the American Chemical Society, the world's largest professional association of chemists. In 1909 Baekeland announced his preparation of the first fully synthetic polymer, a resin he called *Bakelite*. In the following year he founded the Bakelite Corporation to manufacture the material.

Unlike any of the other polymers we've examined so far, Bakelite is a thermosetting plastic rather than a thermoplastic. Bakelite forms as a mixture of *phenol* and *formaldehyde* (Sec. 16.4) polymerizes. As we see in Figure 21.14, each

Left: Leo H. Baekeland, inventor of Bakelite, a condensation polymer of phenol and formaldehyde.

Right: Old phonograph records made of Bakelite. These records rotated at 78 revolutions per minute and were called "78-rpm" records.

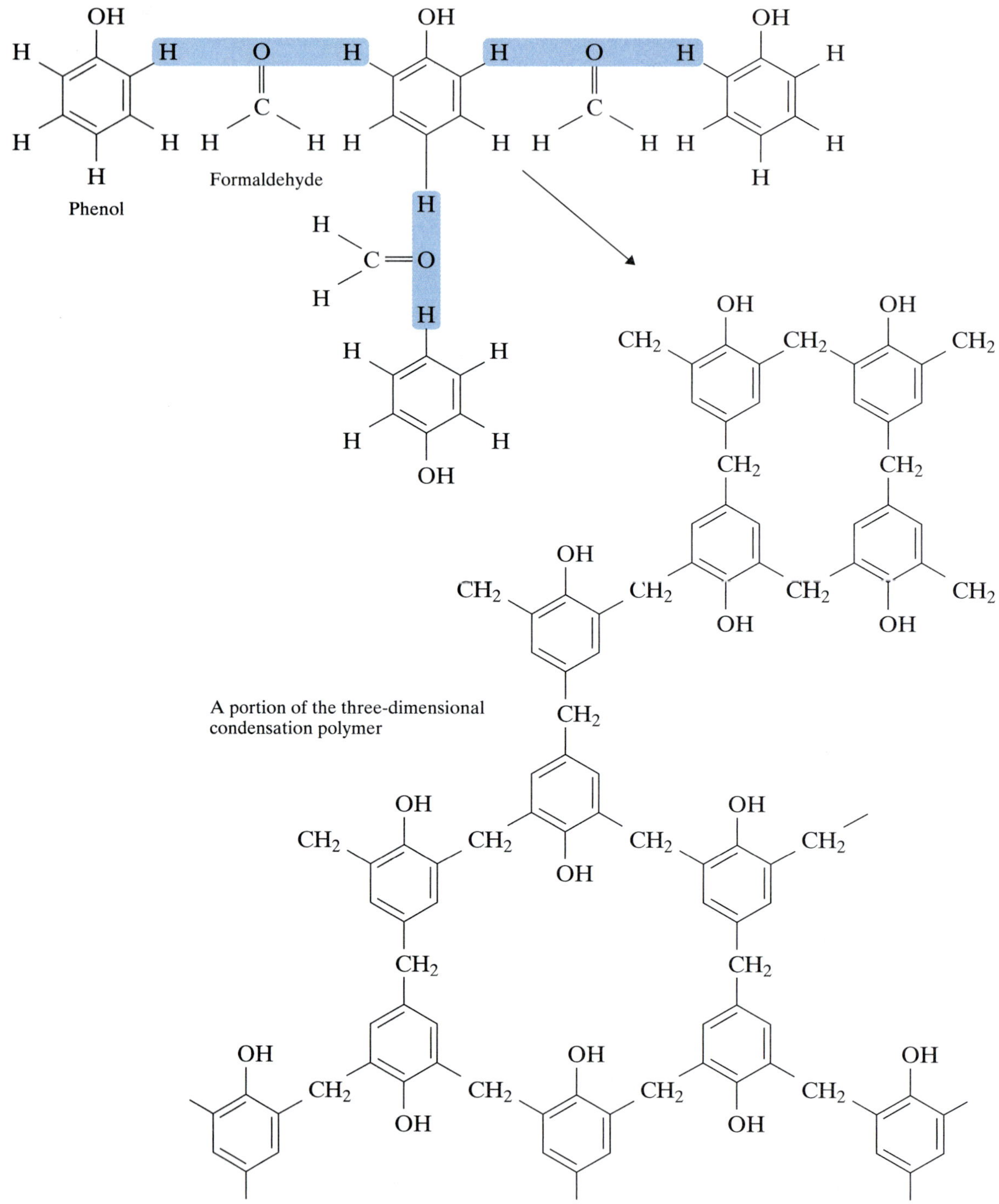

Figure 21.14 Bakelite.

formaldehyde molecule bonds to *two* different phenol molecules, and each phenol ring bonds to *three* different formaldehyde molecules (with a molecule of water lost for each combination of two phenols and one formaldehyde). The geometric possibilities available here produce an intricate, three-dimensional web of resinous polymeric material.

As we've seen (Sec. 21.5), Bakelite is a hard, sturdy material, resistant to heat and electricity and not easily burned or scorched. The use of other aldehydes and of other phenols, bearing groups other than hydrogens here and there on the ring, produces variations on Bakelite and has created an entire class of *phenolic resins*. Today the primary uses of the phenolic resins are as adhesives and fillers in the manufacture of plywood and fiberboard and in the production of insulating materials. A little more than 40% of all thermosetting polymers produced today belong to this class of resin.

■ QUESTION

What is the chemical name of the group that unites the aromatic rings in Bakelite? (See Sec. 6.5 for help.)

21.9 Part IV: From the Kitchen Stove to Nylon

We've seen that the early advances in the development of commercially useful polymers and plastics were made by a few ingenious (and lucky) individuals, sometimes quite by accident, often with little equipment. The days of major advances in polymer and plastic chemistry arriving through small accidents on hot kitchen stoves came to an end in the early years of the 20th century largely because of

- the growing sophistication of scientific equipment and techniques;
- the increasing rigor of research programs carried out in academic, industrial, and institutional laboratories; and
- the development, in the 1920s, of a comprehensive understanding of the molecular structure of polymers.

Wallace H. Carothers, the DuPont chemist who led the research team that invented nylon.

Among the fruits of highly organized, well-directed, and strongly supported research programs is the discovery of the condensation polymer *nylon* by Wallace H. Carothers and his co-workers at DuPont Corporation. In 1928 Carothers (1896–1937), an Iowa-born chemist, left his post as instructor in organic chemistry at Harvard University to lead a research group in DuPont's Wilmington, Delaware, laboratories. There he began a program of fundamental research into polymers, studying how they form and what factors affect their properties. Within a few years he and his co-workers found that by polymerizing a mixture of *adipic acid* and *1,6-diaminohexane*, they could produce a plastic (nylon) that can be drawn out into strong, silky fibers (Fig. 21.15).

Figure 21.15 Nylon, a condensation polymer.

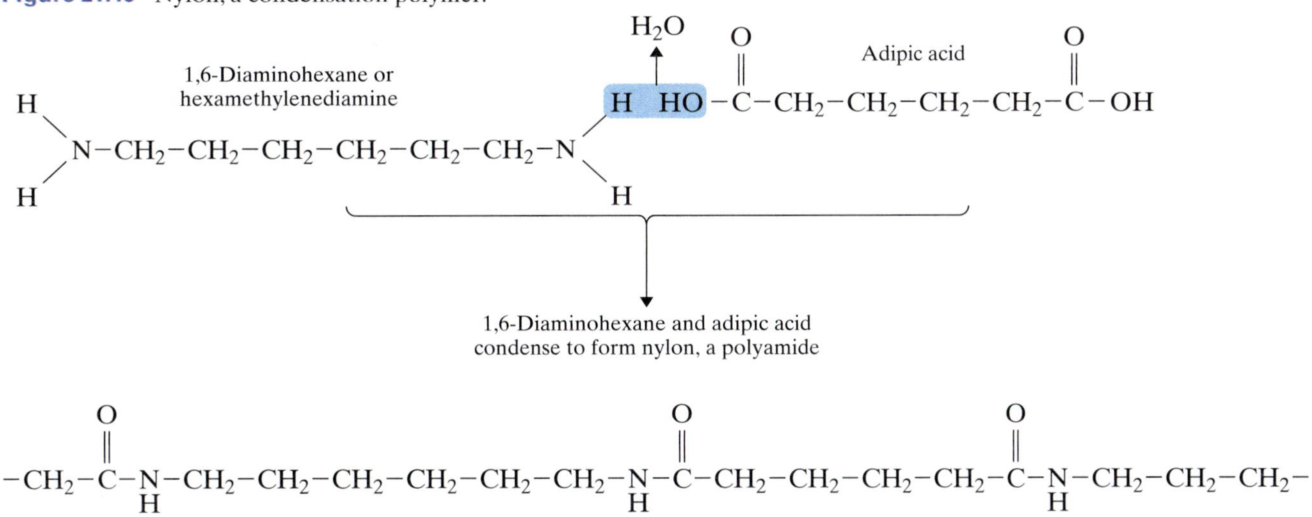

Adipic acid is an example of a *dicarboxylic acid*, one containing two *carboxyl* groups (Sec. 10.8).

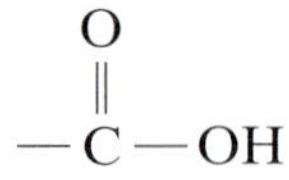

the carboxyl group

The name *adipic* comes from the Latin *adipem*, a fat, and reflects the observation that adipic acid is one of the substances formed when fats are oxidized with nitric acid, HNO_3. Notice, in Figure 21.15, that the adipic acid molecule contains a chain of six carbons, which includes the two carboxyl groups at its ends. We'll refer to this shortly.

The other monomer is another six-carbon compound, 1,6-diaminohexane, with amino groups ($-NH_2$, Sec. 17.2) at the ends of a chain of six methylene groups ($-CH_2-$). Another useful name for this diamine is *hexamethylenediamine*.

With alternating units of the two monomers bonded to each other throughout the length of the chain, the structure of the polymeric molecule itself can be written as

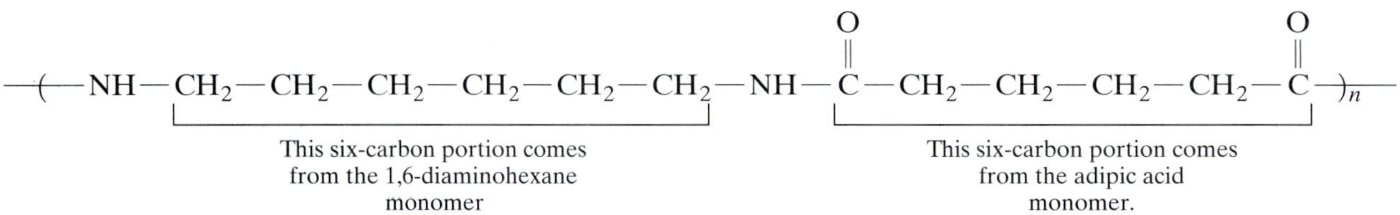

Nylon

Carothers and his group produced several other nylons, each a *polyamide* (Sec. 17.7). To differentiate among all the newly formed nylons, the researchers coded each for the number of carbon atoms in each of its monomers. The one produced from the six-carbon adipic acid and the six-carbon 1,6-diaminohexane became *nylon 6,6*. (In fact, "66 polyamide" was its original name, before the term *nylon* was coined.) Another nylon, *nylon 6*, forms as the ring of *caprolactam* opens with the addition of water and the resulting amino acid undergoes polymerization (Fig. 21.16).

Figure 21.16 Nylon 6.

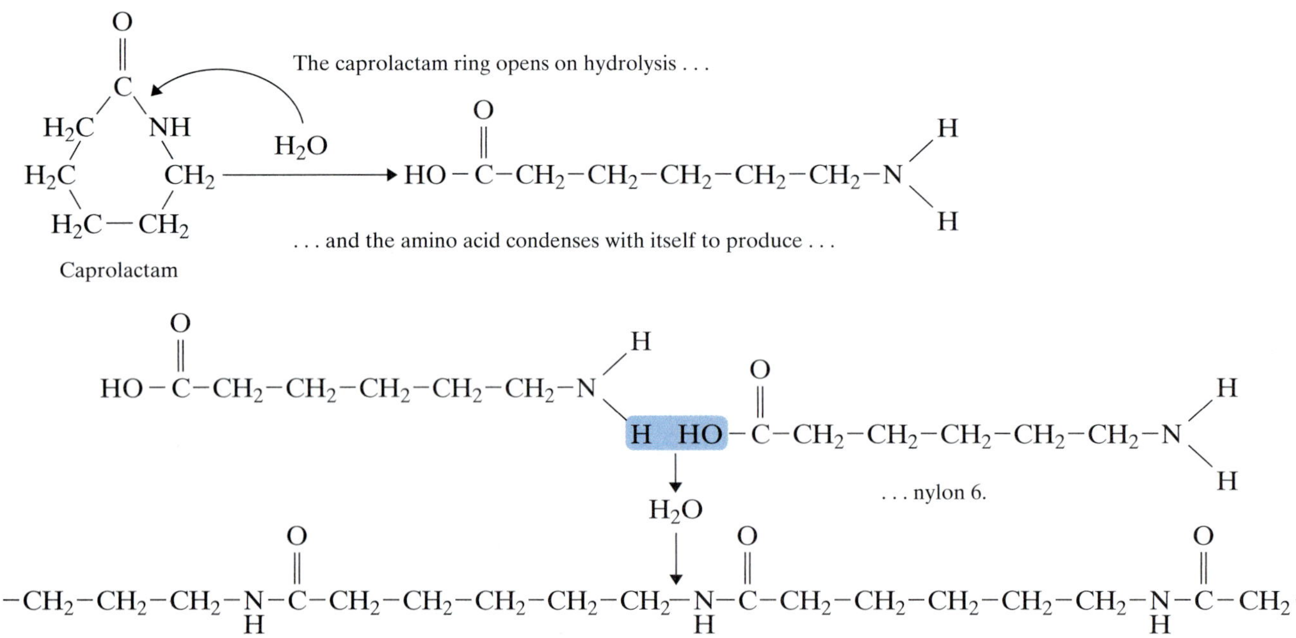

Left: The formation of a strand of nylon 6,6 as two chemicals, dissolved in solvents that do not mix, react with each other. Hexamethylene diamine, $H_2N—(CH_2)_6—NH_2$, dissolved in water and adipoyl chloride, $ClOC—(CH_2)_4—COCl$, dissolved in another solvent, react to form nylon at the surface where the two solvents come into contact. The strand is pulled from the surface of contact. Right: A giant leg near Los Angeles, promoting the sale of nylon stockings after World War II.

Nylon's first practical application to a consumer product came in 1938, when the new polymer was introduced to the public in the form of toothbrush bristles. But it was the polymer's use in stockings, first sold to consumers on a trial basis in October 1939, that made it an overwhelming commercial success. Similar to silk in its properties but far less expensive, nylon became the ideal replacement for the silk of stockings and other fashionable clothing. With the coming of World War II, fashion had to make way for the war effort. The government used most of the nation's limited supplies of nylon for making parachutes, ropes, and other military supplies. Since there wasn't enough nylon for both military and civilian uses, nylon stockings, which had become very popular and were in high demand, were rationed until the end of the war. During the war and shortly afterward, nylon stockings became a valuable item of barter in Europe and achieved the status of an informal currency. Not until the early 1950s was there sufficient production capacity to fill the popular demand for "nylons," as the stockings came to be known, and to provide enough of the plastic for other consumer and commercial uses.

Nylon's name reflects, in a strange and devious way, one of the most appealing characteristics of the stockings made from the polymer: their resistance to snagging and running. An early name, suggested as a more popular replacement for the technical "66 polyamide," was *norun*, referring to the appeal of the stockings. That term was judged unacceptable by those at DuPont who were responsible for giving the polymer a trade name. After spelling it backward as *nuron*, which was still unsatisfactory, they finally agreed on *nylon*.

■ **QUESTION**

Is each of the following a homopolymer or a copolymer? (a) nylon 6; (b) nylon 6,6. Name the monomer(s) of each.

21.10 PET, a Condensation Polymer

The modern, systematic search for commercially useful polymers has led to the development of several valuable condensation and addition polymers, which we'll examine in the next few sections. The commercial leader among condensation polymers is *poly(ethylene terephthalate)*, a *polyester* also known as *PET*. We've already examined

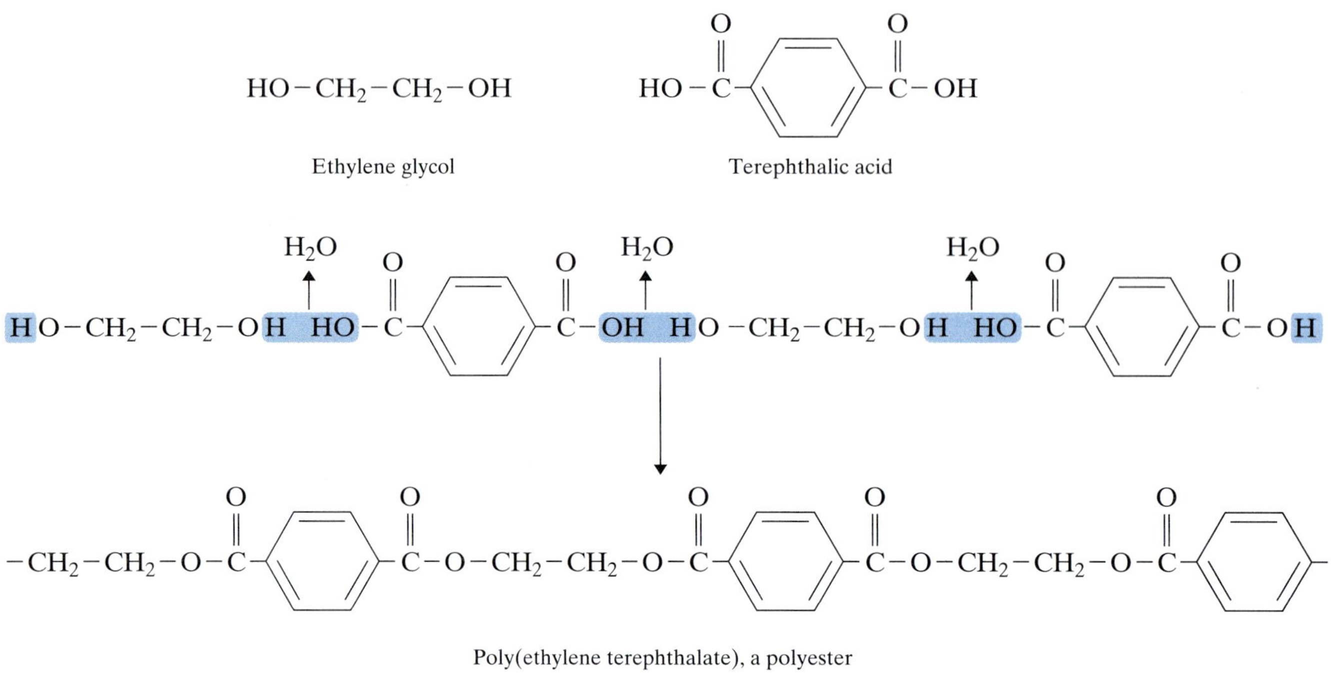

Figure 21.17 Poly(ethylene terephthalate).

a *triester* in the triglycerides that form our fats and oils and that give us our soaps (Sec. 15.1). In those compounds the three hydroxyl groups (—OH) of a single glycerol molecule are all combined with the carboxyl groups (HO_2C—) of the three carboxylic acids that form the triglyceride's side chains. Since each acid can contribute only one reactive group to the process, the reaction stops with the formation of the triglyceride.

In 1941 the British industrial chemist John Rex Whinfield succeeded in producing a polyester, poly(ethylene terephthalate), by polymerizing a mixture of *ethylene glycol* and *terephthalic acid*. As shown in Figure 21.17, ethylene glycol has two —OH groups with which it can form esters. Unlike the carboxylic acids of the triglycerides, though, terephthalic acid also has *two* —CO_2H groups, allowing each one of its molecules to combine with *two different* glycol molecules. With this sort of combining power the two compounds form a copolymer whose monomeric units are linked to each other by ester groups. Chemically the process resembles the condensation of hexamethylenediamine and adipic acid to form the polyamide chain of nylon.

Food storage bags made of poly(ethylene terephthalate), PET.

Drawing out the poly(ethylene terephthalate) into a filament produces the world's leading synthetic polymeric fiber, known as *Dacron* in the United States, *Terylene* in Great Britain, and usually simply referred to as "polyester" when it's woven into a fabric. (Other, related polyester fibers include *Fortrel* and *Kodel*.) PET also forms an extremely thin and extremely tough film, *Mylar*, used as the plastic backing for audio and video tapes and computer diskettes, as well as wrapping material for frozen foods and bags for boil-in foods. In Western Europe PET has grown since the early 1980s into the dominant commercial plastic for packaging consumer goods.

[*Rayon* is a generic term for a group of fibers derived from cellulose, largely by replacing one or more of the hydrogens of the —OH groups on the glucose rings with other chemical groups. Schoenbein's guncotton (Sec. 21.7) is one form of rayon.]

■ **QUESTION**

Fats and oils are triglycerides, with three ester functional groups per molecule. Should they be considered *polyesters*? Explain.

21.11 Polystyrene, an Addition Polymer

For other examples of polymers and plastics—one that both contributed to winning World War II and generated a flash of national mania (Sec. 21.13), and another that produced two Nobel Prizes in chemistry (Sec. 21.14)—we turn now to addition polymers such as the *polyolefins* and closely related polymers. The addition polymers we'll examine in the following several sections dominate commercial plastic production in the United States.

Polyolefins are polymers of *olefins*, which is another and much older name for the unsaturated hydrocarbons we now know as the *alkenes* (Sec. 6.10). "Olefin" is a corruption of a name given to ethylene by several Dutch chemists in 1795 to reflect its character as an "oil-forming" gas in its reaction with chlorine. Since then, the term has been extended to include the entire category of hydrocarbons that, like ethylene, bear a single carbon–carbon double bond. *Olefin* is often used as a synonym for *alkene*. Polymerizing an olefin produces a **polyolefin**.

Polyolefins are polymers produced by the polymerization of alkenes and compounds closely related to them.

We've already seen that polymerizing ethylene, the simplest of all the alkenes, produces polyethylene, a polyolefin we'll examine in more detail in Section 21.13. Replacing one or more of ethylene's hydrogens by other substituents gives monomers useful for preparing polymers closely related to the polyolefins and sharing many of their important properties.

Replacing one of the ethylenic hydrogens by a *phenyl* group, for example, produces *styrene*, which polymerizes to the thermoplastic *polystyrene* (Fig. 21.18). The various techniques available for converting the raw polystyrene polymer into a finished product provide a wide range of useful properties for the resulting plastic. Inexpensive, clear, rigid drinking glasses are made of polystyrene. In a variation known as high-impact polystyrene, the plastic is used to make sturdy furniture, inexpensive tableware, and stereo, television, and computer cabinets. Another form of polystyrene, a solid but lightweight polystyrene foam, is a good thermal insulator and shock absorber, useful for making picnic coolers, egg cartons, clamshell containers for fast foods, disposable

Cups, plates, and containers made from polystyrene foam.

Figure 21.18 Polystyrene.

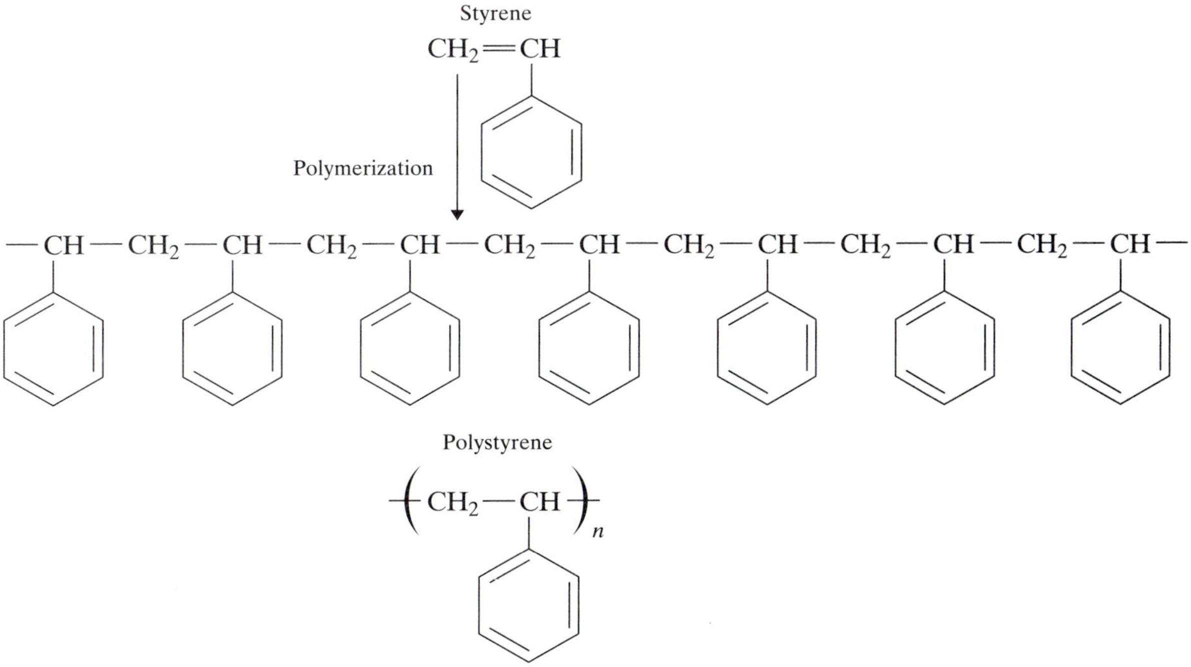

cups for keeping drinks hot or cold, and small polystyrene nuggets used as packing material. These foams, some of which are sold under the name *Styrofoam*, are made by using a gas to generate a foam of liquid polystyrene and allowing the frothy mass to cool and solidify. Chlorofluorocarbons (Sec. 14.9) were once used to generate the froth, but these have been replaced by other gases, including low-boiling alkanes. Today, well over half of all the polystyrene produced goes into inexpensive consumer products.

■ **QUESTION**

What physical characteristics of polystyrene foam make it useful in consumer products such as picnic coolers, packing materials, and cups for both hot and cold drinks?

21.12 Vinyl and Its Chemical Cousins

In the realm of consumer products the word *vinyl* has come to mean a tough, flexible, and often smooth, shiny plastic that serves as an inexpensive substitute for leather. We often find vinyl purses, wallets, and jackets and other vinyl goods for sale at the lower priced counters of stores.

To the chemist, though, vinyl represents a hydrocarbon group ($CH_2{=}CH{-}$) that can be formed by the removal of a hydrogen atom from ethylene, just as a methyl group ($CH_3{-}$) and an ethyl group ($CH_3{-}CH_2{-}$) can be produced by the removal of a hydrogen atom from methane (CH_4) and ethane ($CH_3{-}CH_3$), respectively (Chapter 6). Moreover, just as $CH_3{-}Cl$ and $CH_3{-}CH_2{-}Cl$ represent *methyl* and *ethyl* chloride, $CH_2{=}CH{-}Cl$ is *vinyl* chloride.

The terms *poly(vinyl chloride)*, *polyvinylchloride*, and *PVC* all serve very nicely as names for the thermoplastic formed by the addition polymerization of vinyl chloride (Fig. 21.19). The polymer forms a tough plastic, well suited to pipes, plumbing, electrical conduit, flooring, and both indoor and outdoor wall coverings. Among personal products, it's widely used in toys, garden hoses, and inexpensive wallets, purses, and keyholders. Well over half of all PVC production currently goes into the construction industry, with most of it used in piping, tubing, and similar extruded materials.

A **plasticizer** is a liquid that is added to a plastic to soften it.

Because thin sheets of polyvinylchloride are relatively stiff and crack easily, it's necessary to add a **plasticizer** to give them the same sort of flexibility we expect from leather. Plasticizers are liquids that mix readily with a plastic and soften it. With time a plasticizer can migrate out of the plastic or otherwise deteriorate, allowing aged polyvinylchloride to stiffen and crack.

Figure 21.19 Poly(vinyl chloride).

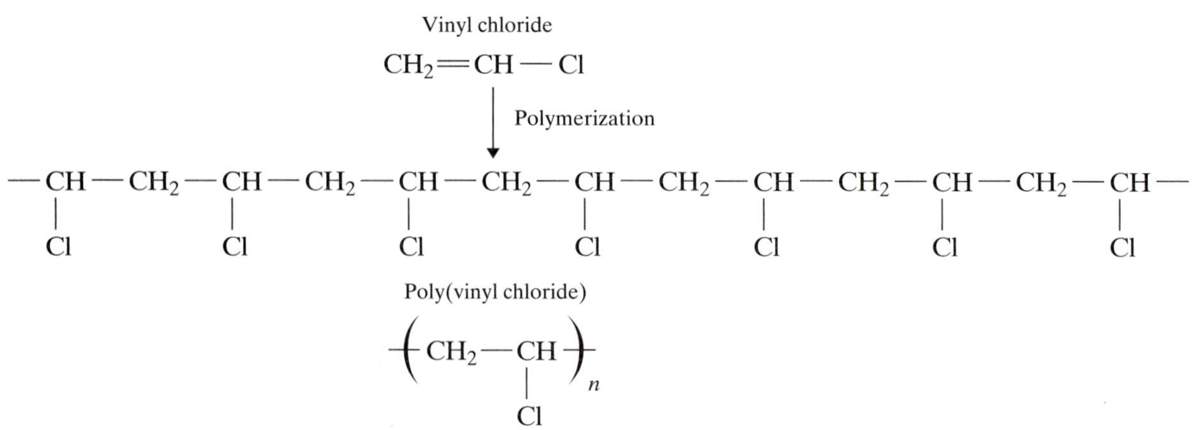

Other useful thermoplastics come from still other monomers closely related to ethylene, including *vinyl acetate*, *acrylonitrile*, *vinylidene chloride*, *tetrafluoroethylene*, *methyl methacrylate* (Table 21.2), and, of course, styrene.

Table 21.2 Addition Polymers

Monomer		Polymer	
Ethylene	$CH_2{=}CH_2$	polyethylene	$+CH_2-CH_2+_n$
Vinyl chloride	$CH_2{=}CH-Cl$	poly(vinyl chloride) polyvinylchloride PVC	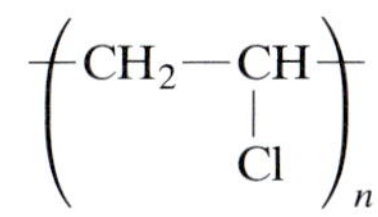
Vinyl acetate	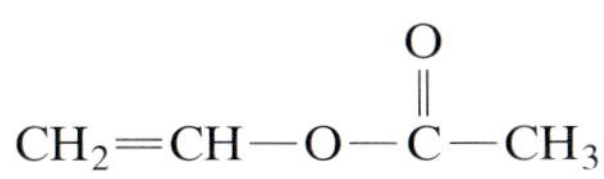	poly(vinyl acetate) polyvinyl acetate PVA	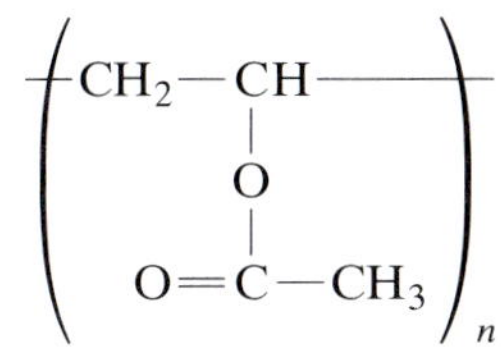
Acrylonitrile	$CH_2{=}CH-C{\equiv}N$	polyacrylonitrile Orlon Acrilan Creslan	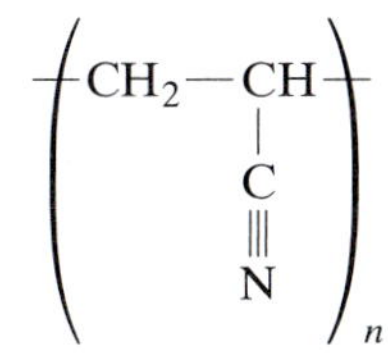
Vinylidine chloride	$CH_2{=}CCl_2$	poly(vinylidine chloride) Saran	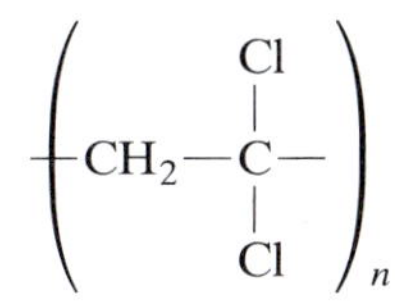
Tetrafluoroethylene	$F_2C{=}CF_2$	polytetrafluoroethylene Teflon	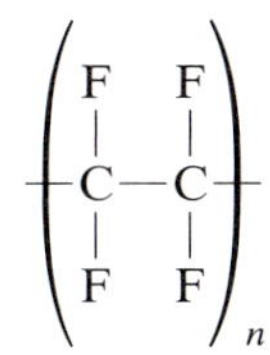
Methyl methacrylate	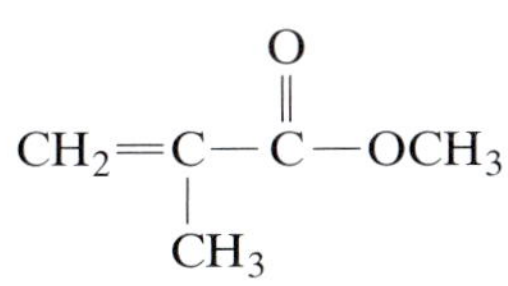	poly(methyl methacrylate) Lucite Plexiglas	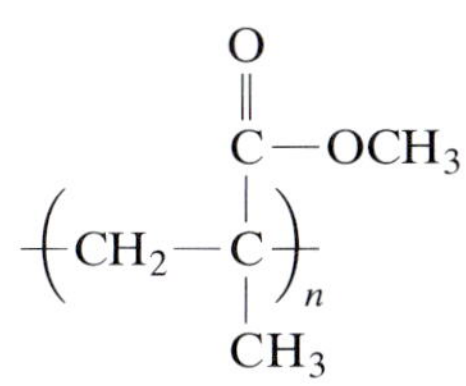
Styrene	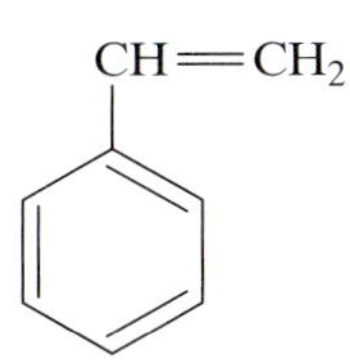	polystyrene Styrofoam	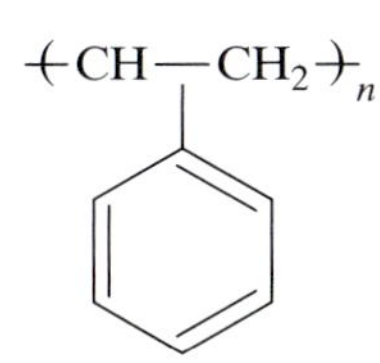

A broken car window made of safety glass. A sheet of polyvinylacetate sandwiched between two sheets of glass prevents the broken glass from disintegrating into flying slivers.

Bubble gum made from polyvinylacetate.

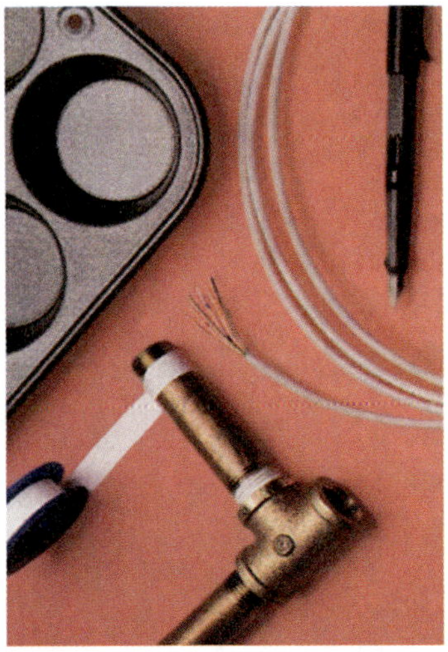

Articles made from or coated with Teflon, a tough, stable, heat-resistant, nearly friction-free polymer of tetrafluoroethylene, $CF_2{=}CF_2$.

Polymerization of vinyl acetate produces *polyvinylacetate*, a plastic with a wide range of properties and an equally wide range of applications. As a thermoplastic with a low softening temperature it's useful in coatings and adhesives. A sandwich of two panes of glass bonded to a central sheet of the material forms a shatterproof or *safety glass*. While the glass can break, it won't shatter into dangerous slivers. Safety glass of this kind is used in car windows.

Mixed with a sweetener, some flavoring, and other ingredients, polyvinylacetate replaces the chewy chicle of chewing gum. (Chicle itself is a rubberlike polymer obtained from the sap of the sapodilla tree.) Partially polymerized vinyl acetate serves as a *binder* in water-based house paints. As the water evaporates, the low-molecular-weight polymer present in the paint polymerizes further, forming a tough sheet that binds the pigment to the coated surface.

Polyacrylonitrile, a thermoplastic that's drawn out into fine threads and woven into synthetic fabrics such as *Orlon*, *Acrilan*, and *Creslan*, comes from the polymerization of acrylonitrile ($CH_2{=}CH{-}CN$).

Putting two chlorines on the single carbon of ethylene converts it into *vinylidene chloride* ($CH_2{=}CCl_2$), the major monomer of *Saran*. Sheets of this polymer form a nearly impregnable barrier to food odors, which makes it useful for wrapping foods that are to be stored near each other in a refrigerator.

Teflon results from the polymerization of *tetrafluoroethylene* ($CF_2{=}CF_2$), a monomer obtained by replacing all of ethylene's hydrogens with fluorines. Teflon's great chemical stability, its resistance to heat, its mechanical toughness, and its nearly friction-free surface make it useful as a coating for bearings, valve seats, gaskets, and other parts of machinery that take heavy wear. Since things don't easily stick to it, Teflon also makes a fine coating for cooking utensils such as pots and pans. Teflon was first prepared at DuPont in 1938; commercial production began 10 years later.

In molecular structure *methyl methacrylate* is a bit further removed from ethylene. Its polymer forms a very hard, clear, colorless plastic that appears in consumer products as Lucite and Plexiglas. It's used in making glasses, camera lenses, and other optical equipment, in costume jewelry, and as windows in aircraft.

■ QUESTION

Of the polymers described in this section, which contain the element: (a) oxygen, (b) chlorine, (c) nitrogen, (d) fluorine?

21.13 Polyethylene, the Plastic That Won the War and Gave Us the Hula-Hoop

Accounting for about 16.6 million tons or over 40% of all plastic production in the United States annually, polyethylene clearly represents the most important polymer of the U.S. plastics industry. In its two principal commercial forms (high-density polyethylene and low-density polyethylene, which we'll examine shortly) it also illustrates very nicely the connection between the structure of a polymeric molecule and the properties of the plastic it forms. As we might imagine, the physical properties of a piece of bulk plastic depend partly on the molecular structure of the monomer that forms it and partly on the average length of its polymeric chains. As we'll see shortly, a third factor also comes into play: the way polymeric chains organize themselves as they constitute the bulk of the material.

Polyethylene itself was first prepared in 1934 in the laboratories of Imperial Chemical Industries, in Great Britain, and went into commercial production there five years later as World War II was about to begin. The first practical use of the plastic was as insulation on the electrical wiring of military radar sets. Considering the critical importance of aircraft radar to the survival of Britain in the early years of the war, we could easily designate polyethylene as the plastic that won the war.

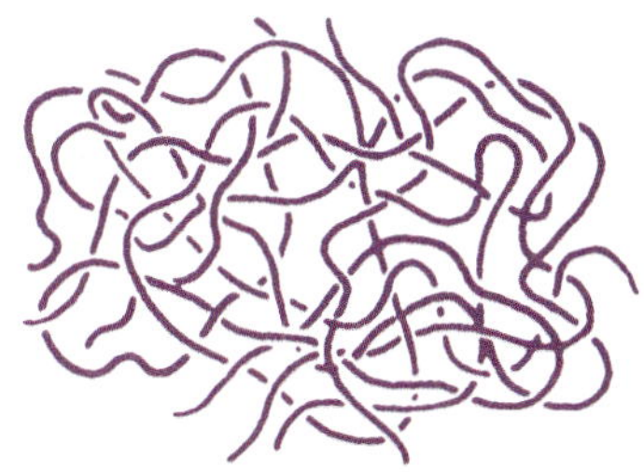

Figure 21.20 Disorganized polymeric strands of low-density polyethylene.

The techniques used for those early polymerizations produce what we now know as *low-density polyethylene (LDPE)*. As the chains of LDPE form, they branch sporadically into short offshoots from the main line of the polymer. These short branches keep the major strands of the polymer from falling into anything resembling a coherent, well-organized pattern. Instead they form a tangled, randomly oriented network of strands, somewhat like a large ball of fuzz (Fig. 21.20). The result is a low-density, soft, waxy, flexible, relatively low melting plastic that accounts for a little over half of all the polyethylene produced in the United States. More of this LDPE goes into producing trash bags than into any other single product. Industrial and commercial packaging, food wrappers, and plastic shopping bags are close behind. As in World War II, the coating of electrical wires and cables remains an important use.

Polymerized by a different method, the polyethylene chains can grow on and on without branching, thereby generating *high-density polyethylene (HDPE)*. This form of the plastic was first produced in Germany in the early 1950s by Karl Ziegler, a German chemist born near the end of 1898. Ziegler prepared HDPE by carrying out the polymerizations in the presence of certain highly specialized *organometallic catalysts*, which consist of combinations of organic molecules and metal atoms joined to each other by covalent bonds. These catalysts control the way the monomers link to each other as they polymerize.

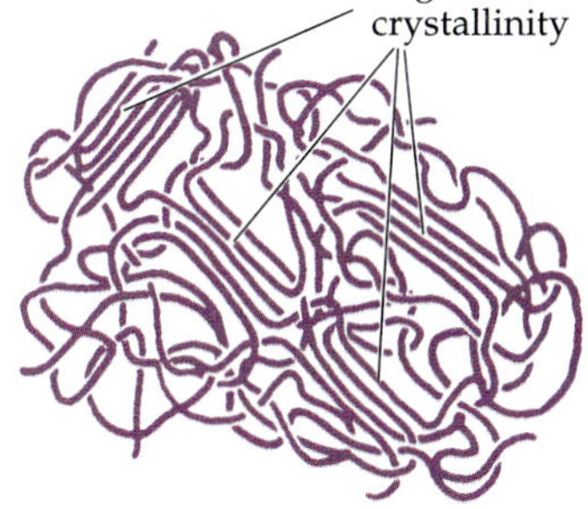

Figure 21.21 Polymeric strands of high-density polyethylene with regions of crystallinity.

In HDPE the long molecular chains aren't fixed into a tangled web as they are in the LDPE, so they can align themselves into localized areas of tightly packed strands, mimicking here and there the orderly structure of crystals such as sodium chloride and sucrose. With many regions of close packing and a high degree of crystallinity, HDPE is a denser, harder, higher melting, and more rigid polymer than LDPE (Fig. 21.21). We saw another example of this same sort of relationship between molecular structure and physical properties in the fats and oils of Chapter 15.

Hula-hoops in action.

Of approximately 7.7 million tons of HDPE produced in the United States in 2000, most went into the manufacture of bottles and other containers designed to hold a variety of liquids, including drinks, bleach, antifreeze, and engine oil. It's also used in fabricating shipping drums and automobile gasoline tanks. Historically, HDPE's introduction into consumer products was as the Hula-Hoop, a large plastic hoop placed at the waist and twirled with hulalike motions. The craze swept the United States in the late 1950s; its memory survives today largely in collections of national trivia.

■ QUESTION

What is the major use of the LDPE produced in the United States?

21.14 A Nobel Prize for Two

We have seen repeatedly that the physical and chemical properties of a substance depend on the structure of its molecules or ions. Among polymers, for example, we saw that the way individual glucose molecules are joined to each other determines whether they form starch or cellulose (Sec. 16.11), and that the geometry of the double bond in the polyisoprene chain determines whether it forms rubber or gutta–percha (Sec. 21.5).

The effects of molecular structure on the properties of a plastic appear again in the polymerization of *propylene*. Figure 21.22 shows a small portion of the polypropylene chain, with emphasis on the orientation of the methyl substituents that appear on every other carbon of the polymer. There are three different ways these methyl groups can be arranged:

1. They can all protrude from the same side of the stretched-out, zig-zag molecular chain (*isotactic* polymer).
2. They can appear on alternating sides (*syndiotactic* polymer).
3. They can be oriented randomly (*atactic* polymer).

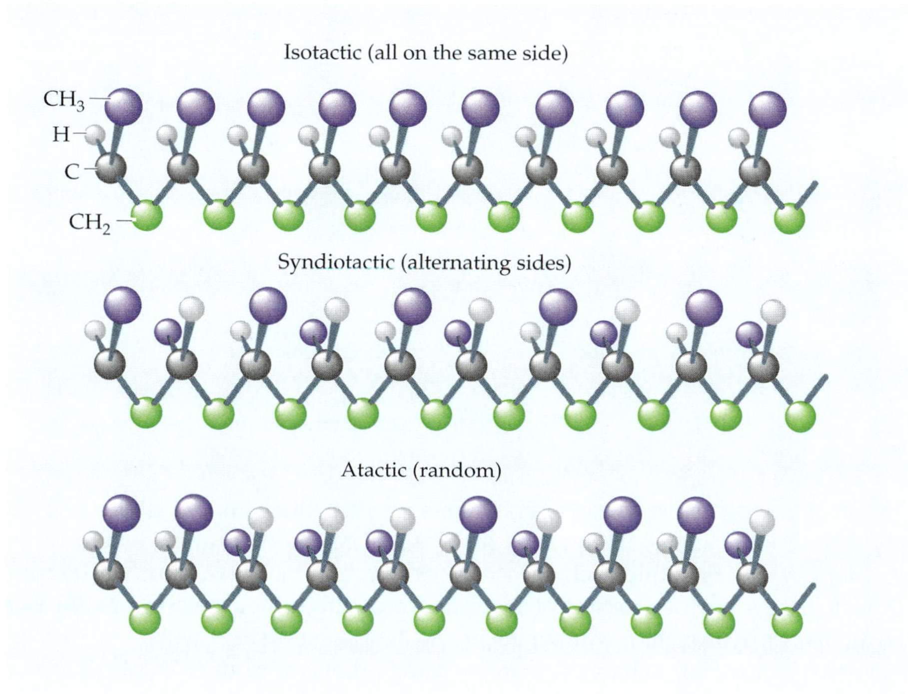

Figure 21.22 The polypropylenes.

The terms for the three orientations come from the Greek *taktos*, meaning ordered, which is modified by the prefixes *iso-* (same), *syndio-* (two together), and *a-* (not).

In 1953 Ziegler (Sec. 21.13) and Giulio Natta, Italian-born (1903) professor of industrial chemistry at the Milan Polytechnic, independently prepared the highly ordered polypropylenes, thereby creating the first *stereochemically ordered* polymers. For their pioneering work Ziegler and Natta shared the 1963 Nobel Prize in chemistry. As we might expect from our knowledge of the effects of molecular order on the melting points of triglycerides and on the hardness of the polyethylenes, the highly ordered isotactic and syndiotactic polypropylenes are higher melting, more crystalline, and harder than the atactic.

Today almost 8 million tons of polypropylene are produced in the United States each year, most of it going into the manufacture of automobile trim and battery cases, carpet filaments and backing, fabrics, small items such as toys and housewares, and the hard plastic caps used on containers made of other materials.

■ QUESTION

Which of the polymers of Table 21.2 could exist in isotactic and syndiotactic forms?

21.15 Silicates, Polymers of the Lithosphere

Not all polymers are organic compounds. Probably the most important and certainly the most abundant of the inorganic polymers are compounds of silicon and oxygen. The clays, rocks, and inorganic soils of the earth's crust are made mostly of these two elements. On average, oxygen constitutes about 46% of the mass of the planet's outer layer of land and ocean bed; silicon, about 28%. Taking into account the difference in their atomic weights, this corresponds to a ratio of almost three atoms of oxygen for every atom of silicon.

Figure 21.23 Tetrahedral carbon and tetrahedral silicon.

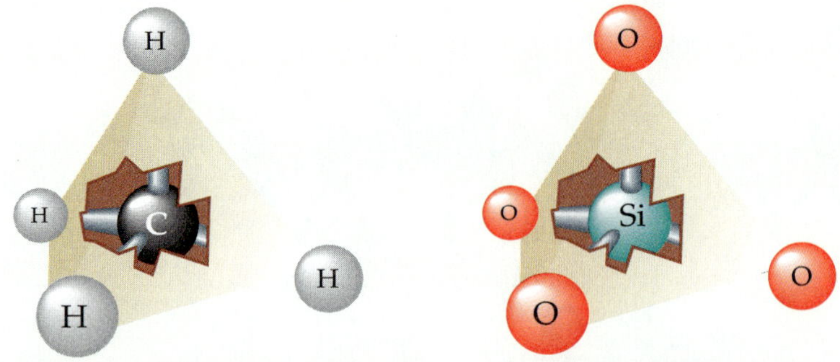

The tetrahedral carbon of methane

The tetrahedral silicon of the silicates (the oxygens either are anionic or they are bonded to another silicon.)

Most of the solid crust consists of polymers with repeating units of silicon dioxide (SiO_2) and of ionic aggregates such as SiO_3, SiO_4, Si_2O_5, Si_4O_{11}, and so forth. In these, four oxygens surround each tetracovalent silicon in much the same way as four hydrogens occupy the four corners of a tetrahedral methane molecule (Fig. 21.23). But while methane's hydrogens are monovalent, the oxygen atoms are divalent, so they either carry a negative ionic charge or bond to a second silicon atom. Bonding of any or all of the oxygens at the corners of the tetrahedra to still other tetrahedral silicons produces threadlike, sheetlike, and three-dimensional, inorganic, silicate polymers.

With the other elements of the crust interspersed among them, these inorganic polymers form much of the substance of the earth's rocks and minerals. One form of the mineral *asbestos* consists of a double strand of the silicate tetrahedra (Fig. 21.24). Asbestos was once used in brake linings and cigarette filters and as insulation in houses, schools, and office buildings. Inhaling asbestos fibers has been shown to produce *mesothelioma*, a cancer of the lungs and the cavity of the chest and the abdomen. Since the mid-1970s the mineral has been abandoned as an insulating material in construction. Its use in consumer products is now prohibited by law.

Two-dimensional sheets of silicate polymers form micas, clays, and talcs, while quartz is a three-dimensional lattice of silicon dioxide (SiO_2). In the quartz lattice (Fig. 21.25) every oxygen is shared by two different silicon atoms, one at the center of each of two adjacent tetrahedra that are connected through the oxygen. Each silicon atom owns, in effect, exactly half of each of the four oxygens bonded to it (with the other half assigned to the other silicon atom bonded to the oxygen). With this arrangement, the molecular formula of quartz becomes SiO_2.

Figure 21.24 A double-stranded asbestos polymer.

Figure 21.25 A portion of the three-dimensional crystal lattice of quartz.

■ QUESTION

What is the most abundant source of polymers of silicon and oxygen?

21.16 Water, Rubbing Alcohol, and the Properties of Plastics

The opening demonstration illustrated one of the most important characteristics of plastics: their range of physical and chemical properties. A little water and rubbing alcohol revealed the difference in densities of two of the plastics that form our consumer goods. Plastics also differ in other important respects, including flexibility, resistance to heat, transmission of light, and tensile strength.

These differences arise not only from differences among the various monomers used in polymerizations, but also from differences that can be generated through the polymerization process itself, as we've seen for low-density and high-density polyethylene (Sec. 21.13) and the isomeric polypropylenes (Sec. 21.14). The properties appropriate to each of our consumer products determine which of the many available plastics go into them. PET, for example, is useful for the bodies of soft drink bottles because gases (such as carbon dioxide) don't pass through it easily. Polypropylene, on the other hand, forms a stronger, more rigid plastic than PET does. Polypropylene, therefore, is more useful than PET for manufacturing large jugs for water, milk, and liquid detergents. In these, strength is a greater concern than the permeability of gases. In the Perspective we'll see how all this affects our disposal of waste plastics and the recycling of plastic trash.

■ QUESTION

Of PET and polypropylene, which plastic is preferred for the manufacture of the bodies of soft drink bottles? Why? Which is preferred for the manufacture of large bottles for water, milk, and liquid detergent? Why?

PERSPECTIVE

The Problem with Plastics

Unlike metal, discarded plastic boxes and bottles don't corrode and decay. Unlike paper and cloth, ordinary plastic bags and wrappings aren't degraded by the weather or by the action of the microorganisms of the soil. Plastics, as a rule, simply aren't **biodegradable**. That is, the microorganisms that inhabit the soil can't degrade plastics to the simpler substances that form our natural environment.

Plastic litter lies by the side of the road and in our parks and beaches, virtually unchanged by weather or microorganisms, until it's gathered up for proper disposal. (An estimated 60% of all the debris collected from our beaches consists of plastics of one sort or another.) Even with proper disposal, though, plastic waste presents a growing societal problem as it accumulates, along with other trash, in expensive landfills. These areas, where urban trash of all sorts is dumped and mixed into the land itself, are quickly filling to capacity and closing. About 80% of all our plastic waste ends up in landfills.

Incineration, an alternative approach used with about 10% of our plastic trash, is not only expensive but potentially hazardous as well. Some plastics, including Teflon, poly(vinyl chloride) and poly(vinylidene chloride), produce irritating or toxic gases when they burn. Added to this is the impact of incineration on the greenhouse effect (Sec. 8.14).

Science and technology offer some hope for solving the problem through two possible routes: recycling plastics so that they don't accumulate in the environment as trash, and decomposing the plastics that do become environmental trash. Recycling waste plastics into new products, much as paper, glass, and metal wastes are recovered and recast into new and useful items, would certainly seem to be a promising approach. Yet, as we've seen, plastics come in a variety of molecular structures, with a variety of properties. Each synthetic plastic possesses its own particular characteristics that suit it to a set of specific applications. For effective recycling, the various plastics of our wastes would have to be sorted into a variety of individual piles, an expensive process, with each pile representing a separate category of properties. Tossing them all together, into a single batch, lowers the cost of the entire operation but produces a low-grade product known informally as "plastic lumber," a satisfactory substitute for wood in some of its rougher uses but hardly suitable for more specialized applications. In all, while a little over a quarter of our residential and commercial aluminum wastes and about 40% of similar waste paper are recycled, only about 6% of our plastics are reused.

One reason for the low level of recycling plastics is the inherent difficulty and cost of separating different kinds of plastic from each other so that each can be reused effectively. Another problem is that some plastic products just don't lend themselves to collection for recycling. The plastic liners of baby diapers probably offer the best example of this sort of problem.

In another approach plastics can be removed from the environment by chemically induced degradation. Impregnating the plastic with a substance that promotes its decomposition without significantly affecting its properties can cause it to decay much like paper or cloth. Incorporating a readily biodegradable form of starch, cellulose, or protein, for example, attracts soil microorganisms into the discarded plastic. As the microorganisms feed on these nutrients they also clip the long molecular chains of the polymers into shorter segments, causing the plastic to decompose. Alternatively, chemical activators can be added to the plastic so that continued exposure to sunlight leads to degradation.

Either one of these treatments leaves the plastic suitable only for decomposition rather than for recycling and, therefore, does nothing to slow the introduction of plastics, or their degradation products, into the environment. What's more, the hazards of all the possible decomposition products produced by a large variety of plastic wastes are still uncertain and could pose a greater long-term threat to the environment than do the intact plastics themselves. Once again, the question of the safe disposal of longlived plastics depends on which risks we as a society are willing to accept. ■

A substance is **biodegradable** if microorganisms within the environment can convert it to the simpler substances that form our natural environment.

EXERCISES

For Review

1. Following are a statement containing a number of blanks and a list of words and phrases. The number of words equals the number of blanks within the statement, and all but two of the words fit correctly into these blanks. Fill in the blanks of the statement with those words that do fit, then complete the statement by filling in the two remaining blanks with correct words (not in the list) in place of the two words that don't fit.

_________ are macromolecules composed of long chains of structural units known as _________. If all of these units are identical in structure, the long chain is a _________; if they represent two or more different monomers, the result is a _________. The natural polymer _________, with its monomeric units of _________, represents a typical homopolymer. The synthetic polymer _________, with alternating units of _________ and _________, represents a typical copolymer. Cellulose and nylon are similar in that they are both _________ polymers. Another class of polymers, the _________ polymers, includes _________, which is produced in two forms. The _________ polymer forms branched chains and is used for manufacturing trash bags and other forms of wrapping, while the _________ product, a _________ polymer, is used in plastic bottles and other containers for soft drinks, antifreeze, bleach, and engine oils.

1,6-diaminohexane	homopolymer
addition	linear
adipic acid	low-density
cellulose	monomers
condensation	poly(ethylene terephthalate)
copolymer	polymers
glucose	polystyrene
high-density	

2. Define, identify, or explain each of the following:

HDPE
isoprene
isotactic
LDPE
PET
plasticizer
SiO_2
thermoplastic
vulcanization

3. Match each of the following with the contribution made to polymer chemistry:

1. discovered the process of vulcanization of rubber
2. first prepared poly(ethylene terephthalate)
3. shared a Nobel Prize with Karl Ziegler for preparing stereochemically ordered polypropylenes
4. prepared the first fully synthetic polymer from phenol and formaldehyde
5. invented celluloid
6. invented nylon
7. first prepared high-density polyethylene
8. accidentally invented nitrocellulose (guncotton)

_____ a. Leo H. Baekeland
_____ b. Wallace H. Carothers
_____ c. Charles Goodyear
_____ d. John and Isaiah Hyatt
_____ e. Giulio Natta
_____ f. Christian Schoenbein
_____ g. John R. Whinfield
_____ h. Karl Ziegler

4. Give the structures of each of the following pairs of compounds, name the polymer they form, and describe its use or importance:

a. adipic acid and 1,6-diaminohexane
b. phenol and formaldehyde
c. styrene and butadiene
d. ethylene glycol and terephthalic acid

5. Identify the monomers used to make each of the following plastics:

Acrilan	Mylar
Bakelite	Nylon
Creslan	Orlon
Dacron	Saran
Fortrel	Teflon
Kodel	Terylene

6. Describe a consumer product in which each of the polymers in Exercise 5 is used.

7. Where possible, describe the properties of each of the polymers in Exercise 5 that make it particularly suitable for its commercial use.

8. What's the difference between a plastic and a polymer?

9. What are three factors, operating at the molecular level, that affect the bulk properties of a plastic?

10. How do high-density and low-density polyethylene differ in (a) the structures of their polymeric molecules; (b) the forms or shapes that their molecules take within the bulk plastic; (c) the physical properties of the bulk plastic?

11. Give two examples of polymers that we eat.

12. What is the difference between a thermoplastic and a thermosetting plastic?

13. What is an *olefin*? How are olefins used in the manufacture of plastics?

14. What are the two commercial forms of polyethylene?

15. Is Bakelite (a) an addition polymer or a condensation polymer? (b) a thermoplastic or a thermoset?

16. Name a commercial plastic that contains (a) phenyl, (b) chlorine, (c) fluorine, (d) nitrogen, (e) acetate groups.

17. What chemical element occurs in fibers of nylon but not in fibers of Orlon?

18. What is the principal environmental concern about the continued use of plastics?

19. In what two ways can science and technology help alleviate this concern?

20. Describe two factors that currently inhibit the effective recycling and reuse of plastics.

21. Name or describe one substance that can be added to a plastic to increase the likelihood that it will be biodegradable.

22. In addition to degradation by microorganisms, plastics can be degraded by another natural phenomenon. What is that phenomenon?

A Little Arithmetic and Other Quantitative Puzzles

23. What commercial plastic is produced in the largest quantities each year?

24. The nitric acid in Schoenbein's accident reacted with free hydroxyl groups on each of the monomeric glucose links of the cellulose polymer. How many free hydroxyl groups does each of these monomeric units have available for this reaction?

25. Figure 21.12 shows a portion of the polymer that forms on polymerization of a mixture of 75% butadiene and 25% styrene. Write a similar segment of the polymer you would expect to form from a mixture of 25% butadiene and 75% styrene.

Think, Speculate, Reflect, and Ponder

26. Why was the discovery of vulcanization so important to the development of the commercial rubber industry?

27. Give an example of the importance of geometric isomerism to the properties of elastomers.

28. Give an example of the importance of stereochemical order to the properties of a plastic.

29. Give two examples of serendipity in the discovery or development of a polymer.

30. Give examples of (a) a natural condensation homopolymer; (b) a natural condensation copolymer; (c) a natural addition homopolymer.

31. Give examples of (a) a synthetic addition homopolymer; (b) a synthetic addition copolymer; (c) a synthetic condensation copolymer.

32. Terephthalic acid is a dicarboxylic acid used as a monomer in the manufacture of a condensation polymer. What is another dicarboxylic acid used successfully in the manufacture of a condensation polymer?

33. What name would you give the plastic made from the addition polymer of $CH_2{=}CH{-}Br$?

34. What polymer would you use if you wanted to manufacture the following items:

a. skis with a coating that would make them slide over snow particularly easily
b. a wrapper for boxes of perfume that would not let the perfume's odor escape into a store or a room
c. automobile windows
d. an inexpensive, imitation silk fabric
e. an anchor for the ends of the heating elements of a toaster
f. an elastic, rubbery lining for gasoline tanks
g. a plastic film that's very thin, very strong, and very resistant to stretching

35. Molecules of low-density polyethylene have the structure $-CH_2-CH_2-CH_2-CH_2-CH_2-CH_2-CH_2-$ yet the shorthand form $\leftarrow\!\!\!\!+ CH_2 \rightarrow\!\!\!\!\!+_n$ does not adequately describe the polymer. Explain why it does not.

36. Write the molecular structure of neoprene in the form $\leftarrow\!\!\!\!+ X \rightarrow\!\!\!\!\!+_n$.

37. In what important way are the molecular structures of nylon polymers and protein polymers similar to each other? In what ways are they different?

38. Name two commercial products made of or containing plastic that you have used within the past 24 hours.

39. What structural characteristic must a molecule have if it is to be used as a monomer in an addition polymerization?

40. What property or combination of properties does each of the following polymers contribute to the consumer product associated with it: (a) poly(methyl methacrylate) as a camera lens; (b) poly(vinyl acetate) as a substitute for chicle in chewing gum; (c) polytetrafluoroethylene as a coating for pots, pans, and skillets; (d) poly(vinylidene chloride) as a food wrapper?

41. High-density polyethylene is used for manufacturing large containers for water and liquid laundry detergent, while low-density polyethylene is used for trash bags. What consequences would result from using high-density polyethylene for trash bags and low-density polyethylene for containers of water and liquid detergents?

42. Would you use high-density or low-density polyethylene to manufacture each of the following: (a) rope, (b) disposable medical syringes, (c) utensils for use in microwave ovens, (d) plastic sheets to protect cars parked outdoors?

43. If you cool a tennis ball to −195°C (with liquid nitrogen) and then throw it against a wall, the ball will shatter instead of bouncing back. Yet if you let the cold ball warm to room temperature, it will once again bounce normally. This cycle of cooling and heating, with corresponding changes in the properties of the tennis ball, can be repeated almost indefinitely. Does the elastomer that forms the ball resemble a *thermoplastic* or a *thermoset* in its behavior?

44. A plastic can be prepared easily by the polymerization of vinyl chloride, $CH_2{=}CH{-}Cl$. Can a plastic be prepared easily by the polymerization of ethyl chloride, $CH_3—CH_2—Cl$? Explain.

45. Describe the environmental problems generated by the production of synthetic plastics, their possible solutions, and the difficulties associated with each possible solution.

46. In an analogy to polypeptides and proteins, the primary structure of a polymer is represented by $\mathrm{-\!(X)\!-}_n$. Which polymers discussed in this chapter, if any, would you expect to show a secondary structure (Sec. 17.10)? Explain.

47. Is the monomer of nylon 6 an amino acid? What term is given to the functional group that links the monomers of nylon 6? Would you classify nylon 6 as a either a polypeptide or a protein? Explain.

48. Does DNA (Sec. 18.10) represent a condensation polymer or an addition polymer? Explain.

49. Does DNA represent a homopolymer or a copolymer? Explain.

50. In what way(s) are asbestos and quartz alike? In what way(s) are they different?

Web Questions

Use the Web to answer or accomplish each of the following. Report your Web source for each response.

1. Write a brief summary of the life of Wallace H. Carothers and the development of nylon.

2. Describe how household trash is disposed of in your community or in a community within your state. Is it placed in a landfill? Incinerated? Transferred to a distant location? Disposed of in some other way?

3. Summarize current procedures for the recycling of plastic wastes in your community or in your state.

4. Polyethylene is manufactured as low-density polyethylene (LDPE) and high-density polyethylene (HDPE). The polymer is also manufactured as linear polyethylene and branched polyethylene. Describe the structures and properties of these two forms of polyethylene and explain how they differ (if they do) from LDPE and HDPE.

5. Write a summary of the connection between asbestos and mesothelioma. Describe mesothelioma itself, how we came to connect mesothelioma with exposure to asbestos, current restrictions on the use of asbestos, and efforts to remove asbestos from the environment.

6. Describe the work and the discoveries of Karl Ziegler and Giulio Natta that led to the award of the Nobel Prize in chemistry.

7. Teflon is widely used as a nonstick surface. It's useful as a coating in pots and pans because foods tend not to stick to the Teflon surface at they are cooked. Why, then, does the Teflon coating stick firmly to the metal surface of the pot or pan? How is the Teflon coating applied so that it adheres strongly to the metal surface?

CHAPTER 22

Cosmetics and Personal Care

LOOKING GOOD AND SMELLING NICE WITH CHEMISTRY

Cosmetics help define our personal image.

The Wave of Chemistry

Figure 22.1 The home permanent uses the chemistry of thioglycolic acid and hydrogen peroxide to change the shapes of the protein molecules of hair.

If you give yourself a home permanent for waves or curls or to add a bit of body to your hair, you carry out some exceptionally interesting protein chemistry on the top of your head. Whether you use a kit that provides only the essentials, or the full treatment with a preperm shampoo and a postperm conditioner, you invariably go through two steps common to all. At some point in the process you use a waving lotion. Then after perhaps 10 or 15 minutes, you rinse out the waving lotion and apply a neutralizing solution.

The next time you have the box in front of you, take a look at the list of ingredients. In addition to lots of chemicals to give the waving lotion a nice odor, a desirable sense of thickness, and the like, you'll probably find that it also contains a chemical with "thioglycolate" (or something resembling that) in its name. The neutralizing solution is likely to contain some hydrogen peroxide along with all of its other ingredients.

These two, the thioglycolate and the hydrogen peroxide, are what do the actual work. With the glycolate of the waving lotion you break enormous numbers of covalent chemical bonds in your hair. But that's OK. Those are the bonds that hold strands of protein firmly in place next to each other. Once you break all those bonds, the proteins that form your hair can shift around to accommodate your wishes, as you express them with the curlers.

After you wait a few minutes to let the molecules loosen up and slide past each other to conform to the curlers, you snap them into their new molecular shapes with the neutralizer. The hydrogen peroxide regenerates just about all the covalent bonds that the thioglycolate broke, but in new locations. Now the protein molecules are just where you want them, and you have the look that you want and that only chemistry can provide (Fig. 22.1). We'll examine the chemistry of the permanent in more detail in Section 22.7.

If a home permanent is one of the ordinary things of life, the chemistry that goes into it is truly extraordinary. The same can be said for the chemistry that goes into our shampoos, toothpastes, colognes, perfumes, skin ointments and cleansers, hand and body soaps, lipsticks, deodorants, mouthwashes, antiperspirants, and related cosmetics and toiletries that are supposed to make us look and smell healthier and more fragrant (or perhaps less fragrant). In this chapter we'll examine some of these personal care products and the chemistry that makes them work.

22.1 The Universal Urge for Adornment

The urge to improve our appearance—to look good and to smell nice, especially by applying chemicals to our bodies—is common to the entire human race, just as it has been throughout the histories of all its civilizations. Peoples of all regions and all times have painted themselves, sometimes with plant extracts, sometimes with powdered metals, to become more vivid and to appeal to the eyes of others. *Henna*, a hair dye extracted from the henna plants of Africa, India, and the Middle East and still in current use, once colored the hair of Egyptians now mummified. Vases of cosmetics, oils, and ointments were sealed into the tombs of the kings of ancient Egypt to make them more presentable in the afterlife. The Bible refers to cosmetics as it tells of Jezebel painting her eyes when she heard of Jehu's arrival in Jezreel (II Kings, 9 : 30) and speaks of those who paint their eyes and bedeck themselves with ornaments (Ezekiel, 23 : 40). The Hebrew *kakhal* (meaning to paint or to stain) and the Arabic *al-koh'l* (a fine powder used in coloring the eyelids) give us our own *alcohol* by a tortuous sequence of transformations.

A **cosmetic** is a substance applied directly to the human body to make it more attractive.

In a subtle sense, the very word **cosmetic** is as universal as the materials it describes. The ancient Greek word *kosmos* served equally to represent order, adornment, and the universe itself. The connection among the three meanings comes from the belief of the Greek philosophers that there is beauty in order and that the universe is an ordered place and therefore beautiful. According to the Federal Food, Drug and Cosmetic Act (Sec. 17.14), a cosmetic is anything intended to be applied directly to the human body for "cleansing, beautifying, promoting attractiveness, or altering the appearance" (presumably for the better). According to the Act, soap isn't legally considered to be a cosmetic. But for our purposes we can consider that anything we apply directly to our bodies to make us more attractive fits the definition of a cosmetic.

The growing appreciation that a healthy body is an attractive body can lead to confusion between a cosmetic and a medicine. We'll examine medicines more closely in Chapter 23, but for now we can say that while cosmetics make us more attractive, medicines are used specifically for treating diseases, illnesses, and injuries. Cosmetic companies that claim their products improve health must be able to verify their claims to the Food and Drug Administration. Manufacturers of skin care products, for example, sometimes claim that their creams and ointments will

Improving personal appearance through the use of cosmetics.

lead to a healthier or healthier looking skin. In fact, unless the substance can actually cure a disease of the skin, it's a cosmetic rather than a medicine.

■ QUESTION

How does the Federal Food, Drug and Cosmetic Act define a *cosmetic*? How do we define it here, in this chapter?

22.2 Rich Suds and Expensive Odors

In the United States, four categories of cosmetics and toiletries account for about half the total sales of all these highly personal consumer chemicals:

- hair care products,
- perfumes and colognes,
- skin care products and emollients (skin softening and smoothing products), and
- hand, face, and body soaps.

The manufacture of these personal care products can be a rewarding activity. Overall, only about a tenth of the retail cost of the average cosmetic or toiletry goes toward its raw materials: chemical dyes, detergents, perfumes, moisturizers, solvents, and the like. The remainder pays for packaging, advertising, research, labor and similar expenses, and, of course, profit.

Of all the chemicals used in these formulations, two classes, aside from water and common solvents, stand out. One consists of synthetic detergents and other surfactants, which constitute the largest single category of chemicals used in their manufacture and which do the work of our shampoos, soaps, toothpastes, and related cleansing toiletries.

The other, significant from both an esthetic and an economic standpoint, comprises the fragrances that add their pleasant aromas to virtually every one of these products. Although these compounds contribute less than 1% to the total weight of all our personal care products combined, their costs make up about a quarter of the total value of all the raw materials used in their manufacture. Gram for gram, fragrances are by far the most expensive of all the ingredients.

■ QUESTION

What kind of chemical makes up the largest category of compounds used in the manufacture of personal care products?

22.3 To Satisfy the Consumer

You wouldn't want to wash your hair with toothpaste or brush your teeth with shampoo, but the active detergent is very likely to be the same in each. This shouldn't surprise us when we recognize that the surfactant that forms suds in our hair can foam up in our mouth or on our hands just about as well.

We don't wash our hair with toothpaste or brush our teeth with shampoo for some very good reasons. Aside from the particular odors and flavors we associate with the two products, we expect a shampoo to do more than simply remove greasy dirt from hair—an ordinary hand soap can do that—and we expect a dentifrice to do more than merely clean our teeth.

To meet our broader expectations, each of our cosmetics and toiletries is formulated to perform a set of at least three different kinds of tasks, each important to the commercial success of the product. To formulate a product that does each

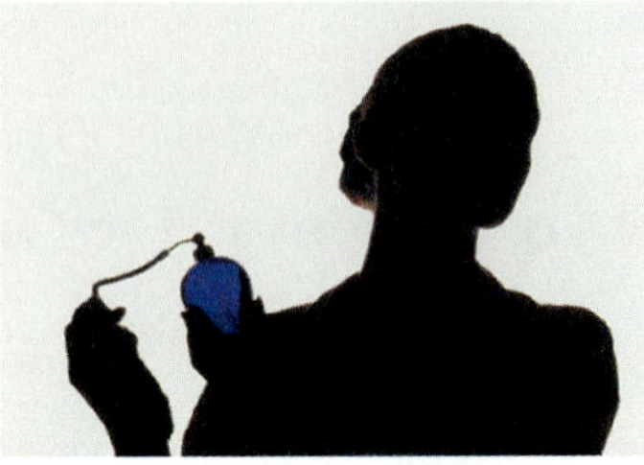

Common to all personal care products are pleasant odors. Gram for gram, fragrances make up the smallest percentage of the weight of our personal care products and the largest percentage of the cost of their ingredients.

one well requires a skillful blending of carefully selected chemicals, each lending its own characteristics to the jobs at hand.

First comes the principal function. Above all, we expect shampoos and toothpastes to clean and to produce suds. For both cleansing and foam production their most important active ingredient is a detergent not much different chemically from those described in Chapter 13. (Actually, the detergent in a good dentifrice isn't nearly as important as the abrasive, but we'll come to that in Sec. 22.8.) Next is a set of secondary activities relating directly to specific applications on our hair, teeth, skin, or nails. We expect our shampoos, for example, to provide body and luster to our hair, as well as to clean it thoroughly without removing all its oils; we expect our toothpastes to give us a fresh-tasting mouth and clean-smelling breath along with clean, cavity-resistant teeth and healthy gums.

Finally there are the more subtle qualities of our personal care products. To produce consumer satisfaction they must be convenient to use, and their odors, colors, and general appearance must convince us that they do their jobs well and effectively. Otherwise they lose out in a competitive marketplace. Hair shampoo that looks like muddy water or toothpaste that reminds us of garlic would hardly be commercial successes no matter how clean and sparkling their chemicals might get our hair or our teeth. Along with all this, these products are also packaged, advertised, and sold in ways designed to reinforce our self-image. But that lies in the realm of marketing rather than chemistry.

■ **QUESTION**

What do you expect as the primary quality of an antiperspirant? What other, secondary qualities do you expect?

22.4 Personal Care Surfactants

As you read the list of ingredients of your favorite shampoo or toothpaste, it's likely you'll find among them a *sodium lauryl sulfate* or an *ammonium lauryl sulfate*. These are surfactants, with hydrophilic heads and hydrophobic tails, not much different in molecular structure from the soaps and synthetic detergents that clean our bodies and our clothes. As they appear in Figure 22.2, all the lauryl sulfates con-

Figure 22.2 Surfactants of toothpastes and shampoos.

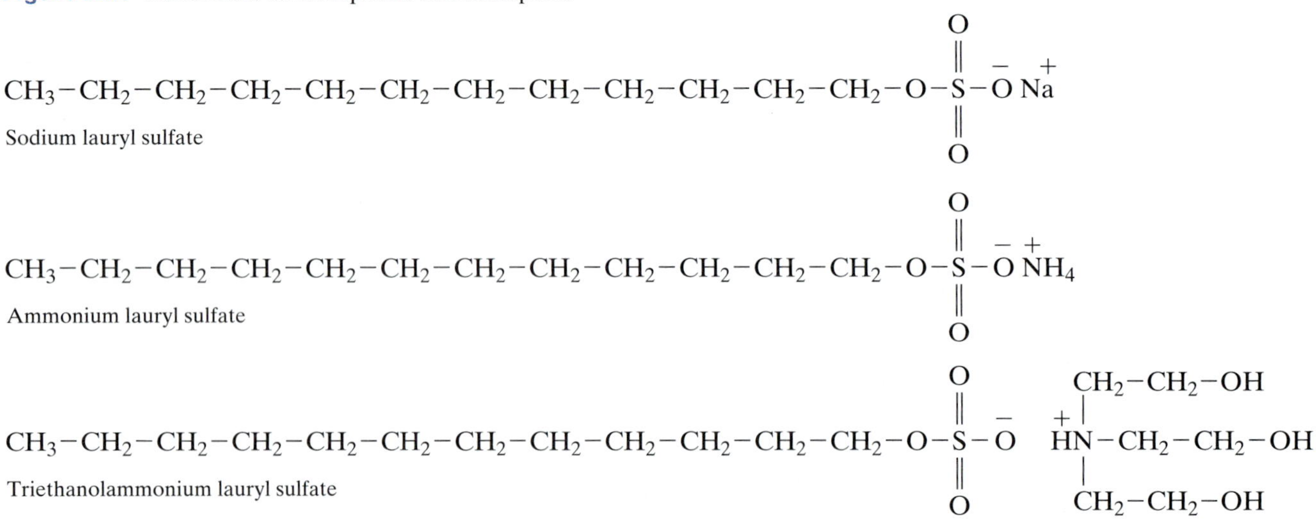

Sodium lauryl sulfate

Ammonium lauryl sulfate

Triethanolammonium lauryl sulfate

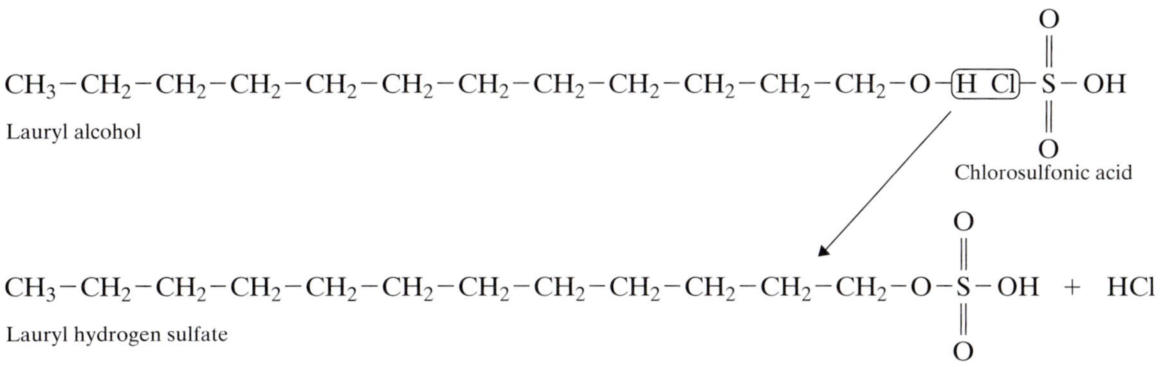

Figure 22.3 Synthesis of lauryl hydrogen sulfate.

tain identical anionic portions and differ only in the cation that balances out the anion's negative charge. Since it's the anion of each that's the active surfactant, with its combination of hydrophilic head and hydrophobic tail, they are all *anionic surfactants* (Sec. 13.11).

Lauryl hydrogen sulfate, the parent compound, is manufactured through several different chemical reactions, principally by the action of *chlorosulfonic acid* on *lauryl alcohol* (Fig. 22.3). In the process the alcohol and the acid combine with the loss of a molecule of HCl. Replacing the acidic proton of the lauryl hydrogen sulfate with another cation, such as the sodium ion or the ammonium ion, generates the particular lauryl sulfate used in the toiletry formulation.

The sodium salt of lauryl hydrogen sulfate, *sodium lauryl sulfate*, the surfactant of many dentifrices, is relatively insoluble in cold water. This decreases its appeal for hair shampoos since in cold weather the salt tends to come out of solution and produce a slight cloudiness. While a bit of turbidity hardly affects the appearance of most toothpastes, watching a bottle of what was once a clear shampoo turn strangely cloudy in cold weather could cause a consumer to switch brands. Because of this property of the sodium salt, you're more likely to find *ammonium lauryl sulfate* or *triethanolammonium (TEA) lauryl sulfate* listed as an ingredient in your shampoo. They' re more soluble in cold water and, as an added benefit, they don't dry out the hair quite as much as the sodium salt.

■ **QUESTION**

What are two properties of ammonium and triethanolammonium lauryl sulfates that make these anionic detergents superior to the sodium salt for use in shampoos?

22.5 Hair: Cleaning the Cuticle

Hair itself is a lifeless structure composed of a protein called *keratin*. A strand of hair contains no more living tissue than do our nails or the keratin-containing claws, horns, hoofs, or feathers of other animals. Like all human proteins, keratin is a polypeptide with monomeric units consisting of a combination of the 20 amino acids of Table 17.1. In this particular protein one amino acid, *cystine*, dominates the polymer and accounts for somewhere between 14% and 18% of keratin's amino acids. The sulfur–sulfur bonds of the cystine give hair much of its strength by connecting the strands of protein and keeping them aligned with one another, as shown in Figure 22.4.

Figure 22.4 Sulfur–sulfur links in a typical segment of keratin.

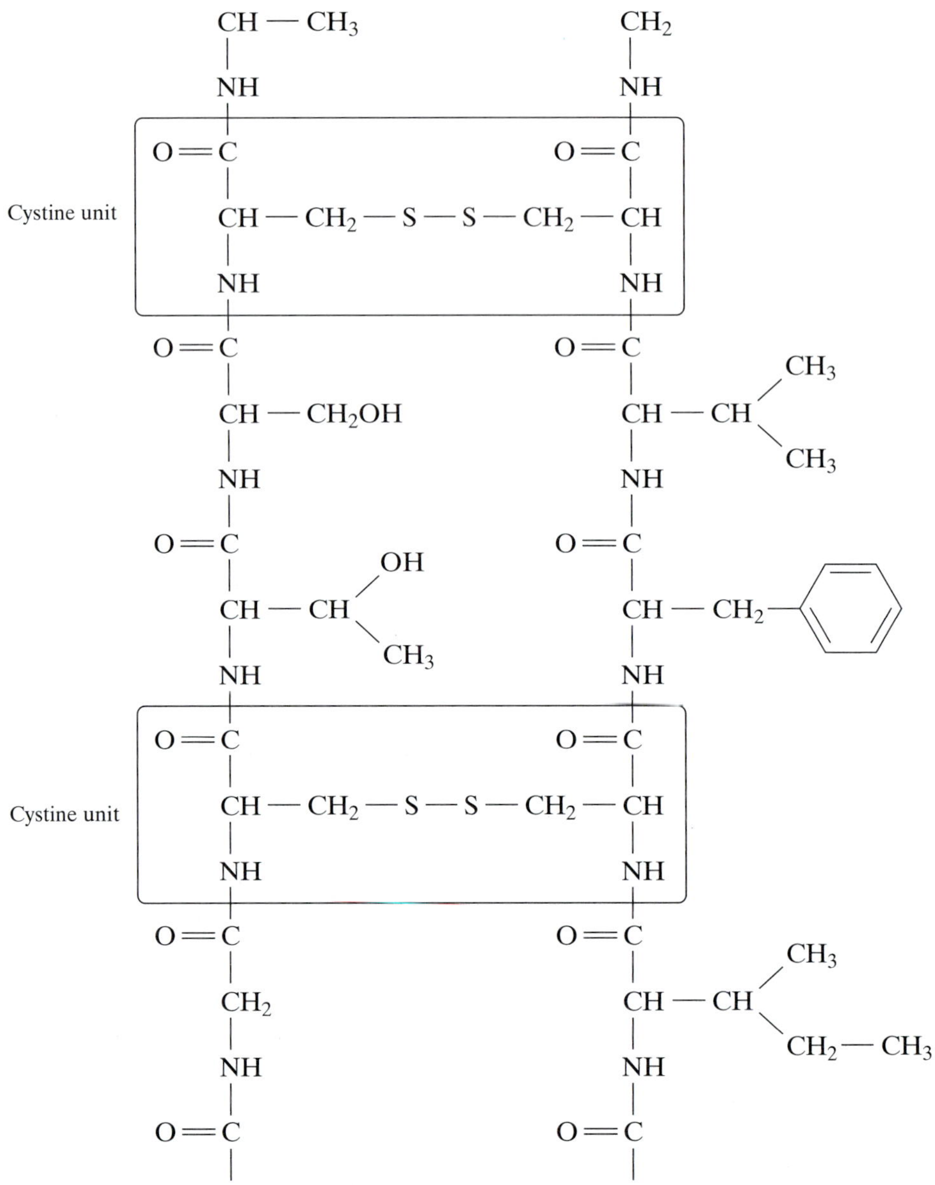

A **follicle** is a microscopic sac, under the surface of the skin, that holds the root of a strand of hair.

The **cortex** is the central core of a strand of hair. The cortex forms the bulk of the strand and contains its pigments. The **cuticle** is the scaly sheath that covers the cortex. **Sebaceous glands** are glands lying in the skin, near the follicle, which secrete **sebum,** an oily substance that lubricates the cuticle and gives it a gloss.

Each strand grows from a living root embedded within microscopic sacks or **follicles**, which are buried under the surface of the skin. The hair of the scalp grows outward at about one centimeter each month, or about a tenth of an inch per week. After 4 to 6 years of steady growth each strand falls out, replaced by a new one as it begins its own growth.

Typically each strand of hair consists of a central core, the **cortex**, which forms the bulk of the fiber and contains its coloring pigments. Enveloping this cortex is a thin, translucent, scaly sheath, the **cuticle**. **Sebaceous glands** lying in the skin near the follicle lubricate the emerging shaft with an oily **sebum**, which gives the hair a gloss, keeps the scales of the cuticle lying flat, and prevents the strand from drying out (Fig. 22.5). Too much sebum and the hair feels greasy and dirty; too little and it's dry, dull, and wild. The detergent of a hair shampoo, then, must remove dirt from the hair and scalp, as well as enough accumulated sebum to keep the hair looking clean, but not so much as to remove all of the oil. The moderate detergent

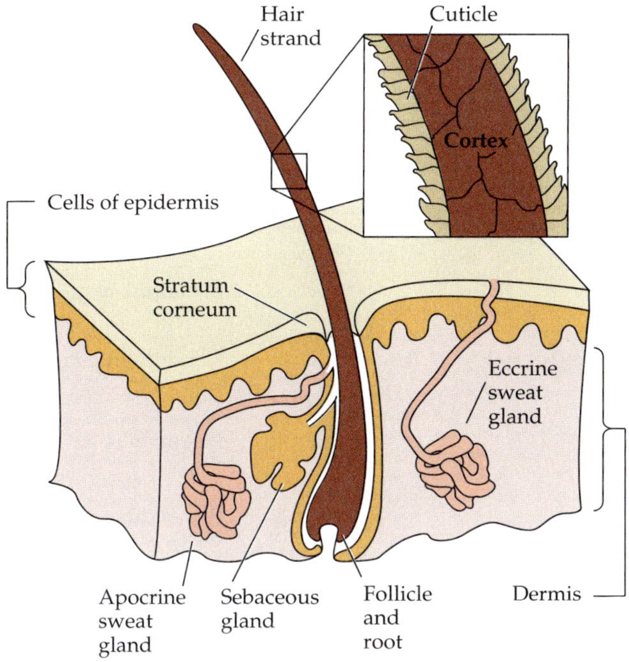

Figure 22.5 A typical section of skin, hair, and glands.

action of the lauryl sulfates and related surfactants accomplishes this tightrope act very nicely. Many shampoo formulations contain added conditioners to replace at least part of the lubricant that might be lost in the washing.

Other shampoo ingredients help stabilize the foam, act as preservatives, give the shampoo itself a pleasing viscosity or body, adjust the pH, chelate (Sec. 19.17) the metal ions of hard water that might degrade the surfactant's action, and add color and fragrance. The acidity of the shampoo has a lot to do with its ability to produce a good luster and resilient hair. Other things being equal, a strand of hair is strongest under slightly acidic conditions, at a pH of about 4 to 6. Moreover, the scales of the cuticle tend to swell up and fluff out under basic conditions. This condition causes reflected light to scatter, making the hair look dull. A tight, slightly acidic cuticle reflects light more coherently, giving the hair a pleasant luster (Fig. 22.6). The ingredients of a typical clear shampoo appear in Table 22.1.

Hair shampoo, a blend of detergents, conditioners, stabilizers, fragrances, and other ingredients that clean your hair and present it in its best light.

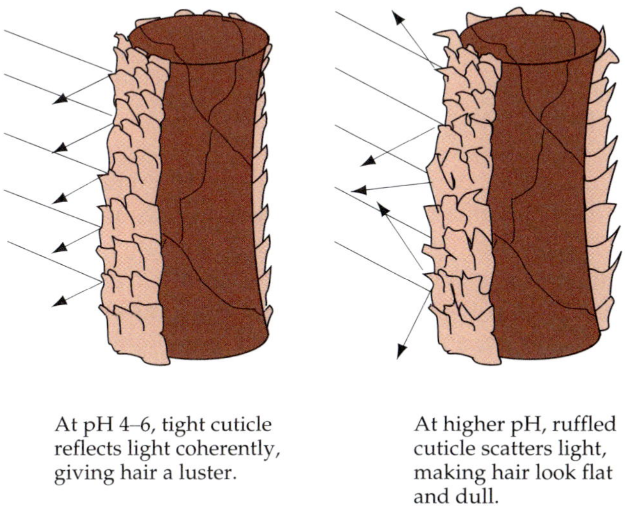

Figure 22.6 pH and the radiance of hair.

Table 22.1 Ingredients of a Typical Clear Liquid Shampoo

Ingredient	Weight (%)	Function
Purified water, H_2O	60	Solvent and filler
Triethanolammonium lauryl sulfate $(HO—CH_2—CH_2)_3N^+H\ ^-O_3SO—CH_2—(CH_2)_{10}—CH_3$	32	Surfactant
Myristic acid, $CH_3—(CH_2)_{12}—CO_2H$	4	pH adjustment
Oleyl alcohol, $CH_3—(CH_2)_7—\underset{H}{C}=\underset{H}{C}—(CH_2)_7—CH_2—OH$	2	Conditioning agent
Fragrance	1	Perfume
Formaldehyde, $H—\overset{O}{\overset{\|\|}{C}}—H$	0.5	Preservative
Other additives	0.5	Sequestrants and colorants

■ **QUESTION** Covalent bonds between two atoms of which element are responsible for keeping strands of hair's keratin aligned with each other?

22.6 Hair: Coloring the Cortex

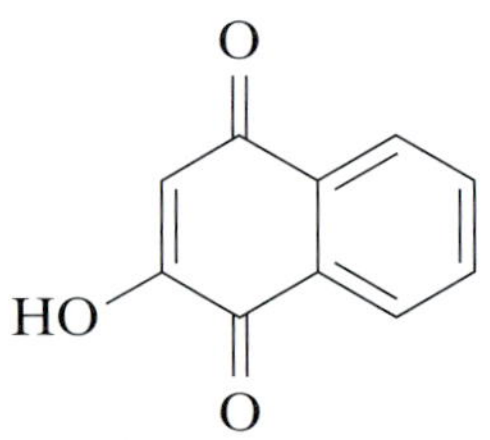

Figure 22.7 Lawsone, the coloring agent of henna.

As we saw in Section 22.1, the use of deeply colored vegetable extracts and powdered metals to color hair goes back to the earliest records of human history. Some of these dyes, such as henna, are still used. *Lawsone* (Fig 22.7), the active component of henna, gives hair a red-orange tint.

While powdered metals themselves are no longer used as commercial hair colorings, some contemporary products darken hair through the action of metal atoms incorporated into organic molecules. One of these is the *lead acetate*, $Pb(O_2C—CH_3)_2$, of products such as Grecian Formula. The lead reacts with the cystine of keratin, converting the amino acid's sulfur into dark lead sulfide, PbS. The process takes place slowly, requiring repeated applications over several days, and results in a gradual darkening of the hair.

While the vegetable dyes of antiquity are still used occasionally, changing the color of hair today is more likely to be accomplished through the reactions of complex synthetic organic and organometallic chemicals devised only within the last hundred years or so.

Melanin is the dark brown pigment of the skin and hair; **phaeomelanin** is a red-brown or yellow-brown pigment chemically similar to melanin.

Hair itself contains two natural pigments: **melanin**, the dark brown pigment of the skin, and **phaeomelanin**, a red-brown or yellow-brown pigment chemically similar to melanin. The color of hair depends on the amounts and physical conditions of these two pigments within the strand. Their absence produces white or gray hair. Converting dark hair to a lighter shade requires bleaching these natural pigments, usually with a solution of hydrogen peroxide, H_2O_2, and replacing them with a synthetic dye. Adding dyes without bleach darkens the color.

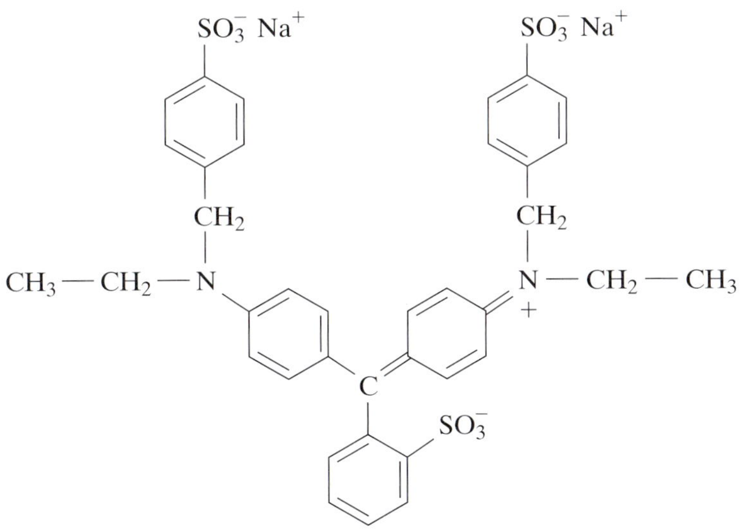

Figure 22.8 FD&C Blue No. 1, a temporary hair dye. It washes out with a single shampooing.

The variety of synthetic dyes now available gives us some flexibility in choosing the permanence of any color change. Temporary dyes and rinses color the hair with chemicals that wash out easily with a single shampooing. The most popular of these contain the salts of large, acidic molecules, such as FD&C Blue No. 1 shown in Figure 22.8. The multiple sulfonate groups, $—SO_3^-$, on the molecule add to its solubility in water, yet because of its large size the colorant doesn't penetrate easily into the cortex. Instead it coats the surface of the cuticle and leaves with the next rinse.

[The *FD&C* of the dye's identification number indicates that the Food and Drug Administration has approved its use in *f*ood, *d*rugs, and *c*osmetics, as specified in the Food, Drug and Cosmetic Act (Sec. 19.13). In an actual commercial formulation the blue dye wouldn't be used alone as the sole coloring agent. It would be blended with others to produce a pleasing shade or tint.]

Smaller molecules, which can penetrate more readily through the cuticle and move into the cortex, make up formulations designed to produce semipermanent dyes, those that last through half a dozen or so washings. Naturally, since the dye molecule is mobile enough to penetrate into the cortex, it can diffuse out, too, during the repeated shampoos. The colors simply last a bit longer than the temporary tints. Typical dyes of this kind appear in Figure 22.9.

Permanent dyes, those that last as long as the strand of hair itself, are made up of molecules with little ability to migrate out of the cortex, either because of their size or because of some structural feature that tends to immobilize them within the keratin.

Organic dyes add color to the cuticle and the cortex of hair.

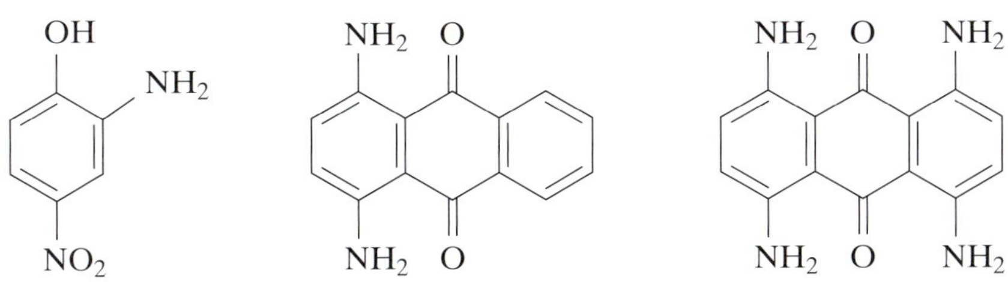

2-Amino-4-nitrophenol (yellow)

1,4-Diaminoanthraquinone (violet)

1,4,5,8-Tetraaminoanthraquinone (blue)

Figure 22.9 Representative semipermanent hair colorants. These last through several shampooings.

Once inside the cortex they remain there for the life of the strand. But if the molecules won't migrate *out*, we might wonder how they travel into the cortex in the first place.

Actually, they don't. The molecules of the permanent dyes don't even exist until they are formed well inside the cortex through a chemical combination of smaller, more mobile molecules that penetrate from the outside. A three-component system consisting of an oxidizing agent, such as hydrogen peroxide, and two organic intermediates—a *primary intermediate* and a *secondary intermediate*, or *coupler* (Fig. 22.10)—does the job by an ingenious bit of chemistry. The small and highly mobile molecules of the intermediates, usually colorless themselves, move into the keratin much like those of the semipermanent dyes. As these relatively small organic molecules travel into the cortex, the one representing the primary intermediate is oxidized by the hydrogen peroxide. The product of this reaction combines with the other component, the secondary intermediate or coupler, to form the immobile, permanent coloring. Shades darker than the natural color of the hair are produced as the newly introduced dye adds to the color of the melanin already present. Lighter shades result as larger amounts of the oxidizing agent bleach the melanin even as they induce formation of the dye.

Dyeing hair with these systems may carry some risks. Many contain ingredients that can generate allergic reactions in sensitive individuals. Used on eyebrows or eyelashes, some could damage the eyes. A few of the intermediates, such as the 2,4-diaminoanisole of Figure 22.10 (also known as MMPD, a shortened form of an alternative name, 4 *m*ethoxy-*m*-*p*henylene-*d*iamine), have been shown to pro-

Figure 22.10 Intermediates in the formation of permanent hair dyes.

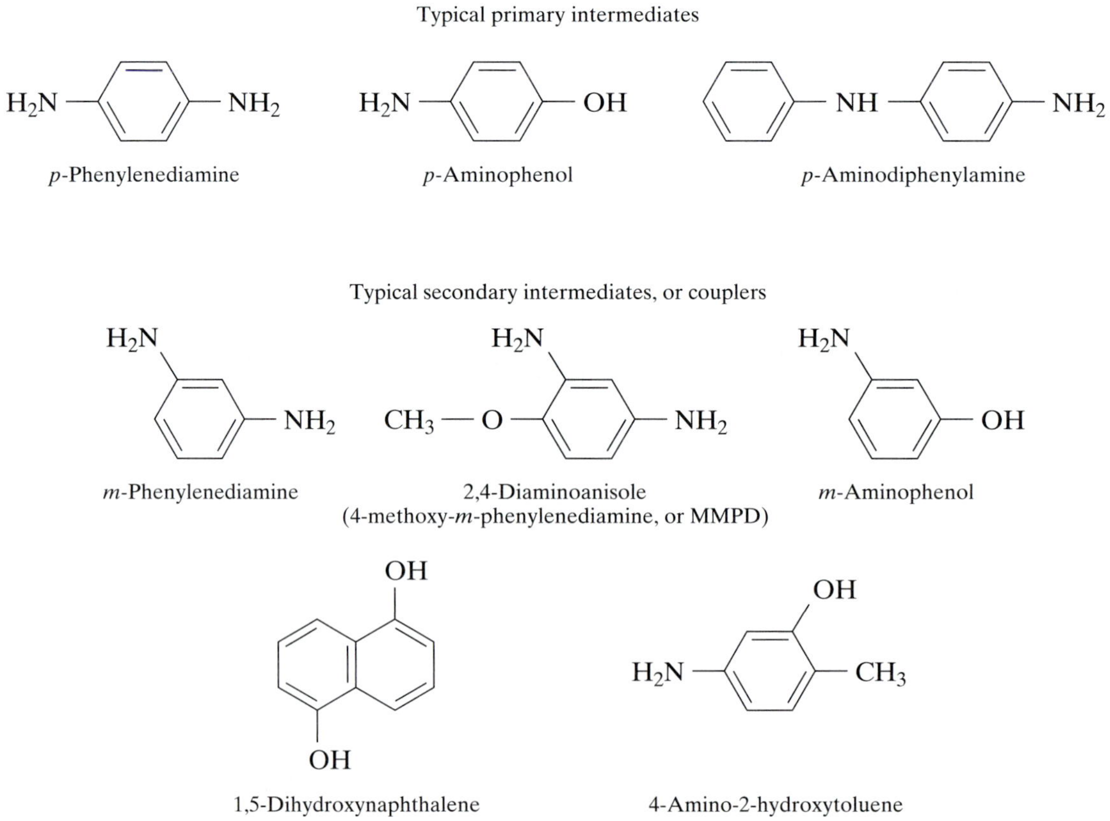

duce cancer in laboratory animals. MMPD itself has been removed from most commercial hair dyes. Those that still contain it must carry a cautionary statement on the label: "Warning—Contains an ingredient that can penetrate your skin and has been determined to cause cancer in laboratory animals." Still other ingredients give positive Ames tests (Perspective, Chapter 20) and are suspect as mutagens or carcinogens. Nevertheless, no authentic connection between these dyes and human cancer or birth defects has been established. As for the question of whether hair dyes are actually safe, the discussion of risks, hazards, and safety in Chapter 20 applies to these ingredients as well as to food and its additives. As before, safety is largely a matter of the acceptability of known risks.

■ QUESTION

If the permanent hair dyes consist of molecules that, because of their size or structural features, are unable to migrate *out* of the cortex, how do they get *into* the cortex in the first place?

22.7 Hair: Curling the Keratin

Several forces combine to keep a strand of hair straight or in loose waves or tight curls. The most important are

- disulfide links—the sulfur–sulfur covalent bonds of the keratin's cystine;
- salt bridges—the ionic bonds that form between the acidic group of one amino acid and a basic (amino) group of another amino acid located somewhere else on the same or an adjacent protein molecule; and
- hydrogen bonds.

We discussed the disulfide links of cystine groups in the keratin earlier (Sec. 22.5). The salt bridges form as the proton of a carboxyl group ($—CO_2H$) on an amino acid, such as aspartic acid or glutamic acid, transfers to a nearby amino acid containing a free amine group ($—NH_2$), such as the lysine or proline of another chain (Table 17.1). This acid–base reaction produces the ionic bonding of a carboxylate anion ($—CO_2^-$) and an ammonium ion ($—NH_3^+$) that holds portions of adjacent keratin chains near each other, just as ionic bonds cement a crystal of sodium chloride.

Hydrogen bonds, which we examined in Section 17.10, are formed by weak electrical attractions between the hydrogen of an NH group or an OH group, on the one hand, and any one of a few other atoms, including oxygen and nitrogen (but not carbon), on the other. Hydrogen bonds are plentiful in water and in the polymeric proteins, where the hydrogen on a nitrogen of one amide link bonds (weakly) to the carbonyl oxygen of another. Figure 22.11 sums up these interactions in the keratin of hair and in water.

While individual hydrogen bonds are weaker than any of the ionic or covalent bonds, their sheer numbers give dry hair most of its strength. In wet hair, though, water molecules intrude between the keratin strands, disrupt the hydrogen bonds that keep them aligned with each other, and allow them to shift a bit. As hair dries, the water molecules leave and the hair retains its new shape, held intact by the combined force of large numbers of hydrogen bonds in the newly realigned polymeric chains. Every time you wash your hair, set it, and then blow-dry it, you use a bit of chemistry to disrupt the keratin's hydrogen bonds, rearrange its strands, then remove the water molecules so that the keratin's own hydrogen bonds hold your hair in its new shape, at least until it gets wet again (Fig. 22.12).

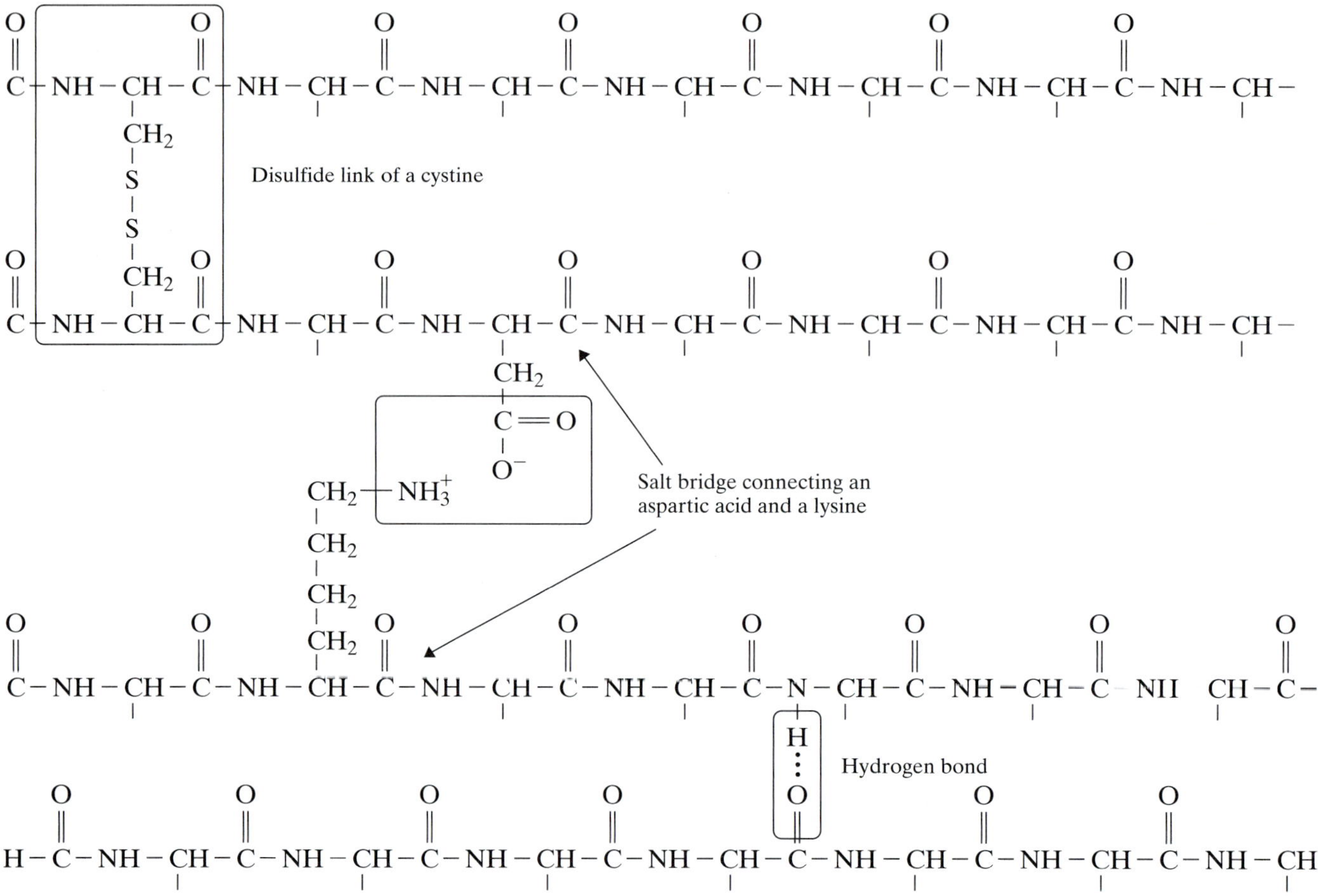

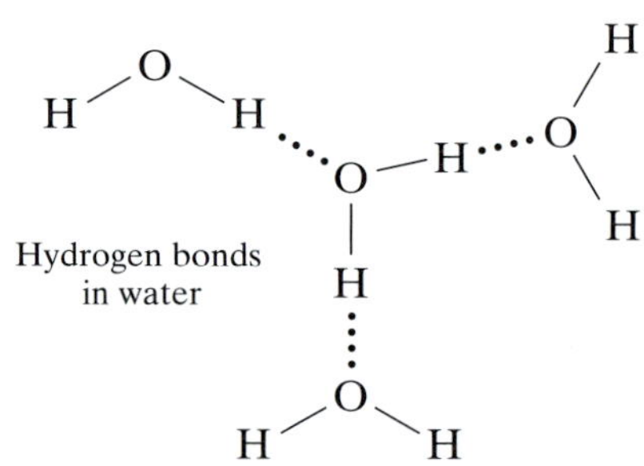

Figure 22.11 Bonding in proteins and in water.

You can produce a longer lasting rearrangement of the keratin strands by using a commercial permanent wave kit and, in this case, a bit of sulfur chemistry. The sulfur–sulfur bonds of the cystine units shared by adjacent keratin strands hold the polymers in a fixed alignment, much as the rungs and steps of a ladder hold its sides parallel. To rearrange the shapes of the strands, you simply break the sulfur–sulfur covalent bonds, reorganize the strands, then regenerate new sulfur-sulfur bonds. Unlike the wet wave, the permanent keeps its shape through the newly formed covalent bonds. These remain firm whether the hair is wet or dry.

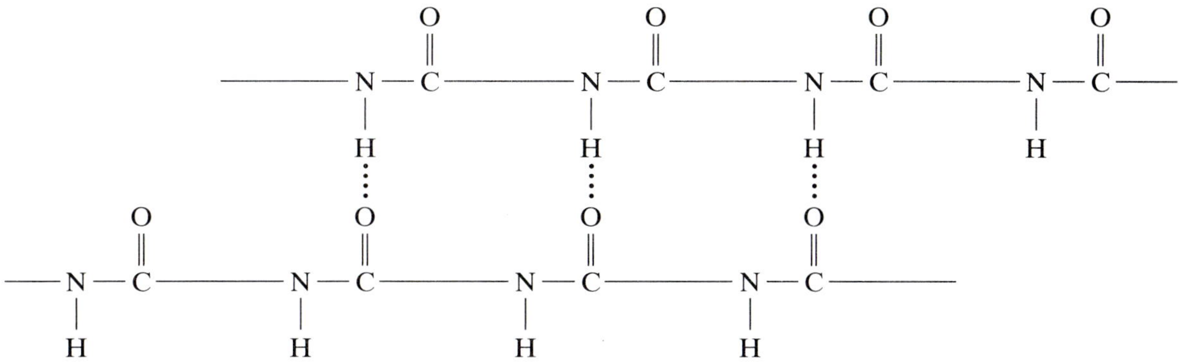

Adjacent protein strands of dry hair are held in place partly by hydrogen bonds.

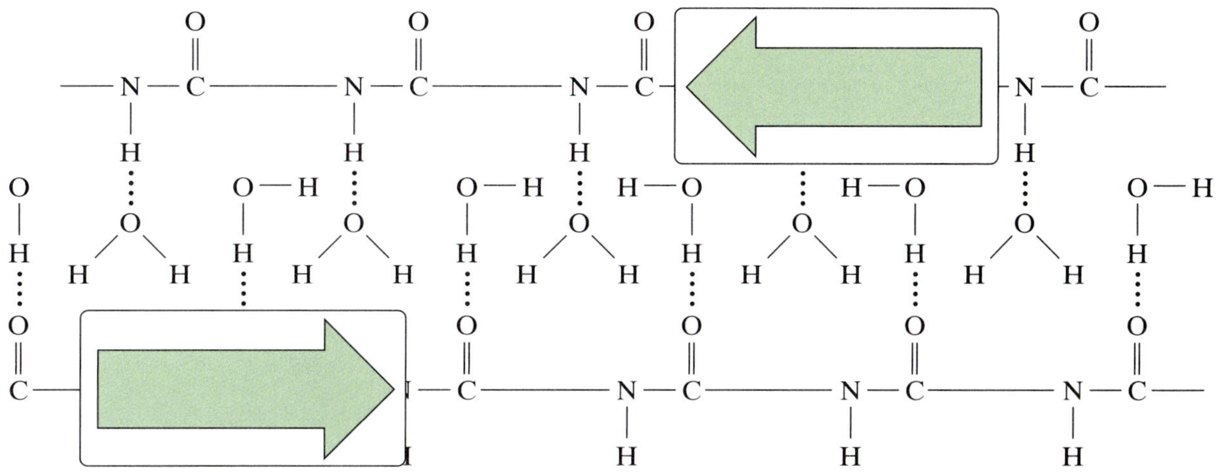

Hydrogen bonding with water allows protein polymers to slide past each other when hair is wet

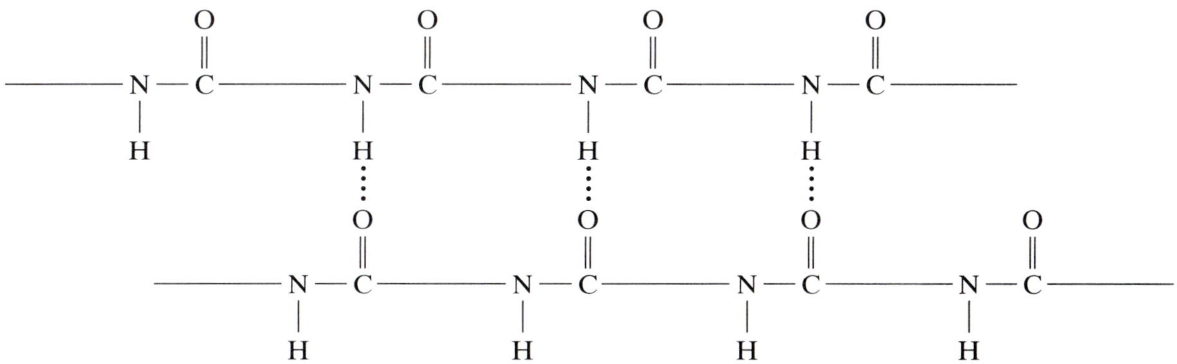

Drying hair removes water and fixes a new orientation of strands.

Figure 22.12 The chemistry of a wet wave.

To accomplish all this, permanent waving kits (such as those described in the opening demonstration) contain two essential parts: a waving lotion and a neutralizer. Most waving lotions are slightly basic solutions of *thioglycolic acid*, a compound that effectively transfers the hydrogen of its own sulfur to the cystine units and cleaves them to cysteines. The result resembles what we'd get if

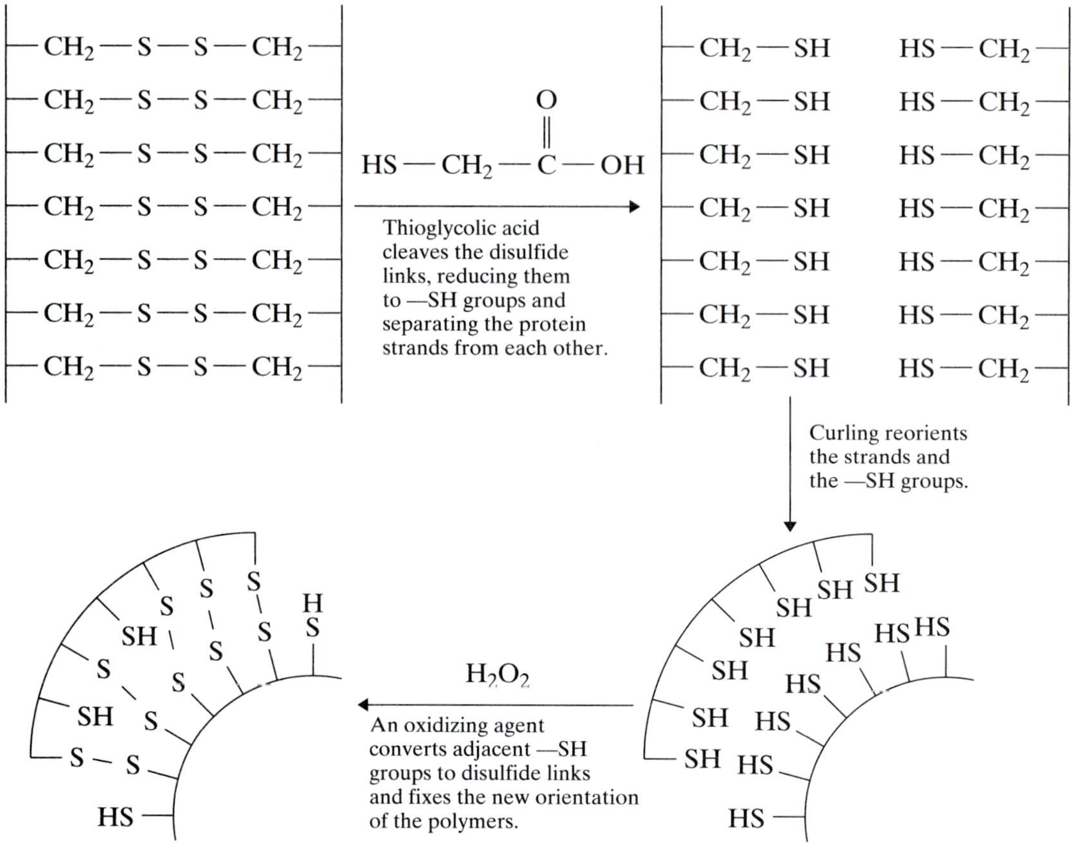

Figure 22.13 The chemistry of the permanent wave.

we sawed straight down the middle of the steps and rungs of the ladder. With this accomplished, straight keratin strands can curl, or curled strands can be straightened. Now the neutralizing part of the kit, usually a dilute solution of hydrogen peroxide, converts the cysteines back to cystine and the new shape takes hold (Fig. 22.13):

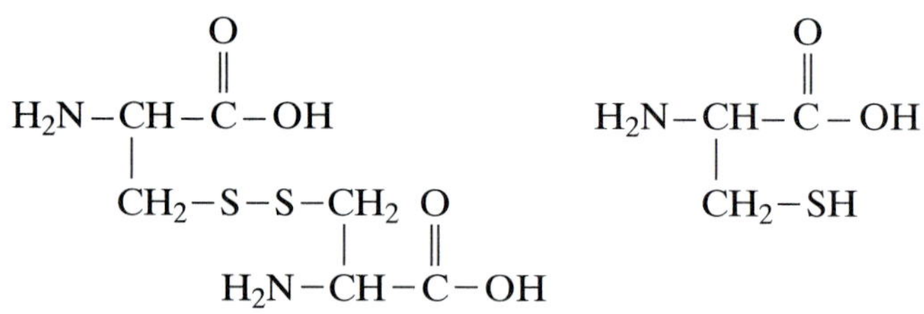

cystine cysteine

■ **QUESTION**

The curl of a permanent wave is produced by breaking covalent bonds between two strands of keratin, reorganizing the strands, then re-forming the covalent bonds. Atoms of which element are connected by the covalent bonds that are broken and re-formed?

22.8 Toothpaste, the Fresh Abrasive

Bright, clean teeth appear to be healthy teeth. But while we see a rich foam as the clearest indication that the toothpaste is doing its job, those same surfactants that produce the foam are among the least important cleansing agents of a good dentifrice. They aren't what keep teeth clean and healthy, and a few powdered dentifrices don't even contain them. In fact, the Food and Drug Administration doesn't consider a surfactant to be an active ingredient of a toothpaste. What's needed for a good dentifrice is an effective abrasive. To understand why takes a bit of dental chemistry.

A healthy tooth is covered by a layer of enamel that serves both to grind food and to protect the interior regions of the tooth itself. As in bone, calcium makes up most of this extremely hard surface, which is composed largely of the mineral *hydroxyapatite*, $Ca_{10}(PO_4)_6(OH)_2$. Despite its toughness, the enamel is susceptible to acids that come from a thin, adhesive, polysaccharide film called **plaque**.

Bacteria of the mouth continuously convert some of the sugars of our food into this plaque, which attaches to the enamel and serves as a home to still other bacteria. These other bacteria, in turn, convert the plaque into an acid that erodes the calcium and the phosphate from the shield of hydroxyapatite that protects the tooth. When enough erosion has occurred microorganisms can pass through the weakened barrier to begin their work on the interior. The result is *dental caries*, better known as tooth decay or cavities.

The key to keeping teeth free of cavities lies in removing the accumulated plaque, and that depends more on grinding it away with a good dental abrasive than on the detergent action of a surfactant. With daily removal of the plaque, the calcium and phosphate normally present in saliva replace any that might have been removed by mouth acids. As long as no bacteria have entered the body of the tooth itself, this reconstitution process, known as *remineralization*, returns the enamel to its original strength. To be useful as a dental abrasive the grinding agent must be harsh enough to remove accumulated plaque, yet not grind away the enamel itself. Table 22.2 presents some of the chemicals used as abrasives in commercial dentifrices.

Toothpaste, a combination of abrasives, detergents, flavorings, and, in many cases, fluorides and other ingredients that protect against decay.

Plaque is a thin layer of a polysaccharide that sticks to the surface of teeth and harbors bacteria.

Table 22.2 Typical Dentifrice Abrasives

Abrasive	Formula
Calcium carbonate	$CaCO_3$
Calcium pyrophosphate	$Ca_2P_2O_7$
Dibasic calcium phosphate	$CaHPO_4$
Hydrated aluminum oxide	$Al(OH)_3$
Magnesium carbonate	$MgCO_3$
Talc	$Mg_3(Si_2O_5)_2(OH)_2$
Titanium dioxide	TiO_2
Tricalcium phosphate	$Ca_3(PO_4)_2$

Fluoride ions also help maintain the strength of the enamel. Fluorides are present in toothpastes largely in the form of stannous fluoride (SnF_2), sodium monofluorophosphate (Na_2PO_3F), and sodium fluoride (NaF). The fluoride seems to act by

- replacing some of the hydroxy groups in the enamel's hydroxyapatite, converting it to a harder mineral, *fluoroapatite*, which is more resistant to erosion by acids, and
- suppressing the bacteria's ability to generate acids.

Although not our principal protection against tooth decay, the surfactants of toothpaste formulations do effectively remove loose debris from the mouth and

Table 22.3 Composition of a Typical Dentifrice

Ingredient	Formula	Weight (%)	Function
Water	H_2O	39	Solvent and filler
Glycerol	$CH_2(OH)-CH(OH)-CH_2(OH)$	32	Humectant, retains moisture
Dibasic calcium phosphate	$CaHPO_4$	27	Abrasive
Carrageenan	A carbohydrate of seaweed	1	Thickening agent and stabilizer
Fluorides and other additives		1	Enamel hardener; sweeteners and preservatives

also give us the sense of cleanliness. Almost all dentifrices contain a bit of saccharin and some flavoring or fragrance to leave us with a sense of sweetness and freshness after brushing. The components of a typical toothpaste appear in Table 22.3.

■ QUESTION

What is the principal function of a good toothpaste in keeping teeth clean and cavity-free?

22.9 To Make Our Skin Moist, Soft, and . . .

On average, about a quarter of every dollar we spend for cosmetics goes toward the care of our skin with moisturizers and emollients (softeners); hand, face, and body soaps; and deodorants and antiperspirants. We've already examined the chemistry of soaps and detergents in Chapter 13; here and in the next section we'll take up the means by which we keep our skin—the organ that envelops our bodies—moist, soft, and inoffensive.

If an organ is a group of tissues that perform one or more specific functions, as the heart pumps blood or the kidneys filter it, then our skin is indeed an organ, the largest one of all. Its functions range from the obvious covering of our bodies and protection from damage by foreign matter and microorganisms, to the more subtle tasks of regulating body temperature, sensing the stimuli provided by the outside world, and even synthesizing compounds such as vitamin D (Sec. 19.5).

The **dermis** is the portion of the skin that contains nerves, blood vessels, sweat glands, and the active portion of the hair follicles, and also supports the upper layer, the **epidermis**.

The **stratum corneum** is the outermost layer of skin, a protective shield consisting of 25 to 30 layers of dead cells.

The skin itself is made up of two major layers, the underlying **dermis**, which contains nerves, blood vessels, sweat glands, and the active portion of the hair follicles, and which also supports the upper layer, the **epidermis** (Fig. 22.5). The epidermis consists of several tiers of cells. At its bottom, resting on top of the dermis, is a single sheet of cells that divide continuously, always pushing upward. As they move toward the outside of the skin, driven along by new cells coming up from beneath, they lose their ability to divide and eventually die. By the time they become the lifeless, outermost layer of the skin, the *stratum corneum*, they have been transformed into keratin, the same protein that forms the hair.

The **stratum corneum** itself is a layer of 25 to 30 tiers of these dead cells whose sole function is to protect us against the outside world, including the world of our cosmetics and toiletries. The condition of this layer depends mostly on its water content. Too much moisture nourishes the growth of fungi and other microorganisms; too little dries it out, producing flaking and cracking. A water content of 10% is just about right for the stratum corneum.

The body supplies protection to the skin with the sebum secreted by the sebaceous glands of hair follicles. This oily substance coats the adjacent skin, lubricating and softening its dead keratin (as well as the hair's), and lowering the rate at which water evaporates from its surface. These sebaceous glands occur only at the hair follicles themselves. No sebum coats the skin of the hairless regions of the body, such as the palms of the hands and the soles of the feet.

Commercial cosmetic lotions and creams are emulsions, or colloidal dispersions (Sec. 13.5) of two or more liquids that are insoluble in each other (usually an oil and water). These personal care products act in much the same way as sebum. A contemporary form of *cold cream*, one of the oldest and most common of the skin moisturizers, consists of an emulsion of about 55% mineral oil, 19% rose water, 13% spermaceti (a wax obtained from the sperm whale), 12% beeswax, and about 1% borax (a mineral that combines sodium borate and water). A typical *vanishing cream*, which fills in wrinkles and appears to make them vanish, incorporates about 70% water, 20% *stearic acid* (partly as its sodium salt; Table 15.1), and 10% *glycerol* (also known as *glycerin*; Sec. 13.8), with traces of potassium hydroxide, preservatives, and perfumes. In the more freely flowing hand and body lotions, water and oils replace some of the waxes of the creams. Since the function of these lotions and creams is to keep the outer layer of skin moist, they are most effective when applied after a bath or shower, while the skin is still damp.

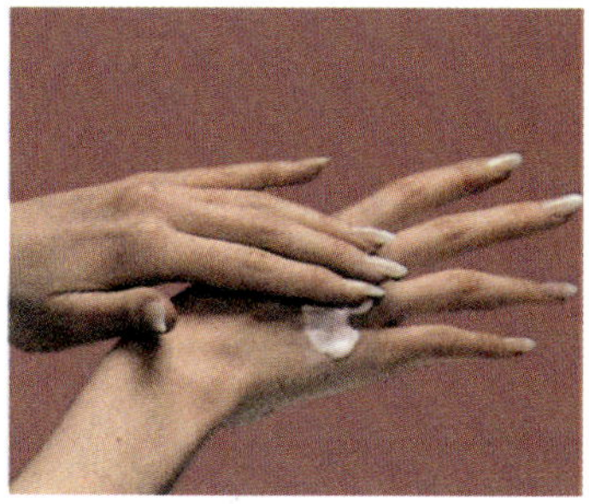

Skin lotions and creams soften the lifeless cells of the stratum corneum.

■ QUESTION

The *dermis*, *epidermis*, and *stratum corneum* are layers of the organ known as the skin. What is their sequence, from top to bottom? What is the composition or structure of each?

22.10 . . . Inoffensive

We control our body's temperature by perspiring (from Latin words meaning "to breathe through") or sweating (an Anglo-Saxon word that means just what it says). When you're warm or tense, your sweat glands begin their work of cooling you off. One set of these glands, the *eccrine* glands (sometimes called the "true" sweat glands), covers most of the skin. They're especially dense on the forehead, face, palms, soles, and armpits, and they secrete a slightly acidic, very dilute solution of inorganic ions (largely sodium, potassium, and chloride), lactic acid ($CH_3—CHOH—CO_2H$), some urea ($H_2N—CO—NH_2$; Sec. 6.1), and a little glucose. The cooling effect of sweating comes from the evaporation of the water from this secretion, which ordinarily has no odor.

Antiperspirants and deodorants help control sweating and associated odors.

Another set, the *apocrine* glands (see Fig. 22.5), releases a different kind of substance, one that can easily become disagreeable. Like the sebaceous glands, these apocrine glands secrete their fluids into the hair follicles. But unlike the sebaceous glands, which occur wherever hair grows, these apocrine glands lie almost exclusively under the arms, in the groin, and in a few other smaller regions of the body. While their secretions produce little or no odors in themselves, bacteria that accumulate in the nearby strands of hair can degrade the contents of the apocrine fluids into foul-smelling products.

Antiperspirants inhibit sweating and keep the body relatively dry. **Deodorants** directly attack odors themselves.

To control the wetness of perspiration and any of its associated odors we have available two personal care products, **antiperspirants** and **deodorants**. The antiperspirants inhibit sweating and keep the body relatively dry. Deodorants, on the other hand, directly attack odors themselves. The most widely used of the antiperspirants are the *aluminum chlorohydrates*, $Al_2(OH)_4Cl_2$ and $Al_2(OH)_5Cl$. These release aluminum cations, which seem to reduce wetness by physically closing the ducts of the eccrine glands for as long as several weeks. They also reduce the odor associated with the apocrine glands, probably by killing the bacteria that decompose the organic portion of the fluid.

Deodorants mask odors with fragrant ingredients that cover the offending aroma and also with antibiotics that eliminate the bacteria. In addition to the aluminum salts, these antibacterial agents include various salts of zinc as well as the broad-spectrum antibiotic *neomycin*. Since the odor itself comes from the action of bacteria on the accumulated residue of perspiration, daily washing alone often solves many of the problems.

■ QUESTION

What's the difference between the action of an antiperspirant and a deodorant?

22.11 Putting a Little Color on Your Skin

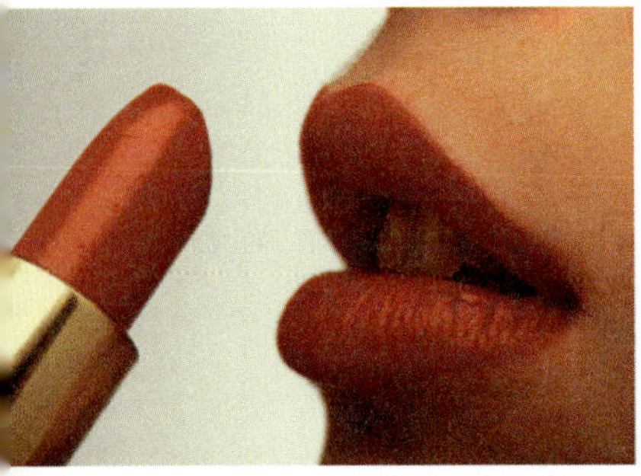

The oils, waxes, polymers, and dyes of lipstick protect, soften and brighten the lips.

With our hair, teeth, and skin clean and in good shape, we'll focus now on cosmetics that add color: lipsticks, eye colorings, nail polish, and face powder. Lipstick provides a softening and protecting film for the lips, as well as colors ranging from soft and unobtrusive to dazzling. About half the weight of a tube of modern lipstick is highly purified castor oil (a mixture of triglycerides obtained from plant seeds). This oil serves to dissolve the dyes and, more importantly, to give the stick the ability to remain a waxy solid in its container yet flow smoothly as it touches the warmer lips. Most of the remainder of the stick is a mixture of oils, waxes, hydrocarbons, esters, *lanolin* (the waxy sebum of sheep), and polymers, all formulated to produce a desirable texture, melting point, ingredient mix, and film flow. In addition, the castor oil and other oils and waxes of lipstick help keep the skin of the lips moist and soft. The remaining ingredients—dyes, perfumes, and preservatives—make up a very small percentage of the weight.

Major dyes of modern lipsticks include D&C Orange No. 5, known chemically as a *dibromofluorescein*, and D&C Red No. 22, a *tetrabromofluorescein* (Fig. 22.14). The D&C here indicates that the Food and Drug Administration has approved these dyes for use in drugs and cosmetics.

Figure 22.14 Lipstick dyes.

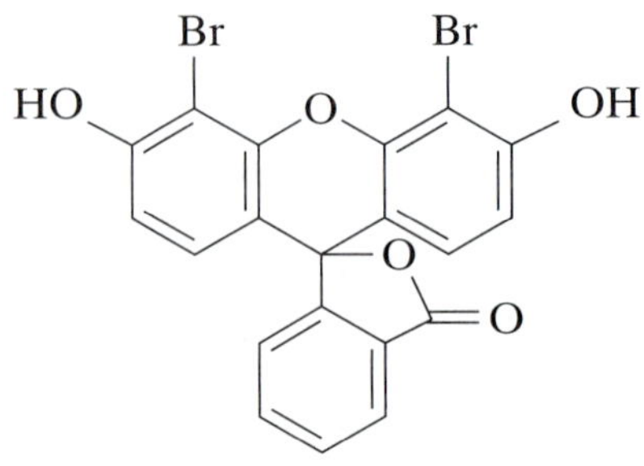

D & C Orange No. 5
4′,5′-dibromofluorescein

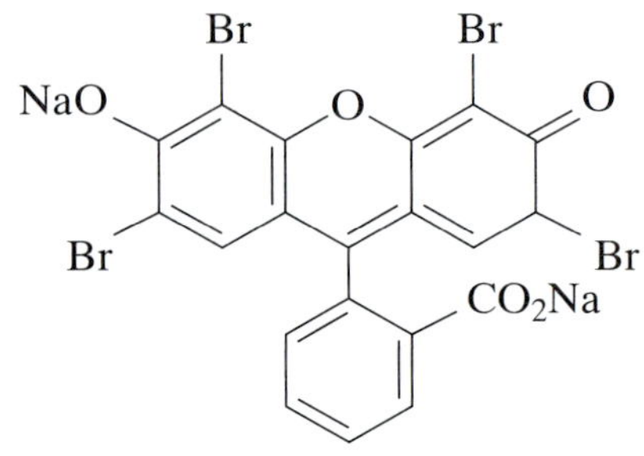

D & C Red No. 22
2′,4′,5′,7′-tetrabromofluorescein

Eye shadows use intensely colored, largely inorganic dyes incorporated into mixtures of beeswax, lanolin, and hydrocarbons. The colorings include *ultramarine blue* (an inorganic polymer containing aluminum, oxygen, silicon, sodium, and sulfur), iron oxides of various shades, *carbon black* (a very finely divided form of carbon resembling powdered charcoal), and *titanium dioxide*, TiO_2. This last ingredient, a white powder, is quite opaque and is useful in hiding or muting the natural color of the skin as the other dyes show through.

Because eyelashes turn pale near their ends, they tend to get lost against the lighter background of the skin and eyes. Mascaras give the lashes a longer look by darkening the ends for more contrast. Despite the term *lengthening mascara*, these cosmetics don't actually lengthen the eyelashes. They just make them seem longer by making their ends more visible against a lighter facial background. To protect the eyes against any potential for allergic reactions to synthetic dyes, the FDA requires that only natural dyes, inorganic pigments, and carbon black be used as mascara colorants. Otherwise, mascaras are prepared in much the same way as other cosmetics.

Nail polish is simply a highly specialized, flexible lacquer that can bend with the nail rather than crack and flake. Its pigments include ultramarine blue, carbon black, organic dyes, and various oxides of iron, chromium, and other metals. These, together with nitrocellulose, plasticizers, and a resin, are all dissolved in a mixture of hydrocarbons and esters, including butyl acetate and ethyl acetate. As the volatile solvents evaporate, they leave the dyes embedded in a polymeric film that grips the nail. Most commercial nail polish removers contain ordinary organic solvents, primarily ethyl acetate (a common solvent for the polish itself) and acetone. Since these remove the natural oils of the nail along with the hardened polish, the removers also contain castor oil, lanolin, or other emollients to keep fingernails and the surrounding skin soft.

Face powders are a bit different from the other cosmetics of this section. Unlike lipsticks, eye shadows, and nail polishes, these powders aren't designed to replace a natural shade with a deep or striking color. Instead, they dull the glossy shine of sebum and perspiration while at the same time adding a pleasant tint, texture, and odor of their own, all without calling attention to their own presence. The composition of a typical cake of face powder appears in Table 22.4.

Table 22.4 Composition of a Caked Face Powder

Ingredient	Weight (%)	Formula	Function
Talc	65	$Mg_3(Si_2O_5)_2(OH)_2$	Forms cosmetic bulk, provides desired texture
Kaolin	10	Al_2SiO_5, hydrated	Absorbs water
Zinc oxide	10	ZnO	Provides hiding power
Magnesium stearate	5	$Mg(O_2C—C_{16}H_{32}—CH_3)_2$	Provides texture
Zinc stearate	5	$Zn(O_2C—C_{16}H_{32}—CH_3)_2$	Provides texture
Mineral oil	2	Hydrocarbons	Emollient
Cetyl alcohol	1	$CH_3—(CH_2)_{14}—CH_2OH$	Binding agent
Lanolin and other additives	2		Softening and coloring agents, perfumes

■ QUESTION

A *lengthening mascara* doesn't actually lengthen eyelashes, but it does make them appear to be longer. How does mascara produce this effect?

22.12 A Few Notes on Perfume

Despite the variety of personal care products we've encountered in this chapter, all share a single characteristic: a pleasant odor or flavor. (Our sense of *taste* is limited to sweet, sour, bitter, and salt. *Flavors* are combinations of these four tastes with the sense of smell. Odors affect our perception of flavor.)

Perfumes are blends of various synthetic chemicals, animal oils, and extracts of fragrant plants, all dissolved as 10 to 25% solutions in alcohol.

Colognes are much more dilute and much less expensive versions of perfumes.

Our perfumes, colognes, and lotions, like the products we use to color our hair and bodies, have their origins in antiquity. The word **perfume** itself comes from the Latin *per* (through) and *fumus* (smoke) and may have applied originally to scents carried by the smoke of incense and odorous plants used in sacred ceremonies. Today's perfumes are the products of a long history of changes in the popularity of different sorts of odors and of the methods of blending them. Modern perfumes are blends of various synthetic chemicals, animal oils, and extracts of fragrant plants, all dissolved as 10% to 25% solutions in alcohol.

Cologne, a shortened form of eau de Cologne (from the French for "water of Cologne"), is a much more dilute and much less expensive version of a perfume, with concentrations of the fragrant oils running about a tenth those used in the perfume. The term itself refers to the city of Cologne, Germany, where an Italian, Giovanni Maria Farina, settled in 1709 and began manufacturing a lotion based on citrus fruit. The product became a very popular toiletry, providing fame to the city and wealth to Farina and his heirs.

For devising the odors of today's cosmetics and toiletries the modern perfumer has available a set of some 4000 aromatic plant and animal substances and another 2000 synthetic organic chemicals. The blended fragrance itself reaches the nose in three phases.

1. The first impact, the *top note*, comes from components that vaporize easily and move to our nose quickly. A typical example is *phenylacetaldehyde*, which brings the odor of hyacinths and lilacs. The top note is the fragrance of the perfume that makes the first impression.
2. The most noticeable odor, the *middle note*, is produced by compounds such as *2-phenylethanol* (also known as *β-phenylethyl alcohol*) with its aroma of roses.
3. The *end note* is a residual, longer lasting scent carried by substances like *civetone*, a cyclic organic compound with a musklike odor. The end note of a perfume is the fragrance that lingers.

Civetone is the odorous component of the secretions of the Ethiopian civet cat and was originally obtained by prodding and scraping the glands of the caged animals. Today a more humane preparation comes from syntheses carried out in chemical laboratories.

Examples of the organic compounds used in blending fragrances appear in Figure 22.15. The choice of specific ingredients depends on the nature of the application, the chemical behavior of the available materials, and safety considerations. A nicely scented substance that oxidizes easily to a foul-smelling product could hardly be used in a face powder, which is exposed to the air for long periods. In addition to these technical factors, though, consumers' perceptions and expectations come into play as well. A deodorant soap, for example, might carry a strong smell suggesting the clean air of a forest, yet the scent of a lipstick must be more subtle and too weak to interfere with the taste of food and drink. These mat-

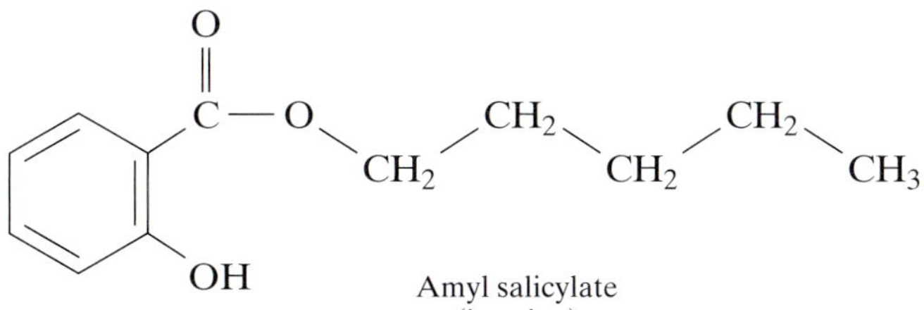

Amyl salicylate
(jasmine)

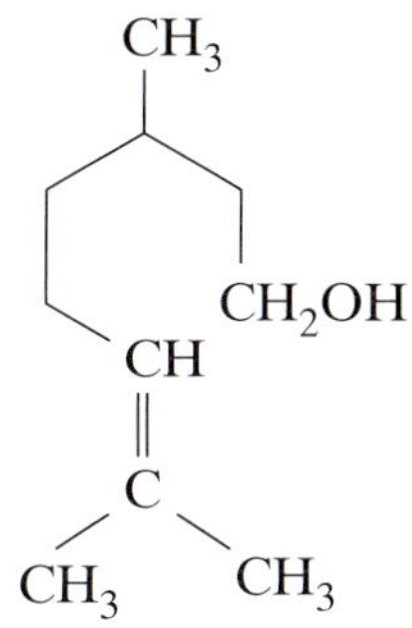

Citronellol
(roses, citrus)

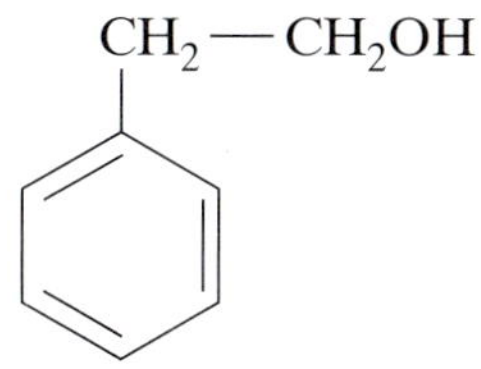

2-Phenylethanol
(β-phenylethyl alcohol)
(roses)

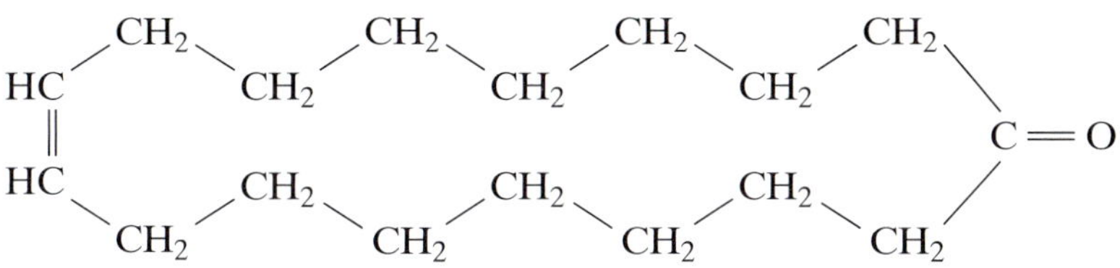

Civetone
(musk)

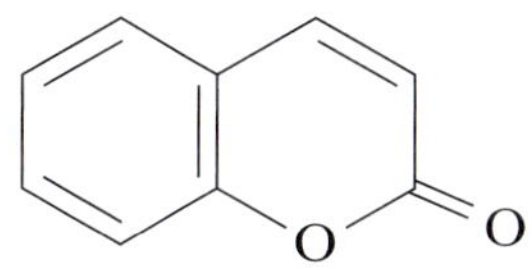

Coumarin
(sweet hay)

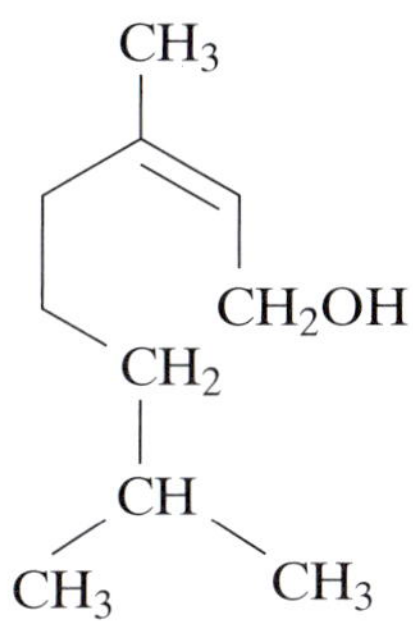

Geraniol
(roses, geraniums)

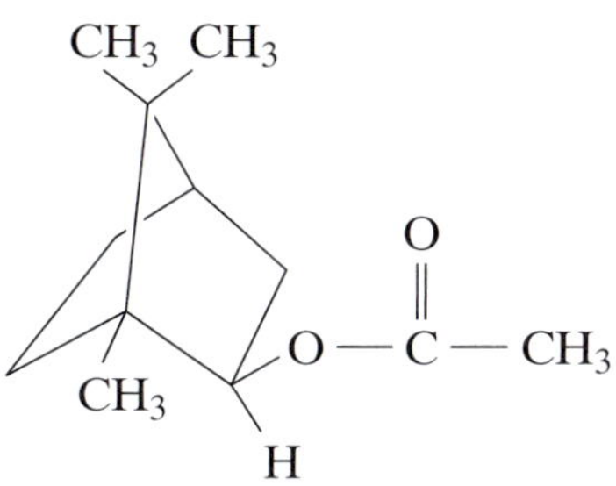

Isobornyl acetate
(pine needles)

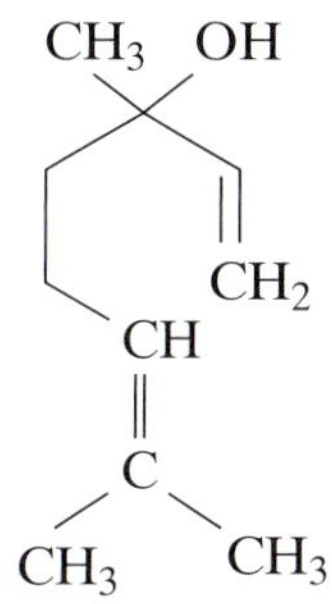

Linaloöl
(lavender)

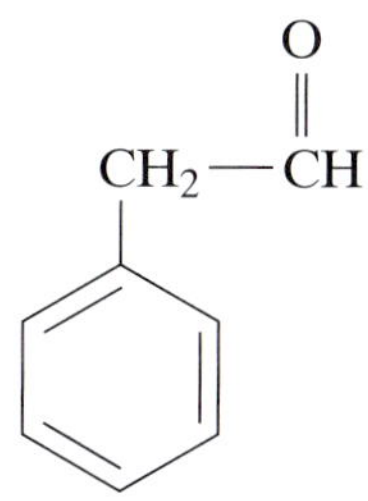

Phenylacetaldehyde
(hyacinths, lilacs)

Figure 22.15 Organic fragrances.

ters of perception, both our perceptions of chemicals and the effects of chemicals on our perceptions, are topics we'll examine more closely in our final chapter.

■ **QUESTION**

What are the functions or actions of the top note, middle note, and end note of a perfume? For each of these categories name a chemical that produces the desired effect.

22.13 Tanning: Waves of Water, Waves of Light

In this section we'll examine still another way you can decorate your body, this time without commercial dyes, detergents, lotions, creams, perfumes, metals, or any other material substance. Despite the absence of chemicals of any kind, this mode of changing the appearance of the human body could pose a hazard far greater than any that might come from the synthetic or natural chemicals of our modern cosmetics. We'll understand why as we examine the science of a suntan.

To understand how our bodies respond to the sun's radiation, why we get a tan by sitting in the sun, and how tanning creams and lotions work, we'll first take a closer look at the sun's rays. The sun's energy reaches the earth as *electromagnetic radiation*, the same sort of phenomenon that carries radio and television signals, radar, microwaves, X-rays, the ultraviolet rays of Section 19.5 and Chapter 14, and the very light and colors that stimulate the retinas of our eyes. The simplest way to discuss this form of energy follows from its resemblance to the waves of water. Both water and electromagnetic radiation move along at a measurable speed, with alternating crests and troughs. As with the waves of water, we can characterize electromagnetic radiation through its **wavelength**, which is simply the distance from one crest to another. What's more, since all forms of electromagnetic radiation travel at a fixed velocity (almost exactly 3.0×10^{10} cm/sec as long as they're moving through a vacuum) we can describe them in terms of their **frequency**, which is the number of waves that pass a fixed point in a specific period (Fig. 22.16).

The **wavelength** of radiation is the distance from one crest to another. The **frequency** of radiation is the number of waves that pass a fixed point in a specific period.

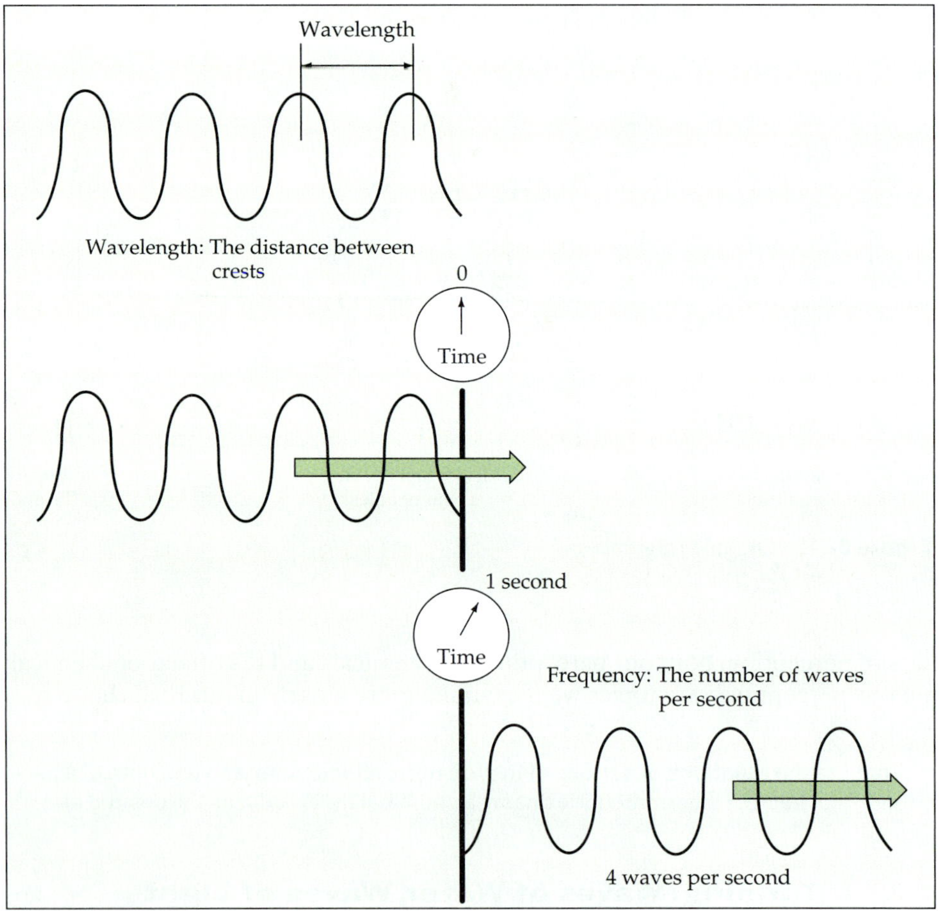

$$\frac{\text{Wave speed}}{\text{Wavelength}} = \text{frequency}$$

Figure 22.16 Wavelength and frequency.

These two characteristics of any wave motion, wavelength and frequency, are mutually reciprocal. That is, the larger (or smaller) the wavelength, the smaller (or larger) the frequency. The shorter the distance between two wave crests, for example, the more waves will pass a fixed point over any particular period, as long as they're moving along at a uniform rate.

At this point radiation takes on its own set of characteristics and our analogy to water loses its value. The energy of electromagnetic radiation, for example, depends *directly* and *only* on its frequency. The greater its frequency (or the shorter its wavelength), the higher its energy. For that matter, frequency is the only characteristic that differentiates one form of electromagnetic radiation from another. X-rays, for example, travel with a shorter wavelength and a higher frequency (and therefore a higher energy as well) than radio or television waves. It's this difference alone that makes them behave as they do.

With one *nanometer* (*nm*) equal to 10^{-9} meter (m), visible light comes to us in a range of wavelengths from about 400 to 700 nm. Red light lies at the longer end of the spectrum, while invisible, heat-bearing infrared radiation lies beyond, at wavelengths a bit longer than 700 nm. At the shorter end of the spectrum visible light carries a blue-violet color. Still shorter wavelengths, running from about 290 to 400 nm, form the portion of the sun's ultraviolet (UV) radiation that penetrates the ozone layer and reaches the earth's surface and our exposed skin (Fig. 22.17; also see Sec. 14.8).

A suntan is simply the visible evidence of the body's attempt to protect itself from the harm this ultraviolet radiation might do. Any exposed skin is bathed in relatively high energy, potentially damaging ultraviolet radiation. In fact, the very absorption of this radiation stimulates the body to take protective measures. As one means of shielding itself, the skin begins producing the dark pigment melanin (Sec. 22.6), which screens out part of the radiation and helps minimize the damage. A tan, then, is the visible evidence of the generation of protective melanin in response to potential harm.

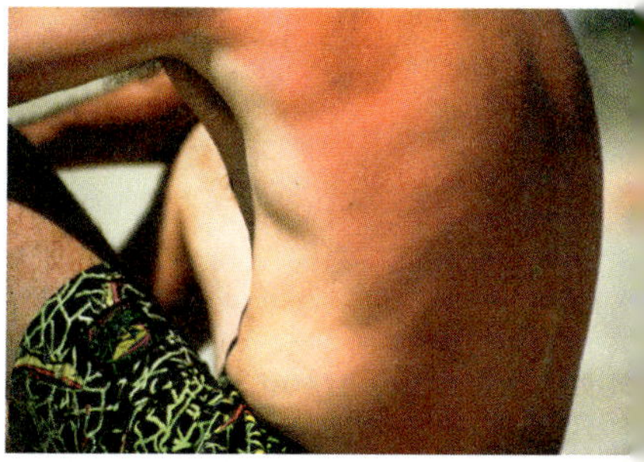

Beginning to burn from too much exposure to the sun's ultraviolet radiation.

The potential for damage comes from two adjoining segments of the sun's UV radiation, its UV-A and UV-B regions, which affect the skin in slightly different ways. UV-B radiation, the more powerful of the two, occupies the shorter wavelength region, from 290 to 320 nm. It does part of its damage quickly, with the remainder coming more slowly, over a longer term. The acute injury lies in the tissue destruction of sunburns. The results—redness, blistered and peeling skin, and pain—are the same as for burns produced by ionizing radiation (Sec. 5.7) and by

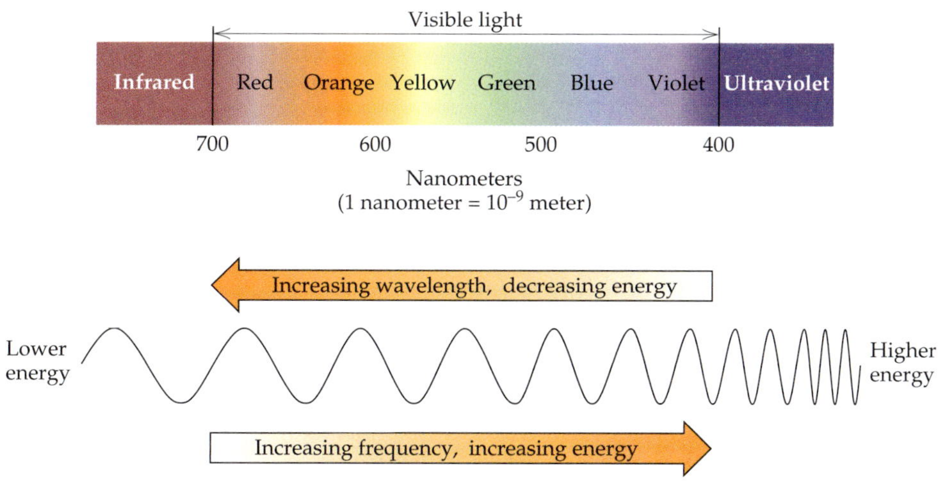

Figure 22.17 Wavelength, frequency, and energy.

intense heat. Repeated exposure to the UV-B region over a longer period produces skin cancers, especially among fairhaired, light-skinned people. Even very brief exposure can begin the formation of wrinkles and sag through the destruction of collagen, one of the proteins responsible for the firmness of skin and similar tissue (Sec. 18.1). One beneficial effect of the absorption of small amounts of UV-B radiation is the generation of vitamin D (Sec. 19.5).

The less energetic zone, the UV-A of longer wavelengths (320 to 400 nm), isn't as effective at tissue destruction. Instead of burning, it tends to produce a slow tan. It's not quite innocuous, though, since it adds to UV-B's ability to do its damage.

■ QUESTION

What single characteristic of electromagnetic radiation directly determines or defines its energy?

22.14 Sunscreen Chemicals

Commercial suntan and sunscreen products act either by blocking the UV-B selectively, letting the UV-A produce its slow tan, or by blocking out both regions, shielding us from the entire solar UV spectrum. The most effective shields are creams containing opaque inorganic oxides, especially zinc oxide and titanium dioxide. These scatter all the radiation, letting none through to the skin. They are usually applied to the nose and the tops of the ears, areas that receive the most direct exposure.

Among the more widely used selective agents are *p-aminobenzoic acid* (PABA or ABA) and its many structural modifications. PABA does a good job of absorbing UV-B radiation while letting UV-A radiation pass through to produce a tan. For this reason lotions containing PABA and similar compounds are considered tanning lotions rather than sunscreens. Other UV-absorbing compounds, such as *benzophenone*, *oxybenzone*, and *dioxybenzone*, absorb throughout the UV spectrum and, sometimes in combination with PABA, provide a wider range of protection. The structures of these absorbers appear in Figure 22.18.

The **sunscreen protection factor (SPF)** of a suntan or sun screen lotion reveals the fraction of ultraviolet radiation it allows to pass through to the skin.

Whatever specific ingredients it may contain, the lotion's **sunscreen protection factor (SPF)**, which appears on its container, serves as a measure of its protecting power. A product with an SPF of 6, for example, reduces the amount of

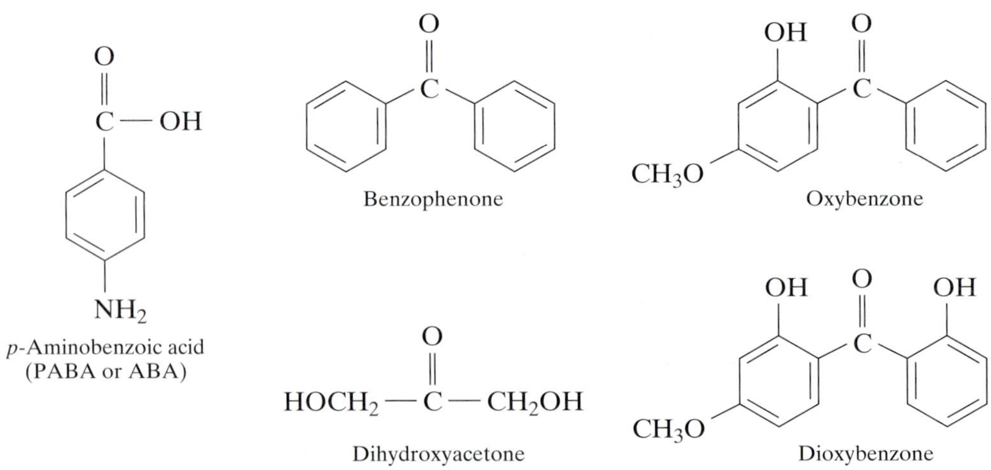

Figure 22.18 Tanning compounds.

UV radiation reaching the skin by a factor of 6. With it you would receive only a sixth of the radiation reaching you without any protection at all. This means that after 6 hours in the sun, protected by a sunscreen lotion with SPF 6, you would have received the same amount of UV radiation as someone who spent only 1 hour in the sun with no UV protection at all. An SPF of 15 or higher effectively blocks out all the UV radiation at sea level in most regions of the United States. Although SPFs much higher than 15 are available, it's doubtful whether they provide any additional, significant protection to the average person at sea level.

To minimize the risk from UV radiation and enhance its benefits, we have a range of degrees of chemical protection available to us, marked by the sunscreen protection factors of lotions and creams. The best way to use them depends, of course, on your own genetic structure. Many authorities recommend that a person with a particularly light skin start slowly, well protected by an SPF of 10 or 15. Only after developing a light, protective tan (through melanin production) is it safe to drop to lower SPFs for a deeper shade of bronze. Because the best protection comes from sunscreen chemicals that have been absorbed into the outer layer of skin, it's best to apply the lotion about half an hour before going out into the sun. As a rough guide, it takes about one ounce of lotion to cover the body of an adult.

It's also possible to develop what appears to be a tan without any exposure to the sun at all. Some chemicals, such as *dihydroxyacetone* (Fig. 22.18), react with the amino acids of the skin to produce a brown pigment. Dihydroxyacetone and other compounds that produce the same effect are components of products sold to produce a tanlike color without exposure to the sun, a "sunless tan." While dihydroxyacetone alone isn't an effective sunscreen, its combination with the lawsone of henna (Sec. 22.6) forms a sunscreen the Food and Drug Administration considers to be safe and effective.

■ QUESTION

What's the difference between a sunscreen and a tanning lotion? Give the full name of a chemical considered to be a tanning lotion rather than a sunscreen. Name a chemical that's a sunscreen.

PERSPECTIVE

Making Choices: Health Versus Beauty

The effects of the sun's ultraviolet radiation on our skin provide us with additional examples of themes we have developed in our study of chemistry:

- The substances and phenomena of our everyday world aren't intrinsically beneficial or harmful, but become so only in the ways we use them.
- In our everyday activities we are continuously faced with choices, many of which are easier to make if they are backed by a knowledge of the composition, properties, and behavior of the substances and phenomena we meet.

In a specific illustration of the first theme, sunlight striking the skin can generate the vitamin D that promotes the formation of healthy bones and teeth. On the other hand, excessive exposure to the intense UV radiation coming from the sun can cause damage to the skin ranging from mild burns to cancer. Sunlight, like so many other substances and effects, provides benefit or harm in proportion to our exposure and use.

As for the second theme, in many regions of the world a well-tanned skin is considered attractive and a mark of someone who has the leisure to lie for many hours in the sun. Desiring this cosmetic effect, yet knowing that UV radiation can damage the skin, many of us choose to expose ourselves to the sun's radiation for long periods, relying on the protection afforded by the chemicals of commercial suntan and sunscreen lotions. An understanding of the effects of sunlight on our skin and how chemicals can help moderate these effects can help us in our choices. Like so many other activities, choosing to lie in the sun for a tan involves balancing a risk, skin damage, against a benefit, the cosmetic effect of a tan. ■

EXERCISES

For Review

1. Following are a statement containing a number of blanks and a list of words and phrases. The number of words equals the number of blanks within the statement, and all but two of the words fit correctly into these blanks. Fill in the blanks of the statement with those words that do fit, then complete the statement by filling in the two remaining blanks with correct words (not in the list) in place of the two words that don't fit.

Hair care products include shampoos, colorants, and curling or straightening agents. Many of the more popular shampoos contain __________ such as sodium and ammonium __________ as well as other agents to stabilize the shampoo, color it, adjust the __________, and __________ the metal ions of hard water. Among the hair colorants are rinses and temporary dyes, which often contain acidic molecules that adhere to the outside of the hair but are too large to penetrate into the __________ and are washed out easily. Many of the permanent dyes operate through a three-component system consisting of two small organic molecules and an oxidizing agent such as __________. The small organic molecules penetrate easily into the hair's cortex where one of them, the primary intermediate, is oxidized and reacts chemically with the second, which is known as a secondary intermediate or __________. The product is an organic molecule too large to migrate out of the cortex easily. Most __________ lotions consist of solutions of __________, which effectively converts the cystine units that unite adjacent protein molecules of the hair into __________. With this conversion accomplished, individual protein strands are separated from each other and the hair can be straightened or curled. The neutralizing agent regenerates the __________ connections, which help fix the new shape.

chelate
cortex
coupler
cysteine
cystine
hydrogen peroxide
keratin
lauryl sulfonates
sulfur
thioglycolic acid
waving

2. Following are a statement containing a number of blanks and a list of words and phrases. The number of words equals the number of blanks within the statement, and all but two of the words fit correctly into these blanks. Fill in the blanks of the statement with those words that do fit, then complete the statement by filling in the two remaining blanks with correct words (not in the list) in place of the two words that don't fit.

The active ingredient of a toothpaste is its __________ rather than the foaming __________ since the prevention of dental cavities, or __________, requires the grinding away of plaque, a thin film of __________ that adheres to the surface of teeth. The caries form as bacteria, which thrive in the plaque and convert it into __________ that erode the hard surface of the teeth. This hard surface consists of the mineral __________, which contains the element __________ along with __________ groups and OH groups.

abrasive
calcium
caries
glycerol
hydroxyapatite
phosphate
sodium fluoride
surfactant

3. Following are a statement containing a number of blanks and a list of words and phrases. The number of words equals the number of blanks within the statement, and all but two of the words fit correctly into these blanks. Fill in the blanks of the statement with those words that do fit, then complete the statement by filling in the two remaining blanks with correct words (not in the list) in place of the two words that don't fit.

Skin care products include lotions and creams, which are __________ of water and oily substances that soften the __________ of the skin in much the same way as the __________ secreted by glands of the skin itself. Other skin care products, the deodorants and antiperspirants, keep the skin free of perspiration and pleasant smelling. While much of the fluid of perspiration comes from secretions of the __________, or the "true" sweat glands, offensive odors are the products of the action of skin bacteria on the secretions of the ________ glands. The effective ingredients of antiperspirants are __________, which physically close the ducts of the eccrine glands, thereby eliminating their secretions. While frequent bathing removes both the residues of perspiration and the bacteria that convert them into offensive substances, deodorants also help by providing fragrances that mask undesirable odors and antibacterial agents such as __________ salts and ________, which help prevent odors by eliminating the bacteria that cause them.

apocrine
eccrine
emulsions
follicle
keratin
sebum
stratum corneum
zinc

4. Following are a statement containing a number of blanks and a list of words and phrases. The number of words equals the number of blanks within the statement, and all but two of the words fit correctly into these blanks. Fill in the blanks of the statement with those words that do fit, then complete the statement by filling in the two remaining blanks with correct words (not in the list) in place of the two words that don't fit.

Lipsticks, eye colorings, nail polish, and face powder add color to the face and nails. A highly purified form of __________, which is a plant __________, forms the major ingredient of most modern lipsticks. In addition to this oil, lipstick also contains other lubricants, including __________, the waxy __________ secreted by sheep. Among the dyes added to lipstick are the red __________ and the orange __________. Coloring agents used in eye shadows include the inorganic polymer __________, blackening agents such as __________, and whiteners such as __________. Nail polish is simply a mixture of a __________, such as __________, dyes, plasticizers, and other substances, all dissolved in organic solvents, including __________, and esters such as __________. As the solvents evaporate they leave the ingredients embedded in the polymer, which adheres to the nail.

carbon black	polymer
castor oil	sebum
dibromofluorescein	tetrabromofluorescein
ethyl acetate	titanium dioxide, TiO_2
hydrocarbons	triglyceride
keratin	ultramarine blue
lawsone	

5. Following are a statement containing a number of blanks and a list of words and phrases. The number of words equals the number of blanks within the statement, and all but two of the words fit correctly into these blanks. Fill in the blanks of the statement with those words that do fit, then complete the statement by filling in the two remaining blanks with correct words (not in the list) in place of the two words that don't fit.

A suntan is the body's defense against the damaging effects of __________. The sun's ultraviolet radiation consists of two segments, the shorter wavelength and more powerful __________ region and the longer wavelength and less powerful __________ region. While the UV-B radiation is responsible for the tissue destruction of sunburn, it also produces the beneficial effect of generating __________ within the skin. The skin's absorption of UV-A, on the other hand, produces a slow tan by stimulating production of the dark pigment __________, which tends to block the absorption of the entire range of ultraviolet radiation. This UV-A can also damage the skin's connective tissue, leading to aging of the skin. Both UV-A and UV-B can generate __________ on prolonged exposure. Among the UV-absorbing agents of commercial suntan and sunscreen products are __________, which selectively blocks absorption of the UV-B, allowing the UV-A to produce a tan, and __________, which absorbs throughout the ultraviolet spectrum, thereby providing a wider range of protection against the harmful radiation. Some commercial products promise a "sunless tan" through the action of __________, which produces a brown pigment through its reaction with the __________ of the skin.

amino acids	skin cancer
benzophenone	sunscreen protection factor
dihydroxyacetone	UV-A
melanin	UV-B
p-aminobenzoic acid, or PABA	zinc oxide

6. Define, explain, or identify each of the following:

civetone	hydrogen bond
cologne	keratin
dermis	manometer
epidermis	sunscreen protection factor
frequency (of radiation)	wavelength

7. Name each of the following compounds and describe its function in personal care products:

a. H_2O_2

b. ZnO

c. $HS{-}CH_2{-}CO_2H$

d. $Al_2(OH)_4Cl_2$

e. Na_2PO_3F

f. $Pb(O_2C{-}CH_3)_2$

g. TiO_2

h. NaF

i. 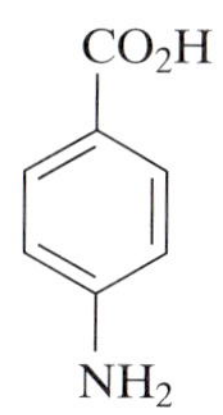

j. 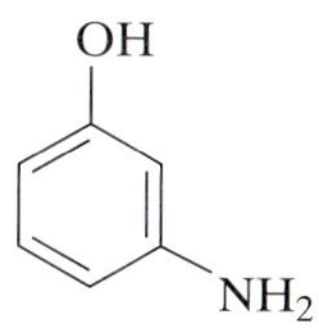

k. 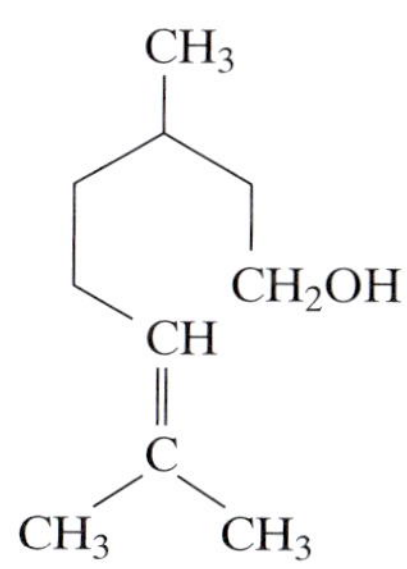

l. 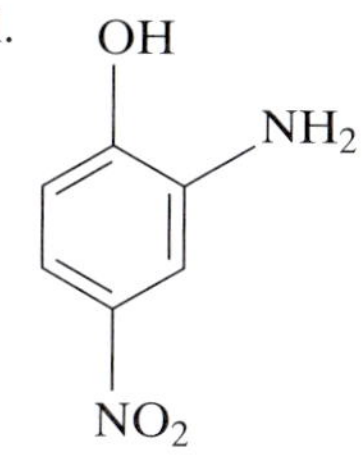

m. 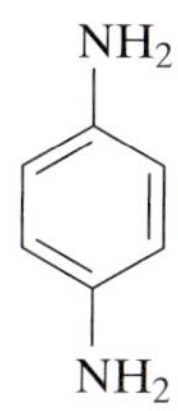

n. 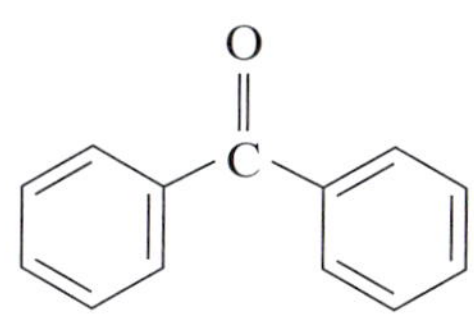

o. 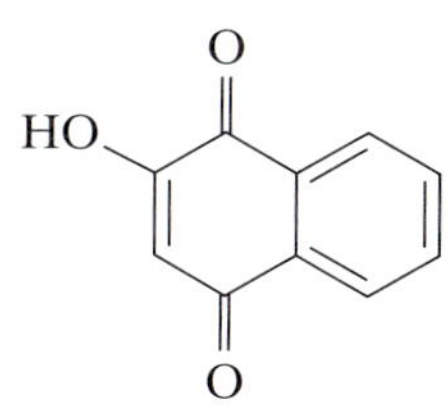

p. 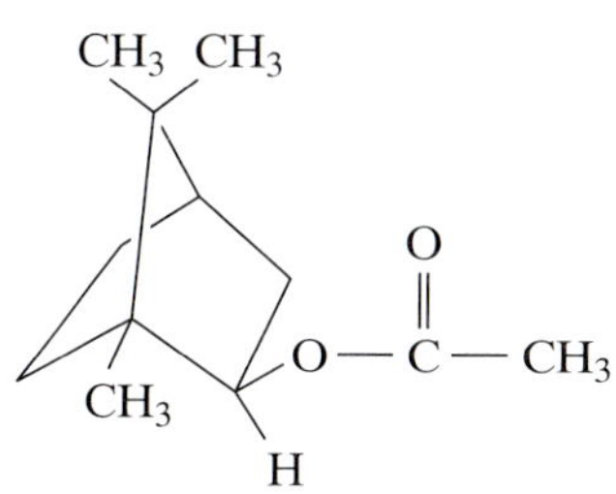

8. Name a hair dye that was used by the ancient Egyptians and is still in use throughout the world today.

9. On a weight basis, what are the most expensive ingredients of cosmetic and personal care products?

10. Describe the chemistry, applications, and effects of each of the following as products related to tans or exposure to the sun: (a) zinc oxide; (b) *p*-aminobenzoic acid; (c) a combination of *p*-aminobenzoic acid and dioxybenzone; (d) dihydroxyacetone.

11. Why is either ammonium lauryl sulfate or triethanolammonium lauryl sulfate better suited than sodium lauryl sulfate for use as a detergent in a hair shampoo?

12. What is the difference between the cortex and the cuticle of a strand of hair?

13. Name three chemical forces that give the hair strength and explain how each works.

14. What kinds of chemical bonds are broken and then re-formed in (a) a wet wave; (b) a permanent wave?

15. What does hydrogen peroxide react with as it lightens the color of hair? What does lead acetate react with as it darkens hair and what is the product of the reaction?

16. What is the difference between UV-A and UV-B radiation?

17. What kind of skin damage results from overexposure to (a) UV-A radiation? (b) UV-B radiation?

18. In what way(s) are light waves and water waves similar?

19. In what way(s) are light waves and water waves different?

20. How do the aluminum chlorohydrates act to reduce perspiration? How do antibiotics act to reduce body odors?

21. What function does titanium dioxide, TiO_2, serve in a cosmetic?

22. What's the difference between a perfume and a cologne?

23. Which carries greater energy: (a) X-rays or television transmissions; (b) red or blue light; (c) ultraviolet or infrared radiation?

24. How do (a) zinc oxide, (b) *p*-aminobenzoic acid, and (c) benzophenone differ in their effects on ultraviolet radiation?

25. Describe the function of the following chemicals:

a. thioglycolic acid in home permanent kits
b. hydrogen peroxide in home permanent kits
c. hydrogen peroxide in permanent hair dyes
d. triethanolammonium sulfate in hair shampoo
e. titanium dioxide in a dentifrice
f. zinc oxide in a face powder
g. castor oil or lanolin in nail polish remover

A Little Arithmetic and Other Quantitative Puzzles

26. Suppose that on your first day of vacation you planned to spend no more than 20 minutes in the sun, without any sunscreen protection. Now you decide you want to spend much more time outside in the sun, on your first day, yet you still want no more than the equivalent of 20 minutes of UV radiation. How much time could you spend in the sun, still getting only the equivalent of 20 minutes of UV radiation, if you use a sunscreen lotion with an SPF of (a) 6? (b) 10? (c) 15? (d) 30?

27. Light with a wavelength of 400 nm lies near the border between ultraviolet light and visible, violet light. If we increase the wavelength by 50%, to 600 nm: (a) Do we increase or decrease the energy of the light? (b) What color light do we obtain?

Think, Speculate, Reflect, and Ponder

28. We have seen repeatedly that a particular chemical can have more than one use and that sometimes these uses are very different from each other. Name two uses for mineral oil that involve the human body. (See Sec. 6.4 for one of them.) With your answer in mind, if your skin is dry and cracked and you have no lotions or creams at hand, what other consumer product could you use in an emergency?

29. Brushing with ordinary baking soda ($NaHCO_3$) has been recommended as an alternative if toothpaste is unavailable. While baking soda doesn't contain either an abrasive or a detergent (or a pleasant flavor or fragrance), it does provide one benefit as a dentifrice. What is the benefit?

30. In what ways are the characteristics of a hair shampoo and a toothpaste similar? In what ways are they different?

31. Even though an SPF of 15 is generally regarded as sufficient at sea level, a person living at a high altitude or on a mountain-climbing expedition might want to use a much higher SPF. Why?

32. Describe the role of dental plaque in tooth decay.

33. Describe the chemical differences between a wet wave of the hair and a permanent wave.

34. Hydrogen peroxide takes part in chemical reactions that make the hair lighter in color and also in reactions that make hair darker. Explain how the hydrogen peroxide acts in each of these systems.

35. Describe the chemical differences among the dyes of hair rinses, temporary dyes, and permanent dyes. Explain how these differences affect the ease with which the dye is washed out of the hair.

36. Explain why the pH of the hair is important to its appearance.

37. The sulfur–sulfur bonds (—S—S—) of a single cystine molecule can be broken by the addition of a molecule of hydrogen to produce a new amino acid, containing an —SH group. What is the name of the amino acid produced by addition of hydrogen to cystine? (See Table 17.1.)

38. Considering the nature of the transformation of cystine to cysteine that is produced by thioglycolic acid in the chemistry of the permanent wave, suggest a possible fate of the thioglycolic acid in this process.

39. Using techniques of nomenclature described in this chapter, name *sarcosine* as a modified amino acid.

Web Questions

Use the Web to answer or accomplish each of the following. Report your Web source for each response.

1. What is *botox*? What is it used for? How does it work? Would you classify botox as a cosmetic?

2. What is Fluoristan? In what consumer product is it found? What is its chemical formula or structure? How does it work?

3. Describe two plant extracts other than henna that have been used as hair dyes. Identify their sources and, if known, identify the specific plant chemicals that produce the colors.

4. Identify three aluminum compounds, in addition to the aluminum chlorhydrates, that are useful ingredients in antiperspirants. Explain how aluminum compounds act to stop or reduce perspiration.

5. What are *lake* pigments and dyes? What function do they serve in cosmetics? Identify two lake pigments or dyes used in cosmetics.

6. Explain what *depilatories* are and how chemical depilatories function. Would you classify depilatories as cosmetics? Explain.

7. Summarize the ability of the Food and Drug Administration (FDA) to regulate the manufacture, sale and use of cosmetics. Describe some of the powers the FDA is permitted to exercise. Identify some areas in which the FDA is not permitted to act or to exercise its authority.

Medicines and Drugs

CHEMICALS AND THE MIND

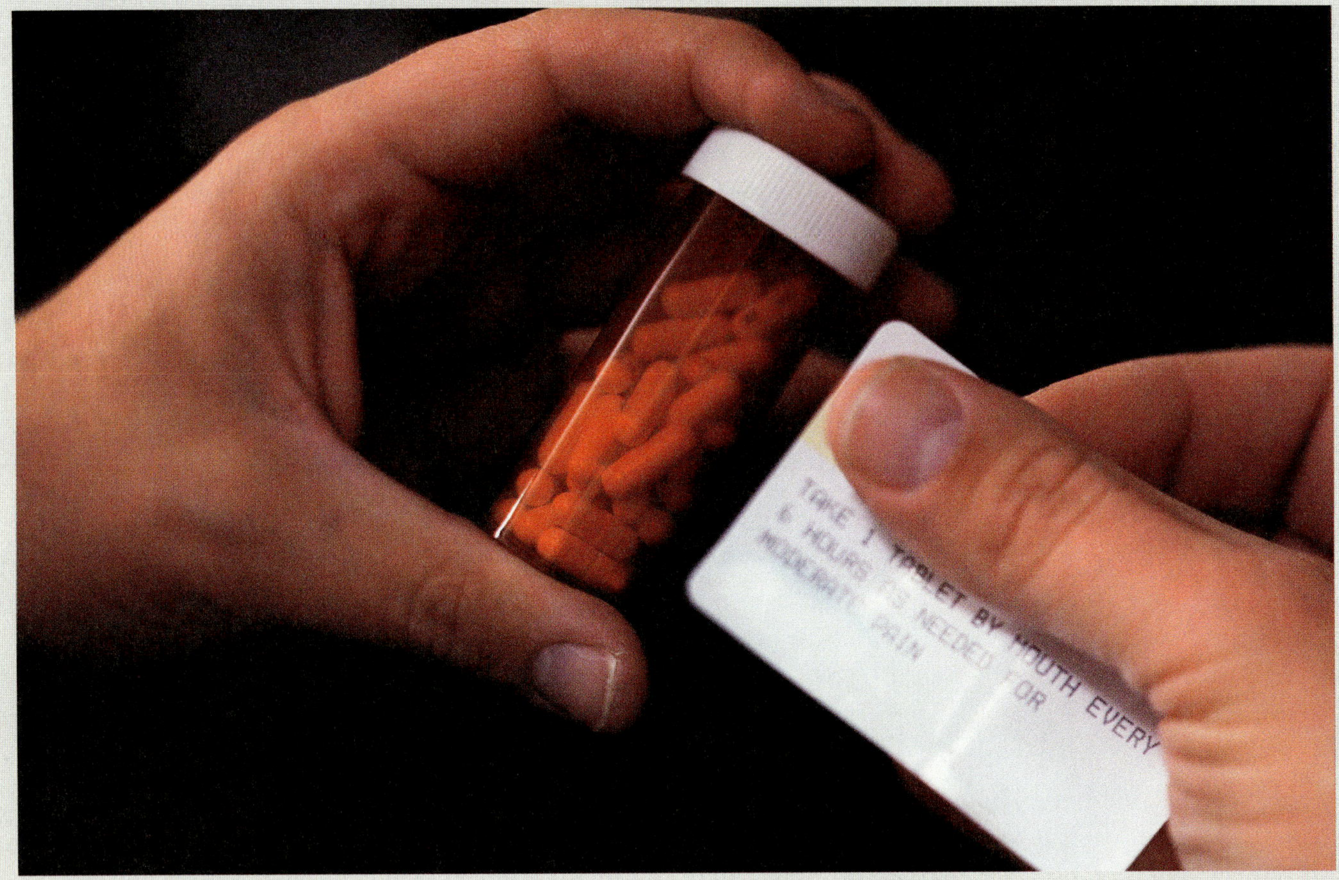

Chemicals can cure illness, reduce pain, inflammation and swelling, and alter states of perception.

Perception, Reality, and Chemicals

The world as we see it isn't necessarily the world as it exists. In Figure 23.1 you see two horizontal lines, each ending in diagonal lines slanting one way or another. Most people see the horizontal line on the right as longer than the one on the left. Measure them, though, and you'll find that the two horizontal lines are of equal length. Then there's the stovepipe hat of Figure 23.2. Most people see the height of the hat as greater than the width of the brim. Again, measure the height and the width and you'll find that they are the same.

Misperceptions like these can be corrected easily by tests of the physical world, such as we use in applications of the scientific method. Here we can test reality with a ruler and learn whether our perceptions about the length of lines are true. The illustration at the top of this page presents a trickier case, one that can cause problems. When you look at it, do you see an oddly shaped vase or the silhouettes of two people facing each other? Some people see one, some see the other. Maybe you can see both. There's no objective test of reality here. One person can call it a vase, another can call it a pair of silhouettes, and both can be right.

As the illusion above demonstrates, two different viewers can perceive the same reality in different ways. Each might agree that someone else could interpret things differently. Then again, they might not. Differences in perceptions of this sort can be troublesome. It can be hard for people to work out their differences over issues they disagree on when they see the same thing differently and neither can view it from the other's perspective. What we call "reality" sometimes depends on just how we view the world we all live in, on what our perceptions are. Living in a society like ours can present problems when the perceptions of some members of our society conflict with the perceptions of others.

Our perceptions of the world and their effects on our actions are important to a discussion of chemistry. Chemicals, including those we classify as medicines and drugs, can affect our views of reality and our actions. The effects can be beneficial, as when we use chemicals to ease pain so that we can carry on with our work and our lives, or they can carry a potential for disaster when they remove us from a normal contact with reality.

In this chapter we'll examine some of the chemicals that affect the mind and the ways they change our perceptions. We'll also examine a phenomenon called the *placebo effect*, in which our own belief about how a specific chemical might affect us can actually influence the way our bodies respond to that same chemical. As a result of the placebo effect, chemicals as innocuous as table sugar sometimes produce or relieve symptoms dramatically, but only because we are led to think they will.

Figure 23.1 Which line is longer?

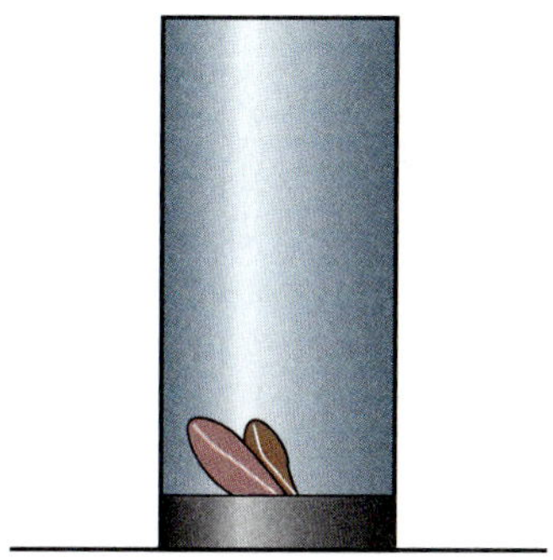

Figure 23.2 Is the hat taller than the brim is wide, or is it the other way around?

We'll also examine a technique called the *double-blind* experiment that allows us to learn whether the action of any particular medication results from a direct chemical influence of that medication, regardless of our expectations, or from the action of those very same expectations through the placebo effect, through nothing more than our faith in the substance's effectiveness. We'll start our examination of the connection between chemicals and perceptions with agents of mercy, the *analgesics* or painkillers.

23.1 Aspirin

Today's leading commercial pain reliever is common *aspirin*. Since its initial synthesis in 1853 and its acceptance into medical practice at the beginning of the 20th century, aspirin has become the most widely used of all drugs for the treatment of illness or injury. More aspirin has been taken over the years by more people throughout the world than any other single medication of any kind.

An **analgesic** reduces or eliminates pain. An **antipyretic** lowers or eliminates fever. An **anti-inflammatory agent** reduces or eliminates inflammation.

Aspirin is the common name or trade name of **acetylsalicylic acid**, an analgesic, antipyretic, and anti-inflammatory medication.

In this examination of the extraordinary chemistry of our ordinary chemicals, aspirin must be recognized as one of the most extraordinary of all, with the power to act as an **analgesic** (to relieve pain), as an **antipyretic** (to lower fever), and as an **anti-inflammatory agent** (to reduce inflammation). It's effective in the treatment of rheumatoid arthritis and in preventing some types of strokes and heart attacks, those resulting from the accumulation of platelets (Sec. 23.4) in the blood vessels. It also shows promise in decreasing the risk of some forms of colon cancer. Yet it's also one of the most common of all medications, inexpensive, and available to virtually everyone, almost everywhere.

Aspirin itself, the common or trade name of the chemical **acetylsalicylic acid**, is prepared easily through the reaction of *acetic anhydride* with *salicylic acid* (Fig. 23.3). (Acetic anhydride results from the dehydration of acetic acid through

Figure 23.3 Aspirin from the acetylation of salicylic acid.

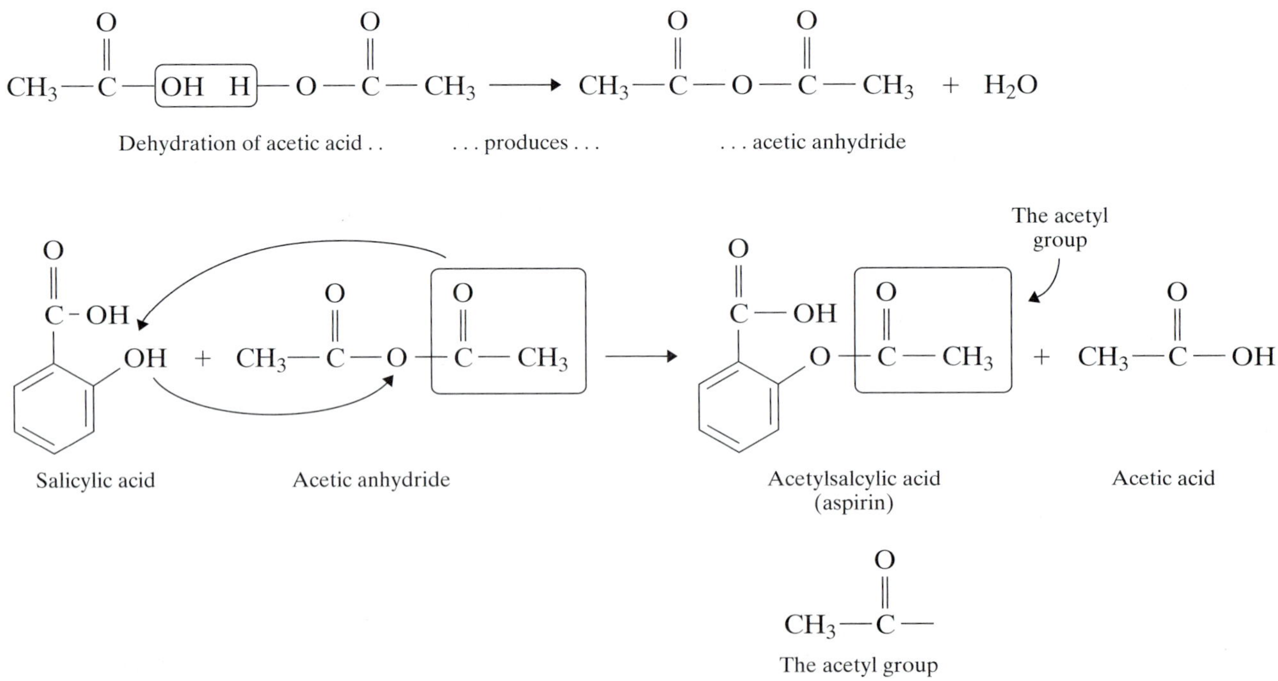

the removal of one water molecule from two molecules of the acid. The product is a good *acetylating agent*, which means that it's useful for adding the *acetyl* group to another molecule, as it does in the synthesis of aspirin.)

■ QUESTION

What function does acetic anhydride serve in the chemical synthesis of aspirin?

23.2 How Aspirin Got Its Name

Although neither salicylic acid (the molecule being acetylated) nor acetylsalicylic acid (the product itself) occurs in nature, several closely related compounds do. Physicians as far back as Hippocrates, the ancient Greek healer known as the Father of Medicine, knew of the curative powers of the bark of the willow tree and closely related plants, especially their antipyretic or fever-reducing properties.

In 1827 willow bark yielded its active agent, *salicin*, to chemists of that era. A few years later this salicin was hydrolyzed to glucose and *salicyl alcohol*, which was in turn oxidized to salicylic acid (Fig. 23.4). The Latin term for willows, *salix*, gives us the name of an entire family of compounds, the *salicylates*, whose molecules resemble those of both the alcohol and the acid. It's also the term for the botanical genus of the tree itself, *Salix*.

At about the same time, salicylic acid was also obtained from the reaction of salicylaldehyde with strong base (Fig. 23.5). This fragrant aldehyde (Sec. 16.4) was itself isolated from meadowsweet flowers, which belong to the genus *Spiraea*. Many years later, with commercial production of acetylated salicylic acid, the term *aspirin* was coined by adding an *a* (for *a*cetylated) to a portion of an older name of the acid, *spir*aeic acid, which was derived from the botanical genus of the meadowsweet.

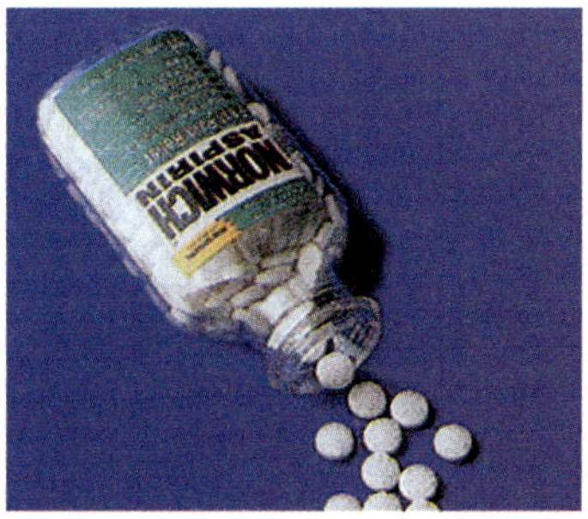

Aspirin is the most widely used of all drugs for the relief of symptoms of illness and injury.

The bark of the willow tree contains salicin, an analgesic that can be converted to salicylic acid and then to aspirin. The Latin name for the tree, salix, gives us both the botanical genus of the tree and the generic term for compounds related to salicin, the salicylates.

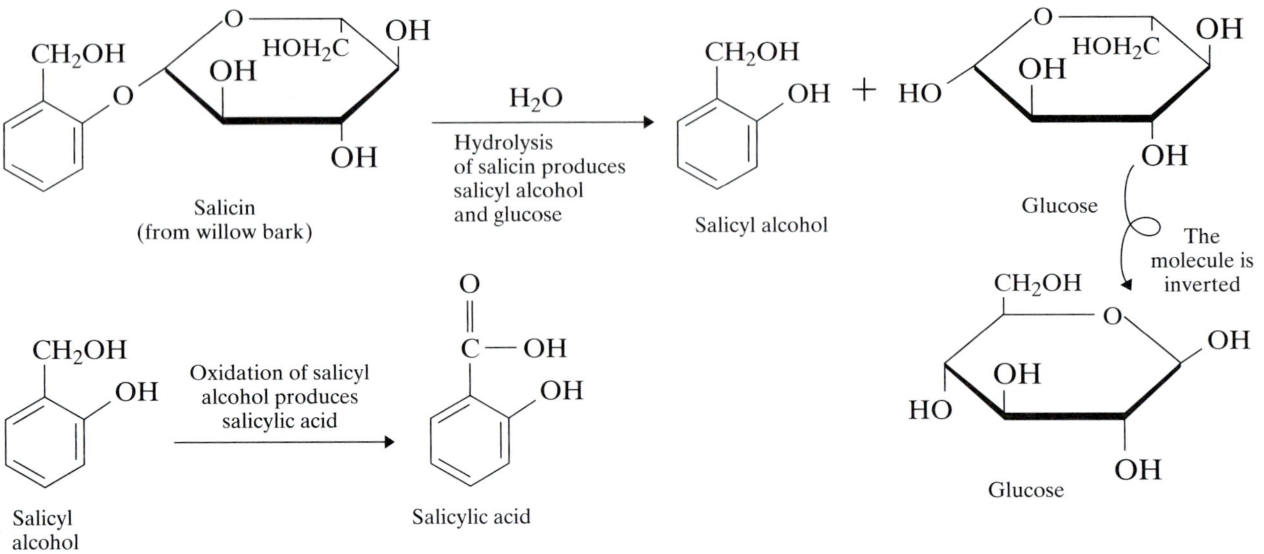

Figure 23.4 Salicylic acid from the bark of the willow tree.

Figure 23.5 Salicylic acid from salicyladehyde.

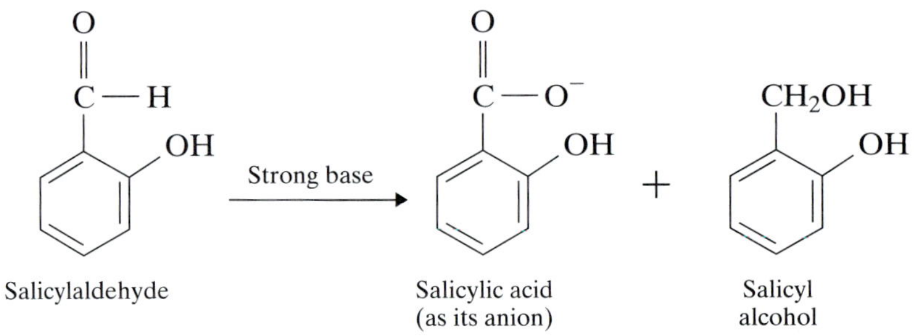

The (unbalanced) reaction of salicylaldehyde with strong base to produce salicylic acid (and salicyl alcohol).

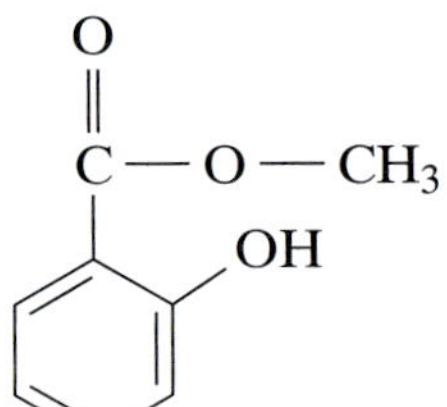

Figure 23.6 Methyl salicylate, oil of wintergreen. This ester of salicylic acid is used in making perfumes; in flavoring candy, chewing gum, and medicines; and in liniments used for the relief of muscular pain.

Another salicylate, *methyl salicylate*, is the fragrant methyl ester of salicylic acid, known more commonly as *oil of wintergreen* (Fig. 23.6). In addition to its use in perfumery and (in very small amounts because of its toxicity) as a flavoring for candy, chewing gum, and medicines, methyl salicylate is also the active ingredient of liniments used for relieving the pain of aching muscles and joints. As it's rubbed into the skin, the ester's irritant action acts as a counterirritant to deeper aches and eases their discomfort. The irritant action of the methyl salicylate also stimulates blood flow into the deeper regions, adding to its analgesic effect.

■ **QUESTION**

In what way is the word *aspirin* related to the meadowsweet plant?

23.3 Salicylates Versus the Prostaglandins

Like many other salicylates, at least those with toxicities low enough to be taken internally in amounts that will do any good, salicylic acid relieves pain, fever, and inflammation. But while both the acid and its sodium salt are effective against all three of these discomforts, the acid itself is far too corrosive to tissues to be taken internally by most people, and its salts are generally unpalatable.

In 1893 Felix Hofmann, a chemist working for the Baeyer firm in Germany, found a way around these problems. To provide relief for his father's rheumatoid arthritis, Hofmann turned from salicin, a weaker but popular analgesic of the time, and from salicylic acid, whose corrosive character produces stomach discomfort, to the acetylated acid that had been prepared 40 years earlier. He started aspirin on its road to success by demonstrating that in its acetylated form the salicylate is easily tolerated and by providing convincing evidence of its potent analgesic effect. Baeyer began selling powdered aspirin in envelopes and capsules in 1899; in 1915 the tablets were introduced. Today's tablets ordinarily contain between 309 and 341 mg of acetylsalicylic acid and an additional *binder*, an inert ingredient that holds the acetylsalicylic acid in the convenient form of a tablet. ("Extra-strength" aspirin contains about 500 mg of acetylsalicylic acid per tablet. The "extra strength" comes simply from the larger amount of the active ingredient in each tablet. The additional strength can be obtained equally well and often at lower cost by taking two of the ordinary tablets.)

As we might expect, aspirin doesn't begin to do its work until it enters the bloodstream. The time it takes for the acetylsalicylic acid bound up in a solid aspirin tablet to enter the blood is governed largely by the rate at which the tablet disintegrates in the stomach. This rate, in turn, depends on pH (Sec. 10.7). The higher the pH, the faster the tablet breaks up and the faster the acetylsalicylic acid gets into the blood. To accelerate the disintegration of the tablet and the absorption of the aspirin itself, some companies produce a **"buffered" aspirin**. (A *buffer* is a combination of an acid or a base and one of its salts. By reacting with both acids and bases, this combination is able to maintain the pH of a solution within a narrow range.)

"Buffered" aspirin is a combination of acetylsalicylic acid and one or more bases. The combination accelerates the disintegration of the tablet.

Despite the use of this term, these tablets aren't truly buffered. Rather than a combination of acetylsalicylic acid and one of its salts, which would constitute a true buffer, the "buffered" aspirins consist of a combination of aspirin and one or more bases, including magnesium carbonate, $MgCO_3$, magnesium hydroxide, $Mg(OH)_2$, aluminum hydroxide, $Al(OH)_3$, and aluminum glycinate, the aluminum salt of the amino acid glycine (Sec. 17.2). Although the addition of these bases to the tablet does increase rates of disintegration and absorption, rigorous clinical studies have failed to produce any evidence of faster or greater pain relief from "buffered" aspirin than from the "nonbuffered" variety. In clinical tests, a simple aspirin tablet of the ordinary variety works as well as the buffered types.

The beneficial effects of aspirin and other salicylates come from their ability to prevent the formation of *prostaglandins* within the body (Fig. 23.7). These compounds, named for the seminal fluid and prostate tissue in which they were discovered, occur in virtually every tissue and fluid of the body and participate, in one way or another, in almost every bodily function. They play particularly important roles in the sensation of pain and its transmission along the nervous system, in the generation of fevers, and in the swelling of inflammations. By interfering with the formation of the prostaglandins, salicylates relieve each of these conditions.

Figure 23.7 Typical prostaglandins. Important in many bodily functions, they play a major role in the sensation of pain. Aspirin relieves pain, fever, and inflammation by inhibiting the synthesis of prostaglandins.

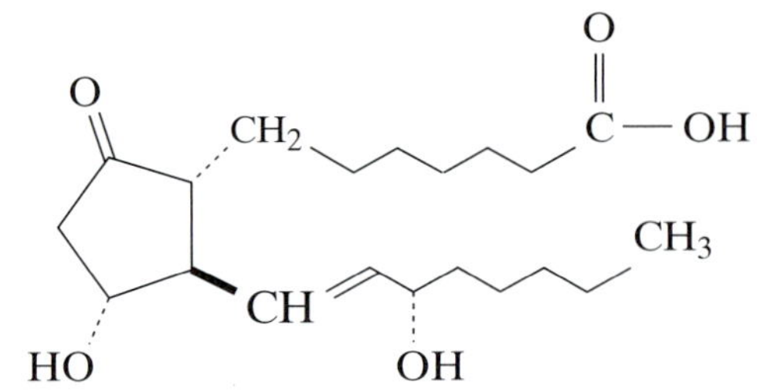

Prostaglandin E_1

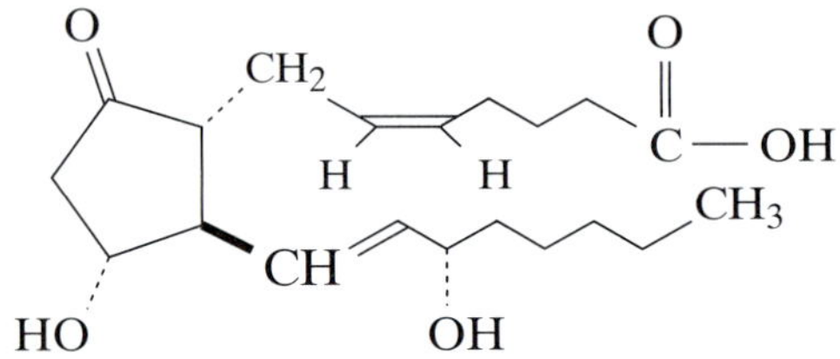

Prostaglandin E_2

(the most abundant and most active of the prostaglandins)

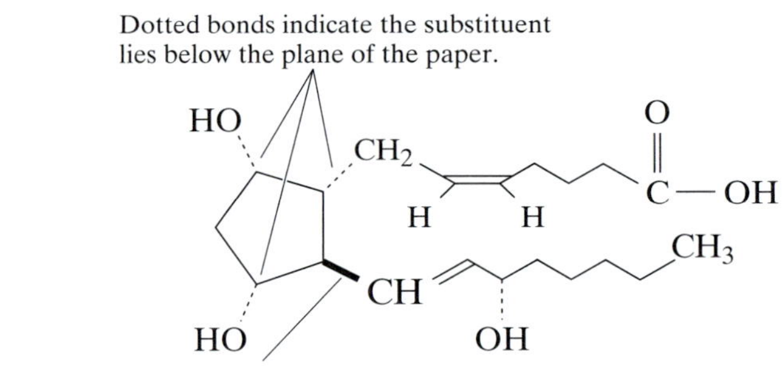

Prostaglandin $F_{2\alpha}$

■ QUESTION

What's the distinction between a true, chemical buffer and the combination of chemicals used in a "buffered" aspirin?

23.4 Aspirin: Risks along with Rewards

While less destructive to body tissues than salicylic acid itself, aspirin can nonetheless produce some nasty side effects, including upset stomach and various allergic reactions. It's estimated that 2% to 10% of all those who take aspirin suffer occasionally from some form of stomach distress or nausea.

Some of the major risks in taking aspirin routinely come from its effect on the blood's ability to clot. Acetylsalicylic acid inhibits the formation of the blood's *platelets*, which are small bodies within the blood serum that initiate clotting. Insufficient concentrations of platelets can lead to gastrointestinal bleeding, susceptibility to bruising, and an increased risk of hemorrhagic stroke, a condition brought about by the rupture of small blood vessels within the brain. To guard against excessive bleeding, surgery patients are often advised not to take aspirin for several days before the scheduled operation.

More serious than these effects is *Reye's syndrome*, a rare and sometimes fatal condition experienced principally by children and adolescents recovering from chicken pox, the flu and other viral infections. For unknown reasons using aspirin and other salicylates to treat these illnesses appears to increase the likelihood of Reye's syndrome, which is accompanied by a swelling of the brain, damage to the liver, and symptoms including confusion, irritability, nausea and lethargy. Aspirin labels in the United States carry warnings to consult a physician before giving the drug to children or adolescents with symptoms of either chicken pox or the flu.

In addition to producing these relatively rare reactions, aspirin is more toxic than we may realize. With an LD_{50} of 1.5 g/kg (orally in mice and rats; Table 20.1), there's a 50% chance that 15 g—about 45 tablets—would be enough to kill a 10-kg (22-lb) child. These, of course, are no more than impersonal statistics that assume humans and rodents are equally susceptible to the toxin. Yet they take on a more human dimension when we recognize that no one is perfectly average and that some of us are more susceptible than others to the lethal effects of aspirin or any other potentially toxic substance. Before enactment of the Poison Prevention Packaging Act of 1970, for example, more than 40 children under the age of five in the United States died each year from accidental aspirin poisoning. With the introduction of childproof caps on aspirin containers, the rate dropped sharply to about 25 per year (Sec. 20.2).

One of aspirin's major benefits—decreasing the risk of heart attacks among certain groups—comes from the same decrease in platelet concentration that produces some of the medication's potential hazards. Because aspirin inhibits the formation of blood clots (by decreasing platelet concentrations), it appears to decrease the incidence of heart attacks resulting from the formation of internal clots that can block the flow of blood to the heart's muscles. Aspirin's protective effect seems to be greatest in people with risk factors including smoking, high serum cholesterol, hypertension, obesity, and family histories of heart attacks, and in those who have already suffered a heart attack. In addition, we've seen in Section 23.1 that aspirin appears to protect as well against some forms of strokes and colon cancer. Once again, particularly with respect to acetylsalicylic acid, we see the need to balance a chemical's risks against its benefits.

■ QUESTION

What are two benefits that can be obtained from the use of aspirin? What are two risks associated with its use?

23.5 Acetaminophen: A Substitute for Aspirin

For those of us who may be allergic to aspirin or who find that it produces stomach disorders, there's an alternative in *acetaminophen*, the active ingredient of Tylenol and Datril. Acetaminophen is the least toxic member of a class of analgesic and antipyretic (but *not* anti-inflammatory) medicines known as the *p-aminophenols* (Fig. 23.8). Compounds of this sort trace back to *acetanilide*, which was used to alleviate fever as early as 1886. Acetanilide proved too toxic for general use, yet it offered promise. Investigations of similar compounds led to trials of *p*-aminophenol (which also proved too toxic) and of compounds related to it.

In the following year, 1887, *phenacetin* (also called *acetophenetidin*) and, in 1893, acetaminophen were introduced. In 1983, almost a century after phenacetin was first

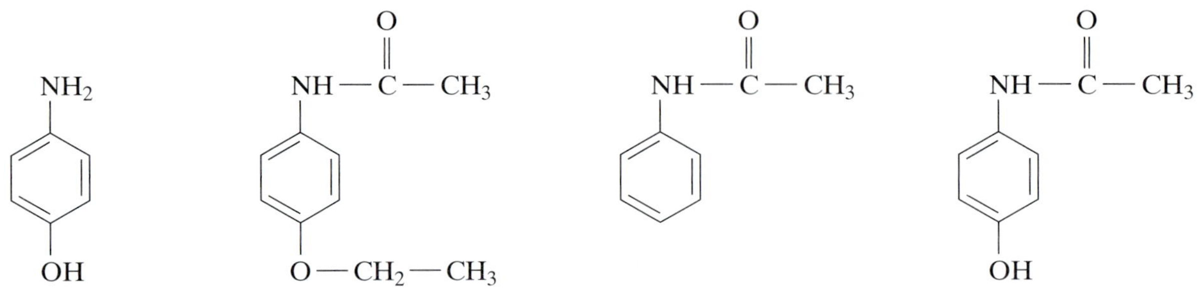

p-Aminophenol | Phenacetin (acetophenetidin) | Acetanilide | Acetaminophen

Figure 23.8 *p*-Aminophenol and related compounds. Acetaminophen is a useful alternative to aspirin.

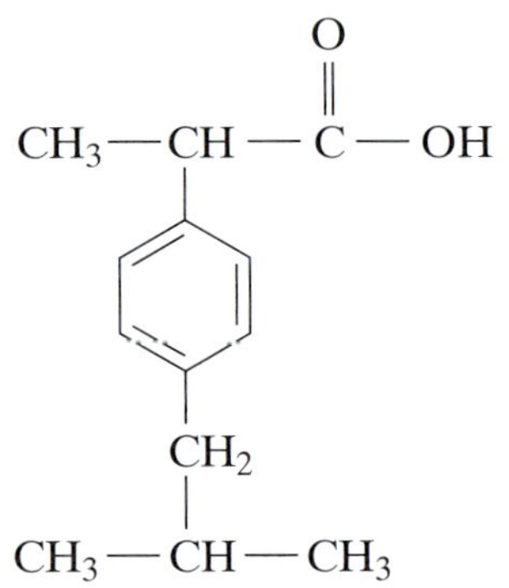

Figure 23.9 Ibuprofen, an alternative to aspirin for the relief of pain, fever, and inflammation.

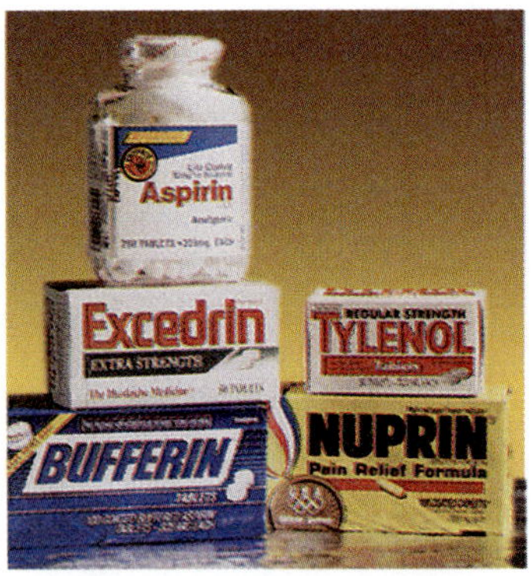

Analgesics containing aspirin, acetaminophen, and ibuprofen.

used as an analgesic, the U.S. Food and Drug Administration banned it from both prescription and nonprescription medications because of its tendency to damage the kidneys and to produce disorders of the blood when used excessively. Despite acetaminophen's relatively low toxicity compared to other *p*-aminophenols, even this comparatively safe medication carries its own risks. Acetaminophen appears able to produce lethal damage to the liver when taken in larger-than-recommended doses with or after several alcoholic drinks.

In 1984, the year after banning phenacetin as both a prescription and nonprescription drug, the FDA approved *ibuprofen* as an over-the-counter (nonprescription) drug for the relief of pain, inflammation, and fever. (Before the FDA's action, ibuprofen had been available in the United States only as a prescription drug.) Currently available in over-the-counter medications such as Advil and Motrin, ibuprofen appears to act much like aspirin in interfering with formation of the body's prostaglandins. What's more, ibuprofen *does* control inflammation (unlike acetaminophen) and is *not* a *p*-aminophenol (again, unlike acetaminophen; Fig. 23.9).

While the occasional use of any of these painkillers by a healthy person who has no allergic reaction to them doesn't seem to result in harmful effects, there is evidence that continuous, daily use of acetaminophen over a long period can produce a risk of kidney damage.

As we move to other chemicals that can change our perceptions of pain and of the world about us, we'll begin with a short look at a few of the analgesics with the longest histories of use, thousands of years in some cases. Since several of these seem to stand at a borderline between what we would now call a medicine and what we might consider to be a drug, we'll first examine some differences between the two categories.

■ **QUESTION**

What hazard does acetaminophen pose when taken in excessively large doses or with alcohol?

23.6 Chemicals, Medicines, and Drugs

As we've seen in earlier chapters, the cry "There are *chemicals* in our foods!" acts to generate suspicion, if not actual fear, in the hearts of some consumers. Something similar may be happening to the language of medicinal chemicals. To say, for example, that someone is "on drugs" could one day cast a shadow on what we might legitimately buy in a "drugstore."

In one sense **drugs** and **medicines** are virtually indistinguishable from each other as chemicals used medically for treating diseases and injuries. Yet the term *drugs*, unlike *medicines*, carries with it the connotation of narcotics, addiction, and crime. All this works to elevate the term *medicines*, despite the valid use of narcotics in the practice of medicine and the possibility of addiction to some of our more readily available, entirely legitimate medications.

Drugs and **medicines** are chemicals used medically for treating diseases and injuries. Drugs carry the added connotation of narcotics and addiction.

We've seen sporadically, in our examination of the chemistry of ordinary things, the subtle connections between the way we use a chemical and the history of its name or of terms connected with it. Something of this sort also works to separate *drugs* from *medicines*. The word *medicine* came into the English language through French, from the Latin *medicina* meaning, according to its context, a medication, the practice of the physician, or the place where the physician works.

The less reputable *drugs* bears an uncertain ancestry. In one form or another it entered into several European languages by obscure routes and with various implications. As with the substances themselves, the word *medicine* seems to have a clearer ancestry, a purer pedigree than does *drug*. It's a bit more respectable. In the next several sections we'll examine several chemicals that fit more neatly into the category of drugs than the category of medicines.

Drugs (medicines) are sold legitimately in drugstores.

■ QUESTION

Name a substance that you would normally consider to be (a) a medicine, (b) a drug. Can you name a substance that might be either a medicine or a drug, depending on the context or circumstances?

23.7 Narcotics and Alkaloids: The Opium Alkaloids

Whether we consider it a medicine or a drug, the essence of a **narcotic** is clear. From the Greek *narkotikos*, meaning numbing or stupefying, a narcotic is a chemical that dulls the mind, induces sleep, and generally numbs the senses. As the origin of the word in the classic Greek language suggests, the use of narcotics is an ancient activity. The Sumerians, one of the earliest of civilizations, were probably familiar with the narcotic effects of **opium**, the dried sap of the poppy, some 6000 years ago. (The Sumerian term for the flower can be translated as "joy plant.") The first clear reference to the sap of the poppy appeared in Greek writings of about 300 B.C. Later, both Arabian and Oriental healers used its medical powers. Thomas Sydenham, the 17th-century English physician known as the English Hippocrates, wrote of the substance's powers to ease pain:

A **narcotic** is a substance that produces a stupefying, dulling effect and induces sleep.

Opium is the dried sap of the poppy. It contains several narcotic compounds.

> Among the remedies which it has pleased Almighty God to give to man to relieve his sufferings, none is so universal and so efficacious as opium.

Opium's power arises from several of its narcotic constituents, all members of a major category of plant compounds known as **alkaloids** for their *alkalinity*, their behavior as bases in the presence of acids. At the molecular level the alkaloids all owe their basicity to one or more amine nitrogens in their molecular structures (Sec. 17.2); most taste bitter and produce physiological reactions of various kinds and intensities.

Alkaloids are basic, bitter-tasting, nitrogen-containing compounds that are found in plants and that produce physiological reactions of various kinds and intensities.

In the dried opium itself, alkaloids of all sorts make up about a quarter of the total weight. **Morphine**, the first alkaloid of any kind ever to be isolated, leads the list at about 10%. In 1803 Friedrich Sertürner, a German pharmacist, obtained this

Morphine is the major alkaloid of opium.

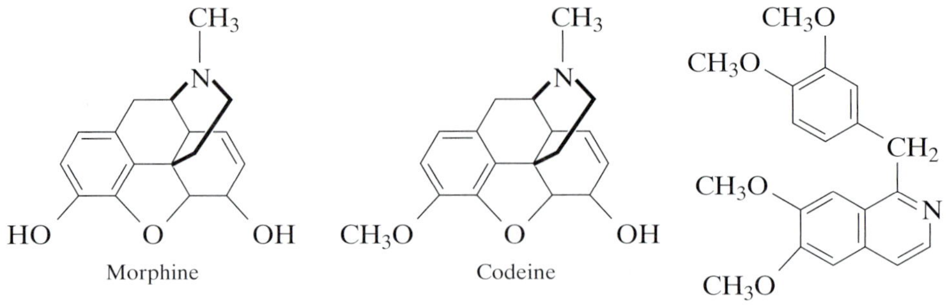

Figure 23.10 Alkaloids of opium. These occur in the sap of the poppy.

alkaloid in pure form, described it, and named it for Morpheus, the Roman god of dreams. Morphine is a powerful narcotic, for many years a valuable tool in the physician's bag. It's a potent analgesic as well as a cough suppressant. Morphine produces a variety of psychological responses, including apathy and euphoria. It's also highly addictive.

Another of the opium alkaloids is *codeine*, isolated in 1832 by a French pharmacist, Pierre-Jean Robiquet. Codeine is a less potent analgesic than morphine, yet it's one of the most powerful cough suppressants known. *Papaverine*, discovered in 1848, is also present in this plant extract. It's useful for relaxing the smooth muscles (such as those of the blood vessels, the stomach, the intestines, the bladder, and the uterus) and for dilating or enlarging blood vessels, especially within the brain.

While they differ in their specific actions on the body and in their potency, these and the other opium alkaloids share two important characteristics: They are all addicting and they are all among the most powerful constipating agents known, with substantial value in controlling diarrhea. Figure 23.10 presents their molecular structures.

■ **QUESTION**

In what way does the molecular structure of codeine differ from the molecular structure of morphine?

23.8 A Good Idea Gone Very, Very Bad

The medical value of morphine in vanquishing pain is counterbalanced by its very great capacity to produce addiction. Just as the search for better, less toxic analgesia through molecular modifications of acetanilide led to the *p*-aminophenols (Sec. 23.5), a search for nonaddictive molecular modifications of morphine soon produced a promising synthetic substance. In 1898 the diacetylation of morphine in the chemical laboratories of a German dye manufacturer led to *diacetylmorphine* through a chemical transformation very much like the acetylation of salicylic acid to aspirin (Sec. 23.1). Immediately on its introduction into medicine at the beginning of the 20th century, this diacetylated molecule proved to be a much more powerful narcotic and cough suppressant than morphine itself. It was so powerful, in fact, and the size of its effective doses were so low that it gave promise, at first, of being nonaddictive.

Heroin is diacetylmorphine, morphine in which two —OH functional groups have been converted to ester functional groups.

The promise fell through, badly. Diacetylmorphine proved to be one of the most addictive drugs known, so powerfully addictive that its use, possession, manufacture, and importation into the United States and several other countries have been banned by law. It's known more commonly now by an early trade name dating from the time of its medical trials, **heroin** (Fig. 23.11).

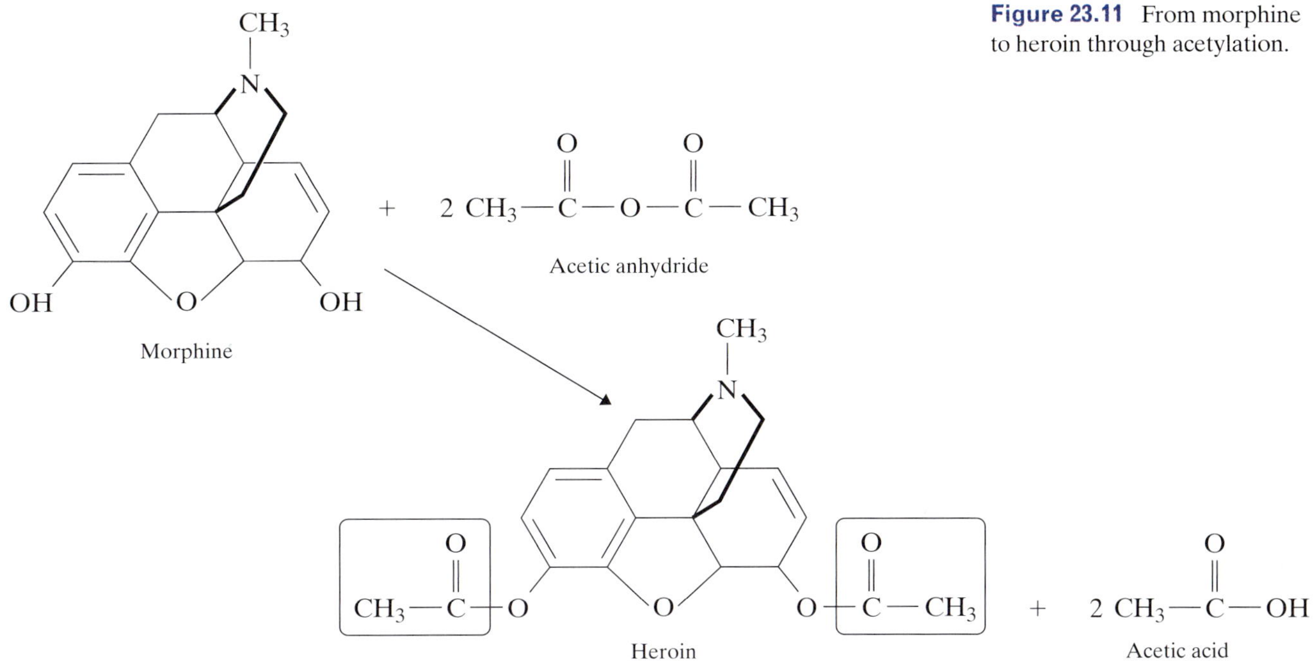

Figure 23.11 From morphine to heroin through acetylation.

(A bit of practical chemistry comes into play in enforcing the ban on its manufacture. Farmhouses in rural France were occasionally used as secret factories to manufacture heroin for international distribution. One of the byproducts of the acetylation with acetic anhydride, shown in Figure 23.11, is acetic acid, the major organic component of vinegar. To locate the heroin factories, French police trained dogs to sniff out the vinegar-like odor and to lead them straight to its source.)

■ QUESTION

When heroin was first synthesized and tested as a narcotic and cough suppressant, it seemed to show promise of being nonaddictive. What property of heroin stimulated this hope?

23.9 Sigmund Freud, Medieval Poisoners, and Beautiful Women: Cocaine and Related Alkaloids

Still another plant, the South American coca bush, produces the alkaloid *cocaine*, used medically as a topical or local anesthetic. Cocaine resembles morphine in its medical value, in this case as a local anesthetic, and in its ability to produce a fast and powerful addiction. Beyond its therapeutic uses, cocaine produces euphoria, a great sense of well-being, and delusions of immense power—all followed by a depression that leads the user to crave another jolt of euphoria, which is followed by the same cycle of euphoria, depression, and a craving for more. Morphine and heroin produce a truly physical addiction in that physiological symptoms—watery eyes and nose, yawning, sweating, goose flesh, and dilated pupils, for example—occur during withdrawal. Cocaine's addiction seems to be purely psychological, without physical symptoms on withdrawal, but its psychological addiction is nonetheless as real and as powerful as any physical addiction.

Sigmund Freud, one of the first to recognize the effects of cocaine.

Sigmund Freud, founder of psychoanalysis, played a major role in the discovery of cocaine's addictive and anesthetic potential. A few years after the compound's isolation from coca leaves, Freud began studying it as an aid to his treatment of patients. He soon persuaded a fellow physician who was addicted to morphine, Karl Koller, to use cocaine as an aid in breaking his morphine habit. Koller succeeded, at the cost of becoming addicted to cocaine instead. During his experimentation with cocaine, Koller recognized its properties as an anesthetic and, in 1884, began using it to anesthetize his patients' eyes during medical procedures. Within the year another physician began using the drug in dentistry as the first local dental anesthetic.

Cocaine provides a fine example of the connection between chemical properties and the use or abuse of a substance. Chemically, cocaine is a nitrogen-containing base, with *chemical* properties similar to those of other amines we have examined (Sec. 17.2). Like other amines, cocaine reacts with acids, including hydrochloric acid, to form salts. Cocaine hydrochloride, the form in which cocaine is usually isolated from coca leaves, is a salt with physical properties similar to those of common sodium chloride. It's readily soluble in water and fairly stable toward heat. It doesn't vaporize readily. On reaction with a base, cocaine hydrochloride is converted to cocaine itself, the pure base, often called the "free base."

This basic form of cocaine has physical and chemical properties far different from those of the hydrochloride. During the chemical conversion of cocaine hydrochloride (the salt) to cocaine (the base), the generated base forms as a white solid sheet that cracks into large lumps or "rocks." This formation of "rocks" as the solid cracks into individual lumps gives the product the name *crack cocaine* or simply *crack*. Unlike the crystalline salt, the free base readily vaporizes. Inhaling cocaine vapors quickly produces a sharper, more intense sensation than the one produced by the hydrochloride salt. The fall into depression that follows is also steeper.

Chemically, cocaine belongs to the *tropane* alkaloids, all of which are molecular variations of the parent, tropane (Fig. 23.12). Other members of the family in-

Figure 23.12 Cocaine and the tropane alkaloids.

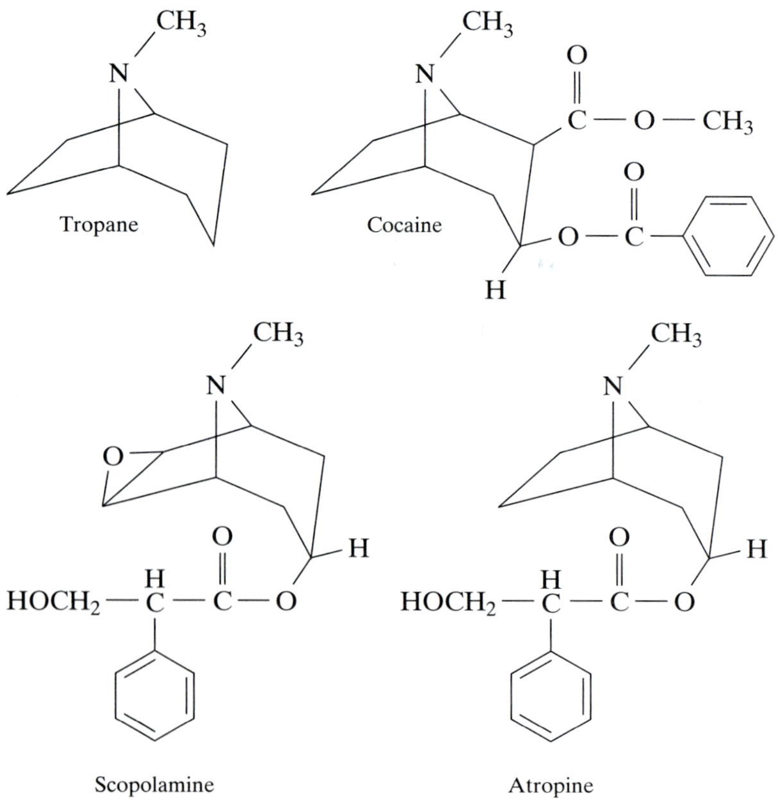

clude *atropine* and *scopolamine*, both of which affect the nervous system in one way or another and occur in plants of the nightshade family, a group that encompasses many poisonous species as well as potatoes, tomatoes, red peppers, and eggplants. One of the more toxic of the nightshade plants, *belladonna*, was a favorite poison in the intrigues of the Middle Ages. Atropine, the major alkaloid of the plant, is the active ingredient in tincture of belladonna, a liquid once popular for its ability to increase the size of the pupils and (presumably) make the eyes more attractive. The name of the plant itself, belladonna, comes from the combination of the Italian words *bella* and *donna*, "beautiful lady."

Scopolamine, which also dilates pupils, is a powerful sedative. In combination with morphine it was once used to induce "twilight sleep" to ease childbirth. (The combination has been replaced by safer drugs.) Currently, it is an ingredient in some over-the-counter sedatives and motion-sickness preparations.

■ QUESTION

In addition to a basic nitrogen atom bearing a methyl group, what molecular structure is characteristic of the tropane alkaloids?

23.10 Caffeine and Nicotine

Probably the most widely used of all the alkaloids are the *caffeine* of the coffee bean and tea leaves and the *nicotine* of the tobacco plant (Fig. 23.13). **Caffeine** stimulates the central nervous system and the heart and heightens a sense of awareness. It's added to cola drinks at a concentration of about 35 to 55 mg per 12-oz can and it occurs naturally in tea and coffee. The amount of caffeine in a cup of coffee depends on the strength of the brew, but the alkaloid is generally present at a concentration of about 100 to 150 mg per cup. (Decaffeinated coffee, of course, contains virtually none.) The lethal dose for an adult is estimated at about 10 g, taken orally. That amounts to 70 to 100 cups of brewed coffee at one sitting, a quantity that would generate plenty of discomfort long before the fatal dose is approached.

Caffeine is an alkaloid that occurs in coffee beans and in tea leaves. It stimulates the central nervous system and heightens a sense of awareness.

Nicotine is a highly poisonous alkaloid that occurs in tobacco leaves.

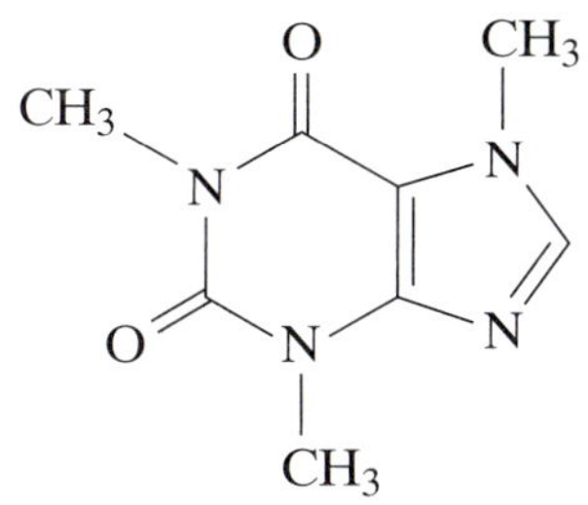

Caffeine

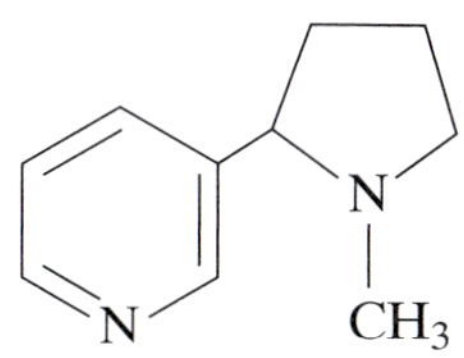

Nicotine

Figure 23.13 Caffeine and nicotine. Caffeine is the major alkaloid of coffee beans. Nicotine is a highly poisonous alkaloid of tobacco.

Nicotine, on the other hand, is a lethal substance that's used as a powerful agricultural insecticide. Absorbing less than 50 mg of nicotine can kill an adult in a few minutes. If it weren't for the oxidation of most of this alkaloid to less toxic products by the high temperatures and the rapidly moving air stream that accompany smoking, no cigarette smoker could possibly last long enough to develop smoking into a habit.

The toxicity of nicotine provides a fine example of the connection between the effects of a poison and the means by which it enters the body. Nicotine administered *orally* to mice is actually slightly less lethal than caffeine (Table 20.1). But with direct absorption into the bloodstream, through the lungs while smoking, for example, nicotine's potency increases by a factor of about 1000. Furthermore, the effects of nicotine aren't limited to its immediate toxicity. Pregnant rats who receive nicotine (in amounts equivalent to moderate to heavy smoking by humans) produce offspring particularly susceptible to the rat equivalent of sudden infant death syndrome.

Several generations ago, cigarette smoking was generally accepted as smart and sophisticated. With growing recognition of the health hazards posed by nicotine and other components of cigarette smoke, both to smokers themselves and to nonsmokers who inhale smoke exhaled by others (passive smoking, Sec. 14.16), smoking in public places is being restricted severely or prohibited entirely. Seating in many restaurants, for example, is separated into smoking

Tea, coffee, and cola: sources of caffeine.

and nonsmoking sections, and smoking is prohibited on all domestic airline flights within the United States.

The lethal effects of smoking aren't due exclusively to nicotine. Cigarette smoke is rich in toxic carbon monoxide (Sec. 6.14 and Chapter 8), which decreases the blood's ability to carry oxygen. The smoke also contains tarry substances capable of generating cancer. It's estimated that about 170,000 people die each year in the United States from smoking-related heart and circulatory diseases. Another 130,000 deaths linked to cigarette smoking result from cancers of the lungs and other major organs. Roughly a third of all deaths from cancer are believed to be smoking related. Smoking during pregnancy is particularly dangerous to the fetus and can lead to spontaneous abortions, low birth weights, birth defects, and infant deaths.

■ QUESTION

In addition to nicotine, what is another lethal substance found in cigarette smoke?

23.11 Natural Versus Synthetic

Our discussion of nicotine, caffeine, aspirin, acetaminophen, morphine, cocaine, and heroin provides us with another opportunity to examine the (far too common) belief that the products of nature must necessarily be good, or at least far less harmful than what we humans manufacture with our intellects and our skills in laboratories and factories. As in Chapter 20, we find once again that the chemicals of this world, both those we isolate from plants and animals and those of the chemical laboratory, just don't fall neatly into the categories of our simpler notions.

The naturally occurring, narcotic alkaloids we've described, for example, are all very useful for the analgesia and other therapeutic effects they provide, but they're also destructive for the addictions they produce. Our laboratories, on the other hand, have produced the wonders of aspirin and acetaminophen, as well as the devastation of heroin. As with the chemicals in our foods, we find again that both benefits and risks come from the fruits of nature as well as from those of the laboratory. To hold that what nature produces is necessarily more benign or less harmful than what humans manufacture is to ignore the difference between acetylsalicylic acid, which does not occur in nature, and cocaine, which does.

■ QUESTION

Name a compound produced by humans that is useful or beneficial and another that is harmful. Repeat for compounds isolated from plants. Explain your choices.

23.12 Licit Versus Illicit

There's another difference between aspirin and cocaine. You can buy a bottle of aspirin easily and freely in any drugstore, convenience store, or supermarket. Cocaine, on the other hand, is one of several federally controlled (regulated) substances whose unauthorized manufacture, possession, importation, and use are rigorously prohibited or regulated by law. Aspirin is a licit drug; cocaine is illicit.

Whether a drug is legal and sanctioned by a society or illegal and condemned depends very much on the outlook of that particular society at that particular time in its history. As with *safety*, any society's tolerance for any particular drug depends (at least partly) on its assessment of the consequences of its use. Today in most Western nations many of the narcotic alkaloids and related synthetic compounds are controlled substances. In the United States this is largely a phenomenon of the 20th century.

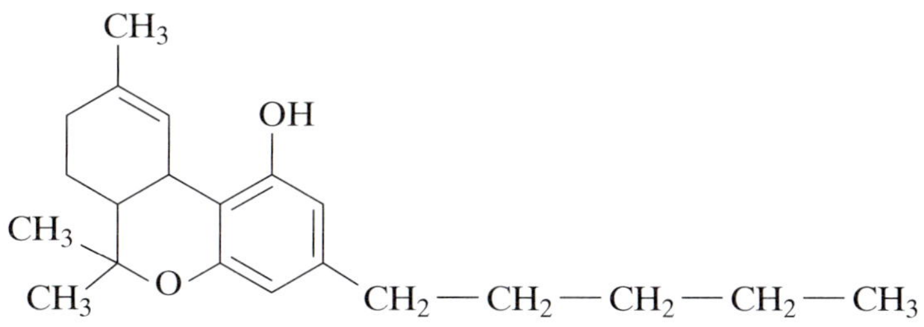

Figure 23.14 Tetrahydrocannabinol, or THC, the active ingredient of marijuana.

Before 1914, habitual opium smoking was both widespread and generally tolerated in the United States. In the first decade of the 20th century, for example, about 150,000 pounds (68,000 kg) of opium were imported *legally* into the United States to support the addiction. The rigorous federal control of addictive substances started with the passage of the Harrison Narcotics Act of 1914. Through that Act and subsequent laws, the use of narcotics to satisfy an addiction became punishable by law. While we now prohibit by law the nonmedical use of virtually all physiologically active, addictive alkaloids, we still generally tolerate or accept a few, especially the caffeine of coffee and the nicotine of tobacco (Fig. 23.13). There's evidence that both these alkaloids can produce addictions.

Our society has had mixed feelings about the nonalkaloid narcotic, *alcohol*, or, more accurately, *ethyl alcohol*, $CH_3—CH_2OH$, which depresses the entire central nervous system and which can produce a powerful psychological addiction. The manufacture, sale and transportation of alcoholic drinks were prohibited with the force of the 18th Amendment to the U.S. Constitution, which became effective January 16, 1920. Almost exactly 14 years later we changed our societal mind and repealed the prohibition with the 21st Amendment. Thus alcoholic drinks, which had been declared unacceptable to U.S. society with the 18th Amendment, became acceptable once again with the 21st.

Marijuana, whose major, active component is *tetrahydrocannabinol*, or THC (Fig. 23.14), has also been the subject of mixed signals. While marijuana seems to have medical value in controlling *glaucoma*, a disease of the eyes in which increasing internal pressure leads to blindness, and in relieving the nausea of chemical treatment of cancers and other diseases, it's a controlled substance with restrictions generally similar to those placed on heroin and morphine. Depending on the decade and on the particular region or locality within the nation, the possession and use of small quantities of marijuana have been subjected to responses ranging from vigorous prosecution to quiet indulgence.

■ QUESTION

To which of the following classes of organic compounds does tetrahydrocannabinol belong: (a) hydrocarbon; (b) carboxylic acid; (c) phenol; (d) alkyl halide; (e) aromatic?

23.13 Synthetic Narcotics: Fruits of Imagination and Skill

Long ago, before the full blossoming of chemistry as a science, medicine could ease human suffering only with drugs that were little more than extracts of readily available plants. Some were potent, some were not. A few were addictive or poisonous or both.

Today, molecular structures can be assembled and modified to increase drug effectiveness and diminish side effects. Although the ideal drug—immensely powerful, perfectly targeted to a specific illness or symptom, and completely harmless and nonaddictive—hasn't yet been developed (and probably never will be),

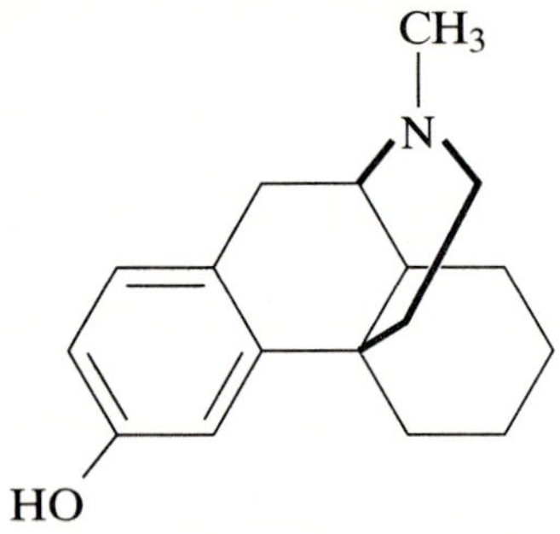

Levorphanol
(Levo-Dromoran)

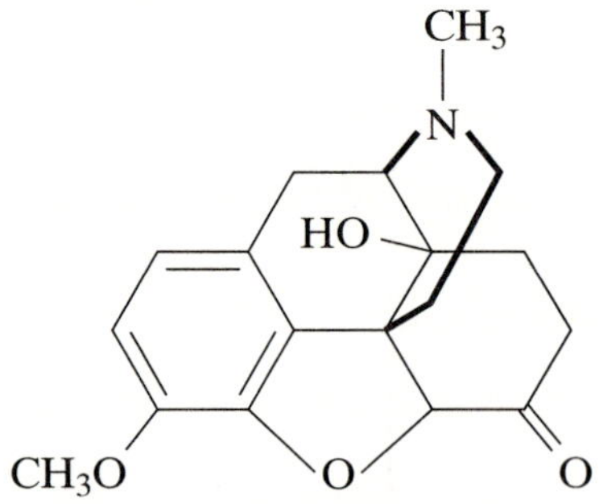

Oxycodone
(Percodan)

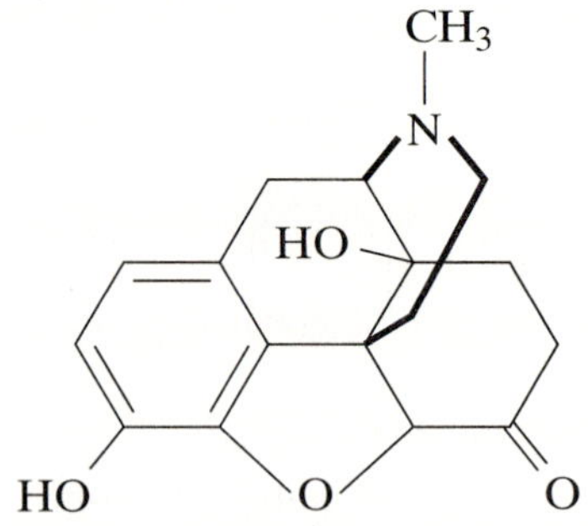

Oxymorphone
(Numorphan)

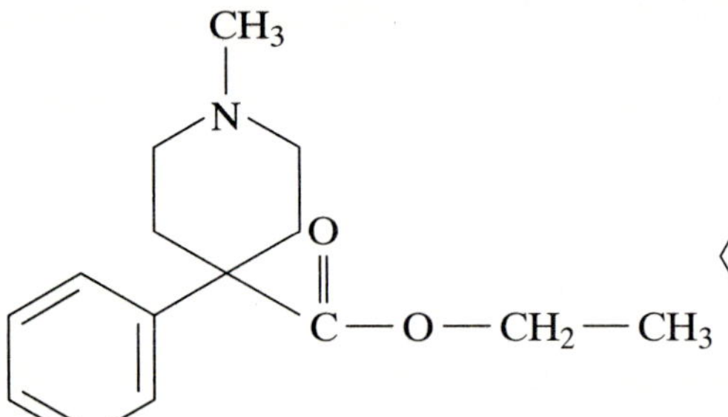

Meperidine
(Demerol)

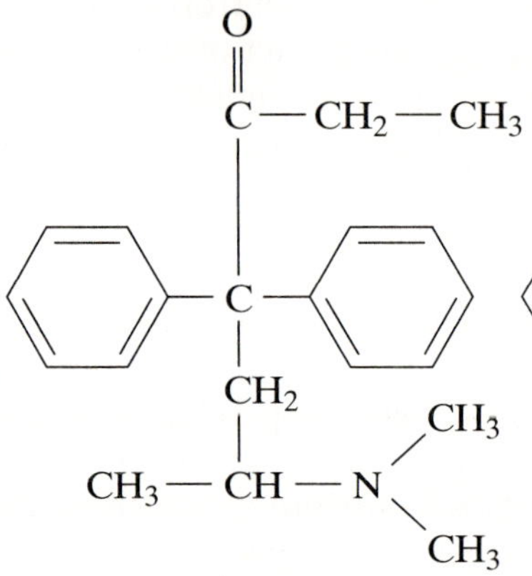

Methadone

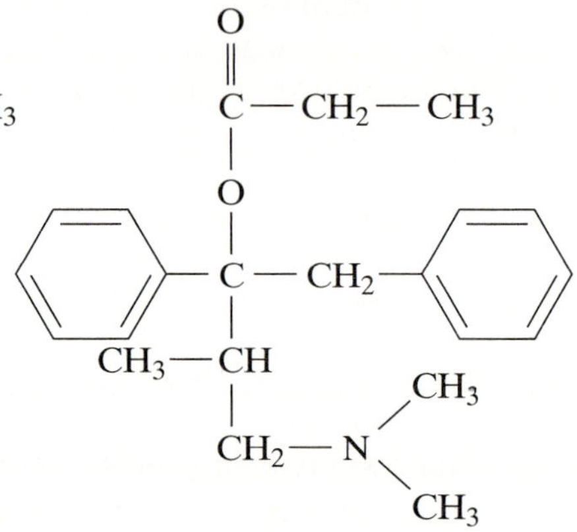

Propoxyphene
(Darvon)

Figure 23.15 Synthetic narcotics.

physicians now have available a collection of synthetic, well-designed, and effective molecules to supplement or replace those that come from plants and animals.

Figure 23.15 presents some of the synthetic narcotics that act much like morphine (and that still present the possibility of addiction). Of all these synthetic compounds, *levorphanol* (Levo-Dromoran) shows the strongest resemblance to morphine in the combination of its molecular structure and actions. *Meperidine* (Demerol) is probably the most widely used synthetic narcotic, while *oxymorphone* (Numorphan) is one of the most powerful, providing about 10 times the analgesia as an equal weight of morphine. Some others, such as *propoxyphene* (Darvon) and *oxycodone* (Percodan), are sometimes combined with aspirin or acetaminophen for more potent relief than either component alone would provide.

Methadone, as its hydrochloride salt (Dolophine hydrochloride), is both an analgesic in its own right and a useful treatment for heroin addiction. While it is itself addictive, methadone doesn't provide the euphoria or other psychological effects of heroin. Heroin addicts can avoid the crippling effects of withdrawal through a methadone maintenance program that permits them to remain productive members of society.

■ **QUESTION**

In what simple way does the molecular structure of oxymorphone (the active ingredient of Numorphan) differ from the molecular structure of oxycodone (the active ingredient of Percodan)?

23.14 For a Pain in the Mouth and Other Places

As with morphine, a search for synthetic analogs to cocaine, the oldest of the local anesthetics (Sec. 23.9), has produced several useful replacements with smaller potential for addiction or other side effects. Like cocaine itself (Fig. 23.12), the synthetic anesthetics of Figure 23.16 consist of three molecular regions:

- an aromatic ring, usually substituted, at one end of the molecule;
- a substituted nitrogen at the other; and
- a section, normally containing an amide or an ester group, connecting the two.

The oldest of these local anesthetics, *procaine*, was synthesized in 1905 and awarded a U.S. patent the following year. As Novocaine it's still used occasionally in

Figure 23.16 Cocaine and related synthetic anesthetics.

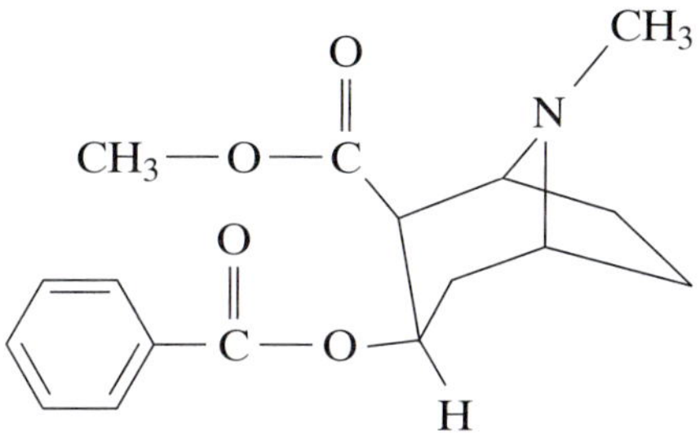

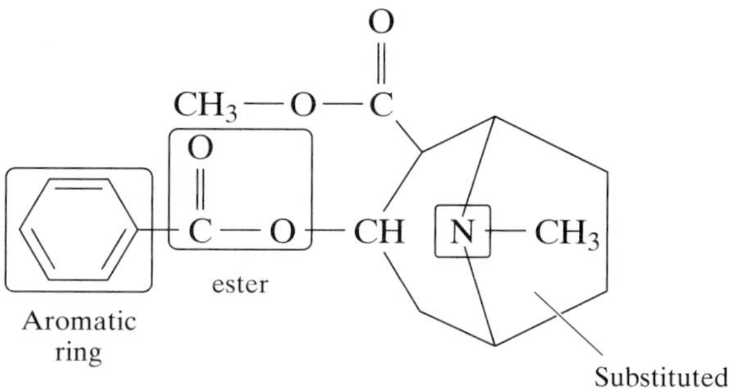

Cocaine
(Two views of the same molecule)

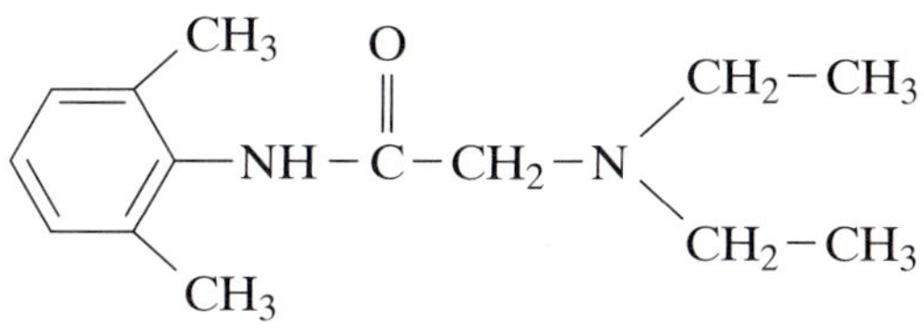

Lidocaine

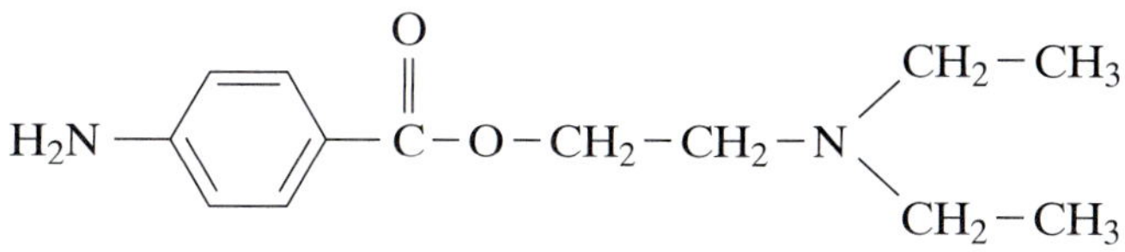

Procaine

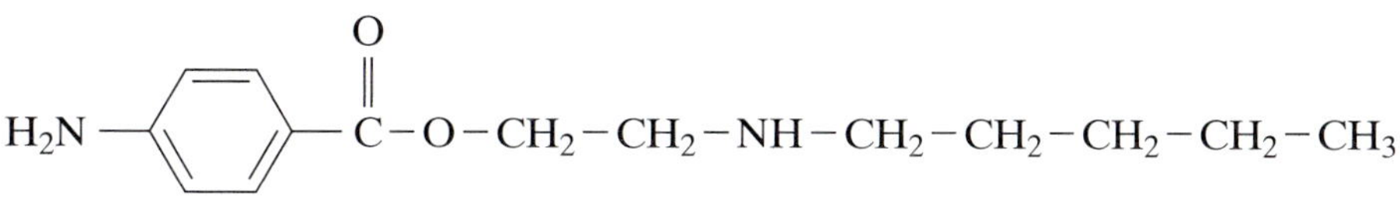

Naepaine

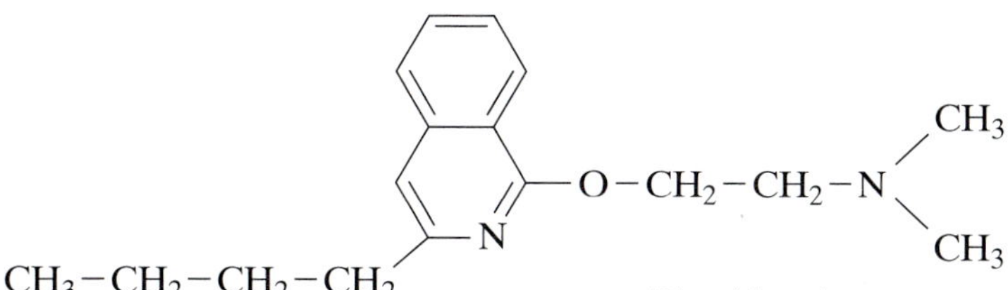

Dimethisoquin

A **vasoconstrictor** is a drug that constricts blood vessels.

dentistry and as a nerve block. Because the body quickly absorbs procaine, it's often injected along with a **vasoconstrictor**, a drug that constricts nearby blood vessels, to keep the anesthetic at the site of the injection as long as possible.

Lidocaine, patented in the United States in 1948 and used as Xylocaine, is a much stronger anesthetic than procaine and can be used without a vasoconstrictor. Lidocaine is also effective at returning an irregular heartbeat to its normal rhythm.

Toxicity or other undesirable properties of an anesthetic can restrict its medical applications. *Naepaine* (Amylsine), for example, is used only on the cornea of the eye, while *dimethisoquin* (Quotane) is most effective on the surface of the skin, especially for the relief of irritations and itchiness, including the discomfort of hemorrhoids.

■ QUESTION

What structural feature is present in molecules of cocaine, lidocaine, naepaine, and procaine, but absent from dimethisoquin?

23.15 Meditations, Fantasies, and Hallucinations

Drugs can also affect a different sort of perception, other than that of pain. Some substances, with no medical value whatever, can stimulate hallucinations ranging from ecstatic to terrifying and can induce psychological disturbances so powerful that they change the perception of reality itself. Two with particularly interesting histories, *mescaline* and *N,N-diethyllysergamide*, also known as *lysergic acid diethylamide* or *LSD* (from its German name, *lysergsaure diethylamid*), appear in Figure 23.17.

Mescaline, another of nature's products, forms in the mescal buttons at the top of the peyote cactus. Eating these buttons produces distortions of reality and states of deep meditation that have been part of religious rituals of Native Americans of the Western Hemisphere since before the time of Columbus. Today the drug is generally illegal in the United States, although in some regions its use is permitted for religious services of the Native American Church of North America, which serves about a quarter of a million Native Americans of the United States and Canada.

Unlike mescaline, LSD came first from the chemical laboratory, although microorganisms were later observed to produce it as well. The drug was initially prepared in 1938 by Albert Hofmann and a co-worker at Sandoz Laboratories, a Swiss pharmaceutical firm. They set it aside after a few tests indicated that it was a thoroughly uninteresting substance. Five years later, while studying the chemistry of *lysergic acid*, one of the alkaloids of the *ergot* fungus that grows on rye and other cereal grains, Hofmann prepared a bit more of the amide, a few milligrams. Shortly

Figure 23.17 The hallucinogens mescaline and LSD. Mescaline occurs in the peyote cactus. LSD is a synthetic compound.

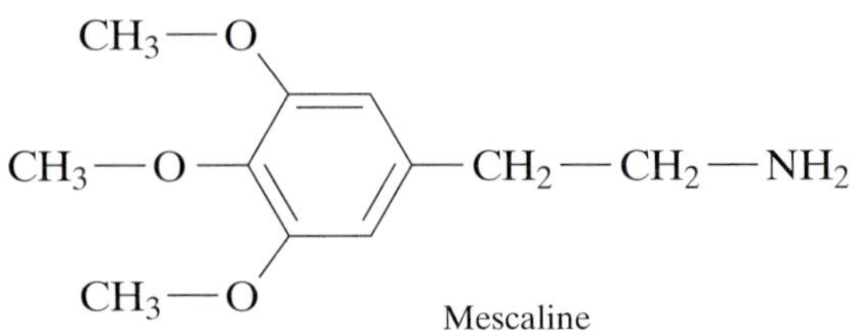

Mescaline

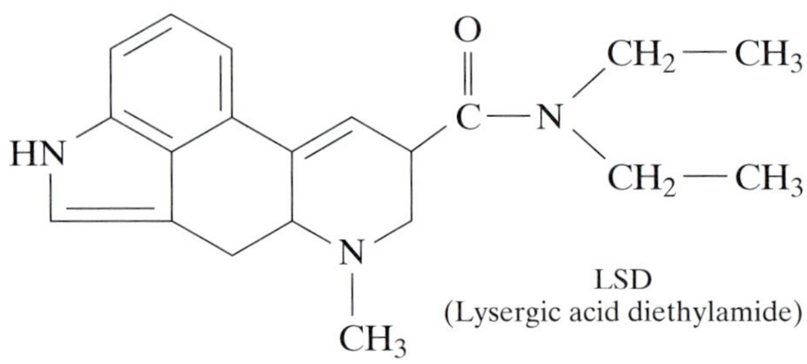

LSD
(Lysergic acid diethylamide)

afterward he developed a restlessness and dizziness followed by a fantasy-filled delirium, including what he later called an "intense kaleidoscopic play of colors."

In tracing the source of the disturbance Hofmann focused on the diethylamide of lysergic acid that he had prepared on the day of the attack. To test his suspicions he swallowed a quarter of a milligram, 0.00025 g, of the substance. The result was more than 6 hours of distortions of vision and of space and time, alternating restlessness and paralysis, dry throat, shouting, babbling, fear of choking, and a sense of existing outside of his body. The worst of the symptoms ended in 6 hours, but the distortions of shapes and colors continued throughout the day. Hofmann recovered completely. The use of LSD has produced far more serious consequences in others, leading some to suicide.

Although neither mescaline nor LSD actually appears to be addictive, each is on the federal list of controlled substances.

■ QUESTION

Molecules of both mescaline and LSD contain a benzene ring. What other functional group is common to both? What functional group is present in the LSD molecule but not in the mescaline molecule?

23.16 The Phenylethylamines

Many of the narcotics and other compounds that alter our perceptions in one way or another, including mescaline, LSD, morphine, heroin, oxycodone, and oxymorphone, are composed of molecules with one bit of chemical architecture in common, a *β-phenylethylamine* (sometimes shortened to *β-phenethylamine*) segment. (See Sec. 17.3 for a review of the use of Greek letters in naming organic compounds.) Molecules containing this structural fragment often affect the way our nerves carry their messages to the brain. Figure 23.18 highlights this structural unit in several of these molecules.

Figure 23.18 Typical β-phenylethylamines. This structure occurs in amphetamines and in many other stimulants and narcotics.

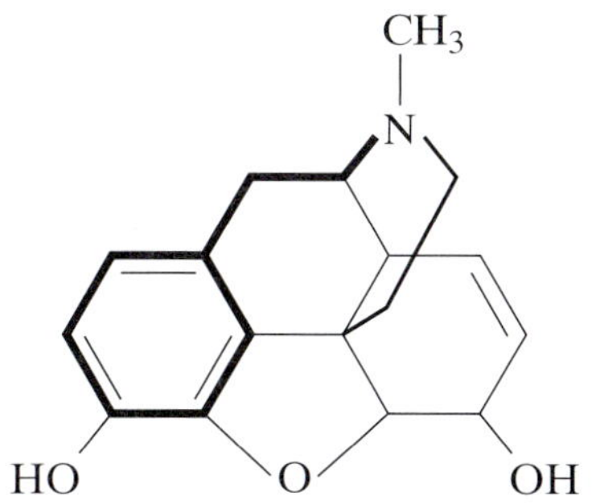

Morphine

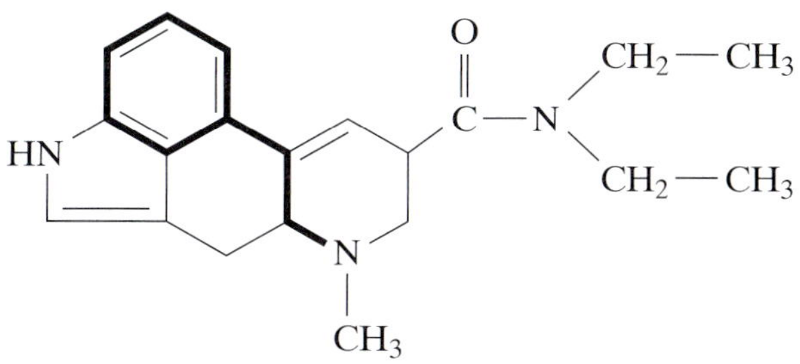

LSD
(lysergic acid diethylamide)

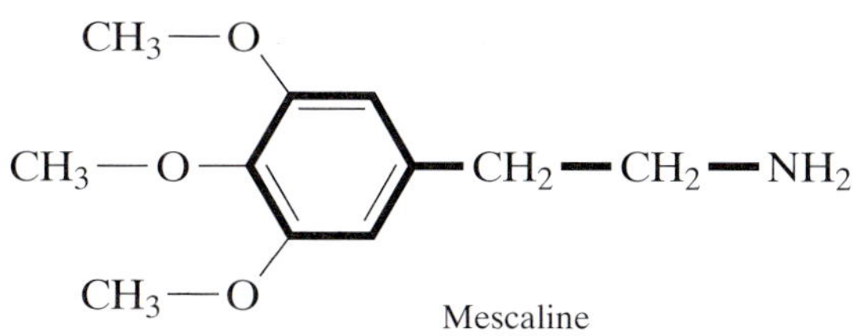

Mescaline

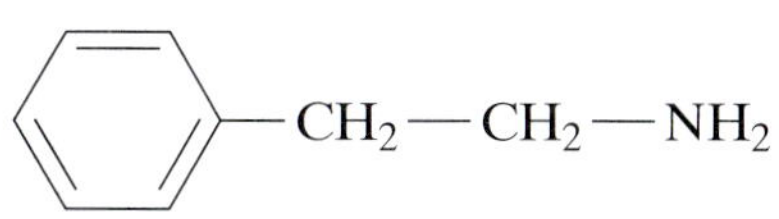

β-Phenylethylamine

Figure 23.19 Amphetamines. These β-phenylethylamines are powerful stimulants.

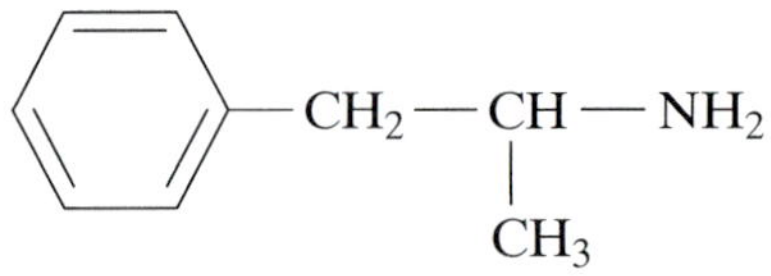

Amphetamine
(Benzedrine)

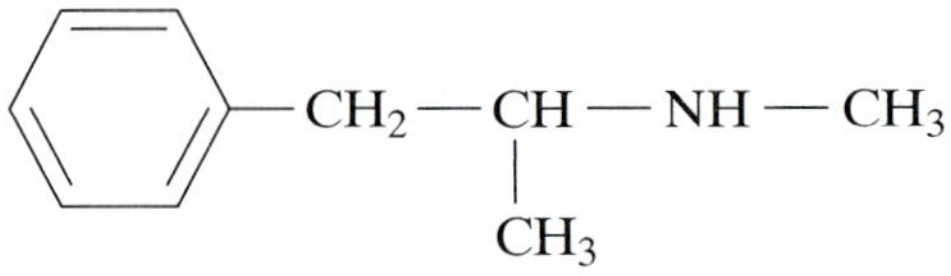

Methamphetamine
(Methedrine)

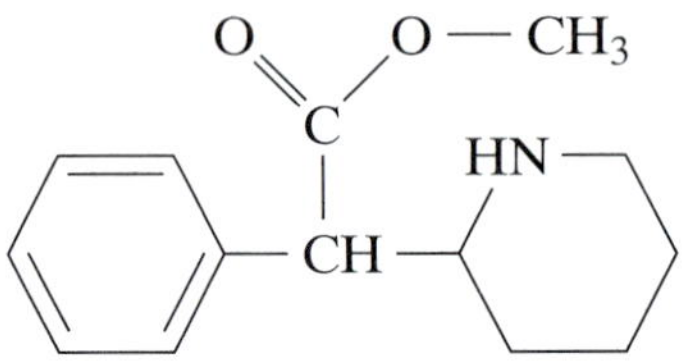

Methylphenidate
(Ritalin)

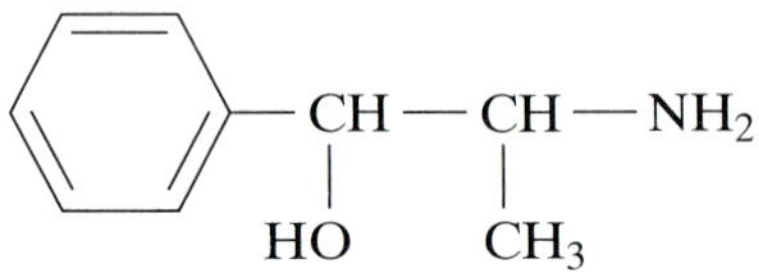

Phenylpropanolamine

Other substances with this unit include the *amphetamines*, the powerful stimulants of Figure 23.19 that raise the pulse rate and blood pressure, reduce fatigue, and suspend temporarily the desire for sleep. Amphetamines are useful medically as decongestants, to control hemorrhaging, to counteract the toxic effects of depressant chemicals, in treating epilepsy, as an aid in weight loss, and for raising blood pressure when it drops to dangerously low levels. They carry a great potential for abuse, though, and can produce or contribute to paranoia and mental illness. Their use, like that of most of the other substances of this chapter, is restricted by federal law.

■ **QUESTION**

Using the Greek letter system of nomenclature, designate the carbon of phenylpropanolamine that (a) bears the methyl group; (b) bears the hydroxyl group.

23.17 For Pain and Suffering of a Different Sort: Mental Illness

Beyond the hurts of raw, physical pain, we are also vulnerable to mental and emotional torments, ranging from debilitating anxieties and irrational fears to mental illnesses, psychoses, and disorders of all descriptions and intensities. Mirroring the scope of mental suffering that can afflict humans is the complexity of chemical substances that can provide cures, or at least relief. These range from the deceptively simple structure of ionic *lithium carbonate*, Li_2CO_3, to the awesome structure of *reserpine* (Fig. 23.20).

Lithium carbonate provides stability to the mood swings of *manic depression*, a condition that produces the extremes of raging exhilaration and intense, paralyzing depression. It was first used to treat manic patients in 1949, in Australia.

Reserpine is the major *rauwolfia* alkaloid produced by rauwolfia shrubs, which are native to India. Ancient Hindus applied extracts of these and similar plants to a variety of disabilities ranging from insomnia to severe mental illness. In modern times reserpine was used to treat mental illness as early as 1954, but it has given way recently to more modern drugs. In 1981 the Environmental Protection Agency registered reserpine as a carcinogen.

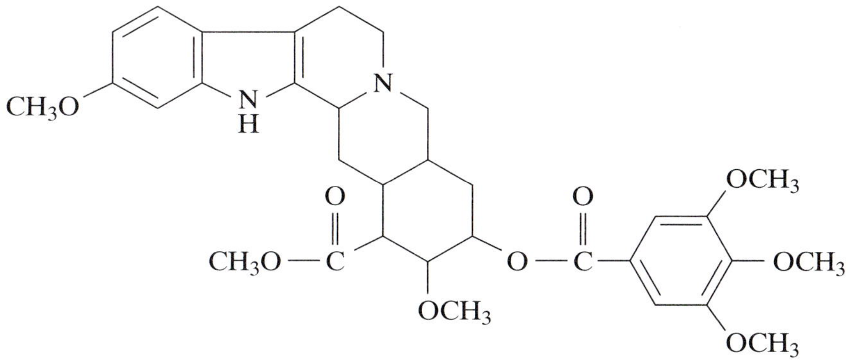

Figure 23.20 Reserpine, an alkaloid of the rauwolfia shrub, has been used to treat various forms of mental and emotional illness.

The kinds of drugs available for the treatment of mental pain (and their effects) are as varied as the symptoms they relieve and their causes. Among the most commonly used are the *tranquilizers* and the *antidepressants*. Physicians write more prescriptions for tranquilizers than for any other class of prescription medicine, with *meprobamate* (Equanil and Miltown) among the most popular. This particular tranquilizer belongs to the chemical class of *carbamates*, the esters of *carbamic acid*, $H_2N—CO_2H$ (Fig. 23.21). *Carisoprodol* (Soma), another carbamate, is an effective muscle relaxant.

Diazepam (Valium), *flurazepam* (Dalmane), *oxazepam* (Serax), *clorazepate* (Tranxene), and *chlordiazepoxide* (Librium) are all tranquilizers of a different structural class (Fig. 23.22). First prepared in 1933, they were soon found to act as muscle relaxants. It wasn't until a generation later, after they had been used to tame violent monkeys and other animals, that compounds of this structure were first used to treat disorders in humans.

Among the antidepressants, *fluoxetine* (Prozac, Fig. 23.23), which was introduced in 1988, holds the lead with annual worldwide sales in excess of $2.5 billion. With relatively low toxicity—roughly the same as acetaminophen (Sec. 23.5)—and potential side effects, such as insomnia, loss of appetite, and nausea, that can seem minor when compared with its benefits, fluoxetine is taken by many people who benefit from its antidepressant and mood- and personality-enhancing effects.

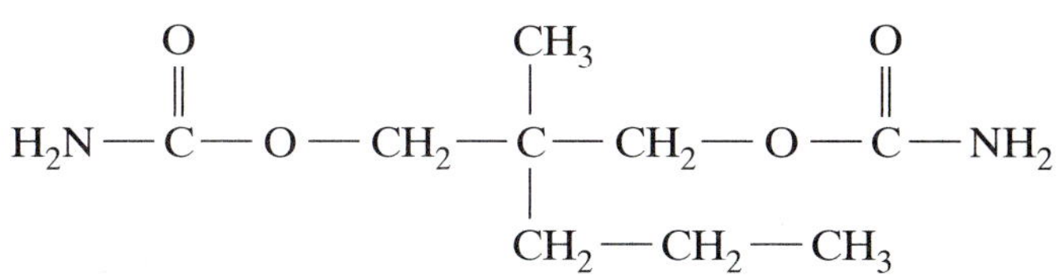

Meprobamate
(Equanil, Miltown)

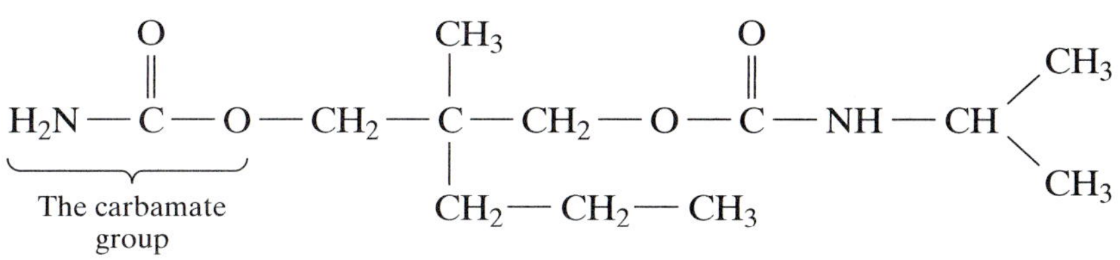

Carisoprodol
(Soma)

Figure 23.21 Carbamate tranquilizers.

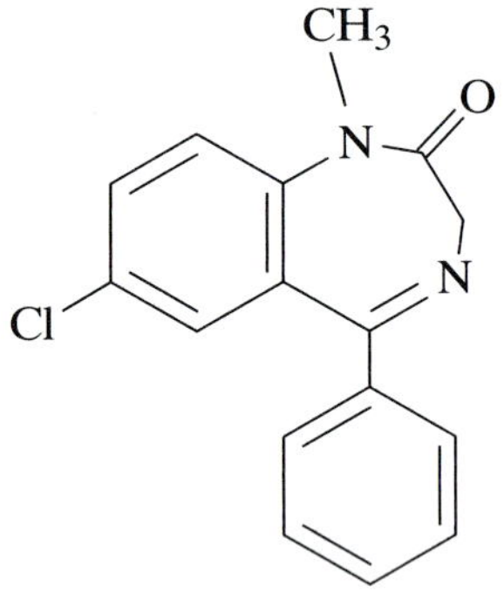

Diazepam
(Valium)

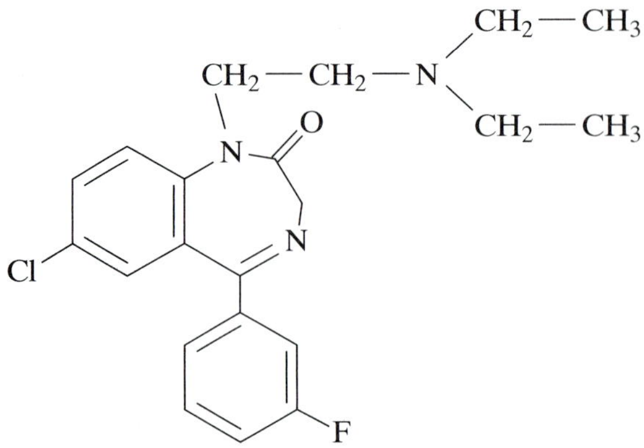

Flurazepam
(Dalmane)

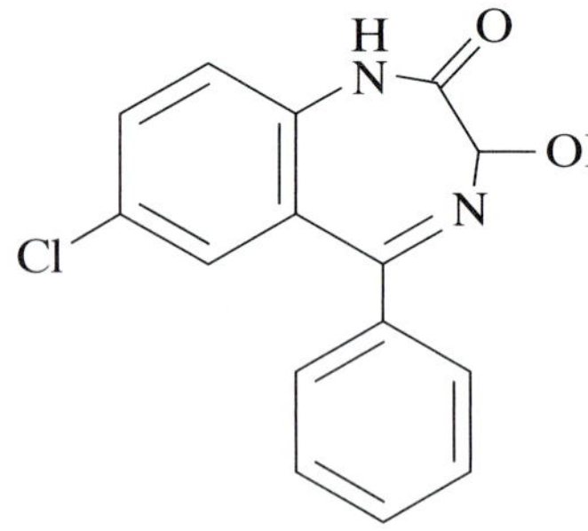

Oxazepam
(Serax)

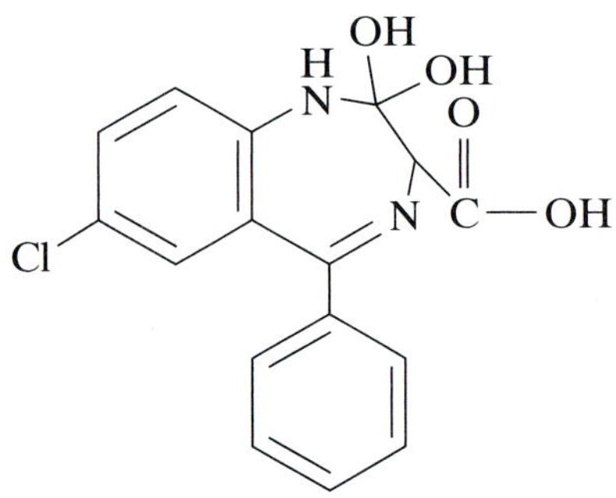

Clorazepate
(Tranxene)

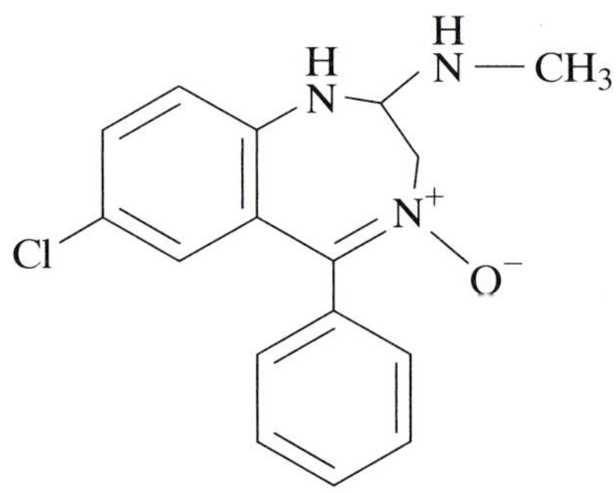

Chlordiazepoxide
(Librium)

Figure 23.22 Muscle relaxant tranquilizers.

Figure 23.23 Fluoxetine (Prozac), an antidepressant.

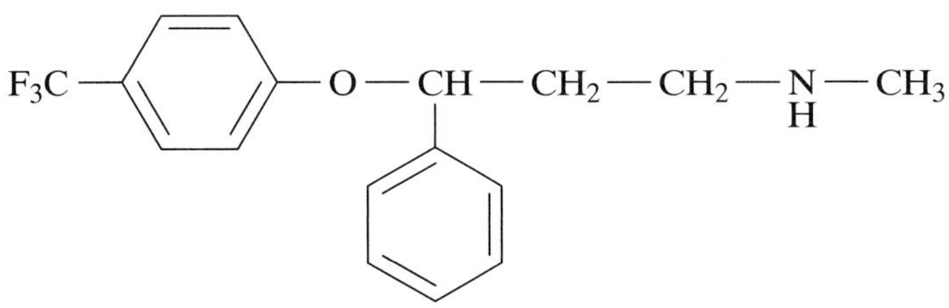

■ **QUESTION**

What chemical element is present in fluoxetine that is not present in any of the other compounds we have examined so far in this chapter?

23.18 Changing the Brain's Tumblers Provokes a Curious Question

Opiate antagonists are compounds that counteract or nullify the effects of opiates, including heroin.

Of all the chemicals that affect our perceptions of pain, a group that could turn out to be the most significant was discovered through a strange connection between human bodies and plants.

The story starts with *naloxone* (Fig. 23.24), a member of a group of **opiate antagonists**, morphine-like compounds that have the power to nullify completely the effects of even the most powerful, most addictive narcotics, including heroin. Among them are several compounds, including naloxone, that not only reverse the effects of

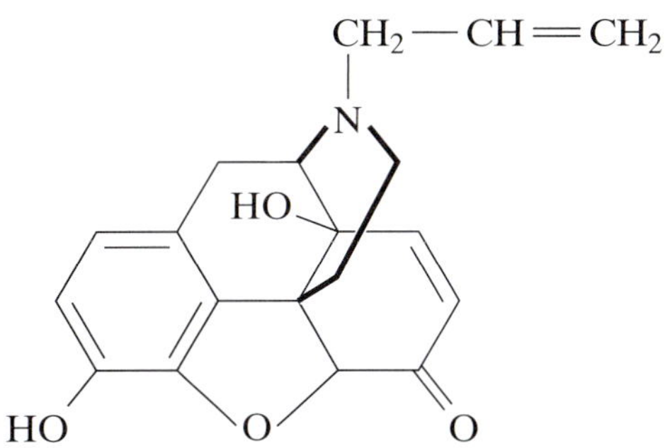

Figure 23.24 Naloxone, a narcotic antagonist, counteracts the effects of narcotics. It is used to treat overdoses of heroin and other narcotics.

the opiates but do so without producing any narcotic effects of their own, not even simple analgesia. Moreover, these *pure antagonists* so effectively counteract the opiates that they even induce withdrawal symptoms in active addicts. Although their greatest value lies in the emergency medical treatment of narcotic overdoses, naloxone and other opiate antagonists have been used in examining and treating eating disorders such as *bulimia* (insatiable hunger) and binge eating.

In studying just how the opiates produce their effects on the body and how naloxone and other antagonists are able to counteract these effects, scientists discovered that certain sites within the nervous system act as **opiate receptors**. The narcotic molecules fit snugly and precisely into these receptor sites, much like a key in a lock, and interfere with the signals flowing through the network (Sec. 16.14). In some circumstances this interference can result in the merciful blocking of a pain signal. Under other conditions, though, when the narcotic molecule enters the brain and occupies one of the receptor sites within its central network, it can garble fundamental perceptions of reality; warp the sense of space, time, and color; generate hallucinations; and produce extremes of emotional states.

Opiate receptors are extremely large molecules within the nervous system. When narcotic molecules interact with these opiate receptors, they block or alter signals flowing through the nervous system.

The antagonists seem to counteract these effects by distorting the opiate receptor site so that the narcotic molecule won't fit. Their action is something like changing the tumblers in the lock, at least for a while, so that the key no longer fits and the signals can no longer be distorted (Fig. 23.25). But the strange question all this provokes is, "Why should *animal* brains have evolved with receptors within them that only molecules produced by *plants* can fit?"

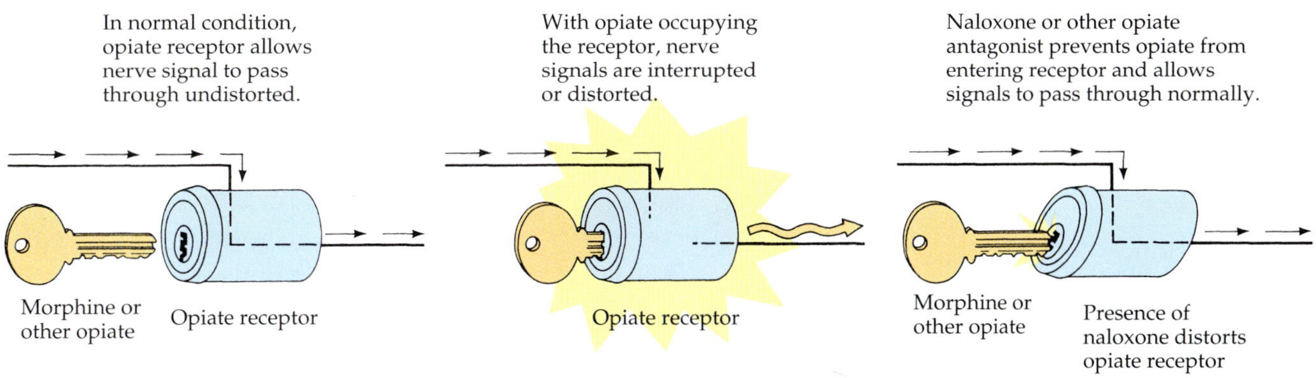

Figure 23.25 The effect of an opiate antagonist.

■ QUESTION

Is naloxone another example of a β-phenethylamine (Sec. 23.16)? If your answer is *yes*, trace the outline of the β-phenethylamine structure in naloxone. If your answer is *no*, explain why it isn't.

23.19 From Antagonists to Enkephalins

The whole idea that animal brains (more specifically, the brains of vertebrates) might contain within their communications networks certain molecular locks that only special molecular keys, produced by certain plants, could fit seemed so absurd that scientists began searching for molecules produced within the animal itself that would also fit these peculiar receptor sites. Among those looking for these special keys were J. Hughes and H. W. Kosterlitz of the University of Aberdeen (Scotland), L. Terenius of the University of Uppsala (Sweden), E. J. Simon of the New York University School of Medicine, and S. H. Snyder and C. B. Pert of the Johns Hopkins University School of Medicine. They found what they were looking for in substances isolated from the brains of pigs and calves.

These molecular keys of the animal kingdom are pentapeptides (peptide chains of five amino acids; Sec. 17.7), which their discoverers named *enkephalins* from Greek words meaning "within the head." Figure 23.26 shows the structures of two enkephalins, *methionine enkephalin* and *leucine enkephalin*. Notice that they differ only in the amino acid bearing the free carboxyl group, the amino acid that gives each its unique name.

As support for the idea of locks and keys in the network of the nervous system, these enkephalins not only produce analgesia but their effects can be blocked by the same naloxone that works against the plant narcotics. What's more, laboratory rats have become physically addicted to enkephalins after repeated injections. (One synthetic enkephalin in which the amino acids of methionine enkephalin have been modified a bit has about 15,000 times the painkilling power of morphine.)

Figure 23.26 Enkephalins, narcotic pentapeptides produced in the bodies of animals.

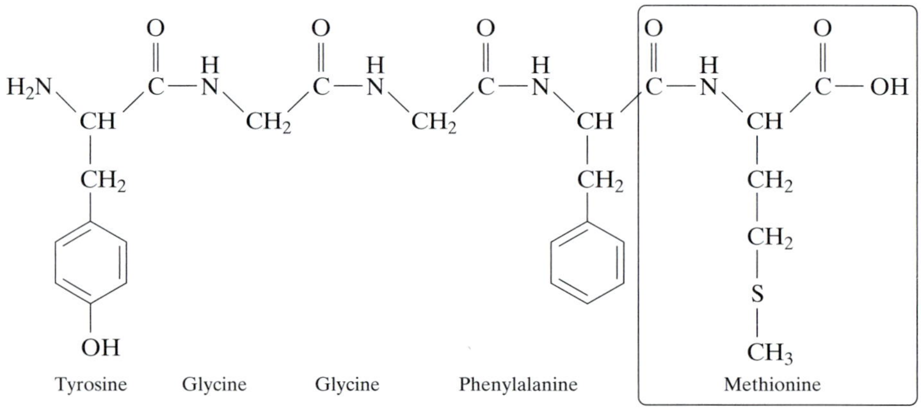

Methionine enkephalin

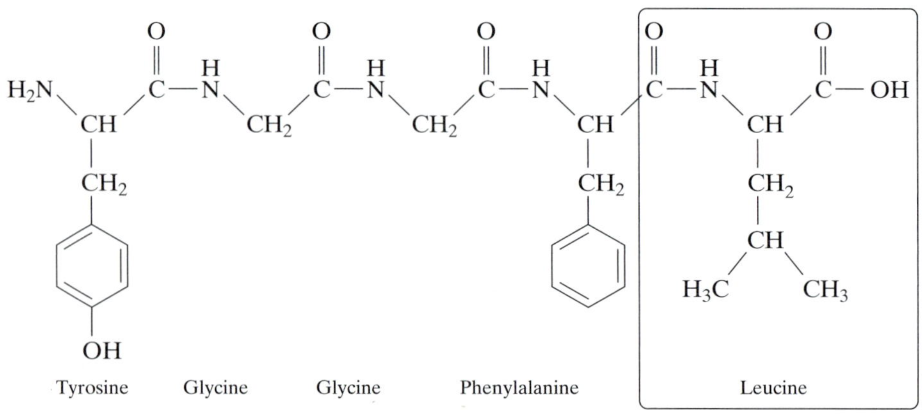

Leucine enkephalin

■ QUESTION

What three amino acids are common to both methionine enkephalin and leucine enkephalin?

23.20 Endorphins, Our Very Own Opiates

To cover the entire field of opiates produced within the animal body—the enkephalins as well as larger and more complex narcotic molecules—the term **endorphins** was coined as a combination of two words: *endogenous* (meaning "formed within") and *morphines* (the kinds of narcotics they mimic). Of these, the major opiate produced by the brain—more exactly by the *pituitary gland*, a small, round gland that hangs down from the base of the brain, roughly at the center of the head—is *β-endorphin*, a polypeptide composed of 31 amino acids. The first five of these amino acids are the same as those that form the methionine enkephalin chain (Fig. 23.27).

Endorphins are opiates produced within the bodies of animals.

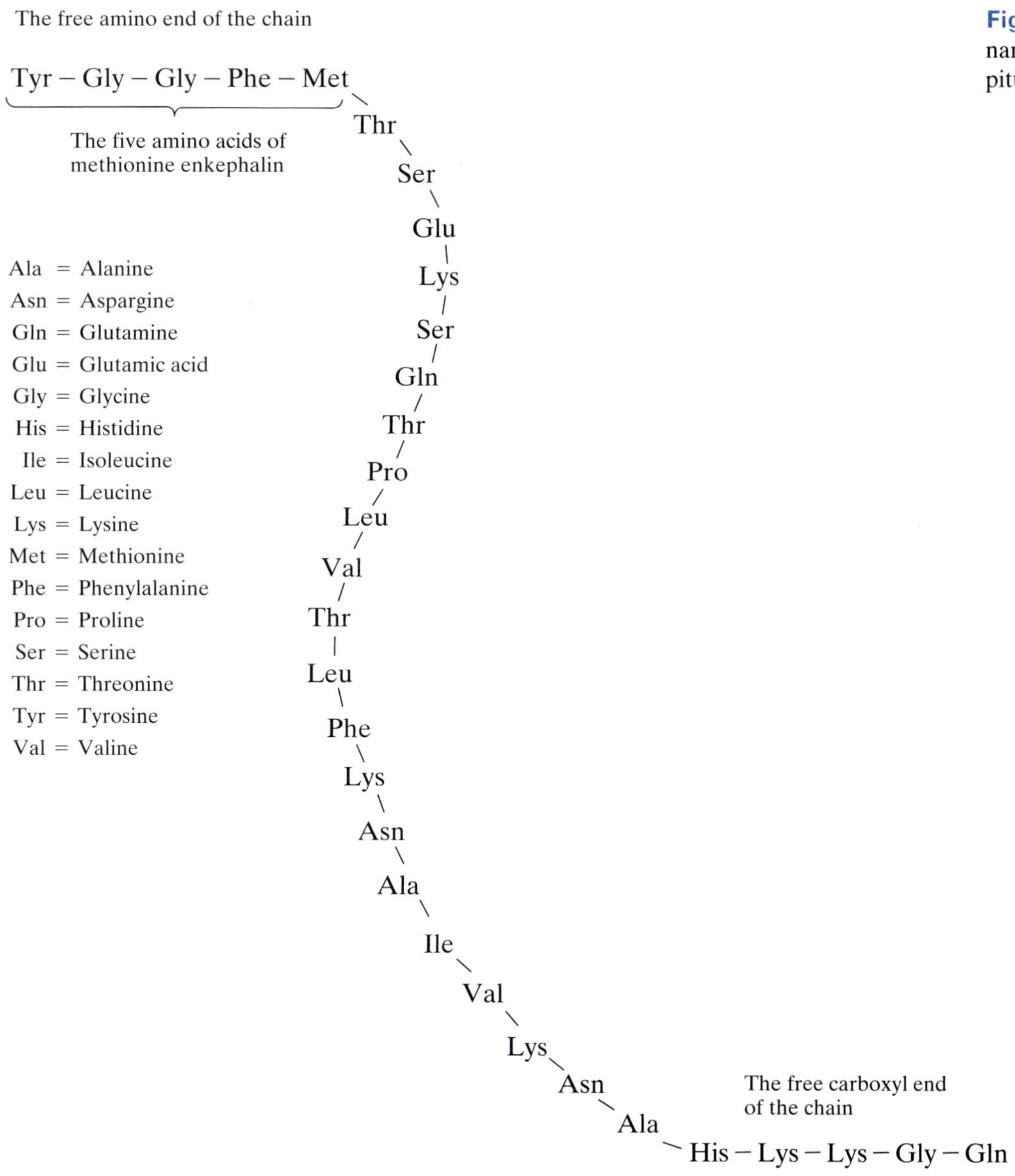

Figure 23.27 β-Endorphin, a narcotic produced by the pituitary gland.

It appears that endorphins are indeed the body's own opiates, generated as needed to reduce pain and perhaps to produce other effects as well. Sharply elevated levels of β-endorphin occur in the blood of women during childbirth and in their newly born infants, generated perhaps to ease the stress of birth on both mother and child. What's more, endorphins appear to produce the analgesia stimulated by some forms of acupuncture. Since naloxone can reverse the effect of acupuncture accompanied by low-frequency electrical currents, this form of the process appears to act by stimulating the formation of endorphins.

In addition to relieving pain, endorphins may also play other roles in the body. Their generation after long periods of vigorous exercise suggests that they may be responsible for the sense of well-being and euphoria that long-distance runners and other athletes experience.

■ QUESTION

What is the name of the amino acid of β-endorphin that bears a free $—NH_2$ group? What is the name of the one that bears a free $—CO_2H$ group?

23.21 The Power of the Placebo

As we've seen throughout this chapter, chemicals can affect profoundly our perceptions of the outside world and especially our sense of pain. In an odd reversal of roles, our perceptions themselves—our beliefs and expectations—can change the ways in which the chemicals and the physical phenomena of the outside world affect us. It all takes place through the *placebo effect.*

A **placebo** is a harmless and normally ineffective substance given to someone who complains of a particular ailment, simply to please and pacify the patient.

Those who practice medicine have long known of the power of the **placebo**, a harmless and (normally) ineffective substance, often a sugar pill or a starch pill in disguise, given to someone complaining of a particular ailment simply to pacify the person. (*Placebo* comes from the Latin and means "I shall please" or "I shall pacify.")

The idea behind the placebo is that the patient's aches and pains are often imaginary, and that since there is (presumably) nothing really wrong, a simple starch pill (a placebo) will make it appear to the patient that something important is being done and will end the complaints. Callous as this may seem, it often works. But not for the original reasoning. The power of the placebo appears to be real, not imaginary.

Simple placebos have eased pains from a variety of causes and a range of intensities. They have stopped coughs, promoted the healing of wounds, relieved depression, lowered blood pressure and pulse rates, and even produced side effects, including dryness of the mouth, headaches, and drowsiness.

It's possible, of course, that chemicals we've long believed to be effective medicines are actually operating through a placebo effect. If a physician says, for example, "This pill will cure you," and if we believe that "This pill will cure me," we may well obtain relief from the placebo effect alone. To examine this possibility, Herbert Benson of the Harvard Medical School studied the reported effectiveness of medicines used during the past two centuries to treat the chest pains of angina. From an examination of medical reports accumulated over that period, he estimated that newly introduced medicines have been effective in 70% to 90% of the patients receiving them, but that their effectiveness dropped to 30% to 40% after even newer ones were found. Perhaps, he concluded, the physician's faith in the newest medicine is transferred to the patient.

What's the origin of the placebo effect? Can faith alone authentically provide analgesia? To find out whether the opiate effect of endorphins is involved, three physicians of the University of California at San Francisco studied a group of dental patients whose wisdom teeth had just been extracted. As part of the study some were given placebos shortly after the extraction while others were given naloxone, the opiate antagonist described in Sec. 23.18. (None of the patients knew what any of their pills contained.)

Some of those who responded to the placebo by reporting a drop in the level of the pain—about one person out of three is just such a *placebo responder*—received naloxone as their next treatment. Their pain shot back up to about the level of those who don't respond to placebos at all. (Those who received naloxone as their only drug found no change in their level of discomfort, so the naloxone itself didn't *cause* any of the pain.) The conclusion reached in this study was that since the opiate antagonist naloxone raises the level of pain in the placebo responders, the analgesic effect of placebos appears to come from the release of endorphins by the body, which were subsequently blocked by the antagonist.

■ QUESTION

What characteristics identify a substance as a placebo?

PERSPECTIVE

The Scientific Method as a Test of Reality

If a simple starch pill can ease pain (for the third of us who are placebo responders) simply because we believe that it will, is there some way to tell whether aspirin, for example, is any more effective than, say, a teaspoon of starch given to us by someone who speaks with authority, wears a white coat with a stethoscope dangling out of the pocket, and occupies a doctor's office—someone, that is, who inspires us to believe?

There is, and it's an elegant example of the scientific method at work. The question we ask is: "Can we tell whether pain relief results from a placebo effect or comes from the inherent analgesic action of a compound, independent of a patient's faith or belief in the healing effect of the medicine?" For the answer we use a simple test of reality that operates through a powerful scientific technique called a *double-blind study*. The procedure involves distributing the medication under examination to a large number of people who have been divided into two groups by random assignment. One set, the *control group*, receives a placebo. The other, the *test group*, receives the authentic medication, aspirin for example. One of the keys to the study's value lies in separating the preparation of the pills from their distribution so that no one taking the pills can possibly know whether it's the medicine or the placebo, thus nullifying any placebo effect.

To avoid any possibility that those conducting the test might pass along a clue, even unconsciously by a facial expression or a bodily attitude, the identity of each individual pill is kept hidden from those distributing them. To achieve this level of secrecy the double-blind test requires two sets of examiners. One identifies each pill with a code number that reveals its contents, but is "blind" to the method of distribution. The other set of examiners, "blind" to the significance of the codes, distributes the pills to the members of both groups. During the double-blind trial itself, no one, neither the investigators nor the participants, knows whether any particular recipient is in the control group or the test group. With this level of concealment, and with the pills appearing identical in all respects except for their coding, the placebo effect operates equally with everyone involved. It's only at the end of the test, through a correlation of the effects of the pills with their coding that any placebo effect comes to light.

If those receiving the authentic medication respond differently from those receiving the placebo, the difference must be due to the real potency of the medicine and not simply to a placebo effect. Here we see the scientific method at work, isolating the narrow question we are asking and removing any possibility that those receiving the pills (placebo or medication) can receive any clues about their contents. With this double-blind procedure, we can be certain that belief and expectation play no part in the answer we obtain.

It's a good test of reality; it's scientific evidence of the effectiveness of any chemical that produces a human response. Use of a double-blind study, for example, proves clearly that aspirin is an effective analgesic, producing relief from pain beyond that due to a placebo effect alone. It's important to treat claims not backed by studies like these with a good deal of skepticism. They may not represent a form of reality beyond the simple release of endorphins from someone's pituitary gland. In the double-blind approach we glimpse a chemical reality untainted by our own expectations. ■

EXERCISES

For Review

1. Following are a statement containing a number of blanks and a list of words and phrases. The number of words equals the number of blanks within the statement, and all but two of the words fit correctly into these blanks. Fill in the blanks of the statement with those words that do fit, then complete the statement by filling in the two remaining blanks with correct words (not in the list) in place of the two words that don't fit.

The world's most widely used __________ is __________, which is known chemically as __________ and is prepared by the reaction of __________ acid with __________. The biological effects of this medication can be traced to its ability to interfere with the body's synthesis of __________, which play a major role in the generation of inflammation and fever and in the sensation of pain. While __________ lacks aspirin's ability to reduce inflammation, it can serve as a substitute for those of us who may be allergic to aspirin.

Among naturally occurring substances, several members of a group of compounds found in __________ and known as __________ produce notable physiological effects. The first member of this class to be isolated was __________, a powerful __________ that is the largest single component of __________. An attempt to produce a less addictive drug by diacetylating morphine led to the powerfully addictive narcotic __________. Another highly addictive alkaloid, __________, is found in the leaves of the coca bush and was once used medically as a __________.

__________, a compound whose molecular structure resembles that of morphine, belongs to a class of compounds, known as __________, that can counteract the effects of narcotics. Studies involving the action of these compounds have revealed the existence of narcotic __________ that are produced within the human body. These compounds, named __________ to reflect their origins and actions, appear to be responsible for the operation of the __________, which allows a simple starch or sugar pill to relieve pain and produce many other remarkable physiological effects.

acetaminophen	local anesthetic
acetic anhydride	morphine
acetylsalicylic acid	naloxone
alkaloids	narcotic
analgesic	opiate antagonists
aspirin	opium
β-phenethylamine	plants
cocaine	polypeptides
heroin	prostaglandins
ibuprofen	salicylic

2. Explain, identify, describe, or define the following:

amphetamines	methadone
β-phenylethylamines	oil of wintergreen
codeine	procaine
ibuprofen	reserpine
lithium carbonate	tetrahydrocannabinol

3. Match each of the following with the corresponding contribution to our understanding of drugs, medicines, and the effects of chemicals on perception:

______ a. Albert Hofmann	1. first to isolate codeine
______ b. Felix Hofmann	2. first to prepare and to feel the effects of LSD
______ c. Karl Koller	3. early workers in the field of endorphins and enkephalins
______ d. Pierre-Jean Robiquet	4. first to isolate morphine
______ e. Friedrich Sertürner	5. first to use cocaine to anesthetize the eye
______ f. J. Hughes, H. W. Kosterlitz, C. B. Pert, E. J. Simon, S. H. Snyder, and L. Terenius	6. provided proof of aspirin's analgesic effect

4. Identify the physiologically active substance that comes from each of the following:

a. the bark of the willow tree
b. the coca bush
c. the sap of the poppy
d. the pituitary gland
e. the belladonna plant
f. climbing shrubs of India

5. Describe the connection between the items in the following pairs:

a. acetylsalicylic acid and Reye's syndrome
b. salicylates and prostaglandins
c. methionine enkephalin and leucine enkephalin
d. methionine enkephalin and β-endorphin
e. naloxone and diacetylmorphine
f. carisoprodol and meprobamate
g. reserpine and Li_2CO_3

6. What compound, substance, or other topic of this chapter owes its name to the following?

a. the Latin name for the willow tree
b. the Roman god of dreams
c. a Latin word for something that pleases
d. the Latin word for a physician's art or practice
e. the Greek word for "stupefying"
f. Greek words meaning "in the head"
g. the Italian term for "beautiful lady"
h. the botanical genus of the meadowsweet plant
i. the bodily organ from which they were first isolated

j. a trade name dating from about 1900

k. words meaning "morphine-like compounds produced within the body"

7. What benefits does aspirin provide? What are the risks in using it?

8. What are platelets? What is aspirin's effect on platelets? What are the risks and benefits of aspirin's effect on platelets?

9. Why was the acetylation of salicylic acid important to the introduction of aspirin for general use as an analgesic?

10. What is the source of the extra strength in "extra-strength" aspirin tablets?

11. What's the difference between an analgesic medication and an antipyretic medication?

12. What three symptoms of illness or injury does aspirin relieve?

13. Name a medication that contains (a) chlorine, but not fluorine; (b) fluorine, but not chlorine; (c) both chlorine and fluorine. Give a trade name for each and describe what each is used for.

14. In what way are methadone and heroin alike in their effects on humans? In what way are they different?

15. To what class of basic, bitter-tasting, nitrogen-containing plant substances do both caffeine and nicotine belong?

16. Name two alkaloids that are used often, in large quantities, and legally throughout most of the world, even though they can be addictive. In what form is each consumed?

17. What family of plants serves as a natural source of the tropane alkaloids?

18. What symptoms are alleviated by (a) aspirin, ibuprofen, *and* acetaminophen? (b) aspirin and ibuprofen, but *not* acetaminophen?

19. Name three mood-altering chemicals that are accepted for general, nonmedical use by adults in many countries of the world. Name three mood-altering chemicals whose general, nonmedical use by adults is prohibited in many countries of the world.

20. Name and give the structure of a compound you could use to convert an alcohol or a phenol into an ester.

21. Chemically, cocaine contains two different functional groups, one of which is an amine. What is the other functional group? (Note that this other functional group occurs twice in the molecule.)

22. Describe the chemical or biological means by which each of the following affects our perception of pain: (a) aspirin; (b) morphine; (c) naloxone; (d) β-endorphin.

23. Why is diacetylmorphine a particularly dangerous chemical?

24. What role did the Harrison Narcotics Act of 1914 play in the history of the use of narcotics in the United States?

25. In what way is the combination of a narcotic and an opiate receptor analogous to a key and a lock?

A Little Arithmetic and Other Quantitative Puzzles

26. As we saw in Section 20.2 and Table 20.1, aspirin has an established LD_{50} and has been fatal to small children. To provide small doses of aspirin for children a "baby" aspirin is available in tablets containing about 80 mg aspirin each. To the nearest 5%, what percentage of an adult's aspirin tablet is found in a "baby" aspirin tablet?

Think, Speculate, Reflect, and Ponder

27. Which one of the muscle-relaxant tranquilizers of Figure 23.22 is a carboxylic acid?

28. Under what conditions or in what circumstances is it dangerous, and possibly fatal, to use aspirin?

29. All acetylations with acetic anhydride always produce one particular organic compound in addition to the desired product. What organic compound is always generated in these reactions?

30. Draw the portion of their molecular structures that is common to all the salicylates.

31. Aspirin tablets kept for many months, especially in hot, humid climates, often smell of vinegar. Why?

32. Under what conditions is acetaminophen a better medication than aspirin? In what circumstances is it inferior to aspirin?

33. What chemical do you think has the longest history of use as a mood-altering drug? Describe your reasoning.

34. The 18th Amendment to the U.S. Constitution prohibited the sale and transportation of alcoholic drinks. The 21st Amendment permitted these activities once again. Why do you think the 18th Amendment was adopted? Why do you think the 21st Amendment was adopted? Discuss the factors you think led to each.

35. Describe the evidence that the analgesic effect of some forms of acupuncture is based on release of endorphins.

36. Describe in your own words how you would distinguish between a "medicine" and a "drug."

37. Do you think that the use of all mood-altering or perception-altering drugs should be prohibited except for medical purposes? Describe your reasoning.

38. In some localities, in the United States and in other countries, establishments commonly known to many of us as "drugstores" are called, instead, "pharmacies" or "apothecaries." Do you think we ought to discontinue the use of the term drugstore because of its connection with the word drug? Describe your reasoning. If you believe we ought to discontinue the use of drugstore, what term would you use to replace it?

39. Describe how a double-blind investigation eliminates the possibility that a placebo effect is operating.

40. Do you think that a person who drinks several cups of decaffeinated coffee, thinking mistakenly that it is caffeine-containing coffee, might show signs of agitation or sleeplessness as a result? If this would occur in some cases, to what would you attribute the phenomenon? Describe how you might go about determining whether agitation or sleeplessness after drinking several cups of decaffeinated coffee might be due to a placebo effect.

Web Questions

Use the Web to answer or accomplish each of the following. Report your Web source for each response.

1. Summarize the history of salicin. Describe its isolation from the bark of the willow tree, its medical applications, and its chemical conversions to salicylates and other substances useful in medicine.

2. Identify Sune K. Bergstrom, Bengt I. Samuelsson and John R. Vane and describe their contributions to one of the topics of this chapter.

3. What is the current legal status of the medical use of marijuana in the United States for relief of side effects of the treatment of cancer, AIDS and other diseases? Are there any states or regions in which its use is legal for medical purposes? What conditions govern any such use? Have Congress or the Supreme Court acted to permit or prohibit such use? Are any proposals currently under consideration to allow or to prohibit such medical use?

4. Identify any societies, countries or regions of the world where alcohol, caffeine and/or nicotine are considered illicit or illegal drugs. What restrictions exist on their use? What penalties are associated with their use?

5. Summarize the current use of fluoxetine, as Prozac and any other brand names. How many people, worldwide and/or in the United States, are estimated to use the drug? How many prescriptions are written for it annually?

6. Summarize the life of Albert Hofmann. Describe the details of his discovery of LSD and his reactions to it.

7. Thomas Sydenham has been called the "Hippocrates of English Medicine" and the "Shakespeare of Medicine." Write a brief biography of Sydenham explaining how and why he earned these descriptions and describing his connections to the topics of this chapter.

8. Describe an actual double-blind test of a newly developed drug. Identify the drug under examination, the number of patients in the study, the conclusions reached, and the principal investigator.

APPENDIX A

A Short Guide to Exponential Notation

COUNTING BEYOND NINE

Counting is easy, recording numbers isn't. The Romans used letters: I for one, V for five, X for ten, C for a hundred, and so on. To Julius Caesar, eight was VIII, nine was IX.

Roman numerals turned out to be too cumbersome. After the Roman Empire fell, Western civilization adopted a simpler scheme for recording numbers, a system based on Arabic numerals: 1, 2, 3, and so on. One of the beauties of this system is that you can multiply any number by ten simply by appending a zero. Three is 3, ten times 3 is 30, and ten times 30 is 300.

With these numerals, 9 is the largest digit. One more than 9, ten, has to be written as 10, which is really no more than ten times 1, just as 30 is ten times 3. Now, long after the Romans, a hundred isn't C any longer. It's ten times ten, 100. A thousand, 1000, is ten times a hundred, or ten times ten times ten if you wish. Each zero we append multiplies everything to its left by ten.

The problem with this system is that very soon, with enormously large numbers, the string of zeros becomes as cumbersome as the Roman letter-numbers used to be, long ago. But ingenuity triumphs again. Just as we simplified the recording of numbers by switching from Roman letters to our modern numerals, we now simplify long strings of zeros by using *exponential notations of ten*.

An exponent itself is just a superscript that tells us how many times the number it's hanging over must be multiplied by itself. The superscript 2 defines 3^2 as 3×3, or 9; the superscript 4 in 3^4 signifies $3 \times 3 \times 3 \times 3$, or 81. In these examples the 2 and 4 are the *exponents* and the 3 is the *base*, which is the number being multiplied by itself any number of times.

(Technically, that paragraph needs a bit of explanation. With 3^2 we surely multiply 3 *by itself* only *once*, not twice. But that's the great strength of mathematics over verbal language. The statement

$$3^2 = 3 \times 3$$

is much more elegant in its simplicity than the dozens of words we'd have to use, in cumbersome constructions, to explain the whole thing in proper sentences.)

In exponential notation to the base 10 we simply replace a string of zeros by a 10 and its exponent. Since 100 is 10×10, it's also 10^2; similarly, 1000 is 10^3, and

10,000 is 10^4. We can plug other numbers into this system as well. For example, 30 is 3×10, 300 is 3×100 or 3×10^2, and 3000 is 3×1000 or 3×10^3. Because of its great utility in expressing the very large numbers of science (and the very small numbers, too, as we will soon see), this form of exponential notation is often called *scientific notation*.

As a practical matter, the exponent of 10 indicates the number of places the decimal point has to be moved to the right to represent the number as it's written in our more common notation. In this way 3×10^3, which can also be written $3. \times 10^3$ (note the decimal behind the 3), becomes 3000 as the decimal moves three places to the right. Again, 3.0173×10^3 translates into 3017.3 with, again, the decimal skipping three places to the right.

As the other side of this coin, *negative* exponents move the decimal to the *left*. The number 3×10^{-1} becomes .3, often written 0.3 to add a bit of clarity. In another example, 3.0173×10^{-5} becomes 0.000030173 as we move the decimal five places to the left.

At the border between mathematics and chemistry, Avogadro's number, 6.02×10^{23}, turns out to be

602,000,000,000,000,000,000,000

Questions: Can you write Avogadro's number in Roman numerals?
Answer: Probably not.

One final thought. In working with exponential notation we can multiply any two numbers together simply by adding their exponents algebraically *as long as their base numbers are identical*. With the base 2 we see that $2^2 \times 2^3$ is the same as 4×8, or 32. Notice that we get the same result by adding the two exponents $(2 + 3)$ to get 2^5, which is still 32. With base 10, the product of 100 (10^2) and 1000 (10^3) is 100,000 (10^5). Here again, the exponent 5 is the sum of the two exponents $(2 + 3)$. By the use of algebraic addition, 0.01 (10^{-2}) × 1,000,000 (10^6) is 10,000 (10^4). Here $6 - 2 = 4$, the exponent of 10 needed to give 10,000.

EXERCISES

1. Write each of the following in common numerals. For example, 3.779×10^2 is 377.9.

a. 2×10^3
b. 4.796×10^3
c. 0.0072553×10^4
d. 142.99×10^{-2}
e. 56.97×10^{-5}

2. Write each of the following in exponential notation to the base 10. For example, 4772.34 is 4.77234×10^3.

a. 47,000
b. 96,723.70
c. 12
d. 0.004195
e. $(5.67 \times 10^3) \times (2 \times 10^{-5})$

3. Write each of the following in common numerals, without using exponents.

a. 10^8
b. 3^4
c. 4^3
d. $2^3 \times 3^2$
e. $24 \times 10^2 \times 2^5 \times 10^{-5}$
f. $2^9 \times 10^{-3}$

The Metric System

THE MEASURE OF ALL THINGS

A Revolution in Measurement

In 1789 the people of France revolted against King Louis XVI. With the success of their great revolution the French not only overthrew their monarch and the system of government he represented, but they began another kind of revolution as well—a revolution in the science and technology of measurement.

Probably since well before recorded history, and surely from the times when the first tribes built complex structures and engaged in trade and commerce, *units* have been vital to any human activity involving measurement: units of length, of area, and of volume; units of mass (or weight) and of time; and units of money. The dollar is a unit, and so are the second, the pound, and the yard. Ask how many miles per gallon a car gets and you are asking about units of distance and volume. Consider the cost of a piece of fabric and you deal with units of area and money.

Until the matter of the Bastille was settled by the French, units of measurement used by the various peoples of the world suffered from two important defects:

- They were based on highly arbitrary standards that were sometimes difficult or impossible to reproduce accurately should the original, physical standard of the unit be lost or destroyed.
- They were converted to other units only by cumbersome and laborious calculations.

To illustrate this second problem, the matter of interconversions, consider a short example. The English *yard*, a measure of length originally established by Edward III in 1353, consists of 3 feet, each of which is made up of 12 inches. Moreover, there are 5280 feet to a mile. Now find the number of yards in 4 miles. It's an easy enough problem, but with the cumbersome English system of units it takes a few moments to solve, even with a calculator. To make matters even worse, before the French Revolution each country of Europe had its own set of units. International trade thus had a double burden. Added to the conversions required within any particular country were those necessary to switch from the units used by one country to those used by another.

In addition to these intricate and time-consuming interconversions, as well as the additional difficulties of the occasional replacement of lost or damaged standards, there was also another important concern in revolutionary France. As in other European countries, the French units of weights and measures were based on arbitrary standards maintained by royal decree. With the revolution they had become anachronistic relics of an overthrown regime, relics that were to be replaced in the conduct of the daily life of France by a more natural and a more rational system of measurement, just as autocratic rule by the king had itself been replaced by a more natural and more rational system of government.

That more rational system, whose units are easily defined and just as easily interconverted, is the *metric system* of weights and measures. Its ease of use is so compelling that the metric system has been adopted as the national system in most countries of the world and has become the universal system of measurement in science. The word itself, *metric*, comes from the Greek word *metron*, to measure.

A Natural Standard of Length: The Meter

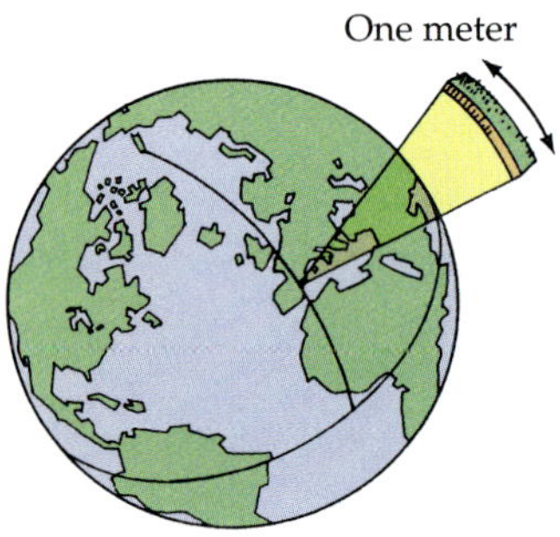

Figure B.1 The original definition of the meter: 1/10,000,000 the distance from the equator to the pole.

In 1791 the French revolutionaries—by now the French government—asked a commission of 12 learned scientists to develop this new set of weights and measures, to be based on readily available standards and ease of interconversion of units. The commission chose the earth itself as the fundamental standard for all measurements of length. The *meter*, the unit of length they proposed, was to be exactly 10^{-7} (1/10,000,000) of the distance from the earth's equator to either of its poles (Fig. B.1).

The meter and all other units of length in the metric system can be interconverted by multiplying or dividing by 10 or by some power of 10. There are, for example, 100 *centimeters* to the meter just as, in monetary terms, there are 100 cents to a dollar. Further, 1000 meters make a *kilometer*. (The prefix *kilo-* is derived from the Greek *khilioi*, a thousand.) With a relationship based purely on powers of 10 it's easy enough to see that there must be 100×1000, or 100,000, centimeters to a kilometer. This decimal relationship makes interconversions among metric units extraordinarily simple.

A Natural Standard of Mass: The Gram

Units of area and of volume are the squares and cubes, respectively, of units of length. A square with a side that's 1 kilometer long has an area of 1 square kilometer (1 km^2). Similarly, a cube whose side is 1 meter has a volume of 1 cubic meter (1 m^3). A cube just 1 centimeter on a side is a cubic centimeter (1 cm^3 or 1 cc). A more convenient metric unit of volume, the one in general scientific use today, is the *liter*, a word derived from an ancient French measure of volume, the *litron*. The liter is the volume of a cube almost exactly (as we will soon see) 10 centimeters on a side.

For the unit of mass the French commission chose the *gram*, which was to be the mass of 1 cubic centimeter of water, measured at the temperature of water's maximum density, 4°C. (The word *gram* comes from the Latin and the Greek *gramma*, a small weight.) For ease of measurement, the actual standard of mass that was adopted was the *kilogram*, the mass of a cube of water 10 cm on edge and therefore occupying a volume of 1000 cm^3. Metric units of mass, like those of length, are all multiples of 10 and are easily interconverted.

On December 10, 1799, near the close of the 18th century, the French Legislative Assembly adopted these natural standards of measurement as the metric system, to be used as the legal and official system of weights and measures of the French people. The idealism of the French Revolution was well on its way to touching every scientific measurement to be made since then.

A Return to Arbitrary Standards: Making Light of the Meter

It wasn't easy, back at the beginning of the 19th century, to measure accurately the arc of the earth from the equator to the pole. But the arc was measured nonetheless, with the best contemporary science and technology available, and a platinum bar was cut and polished to exactly 1 meter. This bar was the standard for the length of the meter during most of the 19th century.

Toward the end of the 19th century came the realization that the platinum bar might be wearing a bit here and there and that its reliability might not last much longer, so a new and sturdier bar was prepared as the standard for the meter. Since it was far easier to duplicate the meter of the old platinum standard than to measure the arc of a quarter of the earth's meridian, and since it had become apparent that the original measurement of the arc—the one that produced the first meter—wasn't very accurate by contemporary standards, the meter was redefined late in the 19th century as the distance between two etched lines on a sturdy metal bar. The distance between these two lines duplicated exactly, at least to the best technology available in the late 19th-century, the length of the original meter. This 19th-century standard, though, was defined with no idealistic reference to the distance between the equator and the pole, and without even a mention of the original platinum bar. As in olden days, the meter was simply and arbitrarily defined as the distance between two scratches on a certain metal bar, and that was that.

Today, with better technology than was available to 19th-century scientists, and perhaps with more wisdom than was possessed by the French revolutionary idealists, science has abandoned concrete, physical standards of length in favor of one that's enormously accurate and reproducible, the speed of light in a vacuum. Today's meter is defined as the distance covered by light as it travels in a vacuum for exactly 1/299,792,458 of a second (Fig. B.2). It's a far cry from the revolutionary 10^{-7} of the arc of the earth stretching from the equator to the pole, or even from the distance between two lines on a metal rod, but it's the same as the length separating those two etched lines, and it's now a meter we can live with.

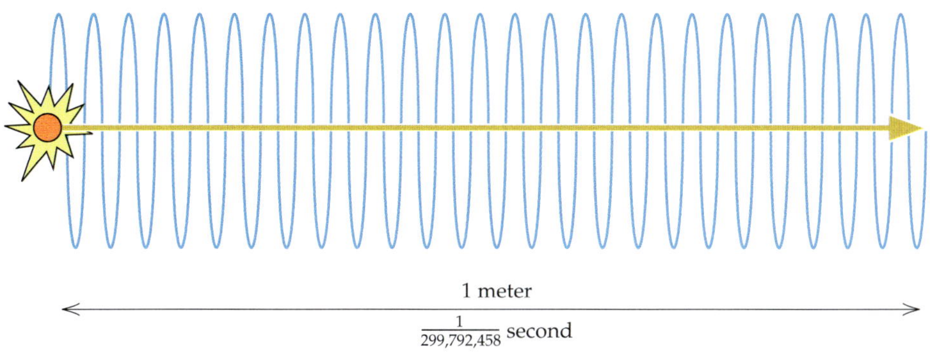

Figure B.2 The modern (1984) meter: the distance light travels, in a vacuum, in 1/299,792,458 second.

Another Arbitrary Standard: The Kilogram

For mass, the original metric ideal was to combine the unit of mass with the unit of length through the density of water. The kilogram was to be the mass of a cube of water exactly a tenth of a meter on edge. It was an elegant combination of cleverness and simplicity, but it didn't work.

As with the measurement of the arc of the earth, the problem lay in the matter of accuracy. Early in the 19th century a platinum cylinder was fabricated with a mass identical to that of exactly 1000 cm^3 of water (as measured in those days) at the temperature of water's greatest density. This mass was declared to be the standard for exactly 1000 g, or 1 kg. Later in the 19th century, with progress in the science of measurement, the volume of exactly 1 kg of water—again, at the temperature of its greatest density—was found to be 1000.028 cm^3, off just a bit from the precise 1000 cm^3 of the metric ideal.

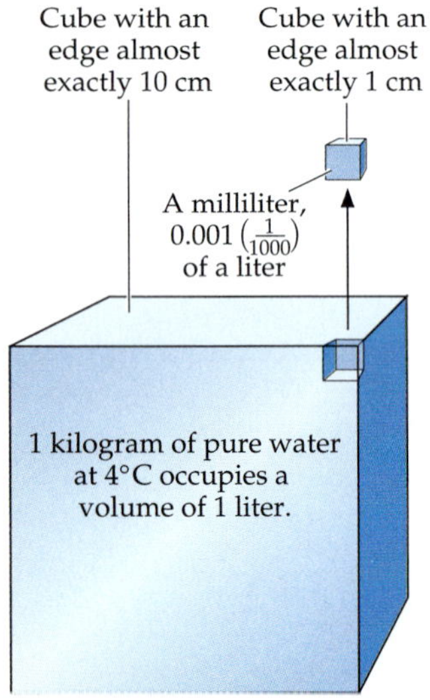

Figure B.3 The liter and the milliliter.

That's certainly not enough of a difference to affect anything in our daily lives, but it's an error large enough to give a scientist sleepless nights. With the mass of the kilogram firmly established and easy to reproduce, the simplest solution was to change the standard unit of volume. In a simple switch of the term that's defined and the basis for the definition, the standard *liter* (abbreviated *L*) is now the volume of exactly 1 kg of pure water at the temperature of its greatest density at 1 atmosphere of pressure. A smaller unit of volume, often more convenient, is the *milliliter* (mL), just one thousandth of a liter (Fig. B.3). There's no noticeable difference between the milliliter and the cubic centimeter in the chemistry we live with daily, but to the theory and practice of a very accurate science the milliliter wins out over the cubic centimeter.

The English System and the Metric System

In the United States the metric system has been a legal system of weights and measures since July 28, 1866. But an older, American adaptation of the English system has been in common use for so long that the two systems, English and metric, now coexist in the United States. Following are some useful comparisons of the two, more detail on the various metric units, and some common and ordinary quantities in both metric and U.S.–English units (Figs. B.4 to B.7).

Metric Units of Length

Unit	Prefix	Abbreviation	Multiple of the Meter
millimeter	milli = one-thousandth	mm	0.001 or 1/1000 of a meter
centimeter	centi = one-hundredth	cm	0.01 or 1/100 of a meter
decimeter	deci = one-tenth	dm	0.1 or 1/10 of a meter
meter		m	
kilometer	kilo = one thousand	km	1000 meters

1 km = 1000 m = 100,000 cm = 1,000,000 mm

1 m = 100 cm = 1000 mm

1 cm = 10 mm

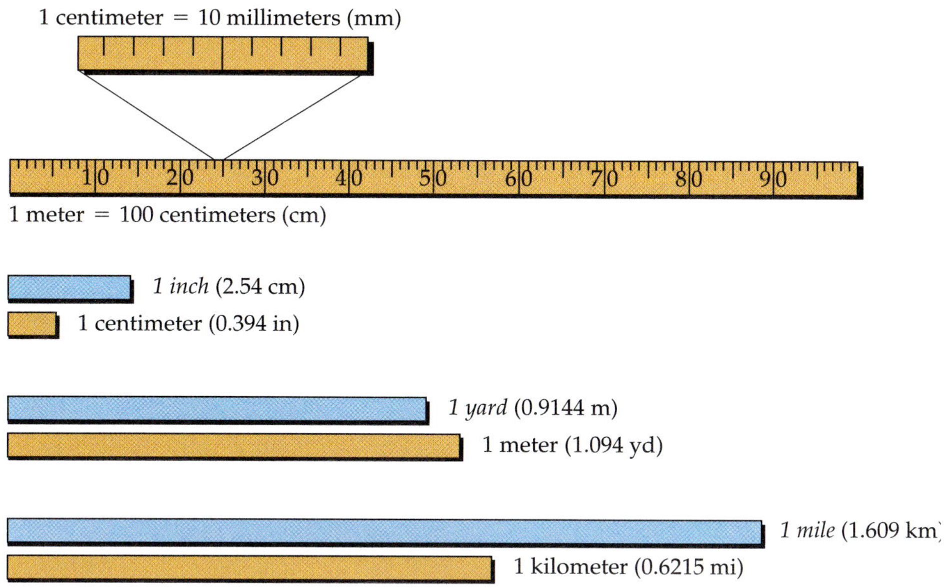

Figure B.4 Measures of length.

Length in the Metric System and in the U.S.–English System

1 inch = 2.54 cm
1 cm = 0.394 inch
1 foot = 30.48 cm
1 yard = 91.44 cm = 0.9144 m
1 m = 39.37 inches = 1.094 yard
1 mile = 1.609 km
1 km = 0.622 mile

Metric Units of Volume

Unit	Prefix	Abbreviation	Multiple of the Liter
milliliter	milli = one-thousandth	mL	0.001 = 1/1000 of a liter
liter		L	

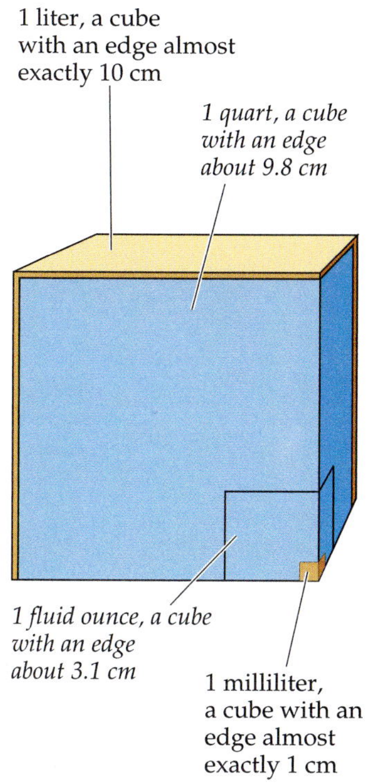

Figure B.5 Measures of volume.

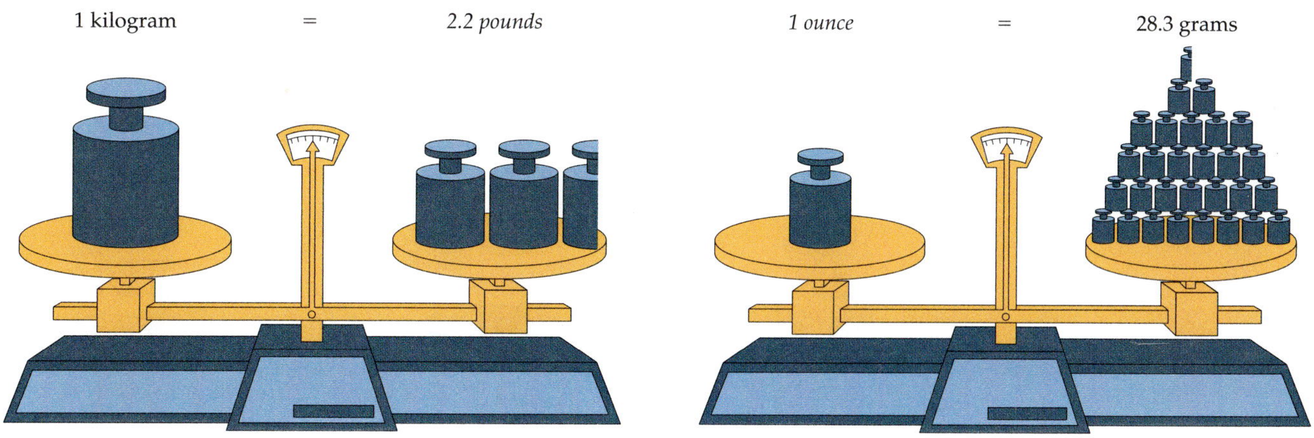

Figure B.6 Measures of mass.

Volume in the Metric System and in the U.S.–English System

1 fluid ounce = 29.6 mL
1 mL = 0.0338 fluid ounce
1 L = 1.057 quart
1 quart = 0.946 liter

Metric Units of Mass

Unit	Prefix	Abbreviation	Multiple of the Gram
milligram	milli = one-thousandth	mg	0.001 or 1/1000 of a gram
gram		g	
kilogram	kilo = one thousand	kg	1000 grams

Mass in the Metric System and in the U.S.–English System

1 ounce = 28.3 g
1 g = 0.0353 ounce
1 pound = 0.454 kg
1 kg = 2.20 pounds

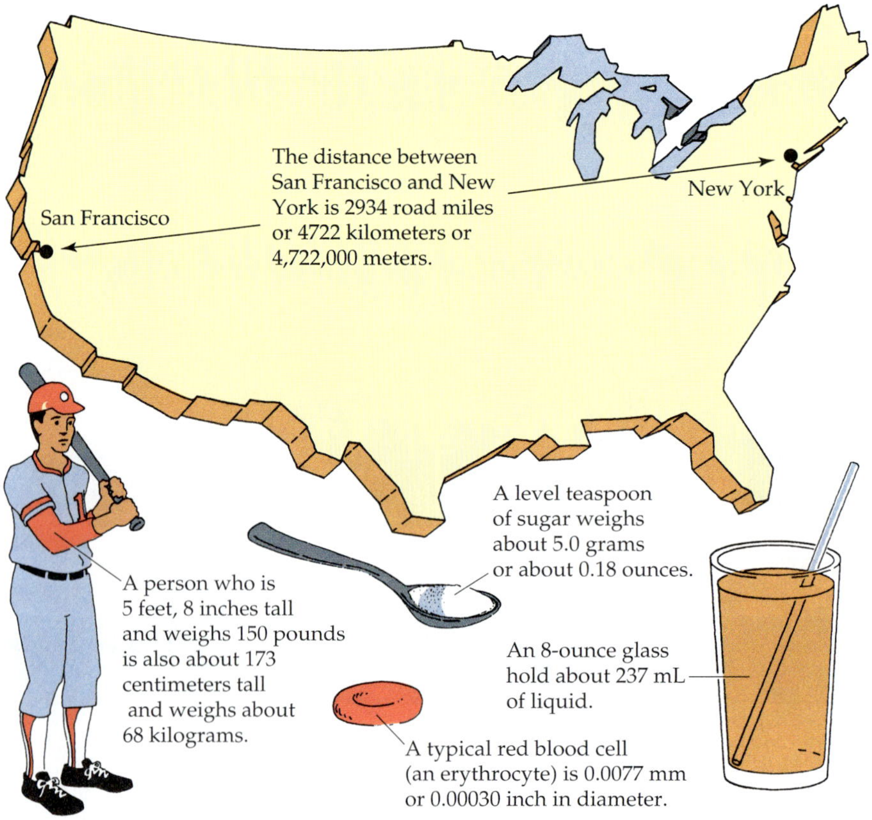

Figure B.7 Typical measurements.

Using Units Cancellation

PETER PIPER PICKED A PECK OF PICKLED PEPPERS

Plenty of mathematical problems sound like that old tongue twister, "If Peter Piper picked a peck of pickled peppers, how many pecks of pickled peppers did Peter Piper pick?" Having the right facts and knowing how to use them make working the simple arithmetic of most chemical problems a lot easier than repeating this or any other tongue twister. The puzzle of deciding what to do with the numbers given in the problem is solved easily by using the technique of *units cancellation*.

Units cancellation is no more than the procedure of canceling out any unit that appears in both the numerator of one measure and the denominator of another. It's a technique that we use, often unconsciously, in working with simple fractions. To return to Peter Piper for an illustration, suppose he picks three-eighths of a peck of pickled peppers in an hour. If Peter works steadily at this rate, how many peppers will he have picked in two-thirds of an hour? We get the answer by multiplying his rate of picking, $\frac{3}{8}$ peck/hour, by the amount of time he works, $\frac{2}{3}$ hour. Writing out the multiplication, we get

$$\frac{3\text{ peck}}{8\text{ hour}} \times \frac{2\text{ hour}}{3}$$

We know we can divide both the numerator and the denominator by 3 without changing the product of the two fractions. In effect, these two divisions cancel the 3s that appear in the numerator of one fraction and in the denominator of the other:

$$\frac{\cancel{3}\text{ peck}}{8\text{ hour}} \times \frac{2\text{ hour}}{\cancel{3}}$$

This gives us

$$\frac{1\text{ peck}}{8\text{ hour}} \times \frac{2\text{ hour}}{1}$$

Now we divide both the numerator and the denominator by 2 to get the answer:

$$\frac{1\text{ peck}}{\cancel{8}\text{ hour}} \times \frac{\cancel{2}\text{ hour}}{1} = \frac{1}{4}\text{ peck}$$

We can use this same principle, one in which we divide both the numerator and the denominator by a common term, in the units cancellation technique. In the problem we're dealing with here, we divide both the numerator and the denominator by the same factor—the unit *hour*—to get the answer in the unit we want, *peck*:

$$\frac{1\text{ peck}}{4\,\cancel{\text{hour}}} \times \frac{\cancel{\text{hour}}}{} = \frac{1}{4}\text{peck}$$

Now let's turn from the world of tongue twisters to the real world. Suppose you're going to drive a car that gets 20 miles/gallon on a trip of 500 miles and you use gasoline that costs \$1.20/gallon. What's the total cost of the gasoline for the trip? What do you multiply and what do you divide to get the answer? It's easy with units cancellation.

First, decide what measurement holds in its numerator the unit you want in your final answer. To get an answer in dollars and cents, we start with \$1.20/gallon. Now cancel out the gallon of the denominator by dividing by the term (20 miles/gallon). Carry out this division by *multiplying* by the term's reciprocal (gallon/20 miles):

$$\frac{\$1.20}{\text{gallon}} \times \frac{\text{gallon}}{20\text{ miles}}$$

With cancellation of the unit *gallon*, you learn that the trip is costing you

$$\frac{\$1.20}{20\text{ miles}}$$

Now multiply by 500 miles to cancel out *miles* and you end up with \$30 for the entire trip:

$$\frac{\$1.20}{\cancel{\text{gallon}}} \times \frac{\cancel{\text{gallon}}}{20\,\cancel{\text{miles}}} \times \frac{500\,\cancel{\text{miles}}}{} = \$30$$

By writing out all the fractions in one long step (with units cancellation always in mind), you can cancel all the units—except the one you want to end up with—in one swoop and have all the arithmetic in front of you. This way we can see that the entire problem becomes simply

$$\frac{\$1.20}{\text{gallon}} \times \frac{\text{gallon}}{20\text{ miles}} \times \frac{500\text{ miles}}{} = \$30$$

With the technique of units cancellation well in hand, we can turn to a problem that involves a chemical principle. Let's look at Exercise 30 of Chapter 9.

■ QUESTION

The sun is 150,000,000 km from the earth and the diameter of a penny is 1.9 cm. If Avogadro's number of pennies were used to build a road from the earth to the sun and the road were just one layer of pennies deep, how many pennies wide would the road be?

To solve this problem we have to find out how many pennies, 1.9 cm in diameter, we could string out from the earth to the sun. Then, knowing the value of Avogadro's number (Sec. 9.3), we can calculate how many of these strings of pennies we could lay out, side by side.

We want an answer to the first part of the problem in units of pennies, so we start with the length of 1.9 cm/penny and use metric factors (Appendix B) in canceling units of length. Notice that in the following we start with penny/1.9 cm so that the unit penny remains in the numerator:

$$\frac{\text{penny}}{1.9\text{ cm}} \times \frac{100\text{ cm}}{\text{m}} \times \frac{1000\text{ m}}{\text{km}} \times \frac{150{,}000{,}000\text{ km}}{}$$

Since it's much easier to work with powers of 10 in this case (Appendix A), we'll rewrite this as

$$\frac{\text{penny}}{1.9\text{ cm}} \times \frac{10^2\text{ cm}}{\text{m}} \times \frac{10^3\text{ m}}{\text{km}} \times \frac{1.5 \times 10^8\text{ km}}{} = 0.789 \times 10^{13}\text{ pennies}$$

This gives us 0.789×10^{13} (or 7.89×10^{12}) pennies in one line from here to the sun. Since Avogadro's number of pennies amounts to 6.02×10^{23} pennies, our next calculation shows how many of these lines Avogadro's number of pennies would form. The road of pennies that would stretch from the earth to the sun would be

$$\frac{\text{line of pennies}}{7.89 \times 10^{12}\text{ pennies}} \times \frac{\text{Avogadro's number of pennies}}{}$$

$$\times \frac{6.02 \times 10^{23}\text{ pennies}}{\text{Avogadro's number of pennies}} = 0.763 \times 10^{11}\text{ lines of pennies}$$

Avogadro's number of pennies would therefore be enough to form a roadway 76,300,000,000 pennies wide from the earth to the sun.

One final example rounds out Appendix C with a problem in the chemistry of combustion. In Section 9.4 we see that methane burns to give carbon dioxide and water according to the equation

$$CH_4 + 2O_2 \longrightarrow CO_2 + 2H_2O$$

Notice that for every molecule (or mole) of CH_4 that burns, two molecules (or moles) of water form. Our question is, "What is the mass of water that forms when 8 g of methane burn in air?" Using units cancellation we get our answer easily and quickly. We want a mass of water as our answer, so we'll start with the fact that there are 18 g of water per mole of water:

$$\frac{18\text{ g water}}{\text{mol water}}$$

Since we know from the chemical equation that we get 2 moles water per mole of methane, or (2 mole water)/(mol methane), we can use this factor to cancel out the unit *mol water*. Then we'll use the molecular weight of methane, 16 amu, and the mass of methane that we start with, 8 g, to arrive at our units-canceled answer:

$$\frac{18\text{ g water}}{\text{mol water}} \times \frac{2\text{ mol water}}{\text{mol methane}} \times \frac{\text{mol methane}}{16\text{ g methane}} \times \frac{8\text{ g methane}}{} = 18\text{ g water}$$

Once again, units cancellation gives us the answer.

EXERCISES

Work the following using the information contained in Appendices A to C and the contents of Chapters 1 to 10.

1. What is the distance from New York to San Francisco, expressed in centimeters? (See Fig. B.7.)

2. Light travels at about 3×10^{10} cm/sec in a vacuum. A light-year is defined as the distance that light travels in a vacuum in 1 year. What is the length of a light-year (a) in meters? (b) in miles?

3. Hydrochloric acid, HCl, forms as hydrogen and chlorine combine according to the chemical equation

$$H_2 + Cl_2 \longrightarrow 2HCl$$

a. What mass of chlorine combines with 40 g of hydrogen to form HCl?

b. What mass of HCl forms when 71 g of chlorine react with an excess of hydrogen?

4. Water forms when hydrogen and oxygen combine according to the chemical equation

$$2H_2 + O_2 \longrightarrow 2H_2O$$

What mass of water forms when 40 g of hydrogen burn in an excess of oxygen?

5. Calculate the number of centuries in the length of time equivalent to Avogadro's number of seconds. Compare your answer with the value given in Section 9.3.

Significant Figures

BUDGETING FOR GASOLINE

Suppose you're going to drive from San Francisco to New York and you want to know how much you should budget for gas. You're driving a car that gets about 19 miles per gallon on the highway, and as a rough estimate you guess that gasoline will cost about \$1.25 per gallon. (That's just a guess, but we'll stick with it for the calculation.) A check of a road atlas shows that the distance from San Francisco to New York is 2934 miles. What's the best estimate of the cost of the gasoline: \$193.02631, \$193.03, \$193, \$190, or about \$200? The first figure, \$193.02631, comes from using 19 miles per gallon, \$1.25 per gallon, 2934 miles, and a hand calculator. The others come from rounding the answer to different degrees of approximation.

The best answer comes from an understanding of *significant figures*, also called *significant digits*. We'll use both terms here. *Digits* are the symbols from 0 to 9 that make up all the numbers we work with. *Significant* digits are the digits in any number that have real meaning, that we can rely on with confidence. We'll examine what significant figures are and how and why we use them in the measurements and calculations of science and everyday life.

The Number of Significant Figures

The key to working with measurements and calculations lies in knowing the number of significant figures in any particular quantity. To determine this, we focus on zeros and on the digits that aren't zero, the nonzero digits from 1 to 9. These are the rules to follow:

Rule 1: All nonzero digits are significant. The number 5692 contains four significant figures. The number 27,477.2941 contains nine significant figures. There's one significant figure in 6.

Rule 2: All zeros that lie between nonzero digits are significant. The number 3904 contains four significant figures. The number 40,028 has five, and so does 30,508. There are seven significant figures in 3,000,002.

Rule 3: None of the zeros that lie to the left of the first nonzero digit are significant, regardless of the position of the decimal. There are four significant figures in 0.002911, and five in 0.000039802. The number 10.0083 has six significant figures by rule 1.

Rule 4: Zeros to the right of the last nonzero digit are significant if they lie to the right of a decimal point. There are four significant figures in 341.0, five in 202.40, and seven in 330.0000.

Rule 5: Zeros to the right of the last nonzero digit but to the left of a decimal point may or may not be significant, depending on the context.

Confusion can come in rounding off numbers for approximations. If someone asks you how far you'll have to drive on your trip from San Francisco to New York and you answer "2934 miles," you're using four significant figures. But if you answer "about 3000 miles," you're using only one. Here, the word *about* gives your friend a cue that this is an approximate answer, probably with only one significant digit. Suppose, on the other hand, that you look up road mileages in an atlas and find that the distances from Cincinnati to Atlanta, Boston, Dallas, and Detroit are, respectively, 440, 840, 920, and 259 miles. Because you're using a reference book and the distance to Detroit is given to three significant figures, you can assume that the other distances are also given to three significant figures.

In scientific discussions, powers of 10 are used to indicate significant zeros. For example, 3×10^2 represents 300 to one significant figure; 3.00×10^2 indicates 300 to three significant figures.

Exact Numbers: Counting and Defining

Numbers we use in counting and in defining other quantities are assumed to be exact and to have an infinite number of significant figures. If you count the number of pages in a book and find exactly 573, you report this as "573." In this case, "573.000 pages" is meaningless. (Naturally, if you *measure* the length of some typed material and find that it runs over three pages, to 3.76 pages for example, that's another matter. That value comes from a measurement, which we'll get to shortly.) This procedure holds with money as well. We might find, for example, that something costs $34.83. We can look at this as an exact count if we recognize that it's equivalent to exactly 3483 cents. A dollar is, after all, *defined* as the equivalent of 100 pennies. It wouldn't make sense to speak of the cost as $34.830 or 3483.0 pennies, for example, since 3483 cents is an exact count of pennies.

Like counting, defining quantities also involves exact numbers, with an infinite number of significant digits. There are exactly 100 pennies to the dollar, 60 seconds to a minute, 100 cm to a meter, 1000 g to a kilogram, and the like. We say that 60 seconds make up 1 minute and understand that both "60" and "1" are exact numbers.

Approximate Numbers: Measuring

Significant figures are vital when we come to measurements, because all measurements involve some degree of inaccuracy. You can see why if you try this. Get a ruler and a dollar bill and have four friends work with you in measuring the width of the bill, edge to edge. You might find that the bill runs between 155 and 156 mm in length. You and your friends would have to estimate down to the tenth of a millimeter the part of the bill that falls between 155 and 156 mm. You'd get a set of five measurements that would be identical in the first three significant figures (155), but that would vary in your estimates of the fourth, the tenths of a millimeter. You might get a set of measurements like this:

Person	Measurement
1	155.7 mm
2	155.3 mm
3	155.5 mm
4	155.6 mm
5	155.3 mm

Knowing that each of you measured the bill to the best of your ability, and that the fourth digit is an estimate, you would naturally take an average of the five measurements. Adding up all five gives 777.4 mm; dividing by 5 gives an average value of 155.48 mm. That's one more significant figure than any of your five measurements. If your measurements are reliable to 0.1 mm, the average is also reliable to 0.1 mm. You can't very well get an average that's trustworthy to 0.01 mm from individual measurements that are good to only 0.1 mm. So you round the average to 155.5 mm. That's your best estimate of the length of the bill, as measured by five different people. The result of any mathematical manipulation of a set of measurements can't be any more reliable than the least reliable measurement. The end result must not contain more significant figures than those in the least reliable measurement.

Rounding

In rounding numbers we use the following rules. (In rounding 155.48 mm to 155.5 mm you used rule 7):

Rule 6: If the digit dropped is less than 5, drop it and all that follow without changing the remaining digits. For four significant figures you would round 1.4033 to 1.403, and 76.584997 to 76.58.

Rule 7: If the digit dropped is greater than 5, or if it's 5 and any nonzero digit follows (also to be dropped), increase the last remaining digit by one. For four significant figures, 0.0029817 becomes 0.002982, and 2848.50003 becomes 2849.

Rule 8: If the digit dropped is 5 and only zeros follow it, do not change the last remaining digit if it's even; raise it by one if it's odd. Thus, for four significant digits, 34.835000 becomes 34.84, and 937.45 becomes 937.4.

Adding and Subtracting

In adding and subtracting, line up the decimal points and round all numbers down to the one with the fewest decimal digits. To add

$$\begin{array}{r} 45.9930 \\ 5.3501 \\ 25.6 \\ \underline{11.45} \end{array}$$

we recognize that 25.6 has only a single significant digit to the right of the decimal so we round the numbers to

$$\begin{array}{r} 46.0 \\ 5.4 \\ 25.6 \\ \underline{11.4} \end{array}$$

for the total

$$88.4$$

For subtraction we have

$$\begin{array}{r} 12{,}983.006 \\ \underline{-\quad 1{,}197.355079} \end{array}$$

which rounds to

$$\begin{array}{r} 12{,}983.006 \\ \underline{-\quad 1{,}197.355} \\ 11{,}785.651 \end{array}$$

Multiplying and Dividing

Regardless of the locations of decimal points, the result must have no more significant digits than those in the factor with the smallest number. Here, two slightly different approaches give the same result. In one, carry out the operation, then round off the result to the same number of significant digits as in the least reliable factor:

$$13.2 \times 242.7 \times 0.023 = 73.68372$$

We round off the answer to 74 since the factor 0.023 has only two significant figures. In another, simpler approach, before you carry out the multiplication or division, round off factors to one digit more than the number in the least reliable factor. With this our multiplication becomes $13.2 \times 243 \times 0.023 = 73.7748$, which once again rounds to 74. This second method gives the same result with a little less work. For division, we have by the first method (and a hand calculator)

$$\frac{143.751}{62} = 2.3185645, \text{ which rounds to } 2.3$$

Using the second, simplified approach we have

$$\frac{143.751}{62}, \text{ which rounds to } \frac{144}{62} = 2.3225806, \text{ or again } 2.3$$

Now back to our original problem. Since the factor with the smallest number of significant figures in your trip from San Francisco to New York is the 19 miles per gallon you estimate for your mileage (with its two significant figures), your best guess of the cost also has two significant figures, $190. But common sense also comes into play here. Since your mileage might drop as you go, and since gasoline prices could change during your travel, you ought to play it safe and assume the trip will cost $200 in gas. It would probably be wise to budget even a few dollars more. Common sense is often far more significant than any figures we work with.

EXERCISES

1. Add the following:

a. 21.0884 + 11.79 + 1.34508 + 14.6003 + 33.0650

b. 0.00330 + 1.002001 + 5.704520 + 0.10445

c. 13.6 + 22.9 + 9.45 + 3.350327 + 4

2. Subtract the following:

a. 14.0409 − 8.723

b. 202.3000 − .0040

c. .0072606 − .000419

3. Multiply the following:

a. 2.78 × 0.0035701 × 4.6

b. 3.7 × 414 × 2.601

c. 2 × 7 × 0.035108

4. Divide the following:

a. 6.4/9.2

b. 5.750/3.6

c. 0.01090/0.003266

Working with Avogadro's Number

A REMARKABLE CALCULATION

In effect, Exercise 26 of Chapter 9 asks this question:

- What is the combined total of all the protons and all the neutrons in the body of a 165-pound person?

Working this problem requires a good grasp of the contents of Chapter 9 and facility with the material of Appendixes A to D. We know from Chapter 9 that one mole of a chemical substance contains Avogadro's number of chemical particles. We also know that one mole of a substance consists of its atomic, molecular, or ionic weight expressed in grams. These masses are ordinarily expressed in atomic mass units (amu). Since the mass of a proton and of a neutron is one amu (Sec. 2.4), one mole of protons must have a mass of one gram and, similarly, one mole of neutrons must also have a mass of one gram.

What we need to do, then, is calculate the total mass of a 165-pound person in units of grams and we will have the combined sum of the moles of protons and the moles of neutrons in the body. (Recall that the mass of the electron is negligible compared with the mass of the proton or the neutron; see Sec. 2.4).

Knowing that each mole of chemical particles contains Avogadro's number of particles gives us the answer we're looking for. The following sequence of fractions makes clear the units cancellation required:

$$\frac{6.02 \times 10^{23}\ \text{particles}}{\text{mol}} \times \frac{\text{mol}}{1\ \text{g particles}} \times \frac{10^3\ \text{g}}{\text{kg}} \times \frac{\text{kg}}{2.20\ \text{pounds}} \times \frac{165\ \text{pounds}}{\text{body}}$$

$$= 452 \times 10^{26}\ \text{particles/body}$$

or a combined total of 4.52×10^{28} protons and neutrons in the body of a 165-pound person.

GLOSSARY

A

Absolute zero The lowest possible temperature; 0 K, or −273°C.

Acceptable daily intake (ADI) One percent of the maximum daily amount of a food additive that produces no observable effect on laboratory animals.

Acetylsalicylic acid The chemical name of aspirin.

Acid By the Arrhenius definition, any substance that can produce a proton (hydrogen ion) in water. By the Brønsted–Lowry definition, any substance that can transfer a proton to another substance. Turns blue litmus red, tastes sour, and reacts with iron, zinc, and certain other metals to generate hydrogen gas.

Acid–base indicator A dye that changes, loses, or acquires color as a solution containing the dye changes in acidity or basicity.

Adipose tissue The fatty tissue of the body. One pound of adipose tissue is equivalent to about 3500 Cal.

Aldehyde A class of organic compounds whose molecules contain a carbonyl group bonded to at least one hydrogen.

Alkaloid A basic, bitter-tasting, nitrogen-containing compound found in plants. Alkaloids produce physiological reactions of various kinds and intensities.

Alkane A hydrocarbon whose molecular formula fits the general formula C_nH_{2n+2}.

Alkene A hydrocarbon that contains a carbon–carbon double bond and fits the general formula C_nH_{2n}.

Alkyne A hydrocarbon that contains a carbon–carbon triple bond and fits the general formula C_nH_{2n-2}.

Alpha (α) particle A high-energy helium nucleus traveling at 5–7% the speed of light.

Amide An organic compound containing a nitrogen bonded to a carbonyl group:

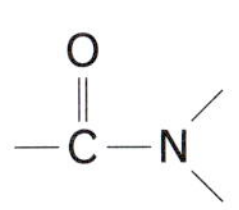

Amine An organic base whose molecules bear the amino group, $—NH_2$.

Amino acid A compound whose molecules carry both a carboxyl group and an amino group. Amino acids have properties of both acids and bases.

Ampere (amp) A unit of electrical current. The rate of flow of electrons.

Analgesic A substance that reduces or eliminates pain.

Anion A negatively charged ion.

Anode The part of a battery that provides the electrons flowing through the external circuit.

Anti-inflammatory agent A substance that reduces or eliminates inflammation.

Antioxidant A chemical so easily oxidized itself that it protects other chemicals from oxidation.

Antiperspirant A substance that inhibits sweating and keeps the skin relatively dry.

Antipyretic A substance that reduces or eliminates fever.

Aquifer A large layer of porous rock that holds fresh water and from which water can be drawn by wells.

Aromatic compound A compound that contains at least one benzene ring as part of its molecular structure.

Aromatization The conversion of a cyclohexane ring into an aromatic ring.

Aspirin The common name or trade name of *acetylsalicylic acid*. An analgesic, antipyretic, and anti-inflammatory medication.

Atmosphere The body of gases that surrounds the earth.

Atmospheric pressure The pressure generated at any location on or above the earth's surface by the combined weight of all the atmospheric gases above that location.

Atom The smallest electrically neutral particle of an element that can be identified as that element.

Atomic mass unit (amu) One-twelfth the mass of the most common isotope of carbon.

Atomic number The sum of all of an atom's protons.

Atomic weight The average mass of all of an element's isotopes, weighted for the abundance of each.

Avogadro's Law Equal volumes of gases (at the same temperature and pressure) contain equal numbers of atoms or molecules.

Avogadro's number The experimentally determined number of chemical particles in a mole: 6.02×10^{23}.

B

Background radiation The low level of natural radiation to which we are all exposed.

Balanced equation An equation containing the same number of atoms of each of the elements among both the reactants and the products.

Base By the Arrhenius definition, any substance that can produce a hydroxide ion in water. By the Brønsted–Lowry definition, any substance that can accept a proton. Turns red litmus blue, tastes bitter, and feels slippery when dissolved in water.

Beta (β) particle A high-energy electron moving at up to 90% the speed of light.

Binary compound A compound composed of two elements.

Binding energy The energy that holds an atom's nucleus together as a coherent whole. The energy equivalent of an atom's mass defect.

Biochemical oxygen demand (BOD) A measure of the aerobic activity of aqueous microorganisms. An indirect measure of organic pollutants.

Biochemistry The branch of science devoted to the study of the chemical reactions taking place within a living organism.

Biodegradable A substance that can be converted, by the action of microorganisms within the environment, into the simpler substances that form our natural environment.

Biological contamination Contamination that results from the presence of disease-causing microorganisms.

Biomass The mass of all living things, including their waste products, residues, and remains.

Boiling point The temperature at which a liquid boils to become a gas and at which the gas condenses to become a liquid. The boiling point is usually determined at normal atmospheric pressure.

Boyle's Law With the temperature and the number of moles of a quantity of gas held constant, the volume of the gas varies inversely with its pressure.

Breeder reactor A nuclear reactor that produces more fissionable fuel than it consumes.

"Buffered" aspirin A combination of acetylsalicylic acid and one or more bases. The combination accelerates the disintegration of the tablet.

C

Caffeine An alkaloid that occurs in coffee beans and in tea leaves. It stimulates the central nervous system and heightens a sense of awareness.

Calorie The amount of heat (or energy) needed to raise the temperature of one gram of water by one degree Celsius. When capitalized, it represents 1000 calories.

Carbon cycle The transfer of carbon among the earth's various natural storage facilities for the element or its compounds. Carbon moves from one region to another largely as carbon dioxide.

Carcinogen A cancer-causing agent.

Catalyst A substance that increases the rate of a reaction but is not itself a reactant.

Catalytic cracking A process by which large hydrocarbon molecules are converted into two or more smaller hydrocarbon molecules.

Catalytic hydrogenation The addition of hydrogen atoms to atoms joined by double or triple bonds, through the use of a catalyst.

Catalytic reforming A process by which hydrocarbon molecules are reorganized into more useful molecular structures of the same carbon content.

Cathode The part of a battery that receives the returning electrons from an external circuit.

Cation A positively charged ion.

Cellobiase An enzyme, absent in humans, needed to digest cellulose.

Chain reaction A continuing series of nuclear fissions that occurs when neutrons released in the fission of one atom cause fission among several additional atoms, which in turn release still more neutrons and produce still more fissions, and so on.

Charles' Law With the pressure and the number of moles of a quantity of gas held constant, the volume of the gas varies directly with its absolute temperature.

Chemical formula A sequence of chemical symbols and subscripts that shows the elements present in a compound and the ratio of their ions in a lattice or the actual number of their atoms in a molecule.

Chemical pollution Pollution resulting from the presence of harmful or undesirable chemicals.

Chemistry The branch of science that studies the composition and properties of matter and the changes that matter undergoes.

Chirality The characteristic "handedness" of right and left hands, as applied to chemical structures.

Cholesterol A steroidal alcohol that contributes to the development of atherosclerosis (hardening of the arteries).

Chromosomes The structures within a cell's nucleus that carry the organism's genetic information.

Circuit The path electrons follow.

Colloid A mixture made up of particles of one substance dispersed throughout another. The particles of the dispersed substance range in diameter from about 10^{-7} to 10^{-4} cm.

Cologne A diluted, inexpensive version of a perfume.

Complementary protein A combination of incomplete or low-quality proteins that, taken together, provide about the same ratio of essential amino acids as do complete or high-quality proteins.

Complete protein (high-quality protein) Protein that contains all the essential amino acids in about the same ratio as they occur, on average, in human proteins.

Compound A pure substance formed by the chemical combination of two or more different elements in a specific ratio.

Compression ratio The ratio of the maximum volume of a cylinder's gasoline–air mixture at the beginning of the compression stroke to the volume of the compressed mixture as the spark plug fires.

Concentration The quantity of solute dissolved in a specific quantity, usually a specific volume, of solution.

Corrosion The erosion and disintegration of a material as a result of chemical reactions.

Cortex The central core of a strand of hair. The cortex forms the bulk of the strand and contains its pigments.

Cosmetic A substance applied directly to the human body to make it more attractive.

Covalent bond A bond consisting of a pair of electrons shared between two atoms.

Critical mass The minimum mass of fissionable material needed to sustain a chain reaction.

Crystal lattice The orderly, three-dimensional arrangement of the chemical particles that make up a crystal.

Cuticle The scaly sheath that covers the cortex of a strand of hair.

Cyclization A chemical process that converts noncyclic molecules into cyclic molecules.

D

Dalton's Law The total pressure of a mixture of gases equals the sum of the partial pressures of each of the gases in the mixture.

Density The mass per unit volume of a substance, normally stated in grams per milliliter or grams per cubic centimeter.

Deodorant A substance that masks or nullifies bodily odors.

Dermis The portion of the skin that contains nerves, blood vessels, sweat glands, and the active portion of the hair follicles and also supports the upper layer of skin, the *epidermis*.

Dipole Results from a separation of positive and negative charges in a molecule.

Disaccharide A molecule formed by the chemical combination of two monosaccharides.

Dissolved oxygen (DO) A measure of the aerobic activity of aqueous microorganisms. An indirect measure of organic pollutants.

Dissolved oxygen deficit (DOD) A measure of the aerobic activity of aqueous microorganisms. An indirect measure of organic pollutants.

Distillation The process of purifying a liquid by boiling it in one container, condensing its vapors again to a liquid, and collecting the separated, condensed liquid in another container.

Drug A chemical used medically for treating a disease, illness or injury, with the added connotation of narcotics and addiction.

E

Elastomer A substance that stretches easily and returns readily to its original shape.

Electric circuit The path electrons follow.

Electric current A flow of electrons.

Electrochemical cell A cell or battery that produces electricity from chemical reactions.

Electrochemical reaction A reaction that can produce a flow of electrons from one location to another. A reaction that is caused by such a flow.

Electrolysis The decomposition of a substance by means of electricity.

Electrolyte A substance that conducts electricity when it is dissolved in water or melted.

Electron A subatomic particle with negligible mass and a charge of 1^-.

Electron dot structure Also called *Lewis structure*. Shows the elemental symbol and the valence electrons, arranged as paired or unpaired dots.

Electronegativity A measure of an atom's ability to attract the electrons of a covalent bond toward itself.

Electron structure The distribution of electrons in an atom's quantum shells.

Element A substance whose atoms all have the same atomic number.

Enantiomers A pair of molecular structures that are nonsuperposable mirror images of each other.

Endorphin An opiate produced within the bodies of animals.

Endothermic reaction A reaction that absorbs heat as it proceeds.

Energy The ability to do work.

Epidermis The layer of the skin that lies above the *dermis*.

Essential amino acid An amino acid that we must obtain from our foods because our bodies cannot synthesize it from other compounds.

Essential fatty acid A fatty acid we must obtain from our foods because our bodies cannot synthesize it from other compounds.

Ether A compound in which an oxygen atom is bonded to the carbons of two organic groups, which may be identical or different in structure.

ETS (environmental tobacco smoke) Tobacco smoke persisting in the atmosphere. An environmental hazard to smokers and nonsmokers alike.

Eutrophication A form of water pollution in which plants, well nourished by pollution, thrive at the expense of aquatic animals.

Exothermic reaction A reaction that releases heat as it proceeds.

F

Fallout Fine radioactive debris that's released into the atmosphere by nuclear explosions and accidents and that settles to the earth.

Family of compounds A group of compounds of similar structure and similar properties.

Family of elements A group of elements all of which lie in the same column of the periodic table and have similar chemical properties.

Fat A solid triester of glycerol and fatty acids. Constitutes one of the macronutrients of our foods and furnishes 9 Cal/g.

Fat-soluble vitamin One of a group of vitamins that are much more soluble in fats, hydrocarbons, and similar solvents than in water. Vitamins A, D, E, and K.

Fatty acid An acid formed on hydrolysis of fats and oils.

Fiber (also known as *bulk* and *roughage*) A component of our foods made up of the carbohydrate *cellulose*. Fiber is indigestible because of our lack of the enzyme *cellobiase*.

Fischer projection A two-dimensional projection used to show the stereochemistry of a three-dimensional, chiral carbon.

Follicle A microscopic sac under the surface of the skin. Holds the root of a strand of hair.

Food additive By a practical definition, anything intentionally added to a food to produce a specific, beneficial result.

Fossil fuel Fuel formed from the partially decayed animal and vegetable matter of living things that inhabited the earth in eras long past.

Free radical A chemical structure containing an unpaired electron.

Frequency The number of waves that pass a fixed point in a specific period.

Functional group A small set of atoms held together by covalent bonds in a specific, characteristic arrangement that is responsible for the principal chemical and physical properties of a compound.

G

Galvanic (bimetallic) corrosion Corrosion that results from the contact of two different metals separated by a thin layer of electrolyte.

Galvanizing A process that provides a protective zinc coating to metals.

Gamma (γ) ray A form of high-energy electromagnetic radiation that has no mass and carries no electrical charge.

Gas A phase or state of matter that takes both the shape and volume of the container that holds it.

Gay-Lussac's Law When gases react with each other at constant temperature and pressure, they combine in volumes that are related to each other as ratios of small, whole numbers.

Gene A segment of a chromosome that carries information necessary for the creation of a specific protein or polypeptide.

Genetic damage Biological damage that is transmitted from one generation to the next.

Genome The complete set of genes responsible for all the hereditary characteristics of an organism or a species.

Geometric isomers Different compounds that have the same four groups bonded to the carbons of a double bond, but with different geometries.

Gluon A subatomic particle that holds quarks in their clusters.

GRAS list A list of food additives that are generally recognized as safe by a panel of experts, without having been subjected to laboratory testing.

Greenhouse effect The warming of the earth by solar radiation trapped as heat through the insulating effect of atmospheric gases.

Greenhouse gas An atmospheric gas that contributes to the greenhouse effect.

Groundwater Water within the earth's crust, lying just below the surface.

H

Half-cell reaction Each of the oxidation and reduction reactions that combine to form a redox reaction.

Half-life The time it takes for exactly half of any given quantity of a radioisotope to decay.

Health food By one definition, a food that supplies a large number of micronutrients compared with its calories. This definition does not take into consideration the food's content of saturated fats, cholesterol, or other factors that can affect health.

Henry's Law At a fixed temperature, the quantity of a gas that dissolves in a liquid depends directly on the pressure of that gas above the liquid.

Heroin *Diacetylmorphine*, a strongly addicting compound in which the two —OH functional groups of *morphine* have been acetylated.

Hydrocarbon A compound composed exclusively of hydrogen and carbon.

Hydrogen bond A weak chemical bond formed through the attraction of a nitrogen, oxygen, or fluorine for a hydrogen bonded to a different nitrogen, oxygen, or fluorine.

Hydrolysis The decomposition of a substance, or its conversion to other substances, through the action of water.

Hydronium ion (H_3O^+) The ion that forms when a proton bonds to a water molecule.

Hydrosphere All the waters of the earth's crust.

I

Incomplete protein (low-quality protein) Protein that is deficient in one or more of the essential amino acids.

Inorganic chemistry The chemistry of compounds that do not contain carbon.

Inorganic compound A compound that does not contain carbon.

Invert sugar A mixture of equal quantities of glucose and fructose.

Iodine number The number of grams of iodine that add to 100 g of a triglyceride.

Ion An atom or a group of atoms that carries an electrical charge.

Ionic bond A chemical bond resulting from the mutual attraction of oppositely charged ions.

Ionic compound A compound composed of ions.

Ionization The transformation of a covalent molecule into two or more ions.

Ionizing radiation Radiation capable of converting electrically neutral matter into ions.

Ionosphere A region of the upper atmosphere filled with ionized gases. Effectively reflects short-wave radio transmissions.

Isomerization The conversion of one compound into another of the same molecular formula.

Isomers Different compounds with the same molecular formula.

Isotope An atom of an element that differs in mass number from other atoms of that same element.

J

Joule The work done by 1 watt of electricity in 1 second. Equivalent to 0.24 cal.

Junk food By one definition, a food that supplies a large number of calories but few micronutrients.

K

Ketone A class of organic compounds whose molecules contain a carbonyl group bonded to two carbon atoms.

Kilocalorie 1000 cal; 1 Cal.

Kinetic energy The energy of motion.

Kinetic-molecular theory of gases A theory that explains the behavior of gases by assuming that they are made up of point-sized, perfectly resilient, constantly moving chemical particles.

Knocking A rapid pinging or knocking sound produced by preignition or irregular combustion of the gasoline–air mixture in the cylinders of an internal combustion engine.

L

Law of Conservation of Energy Energy can be neither created nor destroyed as a result of chemical transformations.

Law of Conservation of Mass Mass can be neither created nor destroyed as a result of chemical transformations.

LD_{50} The amount of a chemical that kills exactly half of a large population of animals.

Le Châtelier's principle When a stress is placed on a system in equilibrium, the system tends to change in a way that relieves the stress.

Lethal dose The amount of a substance that causes death.

Lewis structure Also called *electron dot structure.* Shows the elemental symbol and the valence electrons, arranged as paired or unpaired dots.

Lipoprotein A combination of a protein with various proportions of cholesterol and triglycerides, effective in transporting these substances through the blood.

Lithosphere The earth's hard, rigid outer shell. The surface layer of the lithosphere forms the earth's crust.

Liquid A phase or state of matter that retains its own volume but takes the shape of the container that holds it.

M

Mass A property of matter observed as a body's resistance to acceleration.

Mass defect The difference between the mass of an atom as a whole and the sum of the masses of all the individual protons, neutrons, and electrons that compose it.

Mass number The sum of all the protons and neutrons in an atom's nucleus.

Matter All the substances that form the material things of the universe.

Medicine A chemical used medically for treating a disease, illness or injury.

Melanin The dark brown pigment of the skin and hair.

Melting point The temperature at which a solid melts to become a liquid, and at which the liquid crystallizes to become a solid.

Mesosphere The region of the earth's atmosphere that lies between the stratosphere and the ionosphere.

Metabolism The combination of all the chemical reactions that take place in any living organism.

Micelle A submicroscopic globule or sphere distributed in large numbers throughout another substance, which is usually a liquid.

Micronutrient A component of food, especially a vitamin or a mineral, that we need in very small amounts for life and good health.

Mineral In nutrition, a chemical element we need for life and good health, other than the elements that make up the macronutrients (that is, other than carbon, hydrogen, oxygen, nitrogen, and sulfur).

Molarity The concentration of solute in a solution expressed as the number of moles of solute per liter of solution.

Mole The number of atoms of carbon-12 in exactly 12 g of carbon-12. In practice, the number of chemical particles contained in a quantity of any pure substance equal to its atomic, molecular, or ionic weight expressed in grams.

Molecular formula The chemical formula of a covalent compound.

Molecular weight The sum of the atomic weights of all the atoms in a molecule.

Molecule A discrete chemical structure held together by covalent bonds.

Monomer One of a set of individual molecules that are linked to each other to form a polymer.

Monosaccharide The smallest molecular unit of a carbohydrate, such as glucose, fructose, or galactose.

Monounsaturated fatty acid A fatty acid with a single carbon–carbon double bond in its carbon chain.

Morphine A narcotic. The major alkaloid of opium.

N

Narcotic A substance that produces a stupefying, dulling effect and induces sleep.

Neutralization The combination of an acid and a base to produce a salt (and often water as well).

Neutron A subatomic particle with a mass of 1 amu and no electrical charge. Occurs in the nucleus of every atom except protium.

Nicotine A poisonous alkaloid that occurs in tobacco leaves.

Nonelectrolyte A substance that does not conduct electricity when dissolved in water or when melted.

Nonessential amino acid An amino acid our bodies can synthesize from other chemicals.

Nuclear fission The splitting of an atomic nucleus into two or more large fragments.

Nuclear fusion The combination of two or more nuclei of small mass to form a single nucleus of larger mass.

Nucleus The positively charged central core of an atom. Consists of protons and, except for protium, neutrons.

O

Octane enhancer A gasoline additive used to increase octane ratings.

Octane rating (octane number) A measure of a gasoline's resistance to knocking.

Octet rule Atoms often react so as to obtain eight electrons in their valence shells.

Oil In nutrition, a liquid triester of glycerol and fatty acids. Constitutes one of the macronutrients of our foods and furnishes 9 Cal/g.

Opiate antagonist A compound that counteracts or nullifies the effects of opiates.

Opiate receptors A set of large molecules within the nervous system. When a narcotic molecule connects with an opiate receptor, it blocks or alters signals flowing through the nervous system.

Opium The dried sap of the poppy, composed principally of *morphine*.

Optical activity The ability to rotate the plane of polarized light.

Organic chemistry The chemistry of carbon compounds.

Oxidation The loss of electrons.

Oxidation potential The voltage produced by or required for the removal of electrons from an atom or ion.

Oxidizing agent An agent that acquires electrons from some other substance and thereby causes it to be oxidized.

Oxygenate A gasoline additive that contains oxygen and that improves the efficiency of hydrocarbon combustion.

P

Partial pressure The pressure each gas of a mixture of gases would exert, at the same temperature and in the same volume, in the absence of all the other gases.

Parts per million (ppm) A concentration term referring to the number of units of weight of a solute per million units of weight of solution, such as milligrams of solute per kilogram of solution.

Peptide bond (peptide link) The carbon–nitrogen bond of an amide group that links two amino acids. The resulting compound is a *dipeptide*, a *tripeptide*, a *tetrapeptide*, and so on, depending on the total number of amino acids joined by peptide links.

Perfume A pleasantly fragrant blend of synthetic chemicals, animal oils, and/or extracts of plants, all dissolved as a 10–25% solution in alcohol.

Period The elements in a row of the periodic table.

Petroleum fraction A mixture of components of petroleum with similar boiling points. Obtained through the distillation of petroleum.

pH A measure of a solution's acidity. The negative logarithm of a solution's hydronium ion concentration.

Phaeomelanin A red-brown or yellow-brown pigment chemically similar to melanin.

Photochemical smog The combination of all products resulting from the initial interaction of sunlight and nitrogen dioxide and subsequent reactions involving the bimolecular oxygen of the atmosphere as well as hydrocarbon pollutants.

Placebo A harmless and normally ineffective substance given simply to please and pacify a patient.

Plane of polarization The single plane of electromagnetic oscillations of polarized light.

Plaque A thin layer of a polysaccharide that adheres to the surface of teeth and harbors bacteria.

Plasma A state of matter in which electrons have been stripped out of their atomic shells to produce positively charged nuclei and negatively charged electrons, but without the existence of atomic structure.

Plastic A material capable of being shaped into virtually any form.

Plasticizer A liquid added to a plastic to soften it.

Poison A substance that can cause illness or death when it enters our bodies, usually as a component of our food or drink.

Polar covalent bond Formed when the shared electrons of a covalent bond lie closer to one of the bonded atoms than to the other.

Polarized light Light with electromagnetic oscillations in only one plane, known as the *plane of polarization*.

Pollution An excess of a substance generated by human activity, present in the wrong environmental location.

Polymer A molecule of very large molecular weight formed by the repeated chemical linking of a great many simpler, smaller molecules.

Polymerization The linking of many individual monomers to each other, resulting in the formation of a polymer.

Polyolefin A polymer produced by the polymerization of alkenes or compounds closely related to them.

Polypeptide A polyamide consisting of a chain of 10 or more amino acids.

Polysaccharide A molecule consisting of a chain of hundreds of monosaccharides.

Polyunsaturated fatty acid A fatty acid containing two or more carbon–carbon double bonds.

Positron A subatomic particle that carries a charge of 1+ but is otherwise identical to an electron.

Positron emission tomography (PET) A diagnostic medical technique in which images of planes within the body are generated through the analysis of γ rays emitted by collisions of positrons with electrons.

Potential energy The energy stored in an object or a substance, often as a consequence of its location or composition.

ppm Concentration in *parts per million*, or milligrams of solute per kilogram or liter of solution.

Primary air pollutant A pollutant that enters the air as the direct result of a specific activity.

Primary (1°) carbon A carbon bonded to exactly one other carbon.

Primary structure The sequence of amino acids in a peptide chain.

Protein A polypeptide with a molecular weight greater than 10,000 amu.

Proton A subatomic particle with a mass of 1 amu and a charge of 1+. Protons exist in the nucleus of every atom.

Q

Quantum shell An electron shell surrounding an atomic nucleus.

Quark A fundamental particle that, in combination with other quarks, forms larger subatomic particles such as protons and neutrons.

R

Radical (free radical) A chemical species that carries an unpaired electron.

Radioactivity The spontaneous emission of radiant energy and/or high-energy particles from the nucleus of an atom.

Radioisotope An isotope that emits radioactivity.

Rate-determining step The slowest step in a sequence of steps that, by itself alone, determines the rate of the entire, multistep process.

Redox reaction A reaction that takes place with a transfer of one or more electrons from one chemical species to another.

Reducing agent An agent that transfers electrons to some other substance and thereby causes it to be reduced.

Reduction The gain of electrons.

Reduction potential The voltage produced by or required for the addition of electrons to an atom or ion.

rem A shortened form of roentgen-equivalent-man. A unit of measurement of the biological effect of ionizing radiation on humans.

Rust The corrosion of iron to form various reddish-brown oxides of iron.

S

Safety By one useful definition, the acceptability of risk.

Salt A compound (other than water) produced by the reaction of an acid with a base. Commonly, *sodium chloride*.

Saponification The hydrolysis of an ester carried out in the presence of a base.

Saturated hydrocarbon Contains the maximum number of hydrogens possible and contains only single bonds.

Science A way of knowing and understanding the universe.

Scientific method The process by which science operates, involving the formulation of hypotheses and theories based on experimental tests of the universe.

Sebaceous gland A gland, located in the skin near a hair follicle, that secretes sebum, an oily substance that lubricates the cuticle of the hair and gives it gloss.

Sebum An oily substance that lubricates a hair's cuticle and gives it gloss. Secreted by the sebaceous glands.

Secondary air pollutant A pollutant formed by further reaction of a primary air pollutant.

Secondary (2°) carbon A carbon bonded to exactly two other carbons.

Sedimentary pollution Results from the accumulation of suspended particles in water.

Sequestrant A molecule or ion that bonds with a metal so firmly that it effectively removes the metal from chemical contact with other substances.

Solid A phase or state of matter that retains its own shape and volume regardless of the container that holds it.

Solute A substance that is dissolved in some other substance to form a solution. The minor component of a solution.

Solution A homogeneous mixture of one substance, the solute, dissolved in another, the solvent. In a solution, the solute is present in a smaller proportion than the solvent.

Solvent A substance in which a solute is dissolved to form a solution. The major component of a solution.

Somatic damage Injury to a living body, causing illness or death.

Standard reduction potential The numerical value of a substance's reduction potential, in volts, as compared with the reduction of the hydrogen ion.

Stereochemistry The three-dimensional, spatial arrangement of the atoms of a molecule.

Stratosphere The region of the earth's atmosphere that lies above the troposphere and contains the ozone layer.

Stratum corneum The outermost layer of skin, a protective shield consisting of 25–30 layers of dead cells.

Sublimation The conversion of a solid to a gas without the intermediate formation of a liquid.

Sunscreen protection factor (SPF) The inverse of the fraction of ultraviolet radiation that passes through a suntan or sunscreen lotion to reach the skin.

T

Technology The art, skill, or craft used in applying knowledge provided by science.

Teratogen A substance that produces severe birth defects.

Tertiary (3°) carbon A carbon bonded to exactly three other carbons.

Tetraethyllead ("lead") A chemical originally added to gasoline to inhibit knocking.

Tetrahedral carbon A carbon atom located at the center of a tetrahedron. It forms covalent bonds to the four atoms lying at the apexes or corners of the tetrahedron.

Thermal inversion An atmospheric condition in which a layer of warm air lies above a layer of cooler air, trapping it and any pollutants within it.

Thermal pollution Occurs when environmental harm comes from the warming of a body of water. Can be caused by the discharge of waste coolant.

Thermoplastic A plastic that softens when heated, then hardens again as it cools.

Thermoset A plastic that remains firm and holds its shape, even when heated.

Toxin A harmful substance, usually of biological origin.

Trace element A dietary element needed in our diets at levels of less than 100 mg/day.

Transmutation The conversion of one element into another.

Triglyceride A triester of glycerol and fatty acids. The principal organic component of fats and oils.

Troposphere The region of the earth's atmosphere in contact with the planet's surface. Responsible for the earth's weather.

U

Unsaturated molecule A molecule to which hydrogen can be added.

V

Valence electron An electron located in the outermost electron shell of an atom.

Valence shell The outermost electron shell of an atom.

Vasoconstrictor A drug or medicine that constricts blood vessels.

Vitamin An organic compound of our food that is essential in very small amounts for life and good health.

VOC Volatile organic compounds that can produce air pollution.

Volatility A measure of the ease and speed of a substance's transformation from a liquid to a vapor.

Volt A unit of electrical potential, the pressure that moves electrons from one point to another.

VOS Volatile organic solvents that can produce air pollution.

W

Water-soluble vitamin A vitamin that is more soluble in water than in fats or hydrocarbons. Vitamins B and C.

Wavelength The distance between crests of adjacent waves.

Weight The characteristic of a body of matter that results from the pull of gravity.

ANSWERS

TO SELECTED EVEN-NUMBERED EXERCISES

Chapter 1

4. a. An ion carries either a positive or a negative charge. **b.** A cation carries a positive charge. **c.** An anion carries a negative charge.

6. An element consists of atoms of the same atomic number. A compound is composed of two or more different elements.

8. Ionic bond.

12. Science provides knowledge and understanding of the universe. Technology applies the knowledge and understanding provided by science.

14. The products are sodium chloride and sodium until you use 15.4 g chlorine, which produces only sodium chloride. Thereafter the products are sodium chloride and chlorine.

16. 9 g water and 4 g oxygen.

22. Yes. It contains dissolved salts, which are electrolytes.

Chapter 2

2. a. Nitrogen; **b.** lithium; **c.** fluorine; **d.** hydrogen; **e.** carbon.

4. Elements with atomic numbers 1 to 10.

6. Universe, helium; our bodies, oxygen.

8. No. They are both isotopes of hydrogen.

10. Tritium.

12. 1, Hydrogen, H; 2, helium, He; 3, lithium, Li; 4, beryllium, Be; 5, boron, B; 6, carbon, C; 7, nitrogen, N; 8, oxygen, O; 9, fluorine, F; 10, neon, Ne.

14. No, every atom must contain at least one proton, which has a mass of 1 amu.

16. ${}^{3}_{1}H$ (tritium), ${}^{4}_{2}He$ (helium).

18. Neutrons, protons, protons, neutrons.

20. Both consist of elemental carbon and no other element.

22. The mass of an oxygen atom is 16 times the mass of a hydrogen atom.

24. 12 g; calculated value is 0.62 times measured value.

Chapter 3

4. a. Molecules; **b.** atoms; **c.** ions; **d.** molecules; **e.** ions; **f.** molecules; **g.** atoms.

6. LiF, LiBr, KF, KBr.

8. Elemental carbon: graphite, diamond; contain the element: carbon dioxide.

10. a. Nitrogen; **b.** sulfur; **c.** hydrogen; **d.** neon; **e.** magnesium; **f.** helium; **g.** magnesium; **h.** boron; **i.** carbon.

12. a. Al; **b.** Ar; **c.** Ca; **d.** C; **e.** Cs; **f.** F; **g.** Au; **h.** H; **i.** Pb; **j.** Mn.

16. a. All are inert; all are gases. **b.** All (except hydrogen) are metals; all (except hydrogen) react with water to produce hydrogen gas.

18. The valence shells acquire additional electrons until they become filled.

20. They have the same numbers of valence electrons.

22. O_3.

24. Ionization.

26. 8.

28. 81.

30. 0.

Chapter 4

4. From first to last: e, b, f, c, d, a.

6. a. Uranium; **b.** uranium; **c.** plutonium-239; **d.** to build an atomic bomb; **e.** helium; **f.** lead-206; **g.** nitrogen; **h.** polonium; **i.** nitrogen; **j.** oxygen.

8. A stable isotope.

10. a. Gamma radiation (gamma radiation has no mass); **b.** high-speed neutrons (neutrons have a smaller mass than alpha particles).

12. Quarks.

14. Creates a critical mass of plutonium.

16. Achieved transmutation of one element to another.

18. a. Atomic number, 2; mass number, 4; **b.** atomic number increases by 1.

20. Alpha particle consists of two neutrons and two protons; tritium's nucleus contains only one proton.

22. Because of the atom's mass defect, equivalent to its binding energy.

Chapter 5

4. Heat converts water to steam; steam turns a turbine; turbine turns an electric generator.

6. Falling water turns a turbine, which is connected to an electric generator.

8. (1) Real and imagined fears of an explosion; (2) limited supplies of nuclear fuels; (3) disposal of radioactive wastes; (4) costs of construction of plants and production of power.

10. Critical mass cannot be formed. Nonnuclear explosions can occur, which can release radioactive material. Sufficient heat is released to raise the temperature above the melting or softening point of construction materials and surface rocks.

12. Chernobyl.

14. Neptunium.

16. Health hazard of plutonium; loss of plutonium to terrorist organizations.

18. Technetium; name derived from a Greek word meaning synthetic or artificial.

20. I-131 emits only beta and gamma rays, which do little damage as they leave the body, has a relatively short half-life (8 days), and accumulates in the thyroid gland.

22. Gamma radiation.

24. **a.** Cosmic rays, naturally occurring radioisotopes in rocks and soil, radon gas escaping from the ground, internal radioisotopes such as K-40; **b.** internal radioisotopes such as K-40.

26. 60 hours.

28. Kr-93, U-239, Tc-99*m*, Np-239, I-131, Ba-140: none. All of the remainder: 100.0 g.

30. The utensil is probably more than 57,300 years old.

32. 240,000 years; 9600.

34. **a.** 8 days, as half the I-131 (source of half the radiation) vanishes; **b.** 80 days, as all of the I-131 vanishes; **c.** 1.1×10^{10} years, or 9 half-lives of K-40.

36. Yes; the half-life of Po-210, the source of the final decay step, would have to be used.

Chapter 6

4. Organic: b, c, f, g, h, i; inorganic: a, d, e.

6. A benzene ring.

8. **a.** Hexane; **b.** 3-methylpentane; **c.** 3-ethylpentane; **d.** 2,5-dimethylheptane; **e.** 2,3-dimethylbutane; **f.** 2-hexene; **g.** cyclohexane.

10. 1. Carbon bonds to as many as four other atoms simultaneously. 2. It forms strong bonds to atoms of several other elements. 3. It forms molecules containing many and varied branches as well as molecules composed of long, unbranched strings of carbons.

12. **a.** 4; **b.** 3; **c.** 1; **d.** 2.

14. **a.** All hydrocarbons burn in air. **b.** Alkenes will undergo catalytic hydrogenation.

16. **a.** Hydrocarbon, alkane, not aromatic;
b. hydrocarbon, not alkane, not aromatic;
c. hydrocarbon, alkane, not aromatic;
d. hydrocarbon, not alkane, not aromatic;
e. not hydrocarbon, not alkane, not aromatic;
f. hydrocarbon, not alkane, aromatic;
g. hydrocarbon, alkane, not aromatic;
h. not hydrocarbon, not alkane, aromatic.

18. $CH_3—C≡C—CH_2—CH_3$.

20. **a.** $CH_4 + 2O_2 \longrightarrow CO_2 + 2H_2O$;
b. $CH_2=CH_2 + 3O_2 \longrightarrow 2CO_2 + 2H_2O$;
c. $CH_3—CH_2—CH_3 + 5O_2 \longrightarrow 3CO_2 + 4H_2O$
d. $HC≡C—CH_3 + 4O_2 \longrightarrow 3CO_2 + 2H_2O$;
e. $2C_6H_6 + 15O_2 \longrightarrow 12CO_2 + 6H_2O$.

22. $CH_3—CH_2—CH_2—CH_2—CH_2—CH_2—CH_3$.

24. C_nH_{2n-2}; alkynes.

26. **a.** Alkene or cycloalkane; **b.** alkane; **c.** alkyne; **d.** alkene or cycloalkane; **e.** alkane.

28. It has too many hydrogens. A tetravalent carbon can't bond to enough hydrogens to produce a compound of this formula.

30. $2C_{1398}H_{1278} + 3435O_2 \longrightarrow 2796CO_2 + 1278H_2O$.

Chapter 7

4. H_2O, CO_2, energy; CO.

6. Increased power; octane rating must be increased.

8. CO_2.

10. Separation of petroleum's hydrocarbons.

12. Aromatic hydrocarbons are formed from nonaromatic hydrocarbons through cyclization and aromatization.

14. Cyclization and aromatization.

16. Ethyl alcohol; —OH.

18. 1; 2; 1.5.

20.

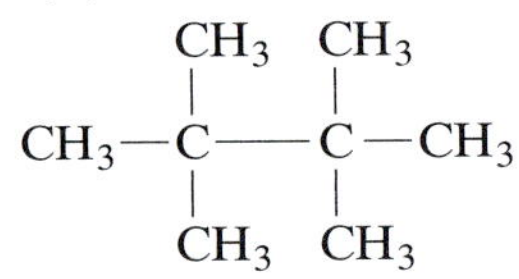

2,2,3,3-tetramethylbutane; 100 or greater.

22. The greater the ratio of 1° to 2°, the greater the ratio of their octane numbers because of the greater degree of branching.

Chapter 8

2. **a.** Proposed that heat consists of "brisk agitation" of the very particles that compose matter; **b.** demonstrated that work produces heat; **c.** established that work of any kind has its exact equivalent in heat and determined the numerical ratio of the amount of work done to the amount of heat produced.

4. **a.** kinetic; **b.** potential; **c.** kinetic; **d.** potential; **e.** kinetic; **f.** potential; **g.** kinetic; **h.** kinetic.

6. **a.** Energy In exceeds Energy Out; **b.** Energy Out exceeds Energy In; **c.** Energy In and Energy Out are equal.

8. The combination of basal metabolism and specific dynamic action.

10. Triglycerides (fats and oils).

12. Exothermic: a, b, c, e; endothermic: d.

14. The extensive burning of fossil fuels, ushered in with the Industrial Revolution, led to an increase in atmospheric CO_2, a greenhouse gas, and an increase in the earth's surface temperature.

16. Methane, biomass activity, cattle farming, coal mining; ozone, burning plant materials, smog, engine exhausts; nitrous oxide, use of nitrogen-containing fertilizers.

18. Would probably have little effect since CH_4 is also a greenhouse gas.

20. Basal metabolism would cause an energy expenditure of about 75 kg × 24 hr × 1 Cal/kg-hr = 1800 Cal. With specific dynamic action amounting to about 10% of our caloric intake, this 165-lb person would need to consume about 2000 Cal/day to maintain body weight.

22. **a.** 57 min; **b.** 34 min; **c.** 33 min; **d.** 104 min; **e.** 138 min; **f.** 129 min; **g.** 132 min.

24. 56 minutes.

26. 250.

28. 50°C.

30. The 100-watt bulb emitted 10,000 cal. 1 joule/0.24 cal × 10,000 cal = 41,667 joules. A 100-watt bulb releases 100 joules/second. The bulb was on 416.67 seconds, or 6.9 minutes.

Chapter 9

4. **a.** Formation of CO. **b.** Outdoors there is sufficient O_2 for CO_2 formation.

6. $CH_4 + 2O_2 \longrightarrow CO_2 + 2H_2O$

8. Percentage, by weight or volume.

10. ppm (parts per million).

12. **a.** 15; **b.** half a dozen; **c.** half a mole.

14. All weights are in grams: **a.** 20.18; **b.** 130.78; **c.** 28.02; **d.** 6.048; **e.** 17.72; **f.** 35.45; **g.** 70.90; **h.** 2.

16. **a.** water, 300 g; isopropyl alcohol, 700 g; **b.** water, 17 mole; isopropyl alcohol, 12 mole; **c.** solvent; **d.** solute; **e.** 1.7 *M*; **f.** 1.2 *M*.

18. 8.67×10^{-4}%.

20. **a.** 7.09 g; **b.** 0.92 g; **c.** sodium would be left over.

22. Selenium, yes; barium, no.

24. 5.2×10^{16} Na^+ cations.

26. See Appendix E.

28. $2CH_4 + 3O_2 \longrightarrow 2CO + 4H_2O$.

30. See Appendix C.

Chapter 10

4. Blue: a, c; red: b, d, e.

6. **a.** Base, aluminum hydroxide, antacid; **b.** base, calcium carbonate, antacid; **c.** acid, acetic acid, vinegar; **d.** acid, sulfuric acid, auto battery fluid; **e.** acid, hydrogen chloride, patio cleaner; **f.** salt, potassium iodide, salt additive; **g.** base, magnesium hydroxide, antacid; **h.** salt, magnesium sulfate, epsom salts; **i.** salt, sodium fluoride, adds fluoride to toothpaste; **j.** base, sodium bicarbonate (sodium hydrogen carbonate), baking soda; **k.** base, sodium hydroxide, drain cleaner; **l.** base, ammonia, household cleaner; **m.** salt, sodium benzoate, food additive.

8. Hydronium ion, H_3O^+.

10. No difference.

12. $HF + NaOH \longrightarrow NaF + H_2O$; sodium fluoride.

14. Lactic acid is the acid of sour milk and accumulates in muscles with exercise.

16. Carbonic acid, lactic acid.

18. **a.** Yes (pH less than 7); **b.** yes (pH less than 7); **c.** no (pH greater than 7).

20. $CH_3—CO_2H$ acid, NaOH base, $CH_3—CO_2^{-+}Na$ salt.

22. $HO_2C—CH(CO_2H)—CO_2H + NaHCO_3 \longrightarrow$
$HO_2C—CH(CO_2H)—CO_2^{-+}Na + H_2O + CO_2$.

24. **a.** 0.01 *M*; **b.** 10^{-12}; **c.** 12.

26. pH will be greater than 2; $[F^-]$ will be less than 0.01 *M*.

28. **a.** Yes, if $[H_3O^+]$ is less than 10^{-14}; Yes, if $[H_3O^+]$ is greater than 1.

30. **a.** 0.1; **b.** 1; **c.** 0.4.

32. 14.

Chapter 11

4. An electric current flows through the external circuit from the anode to the cathode.

6. **a.** Zinc; **b.** to supply electrons; **c.** carbon; **d.** to receive electrons and carry them to the battery's interior; **e.** NH_4Cl, $ZnCl_2$, and MnO_2.

8. From reducing agents to oxidizing agents.

10. Positive: reaction proceeds spontaneously; negative: reaction requires energy from an external source if it is to take place.

12. **a.** Refer to the value in the table. **b.** Reverse the sign of the reduction half-cell that's the reverse of the oxidation half-cell. **c.** Add the two voltage values.

14. **a.** Acquire, Cl_2; **b.** lose, Zn.

16. **a.** Yes; **b.** yes; **c.** Cl_2, F_2.

18. **a.** *Corrosion* is any form of erosion and disintegration of a material resulting from chemical reactions; *oxidation* is a specific form of corrosion resulting from reaction with atmospheric oxygen; **b.** oxidation of iron to a mixture of oxides we call *rust*; **c.** oxidation of aluminum, which forms a tough, corrosion-resistant film on the metal's surface.

20. The absence of exhaust gases that contribute to air pollution. Limited driving range, low speeds, great weight, and long recharge times for batteries.

22. Yes; the redox potential is +0.98 volts.

24. No. If the $PbSO_4$ is reduced to Pb, the reaction is simply the opposite of the oxidation half-cell for the Pb so that the redox potential is zero. Furthermore the $PbSO_4$ cannot be oxidized to PbO_2; there will be no redox reaction since the Pb cannot be reduced.

26. **a.** Converting +2.87 to zero requires that −2.87 be added to it. Therefore −2.87 must be added to all other standard reduction potentials; **b.** −2.87 volts; **c.** yes; **d.** no effect; **e.** yes **f.** in the same way as previously; **g.** (not applicable); **h.** no, because all are used on a relative basis.

Chapter 12

4. **a.** 3; **b.** 1; **c.** 8; **d.** 2; **e.** 7; **f.** 5; **g.** 6; **h.** 4.

6. Temperature.

8. Nitrogen.

10. Rotational, vibrational, translational; rotational; translational; vibrational.

12. Temperature decreases, kinetic energy decreases; cohesive forces overcome repulsive forces.

14. 100 K, or −173°C.

16. 20 liters.

18. $2H_2 + O_2 \longrightarrow 2H_2O$. Since the number of molecules decreases from 3 to 2, the volume decreases.

20. Mt. Everest, 2.6 liters; Mt. McKinley, 1.8 liters; Ben Nevis, 1.1 liters; For each peak, the drop in pressure is the dominant factor.

Chapter 13

4. Detergents lower water's surface tension. Detergents bind grease to water.

6. Carboxyl and hydroxyl groups.

8. All undergo hydrolysis in the presence of NaOH. Both glyceryl tripalmitate and ethyl palmitate would yield a soap.

10. a. Animal fat; **b.** lye and extracts of wood ashes.

12. a. Sodium acetate and ethyl alcohol; **b.** acetic acid and methyl alcohol; **c.** sodium palmitate and methyl alcohol; **d.** $CH_3—CH_2—CO_2—CH_2—CH_3$.

14. They are antiseptic.

16. Fillers, corrosion inhibitors, optical whiteners, fragrances, coloring agents.

18. A and C float in water.

20. Gold.

22. The real thing, gold.

24. The volume of water molecules is about 3×10^{-23} cm^3 per molecule. Since the water molecule has a smaller volume than the smallest colloidal particle, water does not form a colloidal dispersion in ethyl alcohol and it does not show a Tyndall effect in ethyl alcohol.

Chapter 14

4. Primary: NO, lightning, internal combustion engine, SO_2, burning fossil fuels; secondary: NO_2, reaction of NO with oxygen, SO_3, reaction of SO_2 with oxygen.

6. Atmospheric N_2.

8. a. Mechanical removal of particulate matter by filters; **b.** washing water-soluble gases out of exhausts; **c.** placing electrical charges on particulates and collecting the charged particles on an electrode.

10. Nitrogen dioxide, NO_2.

12. a. Highway vehicles; **b.** highway vehicles; **c.** electric utilities; **d.** highway vehicles.

14. $4FeS_2 + 15O_2 + 8H_2O \longrightarrow 2Fe_2O_3 + 8H_2SO_4$.

16. Nitrates and phosphates.

18. Irrigation and other agricultural needs.

20. Biological, through diseases carried by microorganisms; thermal, through depletion of oxygen and by affecting reproductive and other life cycles; sedimentary, by decreasing visibility and by carrying biological and chemical contaminants; chemical, by chemical toxicity.

22. By contributing to excessive growth of selective forms of aquatic life to the exclusion of other forms.

24. Mercury-containing wastes were discharged into a bay near Minimata, Japan, contaminating fish that were part of local diets.

26. a. Fuel combustion; **b.** transportation; **c.** transportation.

28. Industrial waste water: no; sanitary sewers: yes; drinking water: no.

Chapter 15

4. It raises the melting point.

6. After hydrogenation and hydrolysis, each yields stearic acid as the major saturated acid.

8. Cholesterol.

10. Reduces the number of C=C double bonds, which can react with atmospheric oxygen to generate undesirable products.

12. It would be too hard and brittle for use in foods.

14. Omega-9.

16. Foods derived from plants.

18. Each carbon of the double bond must be bonded to two different substituents.

20. Catalytic hydrogenation.

22. Fat is a solid triglyceride. Adipose tissue is composed of body cells containing fat globules as well as additional cellular material.

24. a. Butter; **b.** whole milk; **c.** white meat of chicken; **d.** cheddar cheese.

26. Lowest, glyceryl trilinolenate; highest, glyceryl tristearate.

28. a. Protein, 875 g; **b.** carbohydrate, 875 g; **c.** fat or oil, 388 g.

30. a. 7.7 Cal/g; **b.** 9 Cal/g; **c.** 86%.

Chapter 16

4. a. Acetone; **b.** acetaldehyde; **c.** formaldehyde; **d.** methyl ethyl ketone, MEK, or 2-butanone; **e.** benzaldehyde.

6. Monosaccharide: b, f, h; disaccharide: c, d, g; polysaccharide; a, e.

8. a. Greek, fermenting fruit juice, sweet wine, sweetness; **b.** Latin, *saccharum*, sugar; **c.** Latin, *dextra*, right; **d.** Greek, hand; **e.** Latin, *lac*, milk; **f.** carbon + Greek, water.

10. "Animal starch," provides animals with reservoir of readily available glucose.

12. Their bodies synthesize cellobiase; ours don't.

14. Cellulose/amylose: Both consist of glucose links; differ in the geometry of the glucose-to-glucose bonds. Amylose/amylopectin:

Both consist of glucose links joined by α linkages; differ in chain length and branching.

16. Formaldehyde, biological preservative; benzaldehyde, food flavoring and perfume ingredient.

18. a. 2; **b.** 4; **c.** 8.

20. Four. Both enantiomers contain the same number of chiral carbons.

22. None.

Chapter 17

4. Heating or adding sodium chloride, sodium carbonate, or isopropyl alcohol.

6. Eggs, beef, poultry, seafood, dairy products.

8. Tryptophan.

10. Arginine, histidine.

12. a. Alanine; **b.** glycine; **c.** phenylalanine; **d.** serine; **e.** aspartic acid; **f.** cysteine.

14. Glycine.

16. They contain sulfur.

18. Molecular weight.

20. a. Sickle cell anemia; **b.** kwashiorkor; **c.** low density lipoproteins.

22. Eight: Ala.Ala.Ala, Ala.Ala.Gly, Ala.Gly.Ala, Gly.Ala.Ala, Gly.Gly.Gly, Gly.Gly.Ala, Gly.Ala.Gly, Ala.Gly.Gly.

24. Twenty-seven: Ala.Ala.Ala, Ala.Ala.Gly, Ala.Gly.Ala, Gly.Ala.Ala, Ala.Ala.Leu, Ala.Leu.Ala, Leu.Ala.Ala, Gly.Gly.Gly, Gly.Gly.Ala, Gly.Ala.Gly, Ala.Gly.Gly, Gly.Gly.Leu, Gly.Leu.Gly, Leu.Gly.Gly, Leu.Leu.Leu, Leu.Leu.Ala, Leu.Ala.Leu, Ala.Leu.Leu, Leu.Leu.Gly, Leu.Gly.Leu, Gly.Leu.Leu, Ala.Gly.Leu, Ala.Leu.Gly, Gly.Ala.Leu, Gly.Leu.Ala, Leu.Ala.Gly, Leu.Gly.Ala.

Chapter 18

4. Meat processing; gelatin.

6. Cell nuclei.

8. DNA: adenine, cytosine, guanine, thymine; RNA: adenine, cytosine, guanine, uracil.

10. DNA: a longer molecule, in the shape of a double helix, resides in the cellular nucleus, carries the organism's genetic information, transmits the genetic information from one cellular generation to the next. RNA: a shorter molecule, exists as a simpler molecular strand, travels from the cellular nucleus to the cytoplasm, translates genetic information into molecular structures of polypeptides and proteins.

12. The correspondence between codons and amino acids or biochemical functions.

14. a. Genotype, genetic feature or structure carried by a parent; phenotype, an observable characteristic in offspring.

b. Gene, a segment of a chromosome that carries the information needed to create a specific polypeptide or protein; chromosome, long strand of coiled DNA within the nucleus.

c. Double helix, two strands coiled around each other; triple helix, three strands coiled around one another.

d. DNA, deoxyribonucleic acid; RNA, ribonucleic acid.

e. mRNA, messenger RNA, carries genetic information from the nucleus into the cytoplasm; tRNA, transfer RNA, transfers amino acids to sites where peptide bond formation takes place.

f. Meiosis, sexual cellular division in which half the chromosomes in each new cell come from one parent, half from the other; mitosis, cellular division in which the two new cells are identical copies of the dividing cell.

g. Nucleic acid, a long molecular chain consisting of repeating nucleotide units; nucleotide, a combination of 2-deoxyribose, phosphoric acid, and one of four cyclic amines.

16. The Nobel Prize is not awarded posthumously.

18. Four, one for each of the four possible bases.

20. An estimated 15 billion hydrogen bonds (approximately 6 billion base pairs with an average of 2.5 hydrogen bonds per base pair).

Chapter 19

4. Macronutrients make up the major portion of our food and provide us with energy and the materials of our physical bodies. Micronutrients help to release the energy of the macronutrients and to convert the macronutrients into the substances of our bodies. Macronutrients: carbohydrates, proteins, and fats and oils; micronutrients: minerals such as calcium and phosphorus, and vitamins such as A and C.

6. Vitamin C; vitamin D.

8. One RE of vitamin A is equivalent to 1 μg—1 microgram, 10^{-6} g—of retinol. One IU of vitamin A represents 0.3 μg of retinol. These units include not only the vitamins themselves but also any and all physiological equivalents of the vitamins.

10. The need for dietary sources of vitamin C, ascorbic acid.

12. Calories that are accompanied by few or no micronutrients; by focusing on the relative proportions of calories and micronutrients in foods.

14. a. Nutrition; **b.** nutrition; **c.** appeal; **d.** appeal; **e.** appeal.

16. Fats and oils.

18. Both sought routes to spices that would mask the unpleasant taste of spoiled food.

20. Protects meat against the growth of botulism-producing bacteria.

22. Approximately three Ca atoms for every two P atoms (about 1.55 Ca atoms per P atom).

24. 0.003% retinol equivalent by weight.

26. 4%.

Chapter 20

4. a. Acetylsalicylic acid (aspirin), made it more difficult for children to open bottles of aspirin; **b.** saccharin, allowed continued use of saccharin in food and drinks despite the finding that saccharin is a carcinogen.

6. a. U.S. Public Health Service; **b.** Bureau of Alcohol, Tobacco and Firearms; **c.** Environmental Protection Agency; **d.** Occupational Safety and Health Administration; **e.** Food and Drug Administration.

8. Bufotoxin and solanine; solanine.

10. a. A teratogen produces severe birth defects; a mutagen produces lesser genetic effects. **b.** A carcinogen produces cancer.

12. Its chemical characteristics, how much we use, how we use it, how susceptible we are as humans, how susceptible we are as individuals.

14. Whether it produces mutations in *Salmonella typhimurium* bacteria.

16. Arsenic trioxide, caffeine, nicotine; ethyl alcohol, thiamine hydrochloride (vitamin B_1), trisodium phosphate.

18. 280 liters.

20. Sodium cyanide, aspirin, sodium chloride, water.

22. $5\ g \times 10^3\ mg/g \times kg/(3 \times 10^{-8}\ mg) \times person/70\ kg = 2.4 \times 10^9$ people.

24. 8.3×10^{12} mice.

26. Larger.

28. 0.1 g per person per day.

Chapter 21

4. a. Nylon; **b.** Bakelite; **c.** styrene–butadiene rubber; **d.** Dacron or Terylene.

6. Acrilan, synthetic fibers; Bakelite, electrical outlets, handles for pots and pans; Creslan, Dacron, Fortrel, Kodel, synthetic fibers; Mylar, plastic backing for audio and video tapes and for computer diskettes and food wrappers; nylon, Orlon, synthetic fibers; Saran, food wrapping; Teflon; machine parts and nonstick coatings; Terylene, synthetic fibers.

8. A plastic is a material we can shape into virtually any form. A polymer consists of molecules of very high molecular weight, formed through the repeated chemical linking of a great many smaller molecules.

10. a. High-density polyethylene molecules have few branches; low-density molecules branch sporadically into short offshoots from the main line of the polymer. **b.** High-density polyethylene molecules align themselves into localized areas of tightly packed, highly organized strands; the low-density molecules form a tangled, randomly oriented network of strands. **c.** High-density polyethylene is a denser, harder, higher melting, and more rigid polymer than low-density polyethylene.

12. Thermoplastics soften when heated and become firm again when cooled. Thermosetting plastics are soft enough to be molded when they are first prepared, but on heating they become permanently firm.

14. HDPE, high-density polyethylene, and LDPE, low-density polyethylene.

16. a. Bakelite, polystyrene; **b.** PVC, poly(vinyl chloride); **c.** Teflon; **d.** nylon; **e.** poly(vinyl acetate).

18. Resistance to degradation; persistence in the environment.

20. 1. The inherent difficulty and cost of separating different kinds of plastic from each other so that each can be reused effectively. 2. Some plastic products (e.g., baby diapers) aren't suitable for collection and recycling.

22. Sunlight.

24. 3.

Chapter 22

8. Henna.

10. a. Blocks all solar radiation; **b.** absorbs UV-B radiation while letting UV-A radiation pass through to produce a tan; **c.** the combination absorbs throughout the UV spectrum, providing a wider range of protection than *p*-aminobenzoic acid alone; **d.** reacts with the amino acids of the skin to produce a brown pigment, sometime called a "sunless tan."

12. Cortex forms the bulk of the fiber and contains its coloring pigments; cuticle, a thin, translucent, scaly sheath enveloping the cortex.

14. a. Hydrogen bonds; **b.** S—S bonds.

16. UV-A: less energetic, longer wavelengths (320 to 400 nm) of solar radiation; UV-B, more energetic, shorter wavelength region (290 to 320 nm) of solar radiation.

18. Both have peaks (crests), troughs, wavelength, and amplitude.

20. By physically closing the ducts of the eccrine glands; by killing bacteria that act on the accumulated residue of perspiration, producing odors.

22. A cologne is a dilute, inexpensive version of a perfume.

24. a. Zinc oxide physically blocks ultraviolet radiation, prevents it from reaching the surface of the skin. **b.** *p*-Aminobenzoic acid blocks UV-B radiation while allowing UV-A radiation to pass through to the skin, producing a tan. **c.** Benzophenone absorbs both UV-A and UV-B radiation.

26. a. Two hours; **b.** 3 hours, 20 minutes; **c.** 5 hours; **d.** 10 hours.

Chapter 23

4. a. Salicin; **b.** cocaine; **c.** opium; **d.** β-endorphin; **e.** atropine; **f.** reserpine.

6. **a.** Salicylates; **b.** morphine; **c.** placebo; **d.** medicine; **e.** narcotics; **f.** enkephalins; **g.** belladonna; **h.** aspirin; **i.** prostaglandins; **j.** heroin; **k.** endorphins.

8. Platelets are small bodies within the blood serum that initiate clotting. Aspirin inhibits their formation. Risk: excessive bleeding resulting from blood's decreased ability to clot. Benefit: decreased risk of some types of heart attacks and strokes.

10. Larger amount of active ingredient (acetylsalicylic acid) in each tablet.

12. Pain, fever, inflammation.

14. Alike: Both are addictive. Different: Methadone doesn't produce the euphoria or other psychological effects of heroin.

16. Caffeine, coffee; nicotine, tobacco.

18. **a.** Pain and fever; **b.** inflammation.

20. Acetic anhydride; $CH_3—CO—O—CO—CH_3$.

22. **a.** Aspirin inhibits the formation of prostaglandins. **b.** Morphine inhibits the transmission of signals along the nervous system by occupying opiate receptors within the system. **c.** Naloxone distorts opiate receptors so that they are insensitive to the effects of opiates and narcotics. **d.** β-Endorphin occupies opiate receptors within the nervous system.

24. The Harrison Narcotics Act of 1914 was the first of a series of laws that made the use of narcotics to satisfy an addiction punishable by law.

26. 25%.

Appendix A

2. **a.** 4.7000×10^4; **b.** 9.672370×10^4; **c.** 1.2×10^1; **d.** 4.195×10^{-3}; **e.** 1×10^{-1}.

Appendix C

2. **a.** 9×10^{15} meters; **b.** 6×10^{12} miles.

4. 360 g H_2O.

Appendix D

2. **a.** 5.318; **b.** 202.2960; **c.** 0.006832.

4. **a.** 0.70; **b.** 1.6; **c.** 3.337.

PHOTO CREDITS

Chapter 1 **Page 1:** Stone/Getty Images. **Page 2:** George Chin/DOTS. **Page 4:** Ken Karp. **Page 6:** PhotoDisc, Inc./Getty Images. **Page 6:** Peter Arnold, Inc. **Page 8** (margin): Corbis Stock Market. **Page 8** (left & right): Ken Karp. **Pages 12 & 13:** PhotoDisc, Inc./Getty Images.

Chapter 2 **Page 16:** PhotoDisc, Inc./Getty Images. **Page 17:** Ken Karp. **Page 19:** Courtesy of International Business Machines Corporation. Unauthorized use not permitted. **Page 22:** Courtesy NASA. **Page 24:** Bettmann/Corbis Images. **Page 27:** Vince Streano/Corbis Images. **Page 29:** Andy Washnik.

Chapter 3 **Page 34:** Rich Iwasaki/Stone/Getty Images. **Page 35:** Ken Karp. **Page 37** (right): Yoav Levy/Phototake. **Page 37** (left): Michael Watson. **Page 37** (margin): Courtesy OPC, Inc. **Page 38** (left): Ken Karp. **Page 38** (right): Courtesy USDA. **Page 40:** Courtesy Chemists' Club Library. **Page 41:** Courtesy Chemical Heritage Foundation. **Page 45** (left): Robert Capece. **Page 45** (right): Marc Muench/Stone/Getty Images. **Page 45** (center): Christian Hilsen/Stone/Getty Images.

Chapter 4 **Page 57:** Fermilab/Peter Arnold, Inc. **Page 58** (top): Scott Camazine/Photo Researchers. **Page 58** (margin): Culver Pictures, Inc. **Page 58** (bottom): Courtesy The Center for the History of Chemistry. **Page 61:** Courtesy C.E. Wynn-Williams. **Page 63** (left): Robert Maass/Corbis Images. **Page 63** (right): Courtesy Amersham Technology. **Page 65:** David Parker/Science Photo Library/Photo Researchers. **Page 69:** Donna Coveney/ Courtesy MIT Archives. **Page 70** (top): Bettmann/Corbis Images. **Page 70** (bottom): Courtesy Bibliothek Und Archiv zur Geschichte der Max-Planck-Gesellschaft, Berlin. **Page 71:** Courtesy U.S. Department of Energy. **Page 73:** Bettmann/Corbis Images. **Page 74** (top): Courtesy National Archives, photo no. 77-BT-187. **Page 74** (bottom): Courtesy National Archives, photo no. 77-BT-115. **Page 75:** Shigeo Hayashi. **Page 76:** AP/Wide World Photos. **Page 80:** Courtesy NASA.

Chapter 5 **Page 86:** Roger Ressmeyer/Corbis Images. **Page 87:** Courtesy University of Chicago/AIP Neils Bohr Library. **Page 88:** "Birth of the Atomic Age" by Gary Sheahan/Courtesy Chicago Historical Society. **Page 90:** David Bartruff/Corbis Images. **Page 92:** Sipa Press. **Page 94:** Matthew Neal McVay/Stone/Getty Images. **Page 98:** Courtesy Yucca Mountain Project. **Page 101** (top): Novosti/Science Photo Library/Photo Researchers. **Page 101** (bottom): Custom Medical Stock Photo. **Page 102** (top): David Parker/Science Photo Library/Photo Researchers. **Page 102** (center): CNRI/Phototake. **Page 102** (bottom): Kelley Culpepper/Transparencies, Inc. **Page 106:** Patrick Mesner/Liaison Agency, Inc./Getty Images. **Page 107:** Yoav Levy/Phototake. **Page 108:** AP/Wide World Photos. **Page 110:** PhotoDisc, Inc./Getty Images.

Chapter 6 **Page 115:** Martyn F. Chillmaid/Photo Researchers. **Page 116:** Ken Karp. **Page 117:** The Granger Collection. **Pages 118 & 122:** Andy Washnik. **Page 121:** Courtesy Tripos Associates. **Pages 128 & 129:** Courtesy Michael Pique, The Scripps Research Institute, CA. **Page 134:** Andy Levin/Photo Researchers.

Chapter 7 **Page 147:** PhotoDisc, Inc./Getty Images. **Page 148 & Page 149** (left) & (right): Ken Karp. **Page 157:** Tannen Mauryl/ The Image Works. **Page 160:** Seth Resnick/Liaison/Getty Images. **Page 161:** Photo by Bill Grafton. Courtesy The Reynolds Group. **Page 162:** Brett Frooner/The Image Bank/Getty Images. **Page 163:** Tony Freeman/PhotoEdit. **Page 167** (bottom): Vanessa Vick/Photo Researchers. **Page 167** (top): Paul S. Howell/ Liaison Agency, Inc./Getty Images. **Page 168:** Courtesy Solectria Corporation.

Chapter 8 **Page 172:** DiMaggio/Kalsih/Corbis Stock Market. **Page 173** (left) & (right): Ken Karp. **Page 174** (left, center & right): Dan Helms/Duomo Photography, Inc. **Page 175 & Page 176** (top): Bettmann/Corbis Images. **Page 176** (bottom): Corbis Images. **Page 178:** Custom Medical Stock Photo. **Page 179:** Bettmann/ Corbis Images. **Page 183** (top): Arthur R. Hill/Visuals Unlimited. **Page 183** (bottom): Michael Nichols/Magnum Photos, Inc. **Page 183** (left): TempSport/Corbis Images. **Pages 189 & 191:** PhotoDisc, Inc./Getty Images. **Page 192:** Michael George/Bruce Coleman, Inc. **Page 194** (top): Mark Marten/NASA/Science Source/Photo Researchers. **Page 194** (center): © Philip & Karen Smith/Stone/Getty Images. **Page 194** (bottom): NASA/Phototake. **Page 195:** Peter Newton/Stone/Getty Images.

Chapter 9 **Page 200:** Taxi/Getty Images. **Pages 201 & 202:** Andy Washnik. **Pages 203, 205 & 206:** Ken Karp. **Pages 207 & 208:** Courtesy Edgar Fahs Collection, University of Pennsylvania. **Page 213:** Peter Lerman. **Page 216:** George Chin/DOTS. **Page 221** (top): Corbis Images. **Page 221** (bottom): Lincoln Nutting/Photo Researchers. **Page 222:** Dick Luria/Taxi/Getty Images. **Page 223:** Scott T. Smith/Corbis Images.

Chapter 10 **Page 228:** Susan Johns/Photo Researchers. **Pages 229, 230 & 232:** Andy Washnik. **Page 231** (center) & (bottom): Robert J. Capece. **Page 234** (top): Courtesy Edgar Fahs Smith Collection, University of Pennsylvania. **Page 234** (bottom): Andy Washnik. **Page 235:** Elliot V. Fry, courtesy AIP Emilio Segré Visual Archives. **Page 243:** Charles D. Winters/Photo Researchers. **Page 245:** Ken Karp. **Page 246:** Courtesy Miles Laboratories.

Chapter 11 **Page 255:** Craig Aurness/Corbis Images. **Pages 256, 257 & 259:** Ken Karp. **Page 258:** Courtesy Eveready Battery Company, Inc. **Page 259** (top): Courtesy Skilcraft. **Page 260:** Andy Washnik. **Page 264:** Culver Pictures, Inc. **Page 272:** Courtesy Delco Remy. **Page 273:** Yoav Levy/Phototake.

Chapter 12 **Page 281:** Tim Flach/Stone/Getty Images. **Pages 282 & 298:** Ken Karp. **Page 288:** Pat Lanza Field/Bruce Coleman, Inc. **Pages 289 & 292:** Science Photo Library/Photo Researchers. **Page 299:** Courtesy Miller Brooks, Perfecto Manufacturing. **Page 300:** Frank Viola/Comstock Images.

Chapter 13 **Page 306:** Hermann Eisenbeiss/Photo Researchers. **Pages 307 & 308:** Ken Karp. **Page 310:** Scott Nielson/Bruce Coleman, Inc. **Page 314:** Bettmann/Corbis Images. **Page 315** (top): The Image Bank/Getty Images. **Page 315** (bottom): Richard Megna/Fundamental Photographs. **Page 321:** Ken Karp. **Page 323:** Charles Hamilton. **Page 324:** Andy Washnik.

Chapter 14 **Page 330:** PhotoDisc, Inc./Getty Images. **Page 331:** Andy Washnik. **Page 333:** David Muench Photography. **Page 335:** PhotoDisc, Inc./Getty Images. **Page 336:** Ray Elliot Jr./Stone/Getty Images. **Page 337** (left): Shawn Henry/Corbis-SABA. **Page 337** (top, right): PhotoDisc, Inc./Getty Images. **Page 337** (bottom, right): Y. Hamel/Photo Researchers. **Page 339:** Gerard Fritz/Stone/Getty Images. **Page 342** (left & right): Courtesy Westfalishes Amt fur Denkmalpflege. **Page 348** (top): Courtesy University of California, Irvine. **Page 348** (center): Courtesy Mario J. Molina, Massachusetts Institute of Technology. **Page 348** (bottom): Courtesy Paul Crutzen, Max Planck Institute for Chemistry, Atmospheric Chemistry Division. **Page 349:** Courtesy NASA Goddard Space Flight Center. **Page 350:** Nicole Duplaix/Peter Arnold, Inc. **Page 352:** Eva Demjen/Stock, Boston. **Page 354:** Courtesy NASA. **Page 356:** Craig Newbauer/Peter Arnold, Inc. **Page 357:** Don Mason/Corbis Stock Market. **Page 360:** Joe Sohm/Chromosohm/Stone/Getty Images. **Page 361:** Michael J. Phillippot/Corbis Sygma. **Page 362:** Custom Medical Stock Photo. **Page 363:** Corbis Stock Market.

Chapter 15 **Page 367:** PhotoDisc, Inc./Getty Images. **Pages 368, 369 & 377:** Ken Karp. **Page 374:** Marti Pie/The Image Bank. **Page 375:** Andy Washnik. **Page 377:** Ken Karp. **Page 379:** Courtesy Michael Pique, The Scripps Research Institute, CA. **Page 383:** Jerry Ohlinger's.

Chapter 16 **Page 387:** Digital Stock/Corbis Images. **Pages 388 & 408:** Ken Karp. **Page 389:** Yoav Levy/Phototake. **Page 390:** Courtesy OPC, Inc. **Pages 392, 395 & 400:** Andy Washnik. **Page 403:** Courtesy Dutch Gold Honey, Inc. **Pages 409 & 411:** PhotoDisc, Inc./Getty Images.

Chapter 17 **Page 416:** Photo Researchers. **Page 417** (right): Stephen Frisch/Stock, Boston. **Page 417** (top, right & bottom, right): Andy Washnik. **Page 418:** Corbis Images. **Page 421:** Andy Washnik. **Pages 425, 426 & Page 427** (top): Ken Karp. **Page 427** (bottom): CNRI/Phototake. **Page 434** (top & bottom): Bill Longcore/Photo Researchers. **Page 435:** Leonard Lessin/Peter Arnold, Inc.

Chapter 18 **Page 440:** Alfred Pasieka/Peter Arnold, Inc. **Page 441:** Andy Washnik. **Page 445:** Leslie Holzer/Photo Researchers. **Page 446** (top): David M. Phillips/Visuals Unlimited. **Page 446** (bottom): BioPhoto Assoc./Science Source/Photo Researchers. **Page 447:** J.C. Revy/Copyright ISM/Phototake. **Page 448:** Copyright: A. Barrington/Science Source/Photo Researchers. **Page 456** (top): AP/Wide World Photos. **Page 456** (bottom, right & left): Courtesy Texas A & M University, College of Veterinary Medicine.

Chapter 19 **Page 460:** Stone/Getty Images. **Pages 461 & 462** (left): Andy Washnik. **Page 463** (top, right): Ken Karp. **Page 463** (bottom, right): Digital Stock /Corbis Images. **Page 463** (left): Photo Researchers. **Page 464** (left): CNRI/Phototake. **Page 464** (right): Andy Washnik. **Page 471** (right): Merlin D. Tuttle/Photo Researchers. **Page 471** (top, left): G. Robert Bishop/Stone/Getty Images. **Page 471** (bottom, left): Joseph Van Wormer/Bruce Coleman, Inc. **Page 472** (top): PhotoDisc, Inc./Getty Images. **Page 472** (bottom): Ken Karp. **Page 475:** Shumsky/The Image Works. **Page 478:** Jonathan A. Meyers/Photo Researchers. **Page 479:** Bettmann/Corbis Images. **Page 483:** Ken Karp.

Chapter 20 **Page 493:** Digital Stock/Corbis Images. **Pages 494 & 496:** Ken Karp. **Page 494** (margin): Photo Researchers. **Page 500:** Visuals Unlimited. **Page 505:** Lawrence Manning/Corbis Images. **Page 509:** Andy Washnik. **Page 510:** Tony Freeman/PhotoEdit. **Page 511:** Nino Mascardi/The Image Bank/Getty Images. **Page 512:** Ken Karp. **Page 515:** Bill Foothorap/Black Star.

Chapter 21 **Page 519:** PhotoDisc, Inc./Getty Images. **Page 520:** Ken Karp. **Page 521:** Photo Researchers. **Page 525:** Ben Edwards/Stone/Getty Images. **Page 527** (left): Michael Baytoff/Black Star. **Page 527** (right): Garry Gay/The Image Bank/Getty Images. **Page 530:** Courtesy The Goodyear Tire and Rubber Company. **Page 531:** Eric Sander/Liaison Agency/Getty Images. **Page 532:** Phil Schemeister/Stone/Getty Images. **Page 533** (top): Dr. E.R. Degginger. **Page 533** (bottom left & right): Courtesy Union Carbide. **Page 535:** Courtesy DuPont. **Page 537** (left): Ken Karp/Omni-Photo Communications, Inc. **Page 537** (right): Courtesy DuPont. **Page 538:** Richard Megna/Fundamental Photographs. **Page 539:** Ken Karp. **Page 542** (top, left): Gerard Fritz/Taxi/Getty Images. **Page 542** (top, right): David J. Sams/Stock, Boston. **Page 542** (bottom): Kip & Pat Peticolas/Fundamental Photographs. **Page 544:** PhotoDisc, Inc./Getty Images.

Chapter 22 **Page 552:** PhotoDisc, Inc./Getty Images. **Page 553:** Courtesy L'Oreal. **Page 554:** Corbis Images. **Page 556:** Sunset (MOULU)/Peter Arnold, Inc. **Page 559:** Roy Morsh/Corbis Stock Market. **Page 561:** PhotoDisc, Inc./Getty Images. **Page 567:** Leonard Lessin/Peter Arnold, Inc. **Page 569** (top): Benn Mitchell/The Image Bank/Getty Images. **Page 569** (bottom): Mark D. Phillips/Photo Researchers. **Page 570:** Vision/Photo Researchers. **Page 575:** Sinclair Stammers/Photo Researchers.

Chapter 23 **Page 582:** PhotoDisc, Inc./Getty Images. **Page 585** (top): Ken Karp. **Page 585** (bottom): James Selkin/Liaison Agency/Getty Images. **Page 590:** Andy Washnik. **Page 591:** Will & Demi McIntyre/Photo Researchers. **Page 593:** ARCHIV/Photo Researchers. **Page 596:** Ken Karp.

INDEX

Symbols of the Elements

Symbol	Name	Symbol	Name
Ac	Actinium	Mt	Meitnerium
Ag	Silver	N	Nitrogen
Al	Aluminum	Na	Sodium
Am	Americium	Nb	Niobium
Ar	Argon	Nd	Neodymium
As	Arsenic	Ne	Neon
At	Astatine	Ni	Nickel
Au	Gold	No	Nobelium
B	Boron	Np	Neptunium
Ba	Barium	O	Oxygen
Be	Beryllium	Os	Osmium
Bh	Bohrium	P	Phosphorus
Bi	Bismuth	Pa	Protactinium
Bk	Berkelium	Pb	Lead
Br	Bromine	Pd	Palladium
C	Carbon	Pm	Promethium
Ca	Calcium	Po	Polonium
Cd	Cadmium	Pr	Praseodymium
Ce	Cerium	Pt	Platinum
Cf	Californium	Pu	Plutonium
Cl	Chlorine	Ra	Radium
Cm	Curium	Rb	Rubidium
Co	Cobalt	Re	Rhenium
Cr	Chromium	Rf	Rutherfordium
Cs	Cesium	Rh	Rhodium
Cu	Copper	Rn	Radon
Db	Dubnium	Ru	Ruthenium
Dy	Dysprosium	S	Sulfur
Er	Erbium	Sb	Antimony
Es	Einstenium	Sc	Scandium
Eu	Europium	Se	Selenium
F	Fluorine	Sg	Seaborgium
Fe	Iron	Si	Silicon
Fm	Fermium	Sm	Samarium
Fr	Francium	Sn	Tin
Ga	Gallium	Sr	Strontium
Gd	Gadolinium	Ta	Tantalum
Ge	Germanium	Tb	Terbium
H	Hydrogen	Tc	Technetium
He	Helium	Te	Tellurium
Hf	Hafnium	Th	Thorium
Hg	Mercury	Ti	Titanium
Ho	Holmium	Tl	Thallium
Hs	Hassium	Tm	Thulium
I	Iodine	U	Uranium
In	Indium	Uub	Ununbium
Ir	Iridium	Uun	Ununnilium
K	Potassium	Uuq	Ununquadium
Kr	Krypton	Uuu	Unununium
La	Lanthanum	V	Vanadium
Li	Lithium	W	Tungsten
Lr	Lawrencium	Xe	Xenon
Lu	Lutetium	Y	Yttrium
Md	Mendelevium	Yb	Ytterbium
Mg	Magnesium	Zn	Zinc
Mn	Manganese	Zr	Zirconium
Mo	Molibtenum		